Pots, People, and Politics: A Reconsideration of the Role of Ceramics in Reconstructions of the Iron Age Northern Levant

Matthew R. Whincop

BAR International Series 1902
2009

Published in 2016 by
BAR Publishing, Oxford

BAR International Series 1902

Pots, People, and Politics: A Reconsideration of the Role of Ceramics in Reconstructions of the Iron Age Northern Levant

ISBN 978 1 4073 0384 0

BAR Publishing is the trading name of British Archaeological Reports (Oxford) Ltd. British Archaeological Reports was first incorporated in 1974 to publish the BAR Series, International and British. In 1992 Hadrian Books Ltd became part of the BAR group. This volume was originally published by Archaeopress in conjunction with British Archaeological Reports (Oxford) Ltd / Hadrian Books Ltd, the Series principal publisher, in 2009. This present volume is published by BAR Publishing, 2016.

Printed in England

BAR
PUBLISHING

BAR titles are available from:

BAR Publishing
122 Banbury Rd, Oxford, OX2 7BP, UK
EMAIL info@barpublishing.com
PHONE +44 (0)1865 310431
FAX +44 (0)1865 316916
www.barpublishing.com

CONTENTS

SECTION THREE Analysis of Iron Age Ceramic Data

SECTION FOUR An Alternative Reconstruction for the Iron Age Northern Levant

APPENDICES

Contents of Appended CD

Charts

Clusters

Database

Distribution maps

Seriation

Site Data

Type Data

Please note that the CD referred to above has now been replaced with a download available at www.barpublishing.com/additional-downloads.html

Abstract

Pots, People, and Politics: A Reconsideration of the Role of Ceramics in Reconstructions of the Iron Age Northern Levant

Matthew R WHINCOP

This study aims to reconsider current reconstructions of the Iron Age Northern Levant and the role that ceramics studies have played in these interpretations. It begins with an assessment of the use of the historical narrative in current interpretations. The conventional historical interpretative framework has produced a broad perspective on Iron Age society at the expense of localised behaviours. For this reason, the present study attempts to engage with Iron Age material culture, more specifically pottery, and consider its role within past societies beyond the broad socio-political histories derived from a literal reading of documentary sources. This study presents a regional ceramic typology for the Iron Age (including the Persian period) and undertakes an analysis of the distribution patterns of this typology across the Northern Levant. An alternative interpretation of the ceramic data is offered, before being compared with the conventional historical model. This alternative reconstruction focuses on theories of practice, and foodways, whilst appreciating the dynamic manner by which material culture is used to constantly negotiate and consolidate social structures. In the end, the study offers one perspective on the compatibility of archaeological data and the historical text, and makes some final recommendations for their correlation.

Statement of Copyright

Acknowledgments

I would like to express my gratitude to a number of people for putting at my disposal their excavated material and/or their time; in particular - Graham Philip, but also Peter Parr, Derek Kennet, Stephen Bourke, Tony Sagona, Paul Newson, Rudolph Dornemann, Jesse Casana, Tine Bagh, Michel al-Maqdissi, Wael al-Hafian, Ahmad Taraqji, Abd al Rachman Kusa, Timothy Harrison, Tony Wilkinson, Stefania Mazzoni, Serena Maria Cecchini, Fabrizio Venturi, Jean-Paul Thalmann, Daniele Morandi Bonacossi, Marta Luciani, Marco Iamonu, Guy Bunnens, Arlette Roobaert, Andrew Jamieson, Elisabeth Stone, Paul Zimansky, Martha Sharp Joukowsky, Karin Bartl, Paul Reynolds, Stephen McPhilips, David Ussishkin, Israel Finkelstein, David Ilan, Amihai Mazar, Nava Panitz Cohen, Gunnar Lehmann, Amnon Ben-Tor, Doron Ben-Ami, Jack Green, Charlotte Whiting, Mike Church, Peter Rowley-Conwy, Andrew Millard, Mike Baxter, Prof. Wakita, Sue Sherratt, and Claude Doumet-Serhal. A number of institutions also granted access to archives that were integral to this study; these include the British Museum, the National Museum of Copenhagen, University College London, and Melbourne University.

Financial assistance from a number of sources made this study a feasible undertaking; Council for British Research in the Levant, Palestine Exploration Fund, Rosemary Cramp Fund, Birley Bursary, Ustinov College at Durham University, Haycock Trust, and Katherine Whincop all contributed to the final product.

I would also like to express my gratitude to friends and fellow students who encouraged me throughout the rollercoaster of the past four years: Tom White, Catherine Isaacs, Ed Cork, Andy Shaw(bert), Beccy Scott, Sophie Blain, Abby Antrobus, Tom Moore, Karen Exell, Rachael Dann, Liam Cooney, the 'R' Man, Simon Alderson, Mark Manuel, Rosa Spencer, Jennie Bradbury, Laura Cripps, Karina Croucher, Sam Hemsley, Sally Fletcher, Kat MacRae, Fulya Dedeoglu, Sarah Lynchenhaum, and Martin Makinson. I owe my greatest debt, however, to my family: my wife, Katherine, and parents, Noel and Merran, have always encouraged me to pursue my dreams. I would like to dedicate this study to my inspiring wife, Katherine; for your belief and support, words cannot express.

List of Tables

List of Charts

List of Seriation Matrices

List of Dendrograms

List of Figures

In-Text Figures

Appended Figures

The ceramic typology used in this study is presented in Figures 1-54, with individual vessel numbers representing the CLASS number. For example, Figure 1 001a/1 represents CLASS 001, variant a, first example. Ceramic illustrations without a source reference were produced by the author. All other figures were reproduced from the cited references: see bibliography for full details. All ceramic figures are reproduced at a scale of 1:4, except those included in Figures 10-15 and Figure23 (048a), all of which are reproduced at a scale of 1:10. The scale of Figure 46(128) is unknown.

List of Maps

Not all ceramic distribution maps were included in this volume; only those considered key to the discussion of **Chapters** 7, **8**, and **9**. The complete and exhaustive collection is included on the appended CD. The maps on the CD are coded by colour according to period:

large green points	= Iron I presence
medium red points	= Iron II
small blue points	= Iron III
tiny yellow points	= Persian period

List of Abbreviations

ASOR	American Schools of Oriental Research
AUB	American University in Beirut
BMB	*Bulletin du Musée de Beyrouth*
CA	Correspondence Analysis
CAH	Cambridge Ancient History
CBRL	Council for British Research in the Levant
DGA	Director General of Antiquities, Lebanon
DGAM	Director General of Antiquities and Museums, Syria
IA-NL	Iron Age Northern Levant
IA-SL	Iron Age Southern Levant
OI	Oriental Institute, Chicago University
PEF	Palestine Exploration Fund
PEFQS	Palestine Exploration Fund Quarterly Statement
SAR	Syrian Arab Republic

CHAPTER ONE
Introduction

1.1 Genesis of the Present Study

The present study is, by and large, a development arising from the author's Masters Research on the Iron II pottery from Tell Nebi Mend. In the process of placing the Tell Nebi Mend pottery within its regional context it had become apparent that there were a number of different "regional" influences present in the material. In particular, the Tell Nebi Mend cooking-pots resembled those from the coast, while the large pithoi were linked to the interior. Furthermore, another "inland" form, the Red-Slip pedestal-platter experienced a locally-restricted distribution (Whincop 2007, Fig. 12). The diversity evident at Tell Nebi Mend was difficult to reconcile with the conventional history of the period, which emphasised a bipartite political structure of the Northern Levant; i.e. Phoenicians on the coast, Aramaeans inland. Published studies of Iron Age pottery only seemed to confirm the historical narrative's bisection of the region (e.g. Lehmann 1998, Fig. 14A-B; Mazzoni 1991-1992, Fig. 3).

The mixed influences evident in the Iron Age pottery from Tell Nebi Mend implied that the two-region model did not explain all aspects of material cultural patterning. Within this model, the conventional historical narrative, as derived from a literal reading of the documentary sources, was the main explanation given for the different cultural distribution patterns. In other words, archaeologists have tended to describe the regional patterns and then invoke the historical narrative as "explanation". Ceramic regionalisation, however, is not an explanation of social behaviour, simply its description, and a poor one. The two-region model implied a political and cultural homogeneity within the two broad regions, a homogeneity that was not evident at Tell Nebi Mend. This raises the question of whether Tell Nebi Mend was unique or whether such complex ceramic patterning might be a more widespread feature. The distribution of Iron Age pottery warrants further investigation.

1.2 Aims of the Present Study

The primary aim of the present study is to allow ceramic material culture of the Northern Levant to inform an archaeological reconstruction of Iron Age society. This is aimed at testing whether the discord between the material culture and the conventional historical narrative, as witnessed at Tell Nebi Mend, is symptomatic of much broader interpretative problems. To ascertain whether the two provide harmonious or conflicting reconstructions of Iron Age society, this study will review current reconstructions and undertake a region-wide study of ceramics, before comparing these two corpora of data. The conclusions of this study will provide an important platform from which future research can consider more meaningful interpretations of the archaeological data. In the process, this study will undertake four different tasks:

1. The first task will be to demonstrate how the conventional historical narrative has been used to determine interpretations of the archaeological record. To address these issues, the role that the historical narrative has played in defining archaeological reconstructions of the Iron Age Northern Levant will be explored and current archaeological definitions of the Iron Age will be challenged. While conventional reconstructions will be critiqued through a survey of the material cultural record, they will also be questioned on theoretical grounds. Central to this task is the recognition of alternative approaches to material culture, against which current practices can be measured.

2. The second task is the meaningful ordering of ceramic data. The aim is to develop a comprehensive and reliable ceramic typology for the Iron Age of the Northern Levant, in order that a systematic comparative study can be undertaken. It is important that the categories by which material culture will be measured and analysed are consistent and meaningful. This will be achieved through the collection and categorisation of a large corpus of Iron Age ceramic data from the study area.

3. The third task is essentially one of identification and description. It involves the identification of both broad and specific patterns in the distribution of Iron Age pottery across the Northern Levant. If the broad historical patterns are to be transcended, it is important that the information collected is interrogated in a thorough and systematic fashion. At this stage, the study will only describe these trends.

4. The fourth task is one of comparison. Its aim is to determine whether the observed ceramic patterning is compatible with the conventional historical narrative. This will involve the explanation of patterns in the data from a perspective that emphasises the role of material culture as an active social agent, rather than as a reflection/product of socio-political processes. Different aspects of the resulting interpretation will then be briefly compared with the conventional historical narrative to determine whether there is any correlation between the two.

1.3 Structure of the Present Study

SECTION ONE provides the background for the study as a whole by investigating the historical framework behind current reconstructions of the Iron Age Northern Levant. This section is comprised of three chapters. **Chapter 2** demonstrates that current reconstructions of Iron Age society bear little resemblance to the archaeology. It explores and challenges key concepts behind current interpretations and, in doing so, highlights their fragile foundations. A key component of this exercise will be to demonstrate how the historical narrative has been used to overdetermine interpretations of the archaeological record, thus highlighting the subordinate role of the latter within current reconstructions of Iron Age society. **Chapter 2** will also explore the role that European perspectives on the Near East played in the "entrenchment" of the historical narrative as the dominant interpretative framework. **Chapter 3** explores the impact of the historical interpretive framework on archaeological *practice* in the Northern Levant. This will take the form of a review of Iron Age excavations from within the study area, and an assessment of the methods used and conclusions drawn. While this exercise will emphasise the conclusions of **Chapter 2**, it will also lead into the discussion of **Chapter 4**, which focuses on the study of Iron Age pottery in the Levant. More specifically, **Chapter 4** explores the recurring themes in ceramic studies in the Near East and why these might be questioned on theoretical grounds. Emphasis will be given to the manner by which the historical narrative has penetrated an understanding of material culture. This will be followed by a brief review of ceramic studies in other areas of archaeological research, and how these demonstrate alternative ways for understanding the relationship between people, society, and material culture. **Chapter 4** concludes by addressing the implications that these alternative methods have for archaeological practice in the Levant.

SECTION TWO will present an overview of the data collected and collated for use in the present study. It consists of two chapters. **Chapter 5** recounts the means for the collection and ordering of the ceramic data, and ends with a brief overview of the main general trends and biases in the dataset. Also central to an understanding of the dataset is an appreciation of how the data was conceptualised, categorised, and manipulated ready for analysis. Hence, **Chapter 5** will discuss the computer applications used to store the data, and the structure of the data within those programs. A key consideration of this chapter is the imperfect nature of the data, which required the present study to rely upon presence/absence information. **Chapter 6** presents the ceramic typology used throughout this study and discusses any trends in distribution, decoration, and form, which are immediately apparent within the data. This chapter has a three-tiered structure: presenting the ceramic forms first according to functional categories, second by CLASS, and finally by sub-CLASS. The manner by which this typology was constructed is also discussed.

SECTION THREE consists of two chapters and is concerned with the analysis of the ceramic data. **Chapter 7** contains an exploratory analysis of the data. A number of different analytical techniques are used to explore the development over time and space of different variables in the data; e.g. context type, geography, decoration, and vessel function and form. The results of these simple comparisons and filtered searches will be presented in visual form. **Chapter 8** will continue the exploration of the data through the use of two multivariate analytical techniques well-suited to dealing with presence/absence data. Correspondence Analysis will investigate comparisons between different sub-sets of data based on associations within those sub-sets. Cluster Analysis will be employed to find similarities and dissimilarities between ceramic categories. The results of both techniques will be presented as charts and any patterns will be discussed. The results of **Chapters 7** and **8** will be used to inform the final interpretation of the ceramic data.

SECTION FOUR consists of **Chapter 9**, which discusses the socio-cultural implications of patterns detected in the data. This chapter will present an interpretation of the ceramic data that is not overly reliant on the conventional historical framework. Instead, the author aims to describe the significant patterns in the data and offer socio-cultural explanations for these phenomena, whilst appreciating the dynamic role that material culture played in the construction of social identities. The aim is to transcend the limitations in the data and allow the archaeological record to confirm or challenge the conventional history of the Iron Age Northern Levant. Consequently, **Chapter 9** will present an alternative reconstruction for material culture patterning, one that is not derived directly from the historical narrative, whilst also considering the historical implications of an alternate interpretation of the data.

1.4 Key Definitions in the Present Study

Before continuing, it is necessary to clarify some terms used in the present study. The term "Northern Levant" has been used to designate the northern half of the eastern Mediterranean, an area incorporating the Lebanese Republic; the western half of the Syrian Arab Republic, and the Hatay of the Turkish Republic. The term "Southern Levant" pertains to the southern half of the eastern Mediterranean, and includes the modern nations of Israel, Palestine, and the Hashemite Kingdom of Jordan. The geographical extent of the present study does not map perfectly onto either of these regions, and does not correspond with any modern borders. Instead, the study area was generously conceived so patterns in the data could reveal their own boundaries (**Map 2**).

The way in which the term "Iron Age" is used throughout the present study is in keeping with its conventional use. Hence, the use of the term encompasses two meanings: it designates a broad period of time following the conventional Late Bronze Age; and it describes the material culture complex of this period. While **Chapter 2** demonstrates that conventional definitions of the "Iron Age" are problematic, the abandonment of the term here might render discussion unnecessarily complicated for the reader.

SECTION ONE

Current Theory, Method and Practice in Reconstructions of the Iron Age Northern Levant

CHAPTER TWO

The Imposition of Predetermined Frameworks onto the Archaeological Record

2.1 Introduction

The purpose of this chapter is to "set the scene" for a reconsideration of the ceramic data. If current reconstructions of the Iron Age Northern Levant (hereafter IA-NL) are accurate reflections of the archaeological record then there is little need to progress any further – generally speaking, they are not. This chapter will demonstrate that the archaeological record does not always bear a close resemblance to conventional histories of the IA-NL. More to the point, it is the conventional historical record as derived from a literal interpretation of documentary sources that has had a significant influence on interpretations of the archaeological record. To this end, the chapter will challenge some of the basic interpretative concepts behind archaeological reconstructions of the Iron Age Levant and demonstrate the fragility of many of the interpretations and how they are often due to historiographic accident, European pre-conceptions, and an unquestioned reliance on the historical narrative. At the heart of the problem is the manner by which the historical narrative is frequently used to 'overdetermine' interpretations of material culture. In other words, the archaeological record has rarely been used to construct the chronological framework, but is instead employed to support the pre-existing historical narrative.

Part of the problem lies in the lack of meaningful units of analysis, which is explored in **Section 2.2**. Terms like "Iron Age" and "Syria" are either vague or inappropriate for a study of material culture from this period. **Section 2.3** discusses 'Iron Age' material culture with the view to isolating a more meaningful and appropriate means for defining this period. It shows that current chronological definitions, which are presumed to be based on the archaeological record, are not altogether reliable. **Sections 2.4** and **2.5** then explore some specific examples of how the conventional historical narrative has directly and indirectly impacted the archaeology of the IA-NL. In particular, **Section 2.5** focuses on the significant impact that the biblically-inspired chronology of the Iron Age Southern Levant (hereafter IA-SL) has had on Iron Age reconstructions of the entire eastern Mediterranean. Fault lies not in the text itself but in the particular view of textual histories that has informed archaeological interpretation. **Section 2.6** explores some of the biases and prejudices inherent in European reconstructions of the greater Near East. The chapter will then conclude with **Section 2.7,** which summarises the assumptions upon which the chronology of the IA-NL rests.

By highlighting the weak foundations upon which reconstructions of the IA-NL are based, it will be demonstrated that histories of the Iron Age are not constructed around immutable facts, but instead based on ideas formed within the framework of particular temporal, socio-political, and intellectual contexts. By showing that the historical narrative has long been the means for understanding the material culture of this region, this chapter is an essential precursor to alternative interpretations of the material culture from the IA-NL.

2.2 Arbitrary Units of Analysis

The aim of this section is to demonstrate how the archaeological record has played only a minor, supporting role in defining interpretative frameworks for the Northern Levant. Archaeologists have tended to use terms that hold little intrinsic meaning for the material record of the IA-NL. Indeed, there are a number of important concepts that are treated as self-evident and accepted with no, or very little, discussion. When brought under closer scrutiny, however, terms such as "Iron Age" and "Syria" lose much of their clarity and reveal subtle biases (Bunnens 2000a, 3-4). Neither "Syria" nor "Iron Age", as currently envisaged, is a well-defined unit of analysis. This section briefly explores the manner by which these categories, which are full of hidden meaning, have been clumsily applied to, or rather imposed onto the archaeological record of the IA-NL; i.e. material culture has not been used to help construct the interpretative framework. As a result, conventional reconstructions of the IA-NL do not always fit the archaeological data.

Syria: Land Without Borders

The term "Syria" is commonly used to define a geographic region, yet its precise meaning varies significantly throughout the literature (*cf.* Buccellati 1967, 11-12; Bunnens 2000a, 3; Klengel 2000a, n. 1). Moreover, a number of uses of the term are unclear (see the following titles, none of which define the use of the term: Albright 1934; 1935; 1936; Albright & Glueck 1937; Hogarth 1914b; Mallowan 1937; McEwan 1937; Olmstead 1931; Woolley 1914-1916; 1927). Indeed, because "Syria" does not have a well-defined meaning (except as a modern nation state), it is invested with new meaning every time it is used and, therefore, embodies any number of meanings. An appeal to the origins of the term does not clarify the situation. Etymological studies on the origin of the word have proved inconclusive (*cf.* Frye 1992; Lapointe 1970; Tvedtnes 1981). The original use of "Syria" has been credited to Greek authors of the eighth century BC (e.g. Herodotus *Histories* I.105, II.116, III.5), but its meaning in these texts is ambiguous. According to Frey (1992), the Greeks used "Syria" interchangeably with "Assyria", indicating that it did not incorporate a well-defined notion (Bunnens 2000a, 4). It is unclear whether the Greeks used the term to represent a cultural zone, an ethnic group, a well-defined geographic area, or a loosely-unified region. The lack of political unity and cultural cohesion in the eastern Mediterranean during this period only compounds the ambiguity (*cf.* Bunnens 2000a, 18; Chavalas 1992, 1; Klengel 1992, 17-18; S. Smith 1942, 88). Furthermore, no convenient term known from Near Eastern epigraphic evidence is available when dealing with the Northern Levant. "Syria" was a foreign term and, therefore, one that encompasses an external perspective. We do not know how the ancient inhabitants of the IA-NL referred to themselves or to their region. Whatever the Greeks imagined the term to mean, it is unlikely that it accurately represents the local population. This raises the question of whether "Syria" (whatever that means) is a meaningful category of study.

The use of the term "Syria" to define an ancient land is problematic. Not only is it ill-defined and ambiguous, but Said (1978, 1-6, *passim*) argues that the use of external terms like this embody expressions of power and dominance. For Said (1978, 197, *passim*) this practice is deeply rooted in Orientalist rhetoric, which he suggests incorporates a desire, conscious or not, to disavow the region's inhabitants of their past and present identities. But despite the vague meaning and possibly Orientalist nature, its archaeological use is rarely questioned.

The term "Syria" is largely avoided throughout the present study because it carries strong associations with the modern nation of Syria, as geographically defined by its modern borders. The present study uses the term "Northern Levant" to denote a loosely defined geographical region; an area roughly equivalent to the Republic of Lebanon, the western half of the Syrian Arab Republic, the northern reaches of the modern nation of Israel, and those regions of the Republic of Turkey that lie south of the Taurus Mountains (**Map 1**).

Iron Age: An Imported *Pre*-history

The present study is concernsed with the Iron Age, a period formally classified as the most advanced stage in tool-making technology (Trigger 1989, 73-79). The term "Iron Age" is widely used throughout the eastern Mediterranean, and has come to incorporate a number of meanings beyond its original metallurgical intention; it now holds chronological, cultural, and ceramic associations (Finkelstein 1996c, 107-108). In its original use, however, "Iron Age" was part of a paradigm developed for the classification of artefacts from Scandinavian *pre*history, a period generally devoid of textual data (Bahn 1996, 89; Daniel 1967, 90-109; Maisels 1993, 19; Schnapp 1996, 301; Trigger 1989, 73-79). The paradigm was an effective tool for classifying and ordering broad periods of time from the archaeological record alone.

Before periodisations in the Levant began using the European technological-evolutionary framework (Three-Age System), chronologies were constructed according to "ethnic" categories; such as Amorite, Jewish, Semitic, and Israelite, for the IA-SL (Finkelstein 1996c, 104; Silberman 1993c, 547), and Syro-Hittite/Late-Hittite (Braidwood 1937, 6; Krogman 1949, Tab. I; Woolley 1921a; 1952), Syrian (Weiss 1985), Phoenician (Buhl 1983, 110; Riis 1970, 12, 127), Aramaean (e.g. Pezard 1931; Seton Williams 1961, 70, 75), and Assyrian (Hachmann and Kuschke 1966, 124) for the IA-NL. While such terms should be rejected for defining material culture because they assume a direct correlation between ethnicity and material culture, they were abandoned for another reason. The terminology was too disparate; the variation in terminology prevented inter-site comparison of material culture, which made a systematic regional study impossible (Finkelstein 1996c, 104). An attempt was made to systematise Levantine terminology at a meeting in Jerusalem in 1922, when the European Three-Age System was formally adopted as the local periodisation (*PEFQS* 1923, 54-55), though there are indications that "Iron Age" was already being used by this time (Phythian Adams 1923, 66). The scheme was widely accepted following the publication of Albright's (1933; 1938; 1941-1943) Tell Beit Mirsim stratigraphy.

While "Iron Age" loosely corresponds to a period associated with the manufacture of iron tools, early use of the term in the Levant was concerned with its ability to easily distinguish the Israelite period from that of the Canaanite Bronze Age (e.g. Aharoni 1978, 153; M. Dothan 1985, *passim*). The twelfth and eleventh centuries BC, which were conventionally associated with the settlement of the Israelite tribes, were included within the Iron Age so that the cultural connection between the Israelite settlement and the Davidic Kingdom could be maintained (Finkelstein 1996c, 120). Furthermore, biblical references to the iron working capacity of the Philistines (I Samuel 13:19), who were believed to have settled in the region around the same time as the Israelite tribes, appeared to suggest that the migrating "Sea Peoples" were responsible for the introduction of iron-working technology in the Levant (Aharoni 1978, 156; Dothan 1982, 20; Drews 1993, 73; Ingholt 1942, 472; Lebeau 1983, 21; G.E. Wright 1939). Hence, scholars were able to justify the inclusion of the late second millennium BC within the period associated with the use of iron. In essence, the inclusion of these two centuries within the Iron Age was not decided on archaeological grounds but on historical considerations.

A paradigm that was originally concerned with the *pre*-historic periods of Europe had come to categorise Levantine history. Despite the term "Iron Age" being designed to represent an artefactual category, the prehistoric paradigm was applied through historical data, signalling a clear departure from the Three-Age System in everything but terminology. Only recently has this clumsy application of the paradigm been called into question (Finkelstein 1996c; Strange 2000).

The use of the term "Iron Age", as it was envisioned in the Southern Levant, was eventually adopted in the Northern Levant, though much later (Mesopotamian and Egyptian archaeology never adopted the Three-Age paradigm). The majority of early-twentieth century publications concerned with the IA-NL made no, or only infrequent use of the term (e.g. Braidwood 1937; Fugmann 1958; McEwan 1937, 8; Pezard 1931, 34; Riis 1948, 203-204; Thureau-Dangin *et al.* 1931a; Thureau-Dangin and Dunand 1936a; von Luschan 1902; Woolley 1914, 88; 1938a, 3; 1939b; 1952, 234-235). When "Iron Age" did appear, it was usually in reference to South Levantine sites, and was rarely used to describe North Levantine material culture. For instance, Pezard (1931, 34) used "*âge du fer*" in reference to other sites, but preferred the use of ethnonyms for his Tell Nebi Mend sequence (e.g. *niveau syro-phénicien*; *niveau syro-hitite*). It was not until the second half of the twentieth century that "Iron Age" was commonly used in the Northern Levant (e.g. S.V. Chapman 1972; Hachmann 1966, Abb. 24-25; Moorey, 1975, 108; Poppa 1978; Pritchard 1968; 1975; Saidah 1966; 1977; Thalmann 1978b, 71-89, Figs 46-47). While this survey is not exhaustive, and earlier isolated uses of the term in the Northern Levant are probable, the key point is that the term did not pass into common usage in the Northern Levant until three to four decades after it was adopted in the Southern Levant, and when it did, it was usually via a direct appeal to South Levantine chronology. Hence, the use of "Iron Age" in the Northern Levant holds little relevance to demonstrable patterns in the local archaeological record.

Summary

We have seen from the above discussion that the phrase "Iron Age Syria" is unlikely to be a meaningful unit of analysis. On the one hand, the use of "Syria" within archaeological research is ill-defined, and is applied in a varied and vague manner. On the other hand, "Iron Age" derives from an artefact-based, European prehistoric paradigm that has been clumsily applied to the Southern Levant through historical data and imposed onto the archaeological record of the Northern Levant. Both terms also embody deeper political overtones, as modern populations attempt to claim the region's cultural heritage and locate their own collective identity (Ben Yehuda 1995; 2002; S. Jones 1997, 9; Kletter 2006, 316; Whitelam 1996, 15). Indeed, it is difficult to avoid the identity politics of modern nation states in the region, where the past is often used in highly creative ways to establish continuity between the past, present, and future (Shanks and Tilley 1987a, 195; Silberman 1995, 261; Trigger 1984, 358).

The current study has generally avoided using the value-laden term "Syria", instead opting for the equally vague "Northern Levant" as defined above. For ease of discussion, "Iron Age" is used throughout, but without carrying any of the metallurgical or absolute chronological implications that it might usually encompass. Instead the use of the term "Iron Age" refers to a particular set of cultural horizon conventionally associated with that period.

2.3 Key Elements in Archaeological Definitions of the Iron Age Northern Levant

The term "Iron Age" came to be used in the Northern Levant largely through interregional comparative studies of material culture with the Southern Levant, where the distinction between Bronze and Iron Age was influenced by modern concerns (Whitelam 1996, *passim*). In establishing the parameters of this period, the key issue appears to have been the separation of the history of the Canaanites from that of the Israelites (as depicted in the biblical narrative), whilst maintaining a link between the Israelite tribes and the United Monarchy. In other words, archaeology was used to confirm the already established historical framework. Consequently, Iron Age history of the Northern Levant does not accurately reflect the archaeological record. While the historical narrative implies some distinction between the Bronze and Iron Ages, the archaeological record is not as clear. If one were to consult a standard text on the archaeology of the eastern Mediterranean, a number of phenomena are depicted as indicative of the early Iron Age (e.g. Akkermans and Schwartz 2003, 360-361). Closer scrutiny of the archaeological record, however, reveals that many of these phenomena are not indicative of the early Iron Age; some are clearly a continuation of Late Bronze Age developments while others can only be associated with the later Iron Age. The following section explores these phenomena.

- "Sea Peoples" migration
- New ethnic populations
- Break in material culture
- Change in political-systems
- Development of the alphabet
- Emergence of private enterprise
- Cremation of the dead
- New monumental style of architecture
- Iron as a working metal
- Use of domesticated camel caravans

"Sea Peoples" Migration

Around the close of the thirteenth century BC, when conventional chronologies place the transition of the Late Bronze and Iron Ages (e.g. Buhl 1992, 34; Fugmann 1958, 267; Mazzoni 1990a), the eastern Mediterranean world experienced a period of political and economic instability (Liverani 1987, 69-70). Known as the "Crisis Years", it was during this period that the political systems of the Bronze Age suffered decline and collapse (Klengel 2000a; Ward and Joukowsky 1992). The appearance of "Sea Peoples" in Egyptian texts at around this time has led many scholars to connect the two phenomena (Betancourt 2000, 297; Dothan 1982, 1-13; Lipinski 2000a, 25; Thomas 1967, 65), attributing the collapse of Bronze Age society to the invasion of the "Sea Peoples" (Kuhrt 1995, 386). Indeed, conventional histories tend to emphasise the role of violent migration as the primary cause of collapse (e.g. Courbin 1990b, 503; Klengel 2000a, 23; Lipinski 2000b, 125; Mazzoni 2000a, 31; Pritchard 1968, 99; Woolley 1921a, 48; 1948). The development of the "violent migration" theory coincided with the discovery of destruction levels at a number of sites across the eastern Mediterranean, most of which appeared to coincide with the late-second millennium BC (*ibid*). Widespread destruction, therefore, became a convenient archaeological indicator for the beginning of the Iron Age (e.g. Badre 2006, 93; Hamilton 1934, 77; Ingholt 1942, 472).

The bases for the "violent migration" theory mainly derive from Egypt texts: an inscription of Merneptah in the Temple of Karnak at Luxor, the text of Ramesses III in his mortuary temple at Medinet Habu, and the Papyrus Harris I (Kuhrt 1995, 384-393). Together these texts detail "wars" against a confederation of "Sea Peoples" (Drews 1993, 48-61). The "violent migration" theory also implies that the populations of the Iron Age Levant, at least within the coastal centres, were of a different ethno-political nature to those of the Late Bronze Age (e.g. Fugmann 1958, 275). In particular, the Iron Age population was considered a derivative of Late Bronze Age Aegean cultures (Bell 2006, 15). This has in turn led to great emphasis being placed on the apparent Aegean origin of early Iron Age cultural elements; such as cremation or sub-Mycenaean pottery (e.g. Dothan 1982, 94, 219, 252; 1998; Dothan and Dothan 1992, 159-170; Killebrew 2005, 14; *cf.* Sherratt 1998, 293).

Despite the frequency by which the "Sea Peoples" migration theory has been linked to various destruction levels, the correlation is not straightforward. The "violent migration" theory ignores strong cultural continuity between the two periods (see below), something that is unlikely to occur with the large-scale settlement of a new population, and instead focuses on archaeologically attestable destruction; however one expects that to manifest itself (**§2.4**). For instance, a significant number of Levantine sites have evidence of destruction at the end of the Bronze Age but attest to continuity in material culture despite the destruction levels (e.g. Ras Ibn Hani - Alalakh and Ugarit are obvious exceptions). Along the Lebanese coast, where the effects of invasion by sea would be most acutely felt, continuity in settlement occupation is markedly well-attested (Blaylock 1999, 265; Courbin 1990b, 503; Riis 1970, 40; S. Smith 1942, 90). In fact, the coastal sites of southern Lebanon bear little evidence of destruction (Akkermans and Schwartz 2003, 361; Anderson 1988, 424; Bell 2006, 12; Bikai 1978b, 14-15, 56, 73-75; S. Smith 1942, 90). The archaeological record does not offer indisputable support for the violent 'Sea Peoples' invasion theory; which begs the question – without the "Sea Peoples" texts, would people be searching for twelfth century BC destruction levels?

The thesis that a violent migration of "Sea Peoples" occurred around 1200 BC is supposedly based on Egyptian texts, yet the inscriptions themselves make no direct claim for migration. Hence, we might conclude that the hypothesis is based not on the Egyptian texts but on a particular interpretation of them (Drews 1993, 48-61). The earliest of the three texts comes from the eastern wall of the main Karnak Temple, and commemorates the great victory of Merneptah over Libyan invaders and their allies in the Nile Delta. It reads:

> *Beginning of the victory which his majesty achieved in the land of Libya..., Ekwesh, Teresh, Luka, Sherden, Sheklesh, Northerners coming from all lands (Breasted 1906a, no. 574)*

When this inscription was first read in the nineteenth century, Egyptologists identified the Libyans' allies with regions of the northern Mediterranean: the *Luka* were identified with Lycia, the *Ekwesh* with Achaea, the *Teresh* with Tyrrhenia, *Sheklesh* with Sicily, and the *Sherden* with Sardinia (Drews 1993, 49-50; Kuhrt 1995, 386-393). There is no indication in the text, however, that any of the "northern" contingents, however they are identified, were migrating. On the contrary, a latter part of this text lists very low numbers of casualties amongst the Libyan king's auxiliaries (compared to actual Libyan casualties), suggesting that they were instead mercenary contingents (Kuhrt 1995, 387; Maspero 1896, 432; Schachermeyr 1982, 41-43). There is nothing here to suggest a violent mass migration of northern Mediterranean groups.

A second inscription comes from Ramesses III's mortuary temple at Medinet Habu, where the text accompanies large reliefs depicting Ramesses' "War" with the "Sea Peoples". Before discussing the text, it is worth noting the context of this account. The Medinet Habu temple contains inscriptions and reliefs commemorating every victory with which Ramesses III could conceivably be credited (Drews 1993, 50). Some of these are obviously not true claims of victory, but literary devices for linking this king with those that came before; Ramesses clearly borrows a number of victories from earlier kings (e.g. Ramesses II's victory over the Hittites – *ibid*). For Baines (1996, 347-351) the creation of monuments like Medinet Habu was more about maintaining the "Great Tradition" of propaganda, than recording an accurate history of events. Moreover, Baines (1996, 363-371) suggests that reliefs concerned with foreigners are usually the most propagandistic. This is a particularly important point if we also accept his (Baines *ibid*, 363) thesis that the Egyptian New Kingdom Temple was a model representation of the cosmos. In this model, the temple's interior was associated with the king, Egypt, and order, while its exterior depicted the king's dominion over the foreign, chaotic world. Viewed from this perspective, the Medinet Habu portrayal of the "Sea Peoples" as foreign ethnic groups is characteristic of an Egyptian concern with the rejection and control (through classification) of the non-Egyptian world (Baines 1996, 377). Such a device would also recall the expulsion of the foreign Hyksos rulers from Egypt, who were demonised long after their disappearance (Baines 1996, 378, n. 129). Hence, the Medinet Habu reliefs may not record specific events, but rather an ideological statement of dominance and control, aimed at the consolidation of Egyptian identity.

Having established the ideological context, we may begin to explore the Medinet Habu text. The particular paragraph thought to attest to the violent migration of "Sea Peoples" is the following:

> *The foreign countries made a conspiracy in their islands (rww). All at once the lands removed and scattered in the fray. No land could stand before their arms, from Hatti, Kode, Carchemish, Arzawa, and Alashiya on, being cut off at [one time]. A camp [was set up] in one place in Amor. They desolated its people, and its land was like that which has never come into being. They were coming forward toward Egypt, while the flame was prepared before them. Their confederation was the Peleset, Tjeker, Shekleshesh, Denyen, and Weshesh, lands united. They laid their hands upon the lands as far as the circuit of the earth, their hearts confident and trusting; "Our plans will succeed!" (Wilson 1969a, 262 – emphasis his)*

The basis for the violent migration theory rests on a single sentence within this paragraph, and more specifically one word within this sentence. The word *rww* is usually translated as 'islands' or 'isles' (Edgerton and Wilson 1936, Pl. 37-39), implying that the foreigners originally derived from islands. However, Nibbi (1975, 48, 65) has demonstrated that this word is also frequently used to refer to continental coasts. When this reading of *rww* is combined with the fact that the *peleset* and *tjeker* were frequently referred to as "Asiatics" in Egyptian texts (Edgerton and Wilson 1936, Pl. 31, 43, 44, 46), there is no reason to locate their lands outside of the eastern Mediterranean. Hence, the Medinet Habu text may be recounting a conflict between Egypt and her immediate Levantine neighbours, and may have little to do with a mass migration of northern Mediterranean people groups.

The third and final Egyptian text dealing with the "Sea Peoples" is Papyrus Harris I.

> *I extended all the frontiers of Egypt and overthrew those who had attacked them from their lands. I slew the Denyen in their islands, while the Tjeker and the Philistines were made ashes. The Sherden and the Weshesh of the Sea were made nonexistent, captured altogether and brought in captivity to Egypt like the sands of the shore. I settled them in strongholds, bound in my name. Their military classes were as numerous as hundred-thousands. I assigned portions for them all with clothing and provisions from the treasuries and granaries ... (Wilson 1969f, 262)*

According to Kuhrt (1995, 389) and Van Seters (1997, 177) the text of the Papyrus Harris was composed to illustrate the piety and virtue of Ramesses III. For this purpose, the king is depicted as the bringer of peace and conqueror of Egypt's traditional foes. Key to an understanding of this text is an awareness that it was written during a period of political turmoil (Ramesses III had just died), when Ramesses IV sought to legitimise his claim to the throne (Van Seters 1997, 177-178; J.A. Wilson 1969f, 262, n.18). The historical survey of Ramesses III's "great achievements" is highly-selective and is not presented in any chronological order. Van Seters (1997, 177-178) suggests that rather than accept the historicity of this document, it can be understood as an ideological statement regarding the legitimacy of the royal line. To expound a history of the eastern Mediterranean from this text demonstrates a lack of sensitivity to the kind of message Ramesses IV was trying to convey.

The above Egyptian texts are not sufficiently reliable to form the keystone of historical reconstructions. They do not directly attest to migration, violent or not, at the end of the Late Bronze Age. It seems that the role of the "Sea Peoples" has been over-emphasised (Bauer 1998, 151). Consequently, alternate explanations for the Egyptian "Sea Peoples" texts have been presented by an increasing number of scholars (e.g. Bauer 1998; Drews 1993; Sherratt 1998, 292-293). In my opinion, the most convincing theory is that of Sherratt (1998, 292-293), who has suggested that the "Sea Peoples" concept does not represent a specific group but rather the emergence of decentralised trading following a regional economic crisis. Accordingly, Sherratt (*ibid*) suggests that the Egyptians texts can be read as their efforts to explain the crisis in terms of their own world view; the vague geographical locations in the text may indicate the decentralised nature of this maritime phenomenon, which was probably a difficult concept for highly centralised Egyptian society to understand.

New Ethnic Populations

As we have already seen, the beginning of the Iron Age is conventionally associated with widespread upheaval and political crisis. While the "Sea Peoples" are often cited as the main component within this turmoil, the appearance of other ethnically-distinct peoples have been associated with the "Crisis Years" (Albright 1975, 516-517; Bunnens 1999, 605; Caubet 1992; Klengel 1992, 181; Sader 1992, 161; 2000; Schwartz 1989, Thomas 1967, 68); the few that concern a study of the IA-NL are the "Aramaeans", "Phoenicians", and "Neo-Hittites". The identification of new Iron Age ethnic-groups is based on the recognition of seemingly new languages in the early Iron Age, a time also associated with the abandonment of languages associated with the use of cuneiform. The apparent change in written language has led some scholars to emphasize a break in cultural traditions and equate this with the immigration and settlement of a new ethnic population (Albright 1975, 529-536; Eidem and Ackermann 1999, 315; Klengel 1992, 187; Lipinski 2000a, 25; Peckham 2001, 19, 21; Sader 2000, 64-68). I have three concerns with this approach: the correlation between language and identity is over-simplistic; the association of written language with spoken language cannot be assumed; and these languages and ethnonyms are not new to the Iron Age.

The equation of a newly visible language with a new population is, for a number of reasons, a simplistic view of the relationship between language and identity (S. Jones 1997, 106-110). Language does not always play a role in the construction of group identity (*cf.* S. Schwartz 1995, 3; Sherratt 2003a, 231-233). According to Sherratt (*ibid*), Iron Age Greeks provide the earliest evidence for the elevation of language into a major focus of ethnic identity, implying that other ethnic affiliations at this time were rare. There is no evidence to suggest early Iron Age populations of the Northern Levant sought to establish ethnic identity through the use of written language. On the contrary, many sites show evidence of multiple language use within the literate elements of the population (Aro 2003, 282; Bryce 2003, 124; Bunnens 1999, 614; 2006, 97; Klengel 1992, 187, 193; 2000a, 27; Lipinski 2000a, 234-235, 239; Melchert 2003a, 2-3). At Zincirli, for instance, the local rulers bore Luwian as well as Aramaean names and used both languages in their inscriptions (Bordreuil 1993, 254; Dalley 2000, 80). Written language is not a definitive indicator of ethnicity.

Written language is not always the same language as that being spoken. For instance, Akkadian, used for record-keeping across a vast area of the ancient Near East during the Bronze Age, was used alongside local literary traditions (Liverani 1990, 14). While alphabetic writing is the physical representation of speech and, therefore, is closely related to spoken language, it does not necessarily represent the earliest occurrence of the language being spoken. For instance, Aramaic is generally associated with the Iron Age, but a number of Aramaic personal names are evident within the Bronze Age archives of Mari, Alalakh, and Ugarit (see Gröndahl 1967, 10-202; Malamat 1973, 134). The visibility of a language is not directly linked to the speaking of that language, but is instead affected by the writing systems and media employed by that society (Sherratt 2003a, 232-233). Writing is not something that people automatically embrace just because they have become aware of the possibility and have encountered the technology (*ibid*), the cultural conditions have to be right. In other words, written language is more a cultural choice than an ethnic indicator (Bunnens 1999, 614-615). In the early Iron Age, alphabetic scripts were widely adopted because there was a perceived need; possibly to differentiate emerging sub-elites from Bronze Age palace-elites, or simply to facilitate mercantile activity (*cf.* Riis 1970, 174; Sherratt 2003a, 230). It is important to remember that a large percentage of Iron Age hieroglyphic Luwian texts were inscriptions detailing the exploits of Iron Age rulers, and were inscribed on stone for posterity (Klengel 1992, 187). Such texts were written down for very specific reasons; the choice of written language entails reasons beyond an expression of personal speech or well-defined ethnicity; indeed, language may be one of the means for negotiating or challenging established identities (Bryce 2003, 125; Melchert 2003b, 13). Ultimately, the archaeological record is not a definitive record of the appearance or disappearance of language, but rather the archaeological *visibility* of that language.

Despite Aramaic, Luwian and Phoenician generally being associated with the Iron Age, there is evidence to suggest that they were known in the Late Bronze Age. While the first documented appearance of the Aramaic *language* in written form occurs in the Iron I period (Aramaic names are known from Late Bronze Age Ugarit – Gröndahl 1967, 10-202), there are much earlier references (fourteenth and thirteenth centuries BC) to nomadic people known as Aramaeans (Bunnens 1999, 606, 610-611; Lipinski 2000a, 45-50; 2000b, 132; O'Callaghan 1948, 95; Salvensen 1998, 139). In other words, the archaeological appearance of Aramaic does not coincide with the appearance of Aramaeans, however they were identified (Klengel 2000a, 26). In addition, Hieroglyphic Luwian was already in use in the Northern Levant during the Late Bronze Age (Bryce 2003, 84-88)., and the Phoenician language appears to be directly related to Late Bronze Age Canaanite; neither were not completely new developments within the Iron Age (Isserlin 1982, 804).

Break in Material Culture

Chronological reconstructions of the IA-NL are largely influenced by South Levantine archaeology, where the biblical narrative equates the beginning of the Iron Age with changes in population associated with the "Sea Peoples" and the settlement of the Israelite tribes (Aharoni 1978, 153). Hence conventional chronologies across the Levant place significant emphasis on change at the end of the thirteenth century BC (Ben Ami 2001, 160; Buhl 1992, 34; Sader 2000, 61), but there were also significant continuities between the Bronze and Iron Ages (Fritz 2000, 507; Mazar 1992, 296; Mazzoni 2000a, 31-33; 2000d, 1043). Many aspects of early Iron Age material culture are very similar to those from the Late Bronze Age (e.g. Tell Afis - Bonatz 1993, 134-135; **Table 2.1**). In particular, there is minimal change evident across the conventional Bronze-Iron Age transition in ceramic culture, domestic architectural traditions, and metallurgical technology.

Table 2.1: Summary of LB/Iron Transition in the Northern Levant

Site	LB	Destr.	Iron I	Cont.	Reference
Tille Hoyuk	Yes	Yes	Yes	Yes	Summers 1993, 3
Jerablus	Yes	Yes	Yes	Yes	Hawkins 1974, 70
Tell Rifa'at	Yes	Yes	Yes	Yes	Seton-Williams 1961, 75, 82
Alalakh	Yes	Yes	No	-	Woolley 1955, 398
Ras al Bassit	Yes	Yes	Yes	Yes	Courbin 1983, 122
Ugarit	Yes	Yes	No	-	Yon 1992
Ras Ibn Hani	Yes	Yes	Yes	Yes	Badre 1983, 206
Tell Sukas	Yes	Yes	Yes	Yes	Lund 1986, 41, 188
Tell Kazel	Yes	Yes	Yes	Yes	Badre 2006, 93
Tell Afis	Yes	Yes	Yes	Yes	Venturi 1998a, 135
Hama	Yes	No	Yes	Yes	Fugmann 1958, 267
Sarepta	Yes	No	Yes	Yes	Anderson 1988, 380, 390
Tyre	Yes	No	Yes	Yes	Bikai 1978b, 73
Kamid el Loz	Yes	No	Yes	Yes	Hachmann 1989, 35, 44, 52-54
Tell Keisan	Yes	Yes	Yes	Yes	Dever 1997a, 278
Tell Abu Hawam	Yes	Yes	Yes	Yes	Hamilton 1935, 66
Tel Dan	Yes	No	Yes	Yes	Biran 1994, 126
Hazor	Yes	Yes	Yes	✘	Ben Ami 2001, 160
Megiddo	Yes	No	Yes	Yes	Lamon and Shipton 1939, 7 Finkelstein *et al.* 2006, 848
Pella	Yes	Yes	Yes	Yes	Smith and Potts 1992, 83
Tel Rehov	Yes	No	Yes	Yes	Mazar 1999a, 38

While continuity in material culture has often been noted by excavators, it rarely influences site interpretations. For instance, Fugmann (1958, 267, 274) remarked on the architectural continuity between Hama Stratum G (Bronze Age) and Stratum F (early Iron Age), with no observable difference or interruption evident in the archaeological record, yet he still associated Hama F with the settlement of a new "Sea Peoples" population responsible for the introduction of cremation.

Change in Political-Systems

Associated with the "Crisis Years" is a general acceptance of a collapse of Late Bronze Age culture and, in particular, the political structures of that period (Hallo 1992, 1-3). As a result, archaeological discussions of the Bronze-Iron Age transition have emphasised the disappearance of the Late Bronze Age palaces and ruling elites, and the development in the private sector of activities that were once palace-controlled, such as writing and trade (e.g. Akkermans and Schwartz 2003, 361; Hallo 1992, 3; Liverani 1987, 69; Sader 1992, 158). The archaeological record, however, is not in total agreement with this interpretation. While some sites do provide evidence for the destruction of large, palace-like buildings (e.g. Ugarit, Alalakh, Ras Ibn Hani – Margueron 1985, 152-155), there are sites that instead testify to political continuity. The clearest example of this is the city of Carchemish (modern Jerablus), where a large elite complex survived the Late Bronze Age "crisis" and persisted into the Iron Age (Klengel 1992, 182-183).

Following its conquest by the Hittites (c. 1350 BC), Carchemish was a seat of the Hittite royal family and an administrative control point for the Hittites in the Northern Levant. When Carchemish survived the collapse of the Hittite empire, it retained its dynasty of local rulers with ties to the Hittite royal house, ties that were not only maintained in the Iron Age, but *emphasised* (Akkermans and Schwartz 2003, 360; Klengel 1992, 193). The Carchemish rulers maintained forms of architecture, monumental art, and iconography that were closely related to those of the Hittite empire (Bunnens 1999, 612). Just as significant, however, is the fact that Iron Age Carchemish maintained a dynastic political structure. The names of the Iron Age dynasts emulated the names of Hittite kings, which included the use of the titles 'Great King' and 'Hero', previously the exclusive privilege of Hittite kings (Caubet 2003, 18; Hawkins 1988, 104-108; Klengel 2000a, 27). Foreign powers even referred to Carchemish as the "land of the Hittites" (e.g. Tiglath-Pileser I - Grayson 1991a, texts A.0.87.1, A.0.87.2, A.0.87.3, A.0.87.4, A.0.87.12). The Carchemish dynasts claimed to be direct descendants of true Hittite imperial power, and their dynastic succession was supported by the site's distinctively Hittite artistic iconography and royal titles. According to Woolley (1914, 98), the practice of cremation at Carchemish was maintained in an effort to mirror the Hittite tradition of cremating kings. Even the language used in declarations of dynastic legitimacy (hieroglyphic Luwian) is an indigenous Anatolian construction dating back to the Hittite empire (Akkermans and Schwartz 2003, 360; Klengel 1992, 193; Woolley 1913, 97). Despite whatever the inscriptions of Ramesses III say about the destruction of Carchemish by the "Sea Peoples" (J.A. Wilson 1969a,

262; Woolley 1952, 226, 235), there is clearly a direct link between the Bronze Age and Iron Age political structure at Carchemish (Hawkins 1974, 70; 1982, 372).

In addition to the survival of the political structures at Carchemish, the names of the so-called "Aramaean" kingdoms of the early Iron Age (e.g. *bit Agusi* = house of Agusi) imply a tribal structure; i.e. a political system centred on familial ties or dynasties (Akkermans and Schwartz 2003, 367; Sader 1992, 159-160). Furthermore, the sites on the southern Lebanese coast appear to have suffered little disruption to their trading activities across the Bronze-Iron Age transition (Bell 2006, 95-100). While Albright (1975, 518-519) assumed that these sites were destroyed by the "Sea Peoples" but were the first to recover, Bell (2006, *passim*) has recently demonstrated that there is little evidence for destruction in this region, which instead thrived during the "Crisis Years". This might be due to the presence of a political system there that was no longer centred around the palace, and therefore more flexible and able to overcome the decentralising chaos of the "Crisis Years". Regardless, there is little reason to believe that the early Iron Age political structure along the Lebanese coast was substantially any different to that of the Late Bronze Age. While terminal collapse of a palace-based economy and political administration is evident at a few North Levantine sites (e.g. Alalakh, Ugarit), it is certainly not represented at all sites of the Northern Levant; the "Sea Peoples" invasion model appears to have overdetermined interpretations of the archaeological record.

The Development of the Alphabet

The adoption of the alphabetic-script is conventionally associated with the early Iron Age and the collapse of Late Bronze Age society (e.g. Akkermans and Schwartz 2003, 360-361). This argument implies that the alphabet was simply filling the lacuna left by the abandonment of palatial scribal traditions, a void partly due to the disappearance of the elite patrons of the scribes, and partly to the diplomatic language of Akkadian no longer being necessary (Hawkins 1982, 381; Sader 1992, *passim*). While most scholars would agree that the long established cuneiform writing systems largely disappeared from the Levant around the time of the "Crisis Years", few have addressed the seemingly 'magical' appearance of a fully developed alphabetic writing system.

An explanation for the development of the alphabet instead lies in the archaeological record of the Late Bronze Age (Isserlin 1982). The rich textual record of Late Bronze Age Ugarit contains around 1800 syllabically written texts (Akkadian, Sumerian, Hittite, and Hurrian) and more than 1900 texts of alphabetic Ugaritic (Bordreuil and Pardee 1989; Schloen 2001, 206). About 130 tablets have also been found at the neighbouring Late Bronze Age palatial centre of Ras Ibn Hani, most of which are written in alphabetic Ugaritic (Bordreuil and Caquot 1979; Bordreuil and Pardee 1995, 29; Bounni *et al.* 1998, 91). In fact, alphabetic texts have been recovered from a number of Late Bronze Age contexts throughout the eastern Mediterranean (Albright 1964b; Hillers 1964; Isserlin 1982, 799-804; Millard 1976; Pritchard 1975, 102*ff*; Riis 1970, 174). While Ugarit's archival record confirms that the alphabetic script pre-dates the Iron Age (it appears as early as the fourteenth century BC – Isserlin 1982, 802; Sznycer 1975), it also demonstrates that the two script types (syllabic and alphabetic) co-existed. Clearly, the alphabet is not an Iron Age innovation, but is based upon a principle already present in the mid-second millennium.

From an analysis of subject matter contained within the syllabic and alphabetic texts of Ugarit, Schloen (2001, 206) has demonstrated that alphabetic scripts at this site were used to record economic transactions (**Chart 2.1**). This simplified script, which was essentially a cuneiform shorthand, appears to have developed as a means for facilitating mercantile intercourse. Those merchants operating outside the realm of palatially-administered trade (see below) did not have access to a scribal resource, and consequently employed whatever form of written communication that was available to them. The development of the alphabet appears to be linked to entrepreneurial trade; if so, this would explain the marked increase in the use of alphabetic scripts in the period following the disappearance of the Late Bronze Age palatial scribes (Bell 2006, 17-19). These developments may also explain the presence of a wooden diptych (associated with cursive writing) on the Ulu Burun ship at a time when Akkadian cuneiform was still the "official" language of trade (Bass 1987, 731; Bass *et al.* 1989, Fig. 19). Papasavvas (2003) has suggested that styli found in Late Cypriot urban centres attest to the use of similar waxed wooden tablets as a writing medium.

In contrast to the comparative richness of the Late Bronze Age textual record, the early Iron Age is poorly represented. As a result, some scholars have come to consider the early Iron Age as a type of "dark age" (*cf.* Klengel 1992, 181; 2000a, 21; Lipinski 2000b, 125; Liverani 1987, 71; Pitard 1987, 81; Venturi 2000c). The problem, however, is not one of loss of literacy, but the loss of archaeological visibility of literacy, possibly due to the use of perishable writing materials associated with a more flexible writing system (Dornemann 2003a, 7; Klengel 1992, 181; 2000a, 25). Free from the cumbersome but durable methods of cuneiform, alphabetic writing could be used on a number of new media, some of which - papyrus, wax on wood, leather – are not able to survive the climate of the eastern Mediterranean well (Anastasio *et al.* 2004, 18).

Chart 2.1: Constituents of Ugarit's Alphabetic and Syllabic Texts

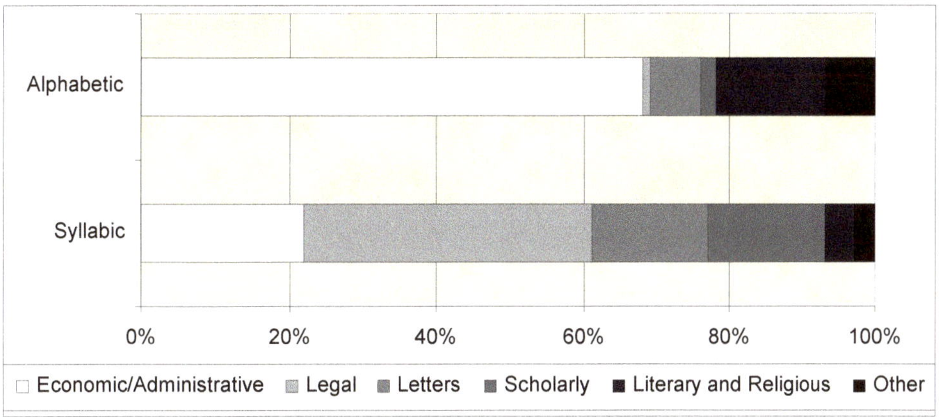

(After Schloen 2001, 206 n.3)

Emergence of Private Enterprise

Another innovation often associated with the Iron Age is the development of private enterprise. For instance, Liverani (1997b, 562) has argued that early Iron Age merchants, who were formerly palace-dependent, were acting for themselves for the first time. In this model, trade had shifted from being an administered process during the Late Bronze Age to an entrepreneurial one in the Iron Age (Liverani 1987, 72; 2003, 128-133). Sherratt and Sherratt (1991, *passim*; 1993, *passim*) viewed this process as one of privatisation, as the people involved in trade changed from being state-controlled to private merchants. There is evidence within the Ugarit archives, however, to suggest that a significant level of trade was undertaken outside of palace controls during the Late Bronze Age (Bell 2006, 19). In fact, the Late Bronze Age texts discuss private enterprise at Ugarit in some detail, with even the names of prosperous individual entrepreneurs known; e.g. Rapanu, Yabninu, Urtenu (Bell 2006, 65-67). Heltzer (1969, 35; 1976; 1978; 1982; 1996) has concluded from the Ugarit archives that both administered and entrepreneurial trade were present in Late Bronze Age Ugarit, with both sectors operating alongside each other.

Sherratt (2003b, 48-50) has also suggested that a number of Levantine coastal cities were already operating outside of palatial controls before the "Crisis Years". In particular, the Lebanese coast had been operating free from Egyptian imperial demands since the death of Ramesses II, and by the beginning of the twelfth century BC had been functioning independently for some decades (*ibid*). At the time of crisis, these cities of southern Lebanon escaped destruction and were able to quickly prosper following Ugarit's demise (Bell 2006, 92; Klengel 2000a, 24; Peckham 2001, 21). At the close of the Bronze Age, those communities with a more decentralised mercantile mechanism were able to continue doing what they were already doing at the end of the Late Bronze Age. This is evident in the continuity of trade between the merchants of Phoenicia and Cyprus, both of which had a long history of engagement in maritime trade (Bell 2006, 95-100; Bikai 1983; 1987; Gilboa and Sharon 2003, 51-55). Private enterprise is not a new phenomenon in the Iron Age, but it does appear to intensify in this period. Free from the monopolising control of the palace economy, the entrepreneurial spirit that was already present in the Late Bronze Age developed and flourished in the Iron Age (Liverani 2003, 128).

Cremation of the Dead

For many scholars, the cremation of the dead in the Levant is unique to the Iron Age, supposedly introduced from the Aegean by the invading "Sea Peoples" (e.g. Barnett 1975, 14; Buhl 1992, 34; Culican 1973, 67; Ingholt 1942, 472; Johns 1938, 121). However, to suggest that a change in burial practices equates to a new population is simplistic, especially when other explanations are possible, such as economy of space, hygiene, changes in ideology (*cf.* Doumet-Serhal 2004b, 73; Courbin 1993a, 104). Indeed, under closer scrutiny the evidence does not support the "Sea Peoples" hypothesis. Contrary to conventional thought (e.g. Doumet-Serhal 2004b, 72-73), there is evidence to suggest that cremation was practiced during the Late Bronze Age at Ugarit, Alalakh, Sukas, Carchemish and Hama, and was, therefore, known in the Levant prior to possible "Sea Peoples" contact (Bienkowski 1982, 80-82; Courbin 1993a, 104-109; Gilmour 1995, 167; Mazzoni 2000a, 35; Prausnitz 1982, 35-36; Riis 1948, 192-203; 1961, 140-141; Woolley 1914, 98; 1952, 225; 1955, 201-203). Furthermore, cremation was not an exclusively Bronze Age Aegean rite, as suggested by some scholars (e.g. Fugmann 1958, 275; Ingholt 1942, 472), but was practised widely in Bronze Age Anatolia (Courbin 1993a, 104; Gaal 1976). Indeed, cremation only became the dominant burial rite in the Aegean during the early Iron Age (Dickinson 2006, 184-195). Moreover, the earliest documented examples of Iron Age cremation in the Levant are not from the coast, as would be expected with a "Sea Peoples" introduction, but were encountered across the inland Northern Levant, at sites such as Carchemish, Hama, and Tell Halaf (Mazzoni 2000a, 35).

While cremation was the distinctive burial practice on the coast in the Iron II period: e.g. Tyre al Bass, Tell Rachidieh, Tambourit, Khalde, Ras al Bassit, and Akhziv (Doumet-Serhal 2004b, 73; E. Mazar 2001, 10; Saidah 1966; 1977), it never became the sole burial practice there; pit, chamber, and tomb inhumations continued to be used alongside cremation for much of the Iron Age. In the end, there is little reason to perpetuate the association of cremation with invading "Sea Peoples" or even as being representative of the Iron Age Levant. Instead, cremation can be said to bear witness to the native character of the Iron Age population.

New Monumental Art and Architecture

The distinctive style of monumental art and architecture of the inland Northern Levant is widely considered characteristic of the Iron Age, as frequently cited as evidence for the presence of "Neo-Hittites" (also "Syro-Hittites") or "Aramaeans" (e.g. Akkermans and Schwartz 2003, 367; Albright 1975, 526-529). Like in other areas of material culture, this style of architecture displays clear continuity with Bronze Age traditions. The heavily fortified sites of Zincirli, Carchemish, Tell Halaf, Ain Dara, Tell Ta'yinat, Aleppo, and Hama, with their monumental gateways and large public buildings (palaces, temples), are often lavishly decorated with an artistic style that draws heavily on the conventions and motifs of Late Bronze Age Imperial Hittite art (Abou Assaf 1985, 347-350; Akkermans and Schwartz 2003, 367; Akurgal 1962, 127-130; Aro 2003, 298-307; Klenegl 1992, 193; Kohlmeyer 2000, 8-11; Mazzoni 2001, 101; Mellink 1974; Pucci 2008b; *contra* Fugmann 1958, 268). The use of guardian figures (lions or sphinxes) at gates, carved orthostats lining the base of walls, and specific iconographic details are all typical of earlier Anatolian traditions (Mellink 1974; Aro 2003, 307-337; Mazzoni 2000d, 1044-1045). While this Hittite-derivative style borrowed heavily from Bronze Age traditions, its appearance across the Northern Levant is only documented during the Iron Age, though the chronology is debated.

The chronology of "Neo-Hittite" art was first outlined by Akurgal (1962; *cf.* Orthmann 1971) who discerned three separate phases in stylistic development (**Table 2.2**). The first phase is defined as the "Traditional Style" ("Early Neo-Hittite Phase"), which he concluded was the perpetuation of "Hittite art which flowered during the second millennium in Anatolia and north Syria" (Akurgal 1962, 127-130). This first phase is represented principally by sculptures at Malatya and Carchemish and has been dated by Akurgal to 1050-850 BC. The second phase is that of the "Slight Assyrian Style" ("Middle Neo-Hittite Phase"), which Akurgal dated to 850-745 BC and is characterised by the first appearance of Assyrian motifs within what is still essentially a "Hittite" style (Akurgal 1962, 130-133). This phase is represented at Carchemish and Zincirli. The third phase is the "Strong Assyrian Style" (or "Late Neo-Hittite Phase"), during which the traditional Hittite element is swamped by Assyrian influence (Akurgal 1962, 133-136). This phase is dated by Akurgal to 745-700 BC and is discernible in the gate lions of Zincirli and Sakçe Gözü.

While Akurgal's sequence outlines the development of different stylistic elements within "Neo Hittite" art, the absolute dates are based on excavations with inadequate stratigraphic control: his reliance on Carchemish or Zincirli chronology, for example, completely undermines his results (**§§3.20, 3.47**). Ignoring absolute chronology, Akurgal's sequence does demonstrate that some early Iron Age populations of the inland Northern Levant maintained a strong cultural link with the stylistic traditions of Bronze Age Anatolia. Nevertheless, "Neo-Hittite" art flourishes during the Iron Age, and might indeed be considered a true Iron Age phenomenon for the Northern Levant. Reconstructions of the Iron Age, which emphasise the political history of the region, seem to correlate well with the political monuments of the Northern Levant.

Iron as a 'Working' Metal

Implicit in the use of the Three-Age System is an acceptance of its inherent structure; in other words, an acceptance that each of the periods is distinct and defined by characteristic developments in metallurgical skill. The term "Iron Age" implies, therefore, that the manufacture of tools and weapons was no longer primarily being undertaken with bronze but had been replaced by iron (Schnapp 1996, 300-301). However, Waldbaum (1978, 17-23; 1980; **Chart 2.2**) has demonstrated that the adoption of iron as a 'working' metal does not coincide with the conventional beginning of the Iron Age. The use of the term "Iron Age" to define the early Iron Age is, accordingly, misleading as the traditional terminology does not fit the contemporary data.

Table 2.2: Stylistic Sequence of Syro-Hittite Monumental Art

Phase	BC	Style
Early Neo-Hittite	1050-850	Traditional Style
Middle Neo-Hittite	850-745	Slight Assyrian Style
Late Neo-Hittite	745-700	Strong Assyrian Style

(After Akurgal 1962, 127-136)

Chart 2.2: Comparative Importance of Iron and Bronze in Iron I Period

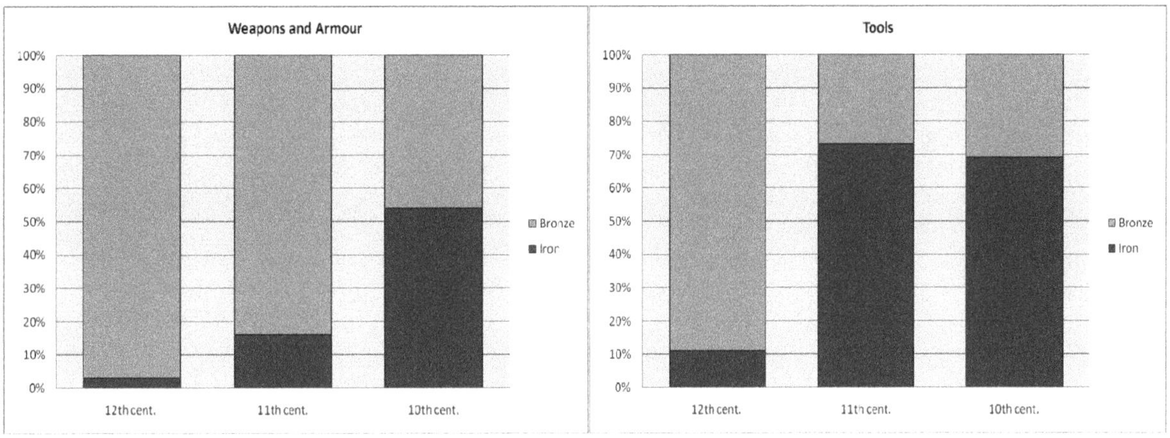

(After Waldbaum 1978, Chart IV.14a-b)

Iron was known and used as a metal in the Bronze Age (Muhly 1980, 34-36), but it was insufficiently understood so as to make it durable for use. The soft low-carbon form of iron, which Wertime (1980, 2) suggests was an unintentional but inevitable by-product of copper- and lead-smelting, was mainly used for jewellery manufacture during the third and second millennia BC. It was during this time that iron was considered a precious metal (Sherratt 1990, 811). Bronze Age texts make frequent reference to the exchange of small iron objects between monarchs and the use and storage of iron objects reserved for ceremonial and ritual use (Muhly 1980, 49-50). This explains why most Bronze Age iron artefacts have been found in contexts associated with elite display and ceremony (Waldbaum 1980, 80-81).

The coming of iron was not a sudden event; nor was its transformation from precious material into 'working' metal. Waldbaum (1980, 83) and Snodgrass (1980a, *passim*) have both suggested that a transitional period of around 200-300 years preceded the adoption of iron for utilitarian purposes. During this transition, bronze continued to be used for utilitarian purposes until iron was viewed, first, as a supplement to and, later, as an acceptable substitute for bronze (Snodgrass 1980a, 337).

Utilitarian iron artefacts, which were present in the early Iron Age but greatly outnumbered by bronze equivalents, do not appear in any real consistency across the eastern Mediterranean until about the tenth century BC, at least two centuries after the conventional beginning of the Iron Age (Waldbaum 1978, 17-23, Tab. IV.1; 1980, 82; 1999). While an unfortunate drawback with Waldbaum's theory was that her dating analysis relied on problematic stratigraphic sequences (e.g. Hama – Waldbaum 1978, 14, 44-46), absolute dates were not needed for Waldbaum to demonstrate that developments in ferrous technology do not coincide with early Iron Age material culture. The relative percentage of early Iron Age bronze and iron weapons and tools, as compiled by Waldbaum (1978, Chart IV.14a-b), are compared in **Chart 2.2**.

Metallurgical evidence demonstrates that bronze continued to be the preferred utilitarian metal until the end of the Iron I period, though the transition did not end the use of bronze altogether. That bronze and iron were used together is attested in the account by the Assyrian king, Assurnasirpal II (883-856 BC) of his difficult passage across Mt Kashiari: "with axes of iron and with picks of bronze, I hewed a path..." (Winter 1988, 194). There is also significant evidence confirming the presence of thriving bronze-working industries across the Levant throughout most of the Iron Age (Biran 1994, 147; Falsone 1988; Tubb 1988; Winter 1988). Despite carburised iron's suitability for tool and weapon manufacture, bronze remained better suited for certain, more specialist types of objects, mainly those produced by casting or the working of metal sheet (Philip *et al.* 2003, 91).

Though Waldbaum (1978, 17-23) has demonstrated that iron only became a key material for the manufacture of tools and weapons toward the end of the Iron I period, some archaeologists continue to associate its appearance with the beginning of the Iron Age (Akkermans and Schwartz 2003, 360; Klengel 1992, 187). These models usually emphasise the terminal *effect* that innovation in ferrous technology had on Late Bronze Age society (see Drews 1993, 73-76). Iron is sometimes considered a primary catalyst in the collapse of the palace-economics for one of two reasons: either the availability of this new 'democratic' metal removed a significant component of the economy from the hands of the elite, and thus broke its monopoly; or the Late Bronze Age palaces suffered directly at the hands of invaders armed with stronger iron weaponry.

According to Snodgrass (1980a, 348), the "Crisis Years" resulted in widespread disruption of trade, which meant bronze-working quickly became unviable. Sources of iron though were abundant, which Snodgrass (*ibid*) suggests made iron generally cheaper, resulting in a wider availability of iron implements. Within Snodgrass' model, control of access to metal tools no longer rested solely in the hands of the elite, which had a terminal effect on palace-based economies. This theory has since been discredited. For instance, Waldbaum (1978, Chart IV.14a) has demonstrated that bronze artefacts continue to appear in significant numbers throughout the early Iron Age, long after the "Crisis Years". In addition, Bell (2006, 95-102, 105-106) has recently highlighted the fact that not all avenues of trade were disrupted, with Cyprus and Phoenicia continuing to trade bronze objects throughout the Late Bronze-Iron Age transition. An alternative model is that of Sherratt (1994a, 61; 2000, 83), who has argued that developments in ferrous technology came about not because of a bronze shortage, or lack of supply, but because there was too much bronze in circulation. According to Sherratt (*ibid*), merchants were simply looking for new commodities to trade. It is also worth noting that despite the abundance of iron ore, the forging of iron objects is a labour intensive operation, while in comparison bronze allows for multiple castings with much less effort (Moorey 1994, 271-273; Tylecote 1980, 209). Hence, iron, like bronze, requires some level of infrastructure, and is, therefore, not as *democratic* as widely suggested (Akkermans and Schwartz 2003, 360; Klengel 1992, 187; Stone and Zimansky 1999, 35). This is reflected in the cost of iron continuing to be greater than bronze well into the Iron Age, which would have prevented iron from penetrating the lower strata of society (Haarer 2001, 264-265; *cf.* Moorey 1994, 263). As we have observed, the conventional beginning of the "Iron Age" does not correlate with the widespread adoption of iron for tool and weapon manufacture.

Use of Domesticated Camel Caravans

Another technological innovation associated with the early Iron Age is the use of domesticated camels for long-distance overland trade (Cline 2003, 364; Klengel 2000a, 24). Retsö (1991) has demonstrated that the camel was instrumental in the establishment of the trans-Arabian incense trade while re-invigorating the local mercantile economies. While generally this is true, the date for this phenomenon is much debated (Artzy 1994, 134-135; Finkelstein 1988b, 246-247; Liverani 1997b, 561; Retsö 1991; Wapnish 1984). Although it is often assumed that the large leap in the number of domesticated camels occurred in the early Iron I period, evidence confirms that numbers increased dramatically in the Levant only in the late Iron II period (Hakker-Orion 1984, 209-210; Wapnish 1984; Wilkens 1998).

According to his reading of the biblical accounts, Albright (1949, 206-207) suggested that camels did not enter the history of the eastern Mediterranean until the eleventh century BC. While archaeologists have often sought to discredit Albright's theory (e.g. Artzy 1994, 134; Ripinsky 1975), there is little evidence to attribute an earlier advent of the domesticated camel; instead, the opposite seems true. Artistic representations in the third and second millennia BC, as well as the appearance of camel bones in second millennium BC archaeological deposits, are not particularly relevant, since they may represent wild camels (Bulliet 1990, 58-65). Indeed, the domestication process of the camel was a long and gradual process, probably taking some centuries, during which there would have been degrees of domestication (*ibid*, 37, 60-64). Bulliet (*ibid*) has suggested that early in the domestication process camels would have been used for meat, milk, leather, and dung, but were unlikely to have been used with a harness, suggesting that the earliest domesticates in the Southern Levant were probably not ridden (Hakkar-Orien 1984). Trying to date the appearance of domesticated camels in the Levant is difficult, and their use as pack animals even more so. Restö (1991, 199, 205) suggests that there is simply no direct evidence to indicate that the animal was used for transport before about 900 BC. Zarins (1978), on the other hand, suggests that unequivocal evidence for domestication of camels does not exist before 500 BC.

The earliest examples of camel faunal material found in the Levant derive from deposits in the Negev (Har Sa'ad, Kadesh Barnea, and Ar'oer) and Gaza strip (Tell Jemmeh) (Hakker-Orion 1984, 210; Wapnish 1984), and have been dated to the tenth century BC. While Artzy (1994, 135) and Bell (2006, 103) both cite Wapnish's study of the faunal assemblage from Tell Jemmeh as evidence of Late Bronze Age camel trade, Wapnish (1984, 171) herself clearly states that "the find locations make it unlikely that any [camel bones] ... pre-date 1100 BC". Moreover, there is a high incidence of butchering marks on the camel bones, suggesting that these animals were primarily used for meat, rather than transportation (Wapnish 1984, 174). This use is mirrored in the camel remains from Tell Afis, where the majority of the Iron Age examples bear butchering marks (Wilkens 1998, 434, 441).

The earliest depictions of camels involved in transportation came from first millennium BC Assyria and inland Northern Levant (Bulliet 1990, 75-86). The appearance of camels on the Black Obelisk of Shalmeneser is problematic, not least because they are depicted as exotic animals, suggesting they were still rare in the ninth century BC (Wapnish 1984, 180). Domesticated camels with harnesses are also depicted on the Balawat gates (Bulliet 1990, 75), and camels being ridden appear in Iron Age art of Tell Halaf and Carchemish, though neither example can be precisely dated (Bulliet 1990, 82; Hogarth 1914a, 186, Pls B16b, B50).

Current evidence for the earliest use of the domesticated camel for trade is not clear. Since it is difficult to distinguish the faunal remains of a domesticated camel from those of a wild specimen, faunal evidence of camels, if found, are not directly able to confirm an animal's use in overland trade in the early Iron Age.

Summary

Key elements in the current conventional definition of the Iron Age have been discussed above, and many have been found to not characterise the beginning of the Iron

Age. Instead, the distinctiveness of the early Iron Age has been overstated, possibly due to the conventional historical narrative being given priority over the archaeological record. If the narrative is instead de-emphasised and focus given to various aspects of material culture, we might no longer consider the current Late Bronze-Iron I division as a significant cultural boundary. The metallurgical evidence, ceramics, domestic architecture, burial practices, mercantile activity, and written language all reflect strong continuity of culture and population. Should the early Iron Age be more appropriately considered a continuation of Late Bronze Age society and culture; a sub-Late Bronze Age?

2.4 Text as Interpretative Framework

The historical narrative has been used to over-determine interpretations of the archaeological record of the IA-NL, resulting in a reconstruction of Iron Age society that is a history of kings, conflicts and peoples – a kind of quasi-politico-military history that considers short-term events more significant than long-term processes (Liverani 1994; Sherratt 1998, 292). These historical "events" are then used to expound wider generalisations about the past, extrapolating histories of complete societies in spite of the fragmentary, biased record. It seems that the archaeological record is often only used as a means for authenticating and informing the already-established historical narrative. In the Northern Levant, classical literature and Assyrian palace inscriptions have had a significant impact on, and influence over interpretations of the archaeological record.

Conventional Histories of the Iron Age Northern Levant

This section will briefly explore the prominence given to the historical narrative in four standard "archaeological-histories" of the Iron Age. This simple exercise was directed toward identifying which assertions in "standard texts" on the subject make no, or little, reference to material culture, and to what extent it is discussed at all. While none of the four books reviewed is solely focused on the Iron Age, each devotes at least one complete chapter to the topic. The texts are:

- Klengel, H. 1992. *Syria: 3000 to 300 B.C. A Handbook of Political History*
- Van De Mieroop, M. 2004. *A History of the Ancient Near East ca. 3000-323 BC*
- Akkermans, P.M.M.G. and G.M. Schwartz. 2003. *The Archaeology of Syria: From Complex Hunter-Gatherers to Early Urban Societies (ca. 16,000-300 BC)*
- Kuhrt, A. 1995. *The Ancient Near East c. 3000-330 BC.*

In his political handbook of Syria, Klengel states that "the scantly epigraphic material offers no reliable basis for a political history of the [early Iron Age]" (Klengel 1992, 182), yet he accepts the 1200 BC date for the start of the Iron Age (*ibid*, 181). For the twelfth and eleventh centuries BC, Klengel focuses on the mauruading "Sea Peoples" and "Aramaeans", and their associated, archaeologically-attested destruction levels (pp. 182-183). He also discusses the role of iron for tools and the use of camels in the transportation of goods (*ibid*, 186-187), which he suggests is "evidenced by texts and archaeological artifacts" (*ibid*, 187), though fails to reference any actual evidence for either of these two developments. Similarly, Klengel's treatment of the Iron II period is focused on the geographical distribution of the different ethnic elements, as divined from the texts, and on the military history of Assyrian armies and anti-Assyrian coalitions. Klengel (*ibid*, 187) once again mentions the paucity of the epigraphic evidence, which he suggest was due to the use of perishable writing materials, but he does not seem concerned about the bias that his reliance on the many political stelae and palace inscriptions might produce; after all, he states in the introduction that the "special concern of this handbook is political history in a restricted sense... [the] social and economic background of these relations cannot be treated *in extenso*" (*ibid*, 15). Furthermore, Klengel does not evaluate the historical reliability of any of the texts. As a result, Klengel's history of Iron Age "Syria" is essentially a history of kings, battles, and ethnic states, with only occasional reference to archaeological evidence; i.e. a political history, as intended.

In his overview of Near Eastern history, van de Mieroop (2004) does not use the term "Iron Age", but instead deals with the first millennium BC under the title of "Empires", which immediately emphasises the fact that this is essentially a political history. Van de Mieroop does not attempt to write a history of the early Iron Age, instead summarising the twelfth and eleventh centuries BC in just a few paragraphs, his reason being a lack of sufficient textual material (*ibid*, 189). Van de Mieroop betrays a heavy reliance on the textual material, as he does not engage with the archaeological record beyond verification purposes. Like Klengel, van de Mieroop has outlined the history of Iron Age "Syria" as one of kings, battles, and ethnic states, with occassional reference to archaeological evidence. It is interesting to note van de Mieroop (*ibid*, 210) critiques the "Hebrew Bible as a historical source", but does not extend his enquiry to other textual corpi.

Akkermans and Schwartz's (2003) book is the only one claiming to be a survey of the "archaeology of Syria", yet Schwartz's (*ibid*, 360-397) chapter on the Iron Age still demonstrates the central role of the historical narrative in the interpretation of the archaeological record. For instance, the main component of his 40–page chapter consists of a 28-page (*ibid*, 366-394) political history, only drawing on archaeological evidence to highlight the veracity of his narrative. In particular, monumental-art, architecture, and urban-planning feature prominently because their "elite" character closely corresponds with the political history being advocated; ceramics and burial practices are dealt with less exhaustively (*ibid*, 366*ff. passim*). Within this lengthy treatise he discusses the archaeological evidence for the "Luwian-Aramaean states" (*ibid*, 366-377), the "Neo-Assyrian empire" (*ibid*, 377-386), the "Phoenicians and Greeks on the Syrian Coast" (*ibid*, 386-388), and the "Neo-Babylonians and Achaemenid Persians" (*ibid*, 389-394); section-headers that highlight an emphasis on the conventional historical narrative and an acceptance of archaeological cultures as "politico-ethnic" units. While "archaeology" is discussed by Schwartz, there remains a heavy reliance on the historical narrative to understand and interpret the archaeological material.

Kuhrt's (1995) 2 volume history of the Ancient Near East devotes a whole volume to the Iron Age, though she refrains from using the term "Iron Age". Kuhrt instead entitles this section "Political Transformation and the Great Empires", which illustrates the fact that this, too, is essentially a political history. Furthermore, Kuhrt's section-headers (e.g. "sea-peoples"; "Aramaeans"; "Phoenicians") betray a reliance on the conventional historical narrative as derived from a literal reading of the ancient documentary sources. Hence, Khurt tends to present archaeological cultures as homogenous political units with a distinct ethno-linguistic character (note her frequent use of "state" to define these past cultures). But while Kuhrt relies on ethno-political terms to define her discussion of the Iron Age, this cannot be said of her discussion of the previous period, which made frequent use of the term "Late Bronze Age" (Kuhrt 1995, section 8c). Kuhrt's history also uses conventional dates for her discussion, dates derived from the historical narrative; e.g. her discussion of the Levant ends in 720 BC with the Assyrian "destruction" of Hama (see below).

The Influence of Classical Texts

While the biblical text has greatly influenced archaeological practice and interpretation in the Southern Levant (**§2.5**), classical texts have had a much more direct impact on reconstructions of the Northern Levant (Klengel 1992, 17). In particular, the identification of many Iron Age sites, peoples, and political and economic structures derives from, or is influenced by, in one way or another, classical texts. This is particularly evident amongst the sites of the Mediterranean coast (e.g. Ras al-Bassit, Tell Kazel, Tyre, Sidon, Sarepta), where the classical and Near Eastern worlds are believed to have come in contact. Ever since Johansen (1923) and Poulsen (1912) insisted that the Phoenicians were important intermediaries between the ancient Near East and the classical world, the IA-NL has been an important area of study for archaeologists interested in following the evolutionary development of European and Greek culture (Gelin 2004, 58-63; Riis 1970, 8-10). This ensured that the classical narrative came to be intricately connected to Near Eastern history, as demonstrated in **Table 2.3**.

Classical literature has various genres of text, including some that are more overtly poetical, mythical, or allegorical, than historical. Nevertheless, the genre of epic, which might be considered closely related to myth, is widely accepted as holding some kernels of historical truth (e.g. Kirk 1975, 820-821). For many years, most of what archaeologists knew of "Phoenician" political history, merchant activity, and craftsmanship derived from Homer's *Iliad* (6.288*ff*; 23.740*ff*) and *Odyssey* (4.614-619; 13.256-286; 14.287-315; 15.403-484). In the absence of contemporary Phoenician historical texts, scholars of the Near East "mined" Homer's few depictions of the Phoenicians for historical value that might illuminate the archaeological record (Muhly 1970, 20-21; Winter 1995, 248-249). Consequently, twentieth century research constructed the whole Phoenician civilisation from external, cursory references, and presented it as a well-defined and unified "nation" of merchant traders and skilled craftsmen (e.g. Kuhrt 1995, 405-407). Little appeal was made to the archaeological record of "Phoenicia". Winter (1995, 261), however, has demonstrated that Homer's "Phoenicians" are foremost a literary device for emphasising the noble character of Odysseus, by contrasting his virtue with Phoenician treachery and deceit (see West 1988, 170). Hence, Homer's "Phoenicians" probably do not accurately depict any real population of the eastern Mediterranean. For this reason, equating the population of the Levantine coast with Homer's "Phoenicians" is problematic.

Once the "Sea Peoples" migration model is set aside, and the continuity in the material culture is emphasised, it is difficult to define a "Phoenician" population: at what point did the so-called "Canaanite" population become "Phoenician"? (Sherratt 1998, 307). This point is complicated by the fact that scholars have accepted that both words, "Canaan" and "Phoenicia" encompass the same meaning; i.e. an association with "red-purple dye" (Albright 1975, 520; Moscati 1988, 24; Muhly 1970, 26-28). Moreover, the culture of the Mediterranean coast of the first millennium BC is not easy to define politically. The general area was not organised as a unified polity, but as a series of only-sometimes-confederated city-states (Sherratt 2005a, 35). Yet scholars pursue studies of "Phoenician" society as if it was a well-defined, homogenous entity (e.g. Aubet 2004b; Aubet *et al.* 1998-1999; Bunnens 1979; Culican 1959; 1970; Falsone 1988; Gubel 1994; Haggi 2006; E. Mazar 2000; 2001; 2004; Moscati 1973; Peckham 2001; Winter 1976). Finally, the term "Phoenician" is a Greek term; we do not know if the local population used this term, or if they maintained the use of *cna'ani* (Moscati 1973, 21-22; Muhly 1970, 26-28; Sherratt 1998, 307; 2005a, 35-36). It is unclear from the Greek texts whether such a term denotes a people, culture, specific region, mercantile class, or even ideological concept. Hence the term "Phoenician" cannot be equated with a specific material culture.

Table 2.3: Classical Texts Cited in Archaeological Interpretations of the Iron Age Northern Levant

Site	Text	Interpretation	Reference
al Mina	Herodotus, III 91	Identification	Woolley 1938a, 28-30
Ras al Bassit	Herodotus, III 91; Strabo VXI 2.8	Site name - Posideion	Boardman 1990, 170
Tell Sukas	Herodotus, III 19.3 Herodotus, V 104-116	Nabonidus destruction End of Greek presence	Riis 1970, 58-59
Tell Kazel	Strabo VXI 2.12 Pliny V 20.17	Site name - Simyra	Bounni 1997, 275
Byblos	Strabo VIII	Origin of name	Frost 2004, 343
Beirut	Excavated to understand absence from texts		Khalifeh 1997b, 294
Sidon	Homer *Il* 23.742-744 Homer *Il* 6.289-292	Phoenician city Traders/Craftsmen	Pritchard 1975, 17
Sarepta	Pseudo-Scylax Lycophron	Site name	Pritchard 1975, 7-9 Khalifeh 1997a, 488
Tyre	Herodotus I 2.1; 2.44 Josephus IX 286-287	Phoenician city Tyrian Fleet	Riis 1970, 138 Pritchard 1975, 19

Herodotus = *Histories*; Homer *Od* = *Odyssey*; Homer *Il* = *Iliad*; Strabo = *Geography*; Josephus = *Antiquities*; Pliny=*Natural History*

Assyrian Political Histories

Since archaeological work began on the Iron Age cities and palaces of Assyria, textual sources relating to the Northern Levant have increased at an exceptional rate; so much so, that the publication of texts has been slow, though this is beginning to change (e.g. Helsinki's State Archives of Assyria Project – Parpola 1987; Toronto's Royal Inscriptions of Mesopotamia series – Grayson 1996). Nevertheless, Assyrian texts have been especially influential within the writing of North Levantine cultural histories (van de Mieroop 2004, 211). The conquest and annexation of the Levant was recounted in many royal inscriptions, the most prominent being those of Ashurnasirpal II (ruled 883-859 BC), Shalmeneser III (858-824 BC), Tiglath-Pileser III (ruled 745-727 BC), and Sargon II (ruled 722-706 BC) (Grayson 1991a; 1996; Luckenbill 1926; 1927). Assyrian interaction with the Levant, however, was primarily through conquest; therefore, Levantine history has inadvertently been linked to Assyrian military history and the supposed archaeological manifestation of military campaigns – destruction layers (*cf.* Ben Tor and Zuckerman 2008; Hawkins 1982, 377; Zuckerman 2007a, 3). Indeed, the correlation of Iron Age "destruction" levels with ninth to seventh century BC campaigns of Assyrian kings has become a common means for imposing an external historical chronology upon the Levantine archaeological record. **Table 2.4** lists a number of Assyrian "destruction" levels that have come to underpin much of the region's Iron Age chronological structure. The validity of associating a layer of ash in the archaeological record with a historical military campaign is rarely assessed. (*cf.* Ben Tor and Zuckerman 2008; Forsberg 1995; Whincop 2007, 186; Zuckerman 2007a, 3; 2007b).

The conflagration of Hama Stratum E (hereafter Hama E) has had a profound impact on the chronological framework of the IA-NL, and has set the precedent for other similar interpretations. The tell of Hama was excavated in the 1930s by Danish archaeologists who exposed a large Iron Age elite-building complex that lay in ruin; the burning was so intense that part of the basalt architecture had melted (Buhl 1992, 35). On historical considerations, the destruction of this complex was attributed to Sargon II (Hawkins 1972-1975, 70; Ingholt 1942, 472). By claiming dominion over the prince and people of the Hamath province, Sargon had supposedly alluded to his destruction of the city of Hama, an event that was assumed to have coincided with his specified destruction of another Hamath city, Qarqar, in 720 BC. Hence, the destruction of the Hama E complex was dated to 720 BC. It followed that the pottery found within the Hama E buildings could also be dated to the eighth century BC (Fugmann 1958, 269). As a result, Red-Slip pedestal platters and monochrome-painted shallow bowls have become the hallmark of eighth century BC material culture in the Northern Levant (Akkermans and Schwartz 2003, 363). The date was also particularly important for Greek chronology, because it provided a date for two Greek *skyphoi* found within the Hama E complex (Francis and Vickers 1985; Hannestad 1996, 48). Indeed, the Hama E "destruction" date was important to a great many scholars, which might explain its almost universal acceptance (Buhl 1992, 35; Coldstream 2003, 248; Mazzoni 2000a, 55; *cf.* Francis and Vickers 1985, 131). Consequently, the interpretation of the Hama E conflagration has led to a number of other "destruction" levels being attributed to Assyrian insurgence (e.g. Makinson 2005, Tab. 2; Moorey 1980, 4; **Table 2.4**). But just how secure is the Hama E "destruction" date?

Table 2.4: Summary of Assyrian "destruction levels" of the Levant

Context	BC	Ruler	Reference
Tille Höyük		Tiglath-Pileser I	Summers 1993, 11
Tell Ahmar	856	Shalmeneser III	Bunnens 1990a, 5
Tell Sukas H2	850	Shalmeneser III	Riis 1960, 123-124
Zincirli	9th	?	Ussishkin 1968, 189
Tell Afis VIII	738?	Tiglath-Pileser III	Cecchini 1998, 296
Tell Arqa 10	738	Tiglath-Pileser III	Thalmann 1983, 217-218
Tel Dan II	732	Tiglath-Pileser III	Biran 2002, Table 1.1
Hazor V	732	Tiglath-Pileser III	Ben Tor 1997, 112-113
Megiddo IVA	732	Tiglath-Pileser III	Finkelstein *et al.* 2006, 856-857
Beth Shan IV	732	Tiglath-Pileser III	Mazar 2001, 289
Tel Rehov 3	732	Tiglath-Pileser III	Mazar 1999a, 30
Samaria VII	722	Sargon II	Mazar 1992, 406
Tell Rifa'at IIb	720	Sargon II	Matthers 1981b, 416
Hama E	720	Sargon II	Ingholt 1942, 472
Tell Kazel I-9	720	Sargon II	Capet and Gubel 2000, 433
Tell Keisan 5	720	Sargon II	Briend and Humbert 1980, 27
Al Mina VIII	720	Sargon II	du Plat Taylor 1959, 87
Al Mina VII	720	Sargon II	Boardman 1980, 44
Carchemish	717	Sargon II	Woolley 1914, 94
Lachish III	701	Sennacherib	Mazar 1992, 432
Al Mina VIII	700	Sennacherib	Riis 1960, 123-125
Al Mina VII	696	Sennacherib	Riis 1970, 159
Tarsus	696	Sennacherib	Coldstream 1968, 385
Zincirli	676	Esarhaddon	Lehmann 1996, 273-274
Tell Sukas	677/671	Esarhaddon	Abou Assaf 1997b, 91

Scholars who accept the historical correlation of the Hama E destruction rarely provide the specific textual source for Sargon's claims to victory of Hama (or Hamath), as if the event is incontestable fact (e.g. Barnett 1963, 81; Gallagher 1999, 155; Otzen 1979, 252). For this reason, it has been particularly difficult for the current author to isolate the specific basis for the interpretation. Nevertheless, the conventional view that written and archaeological evidence provides a reliable and useful date for the destruction of Hama in 720 BC appears to be based on two assumptions, which will be discussed below; the first deriving from textual sources, and the second from the archaeological record.

The first assumption is that textual sources make it clear or at least probable, that the city of Hama was physically destroyed by the Assyrian army in 720 BC in connection with the destruction of Qarqar, located on the edge of the Hamath polity, 65 km to the north. The main text recounting this campaign is Sargon's *Annals* inscribed on stone slabs and the walls of his palace at Khorsabad/Dûr-Sharrukîn (Lie 1929). The key text occurs in lines 33-37 of the so-called *Display Inscription*:

> *Ia'ubidi from Hamath, a commoner without claim to the throne, a cursed Hittite, schemed to become king of Hamath, induced the cities Arvad, Simirra, Damascus and Samaria to desert me, made them collaborate and fitted out an army. I called up the masses of the soldiers of Ashur and besieged him and his warriors in Qarqar, his favourite city. I conquered (it) and burnt (it). Himself I flayed; the rebels I killed in their cities and established (again) peace and harmony (J.A. Wilson 1969d, 285 using the translation of Luckenbill 1927, §55).*[1]

In this text Sargon does not lay direct claim to the destruction of the city of Hamath. Sargon simply boasts of destroying the royal city of Qarqar and killing the king of Hamath (Grayson 1996, 23-24; Hawkins 1982, 417). In later texts Sargon claims to be the "plunderer of the princes of Carchemish, Hamath, ..." (Luckenbill 1927, §92); the one "who blotted out the princes of Hamath, Carchemish..." (*ibid*, §99); the "flayer of Ia'ubidi of Hamath" (*ibid*, §125); the one who "carried off the people of Hamath" (*ibid*); the "uprooter of Hamath" (*ibid*, §137), and the one who "smashed like a flood-storm the country of Hamath (A-ma-at-tu) in its entire [extent]" (J.A. Wilson 1969d, 284; Luckenbill 1927, §183, 186). Despite his boasts, Sargon does not explicitly claim or recount the destruction of the city of Hama, and no other published Sargonid references to Hamath or Ia'ubidi detail the destruction of this city (e.g. Gadd 1954, ND 3411 line 22). Klengel (1992, 226, n. 198) mentions a

[1] While this English translation is 80 years old, there appears to be no modern English revision of it. Current Assyrian archive translation projects have yet to include all of Sargon II's *Annals*. While there is some concern over the use of such an old translation, Luckenbill's work remains the key translation; e.g. Klengel 1992, 220.

stele that Sargon had erected at Hama commemorating his victory at Qarqar, though it seems unlikely Sargon would have erected a stele in front of a destroyed city, especially one "speaking" to the inhabitants of the city. A stele is also known from Tell Acharneh, near Hama, but the fragmentary text of this monument reveals little regarding its intended purpose/audience (Frame 2006, 67). Furthermore, Reade (1976) has concluded that the destruction of Hama was not depicted within the narrative art of Sargon's palace at Khorsabad.

In addition to the inconclusive content, there is some debate regarding the historical veracity of the *Annals* (Tadmor 1958; 1997). Because Assyrian texts primarily recount Assyrian military conquest of foreign lands, it is reasonable to assume that they contain a number of literary devices emphasising Assyrian power and dominion. Indeed, many texts clearly focus on the Assyrian king's subjugation of other kings (or princes), peoples, and lands, presenting this as an idealised concept of dominion over the whole world (Tadmor 1997, *passim*). If we accept that the language used within the Assyrian texts has strong ideological undertones, then it follows that these texts are likely to represent a skewed record of actual events. This is also indicated in the confusing and often contradictory order of events contained within the many *Annals* of Khorsabad and the fragmentary prisms of Nineveh, which led Tadmor (1958, 22-26) to conclude that the *Annals* follow an artificial scheme rather than a strictly historical one. Along similar lines, Gadd (1954, 173, 184) has suggested that the inscribed prisms of Sargon II, found at Nimrud, are arranged in an order that disregards chronology, but which instead emphasises geography.

The second assumption behind the 720 BC destruction date is that the archaeologically-attested burning of the Hama E complex is the result of Assyrian military conquest. When the Hama E complex was excavated, it was found to be covered by a widespread deposit of burnt debris, but there has been no archaeological material published that directly links this conflagration with Assyrian conquest or, indeed, any conquest. One could expect some Assyrian weaponry or armour, or physical damage to the city fortifications to be evident, as was the case at Lachish (Ussishkin 1982; 1990b – for examples of remains of warfare in the archaeological record see Stronach 1997, 317-322; Yon 1992, 117). Following the Hama E conflagration, the majority of the site appears to have been abandoned. While the loss of population appears to reflect the Assyrian policy of mass deportation (*cf.* Buhl 1992, 35; Fugmann 1958, 264-265, 278), the site was supposedly re-settled by 6300 Assyrians there; in other words, the archaeological record would not be characterised by a lack of population, but rather a change in one (Luckenbill 1927, 100*ff*, §183; Oded 1979, 45). Furthermore, there are serious concerns over the stratigraphic integrity of the Hama excavations (Thuesen 1988, 11; **§3.18**). On the current archaeological evidence, other causes for the conflagration, such as earthquake, accidental burning, revolt, civil war, or local feuds, cannot be precluded.

Neither text nor archaeological evidence directly confirms the Assyrian destruction of Hama E. First, the written evidence provides no account of, or claim for, the actual destruction and burning of this city. Surely, the conquest of such an important city as Hamath would have been celebrated in glorious detail by Sargon, but there is no such account. Second, nothing associated with the "destruction" layer can confirm either Assyrian involvement, or even military destruction. The Assyrian interpretation appears to have been favoured because it was a military power considered strong enough to destroy such an impressive and well-fortified citadel. All that can be stated is that the large Hama E complex was destroyed by a massive conflagration, the causes of which are currently unclear. Hence, the Hama E date can no longer be treated as established fact. This has important implications for the chronology of the IA-NL, which is centred on the date of 720 BC (Mazzoni 1990a; 2000a; 2000b).

The fact that neither text nor archaeological evidence confirms the Sargonid destruction of Hama E does not, of course, exclude the possibility that the Hama E destruction date is correct. However, on current historical and archaeological evidence, the common view that Hama E was destroyed by the Assyrians led by Sargon II and that this destruction has been archaeologically identified should be treated with caution. The small archive of texts recovered from the Hama E complex does not clarify the events surrounding the city's destruction, only compounding the problem (Hawkins 1972-1975, 70). This archaeological and historical "rethink" is not unique; Forsberg (1995) has demonstrated the need to revise similar Assyrian destruction dates at Samaria and Tarsus, while Zuckerman (2007a, 3; also Ben Tor and Zuckerman 2008) has recently questioned the Proto-Israelite "destruction" theory for Canaanite Hazor.

Text as Artefact

Far from being independent, objective accounts of actual events, ancient texts were written for a particular purpose, one that can rarely be said to incorporate a desire to provide an accurate record of events (Liverani 1990, 23-26). For instance, Assyrian palace inscriptions focused on display and self-aggrandisement and are as much ideological as historical (Tadmor 1981). With ancient texts, archaeologists are dealing with a world of rhetoric, propaganda and myth-making, wherein abstract truth probably did not exist (Van Seters 1997, 2-6). Moreover, ancient texts often have a formal structure, specific patterns, and recurrent motifs which may appear as "fact" to the modern investigator, but are more concerned with maintaining protocol. Since ancient authors may not have had history (as we understand it) as the aim of their writings, it is important to consider the factors that may actually have shaped their work. Ancient documents need to be understood in terms of function; who was the intended audience and what was the desired effect upon that audience? For example, a royal inscription will depict events differently from royal correspondence; one is aimed at an internal audience, while the other at an external target group (Liverani 1990, 25-26). It is easy to accept that the same event may be narrated in a different way by two different people, but it is not always understood that it will be told differently depending on the audience; this is rarely allowed for in archaeological interpretation (*cf.* Bauer 1998; Sherratt 1998, 292, 307; Stager 1995, 340-341).

While the social context of ancient authors (i.e. the context in which the text was originally conceived) is clearly an important consideration in the construction of ancient histories (Bentley 1999, 127-148; Faust 2006, 6; Silberman 1998b, 268), archaeologists have started to recognise that documentary sources are excavated artefacts (Morris 2000, 25-29; Morrison and Lycett 1997; Thurston 1997; Zettler 1996), and as such need to be considered within their archaeological context (Zettler 1996, 83). Archaeologists routinely consider material culture in light of multiple contexts (cultural, depositional) – historical data must be similarly evaluated, moving from internal considerations of text to the deposition and survival of these documents within an archaeological context (*ibid*). Understanding texts within their context also requires an appreciation of archaeological sampling: which elements of ancient societies kept written records? What kind of information was deemed important enough to record? What type of materials were texts recorded upon? These considerations alone invite caution when dealing with possible biases in ancient documentary sources.

Texts were written for particular reasons, at particular times, by particular people, in particular contexts and, for this reason, are not transparent sources of historical fact. Hence, it is no longer possible to scan historical narratives for the few useful 'facts' that provide the basis for a generalised modern account; any such 'facts' are so embedded in the narrative that they cannot be separated. Instead, texts construct rather than reflect, and invent rather than discover, history. Using text as the basis for a generalised history of the entire Levant, and then linking that history to the archaeological record via vague historical inference is a misguided pursuit.

2.5 Biblical Influence on North Levantine Chronology

During the second half of the twentieth century, archaeologists working on the IA-NL found themselves at a disadvantage; an earlier focus on monumental art and architecture had resulted in a dearth of local ceramics (Eidem and Ackerman 1999, 309; Eidem and Putt 1994, 8; Jamieson 2000, 263; Matthers 1981b, 415; Moorey 1980, 4). Hence, intra-regional comparative studies were impossible. This problem was accentuated by the fact that the few Iron Age ceramic assemblages available for study were mostly non-standard assemblages; the Yunus cemetery at Jerablus consisted of only grave goods (Woolley 1939b), while the appropriate levels at al Mina contained concentrations of imported pottery (Boardman 1959).

Table 2.5: Comparison of North Levantine Pottery with South Levantine Pottery

Northern Levant	Southern Levant	Reference
Tell Abou Danne IId	Hazor VIII-V; Megiddo IV-III Samaria IV-V; Tell Keisan 7-6	Lebeau 1983, 24
Tell Abou Danne IIc	Hazor IV; Megiddo III-II Samaria VI-VII; Tell Keisan 5-4	Lebeau 1983, 24
Tell Afis VII	Gezer XIV; Hazor X-VII Megiddo Tombs; Tell Keisan 9a-c	Venturi 1998a, 128-130 Cecchini 1998, 277
Tell Afis VIII	Gezer VIA; Hazor ? Megiddo?; Tell Keisan 5	Cecchini 1998, 284-285
Tell Afis IX	Gezer VIA-VA; Hazor VI-VA Samaria VII; Tell Keisan 4	Cecchini 1998, 286-287
Hama E	Beth Shemesh IIa; Hazor IX-IV Lachish III; Megiddo VIA Samaria?; Tell Beit Mirsim A	Riis & Buhl 1990, *passim* 168-170, (refs to Amiran 1969)
Al Mina VIII	Hazor IX-X; Megiddo III Samaria VII-VIII	du Plat Taylor 1959, 81
Al Mina VII-VI	Hazor X; Megiddo III-II	du Plat Taylor 1959, 82
Tyre XIV-XIII	Hazor XII; Megiddo VI Tell Abu Hawam IV	Bikai 1978b, 66
Tyre XII-X	Tell Abu Hawam III	Bikai 1978b, 66
Tyre III-II	Ashdod 3; Hazor VI-V Samaria V-VI	Bikai 1978b, 66
Tyre I	Hazor IV	Bikai 1978b, 66

Due to the lack of local ceramics, comparative material was sought in other regions, especially from the abundant ceramic typologies of the Southern Levant (**Table 2.5**). Since ceramic comparison is usually undertaken for chronological purposes, this inadvertently aligned the Northern Levant with the biblically-inspired, chronological framework of the Southern Levant. The acceptance of the "Southern" chronology meant that the Northern Levant effectively lost its own identity, resulting in the entire Levant henceforth being treated as a single region, as evident in the wide use of terms like "Syria-Palestine" (e.g. Dever 1992; Liverani 1983; Perrot 1979).

While pottery from the IA-NL was defined according to pottery from the IA-SL, a number of "Northern" contexts also looked to Cyprus and Greece for comparative purposes. Hence, many North Levantine dates are based on cross-reference of pottery with Greece, Cyprus. Despite comparison with different regions, the results were ultimately the same – the Iron Age ceramic chronologies of the eastern Mediterranean all derived from the IA-SL (see Coldstream and Mazar 2003, 40).

Cypriot Chronology

Cyprus played a pivotal role in the archaeology and chronology of the eastern Mediterranean; its central position and strong trade contacts contributed to its role as a point of cross-reference for different regions. In particular, Cypriot pottery was exported to many areas of the eastern Mediterranean during the second and first millennia BC. As a result, Cypriot pottery has been recovered from a number of Levantine contexts, where its form and style has been used to secure absolute dates. For the Iron Age, reference is usually made to the chronological framework of Gjerstad (1948, 421-427), and occasionally to Birmingham's (1963) revision of Gjerstad's scheme, both of which appear in **Table 2.6**.

Gjerstad's (1948) chronological division of the Cypriot Iron Age was the first undertaken for Cyprus. It was determined at a time before stratified Iron Age sites in Cyprus had been excavated, and was instead based on the abundant tomb material (Birmingham 1963, 23). In the absence of stratigraphy, statistical analysis of mortuary pottery provided Gjerstad with his typological sequence of ceramic forms. The tomb groups, however, lacked absolute dates, and Gjerstad (1948, 184-185) had to turn to external, non-Cypriot data to secure his sequence. For this purpose, Gjerstad (*ibid*, 242-257) turned to the chronology of the IA-SL, which he believed to be the most secure. Through the presence of Cypriot pottery in Palestine, Gjerstad linked his chronological sequence to those contexts he deemed "reliable" (Desborough 1957, 216; Hanfmann 1951, 426-427). The accuracy of Gjerstad's scheme was thus dependent on three factors: the correct identification of pottery development within Cyprus; the correct attribution of sherds in the Southern Levant as Cypriot; and the IA-SL chronology.

Following the publication of his scheme in 1948, Gjerstad's absolute dates aroused considerable criticism. A number of archaeologists called for extensive modifications to the Cypriot chronology, though they, too, based their revisions on the chronology of the IA-SL (e.g. Albright 1950, 175, n. 51; 1953, 22; Desborough 1957, 216-217; Hanfmann 1951, 425; B. Mazar 1951, 24; McFadden 1954, 136; du Plat Taylor 1959, 63, 89; Swift 1958, 159-161; Van Beek 1951, 26-27; 1955, 37-38). The main challenge to Gjerstad's scheme, however, was that of Birmingham (1963) who was trying to account for the growing amount of Cypriot material found in Levantine contexts. Since Gjerstad's publication, excavations at Samaria, Megiddo (IVA-III), Al Mina (X-VIII), Tell Abu Hawam (III), Tell Qasile (VIII-VII) and Hazor (VIII-V) had all provided good amounts of Cypriot pottery from stratified deposits (*ibid*). It was apparent to Birmingham that the increase in material was accompanied by growing discord between the Cypriot and Palestinian datasets. Gjerstad (1953) argued against the revision of his dates based on the new Levantine data, but Gjerstad's absolute dates for the Cypriot Iron Age were based on Levantine sites; hence he could not argue against the Levant dates without undermining his own chronology. But Birmingham (1963, 23, 39-40) was calling for more than a revision of dates, recognising that Gjerstad's actual sequence was in need of revision; i.e. the tomb material needed reordering. According to Birmingham's research, Gjerstad's typological sequence of Cypriot pottery was inaccurate when tested against sites in the Levant. Birmingham (*ibid*, 15, 39) instead proposed a new scheme that incorporated minor revisions to the ceramic sequence and absolute dates, while also introducing to Cyprus the "Iron Age" terminology used in the Southern Levant (**Table 2.6**). Birmingham's scheme, nevertheless, remained as dependent on the South Levantine data as Gjerstad's – she too failed to accomplish a complete revision of the pottery sequence based on well-stratified, independently-dated Cypriot non-mortuary contexts (*ibid*, 15).

While the reliability of the chronology of the IA-SL has recently been called into question (Finkelstein 1996a; Whitelam 1996; Whiting 2007a; Wightman 1990a), the implications of this for Cypriot chronology appear to have been overlooked. Instead, in a reversal of roles, Cypriot chronology, as defined by Gjerstad or Birmingham, has been used to help refine Levantine chronology (e.g. Badre 1998, 83; Bikai 1978b, 66; Gilboa and Sharon 2003, 65-67). Clearly, there is an element of circularity to this practice. Nevertheless, Cypriot imports continue to be used for assigning absolute dates to excavated contexts; consequently, a number of North Levantine contexts are dated via reference to a Cypriot chronology that is derived from South Levantine dates (e.g. Badre 1998, 79-83; Bonatz 1998, 219; Doumet 1982, 133; Doumet-Serhal 2006, 21-25; Hamilton 1934, 75; Saidah 1966, 86-87; 1977, 144).

Table 2.6: Comparison of Cypriot Iron Age Chronology

Gjerstad 1948		Birmingham 1963	
Terminology	BC	Terminology	BC
Cypro-Geometric I	1050-950	Early Iron	1050-900
Cypro-Geometric II	950-850		
Cypro-Geometric III	850-700	Middle Iron I	900-725
Cypro-Archaic I	700-600		
Cypro Archaic II	600-475	Middle Iron II	725-600
Cypro-Classical I	475-400	Late Iron	600-325
Cypro-Classical I	400-325		

Greek Chronology

As with Cypriot pottery, Greek imported pottery frequently appears in Levantine contexts. The distinctive decoration of these vessels has ensured that Greek pottery is rarely missed; nor is its apparent chronological value ignored (e.g. Coldstream 2003; Coldstream and Mazar 2003; Doumet-Serhal 2006, 20-21, fig. 27; Lebeau 1983, 21-26; Mazar 2004; Yasur-Landau 2004). But like with Cypriot pottery, the absolute dating of Greek pottery is problematic. Internal evidence is limited to dates given by classical historians writing about events that occurred centuries before their own time (e.g. Thucydides – Coldstream 1968, 302-327; Desbourough 1957, 217-219). External evidence is based on the appearance of Greek imported pottery in Levantine contexts. While some scholars have questioned the traditional chronology for Greek pottery and tried to lower the dates, the proposed dates too have been based on South Levantine chronology (e.g. Gilboa and Sharon 2003, 67-72; Waldbaum and Magness 1997, 23).

The absolute chronology of Greek pottery is based on the presence of a few pieces of sub-Protogeometric and Geometric pottery in the Levant, usually from contexts that were poorly-defined and historically-dated (Coldstream 1968, 302-311; Cook 1972, 262; Desborough 1952, 293-295). For example, the accepted date for the destruction of Hama E provided scholars with a *terminus ante quem* for the Greek pottery contained therein, though none of the Greek sherds were found in a primary context (Francis and Vickers 1985, 131-134; Fugmann 1958, Figs 310(7B23); 344(L941); Riis 1948, 114; 1965, 80 n. 5-6; 1970, 154-156; Riis and Buhl 1990, 184, Fig. 84). To complicate matters further, a number of Levantine sites also rely on the presence of Greek pottery to date their stratigraphic sequences (e.g. Bikai 1978a, 66; Coldstream and Mazar 2003, 44-45).

Riis (1960, 123-125; *cf.* du Plat Taylor 1959, 91; S. Smith 1942, 94*ff*)) has also suggested that the dating of Greek pottery can be established according to negative evidence. He maintains the absence of Assyrian activity in the Levant during the late ninth and early eight centuries BC (due to trouble with the Urartians in the north) was a favourable time for the arrival of Greek traders. This historical correlation was then used to date the earliest appearance of Greek pottery in the region (at Tell Sukas and al Mina). However, Riis failed to demonstrate that Assyrian presence did sufficiently disrupt commercial activity in other periods.

The Role of the Biblical Text in Iron Age Chronology

As we have seen above, Greek and Cypriot Iron Age chronologies have not been established through scientific means but are based on ceramic comparison with sites from the Levant. In fact, the chronology of the eastern Mediterranean is a network of ceramic correlations that can be traced back to only a handful of supposedly reliable contexts. In the Northern Levant, the one date considered "reliable" is the destruction of Hama E by Sargon II (**§3.18**). For the IA-SL, Finkelstein (1996a, 179-182; *cf.* 2005) has suggested five chronological "anchors" that underpin the chronology: the presence of Philistine pottery; evidence of Shishak's campaign in the Southern Levant; the attribution of Megiddo VA-IVB to Solomon; the construction and destruction of Jezreel; and evidence of Assyrian-conquest (e.g. Ussishkin 1982; 1990b). The underlying theme here is the conventional historical narrative. Each of these archaeological "events" is discussed below, with the exception of Assyrian-conquests which were discussed above.

Philistine Pottery

The chronology of the Iron I period in the Southern Levant is generally based on the presence of distinctive pottery styles that have come to be associated with the biblical Philistines (Finkelstein 1996a, 180; Mazar 1988; Sharon 2001; Singer 1985; Stager 1995). Initially, the Philistines were associated with a distinctive type of bichrome pottery that was found in the coastal plain of the Southern Levant, an area that the biblical narrative had associated with the Philistines (Sharon 2001, *passim*). The Aegean-related forms and relative chronology of the pottery seemed to confirm the "Philistine" label (Faust 2006, 139; Killebrew 2005, 14). Moreover, the appearance of this pottery in the remains of an Egyptian residency at Tell al Far'ah (Petrie 1930) gave credence to a reading of the Papyrus Harris that depicted the *peleset* (later known as philistines) as Egyptian mercenaries (**§2.3**). Hence, this bichrome pottery became an indicator of "Philistine" settlement/presence, whilst simultaneously confirming the historical association of the *peleset* with the "Philistines". However, the whole scheme was turned on its end when the excavations at Ashdod revealed locally-made "Mycenaean IIIC:1b" monochrome pottery underneath the "Philistine" bichrome (Dothan 1982, 36-42). This earlier style was both Aegean-inspired and the supposed pre-cursor to the "Philistine" bichrome, leading some scholars to conclude that "Philistine" settlement in the Southern Levant was a two-phase process, each phase being associated with a ceramic style (e.g. Dothan *ibid*). However, similar monochrome pottery was attested in coastal areas considerably further north – areas not conventionally associated with the "Philistines" – suggesting that this local monochrome style could not be exclusively associated with the "Philistines" (Sherratt 2005a, 33). Furthermore, the monochrome pottery was absent from Egyptian garrison sites, where Egyptian texts placed *peleset* mercenaries. The lack of "Philistine" monochrome pottery at these sites suggested that either the *peleset* of the Egyptian texts could not be equated with the earliest settlement of "Philistines" (in the times of Ramesses III), or that the distinctive style of ceramics could not be associated with the "Philsitines" (Finkelstein 2005, 31-32; 2007, 521; Sherratt 2005a, 34). Were scholars taking a too literal reading of Egyptian texts? The association of the biblical "Philistines" with a specific ceramic style is based on a literal reading of the documentary sources. There are also theoretical grounds for rejecting an identification of a specific people-group on the presence of pottery styles; such an approach has been widely discredited (**§4.3**). Furthermore, archaeologists should be extremely cautious in interpreting changes in ceramic traditions as reflections of historical events and demographic transformations (Adams 1968; 1979).

Shishak/Shoshenq I

Another key archaeological date is Pharaoh Shoshenq I's campaign to the Southern Levant. Conventionally dated to either 926/925 BC or 918/917 BC (e.g. B. Mazar 1957), the event has also been placed in the mid-tenth century BC by some scholars (e.g. Finkelstein 2002, 110; Shortland 2005, 44). Regardless of which chronological scheme is proposed, archaeologists tend to agree that Shoshenq's raid is easily correlated with the archaeological record (usually associated with a destruction layer), since it is documented by both Egyptian (Karnak) and biblical texts (Åhlström 1993; Finkelstein 1996a; 2002; Mazar 1992, 373; 1997b, 157; B. Mazar 1957; Na'aman 1985; 1992, 81). Shoshenq's campaign is important for Iron Age specialists because it helps date tenth century BC strata, whilst also confirming the accuracy of the biblical account. While scholars might debate the specific date of the campaign, it is generally accepted as both historically- and archaeologically-attested fact. In doing so, archaeologists have had to gloss over assumptions that have the potential to undermine the absolute chronology of the IA-SL.

The first assumption is that Shoshenq's campaign to the Southern Levant can be associated with archaeological remains; i.e. "destruction levels". The campaign was recorded in an inscription at Karnak in the Nile Valley that included a topographic list of 154 places in the Levant, (Finkelstein 2002, 109-111). A lot depends upon scholars' ability to link place names listed on this inscription with those of actual sites. Nevertheless, a number of archaeological "destruction" levels have been accepted as resulting from this campaign, and the accompanying ceramic assemblages dated to the tenth century BC (K.A. Wilson 2005, 2). For instance, ash layers sealing Tell Abu Hawam III (Hamilton 1935, 67), Megiddo VA-IVB (Lamon & Shipton 1939, 61), Taanach IIB (Rast 1978, 26-27) and Arad XII (Aharoni 1978, 245; Herzog and Singer-Avitz 2004, 229; B. Mazar 1957, 64; Na'aman 1985; *cf.* Fantalkin and Finkelstein 2006, 18-21) have all been directly linked to Shoshenq's campaign. The association of a historically-attested military campaign with ash layers in the archaeological record certainly seems to oversimplify the relationship between history and material culture. Furthermore, the manner by which destruction levels are attributed to historical campaigns appears arbitrary; for instance, there were two destruction levels at Megiddo that were possible candidates for Shoshenq's campaign (Strata VIA and VA-IVB), (Finkelstein 1996a, 180; Harrison 2004c, 107-108; Mazar 1997b, 157; Watzinger 1929, 56-59).

A second assumption is that Shoshenq's campaign can be decisively dated. The seemingly impressive agreement between the biblical and Egyptian dates is, according to Shortland (2005, 44) and Hughes (1990, 192), the result of Egyptian chronology having been manipulated, consciously or unconsciously, to fit biblical chronology. New Kingdom chronology has, to some degree, been based on the biblical date for Shoshenq's invasion (e.g. Hornung 1964, 24-29; *cf.* Kitchen 1991; 2003, 121-124; 2007, §§8-9, 166-167). The dates of this campaign, however, are by no means secure; the use of Egyptian chronology for dating Palestinian destruction layers to Shoshenq's campaign introduces a real danger of circularity. Also, it has not been conclusively proven that Shoshenq's campaign, recorded only fragmentarily on Egyptian reliefs (J.A. Wilson 1969b), can be equated with biblical references (*I Kings 14:25; II Chronicles 12:2-9*) to the destruction of Jerusalem by an Egyptian king called Shishak (*cf.* Finkelstein 2002, 110; P. James *et al.* 1991, 229-231; Schreiber 2003, 85-89; Shortland 2005, 44).

A third assumption is that the Egyptian inscription relating to the campaign is, or was ever designed to be, an accurate history. The details of the campaign are not explicitly mentioned in the Egyptian text (K.A. Wilson 2005, 64). Discussions of Shoshenq's campaign address a number of issues (its purpose, the exact route of the armies, the order in which the cities were conquered within the topographical list) but rarely is the actual historicity of the text considered (Aharoni 1978, 200-203; Ahituv 1984; Ahlström 1993; Finkelstein 2002; B. Mazar 1957; K.A. Wilson 2005, 1-14; www.reshafim.org.il/ed/egypt/shoshenqi.htm). Scholars have tended to assume what they instead should be demonstrating; that the campaign actually happened as recorded. Moreover, it may not be appropriate to try and link place names in the list with actual destructions because the account is simply an itinerary (J.A. Wilson 1969b, 263; K.A. Wilson 2005, 64). The general style of Shoshenq's inscription is derivative, and draws heavily on earlier Egyptian campaign reliefs; much of the content is stereotypical, and appears to re-use old formulae and phrases in an attempt to copy the military records of the great pharaohs centuries earlier (*ibid*). This raises doubt as to its reliability as a historical account of the campaign at all. K.A. Wilson (*ibid*) has recently demonstrated that itinerary lists are a genre with a specific purpose and are generally not concerned with history. Wilson (*ibid*) points to the Karnak relief scene of the campaign, and notes that it does not portray a specific battle: no town is besieged and no fighting is taking place around the king. Instead Shoshenq is shown in the process of smiting a mixed group of people from foreign lands. The absence of the depiction of a specific battle and the presence of foreigners gives a non-historical and idealised character to the relief. Interpretation should, therefore, view the relief as a depiction of the pharaoh defeating the inhabitants of all foreign lands not as historical fact, but as idealized concept. According to K.A. Wilson (*ibid*), the topographical list should be understood in the same manner.

While Shoshenq's campaign is a key archaeological date for the IA-SL, recent scholarship, as outlined above, has demonstrated that it is inadvisable to use this event for chronological purposes. Nevertheless, archaeologists continue to accept the historicity of Shoshenq's inscriptions (e.g. Finkelstein and Piasetzky 2006). Moreover, scholars declare Shoshenq as legitimate proof for biblical accuracy, but ignore the diversity of population presented in the text which is at odds with the biblical picture of a united Israelite state (*cf.* Aharoni 1978, 192; Liverani 2005b, *passim*).

Solomonic Megiddo

Probably the most important chronological "anchor" for the IA-SL is the attribution of Megiddo Stratum VA-IVB to Solomon and the tenth century BC. This is important for a number of reasons; not the least because the recognition of material remains of the United Monarchy, a seminal period in the construction of the Hebrew people as a nation, was seen as an important means for authenticating modern Israel's claim for legitimacy (Faust 2006, 170*ff*; Kletter 2006, *passim*; Whitelam 1996, *passim*).

The Solomonic identification of Megiddo VA-IVB was based on a form of public architecture (large fortifications with a six-chamber gate) that appeared to fit the biblical description of Solomon's building program mentioned in I Kings 9:15 ("Here is the account of the forced labor King Solomon conscripted to build the Lord's temple, his own palace, the supporting terraces, the wall of Jerusalem, and Hazor, Megiddo and Gezer.") (Guy 1931, 44-48; Lamon and Shipton 1939, 59; *cf.* Mazar 1997b, 159). When the same stratum at Megiddo also revealed a long pillared building, which the excavators interpreted as stables for Solomon's chariot forces mentioned in I Kings 4:26 ("Solomon had four thousand stalls for chariot horses, and twelve thousand horses."), the Solomonic nature of Megiddo VA-IVB appeared to be confirmed. This was followed by the excavation of similar fortifications and gates at Hazor X and Gezer VIII (two cities mentioned alongside Megiddo in I Kings 9:15), which led to the six-chamber gate becoming a hallmark of Solomonic material culture (Aharoni 1978, 192-239; Barkay 1992, 306-308; Mazar 1992, 380-387; Yadin 1970, 66). While other chronological tools were employed to support the Solomonic interpretation (e.g. the Shoshenq stele from Megiddo), the crux of the argument rested on the references of I Kings 9:15 (Yadin 1970, 67). This text, however, does not explicitly detail the nature of Solomon's building program (Finkelstein 1996a, 178). Recent discoveries of six-chamber gates in contexts that clearly post-date the tenth century BC undermines the exclusively-Solomonic association of these features (Herzog 1992, 272-274). Furthermore, there is no extra-biblical evidence confirming the historicity of King Solomon, not least his building program at these sites. The identification of Solomonic Megiddo was based on biblical testimony (Finkelstein 1996a, 178-179). Instead, if biblical accuracy is not assumed, the evidence does not conclusively support the tenth century BC date of Megiddo VA-IVB (and other conventional tenth century BC strata).

As mentioned above, a fragmentary stele of Shoshenq I was found at Megiddo and was used to support the Solomonic interpretation of Stratum VA-IVB. This stele, however, was not found in a context that can be directly associated with the Solomonic stratum – it was found within the spoil heap of the German excavations, which had penetrated into what would later be identified as Stratum VIA (Finkelstein 1996a, 178; Lamon and Shipton 1939, 61; Schumacher 1908). It is unclear from which stratum the stele derived. Moreover, the fragmentary nature of the stele's inscription means it cannot be determined what exactly the stele commemorated (K.A. Wilson 2005, 71); the stele mentions Pharaoh Shoshenq's name and little else. Ussishkin's (1990a, 71) suggestion that the presence of a stele signifies conquest should be rejected, since stelae probably better reflect dominion rather than conquest (K.A. Wilson 2005, 72-73). Hence, the Shoshenq stele from Megiddo cannot be conclusively connected to one specific level in the archaeology of the site, nor to a historical destruction of the site.

Since its original excavation, the Megiddo stratigraphy has undergone a number of revisions, alterations and corrections (e.g. Aharoni 1972; Albright 1941-43, 18; Ussishkin 1990a; G. E. Wright 1950; Yadin 1960; 1970), yet the association of Megiddo VA-IVB with Solomon has generally persisted. Growing criticism of this interpretation in recent years (Davies 1992; Finkelstein 1996a, 178-179; Liverani 2005b, *passim*; Whiting 2007a,

27; Wightman 1990a) has resulted in a widespread debate over absolute dates for the Iron Age of the whole Southern Levant, which of course holds significant implications for the chronology of the entire eastern Mediterranean (Ben Tor 2000; Ben Tor and Ben Ami 1998; Dever 1997b; 2000; 2003; Finkelstein 1996a; 1996c; 1998c; Finkelstein and Piasetzky 2003a; 2003b; 2006; Gal 2003; Herzog and Singer-Avitz 2006; Mazar 1997b; 2000; 2004; 2005; Strange 2000).

Dating of Jezreel

Finkelstein's fourth chronological "anchor" is the dating of the site of Jezreel, located near Megiddo. Jezreel is mentioned several times in biblical texts, with two specific references of particular interest to the excavators of the site (Williamson 1991). The first (I Kings 21) makes reference to the "palace" of King Ahab at Jezreel, while the second (2 Kings 9-10; recounted in Hosea 1:4) makes a vague reference to the destruction of Jezreel during Jehu's *coup d'etat*. The biblical texts imply there was a large ninth century BC palatial structure at Jezreel that was destroyed by Jehu, and therefore the presence of a ninth century BC ceramic assemblage (Na'aman 1997, 125-127). Indeed, the large casemate enclosure found at the site was interpreted by the excavators as the ninth century BC "Palace of Ahab" (Ussishkin and Woodhead 1992, 53). While this correlation was seen as an important means for confirming biblical accuracy, it was also an attempt at resolving some of the concerns over absolute chronology for the IA-SL (Ussishkin and Woodhead 1992, 153). This interpretation, however, is not based on archaeological evidence but on an expectation of how the biblical narrative would manifest itself in the archaeological record; this is despite Williamson's (1991, 89) assertion that it is inadvisable to draw conclusions about the site from biblical references. Nevertheless, Finkelstein (1996a, 183) accepts Jezreel "as an extremely important chronological clue". He came to this conclusion because he found it difficult to understand the large casemate enclosure at Jezreel in any other light than that depicted in the biblical narrative (*ibid*). Hence, Finkelstein (1996a, 179; 1999a; 1999b; 2005) accepted the ninth century BC date, but in doing so committed the same error he had earlier condemned – an interpretation based on biblical testimony.

In addition to the fact that Jezreel's ninth century BC date was not obtained via systematic and scientific method, there is some contradiction and confusion in the evidence. In particular, Zimhoni (1997a, 25-26) considered the ceramic material as comparable with that from Megiddo VA-IVB, which was dated on biblical testimony to the tenth century BC, yet the biblical texts for Jezreel suggest a ninth century BC date - clearly both interpretations cannot be right; or the pottery is not reliable for chronological purposes. Either Jezreel dates to the tenth century BC, or Megiddo VA-IVB belongs in the ninth century BC, or the biblical narrative cannot be considered a reliable chronological witness. Furthermore, there were significant practical concerns with the excavation of Jezreel: the site was greatly disturbed, the recording of *loci* was haphazard and inconsistent, and the strategy was heavily influenced by expected biblical connections (Whiting *personal communication*; Zimhoni 1992, 57-58, 61; 1997c, 89). The date of the Iron Age casemate enclosure at Jezreel is archaeologically unsupported and, therefore, of little benefit for the chronology of the IA-SL. The interpretation of the site is drawn from a literal reading of the biblical narrative.

Scientific Dating Techniques

With the growing criticism of the established chronological framework in the Southern Levant, archaeologists have begun to implement more scientific dating techniques (e.g. Bruins *et al.* 2003; 2005; Finkelstein and Piasetzky 2003a; 2003b; 2006; Mazar and Carmi 2001). This is a conspicuously late development within Near Eastern Iron Age archaeology compared to other archaeological disciplines. While the importance of a sequence with scientifically-determined absolute dates cannot be underestimated, a number of recent radiocarbon programs have attempted to incorporate radiocarbon data into existing traditional frameworks. As a result, little attempt has been made to grapple with the complex methodological problems concerning the way chronological information is used to interpret the archaeological record (e.g. Mazar *et al.* 2005; Bruins *et al.* 2005). For instance, in their 2005 paper on the "Ladder of Time at Tel Rehov", Mazar *et al.* (2005, 193, 253) states *passim* that their radiocarbon dates "fit" the conventional chronologies, and conclude their paper by linking the Rehov Stratum V "destruction" with Shoshenq's raid. Mazar (*ibid*, 254) also links his radiocarbon dates with biblical events without demonstrating the historical veracity of the biblical account. The chronological data alone appears to be the motivation, as if this would support the historical narrative.

The historical narrative remains the key interpretive framework for the IA-SL, despite the application of scientific dating techniques. The so-called precision dating is instead being used to slot the archaeological record into the pre-existing conventional historical framework. The resulting dates could instead be used to construct an alternative, more–meaningful framework from the archaeological record, but they have not (Whiting 2007b). As a result, the chronology of the IA-SL remains structured according to biblical/historical testimony. This inability to break-away from existing methodologies is the reason the results of radiocarbon dating in the Southern Levant have been inconclusive – advocates of the "High Chronology" (Bruins et al. 2003a; 2003b; 2005; Mazar and Carmi 2001) and of the "Low Chronology" (Finkelstein and Piasetzky 2003a; 2003b; 2003c; 2006) have both interpreted the same results as supporting each of their views. Instead I echo Sherratt's (2005b, 119) call for "chronological flexibility" within the Iron Age, as the scientific dating techniques now being employed are evidently unable to give the chronological clarity originally expected.

Summary and Implications

Amongst the many textual sources that have influenced Near Eastern Iron Age archaeology, the biblical text holds a significant place. No other document has contributed so profoundly to excavation techniques, sampling methodologies and interpretative frameworks as the Hebrew Bible. The influence of the biblical narrative is spread across the wider eastern Mediterranean region through the wholesale adoption of South Levantine chronology; this is despite significant geographical, cultural and historical differences between regions. The above discussion has demonstrated that reconstructions of the IA-SL are not overly sensitive to the witness of the archaeological record, and have instead forced the archaeological evidence to fit into a pre-determined framework of established biblical history. The recent chronology debate has avoided addressing the fundamental interpretative framework and has instead focused on providing a temporally more-precise correlation between the biblical narrative and the archaeological record. Nevertheless, the debate in the Southern Levant has important implications for the re-writing of Israel's history (Liverani 2005b, 308-323; Whitelam 1996; Whiting 2007a). For instance, by lowering the dates of conventional tenth century BC strata, Finkelstein's (2005, 39) "Low Chronology" revokes the glorious past of the United Monarchy and undermines biblical veracity; the first Israelite nation-state is instead found in the Northern Kingdom of the Omride Dynasty.

The High/Low chronology debate in the Southern Levant also has important implications for the IA-NL. By bringing into question the bases for South Levantine chronology, researchers like Finkelstein (2005, 39), have effectively undermined not only the chronological framework for the entire eastern Mediterranean, but also current interpretative methodologies based on a literal reading of ancient historical sources. Regardless of whether the "High" or "Low" chronology eventually gains wider acceptance, one thing is clear - Iron Age chronology of the Northern Levant awaits the provision of a chronological framework independent of other regional chronologies (Akkermans and Schwartz 2003, 13; Cichocki 2000, 64-65). A good, well-controlled sampling program for charcoal and short-life carbonised organic matter during excavation is a simple, yet effective means for providing scientific data to supplement and inform the archaeological record.

2.6 Near Eastern Archaeology as a Product of Western Ideas

Reconstructions of the IA-NL have been formed within the framework of particular temporal, socio-political, and intellectual contexts (Silberman 1993c, 546). This section explores how European ideas, assumptions, and priorities have framed discussions of the ancient Near East. This is particularly clear for the Southern Levant, where European views were influenced by a biblical perspective of the "Holy Land" formed through centuries of pilgrimage and crusades (Grabois 1988, 66; Silberman 1982, 8; Whiting 2007a, 5-7; Wilken 1992, 102). The European image of the Northern Levant, however, has developed through slightly different stimuli.

Post-medieval Europe considered itself to be the most morally and intellectually advanced culture in the world; a view intricately connected to the belief that Europe was heir to the superior form of culture developed by the Greeks and Romans (Maisels 1999, 5; McCall 1998, 183). As a result, Renaissance Europe became obsessed with all aspects of the classical world (Lowenthal 1985, 75-80; Trigger 1989, 35). Accordingly, classical texts were seen as a cultural guide for all manner of study: politics, philosophy, art, literature, and the ancient Near East (*ibid*). One side-effect of this perspective was the imposition of a classically-inspired map onto the modern landscape of the Northern Levant, despite the formulaic and stereotypical views presented by the classical authors (see Said 1978, 56-58 for a brief discussion of a few classical authors). The archaeological identification of sites mentioned in classical literature, therefore, became an important motivation for the study of the ancient Northern Levant: e.g. Ras al Bassit was excavated because it was identified with classical Posidaion (Courbin 1990b, *passim*); the excavation of Tell Kazel sought to confirm its identification with classical Sumur (Badre 1990a, 14). Classical texts became the interpretative lens through which archaeologists viewed material culture of the region.

Linked to Europe's classical view of the past was a desire to illustrate the superiority of Greek culture and progression of human civilisation (Jenkins 1992, 56-74; Reade 1987, 48; Waterfield 1963, 138). This was especially evident in the evolutionary theory of art, which suggested that a line could be drawn linking the most primitive attempts at art with that of the Greeks, whose Parthenon sculptures were considered the absolute pinnacle of artistic tradition (McCall 1998, 198). As a result, the artistic traditions of the ancient Near East were only understood relative to the Greek model and were thus considered primitive (e.g. Canby 1985; Ingholt 1942, 474). Western archaeological literature tended to portray the Northern Levant as a land of origins and early civilisation, but one that fell off the main trajectory at some point, when the "torch" of civilisation passed to Greece (Bahrani 2000, 6).

By emphasising the transmission of "civilisation" from the Levant to Greece, European scholars were effectively dispossessing the modern population of its past (Bahrani 2000, 6; Shank and Tilley 1987a, 195). This was achieved through the perpetuation of apolitical geographical terms that held no relevance for the modern inhabitants of the Levant; e.g. "Mesopotamia", "Assyria",

"Phoenicia", "Levant", "Syria" (Bahrani 1998; 2000, 7; Lowenthal 1985 Wengrow 2006). Designations like "Holy Land", "Levant", "Near East", and "Middle East", also betrayed a Eurocentric conception of the world (McCall 1998, 211; Whitelam 1996, 40). In other words, the ancient Northern Levant was conceived, performed and interpreted via a distinctly Western European perspective; what Said (1978, 1) has called an "Orientalist" perspective. Indeed, recent research has demonstrated the strong vein of Orientalist thought in archaeological work across the region (Bahrani 1998, 2000; Kohl 1989; Larsen 1987a; 1989; Liverani 1999; 2005a). But for Said, Orientalism is not only about Europe trying to make sense of the "Orient" (Said 1978, 166), it was about control. Indeed, Said (1996, 28) suggests that the naming of the land implies *control* of that land. Hence, the archaeological "map" of the Northern Levant played a role in the conceptualisation and control of European colonial territories (Bahrani 1998, 171; Liverani 1994; Trigger 1984, 360-363).

During the French Mandate period, French colonial interests ensured the persistence of the European Orientalist perspective. At the same time, Islamic archaeology, which was more closely aligned with the modern population, was relatively ignored (Strika 2000, 1583). The antiquities departments of Syria and Lebanon were administered not by the French Mandate authority but by French national academic and government bodies based in France; the *Académie des Inscriptions et Belles-lettres*, the *Musées Nationaux*, the *Ministère de l'Instruction Publique*, and the *Ministère des Affaires Étrangères* (Chevalier 2002, 308-309; Matthiae 1981, 32). With the establishment of the *Services des Antiquités* and the enactment of laws protecting a recognised list of ancient sites, the tells and ruins of the Northern Levant acquired a legal status that superseded any local meaning they might have possessed; the link between the indigenous population and the region's history was further fractured (Gelin 2004, 28).

In the post-colonial period, European interpretative models have persisted. Avowedly nationalist in outlook, indigenous archaeology's rejection of terms such as "Near East" or "Orient" confirms a strong anti-Orientalist perspective, yet archaeological practice remains squarely rooted in European traditions (Bahrani 2000, 6; Masry 1981, 239; Matthiae 1981, 31-32). This is one of the ironies of the post-colonial situation; i.e. that colonial thought has shaped nationalist discourse that is avowedly anti-colonial. Nevertheless, indigenous archaeology has sought to reclaim its past through the promotion of the Northern Levant as a land of origins, emphasising its role in the universal development of humanity (e.g. the shift from hunter-gathers to sedentary farming communities; the development of the alphabet – www.syriatourism.org).

A number of deeply-embedded European traditions have had an impact, positive or negative, on the practice of Iron Age archaeology in the Northern Levant. The multiple threads of biblical and classical traditions, as well as Orientalist, Colonialist and Nationalist perspectives, have all played important roles in shaping the intellectual climate within which the discipline has developed. The result is an archaeological paradigm that continues to emphasise European images of the Levant.

2.7 Reassessing Chronologies of the Iron Age Northern Levant

The above discussion has highlighted the fact that current reconstructions of the IA-NL are inadequate. The undue prominence given to the conventional historical narrative has resulted in interpretations of the archaeological record that overemphasise political histories, military campaigns, people movement, and the accession and succession of kings. Furthermore, the ascription of "absolute" dates to material culture is often based on circular reasoning and unreliable correlations between the archaeological record and conventional history. In fact, the foundations are remarkably weak, and the basis for maintaining the current paradigm insufficient.

The Beginning of the Iron Age

Conventional archaeological practice dates the beginning of the Iron Age to around 1200 BC in the eastern Mediterranean (except Cyprus, which Gjerstad [1948, 421-427] and Birmingham [1963, 39] both date to the mid-eleventh century BC). This interpretation is based upon the conventional dating of the political crisis responsible for the collapse of the Late Bronze Age palace-economies. While the cause of this crisis is a much-debated topic (*cf.* Liverani 1987; McClellan 1992; Sader 1992; Sherratt 1998, 292-293), its date is widely accepted. However, neither the date nor cause for this crisis appears to be connected with the development and widespread adoption of ferrous technology; a phenomenon generally meant to define the "Iron Age". Indeed, many of the conventional archaeological indicators for the beginning of the Iron Age are not characteristic of this period. For instance, the development of the alphabet and private enterprise are Late Bronze Age innovations, while the domestication of the camel and adoption of iron as "working metal" come some centuries later (**§2.3**). Furthermore, the material culture of the Iron I period is remarkably similar to that of the Late Bronze Age. Any fixing of the Bronze/Iron Age division should acknowledge the cultural continuity that exists between the conventional Late Bronze and early Iron Ages (Mazzoni 2000d, 1043; R. H. Smith and Potts 1992, 83).

The "Crisis Years" are generally associated with economic and political decline and/or collapse, followed by a period of stunted development (e.g. Burdajewicz 1990, 1-23). This 300 year 'period of recovery' has been called a "dark age" (Morris 2000, 33; Muhly 1992, 20-

21); a theory supposedly supported by a general lack of surviving textual and cultural material (Klengel 1992, 182). This "dark age" theory has encouraged some scholars to try to redefine early Iron Age chronology. For instance, James (*et al.* 1991) accounts for the apparent gap by lowering the Late Bronze-Iron Age transition by 300 years. However, in doing so, James also removes a large block of archaeological material (i.e. all pottery conventionally dated to the Iron I period is effectively ignored and lost). Instead of simply removing the 300 year gap from chronologies, an alternative approach would be to explore possible explanations for the apparent gap in occupation. For instance, is the absence of textual evidence really only the loss of archaeologically-visible texts?

The sweeping manner by which James applies his "Low Chronology" suggests that a "dark age" has been proposed for every site in the Mediterranean, but this is not the case. A number of sites display clear evidence of occupation between the conventional Late Bronze Age and Iron II period (e.g. Tell Afis Phase VII; Megiddo VI). This is the main fault with the "dark age" theory; not all sites display an occupation gap for the period in question. I would instead suggest that the "dark age" has been artificially created by the inability of early Iron Age written documents to survive, and is compounded by a widespread inability to recognise early Iron Age material culture. This might be due to the fact that material culture of the conventional Iron I period is not always easily discernible from that of the Late Bronze Age (Anderson 1988, 390), especially away from coastal regions. This may also explain the lack of Iron I data from recent surveys around Hama (Bartl *personal communication*), Tell Mishrifeh (Morandi Bonacossi *personal communication*), and Homs (Philip *et al.* 2005, 40). Consequently, if Iron I pottery was mistakenly attributed to the Late Bronze Age, a gap in occupation would be apparent. For example, at Carchemish Woolley (1952, 235) assigned the Amarna pottery to the Late Bronze Age and the Yunus pottery to the Iron II period, yet he comments that there is no abandonment of the site discernible in the stratigraphy between these two ceramic styles. Moreover, the city architecture displays close links with Hittite Imperial art, suggesting no gap in occupation during the early Iron Age. Accordingly, while James (1987; *et al.* 1996, 318) saw this as a reason for lowering the chronology by 300 years, I would suggest that it implies Woolley was unable to identify Iron I pottery from within the Amarna–styled assemblage.

The archaeological evidence supports neither the "dark age" theory, in the sense of an occupational gap, nor James' (*et al.* 1991) "Low Chronology". Nevertheless, the current chronological framework is inadequate. Current terminology emphasises rapid change between the conventional Late Bronze and Iron Ages that is not always evident in the archaeological record of the Northern Levant. Instead, continuities in material culture invite the extension of the Late Bronze Age to include what is currently called the Iron I period (Anderson 1988, 390).

Sub-dividing the Iron Age

Assigning an absolute date to the Iron I-II transition has proven difficult, as attested by variation in the chronological periodisations of the IA-NL (Cecchini and Mazzoni 1998; Jamieson 2000; Lebeau 1983; Makinson 2005; Mazzoni 1990a; 2000a; Moorey 1980 – see **Table 2.7**). While archaeologists are unlikely to declare that Iron Age chronology is finalised and definite, most agree on some key points: that the Iron Age began in the early twelfth century BC; the Iron I-II transition corresponded with the advent of Red-Slip pottery in the Northern Levant; and the Iron II-III transition coincided with Neo-Assyrian ascendancy in the region.

The most widely-accepted periodisation of the IA-NL is that of Mazzoni (2000a; 2001), who has identified a number of phases within the conventional Iron Age marked by rapid change in material culture. Foremost of these changes is the appearance of burnished Red-Slip pottery. According to Mazzoni (1990a, 79), the advent of this distinctive ceramic style coincided with a decrease in monochrome painted pottery and an improvement and standardisation in firing and throwing technologies (also Cecchini 1998, 277; *contra* Fugmann 1958, 267-268). Mazzoni (2000a, 41-42; 2000d, 1050) has tried to date this transition to the end of the tenth century BC by the presence of Greek pottery in Iron I contexts at Tell Afis (Bonatz 1998; Mazzoni 2000a, 41-42) and through an appeal to Braemer's (1986, 222, 246) Red-Slip chronology, but neither of these chronological "tools" can provide a secure date. Braemer's Red-Slip sequence dates derive from the Ras al Bassit stratigraphy which has never been published and cannot be truly assessed, and the dating of Greek imported pottery is based on circular reasoning (**§2.5**). Indeed, Mazzoni (2000a, 41-42; 2001, 101) admits that her dating of the Iron I-II transition cannot, on present evidence, be defended archaeologically. In the end, the chronological date for the Iron I/II transition has not been independently established in the Northern Levant, but has been loosely based on the same transition in the Southern Levant; i.e. sometime in the tenth century BC.

A key date for the IA-NL is the 720 BC destruction of Hama (**§2.4**). This event is usually associated with a break in Iron Age cultural traditions, and has become the hallmark of the Iron II-III transition (e.g. Matthers 1981b, 415; Mazzoni 2000a). But the change in material culture associated with the end of this period is connected with the Assyrian conquest only because the latter is considered significant enough to bring cultural change. There is little direct evidence to link the two. The dating of the shift in material culture is based on the expected manifestation of conventional history in the archaeological record.

Table 2.7: Comparative periodisations for the Iron Age Northern Levant

BC	Mazzoni 1990a	Mazzoni 2000a	Makinson 2005	Jamieson 2000	Lebeau 1983	Moorey 1980	Woolley 1914	Hachmann 1983	Haines 1971
1200			LB II						
1150									Amuq N Iron I
1100					Iron I	Early Iron			
	Iron IA	Iron IA-C	Iron I						
1050				Iron I				Ältere Eisenzeit	
1000			(1150-1000)						(1200-1000)
950	Iron IB		Iron IIA				Late Hittite		
900			(1000-900)	(12th-10th)					
850					Iron II				
	Iron IIA		Iron IIB			Middle Iron			Amuq O Iron II
800		Iron II		Iron II					
750	Iron IIB		(900-750)						
							(1100-718)		
700				(9th-8th)				Mittlere Eisenzeit	
			Iron III						
650	Iron IIIA	Iron III		Iron III	Iron IIIA	Late Iron	Late Hittite		
600		(720-586)	(750-600)	(7th)			(718-605)		
	Iron IIIB								(1000-500)
550		Persian	Persian	Persian	Iron IIIB	Persian	Persian	(800-500)	Amuq P
	Persian							Jüngere EZ	
350		(586-330)	(600-330)				(605-4th)	(500-0)	

The End of the Iron Age

Absolute dates for the Iron Age rely on historical data, and the end of the Iron Age is no different. Following conventions in South Levantine archaeology (e.g. Aharoni 1978, xviii), many North Levantine scholars use the end of the Neo-Babylonian empire as a terminal date for the Iron Age (e.g. Jamieson 2000; Makinson 2005; Mazzoni 2000a; Moorey 1980). However, Lebeau (1983, Fig. 6, 22) and Lehmann (1996, Tab. 4.9) have proposed the inclusion of the Persian period in the Iron Age, preferring to see the end of Iron Age culture (however that is defined) with the arrival of Alexander's Hellenism in the fourth century BC (Hachmann 1983, 186 is alone in including both the Persian and Hellenistic periods in the Iron Age). Whether the Persian period is included in the Iron Age or not, Lebeau (1983), Lehmann (1996, Tab. 4.8) and Mazzoni (1990b) have each identified a significant break in material culture between the Iron III period and the following, so-called Persian period (Lehmann 1996, 86-87). The argument for attributing a mid-sixth century break in culture to the growth of the Persian Empire, however, is based on an unconvincing premise; i.e. Persian ascendancy is expected to have a significant effect on material culture. Once again, absolute dates are not based on scientific data, but historical convention. Regardless of which date/event is preferred as the end of the Iron Age, it is clear that significant historical events provide the impetus for assigning absolute dates (Bunnens 2000a, 19).

2.8 Concluding Summary

This chapter has explored some of the key assumptions behind conventional interpretations of the IA-NL. It has demonstrated that reconstructions are based on the conventional historical narrative, as derived from a literal reading of ancient documentary sources. Archaeologists have also been influenced by their own modern intellectual environments. In contrast, the archaeological record has played only a minor role.

Historical sources play a central role in interpretations of material culture proposed by archaeologists working on the IA-NL. By implication this approach assumes a direct link between historical events and the archaeological record, as if all significant events, and only significant events, had an immediate and lasting effect on material culture. As a result, archaeological reconstructions have tended to emphasise change (e.g. conquest, migration). However, if the archaeological record is emphasised over historical events, an archaeological account of the IA-NL might look significantly different. For instance, there is significant evidence to suggest that many early "Iron Age" societies were continuations of Late Bronze Age cultures, and that rapid cultural change only occurred toward the end of the Iron I period. This cultural "rethink" has implications for the use of terms such as "Phoenician", "Aramaean" and "Sea Peoples". It would also remove the chronological anchorage of destruction layers (e.g. "Sea Peoples", "Assyrians") and open the way for under-utilised scientific dating methods. Moreover, historical events could no longer be used to "explain" cultural change (e.g. Aegean influence of "Sea Peoples"), which would encourage an exploration of internal agents for cultural transformation.

This chapter has also highlighted the fact that archaeologists are products of their environment. Emphasis on the historical narrative reveals a deeply-embedded European view of texts and their historical primacy. The adoption, conscious or not, of European traditions has also resulted in an archaeological paradigm that has emphasised a particular view of the Levant; a view that is value-laden and to some extent political in nature.

Despite the flawed nature of the current dataset, archaeologists can access an understanding of Iron Age society. This thesis will not argue for the abandonment of history; it instead suggests that the historical narrative is not the whole story. The historical narrative provides a somewhat more generalised reconstruction of past communities. A close engagement with the archaeological record will instead present a reconstruction of the Iron Age Northern Levant that accounts for the complex and diverse nature of ancient communities. But while the "limited" dataset used for this study will be given "a new voice", better data would "speak volumes". Hence, the present study has implications for the future study of ceramics in the Levant. Through more systematic and thorough recording of ceramics, combined with a transparency of sampling strategies, quantified data will become available and a more detailed investigation of Iron Age communities in the Northern Levant will be possible.

The conclusion that historical texts do not provide a single, "factual" history of the Iron Age has important implications for archaeology. Since such histories have formed the main interpretative framework for the archaeology of the IA-NL, current approaches to interpreting material culture (e.g. pottery) need to be reconsidered (**Chapter 4**). To do so it is important first to survey specific cases of archaeological practice in the Northern Levant and isolate the manner by which material culture is being interpreted; this is explored in the following chapter.

CHAPTER THREE

Critical Review of Iron Age Excavations

Chapter 2 has shown that current reconstructions of the Iron Age Northern Levant are not the result of an engagement with material culture but the imposition of a pre-determined historical framework onto the archaeological record. The aim of this chapter is to explore the excavation, recording, interpretation, and publication of North Levantine Iron Age sites through a review of chronological and historical conclusions from each site. Apart from presenting a history of research, the objective is to reveal recurring themes and assumptions within this field of research. While the exercise will reveal limitations in current reconstructions of the IA-NL, it is an important tool for identifying possible directions for future research.

The following review is directed toward all "systematic" excavations of Iron Age sites in the Northern Levant and surrounding areas, as defined in **Chapter 1**. Consequently, it will include a number of excavations that could hardly qualify as "systematic" in their approaches, as well as a number of more meticulous investigations. In both situations, however, it is usually apparent that the conventional historical narrative has been used to interpret the archaeological record.

3.1 Abou Danne, Tell (Syria)

Summary of Excavations

The main strategy of research at Tell Abou Danne was to illuminate the Iron Age of the Aleppo area, a period poorly represented by ceramic material at the time of excavation (Tefnin 1983, 141). To this end, a large sondage was begun on the northern edge of the mound, where it was expected to reveal a complete stratigraphic history of the site; the earliest occupational deposits dated to the beginning of the third millennium BC (*ibid*). An Early Bronze Age casemate wall was exposed in Area B, on the western slope of the tell, while Area A, situated on the south-east corner of the acropolis, revealed the Iron Age occupation of the site. No cohesive architectural plans were immediately evident (*ibid*).

Review

At the time of publication, Lebeau's (1983) study of the Tell Abou Danne ceramics was one of the first typologies for the inland Northern Levant – only the limited typologies of Hama (Riis 1948) and Jerablus (Woolley 1939b) had been published. Consequently, Lebeau had little local material available for a comparative study. He was instead required to look to the Southern Levant, with its abundance of published typologies. Lebeau considered Tell Abou Danne level IId contemporary with Samaria IV-V, Megiddo IV-III, Hazor VIII-V, Hama E, Amuq Ob-c, al Mina X-VIII, Tell Arqa 10B-D, Sarepta D, Tyre X-IV and Tell Keisan 7-6, while level IIc was comparable with Samaria VI-VII, Megiddo III-II, Hazor IV, Amuq Od, al Mina VI-IV, Tell Arqa 9C, Sarepta C, and Tell Keisan 5-4 (Lebeau 1983, 24). The far-reaching comparisons suggest that Lebeau was hoping to position Tell Abou Danne within the greater narrative of the eastern Mediterranean. The main output of his study was a ceramic chronology for the later Iron Age. Lebeau's proposed chronology, however, differed from that of Tefnin (1980b), the project director, by as much as two centuries (**Table 3.1**). Nevertheless, it is Lebeau's ceramic chronology that is more widely accepted over Tefnin's (e.g. Makinson 2005, Tab. 2, p. 486).

Lebeau's ceramic study has been criticised in recent years (Lehmann 1996, 99). Not only does Lebeau's comparative analysis include a number of distant parallels, he includes comparative material without any critique of excavation method at other sites. The result is a chronology that relies upon a number of problematic sequences (e.g. Hazor, Hama) and unsupported dating schemes. The use of Greek imports to support his proposed sequence only compounds the problem, since Greek dates are based on the same South Levantine data upon which Lebeau (1983, 21-26, 134-135) also relied. While this circularity might account for the discord between Lebeau's and Tefnin's dates, problems with the chronology of Tell Abou Danne remain unresolved.

Table 3.1: Comparative Stratigraphy of Iron Age Tell Abou Danne

Level	Tefnin 1980b	Lebeau 1983
IIb (A4)	400-200/150 BC	6th & 5th cent. BC
IIc (A5)	500-400 BC	7th cent. BC
IId (A6)	650-500 BC	875-750/700 BC

(After Lehmann1996, 99)

3.2 Abu Hawam, Tell (Israel)

Summary of Excavations

In 1932, the British Mandatory Department of Antiquities began systematic excavation of Tell Abu Hawam in response to unhindered looting at the site. Work progressed quickly and a sequence of five main architectural phases (strata) belonging to the Late Bronze and Iron Ages were identified. The earliest stratum (V) revealed large orthostat walls that were laid directly onto low sand dunes, suggesting that the settlement was founded during the latter stages of the Late Bronze Age. The material from Stratum V bears strong Cypriot affinities (Hamilton 1934, 75). The occupation of the town seems to have continued without serious interruption into the early Iron Age (Stratum IV), with the pottery of the two strata merging into one another almost imperceptibly (Hamilton 1935, 66). These two strata were distinguished by a thin layer of ash separating the architecture. A more significant ash layer within Stratum IV was interpreted by Hamilton (1934, 77) as a violent conflagration, which he dated to the twelfth century BC and attributed to the "Sea Peoples". This "destruction" layer effectively split Stratum IV into two distinct sub-strata; Strata IVa and IVb (**Table 3.2**). A small ash layer also separated Stratum IV from Stratum III, with the foundations of the latter being placed directly on the ashes of the former. In this case, however, Hamilton chose not to attribute the ash layer to any historical conquest. Stratum III yielded significant architectural remains of buildings and a city-wall, the quality of which Hamilton (1935, 6) considered "not remarkable either for stability or design". The Stratum III pottery holds clear parallels with pottery from the mid-Levantine coast, as one would expect considering its proximity. Stratum III was also sealed by a layer of ash. This ash layer and the clear discontinuity between Strata III and II – in both alignment and construction – implied a break in habitation of the site; a break Hamilton attributed to the violent campaign of the Egyptian Pharaoh Shishak (Shoshenq I?), conventionally dated to the tenth century BC (**§2.5**). The large amount of late Iron Age-Persian period pottery in Stratum II suggested to Hamilton (1934, 77) a lengthy abandonment of the site following Shishak's destruction, before significant re-occupation in the late sixth century BC (Hamilton 1935, Pls I, III).

Since the 1930s, two excavation projects have returned to the site. A French-Israeli project (1980s) reinitiated excavation with the primary goal of assessing Hamilton's much-maligned stratigraphic sequence (e.g. B. Mazar 1951; Van Beek 1951; 1955) and in 2001, a salvage project, led by Michal Artzy (2007), returned to the site and excavated Persian and Late Bronze Age levels.

Review

There are a number of problems plaguing Hamilton's Tell Abu Hawam chronology. First, there was a conspicuous absence of what Hamilton (1935, 67) considered "datable" material at Tell Abu Hawam; two scarabs and a bead bearing the cartouche of Amenophis III were the only examples discussed in his two published reports. Second, Hamilton (*ibid*) imposed absolute dates upon the site by associating arbitrarily-chosen ash deposits with historically-attested military campaigns (e.g. Shishak; "Sea Peoples") (**§2.3**; **§2.5**). Apparently, Hamilton's interpretive method was strongly influenced by the historical record and his chronology constructed to fit the historical data. Third, Hamilton (*ibid*, 8) held a limited understanding of Iron Age material culture; his proposal for a lengthy abandonment of the site (between the ninth and sixth century BC) was based on the absence of what he considered to be characteristic "Israelite" pottery. Hamilton did not consider possible influences on material culture outside those depicted in the biblical account. Finally, Hamilton's practical method has come under close scrutiny (Artzy 2007, 357-358; Balensi *et al.* 1993). When archaeologists returned to the site, the aim was to refine Hamilton's absolute dating and evaluate and correct his stratigraphic sequence (Balensi 1985, *passim*).

Table 3.2: Hamilton's Summary of Tell Abu Hawam Strata

Str.	Period	Date	Pottery
I	Mixed topsoil	Hellenistic/Roman	Mixed
II	Graeco-Persian	Late 6th - early 4th cent.	Greek glazed Basket-handled
	destruction by Shishak		
III	Iron Age I - II	1100-925 BC	Red-Slip Black-on-Red Bichrome painted
IVb	Early Iron Age	1195-1000 BC	No Late Bronze No Late Helladic Bichrome painted
	destruction by "Sea Peoples"		
IVa	LB-Iron trans	1230-1195 BC	Late Helladic III Bichrome painted
V	Late Bronze Age	1400-1230 BC	Late Helladic III Cypriot LB

(After Hamilton 1935, 66)

Table 3.3: Balensi's Revised Chronology of Tell Abu Hawam

Revised	Original	Period	Cent. BC
IIA-B	II	Persian	Fifth-fourth
IIIA-B	None	Iron II	Tenth – eighth
destruction		-	*Late tenth*
IVA-B	IVb-III	Iron I	Eleventh – tenth
destruction		-	*c. 1200*
VA-C	IVa	LB	Fifteenth-twelfth
VI	V	MB	Sixteenth-fifteenth

(After Balensi *et al.* 1993)

The exposure in 1984 of "Middle Iron Age" material led archaeologists to question the validity of Hamilton's proposed abandonment of the site (between Strata III and II) and, consequently, his whole chronological scheme (*ibid*). But while the revised sequence redefined the site's strata, the absolute dates continued to be derived from the conventional historical narrative (**Table 3.3**). While not explicitly named, the historical conquests of the "Sea Peoples" and Shoshenq I (Aramaean conquest has also been suggested – Negev 1972, 10) were key chronological anchors for Balensi's (*et al.* 1993) sequence. Moreover, the ash layers associated with these events were different to those used by Hamilton (e.g. Balensi associated the "Sea Peoples" with the ash layer between Stratum V and Stratum IV, while Hamilton preferred the ash layer within Stratum IV). Dating the different ash layers in the Tell Abu Hawam sequence appears to have been a case of historical "best fit". There was no attempt to demonstrate that the selected ash layers represent violent conquest. There are numerous candidates for destruction other than an invading foreign power; e.g. revolt, earthquake, drought and abandonment, accidental fire.

Balensi's sequence was able to draw on a much larger corpus of Iron Age pottery than Hamilton and, as a result, possibly represents a more accurate ceramic sequence – Balensi's limited publication prevents confirmation of this. Balensi also had at her disposal much-improved ceramic theory, yet she employed a historical interpretative framework unchanged from that of Hamilton, fifty years earlier. The result is an absolute chronology for the site that is based on arbitrary associations of history with the archaeological record. In the end, Balensi's chronology is a revision of Hamilton's conclusions, rather than a complete reworking of the site's chronology.

While Artzy (2006; 2007) has published limited results of the recent salvage project, her conclusions appear to support Hamilton's dates for Stratum V; i.e. Late Bronze Age. Artzy's Stratum V, however, was directly overlaid with Stratum II material, implying a very limited Iron Age occupation (Strata IV-III) of Tell Abu Hawam (*ibid*, 365), which is contrary to the conclusions of both Hamilton and Balensi. There is still much work to be done before the Tell Abu Hawam sequence is clarified.

3.3 Afis, Tell (Syria)

Summary of Excavations

The first excavation of Tell Afis, as part of the Tell Mardikh/Ebla project, was a trench situated on the acropolis (Matthiae 1979; 1985). An earlier survey of site by Albright yielded significant amounts of Iron Age pottery, which seemed to confirm its identification with ancient Hazrek (Mazzoni 1998b, 8). Matthiae's investigative probe revealed the fragmentary remains of a sizeable *bit hilani* palace and densely-packed domestic units indicative of an extensive Iron Age settlement.

Another Italian project returned to the site in 1986 with the intention of classifying and defining the phases of the Iron Age ceramic sequence (Cecchini and Mazzoni 1998, 1). To this end, a number of trenches were excavated across the site's upper and lower mounds (**Table 3.4**) The investigations in Area D of the southern Lower City successfully provided an uninterrupted sequence of occupation for the eighth and seventh centuries BC (*ibid*; Mazzoni 1988a; 1998b, 23), while Area B, along the northern slope of the lower mound, explored the Iron Age outer city-wall (Virgilio 2005). The discovery of other segments of the city-wall in Area M and of the foundations of a dismantled gate have helped confirm the extent of the site in the Iron Age (*ibid*; Del Vesco 2002).

In addition to trenches on the lower mound, work resumed on the western half of the acropolis (Area A), where the bit-hilani palace had been uncovered. The renewed work revealed fragments of mud-brick architecture from the Iron II and Iron III periods (D'Amore 2002; 2005; Soldi 2005). In 1988 Area E_1 was opened on the western slope of the acropolis with the aim of obtaining a complete stratigraphic history of the site; twenty-six levels have been exposed so far, dating from the Late Chalcolithic through to the Iron Age, when Tell Afis reached its greatest size (Mazzoni 1998a; Venturi 1998a). The sequence included the elusive and poorly-understood Iron I period (Mazzoni 1998b, 17; Venturi 1998a, 124). Excavation in Area E_1 also revealed a Late Bronze II 'Residency' and "Pillared Building".

Table 3.4: Location of Tell Afis Trenches

Area	Location
A	W area of acropolis
B	N slope of lower mound
C	S slope of acropolis
D	S area of lower mound
E	W slope of acropolis
F	N outer city wall
G	E summit of tell
H	NW edge of lower mound
J	S edge of acropolis
L	SE edge of acropolis
M	NW area of tell
N	E side of acropolis

In Area G, on the eastern summit of the acropolis, the 1992 discovery of a large sunken courtyard, dated to the Iron II period, persuaded the excavators to intensify work there. The perimeter of this enigmatic structure was originally surrounded by mud-brick walls eight metres high, the collapse of which effectively buried the whole square (Cecchini 2000a, Fig. 1; Mazzoni 1998b, 21). The almost complete Iron Age sequence from Area G, with its abundance of Red-Slip bowls (tablewares), provides an important comparison for the domestic material from Area D (Cecchini 1998; Oggiano 1997, 186). Areas L and N on the acropolis also yielded a significant amount of Iron Age material (Cecchini 2002; 2005; D'Amore 1998a; Magazzu 2002).

Table 3.5: Relative Iron Age Chronology of Tell Afis Excavation Areas

c. BC	Phases	Strata	Area E		Area G			Area D
			E_1	E_2	East	Central	North	
1300		Afis VI						
	LB II		10					
1200			9c					
			9b					
1150			9a					
1100	IA IA		8					
1050		Afis VII						
1000			7					
	IA IB		6					
950								
			5	5	5			
900	IA IC		4	4	4			
			3	3	3			
850	IA IIA		2		2			8
			1					
800		Afis VIII						7 6
	IA IIB			2				
750						8b		5 4
				1	1	8a	5	3
700						↑	↑	
								2
650								
	IA III	Afis IX						1
600						↓	↓	
						1	1	
550								

(After Mazzoni and Cecchini eds 1998, 4)

Review

Like other sites bearing evidence of both the Late Bronze and Iron I periods, Tell Afis Area E_1 bears witness to remarkable cultural continuity across the two periods (Oggiano 1997, 186; Venturi 1998a, 135-136). Nevertheless, the two periods were considered distinct enough to maintain the chronological division. The Late Bronze II period is represented by Levels 10-9c, which are characterised by the so-called "Residency" building that was destroyed by fire (*ibid*, 134). This "destruction" was followed by a lengthy period of sporadic occupation (Level 9b), before Levels 9a-8 witnessed a full reoccupation of the site in the early Iron Age. While Venturi (*ibid*, 134-137) suggests that the idea of a total regional collapse as a result of an invasion by "Sea Peoples" is not a good explanation for archaeological "events" across the inland Northern Levant, he did look to "the widespread crisis at the end of the thirteenth century BC" (*ibid*, 135) as the key chronological indicator for the "destruction" of the Tell Afis "Residency". In other words, the dating of the Tell Afis Late Bronze-Iron Age transition is based on the conventional historical narrative surrounding the "Crisis Years" and the expected effect these events would have on material culture (Mazzoni 2000a, 31). The appearance in Level 9b of Aegean-style Monochrome vessels alongside the local pottery tradition appears to steel Venturi's resolve regarding population movement and a late thirteenth century BC date (Venturi 1998a, 135). If the assumption of "invading peoples" or "settling tribes" is removed, would the beginning of the Iron Age at Tell Afis still be dated to 1200 BC (Bonatz 1998, 219; Venturi 1998a, 135)? While the destruction of the "Residency" is a likely indicator of cultural change, this "event" has not been definitively dated, except through references to the historical narrative. Indeed, Venturi (1998a, 134) laments the general lack of imported wares because it "prevents the establishment of a firm date". Even those ceramic vessels that are presented as imports, which Bonatz (1998, 219) considers unusual for a site so far removed from the coast, have not been conclusively proven to be non-local. Furthermore, when Venturi (1998a, 135) does discuss the chronological value of certain ceramic types, he fails to provide a reference to support the dates (e.g. the bell-shaped bowl with antithetical spiral design; the evolution of Kamid el Loz cooking-pot types). It seems that the chronology of the early Iron Age at Tell Afis has largely been derived from the conventional historical narrative, and therefore remains open to revision.

Like with the start of the Iron Age, the date for the Iron I–II transition at Tell Afis has not been decisively established. Level 2 in Area E_1 witnessed a marked decrease in painted pottery, which give way to the Red-Slip and "Orange Simple Wares" associated with the Iron II period (Mazzoni 1998a, 169; Oggiano 1997, 186). This shift in ceramic horizon is also marked by a remodelling of the town's defensive system and expansion of the Lower City (*ibid*). While the occurrence of this cultural transformation is not disputed here, the assigning of a date to this "event" by comparison with South Levantine sites introduces some chronological ambiguity (Mazzoni 1998a, 169). Comparison is also made with the similar assemblage of Levels 6-4 in Area D in the Lower City of Tell Afis, which is dated to the eighth and seventh century BC by the presence of four sherds of imported pottery (Mazzoni 1998a, 169; Oggiano 1997, 287). The Area D date is confirmed on historical considerations; the corresponding expansion of the Lower City accords well with the situation described in the stele of Zakkur, which describes Hazrek as prosperous and strong enough to defend itself against the coalition of Damascus (Mazzoni 1998b, 7-8). Another factor in assigning the mid-ninth century BC date for the Iron I-II transition at Tell Afis appears to derive from Braemer's (1986, 222) inconclusive date for the appearance of Red-Slip in the Northern Levant (**§2.7**; Cecchini 1998, 277; Degli Esposti 1998, 231; Mazzoni 1990a). Further external support is gained through reference to Hazor, which, putting its remoteness aside, is not without its own stratigraphic and chronological problems (**§3.19**; Cecchini 1998, 296; Degli Esposti 1998, 231; Mazzoni 1998a, 169). In the end, the assigning of a mid-ninth century BC date for the Iron I-II transition is not based directly on the archaeological data, and Mazzoni (2000a, 41) herself admits the date cannot be defended archaeologically.

Like previous transitional periods, the shift from the Iron II to Iron III horizon was assigned an absolute date based on ceramic parallels and historical correlations. The large sunken courtyard of Area G (the '*batiment mysterieux*' of Cecchini 2000a) was filled with the debris of the structure's large walls. While there is little evidence to determine exactly how the walls collapsed, the excavators see the likely catalyst for this event as the marauding Assyrian army (Cecchini 1998, 296). There is, however, no direct archaeological evidence for destruction by the Assyrians. This interpretation is instead based on the historical inclusion of Hazrek, identified with Tell Afis (Amadasi 2001; 2005; Dion 1997, 139-143; Lipinski 2000a, 255-258; Mazzoni 2005b, 12-13), within the sphere of Tiglath-Pileser III's empire in 738 BC.

Tell Afis boasts one of the few complete ceramic sequences for the IA-NL, and the most extensively published to date. One cannot overestimate the influence that the chronology of the Tell Afis sequence holds within the discipline (Akkermans and Schwartz 2003, 363). The periodisation of the Iron Age as encountered, or at least interpreted, at Tell Afis has become the keystone of Iron Age archaeology of the Northern Levant. Under close scrutiny, however, the foundations for this chronological framework are not as secure as expected. While the ceramic sequence at Tell Afis has been meticulously constructed, its correlation with distant sites, combined with a reliance on general historical trends and specific historical events, and the uncritical use of imported pottery, warrants a cautious adoption of its chronological framework.

3.4 Ahmar, Tell (Syria)

Summary of Excavations

French archaeologists excavated a sounding at Tell Ahmar in 1928, which was followed by more extensive investigation of the tell in the following three years (E. Dhorme 1938). Near the summit of the tell, excavation revealed the plan of a large palace that the excavators assigned to the Neo-Assyrian period (eighth and seventh centuries BC) based on the evidence of two large stelae of the Assyrian king Esarhaddon (680-669 BC) found therein (Thureau-Dangin and Dunand 1936b, Pls XII, XIII). Though the Assyrian palace was only partially preserved, it closely resembled the plan of an Assyrian palace at Arslan Tash (Akkermans and Schwartz 2003, 382-383). Tell Ahmar's palace is renowned for its beautiful wall paintings, with figurative compositions that are considered significantly richer than the purely-ornamental friezes from Arslan Tash (Strommenger 1985b, 330). Tell Ahmar's paintings are considered truly Assyrian in composition and execution, there is little trace of local influence, leading to the conclusion that Tell Ahmar was part of the Assyrian heartland in the eighth century BC (*ibid*).

By the close of French excavations, a number of periods were reportedly attested at the site; Ubaid, Early Bronze Age IV, early Iron Age, "Assyrian" Iron Age, and the Persian (burials), Hellenistic, Islamic and modern periods. Evidence of the Early Bronze Age IV included *l'hypogee* of Thureau-Dangin and Dunand (1936a, 96-108), a large underground chamber tomb that yielded an abundance (over one thousand) of ceramic and bronze artefacts (Thureau-Dangin and Dunand 1936a, figs 28-32). Though a plan of the "Assyrian" palace and wall paintings appeared in the two-volume report, no Iron Age pottery was published from this expedition (Jamieson 2005, 749).

Fifty-eight years later, when Melbourne University renewed excavation of the site, the archaeological landscape had changed dramatically. The river, which had flowed past the very foot of the tell during the 1920s (Thureau-Dangin 1929, 185), had shifted some hundred metres to the south (Bunnens 1990a; 2006, 5). Furthermore, the tell had become heavily occupied by the modern village, and the Lower City had been intensively bulldozed to optimize agricultural production - the outer city-wall was all but destroyed (Bunnens 1990a, 3). The site would soon undergo more extensive and permanent damage; the construction of the Tishrin Dam downstream was expected to completely inundate the site – this being the reason for a return to the site (*ibid*, 1). The stated aims of the new project were threefold: to isolate a complete stratigraphic sequence for the site; to study the interaction between the Assyrian and local cultures; and to study the urban layout of the Iron Age settlement (Roobaert and Bunnens 1999, 163). The Iron Age of Tel Ahmar was apparently a key focus of the new project.

The renewed excavations have revealed a stratigraphic sequence of the site extending from the Chalcolithic Period down to the Assyrian occupation (Bunnens 2002-2003). Highlights of this sequence include the Early Bronze Age IV, when Tell Ahmar bears evidence of a sophisticated urban centre (*ibid*, 165-168; Jamieson 1990, 25-26). The Middle and Late Bronze Age were also identified in the sondage on the south-east corner of the tell, though the latter was heavily disturbed by construction dated to the Iron Age (Bunnens 2002-2003, 168). According to textual evidence, the site of Tell Ahmar took on regional significance in the Iron Age (Hawkins 1983), yet the renewed excavations have isolated very small amounts of pre-Assyrian Iron Age material (*ibid*, 169). Nevertheless, there is an abundance of Assyrian material at the site; ceramic, architectural, artistic (Roobaert and Bunnens 1999, 171*ff*). A significant corpus of Assyrian pottery has been recovered from the renewed excavations, the bulk of which was excavated from domestic contexts in Area C in the Middle City. Smaller samples derive from Areas D, E and F, also located off the main tell. Pottery from Area C has been dated to the seventh century BC and, according to Jamieson (1999, 287), holds close parallels to the Assyrian pottery from Fort Shalmeneser at Nimrud, published by Oates (1959).

Excavation at Tell Ahmar has been supplemented by a short archaeological reconnaissance of the Lower City (Green and Hausleiter 2000; 2001). The majority of the unstratified pottery collected from the fields has been dated to the Iron II period. Once again, very little pre-Assyrian Iron Age material was identified (Green and Hausleiter 2001, 105).

Table 3.6: Stratigraphy of Tell Ahmar Area C

Stratum	Period	Date
1	Roman	
2A-B	Iron III	1st half sixth century BC
2C	Iron III	2nd half seventh century BC
3		

(After Jamieson 2000, 264-269)

Review

Much has been written about Iron Age Tell Ahmar (Bunnens 1990a, 3; 1999; 2000a; 2002-2003; Jamieson 1999; 2000; Thureau-Dangin and Dunand 1936a, 134), with discussion focusing on the early Iron Age population; i.e. whether they were Aramaean or Luwian. The site is widely recognised as the capital of the small "Syro-Hittite" kingdom of Masuwari (Hawkins 1983), before being transformed into the headquarters of the "Aramaean" state of Bit-Adini, and then becoming the Assyrian Kar-Shalmeneser of the eighth and seventh centuries BC (G. Schwartz 1989, 278). Ussishkin (1971), however, has reversed the sequence and prefers to see the early Iron Age site as Aramaean before becoming "Syro-Hittite" about a century before Assyrian conquest. Hawkins (1980; 1983) has suggested an alternative interpretation; that Tell Ahmar was a small kingdom of Hittite (Luwian) descent that was under regional authority of an Aramaean tribal leader. This, according to Hawkins (*ibid*), explains the "Syro-Hittite" character of the local art and the mention of Tell Ahmar as part of the Aramaean coalition against the Assyrians.

Discussion over the ethnic affiliation of the Iron Age population of Tell Ahmar has largely ignored the difficulty in associating an historical people with material culture (**§4.3**). Furthermore, the early Iron Age "is not very well attested so far" within the renewed excavations, which means most of the discussion about this period has been concerned with different interpretations of the historical narrative (Bunnens 2002-2003, 169). The archaeological record has not been able to contribute to the discussion precisely because there is little definitive early Iron Age material culture known at the site. A few sculptural pieces reminiscent of those from nearby Carchemish and bearing hieroglyphic Luwian and/or alphabetic Aramaic inscriptions indicate that the site was indeed occupied during the early Iron Age, but the items are usually isolated finds from secondary contexts (e.g. a Luwian inscription re-used in the Late Assyrian building – *ibid*) and not always found on the actual site (Bunnens 2006; Hawkins 1980; L.W. King 1909; Roobaert 1990). The claim of French archaeologists to have uncovered pre-Assyrian Iron Age levels has not been decisively established by the current project (Thureau-Dangin and Dunand 1936a, 84-96). Thureau-Dangin and Dunand's (1936a, 84-96) "Aramaean" levels appear to have been so-designated simply because they lay under the Assyrian palace. The fact that material culture from the early Iron Age is poorly-attested is problematic, considering the prominence given to the site in the historical narrative for the corresponding period. It seems that the historical narrative has driven much of the discussion on this period, which may account for a pre-occupation with the ethnicity of the population.

An abundance of Assyrian pottery and architecture has been exposed in all areas of the site; Tell Ahmar apparently reached the zenith of its size and influence as the Assyrian city of Kar-Shalmeneser. There is little doubt regarding the Assyrian nature of the material culture during this phase, though the absolute dates are less certain. While the two Assyrian stelae found within the palace complex were attributed to Esarhaddon of the seventh century BC, the construction of this building has been dated to the eighth century BC on historical grounds (Lipinski 2000a, 185). The cross-comparison of Area C pottery (Jamieson 1999, 287) with that from distant Nimrud (Oates 1959), even if relevant, is unable to provide refinement of the dating scheme.

The chronological and interpretative framework at Tell Ahmar places a strong emphasis on the conventional historical narrative. As a result, the chronology of Tell Ahmar remains more reliant on the unsubstantiated witness of ancient texts than on the archaeological record. Recently published results indicate that a radiocarbon sampling program is being employed at the site, which should provide a more decisive chronology, though this can tell us little regarding the ethnic identity of the site's population (Bunnens 2002-2003, ns 11, 21).

3.5 Ain Dara (Syria)

Summary of Excavations

The site of Ain Dara was already known in 1955, when the discovery of a monumental basalt lion prompted excavation (Abou Assaf 1997a). When excavation began four more "unfinished" lions were uncovered on the acropolis, supposedly all belonging to an unfinished monumental gate complex. Excavation also revealed the significant remains of a large "Syro-Hittite" temple on the eastern acropolis (Abou Assaf 1990). In contrast to similar buildings at Tell Halaf (Guzana), Zincirli (Sam'al), Tell Ta'yinat, and Sakçe Gözü, all of which were internally decorated, the *outer* walls of the Ain Dara temple were heavily-decorated with animal (lions and sphinxes) sculptures (*ibid*, Abb. 16-17). The two step-like thresholds leading in to the temple bore three large footprints; the outer threshold bearing a pair of imprints, while the inner threshold preserved an image of only the left foot (About Assaf 1990, Tf. 11). Once the disused temple fell into disrepair, the later Iron Age settlement (Levels 6 and 5) reused the dressed stone, in particular the large stone lions (Abou Assaf 1985, 347-350).

Though initially the objective of the project was to reveal the nature of occupation of the whole site throughout all periods (early seasons excavated the northern gate of the lower city), the focus of the Syrian expedition quickly shifted to the temple and its ornate orthostats and sculpture (**Table 3.7**; Stone and Zimansky 1999, 2). Indeed, scholarly interest in the site rarely extends beyond a discussion of the temple's artistic style and date (e.g. Abou Assaf 1997a; Orthmann 1971, 198).

Table 3.7: Ain Dara Acropolis Stratigraphy

Level	Period	Dates	Comments
1	Seljuk	1100–1400 CE	Restricted to Acropolis
2	Byzantine	969-1075 CE	City-wall; significant occupation
3	Abbasid Ummayyad	640-969 CE	Similar to level 2
	hiatus ------------------------------------		1 metre debris layer
4	Seleucid	330–75 BC	Seleucid coins, eastern sigillata
5	Persian	530–330 BC	Achaemenid horsemen figurines
6	Iron III	7th–6th cent. BC	Poorly preserved, Attic wares
7	Iron II	10th–9th cent. BC	Temple and Lion Orthostats

(After Abou Assaf 1990, 1-10)

An obvious exception to scholarly fascination with Ain Dara's temple was the American project at the site. Stone and Zimansky (1999, 1-7) recognised that the temple, as impressive as it was, did not stand in isolation, but was erected amidst a substantial population overlooking a large thriving community. Consequently, the American project focused fieldwork on the lower town. Their first season (1982) was devoted to understanding what periods were represented in the lower city, with excavation and surface survey producing material from the Late Bronze Age, and the Iron I, Iron II and Hellenistic periods (*ibid*, 6). During 1983 and 1984 more extensive excavation was undertaken in the north-eastern area of the lower mound with the aim of investigating the Late Bronze-Iron Age transition. After only two seasons of excavation, when the Late Bronze Age levels were beginning to be exposed, and the associated transition between the Bronze and Iron Ages, the expedition was brought to an abrupt halt (*ibid*).

Review

Despite evidence for much earlier occupation being found at the site, the priorities of the earlier expedition to Ain Dara were centred on the Iron Age levels of the citadel and the impressive temple found therein. A decisive dating scheme for the temple, however, has not been established. The temple's embellishments were executed in a style that clearly owed much to the artistic traditions of the Hittite Empire, most unambiguously in the row of mountain god and demon figures decorating the innermost courtyard (Seirafi *et al.* 1965, Pl. ix a-b). To some scholars, Ain Dara is a clear example of Orthmann's (1971, 136-138) "Late Hittite I" artistic style associated with the early Iron Age (Seirafi *et al.* 1965, 19). Abou Assaf (1990, 39-41), however, argues that three stylistic phases were evident in the building, and presented a chronological sequence spanning more than five centuries (**Table 3.8**). Phase I is dated to the last quarter of the second millennium via comparison with a temple from Hazor that has conventionally been dated to the thirteenth century BC (Abou Assaf 1990, 39): issues of remoteness aside, absolute dates for the sequence at Hazor are unreliable (**§3.19**).

Temple Phases II and III are dated to Orthmann's (1971, 136-138) "Late Hittite I/II" and "Late Hittite II/III", respectively, based on the relative presence of Assyrian motifs within the ornamentation (Abou Assaf 1990, 40-41). Orthmann (1971, 136-138), however, does not agree with Abou Assaf's phasing. Regardless, even if scholars agreed that the 'Ain Dara sculptures are of "Late Hittite I" style, this would not provide a precise date for the temple's use since none of the parallels cited by Abou Assaf are stratigraphically reliable (e.g. Arslan Tash, Carchemish, Zincirli). Furthermore, there is little consensus over the dates of Orthmann's scheme (**§2.3**). There is also some confusion regarding the relationship between the sculptures and the temple building – there are indications that the adornments were never finished (lions in varying degrees of completeness), and signs of rebuilding and discard (Abou Assaf 1990, 61). Moreover, some reliefs were found in secondary contexts (*ibid*, 61, Tf. 51). Indeed, the stratigraphy surrounding the temple is complex, compounded by the fact that the temple stood exposed for some time after it was abandoned (Abou Assaf 1990, 10; Stone and Zimansky 1999, 3). Any possibility of further examination of stratigraphic connection between the temple and its surroundings has been eliminated by a large trench dug around the building for the construction of a protecting roof (*ibid*).

Table 3.8: 'Ain Dara Temple - Stylistic Phasing

Phase	BC	Parallels
I	1300-1000	Hazor
II	1000-900	Carchemish
III	900-740	Arslan Tash, Zincirli, Carchemish, Sakçe Gözü, Tell Halaf

(After Abou Assaf 1990, 39-41)

Table 3.9: Phases of Akhziv Cemeteries

BC	Southern Tombs	Northern	Eastern
end 11th	Cist – rough stones		Not Used
end 11th – early 10th	Chamber – rough stones		
10th	Chamber – ashlar		
9th	Shaft	Cremation	Rock Cut Tombs
8th			
7th			
6th			
5th			

(After E. Mazar 2001, 159)

3.6 Akhziv (Israel)

Summary of Excavations

Prior to the commencement of archaeological work at the site, looting was rampant amongst the ancient tombs of Akhziv. Consequently, the first excavations focused solely on recording and protecting the cemeteries (Dayagi-Mendels 2002, 2; E. Mazar 2001, 9). Work on the tell was restricted to only two seasons of excavation by Prausnitz in 1963 and 1964 (Prausnitz 1963; 1965). Nevertheless, Prausnitz' work on the northern part of the city-mound revealed Middle Bronze fortifications, which were in ruin by the end of the Late Bronze Age, and significant Iron Age deposits (Dayagi-Mendels 2002, 1-2). It was clear from both the tell and the cemeteries that the settlement enjoyed its greatest expansion (c. 8 ha.) during the Iron Age (*ibid*). Publication of the Iron Age material has focused on the cemeteries and the accompanying assemblages of decorated pottery (*ibid*).

The tombs recorded by Ben Dor in the southern and eastern cemeteries were mostly shaft-tombs with a single rock-cut chamber (Dayagi-Mendels 2002, 3-4). The later excavations, by Eilat Mazar, uncovered additional grave types: round-graves, pit-graves, cist-tombs, and jar-burials, as well as a cremation cemetery to the north of the tell (E. Mazar 2001, 10, 157). The many burial types appeared to reflect chronological developments in burial customs at Akhziv, which Mazar was able to order into a general phasing for the cemeteries (**Table 3.9**). One particular tomb in the southern group (TC4), which was a chamber tomb built of rough stones, contained the remains of approximately 50 individuals (men, women and children), most of which were secondary burials, and abundant pottery. While this may be evidence for this tomb's continued use over a long period of time involving specific funerary rituals (dining with the dead, mixing the remains of the dead), Mazar (2001, 157) has interpreted this as the relocation of an earlier northern cemetery that was cleared to make way for the cremation burials.

The large ashlar-built chamber-tomb in the northern cemetery bears witness to four phases of use extending from the tenth through to the mid-sixth century BC (**Table 3.10**). The long period of continuous use and extensive collection of ceramics from the Lebanese coast, led to its interpretation as a "Phoenician Family Tomb".

Review

A large corpus of ceramics has been published from the Akhziv cemeteries, though there is some concern over the presentation of the material. While Dayagi-Mendels (2002, 2) recounts the significant difficulties encountered in the study and publication of the material from Ben Dor's excavations (excavation during time of war – WWII; the long period that elapsed since excavation; inadequate methods of recording; incomplete surveying of tombs which can no longer be located; damage caused by human and natural activities), it is not surprising to read Mazar's (2001, 4) criticism of Dayagi-Mendels' publication. According to E. Mazar (2001, 10), inconsistencies in Ben Dor's records led to Dayagi-Mendels assigning some vessels (e.g. a Cypriot White-Painted barrel jug) to wrong tomb assemblages. Moreover, the aim of Dayagi-Mendels' (2002, 2) publication was the presentation of the rich repertory of pottery – chronology was not important.

Table 3.10: Akhziv "Phoenician Family Tomb"

Phase	Date
1	10th – early 9th cent.
2	Mix of phases 1 and 3
3	Late 9th – 7th cent.
4	Late 7th – mid 6th cent.

(After E. Mazar 2004, 21-23)

For E. Mazar, however, chronology was a key consideration. The Akhziv shaft-tombs were dated on ceramic parallels and the presence of datable scarabs (ignoring the heirloom factor) in some tombs (E. Mazar 2001, 77-146). The dates for other Akhziv tomb types, including the "Phoenician Family Tomb", rely on the presence of datable ceramic data and architectural parallels (E. Mazar 2004, 21-23). Hence, Mazar's chronological framework ignores the many inherent problems with current Iron Age chronology of the Southern Levant. Furthermore, E. Mazar's interpretation of the cemeteries draws heavily from the conventional historical narrative. For instance, the cist-tombs of the southern cemetery are attributed to a "Sea Peoples" group (the *Sherden*), based on similarities between Akhziv cist-tombs and examples from Tel Zeror and Tel Far'ah(S) (E. Mazar 2001, 75). Petrie's (1930, 11-12) assigning of the Tell Far'ah(S) cist-tombs to the "Philistines" was based on the biblical narrative, presence of "Philistine" pottery (which is based on a literal reading of Egyptian and biblical texts), and the proximity of the site to the traditional location of "Philistia" (as derived from the biblical account); none of which have proved to be decisive interpretative tools (**§2.5**). In the end, the Akhziv chronology and "Sea Peoples"/"Phoenician" association are the product of a literal reading of historical sources.

3.7 Aleppo (Syria)

Summary of Excavations

Several soundings were made on the Aleppo Citadel following World War I (Baurin 1923; Dussaud 1931; 1934, 300ff.; Ploix de Rotrou 1931a; 1932b), recovering various basalt and limestone slabs with inscriptions and geometric decoration, most of which have close parallels to orthostats from the 'Ain Dara temple (Gonnella *et al.* 2005, Figs 86, 87; Ploix de Rotrou 1932a). A number of basalt lion orthostats also came to light but have since been lost or damaged (Shaath 1996, Fig. 63). An early Iron Age relief depicting two winged genii was recovered and now resides in the forecourt of the National Museum of Aleppo (G. Miller 1958, Fig. 69; Orthmann 1971, 54; van Loon 1995). Very little was ever published regarding any of these finds, and the pottery from these early expeditions has been lost.

In 1995, a Syro-German project returned to Aleppo for the purpose of investigating "the pre-Hellenistic layers on the citadel of Aleppo" (Khayyata and Kohlmeyer 2000, 733). The location for the renewed work took into account the find-spots for the Iron Age orthostats, as well as the restricting nature of the significant (both in size and importance) Islamic overburden. It is little wonder, therefore, that the new trenches reopened the old sounding (*ibid*, 734). It was soon revealed that the various basalt and limestone slabs exposed in the 1930s turned out to be part of an ancient wall, one which belonged to a large building with a floor plan not unlike the Iron Age temple at 'Ain Dara (*ibid*, 734-735). The main component of the plan is a double wall of orthostats surrounding what appears to be an elongated *cella.* Furthermore, the interior wall is decorated with a long series of divine and mythical reliefs in a typical "Syro-Hittite" compositional style, similar to Zincirli and Carchemish (*ibid*, 736-737). More importantly, the reliefs were *in situ.*

Review

The Syro-German project has brought the presence of significant Iron Age occupation on the tell beyond doubt, but dating the temple more precisely remains elusive. Surprisingly, no pottery has been reported from the recent excavations, which in turn raises concerns over the collection of faunal and botanical samples important for scientific dating and analysis (Khayyata and Kohlmeyer 2000; Kohlmeyer 2000). Like other "Syro-Hittite" projects (e.g. Ain Dara, Carchemish, Zincirli, Tell Halaf), the architecture/reliefs appear to have been a primary concern. As a result, the only means available to date the temple is the unreliable analysis of artistic style. While there exist chronological schemes for the development of "Syro-Hittite" architecture and art, debate surrounds the accuracy of the sequence and absolute dates (**§2.3**). Hence, architectural and artistic styles are unable to provide a solution. Dating the epigraphic style of the Luwian inscriptions is reportedly being undertaken, though as yet no results have been published; they have been assigned a preliminary eleventh century BC date based on conventional dating of the associated reliefs (Hawkins, *4ICAANE conference presentation April 2, 2004*). This important temple is a lost opportunity for modern scientific methods to interpret and independently date the distinctive architecture of the early Iron Age.

3.8 Amuq Plain (Turkey)

Summary of Excavations

Between 1932 and 1938 the Oriental Institute (hereafter OI) of the University of Chicago undertook a survey of 178 mound sites throughout the Amuq Plain, including the adjacent Afrin and Kara Su Valleys (Braidwood 1937, 1). The survey lacked a comparative local ceramic sequence, which could only be obtained through the excavation of carefully selected sites. While the result of this project was a long regional ceramic sequence, its primary design was for the prospection of archaeological sites, more specifically (as the project name "Syro-Hittite Expedition" suggests) the search for monumental Syro-Hittite architecture (Breasted 1933; McEwan 1937, 8; Yener 2000a, 1801; Yener *et al.* 2000a, 163).

Table 3.11: Correlation of OI Iron Age Excavations in Amuq Plain

Amuq Phase	Description	Dates	Tell Judeideh	Chatal Höyük	Tell Ta'yinat
U	Medieval Arab			I	
T	Byzantine			I	
S	Early Christian		I	?	
R	Roman		II	II	
Q	Hellenistic	300-64 BC	III	II	?
P	Syro-Hellenic	500-300 BC	III	II	?
O	Syro-Hittite	1000-500 BC	IV	III	I
N	Levanto-Helladic IV (Sub-Mycenaean)	1200-1000 BC	V	IV	II
M	Levanto-Mycenaean	1600-1200 BC	VI	V	
L	Qatna affinities	1800-1600 BC	VII		
K	Hama affinities	2000-1800 BC	VIII		
J	Chalciform pot series	2400-2000 BC	IX		
I	Smeared-wash pot series	2600-2400 BC	X		
H	Red-Black burnished pots (Early Dynastic affinities)	3100-2600 BC	XI		
G	Jemdet Nasr affinities	3500-3100 BC	XII		
F	Uruk affinities	4000-3500 BC	XIII		
E	Ubaid affinities	4500-4000 BC	?		
D	Derived Halaf and earliest Obeid affinities	4500 BC	?		
C	Developed primitive and true Halaf affinities	5000-4500 BC	XIV		
B	Developed primitive and first painted wares, etc...	pre-5000 BC	XIV		
A	Primitive burnished ware	?	XIV		

(After Haines 1971, 1; Cf. Braidwood 1937, 6-7; McEwan 1937, 10-11)

The first site excavated within the OI Amuq Project was Chatal Höyük (Braidwood 1937, 37 - AS 167), located on the River Afrin in the north-east of the plain. In keeping with the original scope of the project, excavation here focused on the Iron Age strata (Albright 1935, 146). The second site excavated within the project was Tell Judeideh (Braidwood 1937, 37 - AS 176), which yielded an almost complete stratigraphic sequence from the Byzantine period back to the sixth millennium BC and provided the main framework for the Amuq sequence (**Table 3.11**; Braidwood 1937, 4-8). In 1935 excavations were broadened to include a site that had already provided evidence of important "Syro-Hittite" architecture (Albright 1936, 165; Haines 1971, 37). Tell Ta'yinat (Braidwood 1937, 33 - AS 126) incorporated a large low-lying mound surrounded by an extensive (c. 35 ha) Lower City (Batiuk *et al.* 2005, 171). Excavations focused on the "West Central Area" of the upper mound, with additional areas opened on the tell edge and Lower City. Four years of excavation at Tell Ta'yinat resulting in large horizontal exposures of (at least) five Iron Age "Building Periods" (**Table 3.12**; Batiuk *et al.* 2005, Fig. 7.3; Haines 1971, 64-66). The significant corpus of "Syro-Hittite" architectural elements exposed includes large *bit-hilani* buildings (palaces?), a *megaron* temple, and massive city walls and gates, as well as a palatial structure of Assyrian provincial style, (Batiuk *et al.* 2005, Figs 7.2-3).

Table 3.12: Tell Ta'yinat Building Periods (OI excavations)

Period	Buildings	BC
First	XIII (*hilani*); XIV	875-825
Levelling		
Second	I & IV (2 *hilani*); II (*megaron*); VI; Gate XII	825-720
Ta'yinat greatest extent (35 ha)		
Third	IX (Assyrian palace); VI levelled	720-680
Fourth	I rebuilt; II abandoned	7th cent.
Fifth	X; fragmentary	6th cent.

(After Haines 1971, 64-66; cf. Harrison 2001a, 125-126)

Soundings under the Iron Age levels at Tell Ta'yinat revealed Early Bronze Age material, indicating a lengthy abandonment (a millennium?) of the site between the two periods (Harrison *et al.* 2004, 122). According to Woolley (1955, 398; also Yener 2005, 3), nearby Alalakh was occupied from the end of the Early Bronze Age until its destruction at the end of the Late Bronze Age, when regional political power shifted back to Tell Ta'yinat.

The OI excavations were prematurely terminated by the hostilities following withdrawal of the French Mandate and the creation of the Republic of Hatay, but were resumed in 1995 after a hiatus of fifty-seven years. The first four seasons of the Amuq Valley Regional Projects (AVRP) were devoted to a geo-archaeological study and surface survey of the basin area. In addition, salvage excavations were undertaken at a number of sites heavily damaged by modern farming, all of which exposed levels that pre-dated the Iron Age (e.g. Tell Kurdu AS94, Tell Atchana/Alalakh AS136). In co-operation with the AVRP, the University of Toronto has, in recent years, returned to Tell Ta'yinat to undertake an extensive surface and geo-magnetic survey of the site (1999-2002), as well as reinitiate excavation of the upper mound in 2004 (Batiuk *et al.* 2005). Whilst it is still too early for the results of the renewed excavation to be available, some preliminary reports have appeared in Turkish publications (e.g. Harrison *et al.* 2004; 2005; 2006).

Review

Since the Braidwoods' (1960) publication of the prehistoric phases, the Amuq sequence has become a standard reference point for chronologies of the eastern Mediterranean, despite the conspicuous lack of published Iron Age ceramics (Batiuk *et al.* 2005, 171; Yener *et al.* 2000a, 165). According to the established sequence (**Table 3.11**), Phases N-O relate to the Iron Age (Haines 1971, 1-2). The earliest of the two, Phase N, is characterised by an abundance of painted ceramics of Mycenaean IIIC:1 tradition, which led Haines (*ibid*) to assign an Iron I date – the sub-Mycenaean period is conventionally dated to the end of the second millennium BC (McEwan 1937, 10). The Phase N-O transition is associated with a marked decrease in painted wares and the advent of Red-Slip, which came to characterise the Amuq O period (Batiuk *et al.* 2005, 172; Haines 1971, 1, 64; Janeway 2006-2007, 129-137; Swift 1958, 124-126).

Drawing primarily on artefactual data from Tell Judeideh, Chatal Höyük and Tell Ta'yinat, Swift (1958, 139-141) proposed subdividing the Amuq Phase O into four stages, each coinciding with changes in surface treatment of Red Slip pottery (**Table 3.13**). Greek imports (which were abundant at Tell Ta'yinat during Amuq O – Boardman 1990, 174) and key historical events provided Swift with absolute dates, though neither are reliable chronological tools (**§2.5**). Amuq P was associated with the Persian period via comparison with Nayrab and Deve Höyük II; two poorly recorded and published sites conventionally dated to the sixth century BC.

The structural remains at Tell Ta'yinat have been dated, through the presence of Red-Slip, to Phase O of the broader Amuq sequence, largely based on the Tell Judeideh sequence. The Judeideh sequence, however, has only been published as a generalised chronology and final publication of the pottery is still awaited (Braidwood 1937, 4-8). This sequence was "based on field observations only" (*ibid*, 4) and was not intended to be definitive, yet it has persisted.

More than 100 epigraphic finds at Tell Ta'yinat have been used to establish dates for the building periods. For instance, one Luwian text refers to *Halpa^pa-runta-a-s(a)*, possibly the same ruler who paid tribute to Shalmeneser III in the mid-ninth century BC (Batiuk *et al.* 2005, 173). Most of the Luwian inscriptions were found in the fill or foundation trenches of structures dating to the Second Building Period (Gelb 1939, 39-40; Haines 1971, 66); while this suggests a ninth century date for the First Building Period the secondary contexts are not reliable (Haines 1971, 66; Harrison *et al.* 2004, 122-123). The Third Building Period was assigned to the late-eighth/early seventh century BC based on the presence of an inscribed Aramaic sherd from a floor deposit. This, in turn, suggested that the Second Building Period belonged to the eighth century BC (Harrison 2001a, 129). The Fourth and Fifth Building Periods were dated to the sixth and seventh centuries BC, respectively (*ibid*). While the Tell Ta'yinat sequence cannot yet be considered definitive, progress is being made toward providing a reliably dated ceramic sequence, offering the excavators an important opportunity to correlate archaeological remains with the historical record (Batiuk *et al.* 2005, 172). The current Amuq sequence, however, is largely dependent on the conventional historical narrative.

Table 3.13: Swift's Division of Amuq Phase O

Phase	Period	BC	Surface	Imports
Oa	Iron IB	950-900	Hand burnish	
Ob	Iron IIA	900-800	Hand and wheel burnish	
Oc	Iron IIB	800-725	Wheel burnish	Attic Geometric
Od	Iron III	725-550	Wheel burnish	Corinthian Black-Figure Assyrian 'Palace Ware'

(After Swift 1958, 139-141, Tab. 11)

Table 3.14: Stratigraphic sequence of Tell Arqa

Phase	Period	Level
A	Mamelouk	1
		2
B	Crusader	3
		4
C	Early Islamic	
D	Byzantine	5
		6
E	Roman Empire	
F	Hellenistic	7
		8
G	Iron III	9
H	Iron II	10
J	Iron I	
K	Late Bronze II-III	11
L	Late Bronze I	12
M	Middle Bronze II	13
N	Middle Bronze I	14
P	Early Bronze IV	15
		16
R	Early Bronze III	17

(After Thalmann 2006a, Fig. 3)

3.9 Arqa, Tell (Lebanon)

Summary of Excavations

During his nineteenth century *Mission to Phénicie*, Renan (1864) recognised the importance of Tell Arqa, but did not excavate the site. In fact, the site was not the subject of serious archaeological prospection until Thalmann and Will's expedition began in 1972. This French project managed seven seasons of excavation in the following decade before work was postponed during the Lebanese civil war. Work finally recommenced in 1992 and has continued unabated (Thalmann 2006a, 2-3).

Tell Arqa has revealed evidence of sporadic occupation dating back to the sixth millennium BC, with the earliest period of prosperity associated with the Early Bronze IV (**Table 3.14**). The Middle Bronze I is characterised by a warrior's tomb and large pottery workshop, and the Middle Bronze II by a large fortification wall (Thalmann 2006a, 44-45, 56). Following the urban prosperity of the Middle Bronze Age, Tell Arqa was little more than a small village for much of the Late Bronze and early Iron Ages (Chaaya 2000, 215*ff*).

The Iron II period (Level 10) is characterised by a sanctuary and evidence of a cremation necropolis (*ibid*). Thalmann identified four main architectural sub-phases within Level 10 (**Table 3.15**). The Level 10 ceramics are essentially local forms; fine wares and imports are both rare (Chaaya 2000). The Iron II assemblage is characterised by a large quantity of standardised amphorae, some of which bear a painted *lmlk* inscription (Thalmann 1978b, Fig. 23). There is a clear break in ceramic traditions between Levels 10 and 9, with the latter being indicative of less standardised production.

Table 3.15: Summary of Iron Age levels at Tell Arqa

Level	Period	Notes
10F	Iron II	Much fill
10E	Iron II	Inferior level, habitation
10CD	Iron II	Rampart constructed; houses re-used
10AB	Iron II	Superior level, funerary structures
destruction layer attributed to Assyrian conquest (738 BC)		
9	Iron III	Poor architecture; many pits

(After Thalmann 1978a, 69)

Table 3.16: Summary of Archaeological Contexts on mid-Levantine littoral

Akhziv	Mortuary
Khirbet Silm	Mortuary
Tell Rachidieh	Mortuary
Tyre - Bikai	Limited exposure
Tyre - Chéhab	Mixed deposits
Tyre Al-Bass	Mortuary
Sidon	Limited exposure
Joya	Mortuary
Qraye	Mortuary
Qasmieh	Mortuary
Tambourit	Mortuary
Khalde	Mortuary
Byblos	Poor stratigraphy
Beirut	Limited Exposure

Review

The excavations at Tell Arqa are particularly important for the study of the IA-NL, because the site is one of only a few Lebanese sites that have produced a significant amount of Iron Age pottery from mortuary and non-mortuary contexts alike (*ibid*, 71-89). This is contrasted with the high number of Lebanese excavations that have uncovered poorly stratified mortuary contexts (e.g. Khirbet Silm; Joya – S.V. Chapman 1972). When non-mortuary contexts have been encountered, they have been either excavated with poor stratigraphic control or with minimal exposure (e.g. Byblos, Bikai's Tyre trench – **Table 3.16**). The lack of a well-controlled stratigraphic sequence is keenly felt within this discipline.

The Iron Age levels of Tell Arqa have not yet been fully published, though a few preliminary reports have appeared. From reading these reports, it seems Levels 10 and 9 (the Iron Age) have been dated on historical considerations. The ash layer sealing Level 10 has been associated by Thalmann (1983, 217-218) and Chaaya (2000, 215) with the 738 BC campaign of Tiglath-Pileser III to the Akkar Plain, despite textual accounts not explicitly mentioning the site. In the end, evidence for this correlation is not archaeological, but historical. An epigraphic study into the *lmlk* scripts has confirmed a mid-eighth century BC date for the Iron II amphorae, but the results are not yet published (Thalmann *personal communication*).

3.10 Arslan Tash (Syria)

Summary of Excavations

A number of stone bas reliefs were removed from the site of Arslan Tash in the late nineteenth century and taken to the Istanbul Museum (Strommenger 1985a). A French expedition excavated the site from 1927 to 1929 exposing a number of Iron Age levels that have been dated between the ninth and fourth century BC (*ibid*; Thureau-Dangin *et al.* 1931a; 1931b). In particular, the French exposed a large Iron Age palace, temple, small residential building (*bâtiment aux ivoires*), and city-wall (Strommenger 1985a). Three gates in the wall were guarded by large basalt lions, which have been dated on stylistic grounds to the reign of Tiglath-pileser III (744-727 BC) (*ibid*). The lions are considered examples of the less-grandiose Assyrian provincial style (F. Thureau-Dangin *et al.* 1931b, Pls III-VI). The palace follows a typical Assyrian plan, with courtyard, reception rooms, and an inner court for the royal family (Turner 1968, Pl. XVII). In a few rooms the excavators found the remains of horizontal friezes set two metres above the floor. The entrance to the Assyrian temple was protected by two large bull statues with inscriptions stating that Tiglath-Pileser III was lord of this temple of Ishtar (Strommenger 1985a, 330-332). A small building to the east of the palace has become known as '*le batiment aux ivoires*' after the discovery of a hoard of ivories therein.

Review

The publication of the Arslan Tash excavations did not include any ceramic material, plausibly because none was collected or recorded (Thureau-Dangin *et al.* 1931a; 1931b). Instead, the French archaeologists apparently focused on a broad exposure of the site's elite architecture and a study of the "provincial" Assyrian art. The exact manner by which the remains were dated is not clear, though Assyrian "history" appears to have played an important part. Once dated, the sculptures were arranged in a chronological sequence aimed at revealing the development of Assyrian art over time, which could then be used as a chronological tool. The final result, however, was a chronological sequence that was self-referencing. While Arslan Tash provides an important insight into Assyrian architecture and urban planning, little information regarding the wider Iron Age can be gleaned from the confusing and inadequate excavation volumes (Thureau-Dangin *et al.* 1931a; 1931b).

3.11 'Athlit (Israel)

Summary of Excavations

Rock-hewn shaft-graves belonging to the Iron Age were exposed during excavation of the medieval seaport at 'Athlit in the 1920s (Johns 1932, 41). Though containing mainly Persian period material, the graves also contained artefacts that the excavators dated between the ninth and fourth centuries BC, suggesting a long-period of use (*ibid*). No early Iron Age graves were found intact, nor any pots found complete, yet some fragments of Black-on-Red (*ibid*, 63, 82, 104), Red-Slip (*ibid*, 63, 84, 100) and Bichrome (*ibid*, 100) bear witness to the presence of an early Iron Age cemetery. The cemetery was greatly disturbed and partially destroyed by the medieval builders who, themselves, remarked on strange artefacts and "a coinage unknown to us today" (*ibid*).

In all, fourteen chamber tombs were excavated (L7, L12, L13, L14, L16, L19, L20, L21, L21b, L22, L23, L24, L34, L35), incorporating over one hundred burials. Common material within the tombs included pottery, jewellery, and weapons (*ibid*, 49). While the cemetery suffered significant damage at the hands of the medieval castle-builders, most of the disturbance at the site has been attributed to the original users of the cemetery. Johns (*ibid*, 42-43) suggests that at certain, indeterminate intervals, tombs were re-opened for subsequent burials, which would disturb and mix all previous deposits. Furthermore, once a burial chamber was full or had collapsed, "burials were made in the shaft" (*ibid*, 58). Hence, some tombs displayed evidence of continued use spanning a number of centuries.

In 1932, the limits of excavation were extended inland to include the medieval town. It was then that the remains of an ancient settlement were exposed, the earliest of which, according to Johns (1938, 137), pointed to a foundation within the seventh century BC. Evidence for Iron Age occupation is somewhat limited and is represented by a few sherds of Black-on-Red, Red-Slip and Iron Age cooking-pots (Johns 1934, 149-151).

In addition to the Persian period shaft tombs, Johns (1938, 135-137) also exposed a small Iron Age cremation cemetery to the south-east of the settlement. Situated on a sandstone ridge overlooking the promontory, the cremation cemetery contained at least eighteen simple cremation burials, one grave with cremated remains in a cinerary urn, and a single inhumation burial (*ibid*, 124). Johns (*ibid*, 134-135) attributed the inhumation burial to the Persian period, though a number of mixed use cemeteries (inhumation and cremation) have since been excavated (e.g. Tell Rachidieh), suggesting that the two might have been contemporaneous. The cremation cemetery was dated by Johns (*ibid*) to the seventh century BC.

Review

Considering the long period of use, the dating of individual burials within the "Persian" cemetery is particularly difficult. The excavator had to rely on superposition of materials, a complicated task considering the high level of disturbance. For instance, sherds from a single Red-Figure lekythos (Johns 1932, Pl. XXII.307) were found in the fill of three neighbouring shafts (L21, L21b, L23), while two coins, centuries apart in date, were found within centimetres of each other. As a result, the so-called "Persian" cemetery cannot be precisely dated.

The cremation cemetery was dated by Johns (1938, 134) to the seventh century BC on a review of the pottery and the presence of a single Egyptian steatite scarab. Haggi (2006, 48), on the other hand, contends that many of the vessels are indicative of a ninth-eighth century BC date, suggesting a slightly earlier beginning for cremation at 'Athlit. Haggi's (2006, 48-49) argument was based on the presence of similar pottery at other cremation sites along the coast. However, Haggi's comparison with Akhziv, al Mina, and Khalde was made with no assessment of their chronologies (none could be considered reliable).

Recent underwater excavations have dated the artificial harbour at Athlit to the late-ninth or early-eighth century BC, based on radiocarbon testing of a number of wood samples (Haggi 2006, 52). This suggests that the harbour was constructed at least 100 years earlier than the seventh century BC date that Johns (1938, 137) proposed for the settlement's foundation. The dates from the harbour, though, are essentially the dates for the felling of timber, which could have occurred many years before their use in construction. Regardless, dating the harbour does not help date the cremation cemetery; Haggi's (2006) criticism of Johns' chronology appears to be based on her dating of the harbour. Hence, the revised chronology is no more secure than that of Johns.

3.12 Beirut (Lebanon)

Summary of Excavations

Over the past century a number of excavations have explored Beirut's past, with more recent occupation greatly hindering access; e.g. the tell has long been covered by a large Crusader Castle. Widespread destruction and urban renewal associated with the Lebanese Civil War, however, has provided an opportunity to explore previously inaccessible areas (Badre 1997b, 1-45). Beginning in 1993, the Beirut Central District Archaeology Project (BCD) initiated the excavation of over 100 areas across Beirut.

Table 3.17: Summary of Published BCD Excavation Areas

Area	Location	Institution	Periods
Bey 001	S of Martyrs' Sq.	Lebanese University	H, RB
Bey 002	N of Martyrs' Sq.	IFAPO, Beirut	P, H, RB
Bey 003	Tell	AUB Museum	EB thru H
Bey 004	'Zone des Eglises'	Lebanese University	H, RB
Bey 006	Souks Area	AUB	H, RB
Bey 007	Souks Area	AUB/ACRE	LB, H, RB
Bey 008	Medieval Rampart	University of Amsterdam	H, RB
Bey 009	Banco di Roma	UNESCO/ DGA	RB
Bey 010	NE Souks Area	Université Libanaise	P, H, RB
Bey 011	Souks Area	Leiden University	H, RB
Bey 013	E edge of tell	Université Libanaise	LB thru H
Bey 020	E edge of tell	University of Tubingen AUB	LB thru H
Bey 024	'*Place Debbas*'	A-L Universität Freiburg Freie Universität Berlin	RB
Bey 027	S of Martyrs' Sq.	Université de Nice	EB, H, RB
Bey 045	SE of Souks Area	AUB/ACRE	H, RB
Bey 048	Martyrs' Sq.	IFAPO l'Université de Nice	H, RB
Bey 069	SW of Bey 027	Charles University, Czech	H, RB

Key: EB=Early Bronze; LB=Late Bronze; P=Persian; H=Hellenistic; RB=Roman/Byz.

Prior to the BCD project, little was known about Iron Age Beirut – the few related discoveries were restricted to the Persian period (Finkbeiner and Sader 1997, 118). While the majority of BCD areas revealed an extensive classical and medieval city plan, Bronze and Iron Age remains were exposed in only a few areas, mostly those associated with the tell (**Table 3.17** - Curvers and Stuart 1997, 176; 1998-1999).

Excavation of the tell was undertaken for the purpose of answering the absence of pre-classical material (**Table 3.18**; Badre 1997b, 12). The earliest human settlement on the tell dated to the Early Bronze III period, though architecturally the remains were fragmentary (Badre 1997b, 20). The Middle Bronze Age settlement is better represented, with its large defensive wall, monumental gateway and urban architecture. Of particular note is the Middle Bronze Age "silo" containing a small hoard of Middle Bronze Age objects reminiscent of the "*dépôt d'offrandes*" of the Obelisks Temple at Byblos (Badre 1997b, 40); the associated architecture has since been interpreted as the remains of a temple (Badre 2000a; Badre 2001-2002, 5-9). Sometime around the transition from the Middle to the Late Bronze Age, the fortifications of Beirut were renewed by the construction of a glacis, a feature that continued to be used and renewed well into the Iron Age and Persian period. The Late Bronze Age is represented by a rock-cut chamber and a significant amount of local and imported pottery overlying the first phase of the glacis (Badre 1998, 73-79). The Late Bronze Age gate of the city, the outer gate, and stairway entrance to the upper city have also been preserved (Badre 1997b; Curvers 2001-2002; Curvers and Stuart 1997; Finkbeiner and Sader 1997; Karam 1997). A second, larger glacis (glacis II) was constructed sometime around the Bronze-Iron Age transition (Badre 1998, 79). This enormous mound of earth, which was heaped up in front of and above the earlier glacis, and coated with unhewn stones, was exposed in three excavation areas; Bey 032 (Curvers and Stuart 1997, 178-180), Bey 013 (Karam 1997), and Bey 020 (Finkbeiner 2001-2002, Finkbeiner and Sader 1997), mostly to the east (Curvers 2001-2002, Fig. 1). Overlying this glacis were various Iron Age occupational and destruction deposits, all containing significant amounts of pottery (Badre 1998). Sealing the burning and collapse levels of the last glacis, a large complete building was exposed. This building, with its well-dressed limestone blocks and series of storage rooms, forms part of a casemate wall protecting the Iron Age settlement (Badre 1997b, 76, Fig. 40b). The preliminary synthesis suggests that Iron Age Beirut consisted of a stronghold and Lower City surrounding the ancient harbour, with cemeteries to the south and west and an industrial area to the south-west (Curvers 2001-2002, Fig. 6). The Persian settlement of the fifth and fourth centuries BC was more extensive and covered large areas of the promontory to the west and south of the tell (Curvers and Stuart 2004, 252; Finkbeiner 2001-2002, 29-30; Marquis 2004, 272).

On the whole, Bronze Age Beirut followed the general development of the region, namely the building of palaces, temples, and city-walls, all meant to express and protect the power and wealth that resided with the settlement's elite. The Amarna letters depict *biruta* (Beirut?) as a city actively involved in trade relations of the eastern Mediterranean. Beirut's role in the Iron Age, however, seems to have diminished somewhat, as it was overshadowed by the prosperity of its neighbours; Tyre, Sidon, and Byblos.

Table 3.18: Stratigraphy Bey 003 (tell)

Period	Bey 003
EB	Fragmentary remains
MB	Fortification wall Monumental gate Urban settlement
MB-LB	Silo Foundation deposit Well glacis I
LB	4 ash lenses on glacis Rock-cut chamber
LB-Iron	glacis II
Iron Age	Ash layers Collapse Casemate wall
Persian	glacis continues Urban area expands S & W

(After Badre 1997b)

Review

According to Badre (1998), the dates for Beirut's Late Bronze and Iron Age strata were based on ceramic data. The *terminus ante quem* date for the large Iron Age glacis II was established from pottery collected from the three destruction layers covering this feature (*ibid,* 79). The two earliest destruction deposits did not contain any imported material, and the local ceramic horizon was dated by the absence (negative evidence!) of Red-Slip (Badre 1997b, 32; 1998, 79). The third and final destruction level, however, was characterised by the appearance of local Red-Slip and imported pottery, primarily Cypriot Archaic I (*ibid*, 79). The following Iron Age phase, Badre's (1998, 80-81) "level of abandonment", was rich in ceramic material that offered a relatively precise provisional date for this level (750-700 BC), though only the imported pottery was considered in detail. The final Iron Age phase, with its complete seven-room house plan and casemate wall, was again dated by Badre (1998, 82-83) on ceramic evidence: the local storage jars (which have a very wide distribution across the eastern Mediterranean and are not truly "local"), Red-Slip, and imported material suggested a mid-seventh century BC date. Despite her call to date local pottery by more independent means, such as radiocarbon, Badre (1998, 83) has established a chronological framework dependent on the chronology of Cypriot pottery (**§2.5**).

3.13 Beth Shan, Tel (Israel)

Summary of Excavations

The American expedition to Beth Shan (1921-1933) was the first large-scale excavation in the Levant after World War I. The early seasons encountered mainly Medieval and Byzantine levels, and it was not until the third season of excavation that Iron Age deposits were reached. Three strata were attributed to the Iron Age (VI-IV), and incorporated a number of public buildings, including the so-called "Egyptian governor's residence", temples of Ashtaroth and Dagan, and the Stratum V administrative complex (Mazar 1993a, 216). Later seasons worked mainly in the Middle and Early Bronze Age levels. A number of tombs were discovered to the north of the site and were considered contemporary with Strata VII and VI (Mazar 1993a, 218; Oren 1973). Eleven tombs yielded anthropoid coffins of Egyptian tradition, though some archaeologists have connected these with "Sea Peoples" (e.g. Mazar 1993a, 218; Wright 1959; *contra* Oren 1973, 142). The interpretation of the Northern cemetery is complicated by the fact that the tombs were constructed in the Middle Bronze Age and were in secondary use when the anthropoid coffins were placed therein.

A team from Hebrew University initiated further work on the tell in 1983 with the aim of clarifying the early Iron Age sequence (Geva 1971). For the directors, Yadin and Geva (1986, 1), Beth Shan held special importance for the study of early Israelite history. Consequently, they were primarily interested in the excavation of Strata VI-V (**Table 3.19**), conventionally associated with the settlement of "Proto-Israelites" and the foundation of the United Monarchy (*ibid*).

A second expedition from Hebrew University (directed by A. Mazar) returned to the site in 1989 and renewed investigation of the Bronze and Iron Age settlements (Mazar and Mullins 2006, 10-14). Despite opening eight trenches across the mound, Iron Age material was isolated in only Area S (Iron I), on the south-eastern area of the summit, and Area P (Iron II), on the western edge of the summit (*ibid*, Tables 1.1, 1.2). Apparently, the American project had removed the majority of Iron Age deposits (Mazar and Panitz-Cohen 2006, 173). Areas G and L held evidence for Byzantine levels directly overlying Middle Bronze Age material; i.e. no Iron Age (Mazar 2006, 32). While this phenomenon could be interpreted as evidence for Byzantine builders clearing the top of the mound, Mazar (2001, 290-292) suggests it is indicative of the parochial nature of Beth Shan during the Late Bronze and Iron Ages.

Review

Despite considerable undertaking, the published results of the American excavations were disappointing: the Iron Age levels and Northern Cemetery were published after a lengthy interval; only a small portion of material was included; the architectural plans, being schematic, are disjointed; the relationship between architectural features is not always clear; and the ceramic material from some of the strata appears mixed (James 1966, 3, 21, 30, Fig. 73; Oren 1973). Consequently, the stratigraphic and chronological conclusions of James (1966) have been largely rejected and a call for re-examination issued; the primary point of contention was the dating of Strata V and IV (Geva 1971; Mazar 1993a, 219).

The monumental buildings of Stratum V-upper were destroyed and covered by a layer of ash, dated to the late-ninth century BC by the American archaeologists (James 1966, 44). This date was a product of James' (*ibid*, 132, 154) eighth century BC dating of Stratum IV, which relied on ceramic comparisons with Samaria and Megiddo, two sites not without their own stratigraphic problems (**§2.5**). Stratum IV was also sealed by an ash layer, which was associated with Tiglath-Pileser III and dated to 732 BC (Mazar 2001, 289). Geva (1971, 7) points out that many of the ceramic forms that James used for comparison were poor chronological tools; e.g. some bowl forms experienced a long period of use at Megiddo and Samaria and were not indicative of the eighth century BC. Geva (1971, 9) instead suggested a seventh century BC date for Stratum IV, implying that the Assyrian conquest had to be shifted to the Stratum V ash layer. Geva's (*ibid*) dates relied on comparisons with Samaria and Megiddo. At both these sites a heavily fortified city was supposedly destroyed by Assyrians followed by modest reconstruction. This pattern was accepted because it correlated with how Geva *expected* an Assyrian conquest to manifest itself archaeologically; i.e. prosperous city attracted Assyrian interest, leading to destruction, followed by impoverished habitation. In the end, the chronologies of James and Geva both rely upon the historical correlation of an archaeological ash layer.

Yadin and Geva's expedition to Beth Shan focused on the early Iron Age with the hope of clarifying the site's role in the establishment of an Israelite nation-state (Mazar 1993a, 215; Yadin and Geva 1986, 1). Indeed, Israeli nationalist sentiment was a significant driving force for much of Yadin's archaeological method and practice, leaving many of his conclusions influenced by uncritical use of the conventional historical narrative (Kletter 2006, 316). For example, the "destruction" of Stratum V-lower was dated to the late-tenth century BC due to parallels with Megiddo VA-IVB, which Yadin interpreted as being conquered by Shishak in 926 BC (Yadin and Geva 1986, 7). Yadin and Geva's short campaign excavated only a small area and revealed very little architecture. The few conclusions that were offered appear to have been extrapolated from small amounts of data.

Table 3.19: Correlation of Beth Shan strata

Period	American	Yadin/Geva	Mazar
Early Mediaeval	I		
Byzantine	II		
Roman	III		
Iron IIB	IV		P6-P7
Iron IIA	V-upper		P9-P8
Iron IB	V-lower	1-2	S1/P10
Iron IB	VI-upper	3	S2
Iron IA	VI-lower	4	S3
?	VII		
LB IIB	VIII		
LB IIA	IX		R1
LB IB	IX		R2
LB IA			R3
?	XA		R4
MB II	XB		R5
MB I	XI		R6
?	XII		R7
EB	XVI		
Chalcolithic	XVII		
Neolithic	XVIII		

(After Mazar 1993a, 215)

Amongst the objectives for Mazar's (1993b, 202) Beth Shan project was a desire to know the ethnic identity of those buried in the anthropoid coffins from the Northern Cemetery, and to determine whether there was "any evidence for the presence of Philistines or other Sea Peoples at Beth Shan". It is clear from such statements that the results of the latest expedition too were influenced by the conventional historical narrative. Indeed, despite detailed work on the Iron Age stratigraphy, preliminary results have emphasised significant historical events in the life of the site: e.g. David's tenth century BC conquest; the Assyrian destruction in 732 BC (Mazar 1993a, 221; 2001, 289). Furthermore, these important Iron Age events are largely restricted to those that are biblically-attested, resulting in a history of the site that cannot be considered outside of, or be separated from, the biblical narrative (see Mazar 2006, 32).

3.14 Byblos (Lebanon)

Summary of Excavations

Investigations started at Byblos as long ago as 1860 when Renan (1864) began his meta-study of ancient Phoenicia. Renan relocated ancient Byblos from its description in classical sources and made several small soundings at the site (even though the site was largely occupied by a modern village) (Joukowsky 1997, 391).

The first large scale excavation of the site was undertaken by the Egyptologist P. Montet, who had been attracted to the site by reports of Egyptian inscriptions and mythological legends. Montet's excavations focused on the so-called "tombs of the kings", where he found a decorated sarcophagus that has since been linked to Ahiram, King of Byblos (Montet 1929, Pls 128-141). Once again, however, the overlying village restricted progress and Montet left after just five seasons. In 1926 Montet's former assistant, Maurice Dunand returned to the site to recommence excavation. This time, however, excavation was unhindered by land ownership disputes: the enlightened Director of Antiquities, the Emir Maurice Chéhab had resolved problems of private ownership and secured a large portion of the site for the explicit purpose of archaeological investigation (Joukowsky 1997, 391). Dunand would continue to excavate at Byblos for nearly 50 years, exposing evidence of almost continuous occupation of the site since the Neolithic (Dunand 1973). Despite the long history of French works at the site, and the many periods exposed, little has ever come to light regarding the Iron Age (Jidejian 1971, 57-59; Pritchard 1978, 10-11). In many areas of the site, the classical periods were directly superimposed upon Bronze Age levels (Dunand 1939a, 64, 79).

Located off the tell, to the immediate east of the monumental Achaemenid fortress, "Necropolis K" has yielded a significant amount of Middle and Late Bronze Age pottery (Dunand 1973, 75). The small amounts of locally-made Black-on-Red and Bichrome wares (Salles 1980, Pl. VIII.1-5) imply some continuation of use into the early first millennium BC, though Salles (1980, 20-21) considers the necropolis was unused from the end of the Late Bronze Age until the Hellenistic period.

Review

The publication of Montet's (1928; 1929) expedition to Byblos is primarily a catalogue of finds; no attempt was made to relate them to the architecture or stratigraphy. The results of Dunand's expedition are only a minor improvement. In an effort to record precisely the geometrical location of each object and architectural feature across the whole site, Dunand covered the entire headland with a grid of ten metre squares (though not all grid units were of uniform size!). Each square was dug by means of rigid horizontal layers (*levées*) 0.20 metres thick (Dunand 1973, 100). As logical as this method seemed to Dunand, it imposed an artificial stratigraphy onto the site and completely ignored the depositional formation processes of the site. Consequently, Byblian stratigraphy should be disregarded and all excavated material treated as unstratified. Without the provision of a meaningful stratigraphic sequence the objects are all isolated. Furthermore, Dunand's publication is limited to only five volumes (Dunand 1939a; 1939b; 1950; 1954; Cauvin 1968; see Lehmann 2002a, 122), each presented as a catalogue of finds. Dunand's reports require a certain level of "re-excavation" to be of any value.

Chatal Hoyük (Turkey) - see 3.8 Amuq

3.15 Dan, Tel (Israel)

Summary of Excavations

Late in 1963, Yeivin conducted a brief exploratory excavation at Tel Dan, uncovering Early Bronze, Middle Bronze and Iron Age levels (Biran 1996c, 1). More extensive and permanent work began in 1966, when the site was threatened by military construction, and has since become the longest ongoing archaeological investigation in Israel (Biran 1994, 7). This project has exposed Iron Age remains across much of the site (**Table 3.20**), including a sacred precinct, city walls and gate, and metallurgical industrial installations. The earliest Iron

Age settlement was Dan VI, which was characterised by meagre architectural remains and stone-lined pits (Biran 1994, 126-128). Dan VI pottery was characterised by storage vessels and cooking-pots, all of which displayed strong continuity with Bronze Age ceramic traditions (*ibid*, 126), though strong Cypro-Aegean influences were no longer evident (*ibid*, 128). Biran (*ibid*) interpreted the Dan VI "collared-rim" pithoi as representative of Israelite settlement, more specifically the tribe of Dan, which was conventionally dated to the early-twelfth century BC. Hence, the ash layer upon which Dan VI was built was accepted as confirmation of the biblical account of the tribe of Dan's conquest of Canaanite Laish (Judges 18:27). Also belonging to the early Iron Age, Dan V was characterised by permanent settlement (stone-walls, plaster-floors, metallurgical activity) and was sealed by thick ash and rubble, which Biran (*ibid*, 132) accepted as evidence of destruction. While the rich ceramic assemblage of Dan V was a continuation of Dan VI culture, there was a marked shift towards smaller vessels (e.g. jugs, juglets, bowls, flasks, pyxides). Biran (*ibid*, 138, 141) used pottery to date the "destruction" of Dan V to the mid-eleventh century BC. Despite the supposed destruction of Dan V, there was significant cultural continuity into Dan IV (*ibid*, 142). The second of the two Dan IV sub-phases (IVA) witnessed the appearance of Red-Slip, Bichrome, and Black-on-Red pottery.

On the southern slopes of the tell (Area A), two phases of Iron Age fortifications were exposed (Dan IV and III), (Biran 1994, 235-254). In 1993 a stele written in Aramaic was discovered in a secondary context near the city-gate (*ibid*, 275-278), and included the phrase "*byt dwd*" (house of David), the first extra-biblical reference to King David (Biran 2002, 6). Also uncovered in Area A were three successive buildings associated with the Iron Age *hussot* ("piazza"), all of which belonged to the Iron II ceramic horizon (Biran 1994; 235-254; 1999). On the northern side of the mound a "sacred precinct" was exposed, the main feature being the large *bamah*, or high place (Biran 1982; 1994, 159-234). A horned altar and three storerooms full of decorated jugs, bowls, and pithoi, were found in association with this structure.

Review

According to Biran (1994, 128) the destruction of the Late Bronze Age settlement at Dan was attributable to the Israelite tribe of Dan. This interpretation ignores the strong continuity of culture that persisted between Dan VII and VI. Instead, the appearance of "collared-rim" pithoi suggested to Biran (1989a) the settlement of the Israelites. While these vessels have traditionally been associated with the Israelites (Albright 1937, 25; Biran 1989a; 1993; Esse 1992), recent finds of "collared-rim" pithoi outside conventional Israelite areas (e.g. Tel Nami – Artzy 1994) has brought the association into question (London 2003, 148-149). The archaeological record does not provide strong evidence for the presence of a new population in Dan VI. In a reversal of this methodology, Biran (1994, 125) accepts the Dan V-IV continuity as representing no change in population, despite the presence of a "destruction" level in-between. It seems that a literal reading of the biblical narrative was the key interpretative tool used by Biran to decipher the history of the site, as he himself indicates:

> *"When we began the rescue excavation of Tel Dan, we wondered whether archaeological evidence would introduce a new objective element into the discussion [on when the tribe of Dan conquered and settled Laish], and help resolve the issue. A datable conflagration layer could, for example, relate to the account in Judges 18:27 if taken literally... We hoped for tangible evidence from the excavation to determine a date for the settlement of the tribe." (Biran 1994, 125).*

Table 3.20: Tel Dan Iron Age Chronology

Strata	Period	BC
I	Iron II	late 8th–early 6th
Assyrian conquest		*732 BC*
II	Iron II	mid 8th
III	Iron II	9th–early 8th
IVA	Iron II	2nd half 10th–early 9th
IVB	Iron I/II	2nd half 11th–1st half 10th
ash layer = destruction?		
V	Iron I	12th–1st half 11th
VI	Iron I	12th
ash layer – destruction by tribe of Dan		
VII	LB II	14th–13th

(After Biran 2002, Table 1.1)

The circularity of Biran's approach is clear: he hopes to find a datable destruction to support the biblical narrative, but the means he uses for dating that destruction is the biblical narrative – i.e. support for the Bible rests within the Bible. Disregard the biblical reference, and the destruction of Dan VI cannot be conclusively associated with a 1200 BC conquest. Furthermore, Biran's presentation of the material from the sacred precinct makes almost no reference to the site's strata, instead referring to the phases via biblical reference: e.g. time of Jereboam I, the Assyrian conflagration level, etc… (Biran 1994, 159-233). Also, it is not clear from the publications exactly how the sacred precinct was dated, but it does appear to be based on *likely* correlations with the biblical narrative.

In addition to the two "destruction" layers already mentioned (Dan VII and V), Biran (1994, 260) identified another Iron Age "destruction"; a thick layer of ash sealing Dan II was attributed to the Assyrian campaigns of 732 BC. Once again, Biran's interpretation was based on the conventional historical narrative.

3.16 Deve Höyük (Turkey)

Summary of Excavations

Surviving artefacts from Deve Höyük were purchased directly from the looters of the site, and in competition with the dealers of Aleppo (Moorey 1980, 3-4). Woolley and Lawrence were occasionally present to observe the looting, and did make some effort to record the general context of the burials, but the material was being removed so quickly that they were unable to make more than passing observations (*ibid*). From the information at hand, Woolley (1914-1916, 116) was able to classify the pottery and objects into two distinct periods, both associated with mortuary remains. The earlier material appears to represent an Iron Age cremation cemetery (eighth or seventh century BC), while the second phase is associated with a Persian period inhumation cemetery (Moorey 1975; 1980, 4-10). Woolley and Lawrence were greatly interested in the Persian period material, at the expense of the cremation cemetery.

Review

Considering how little information was available to Woolley, his conclusions, however tenuous, are admirable. Nevertheless, the mixed nature of the material and the unscientific manner by which it was "excavated" means that Deve Höyük holds poor implications for the chronology of the region. The broad dates that Woolley (1914-1916, 127) does offer are based on Greek imports, and are therefore only indicative (**§2.5**).

3.17 Ghassil, Tell el- (Lebanon)

Summary of Excavations

In 1956, the American University of Beirut began excavating Tell el-Ghassil – the first systematic excavation within the Beqa' Valley (Baramki 1961, 1964, 1966). The first four seasons focused on Area I, which uncovered the remains of a large temple complex. The next three seasons added Areas II and III; habitation areas to the west and north-west of the temple, respectively (Joukowsky 1972, 42). During the 1970s excavation focused on Area III (Doumet-Serhal 1996, Fig. 7). While isolated, unstratified finds from the Chalcolithic and Early Bronze Age were found, it is only the Middle Bronze Age through to the Iron Age that was represented by stratified occupational deposits (Joukowsky 1972, vii).

One of the key objectives of this project was to study the development of the ceramic industry at the site throughout the millennium and a half of stratified occupation (*ibid*). A limited corpus of Iron I pottery was found in Area III (Levels 7-4/5) and Area I (Level 5). Joukowsky (*ibid*, 218-221) identified strong parallels for this material in the Southern Levant, especially within the Iron I levels of Hazor and Megiddo. Tell el-Ghassil is the Beqa' Valley's sole representative of the Iron II period: Areas III/II (Levels 4-1) and Area I (Levels 4-1). The Iron II pottery was compared with Sarepta and Tyre, with some parallels in cemetery I at Hama and Amuq O (*ibid*). Once again the closest parallels were with the Southern Levant, at Hazor and Megiddo; the principal indicators being cooking-pots, storage jars, Red-Slip bowls, and jugs (Marfoe 1998, 218).

Review

The chronological framework at Tell el-Ghassil was constructed from "a thorough study of architectural deposits, pottery and other objects" (Joukowsky 1972, 197). In other words, absolute dates were assigned by comparison with other assemblages, with some effort being made to distinguish between misleading or debated chronologies and more-reliable stratified deposits. The site's affinity with the Southern Levant meant the closest parallels lay with the debated chronologies of Hazor, Megiddo and Samaria (**§3.19**; **§3.28**). Consequently, the chronology of Tell al Ghassil does not stand alone, and cannot be used for further chronological comparison.

3.18 Hama (Syria)

Summary of Excavations

The tell of Hama was excavated in the 1930s by a Danish expedition, which exposed a large area of the mound and recorded a sequence from the Neolithic through to the Medieval Period (Hama M-A) (Buhl 1992). Hama F and E relate to the Iron Age. While the remains of Hama F were only fragmentary, Hama E contained a large "Royal Quarter", which the excavators equated with Aramaean kingship (Ingholt 1942, 472; for plan see Buhl 1992, Fig. HAM 01; Fugmann 1958, Fig. 185). This large complex incorporated a towered gate (*Batiment* I), two "palaces" (II and V), a temple (III), and the enigmatic *Batiment* IV, originally interpreted by Ingholt (1940, 91) as a harem, then later as an official's residence (Ingholt 1942, 472), and afterwards by Fugmann (1958, 237-245) and Buhl (1992, 35) as a small gate, and by Ussishkin (1966) as a temple. These buildings surrounded a large, open central-space with a small sanctuary (Fugmann 1958, Pl. Ib). While the architecture borrows some features from the "Syro-Hittite" tradition (e.g. basalt lion orthostats), the Hama buildings are a less-decorative example of the tradition (Fugmann 1958, 267-268). These elite buildings were destroyed in a massive conflagration, which the excavators attributed to the 720 BC campaign of Sargon II (Ingholt 1942, 472), a date that has greatly influenced regional Iron Age chronologies (e.g. Francis and Vickers 1985, 131).

In addition to excavations on the tell, smaller excavations throughout the modern town recovered over 1100 cinerary urns (Riis 1948, 1-26). Grave goods were abundant, though not particularly "rich" (Riis 1948, Fig. 230). This large cemetery was divided into four distinct phases of use throughout the Iron I and Iron II periods (**Table 3.21**; Riis 1948, 202).

Review

Throughout his discussion of the Hama Iron Age chronology, Fugmann (1958, 267-269) uses a number of terms that suggested a cautionary approach to the material (e.g. *vraisemblablement*, *semblent*, *peut-être* and *environ*), yet one date he considered beyond doubt is the Hama E "destruction".

> *Tandis qu'il n'a pas été possible de fixer avec certitude la date de la construction des différents bâtiments sur la base de la documentation qui nous a été fournie ici, il n'y a pas de doute que la destruction définitive de la citadelle a eu lieu en 720 BC pour punir la ville de sa résistance contre Sargon II. (Fugmann 1958, 269)*
> (While it was not possible to fix with certainty the date of the construction of the various buildings on the basis of documentation, there is no doubt that the final destruction of the citadel took place in 720 BC to punish the town for its resistance against Sargon II - translation mine)

The Hama E date was essentially an exercise in historical 'best-fit' (**§2.4**), one which also had to ignore inconsistencies in the conventional historical narrative. For instance, by the start of the eighth century BC, Hama was considerably weakened by ongoing Assyrian aggression, yet the king of Hamath and Lu'ash was strong enough to repel an attack by a local coalition including the kings of Damascus and Zincirli (as depicted in the Zakkur Stele found at Tell Afis – Klengel 1992, 210-215).

Another important date at Hama was that associated with the beginning of the Iron Age. Despite being unable to observe any difference in architectural traditions between Strata G (Late Bronze Age) and F (Iron Age), Fugmann (1958, 267, 274) believed, "*sans aucun doute*" (another definite term!), that Hama had been conquered in 1200 BC and settled by a new population. The new inhabitants were said to be characterised by the use of iron weapons and tools, fibulae, painted ceramics, and the practice of cremation (Fugmann 1958, 275). Indeed, the appearance of what Ingholt (1942, 472) considered an essentially Aegean burial rite (cremation) confirmed for him the association of Hama F with the arrival of "Sea Peoples" and thus an early twelfth century BC date for Hama F. These conclusions can be dismissed on a number of considerations. First, cremation is known in the Levant during the Late Bronze Age (**§2.3**). Second, painted ceramics, while present in Hama F, were much more common in Hama G and E. Third, while fibulae are associated with the Iron Age, they were not associated with the adoption of iron for utilitarian items; iron had been used for ornamental reasons since the third millennium BC (Muhly 1980, 34-36). Finally, the presence of iron weapons and tools in Hama F is less significant than what Fugmann (1958, 275) suggests: the 25 weapons and tools recovered from Iron I contexts, were swamped by the presence of bronze weapons and tools (Waldbaum 1978, 27-28, 44, Figs IV.3-4; Tab. IV.4). Moreover, the presence of iron tools and weapons in Hama F is exceptional; no other North Levantine site has produced significant examples. In the end, Fugmann and Ingholt dated the Hama G-F transition to 1200 BC based on the conventional historical narrative, and then proceeded to superimpose the evidence onto their interpretation. In addition, the presence of Cypriot and Greek imports in Hama burials were used to date these contexts (Riis 1948, 114-115), though the same vessels were used to refine Cypriot and Greek chronologies (**§2.5**). There is also some doubt regarding practical aspects of excavation and recording at Hama (Dornemann 1997, 467): for instance, finds were sometimes recorded weekly, leaving much of the material without accurate contextual data (Thuesen 1988, 11). Evidently, the Hama sequence is not without significant problems.

Table 3.21: Correlation of Hama phases

Cem.	Tell	Features	BC
I	F2		1200-1075
II	F1		1075-925
III	E2	Building I, II, III & V	925-800
IV	E1	Restored I-III, V	800-720
Assyrian destruction			720
-	Area 10	Assyrian garrison	Ca. 715
	Hiatus		

(After Fugmann 1958, 275-277; Riis 1948)

3.19 Hazor (Israel)

Summary of Excavations

Garstang's short-lived work at Hazor included soundings on both the tell and lower mound, but his results were not published in detail (Ben Tor 1993a, 595). In 1955, Yadin recommenced work on the lower and upper mounds where he revealed the existence of 21 strata spanning 3000 years (Ben Tor 1997, 108). The lower mound revealed no evidence of Iron Age occupation and was completely destroyed or abandoned by the end of the Bronze Age (Yadin 1975, 129-145; *cf.* Ben Tor and Zuckerman 2008; Zuckerman 2007a, 17, 23). In contrast, the upper mound, which, according to Yadin (1975) and Ben Tor (1993a, 600), was also destroyed at the end of the Bronze Age, preserved Iron Age occupation (Strata XII-III; **Table 3.22**). Yadin dated this sequence by isolating key historical events within the archaeological record: e.g. the end of the Late Bronze Age was associated with the presence of the "Israelite" destruction of the Canaanite city and dated on the presence of imported Mycenaean pottery (Ben Ami 2001, 148-150; Ben Tor 1993a, 600; Zuckerman 2007b, 621).

Ben Tor, one of Yadin's former area supervisors, returned to the site in 1990 with two primary objectives; to examine and reassess Yadin's stratigraphy and chronology, and to confront any problems left unresolved by the previous expedition (Ben Tor 1997, 110; Ben Tor and Ben Ami 1998, 2-3).

Review

Since Yadin's (1975, 143; *et al.* 1958, xix) self-proclaimed objective in excavation was the authentication of the biblical account, it is little wonder that he fulfilled his goal. According to Yadin (1975, 145), the thick ash deposit sealing the final phase of the impressive Late Bronze Age city was clear evidence for the violent destruction of Canaanite Hazor as depicted in the biblical text (Joshua 11) (*cf.* Ben Ami 2001, 148-150; Ben Tor 1993a, 603; Ben Tor and Zuckerman 2008; Zuckerman 2007a). Yadin believed that the meagre Iron I re-occupation (Strata XII-XI) reflected a seasonal settlement of a semi-nomadic population, whom he identified with the Proto-Israelite tribes (Ben Ami 2001, 151). This historical correlation also provided Yadin with absolute dates for the beginning of the Iron Age. There are, however, problems with Yadin's interpretation. First, Yadin's theory ignores significant continuity in material culture across the Bronze-Iron Age transition, and which implies no change in population. Second, Yadin's reasoning is circular – archaeological data is used to support the biblical narrative, yet it is the biblical narrative that is used to interpret and date the archaeological record. Finally, Zuckerman (2007a, 17, 23) has pointed out that the Late Bronze Age destruction deposits were isolated, and more indicative of only a *partial* destruction of the site. Zuckerman (*ibid*) instead views the "destruction" of Canaanite Hazor as the end product of a long, drawn-out process of social and economic decline, and bears little resemblance to a mighty conquest as depicted in the biblical narrative.

In addition to the Late Bronze-Iron Age transition, Yadin distinguished three Iron Age "destruction" layers which he believed were caused by enemy conquest: Stratum IX by Ben-Hadad of Aram; Stratum VII by another Aramaean campaign under the leadership of Hazael; and Stratum VA by Tiglath-Pileser III of Assyria (Ben Tor 1993a, 601; 1997, 112-113). By isolating the archaeological occurrence of historical events, Yadin (*et al.* 1960, xxii) was also assigning absolute dates to his stratigraphic sequence. For confirmation, Yadin drew upon comparative ceramic and architectural data from other sites, though his cross-referencing did not include an appraisal of those sites' chronology. Of particular interest for Yadin was the six-chambered gate and casemate wall of Megiddo IV which had been attributed to Solomon and appeared to confirm the date of a similar gate and fortification at Hazor (Ben Tor 1993a, 601). Yadin seems to miss the fact that the Megiddo excavators relied on the very same biblical passage (1 Kings 9:15), as vague as it is, that Yadin employed at Hazor (**§2.5**).

His methodology aside, Yadin's practice has also come under scrutiny. In a comparison between Yadin's architectural and ceramic data, Ben Tor (1997) has identified a lack of correlation between these two datasets, bringing Yadin's whole chronological sequence into question.

Table 3.22: Stratigraphy of Iron Age Hazor

Str.	Period	Suggested Dates
II	Persian	4th cent. BC
III	Iron III	7th
IV	Iron II	8th
destruction		732 BC
V	Iron II	8th
VI	Iron II	8th
destruction		810 BC
VII	Iron II	9th
VIII	Iron II	9th
destruction		880 BC
IX	Iron I	Late-10th/early-9th
X	Iron I	Mid-10th
XI	Iron I	11th
XII	Iron I	12th
destruction		

(After Ben Tor 1993a, 606; 1997, 112)

Despite shortcomings in Yadin's method and practice, the excavations at Hazor have significantly influenced contemporary Israeli archaeology, both through the dispersal of Yadin-trained archaeologists (e.g. A. Ben Tor, I. Dunayevsky, T. and M. Dothan, Y. Aharoni), and by way of the conclusions reached there. Yadin's interpretation was widely-accepted and today remains an oft-celebrated correlation between biblical history and archaeology (e.g. Aharoni 1978, 178).

While the renewed excavations at Hazor began in 1990, a definitive interpretation of the site's Iron Age history has yet to be produced. A few conclusions have been offered by way of preliminary remarks, which have largely sought to verify Yadin's conclusions, though slightly modified (Ben Tor and Ben Ami 1998, 2; Ben Tor and Zuckerman 2008, 4). For instance, Stratum IXa was dated to the tenth century BC based on Yadin's dates for the surrounding strata; the underlying level was dated to the twelfth and eleventh centuries BC because it contained pits that were equated with the settlement of Proto-Israelites, while the overlying levels were associated with the Stratum VIII pillared building (ninth century BC) (Ben Tor and Ben Ami 1998, 3-4). In the end, Ben Tor *et al* (*ibid*) offer confirmation for Yadin's tenth century BC chronology through reliance on Yadin's chronology. Once the tenth century BC date was confirmed for the Solomonic gate, other anchor points in Yadin's chronology were also supported; e.g. the 880 BC destruction of Stratum IX by the Aramaeans (Ben Tor and Ben Ami 1998, 11). More recently, however, one of the project co-directors at Hazor has begun to question some of Yadin's conclusions. Sharon Zuckerman (2007a; 2007b) has re-evaluated both the date and agent for the destruction of Late Bronze Age Hazor in an earnest attempt to move the reconstruction of the site out from the constraints imposed by Yadin's interpretation and the conventional historical narrative (*cf.* Ben Tor 1998; Ben Tor and Zuckerman 2008, 4).

3.20 Jerablus (Carchemish), Kefrik, Merj Khamis (Turkey)

Summary of Excavations

In 1876, the architect A. E. Henderson was the first to excavate the extensive ruins at Jerablus, undertaking the work on behalf of the British Museum while employed as Consul at Aleppo (Hogarth 1909, 171). Six years prior, George Smith had visited the site and reported a fine "Hittite" sculpture (Winstone 1990, 26-7). Henderson's brief expedition recovered a small group of four *in situ* monuments – orthostats carved with Hittite-looking figures and motifs reminiscent of monuments exposed at Zincirli, where they were part of a ceremonial approach to a large palace (Hogarth 1909, 171). The abundance of the "Hittite" monuments at Zincirli was incentive enough for the British Museum to initiate a return to the site in 1911 for more intensive excavation. Unfortunately, the first season, led by Hogarth, was noted for its confusion, squabbling, and paucity of finds, as well as bringing the ownership of the mound into question (Winstone 1990, 26-32). The situation improved only marginally with a change of director: Woolley and Lawrence appear to have waged personal battles with their workers, servants, German neighbours, and the local Turkish officials (*ibid*).

Evidence from the Jerablus mound bears witness to occupation that extends back into the Neolithic (Hawkins 1976-1980, 435). It was during the Late Bronze Age, however, that the city became an important centre of the Hittite Empire, before reaching its apogee in the Iron Age. The British expedition uncovered substantial remains of the Iron Age at Jerablus, including defensive structures, temples, palaces, a monumental processional way, and numerous basalt statues and reliefs with Luwian hieroglyphic inscriptions (Woolley 1921a; 1952).

The Iron Age pottery from cemeteries associated with Jerablus (Carchemish) is characterised by painted kraters and urns. The majority of these vessels, which show an astonishing variety of decoration, were used as cinerary urns in the Yunus cemetery (Woolley 1939b). Decorative motifs were usually geometric in form; the few bird and animal motifs are similar to late Iron Age Phrygian designs (*cf.* Akurgal 1955, Figs. 1-9; Woolley 1952, 234*ff*). Red-Slip bowls and trefoil-lip jugs were also prominent (*ibid*). The Yunus cremation cemetery was located to the immediate north of the site, where scattered Iron Age sherds drew attention to the presence of over 150 graves (Woolley 1939b, 13). Another contemporary cemetery was identified outside the west gate of the outer city-wall, but it lay beyond the expedition permit and remained unexplored. According to Woolley (1939b, 14), the Yunus cinerary urns were almost always "of a uniform general type, differing from one another only in such details as whether they [had] handles or no handles, ring bases or flat bottoms". The cinerary urns were also covered by either an upturned bath or upturned pot (bowl or krater), presumably to prevent the infiltration of soil.

An additional two Iron Age cemeteries were identified in the vicinity of Jerablus. Little has been recorded regarding the specific context of these sites, but they appear to closely parallel the Yunus material (Moorey 1980, 146). The first is a cremation "tomb group" from Kefrik, a small village 15 km west of Jerablus (*ibid*). This tomb was not systematically investigated and the material only assigned a general Iron Age date, though Moorey (*ibid*) suggests it is remarkably similar to the Deve Höyük cremation cemetery. The second, Merj Khamis, is a small cemetery 6 km to the north that was excavated as part of the British Expedition (Woolley 1939b, 12), though the results were disappointing. By the time the British team began work, over 30 Iron Age burials had been plundered (*ibid*). Woolley and Lawrence excavated eight intact graves, of which only one produced a complete vessel. It was clear from what did remain, however, that this was another cremation cemetery of the Iron Age, more or less contemporary with Deve Höyük I and the Yunus cemetery (Woolley 1939b, 20).

Review

The publication of results, which was completed a number of years after excavation, focused on the architecture and sculptural remains of the site (Woolley 1921a; 1952). Little attention was given to the pottery and small finds, while the inadequately published stratigraphy appears poorly understood (Hawkins 1976-1980, 435). The published plans of the site are neither thorough nor cohesive as the British project evidently lacked sufficient architectural experience (Winstone 1990, 33). Furthermore, archaeological method and practice may not have been of primary concern for Woolley and Lawrence; a number of theories still circulate suggesting the British presence at Jerablus was for military purposes (Pucci 2008b, 3; Winstone 1990, 48, 56).

When Hogarth and Woolley excavated the site, they were particularly interested in finding archaeological evidence confirming "Hittite" presence in the area (Woolley 1952, 227). To this end, Woolley accepted Hogarth's preliminary dating of the site and tried to associate a number of reliefs with the Late Bronze Age. For instance, the reliefs of the "Water Gate" were assigned to the "Middle Hittite Period", "not later in date than the thirteenth century B.C., and, perhaps, considerably earlier than that" (Woolley 1921a, 110). Later scholarship has firmly disagreed, as summarised by Mallowan:

> *The chronology and sequence dating of the rich series of sculptures discovered at Carchemish remains a problem, even after 60 years of investigation, but it is generally recognised that Leonard Woolley exaggerated the antiquity of some of the orthostats and it is no longer possible to assign any of them to the second millennium B.C. On the contrary, many critics will now support Frankfort's view that none of this particular series of sculptures could have been executed without an awareness of Neo-Assyrian art (Mallowan 1972, 63).*

In the end, the site's stratigraphy and architectural sequence has been questioned and cannot be relied upon to aid interpretation of the site's material culture.

Woolley also perceived two distinct ceramic phases associated with the city of Carchemish; the styles were named after two type sites, Amarna and Yunus. The Yunus pottery was associated with the construction of the temple complex and was dated by Woolley (1952, 167-175) to the "Late Hittite Period", or Iron Age. The Amarna style was dated to the Late Bronze Age, via reference to the "Sea Peoples" destruction of Carchemish mentioned in Egyptian texts (J.A. Wilson 1969a, 262; Woolley 1952, 226, 235). Working from the conventional historical narrative, which emphasised change between the Late Bronze and Iron Ages, Woolley (1952, 235) then placed the main cultural break (the shift from Amarna to Yunus pottery) around 1200 BC, though he did admit that the two ceramic styles overlapped. It seems that Woolley's interpretation of the Carchemish pottery was influenced by the historical narrative rather than a close analysis of the site's stratigraphy. Is it possible that Woolley was unable to distinguish early Iron Age pottery from the Amarna ceramic horizon? This would account for the apparent chronological gap ("dark age") between the Late Bronze Age and Iron II pottery despite the overlap of ceramic horizons.

3.21 Jezreel, Tel (Israel)

Summary of Excavations

The earliest excavations at Tel Jezreel, ignoring those inadvertently undertaken by an overly-keen construction company, were two seasons of salvage work by the Israeli Department of Antiquities (Yogev 1988-1989). While it is clear that Iron Age remains were reached, little else regarding this campaign is known. The brief publication discusses only a few Iron Age sherds; notably a Judean handle stamped with a royal *lmlk* impression (*ibid*). Work recommenced at Tel Jezreel in 1990, with a work being undertaken jointly by the Institute of Archaeology of Tel Aviv University and the British School of Archaeology in Jerusalem (Ussishkin and Woodhead 1992, 11; Williamson 1991).

The main feature of the Iron Age strata at Tel Jezreel was a large, rectangular casemate enclosure (c. 145 x 250 m) with fortified gates, corner towers and excavated moat (Ussishkin and Woodhead 1997, Fig. 4). The walls were founded on bedrock suggesting the site was established in the Iron Age, though small amounts of Bronze Age pottery were located in the construction fills of the walls (Zimhoni 1997b, 29). In addition to the predominantly Iron Age pottery, very small amounts of Persian material were also present, though greatly disturbed by later construction.

Review

The interpretation of Jezreel is reviewed in **§2.5**.

Judeideh, Tell (Turkey) - see 3.8 Amuq

3.22 Jurn Kabir, Tell (Syria)

Summary of Excavations

Work at Tell Jurn Kabir began as part of the Tishrin Dam salvage project; the site was chosen because of its known Iron Age material (Eidem and Putt 1999, 193). Excavation was aimed at supplementing earlier excavations in the region that had focused on monumental art and architecture to the detriment of ceramics; e.g. Arslan Tash (Eidem and Putt 1994; Eidem and Ackermann 1999, 309). Two additional Iron Age sites were excavated within this project (Tell Qadahiye and Sandaliye) with the material from all three very similar (Eidem and Putt 1999, 193).

Phase IV represents the earliest occupation at Tell Jurn Kabir. Consisting of flimsy architectural remains, this phase is present only on the acropolis. The few sherds found in this layer are similar to those from Phase III, which covers the entire site (Eidem and Putt 1999, 194). The main feature of Phase III is a round enclosure wall surrounding the tell's summit (*ibid*, Fig. 2). The Phase III settlement was abandoned (*ibid*, 194-195). During Phase II, the foundations for two large buildings were laid on the acropolis, but were never fully completed (Eidem and Ackermann 1999, Fig. 2): Building II bears no traces of any superstructure, while the *bit hilani* Building I was partially back-filled with mud-brick. The casemate enclosure was rebuilt in Phase I, when a mud-brick fort-like structure was erected on top of Building II; the "fort" contained large quantities of "Neo-Assyrian" pottery.

Three Iron Age ceramic assemblages were identified at Tell Jurn Kabir (**Table 3.23**). Assemblage C includes Assyrian ceramic forms that are well-represented at other sites and dated to the seventh and sixth centuries BC (e.g. Tell Ahmar). Assemblage B was dated to the ninth and eighth centuries BC based on parallels with Iron II levels at Tell Abou Danne, Hama and Tell Afis. Assemblage A, dated to the eleventh and tenth centuries BC, was poorly represented elsewhere in the Northern Levant.

Table 3.23: Tell Jurn Kabir Stratigraphic Phasing

Phase	Features	BC	Ceramics
I	'Fort'	7th century	Group C Assyrian
II	Ambitious/unfinished Buildings	9th–8th centuries	Group B
abandonment			
III	Large enclosure wall	11th–10th centuries	Group A
IV	Flimsy architecture	? - oldest	Little pottery

(After Eidem and Ackermann 1999)

Review

The few published reports on Tell Jurn Kabir indicate that the stratigraphic sequence has been assigned broad dates. While the pottery holds parallels with the Northern Levant, a comparative study of pottery was unable to refine the chronology. Nevertheless, the conventional historical narrative had an impact (e.g. Eidem (1999a, 153). For instance, Assemblage C, with its "Assyrian" cups and bottles, was consigned to the Iron III period because it correlated with a strong Assyrian presence along the Euphrates – anything more precise was not possible (Eidem 1999a, 153; Eidem and Putt 1999, 195). A series of radiocarbon samples from Tell Jurn Kabir was analysed, but did not yield a coherent chronological scheme (Eidem and Putt 1999, 196).

3.23 Kamid el Loz (Lebanon)

Summary of Excavations

The first major excavation of Kamid el Loz began in 1963 under the direction of two German professors, R. Hachmann and A. Kuschke (1966, 7). The initial goal of the project was to investigate the pre-Hellenistic settlement of the entire Beqa' Valley; consequently, survey was an important focus (Kuschke 1966; 1978; Marfoe 1995; 1998). Though Kuschke withdrew after two years, Hachmann continued at the site until 1982. Unfortunately, the close of excavations was followed by the illicit looting. Bulldozers turned the site upside-down in search of "treasure", destroying much of the early Iron Age deposits in the process (Seeden 1989). It was not until 1997 that archaeological work at the site resumed.

The period best-represented in Hachmann's (1989, 54-68) excavation was the Late Bronze Age, when an elite building complex, characterised by a large temple adjacent to a palace, crowned the site (**Table 3.24**). The palace was not fully excavated because much of it lay under the modern cemetery. The 1973 campaign also exposed elements of a third elite building, the so-called "*Schatzhaus*", or "Treasury", though the building was not really a treasury but received its name from its rich contents (treasure house?) (Hachmann 1989, 97). The so-called "treasure" was really a rich collection of grave goods from burials placed within the building. The richness of the grave goods and the close proximity to the palace were indicative of elite burials. When children were identified amongst the burials, Hachmann (1989, 37) concluded that these were "royal tombs" (i.e. inherited status). There is also a strong Egyptian influence in the objects from these burials.

Hachmann (1989, 35) distinguished three Iron Age building phases on the tell, all of which were dated to the Iron I period. The Iron Age was separated from the earlier elite buildings by an alluvial and sedimentary level, which suggested a period of abandonment (*ibid*). In comparison with the Late Bronze Age structures, the early Iron Age settlement was characterised by what Hachmann (1989, 54) called rural architecture. After the early Iron Age, Kamid el Loz was supposedly abandoned until the Persian period; when a cemetery occupied part of the tell (Hachmann 1970b, 136). Excavation of the Persian cemetery uncovered 94 graves, dated between the mid-fifth and early-fourth century BC (Poppa 1978, 70). The Persian period settlement was not located.

Table 3.24: Hachmann's Stratigraphy of Kamid el Loz

Phase	Level	Temple	Palace	Period
	1			
1	2			
	3			
	4			Iron Age I
2	5			
	6			
3	7			
	8			
4	9	T1	P1	
	10		P2	
	11a			
	11b			
	11c	T2		
5	11/12			LB I-IIB
	12a		P3	
	12b		P4	
	13a	T3		
	13b		P5	
	13c			
6	14	T4		MB IIB

(After Hachmann 1989, 44)

Heinz (2004; *et al.* 2001, 2004) resumed excavation at Kamid el Loz in 1997. Unlike the previous project, Heinz was interested in the entire site (not just the tell) and expanded the area of excavation to include the Hellenistic and Roman settlements. Work also resumed on the Late Bronze Age palace and temple, yielding evidence of occupation back to the Early Bronze Age. Work on the pre-Roman levels on the east slope of the tell (Heinz *et al.* 2004, Fig. 3) documented evidence for the Iron Age; an Iron II residential area and a fortifying wall of the Late Bronze-Iron Age transition (Heinz *et al.* 2004, 102-105).

Review

It had been decided beforehand that the excavation of Kamid el Loz by the Saarbrüken expedition was not to follow the "*Deutsche Bauschule*" method of broad exposures of architecture. Instead, Hachmann (1989, 27) envisaged the excavation at Kamid el Loz "as an ongoing discussion of the essence of culture-history and the character of cultural and historical processes". While excavation technique was based on the Wheeler-Kenyon method, Hachmann's interpretative emphasis was on broad cultural phases. This mixed method has probably contributed to the discrepancies between his and Heinz's interpretation. For instance, Hachmann (1989, 52) concluded that Kamid el Loz was a small unfortified rural settlement in the early Iron Age, yet Heinz (2004, 579-581) exposed a large fortification-wall from this period. Moreover, Heinz found evidence for Iron II occupation on the mound, contradicting Hachmann's conclusion that Iron Age occupation was limited to the Iron I period. While the results of Heinz's (*et al.* 2004) project are of, as yet, only a preliminary nature, the growing discord between the results of the two projects warrants a cautious approach to Hachmann's conclusions.

3.24 Kazel, Tell (Syria)

Summary of Excavations

Tell Kazel was surveyed in 1956, followed by limited excavation in the 1960s (Badre 1990a, 13). Preliminary results include a chronological sequence from the Middle Bronze Age to the Hellenistic period; the Late Bronze and Persian periods were considered the most significant (Dunand and Saliby 1957; Dunand *et al.* 1964).

After an interval of over 20 years, a joint Syro-Lebanese project resumed work in 1985; one purpose was to train Lebanese archaeologists unable to excavate in Lebanon due to civil war (Seeden 1990, 5). Four trenches were opened across the mound, three of which (Areas I, II and IV) yielded significant evidence for Iron Age occupation (Area III exposed mixed deposits).

Expectations were that Area I, on the western half of the acropolis, would provide a complete stratigraphic sequence of the site (Badre 1990a, 14). Excavation recovered evidence for occupation from the Mamluk period back to the Late Bronze Age (**Table 3.25**).

Area II, on the tell's south-east corner, exposed a Hellenistic cemetery with Iron Age occupation underneath (**Table 3.26**). Work continued here until Late Bronze Age levels were reached, though many of Iron Age deposits were characterised by fragmentary and insubstantial architecture.

Table 3.25: Tell Kazel Area I Stratigraphy

Str.	Period
1-1b	Islamic Medieval
2	Byzantine
3	Hellenistic
4	Late Persian
5	Early Persian/Iron III
6-13	Iron II
14	Iron I-II transition
15-16	Iron I

(After Badre 1990b)

Area IV, on the western side of the tell (**Table 3.27**), sought to clarify the stratigraphy of the earlier excavation (Badre and Gubel 1999-2000, 136). In the process, a series of Late Bronze and early Iron Age structures was uncovered belonging to a temple complex (*ibid*, 136-198). The cultic interpretation was based on the plan (5 x 15 m *cella*) and recovered finds (Badre 2000a, 39-42).

Table 3.26: Tell Kazel Area II Stratigraphy

Str.	Period
1-2	Hellenistic-Roman cemetery
3	Iron III: pits
4	Iron II: pits and rural habitation
5	Iron I: occupation revival, solid mud-brick architecture
ash layer	
6a	Iron – LB transition: poor architectural remains
6b	Iron – LB transition: *residence de luxe*
7	Late Bronze (14th century BC)

(After Capet 2003, 117)

Table 3.27: Tell Kazel Area IV (temple) Stratigraphy

Str.	Period	Comments
1-2	Mixed - Topsoil	No Iron II - little Iron III
	area abandoned	
3	Iron I	Revival of *cella*
4	Iron I	Poorly defined *cella*
	70 cm ash 'destruction' deposit	
5	Late Bronze II	Two-room temple
6	Late Bronze Age	Earliest *cella*, poorly defined

(After Badre and Gubel 1999-2000)

Review

Apart from training purposes, the objectives of the Lebanese excavations were to test the identification of Tell Kazel with historical Simyra, and to illuminate the Late Bronze-Iron Age transition in light of understanding the invasion of the "Sea Peoples" (Badre 1990a, 14). Both of these objectives betray a preoccupation with historical considerations. Indeed, the ash layer overlying the Late Bronze Age deposits was associated with a "Sea Peoples" destruction based primarily the conventional historical narrative (Badre 2006, 93; Badre and Gubel 1999-2000, 127). If the historical events are downplayed, the excavators have not demonstrated how this particular ash layer needs to be a) interpreted as violent conquest, or b) associated with the historical "Sea Peoples". First and foremost, an independent date for this destruction level needs to be established using scientific methods – cross-dating of imported pottery is not an accurate chronological tool, but just adds to the circularity of eastern Mediterranean chronological arguments. Second, Aegean style pottery of the early Iron Age is not indicative of "Sea Peoples"; similar pottery appears at sites without destruction (Badre 1990a, 14; 2006, 93; Badre *et al.* 2005, 16, 36). Moreover, the historicity of the "Sea Peoples" has not been conclusively established. The conventional historical narrative was also used to date a second "destruction" level at the site; the ash layer of Area I Level 9 was attributed to Sargon II, who claimed victory over Simyra (Capet and Gubel 2000, 433). Clearly, the conventional historical narrative was an important interpretative tool for the excavators of Tell Kazel.

3.25 Keisan, Tell (Israel)

Summary of Excavations

Garstang excavated the southeast slope of Tell Keisan in 1935 with a view to obtaining a complete stratigraphic sequence of the mound. Sixteen levels were recorded, extending from the Early Bronze Age to the Hellenistic period, including eight Iron Age levels (**Table 3.28**; Seton-Williams 1980, 382). After only two seasons the outbreak of war brought the project to a close and the results have only been published as a minor appendix to the French expedition's results (Seton-Williams 1980). There is only a cursory discussion of stratigraphy and a total of four plates of ceramics in the report.

Table 3.28: Stratigraphy of Tell Keisan - British Expedition

Level	Period	Remains
I	Early Hellenistic	Rubble floor, plaster
II	Persian/Hellenistic	Plaster floor
III	Iron I	Floor and masonry wall
IV	Iron I	Floor and masonry wall
V	Iron I	Rubble floor
VI	Iron I	Plaster floor with pottery sherds
VII	LB-Iron I	Earth and lime floor
VIII	LB-Iron I	Rubble floor
IX	LB-Iron I	Possible floor?
X	LB-Iron I	Floor and kiln
XI	Late Bronze	Occupation level
XII	Late Bronze	Floor and wall
XIII	Late Bronze	Floor and masonry wall
XIV	Late Bronze	Floor and masonry wall
XV	Middle Bronze I & II	Great stone fortification wall

(After Seton-Williams 1980, 382; *cf.* Briend and Humbert 1980, Tab. 1)

Table 3.29: Stratigraphy of Tell Keisan - French Expedition

Period	Level	Dates	Structures	Ceramics
Persian	3a	450-380	Much destruction	Greek imports
	3b	580-450	Houses with reinforced corners	Ionian & Cypro-Archaic II imports
		destruction		
Iron IIC	4a	600-580	Houses with silos	Basket-handle amphorae
		destruction		
	4b	650-600	Large dry-brick constructions	Phoenician traditions
	5	720-650	Much destruction	Assyrian types (little Southern influence)
		abandoned		
Iron IIB	6	850-800	Occupation continues but short	Little material, reduced occupation
	7	900-850	Occupation continues	End of Bichrome
Iron IIA	8a			
	8b	980-900	Modest reoccupation,	Black-on-Red
	8c		then more significant	
		destruction		
Iron IB	9a	980	Declining house repair	Pithoi & 'Philistine Bichrome'
	9b		Ashlar-masonry	Appearance of Bichrome
	9c	1075-1050		Continuation of 10a
Iron IA	10a 10b	1100-1075	Modest occupation Dry-brick houses	'Philistine' & Mycenaean
		destruction?		
	11	?-1100	Massive construction	No material
	12			

(After Briend and Humbert 1980, Tab. I, p. 27)

The later, French expedition to Tell Keisan (1971-1976) confirmed that settlement on the tell began sometime during the Early Bronze Age and continued into the Hellenistic period, though the French concentrated their efforts on the Late Bronze and Iron Ages (**Table 3.29**; Humbert 1993, 862-864). While the Late Bronze Age city was poorly attested by Garstang, the French delineated an important Late Bronze-Iron Age transitional horizon (British Strata X-VII) characterised by Egyptian imports and locally-made Mycenaean IIIC pottery (Humbert 1993, 864). A large amount of carbonised material was found covering the final Late Bronze Age level, which the French associated with violent destruction by the "Sea Peoples" (Humbert 1993, 864). The Iron I levels (French Strata 12-9) at Tell Keisan are exceptionally thick (c. 3 m), and are culturally similar to Late Bronze Age traditions (Seton-Williams 1980, 385). A significant cultural break instead followed the "destruction" of Stratum 9, ushering in the poor and sporadic architecture of an impoverished Iron II settlement (Strata 8-6). The Iron I-II transition was dated by Briend and Humbert (1980, 27) to the beginning of the tenth century BC, while Assyrian Palace-Ware was used to date Strata 5-4 to the seventh century BC. The presence of Ionic and Rhodian imports of the sixth to second centuries BC was the indicator for Persian and Hellenistic periods.

Review

The construction and interpretation of the Iron Age sequence at Tell Keisan relied upon a number of assumptions: the literal reading of the ancient historical sources; the acceptance of conventional chronologies derived from the conventional historical narrative; and uncritical comparisons with unreliable regional sequences. The result is a stratigraphic sequence dependent upon a literal reading of documentary sources, rather than archaeological evidence. Consequently, the French excavators admit that the chronology is difficult to fix with any precision and is prone to revision (Briend and Humbert 1980, 189, 229).

The end of the Late Bronze Age at Tell Keisan was dated according to conventional history. Despite strong cultural continuity across the Strata 13-12 transition, the French excavators declared that Stratum 12 marked the settlement of a new Aegean-derived population; i.e. the "Sea Peoples" (Dever 1997a, 278; Humbert 1993, 864-866). Archaeologically there is no direct evidence to decisively confirm a change in population, but conventional chronologies emphasise change at this point (Dever 1997a). The presence of a "destruction" layer between the Late Bronze and Iron Age strata invited a conventional incursionary interpretation, and was supported by the presence of "Philistine" Bichrome and locally-made Mycenaean IIIC pottery. In contrast, the Late Bronze Age Stratum 13 bore witness to significant quantities of Aegean pottery, yet this influence was linked only to trade. The conventional historical narrative was a significant influence on the 1200 BC date, and remained an important means for interpreting the whole Iron Age chronological sequence (e.g. Assyrian influence in the mid-eighth century BC; conquest of the Akkar in 643 BC – Humbert 1993, 866).

In addition to direct historical influences, the Tell Keisan stratigraphy was dated via cross-comparison of pottery with other sites. While Briend and Humbert (1980, 177, 214) insisted that only reliable sequences were used, many are not without significant problems. Comparison was also made with Cypriot chronology, though it is not clear whether Gjerstad's or Birmingham's scheme was used (Humbert 1993, 867 - **§2.5**). Consequently, the Tell Keisan sequence is open to revision.

3.26 Khalde (Lebanon)

Summary of Excavations

Ruins of a Roman-Byzantine settlement have been known at Khalde since the nineteenth century, but it was not until 1960 that earlier material was identified (Saidah 1966, 53). It was during the measurement of the classical mosaics, being threatened by modern construction, that Kalayan uncovered a small collection of Iron Age pottery. Understanding the importance of finding *in situ* Iron Age pottery, a salvage excavation was organised (*ibid*). The ensuing excavation uncovered an extensive Iron Age cemetery. Archaeologists exposed two different types of Iron Age burial at Khalde, inhumation and cremation. The funerary remains of 422 inhumations were exposed, with the body usually placed directly on the ground between two rows of stones and surrounded by Iron Age pottery (Badre 1997a). Two cremation burials were also found in cinerary urns (Courbin 1993a, 105; Saidah 1966, 66-67). Two distinct phases of use were discerned within the mortuary material, which the excavator called Levels III and IV (Saidah 1966, 90; **Table 3.30**). The earlier level (IV) was dated to the tenth and ninth centuries BC, while Level III was assigned an eighth century BC date (S.V. Chapman 1972, 181; Pritchard 1978, 33; Saidah 1966, 90; 1969, 130).

In addition to the Iron Age cemetery, excavation also revealed evidence of settlement in the Late Bronze Age and Late Chalcolithic periods (Saidah 1969, 130), though the material consisted of isolated, secondary finds.

Table 3.30: Phases of Khalde Cemetery

Level	Tombs	Cent. BC
III	1, 2, 3, 4, 121	8th
IV	21, 22, 23, 165, 166, 167	10th & 9th

(After Saidah 1966, 90)

Review

The Khalde publication was not an extensive undertaking. The few preliminary reports that have appeared were little more than a catalogue of finds, with little contextual information provided for the burials (e.g. Saidah 1966). Furthermore, the dating of the cemetery's two phases was apparently based on three Egyptian scarabs (Saidah 1966, Nos 3, 35, 36); while the scarabs themselves are indeed datable, their secondary use within non-Egyptian mortuary contexts introduces a level of chronological ambiguity. No ceramic typology was presented, and very few parallels were discussed. The Red-Slip jugs of Phase III were dated by comparison with Gjerstad's Cypriot chronology, the al Mina material, Megiddo, Hazor, and Athlit (Saidah 1966, 86-87): none of which can boast a securely dated sequence. Similarly, Phase IV is compared with Megiddo, Tell Abu Hawam, and Cyprus (Saidah 1966, 88-89). The proposed chronology for Khalde is not based on independent data and can, therefore, only offer a broad indication of date.

3.27 Mastuma, Tell (Syria)

Summary of Excavations

A Japanese project began excavations at Tell Mastuma in 1980. The director, Egami, approached the site with two primary goals: to understand the complete stratigraphic sequence of the site; and to understand the extent and nature of the Iron Age settlement crowning the tell (Egami and Masuda 1982, 26; Egami 1988, 51). Excavation at Tell Mastuma was undertaken in unison with a regional survey of the site's environs (Egami 1983), though the survey identified only a few Iron Age sites not previously identified (e.g. Tell Afis, Tell Tuqan) (Egami and Masuda 1984, 34, Pl. 1).

Table 3.31: Tell Mastuma Central Trench

Levels	Period	BC
O	Persian	6th–4th cent
I-1 I-2	Iron II/III	9th–6th cent.
I-3	Iron I	12th–10th cent.

(After Wakita *et al.* 1995, 2)

In the Central Area a large trench exposed the layout of the Iron Age city, revealing domestic buildings transected by a large street running along the edge of the tell

(Wakita *et al.* 1995, Fig. 4; 2000, Fig. 4). "[B]uildings situated on the outside perimeter of the street were constructed as if longitudinal walls had radiated from the centre of the *tell*, while the walls within the street perimeter area oriented east to west" (Egami 1988, 52); i.e. the perimeter buildings formed a defensive barrier for the upper mound. The extensive work in this area helped refine the Iron Age sequence and recognise some Persian material above Level I, which was not evident in the North Trench (Wakita *et al.* 1995, 2 –**Table 3.31**; *cf.* **Table 3.32**). The Iron Age was subsequently split into three sub-phases, the lowest dated to the Iron I period.

The North Trench, situated at the highest point on the site, was designed to reveal a complete stratigraphic account of the site's occupation. It was excavated down to bedrock, revealing fourteen levels covering three main periods: the Early Bronze IV (A), Middle Bronze (B), and Iron Ages (C) (**Table 3.32**; Egami 1988, 51).

Review

In the course of excavation, 4000 sq. m. of the Iron II settlement was exposed; i.e. about 40% of its total size (Wakita *et al.* 2000, 538). All of the buildings exposed were domestic in nature. Prior to this, domestic architecture of the IA-NL was a largely unexplored phenomenon (Braemer 1982 has a lack of North Levantine data). Excavation of major sites had tended to focus on the elite buildings of important regional centres (e.g. Zincirli, Hama, Tell Ahmar, Jerablus), or simply omitted information on domestic structures. The study of the Tell Mastuma domestic architecture was a significant achievement, made all the more remarkable considering the scarcity of comparable data (an exception being Chatal Höyük – Haines 1971).

In addition to a general lack of domestic parallels, little Iron Age pottery for the Aleppo region had been published. Unfortunately, what would have been an important initiative for the study of the region's ceramics was pre-empted by the publication of a number of comparable assemblages (e.g. Tell Afis). Moreover, the Tell Mastuma sequence was not securely dated, with imported pottery being the main tool for applying chronological definition for the Iron Age settlement. Level I-1 was dated on the evidence of just two vessels (Wakita *et al.* 2000, 552): the first was a Cypriot-White-Painted III juglet dated according to Gjerstad's (1948) Cypriot chronology; the second was a "Phoenician" Red-Slip jug dated according to Lehmann's (1998) Assemblages 1 and 2 dates. Level I-2 was dated via two imported Greek *skyphoi*. Cypriot and Greek ceramic are unable to provide sufficient chronological definition (**§2.5**).

The dating of Level I-3 to the Iron I period is also problematic. Only nine ceramic vessels from this level have been published (Egami and Masuda 1984, Pls. 6.7; 7.3; 8.2-4, 9, 13-14; Wakita *et al.* 1995, Fig. 7.9), none of which are characteristic of the Iron I period. Instead, the presence of a hole-mouth cooking-pot and deep pithoi with heavy rolled rims suggests a date in the Iron II period. The identification of an Iron I level was based on an increase in painted pottery compared to Level I-2, which is not an infallible guide; e.g. Hama E contained significant quantities of Monochrome-painted shallow-bowls (Wakita *et al.* 1995, 19). Furthermore, statistics for the painted pottery have not been published.

While the excavation of Tell Mastuma held much promise for the study of less grandiose aspects of Iron Age life, the results are only partially available. Despite significant contributions to our understanding of domestic architecture, the ceramic material has not been published in any detail, and the Iron Age sequence was dated on comparison with unreliable regional chronologies. Moreover, many of the project's publications have appeared in a Japanese journal (sometimes in Japanese), with a somewhat restricted distribution.

Table 3.32: Stratigraphy of North Trench, Tell Mastuma

Level	Phase	Period	Comparative
I	A	Iron Age	Hama E Mardikh VB
II III IV V	B	Middle Bronze I-II	Hama H Mardikh III
VI VII VIII IX X XI XII XIII XIV	C	Early Bronze IV A-B	Hama J Mardkih IIB

(After Egami 1988)

Table 3.33: Summary of OI Megiddo Sequence - Later Periods

Strata	BC	Period	Character
I	600-350	Persian	
II	650-600	Iron III (Iron IIC)	Destroyed by Necho
III	780-650	Iron III (Iron IIC)	Domestic
IV	1000-800	Iron II	"Solomonic"
VA	1050-1000	Iron I	
VB			Fragmentary, new orientation
VIA	1150-1100	early Iron I	Destroyed by David
VIB			Fragmentary, domestic
VII	1350-1150	Late Bronze II	
VIII	1350-1150	Late Bronze I	

(After Loud 1948, 5)

3.28 Megiddo (Israel)

Summary of Excavations

Megiddo was first excavated by a German expedition, led by Schumacher, at the beginning of the twentieth century. In just three seasons, the tell was surveyed, a topographic map produced, and a 20 metre trench excavated from north to south through the centre of the tell (Ussishkin 1997, 461). The publication of Schumacher's (1908) excavations and final conclusions have been widely criticised (e.g. Finkelstein 1996a; Whiting 2007a, 27).

Excavation recommenced in 1925 with the OI commencing what was then the largest excavation in the Southern Levant (Finkelstein *et al.* 2000a, 1). By the end of the campaign, some 14 years later, an almost continuous occupational sequence, from the Pre-Pottery Neolithic to the Persian period, had been compiled for the site (Aharoni *et al.* 1993, 1023). In the process, significant deposits of Iron Age occupation were exposed (**Table 3.33**), including a palace, extensive fortifications, large gate complex and other enigmatic public buildings. Among the most important (and controversial) Iron Age structures are those that have been associated with Solomon (e.g. "stables", six-chamber gate), and which has become a celebrated case for the archaeological authentication of the biblical text (e.g. Aharoni 1978, 197). The early seasons also uncovered a considerable number of Bronze Age rock-cut tombs to the immediate east and south of the tell, some of which were dated to the Iron Age (Guy 1938, 159-160).

A third expedition to Megiddo was undertaken by Yadin during the 1960s and 1970s. Having discovered a fortification system at Hazor similar to that from Megiddo, Yadin (1975, 207) sought to unravel the problems with Megiddo's tenth century BC (Solomonic) stratigraphy. Yadin (1960, 64) hoped to find an indisputable example of Solomonic architecture, which would confirm his Hazor interpretation and the veracity of the biblical text, though the majority of his Megiddo results remain unpublished and, therefore, difficult to assess (Yadin 1972).

A fourth expedition to Megiddo has been underway since 1992, when Finkelstein and Ussishkin initiated the Tel Aviv University Megiddo Expedition with three primary objectives: to resume excavation in previously dug areas intending to clarify stratigraphical problems; to open new areas for excavation using modern archaeological technique; to supplement excavation with a comprehensive survey of the western Jezreel Valley, thus placing the site within its wider cultural context (Finkelstein *et al.* 2000a, 5). To this end, five areas (F, G, H, J and K) were begun, with three yielding significant Iron Age deposits (Areas F, H and K) (**Table 3.34**).

Table 3.34: Iron Age Megiddo - Correlation of 'OI' and Tel Aviv Strata

Tel Aviv Excavations			*OI Project*
	H3/H4	K1	IVA
	H5	K2	VA-IVB
		K3	VB
F5		K4	VIA
F6		K5	VIB
		K6	VIIA

(After Finkelstein *et al.* 2000, Table 11.1; Finkelstein *et al.* 2006)

Review

The stratigraphy of Iron Age Megiddo is the focus of fierce scholarly dispute. This is due to the importance that these strata hold for some scholars for authenticating modern Israel's identity, and for the archaeological dating of Iron Age sequences across the eastern Mediterranean (**§2.5**). Consequently, the debate has maintained a high-profile and the issues are well-known (for a review of current positions see Finkelstein 2005; Mazar 2005).

The stratigraphic phasing of Megiddo by the OI expedition was based largely on architectural remains (Loud 1948, 1). In fact, the original goal of the expedition was to expose each stratum in its entirety, in the hope that it would produce a complete and exhaustive history of occupation. Limited resources, however, necessitated the use of a large sondage trench, resulting in the insufficient exposure of architecture for the chosen stratigraphic method (*ibid*). Consequently, buildings and structures from the same stratum but different trenches could not be interrelated with one another. Furthermore, no attempt was made to understand the development of material culture diachronically. No ceramic typology was prepared, and each publication was intended to be "no more than a catalogue of the architecture and artefacts recovered" (Loud 1948, vii). Insufficiencies in method and practice were compounded by the frequent change of project director, who was expected to manage the large scale excavation and prepare publication proofs simultaneously (Esse and Harrison 2004, 3-5). Consequently, publication was slow and the application of archaeological practice inconsistent. Furthermore, new directors often revised the interpretation of previous directors (e.g. the Stratum III city gate was originally attributed to Stratum IV; Lamon 1948, 46).

Material culture from Strata VIA and VIB maintained a number of cultural conventions that had their origins in the Late Bronze Age, suggesting continuity of a Canaanite population in the early Iron Age (Lamon and Shipton 1939, 7). Contrary to the conventional historical narrative, rapid change in material culture only came in Stratum VB following the "destruction" of Stratum VIA. This led the excavators to attribute the "destruction" of Megiddo VIA to the campaigns of David as recounted in the biblical narrative. This interpretation, however, was not based on clear archaeological evidence, but on the archaeological record's *apparent* confirmation of the biblical narrative; the more the archaeological record appeared to support the biblical narrative, the more the same narrative could be "securely" applied to the archaeological record, despite the inherent circularity in this approach. There was no archaeological evidence to associate the Megiddo VIA "destruction" with David, except for a desire to link the biblical narrative with the archaeological data. Yadin (Aharoni *et al.* 1993, 1016) dated Stratum VB to the early-tenth century BC. According to Yadin (1970), the cultural break between Strata VIA and VB signalled the beginning of the Israelite occupation of Megiddo. Stratum VA-IVB was then dated to Solomon by its stratigraphic position above (or chronologically after) the early Israelite occupation (VB) (Franklin 2006, 95; Lamon & Shipton 1939, 59). Hence, the attribution of Megiddo VA-IVB to Solomon was based largely on biblical testimony (**§2.5**), with confirmation provided by the unprovenienced tenth century BC Shoshenq stele found in the spoil heap of the earlier German expedition (Lamon & Shipton 1939, 61).

Stratum II was also dated according to a correlation of conventional history with expected archaeological "events"; the ash layer sealing Stratum II was attributed to Pharaoh Necho (c. 605 BC) (*ibid*, 87).

3.29 Mina, al (Turkey)

Summary of Excavations

Woolley's (1938a, 6; 1948) excavation of al Mina uncovered a series of occupational phases that he dated to the eighth century BC and later (**Table 3.35**). Although little architecture was discernible in the heavily-eroded tell, the site produced an uncharacteristically rich assemblage of Greek pottery (Woolley 1938a). The abundance of Geometric pottery was accepted by some as an indication of a Greek "colony", or *apoikia* (Boardman 1959; Riis 1970, 159), which in turn helped to explain the strong "Orientalising" nature of eighth century BC Greek art (Niemeier 2001, 12-16). Woolley, however, was not able to locate the residential area of al Mina and preferred to interpret the exposed architecture as the remains of successive warehouses attached to a thriving trading post (du Plat Taylor 1959, 91).

Table 3.35: Original al Mina Stratigraphy

Level	Dates	Ceramics
X-IX	750-700	Greek Sub-Geometric
VIII	700-675	Cypriot
VII	675-650	Cypriot & Rhodian Greek
VI-V	650-550	Rhodian "Orientalising"
	hiatus - clearing of site	
IV	520-430	Black-figure; Red-figure

(After Woolley 1938a, 16*ff*)

Table 3.36: Smith's Revised al Mina Stratigraphy

Level	Dates BC	Imported Material
X	800?-760?	Sub-Geometric
IX-VIII	760?-680?	Cycladic. 'Early-Proto-Attic'
VII-V	680?-580?	Cycladic. Rhodian A. Proto-Corinthian
	hiatus?	
IV	520?-430	Black-figure, Red-figure.
III	430-375	Bell-kraters. Calyx-kraters
II	375-312	Macedonian, Seleucid, Ptolemaic coins

(After S. Smith 1942, 91)

Few architectural elements were identified in the earliest levels at al Mina (Woolley 1938a, 12). Level X, which rested on virgin soil, included an abundance of imported sub-Geometric and Geometric pottery (*ibid*, 16). While it was difficult to isolate architectural features of Level IX, there was a noticeable shift in ceramic horizon; the pottery was primarily of Cypriot influence, with few Greek imports (*ibid*). Change in ceramic culture between Levels IX and VIII was associated with violent conquest attributed to either the Assyrians (du Plat Taylor 1959, 87) or invaders from Asia Minor (Woolley 1938a, 17-18). While Level VII witnessed another slight shift in ceramic horizon, it represented the reconstruction and continuation of Level VIII buildings (*ibid*, 18). The "destruction" of Level VII has been linked to the Assyrian campaign to Tarsus in 696 BC (Riis 1970, 159). Level VI was a replacement of the decayed Level VII buildings, though the two ceramic assemblages were difficult to separate. Level V was a continuation of Level VI, and coincided with the appearance of true Corinthian wares and the disappearance of Cypriot imports.

Review

The absolute dating of the al Mina sequence is problematic. While a number of scholars have revised Woolley's chronology (e.g. du Plat Taylor 1959 – **Table 3.37**; Robertson 1940; S. Smith 1942 – **Table 3.36**), there remains no single definitive scheme for the site. Much of the debate arises over the earlier levels (X-V) and the dating of specific imported pottery styles (Woolley 1938a, 16*ff*). For instance, the foundation of Level X is associated with the presence of sub-Geometric pottery, but the resulting dates for this style vary greatly: 750 BC (Woolley 1938a, 16); 825 BC (du Plat Taylor 1959, 85-86, 91-92), 800 BC (S. Smith 1942, 91). At the time of both Woolley's (1938a) and Smith's (1942) publications, the Iron Age chronology of Greece and Cyprus had not yet been systematised. In fact, scholars had held some hope that the al Mina sequence would establish a more definitive dating of Greek pottery (du Plat Taylor 1959, 62), the same pottery that has been invariably used to date the al Mina sequence. We might wonder how Woolley and Smith arrived at their absolute dates, other than arbitrarily. Du Plat Taylor's (1959) discussion of Levantine parallels for al Mina's Red-Slip and Bichrome pottery only introduces further circularity into the problem, since Gjerstad's Cypriot chronology is based on the same Levantine contexts du Plat Taylor cites (**§2.5**).

Table 3.37: du Plat Taylor's al Mina Phasing

Level	BC
X-VIII	825-720
Assyrian destruction c. 720	
VII	Late 8th cent.
VI-V	7th cent

(After du Plat Taylor 1959, 85-86, 91-92)

Although an exact number of vessels recovered from al Mina is not known (the material has been dispersed throughout various collections), a minimum figure of 820 items of Greek provenance has been suggested by Boardman (1990, 172). This number, which according to Boardman (1990, 175) is a conservative estimate, leaves al Mina with the most significant presence of Greek pottery in the region. Although neighbouring sites also produced collections of Greek pottery, their numbers constituted less than 5% of their overall ceramic corpus: at al Mina the figure has been suggested to be closer to 50% (*ibid*; sampling strategies might account for the significant difference). Al Mina and its hinterland clearly had an exceptional record of Greek imports in the Geometric period (Levels X-VIII). However, this does not automatically indicate the presence of a Greek colony (as has been suggested – e.g. Boardman 1959), rather just very fluent Greek trade (Boardman, 1980, 42-43, 66-67). Woolley (1938a, 11) regarded Levels X-VII as essentially the same town, and one that had no specific Greek character. Indeed, it is more natural to compare al Mina's irregular buildings and small blind alleys with those of neighbouring Ras Shamra/Ugarit (Lund 1986, Fig. 160; Riis 1970, Fig. 57; Schaeffer 1938, Fig. 2), than with Greece. In other words, al Mina was at home amongst local traditions of the North Levantine coast.

There are significant problems with the al Mina sequence, accentuated by a heavy reliance on regional historical trends for chronological definition. For instance, the destruction of Level VIII at al Mina was attributed to the later-eight century BC to coincide with the period of Assyrian campaigning in the region. There is no archaeological evidence for conquest, or even Assyrian claim to victory. In the end, the chronology of al Mina was dated in an inconsistent and unreliable manner, and should be used for ceramic comparison with caution.

3.30 Mishrifeh, Tell (Syria)

Summary of Excavations

During the French Mandate, du Mesnil du Buisson investigated seven large excavation areas across Tell Mishrifeh, a large, square ramparted site east of Homs (du Mesnil du Buisson 1926; 1927a; 1927b; 1928; 1930; 1935). The greatest exposure was on the northern part of the acropolis where a large Bronze Age palace, with an internal temple and high-place, was revealed (al Maqdissi *et al.* 2002a, 10). Excavation ceased after four seasons due to land ownership issues (*ibid*); a modern village covered much of the site. Following the re-settlement of the village in the 1980s, the DGAM resumed excavations in 1994 by opening six trenches across the site: with the Iron II and the Late, Middle and Early Bronze Ages attested (*ibid*, 11). In 1999, this project was enlarged to incorporate Italian and German teams, with the overall aim being to reconstruct "the history, cultural relations, and natural environment of this significant urban centre of Inner Syria" (*ibid*). To this end, a number of areas were opened (and re-opened) across the site. Level II of Area C, situated on the western slope of the central mound, exposed a large building, identified as an "Aramaean" palace and dated to the eighth century BC (al Maqdissi 2003, 225-235; al Maqdissi and Badawi 2002).

Areas G and H were located within the western and eastern parts of the extensive Bronze Age palace, respectively, and were intended to re-evaluate du Mesnil du Buisson's plan of this building (Novák and Pfälzner 2002, 65-69). The two excavation areas revealed a small Iron Age re-occupation of the palace area following its abandonment in the Late Bronze Age (Barro 2002, 119). Situated on the acropolis, Area J was aimed at producing a complete stratigraphic history of the site (Morandi Bonacossi 2002, 123). This sequence extends from the Early Bronze III to the Iron III, with a considerable hiatus in occupation corresponding to the Late Bronze II and Iron Age I (Morandi Bonacossi 2008b, 57). Area K was located in the northern Lower City, where the large Building 1 (Levels K8-4) was used for metallurgical, culinary, weaving, cultic, and domestic activities during the late Iron I (Luciani 2002, 167).

In 2003, the German team at Tell Mishrifeh, while excavating the Bronze Age palace, discovered an in-tact royal tomb within the palace's foundations (www.qatna.org/en-index.html). This spectacular, once-in-a-lifetime find has since raised the profile of the site, and is set to greatly increase our understanding of Late Bronze Age mortuary practice and belief systems.

Review

The results from du Mesnil du Buisson's excavation are unreliable, to say the least. His preliminary reports are difficult to follow, the pottery is only summarily presented, and no stratigraphy is apparent. The broad exposure of monumental architecture (e.g. the large Bronze Age palace) appears to have been an important motivation behind his work at Tell Mishrifeh (du Mesnil du Buisson 1926, 311; 1927b, 298).

In the recent project, the Iron Age sequence of Area C (Level II) was dated to the late-eighth century BC based on ceramic data. According to al Maqdissi and Badawi (2002, 34-35), the pottery from Level II was characteristic of local inland production of the Iron IIB period. While the pottery closely parallels that from Hama E, Tell Afis VIII, Tell Abou Danne IId-c, and Tell Mastuma I, the dates for these external assemblages derive from one historical event – Sargon's destruction of Hama in 720 BC (**§2.4**). The Hama E comparison also led al Maqdissi (2003) to label the Iron II palace as "Aramaean", despite the fact that Qatna is not mentioned in known Iron Age texts. The association of this building with an ethnic identity exemplifies a reliance on the conventional historical narrative (**§4.3**).

Table 3.38: Summary of Tell Mishrifeh Phases

Period	Area			
	C	G/H	J	K
Modern	I	1-4		0-1
hiatus				
Iron III			1-4	
Iron II	II	5	5-9	2-3
Iron I		6		4-9
hiatus				
Late Bronze	III	7		10-13
Middle Bronze	IV	8-9	10-16	
Early Bronze		10		

(After al Maqdissi *et al.* 2002b)

In Area J, the Iron Age sequence was originally divided into nine phases, broadly dated to the mid-ninth to early-seventh centuries BC, but has since been modified to six phases (*cf.* Morandi Bonacossi 2002, 123-128, 141; 2008b, 60). A series of 18 radiocarbon determinations was obtained from the Tell Mishrifeh Area J sequence, which provided a reliably dated sequence for the Bronze Age levels, though only three Iron Age samples were included (2008b, 60-64). The combined testimony of two radiocarbon determinations from Phase J5, the local ceramic assemblage (Red Slip, Pithoi, Hole-mouth cooking-pots), and import Black-on-Red II juglets, have established a date for this phase in the late Iron II period (*ibid*, 118). Unfortunately, the flatness of the Iron Age calibration curve has meant that the two J5 AMS dates span a 2-sigma interval of c. 300 years, hindering a precise absolute dating of this phase (*ibid*, Tables 2 & 3). In the end, the Iron II sequence must once again rely upon ceramic comparisons with other inland sequences and the witness of Cypriot imports. The Area K sequence is also only broadly according to Cypriot imports (Luciani 2002, 151). Despite limited success in the scientific dating of the Iron Age at Tell Mishrifeh, this sequence sets an important precedence for the construction of a reliable, absolute chronology through meticulous technique and scientific data.

3.31 Nayrab (Syria)

Summary of Excavations

Two seasons of excavation at Nayrab, in the suburbs of modern Aleppo, yielded finds from the Iron Age and Late Antiquity. The most significant finds (two funerary stelae with Aramaic inscriptions) (**Figure 64**), however, were looted in 1891; the French project was a response to the damage being caused by illicit digging and encroaching dwellings. Much of the Iron Age material from Nayrab derives from a cemetery used during the late Iron Age, as suggested by a corpus of twenty-five Neo-Babylonian tablets (Abel and Barrois 1928, 187; P. Dhorme 1927). A remarkable variety of burial-type was attested, implying a period of use longer than the tablets suggest. Within one published plan (Abel and Barrois 1928, Pl. LII) five different types of inhumation are discernible: sarcophagus (Nos S1, 53); single pithos (Nos 12, 29); double-pithos (Nos 58, 67, 70); pit covered with torpedo-amphorae (Nos 4, 19, 40, 64, 68; see also Carrière & Barrois 1927, Fig. 3); and simple pit burial.

Review

While the two preliminary reports do not provide a clear picture of the specific context for each burial, the few published section-drawings and plans offer some basic information on stratigraphy. For instance, some section-drawings depict significant height (and time?) differences between certain burials, which might indicate an extended period of use for the cemetery (e.g. Carriere & Barrois 1927, Pls 33 & 34). Although the excavators dated some tomb groups to the Iron I period, no explanation was given as to how these dates were determined; possibly only because they predated the context of the Neo-Babylonian tablets (Abel and Barrois 1928, 187; Carrière and Barrois 1927, 129). Underneath the burials, the vestiges of mud-brick buildings testified to older installations, though they were left uninvestigated. The Nayrab chronology is confused and unreliable.

3.32 Nebi Mend, Tell (Syria)

Summary of Excavations

Pézard initiated the first full-scale excavations of Tell Nebi Mend in the early 1920's (Pézard 1922; 1931). Following his death in 1923, the project was brought to an abrupt close. Nevertheless, in just two seasons, Pézard's legion of workmen had cut an enormous sondage into the tell's north-east corner; a feature clearly discernible on satellite imagery (Whincop 2007, Fig. 2).

A second expedition to the site was initiated in 1975 by Peter Parr from University College London. The UCL project was directed towards providing a reliably excavated stratigraphic sequence for the inland Northern Levant (Parr 1983), though progress was significantly hindered by a modern village and cemetery. As a result, trenches were situated within "Pézard's Cutting" and atop terracing on the tell's northeast corner (Mathias and Parr 1989, Fig. 2).

Review

In order to reveal a complete occupational history of the site, Pézard's (1931) sondage was excavated the depth of the tell. While Pézard established the presence of second and first millennia BC levels (*ibid*), no coherent plans were produced, nor were any detailed stratigraphic records kept. Furthermore, broad phases were labelled according to historical ethnonyms (e.g. *niveau syro-phénicien*; *niveau syro-hitite*) and no absolute dates were imposed on the ill-defined sequence. Of the abundant ceramics and objects recovered, the stele of Seti I was the only item published in any detail (Pézard 1931, 18-22). A ceramic typology was not established and inter-site comparisons not undertaken. Considering the limited area of excavation, disappointing publication, and unspectacular results, it is not surprising that interest in the site waned. Only preliminary reports have been published from the UCL project so far (Whincop 2007).

3.33 Pella (Jordan)

Summary of Excavations

Apart from some preliminary expeditions to Pella in the 1950s and 1960s (Funk and Richardson 1958; R.H. Smith 1973), focused investigation of the site began in 1985. The American-Australian expedition (College of Wooster and University of Sydney) recovered post-Iron Age material across most of the site, though Iron Age material appeared in Areas III and VIII (McNicoll *et al.* 1982a, 14). The earlier seasons (1979-1981) dated the Iron Age phases according to pottery form and ware (McNicoll *et al.* 1982a, 63). Radiocarbon samples were collected but the results never sufficiently published (McNicoll 1982a, 15). According to Smith and Potts (1992, 83), Pella was destroyed during the Late Bronze Age, which initiated a lengthy abandonment of the site, before it was eventually resettled in the Iron I period. The Late Bronze-Iron Age transition was not marked by a distinctive break in material culture: architecturally or ceramically (*ibid*). The quality of architecture for the Iron Age was poor, implying that Iron Age Pella was only of secondary importance within the regional political power.

Table 3.39: Pella Iron Age Phases 1979-1985

Area		BC	Period
VIII	III		
9			
	I		Iron IA
8-7		1200-1000	Iron I
	O		Iron I-IIA
6		1000-900	Iron IIA
5			Iron IIC

(After R. H. Smith and Potts 1992)

Following the dissolution of the American-Australian partnership in 1985, the University of Sydney continued to excavate the site (Bourke 1997). New areas were opened, and Iron Age occupation was identified in four trenches (III, IV, XXIII, and XXXII), including the large Iron Age "Fortress Temple" built directly atop a larger Bronze Age equivalent (Bourke *et al.* 2003, 344-353).

Review

Despite comments regarding limited parallels for the Iron Age pottery at Pella (R. H. Smith and Potts 1992, 85), close parallels are evident at Beth Shan and Tel Rehov (Bourke *pers. comm.*): in particular, the cooking-wares, cult-stands, and storage jars. The earlier comment was due to a general lack in Iron Age ceramic material, a product of limited Iron Age exposure. The Sydney expedition has since recovered Iron I and Iron II pottery, leading to a re-interpretation of the Late Bronze and Iron Age history of the site (**Table 3.40**).

In particular, Phase Ia of Area III, which was originally dated to the Iron I period, because it post-dated a major destruction level ("Sea Peoples" according to Potts *et al.* 1988, 136), has been reassigned to the Late Bronze Age by Bourke (1997, 113). Bourke's revision is based on a "best fit" correlation between the conventional historical narrative and the archaeological data. In the end, Bourke's preliminary discussion of the Late Bronze and Iron Age levels rely on events in the conventional historical narrative (Bourke *et al.* 2003, 344-353).

Preliminary reports have been published from the Sydney project, with little Iron Age pottery presented in detail. While recent campaigns have extended investigations to include faunal and floral analyses, scientific analysis is broadly restricted to the description of what varieties of animal and plant are present in the archaeological record (Bourke *et al.* 1994, 1998, 2003).

Table 3.40: Pella Archaeological Horizons as Adjusted Post-1985

Stratum	Period	Excavation Area	Dates
I	Mameluke/Ottoman	IV, XXIII	AD 1500-1600
II	Ayyubid/Mameluke	III, IV, XVII, XXIII	AD 1200-1400
III	Abbasid	XXIX, XXIII	AD 750-1050
IV	Umayyad	III, IV, XXIII	AD 660-750
V	Byzantine	III, IV, XXII, XXXIV	AD 320-660
VI	Roman	III, IV, XXIII, XXXIV	60 BC-AD 320
VII	Hellenistic	III, IV, XXIII, XXVII	200-60 BC
VIII	Late Iron	III, IV, XXIII, XXXII	1000-600 BC
IX	Early Iron	III, IV, XXIII, XXXII	1200-1000 BC
X	Late Bronze Age	III, IV, XXXIV	1500-1200 BC
XI	Middle Bronze Age	III, IV, XXVIII	2000-1500 BC
XII	Early Bronze Age	III, IV, XXXII, XXXIV	3500-2000 BC
XIII	Chalcolithic	IV, XIV, XXXII	4500-3500 BC
XIV	Pottery Neolithic	IV, XXXII	5500-4500 BC
XV	Aceramic Neolithic	IV	-6500 BC

(After Bourke 1997, Tab. 1)

3.34 Qarqur, Tell (Syria)

Summary of Excavations

Tell Qarqur was first excavated during the 1980s (Lundquist 1983). Work exposed Iron Age layers across the mound, including a large gateway on the southern slope (Dornemann 2003a, 10). The gate sealed deposits associated with Bronze Age city-walls. Following only two seasons of excavation, work was stopped, and a break of nine years passed before Dornemann resumed investigations. The primary objectives for the renewed excavations were fourfold: the development of the site's settlement sequence; the size and nature of the settlement in different periods; the acquisition of a good sample of cultural materials; the collection of as complete a record as possible of palaeo-zoological and palaeo-botanical remains (*ibid*, 3).

The renewed excavations made prudent use of earlier trenches, with work continuing on the Area A gate, exposing an internal street running north to the tell (*ibid*, 10-29). Though the gate was expected to be part of fortifications, no attached casemate wall was found (*ibid*, 20). Earlier excavations, however, revealed a portion of an Iron Age casemate wall in Area C, located on the western acropolis (*ibid*). Dornemann suspects the gateway might lead to an Iron Age citadel complex, like those at Hama (Fugmann 1958, Fig. 186), Tell Ta'yinat (Haines 1971, Pl. 109) and Zincirli (Koldewey 1898, Tf. 28), though there has not yet been any direct evidence for this. The Iron II period is well-represented at Tell Qarqur; in addition to the Area A gateway, contemporary material has been excavated in Areas B, C, D and E, spanning both the upper and lower tells (Dornemann 2003a, 29). The stratigraphic sequence in Area B, on the eastern acropolis, provided an especially good sequence of Iron II pottery. The presence of Iron II pottery in the lower mound implies that Tell Qarqur was a large city in that period. Iron I material is also well-represented, though the architecture is fragmentary (*ibid*, 59; 2008, 85). According to Dornemann (*ibid*), all but a few sherds of Iron I pottery at Tell Qarqur appear to be locally produced. Apart from the important Iron Age settlement, Tell Qarqur boasts a long history of occupation (**Table 3.41**).

Review

Iron II pottery is abundant at Tell Qarqur, and is characterised by Red-Slip platters and bowls (*ibid*, 41-47). Dornemann (*ibid*, 7), who has studied the OI Amuq sequence, has suggested that pottery from Tell Ta'yinat's First and Second Building Periods closely parallels that from the Tell Qarqur gateway. In Area B, levels were reached that appear to belong to the early development of Red-Slip, tentatively dated by Dornemann (1999, 139) to the early-tenth century BC (late Iron I). While some of the early Red-Slip forms continue into the ninth and eighth centuries BC, they generally boast slips that are dark-brown and reddish-brown in colour, and only sometimes burnished. Dornemann (*ibid*) suggests these early slips are characteristic of Amuq Phase Oa, as defined by Swift (1958). This led Dornemann (2003a, 43) to conclude, although tentatively, that the tenth century BC was an Iron I-II transitional period at Tell Qarqur, and one which he terms the Iron IC. The tenth century BC pottery at Tell Qarqur has the potential to refine the internal divisions of the Iron Age in the Northern Levant. There was also a limited amount of Persian period pottery recovered, though the Iron III period is not represented (Dornemann's lack of an Iron III periodisation may complicate this point – *pers. comm*).

Despite 14 seasons of excavation at Tell Qarqur, few conclusions have been published in detail. Indeed, Dornemann displays a level of measured caution that is uncustomary within the discipline (but not unwelcome). The introductory material in Tell Qarqur's lengthy preliminary reports ask many questions, but offer few definitive answers (Dornemann 1999; 2003a; 2008). But while some might view this cautious approach in a negative light, it is clear that Dornemann is asking all the right questions. Obviously understanding the problems with the current periodisation, there appears to be an earnest desire on Dornemann's part to contribute something meaningful to the discipline, something the current data from Tell Qarqur has yet been able to offer.

Table 3.41: Summary of Tell Qarqur Stratigraphy

Str.	Period
1	Mamluk
2	Ayyubid
3	Early Islamic
4	Byzantine
5	Roman
6	Hellenistic
7	Persian
8	Iron II
9	Iron I
10	Late Bronze
11	Middle Bronze II
12	End Early Bronze IV
13	Early Bronze IVB
14	Early Bronze IVA
15	Early Bronze III
16	Early Bronze II
17	Early Bronze I
18	Uruk
19	Chalcolithic
20	Neolithic

(After Dornemann 2003a, 10)

3.35 Rachidieh, Tell (Lebanon)

Summary of Excavations

At the beginning of the twentieth century, the curator of the Imperial Museum of Constantinople, Macridi-Bey, excavated the eastern slope of Tell Rachidieh where the weak growth of mulberry trees implied the presence of significant sub-surface structures (Macridi Bey 1904a, 564-571). Seven tombs were excavated, yielding numerous cinerary urns belonging to the Iron Age (Pierre Bikai 1992b, 29). When French soldiers discovered four more tombs in 1942, Maurice Chéhab of the Lebanese Department of Antiquities returned for two seasons of excavation (BMB 1942-1943, 86). By 1943, over 100 cinerary urns had been recovered from Tell Rachidieh, with few making it into publication. In 1974, a salvage project returned to the site to excavate five more Iron Age tombs; these constitute the main corpus of published material to date (Doumet 1982; Doumet-Serhal 2004b, 72). According to Doumet-Serhal (*ibid*), cremation was the rule at Tell Rachidieh; nearly all the jars contained cremated human remains. However, Tell Rachidieh was not a true cremation *tophet*, like Tyre-al Bass, but the site of cremation-burials within an inhumation context (tomb).

Review

Five tombs were excavated in 1974, but only two have been dated. Tomb IV was dated to the second quarter of the eighth century BC based on the presence of Cypro-Geometric III and Cypro-Archaic I styles (Doumet 1982, 133). Hence, Doumet relies on Gjerstad's Cypriot chronology, but with no consideration of its value (**§2.5**). Tomb V was dated to the fifth century BC, based on comparisons with Lapp's Persian assemblage from Ta'anach (Doumet 1982, 135). Stratigraphic problems aside, Lapp's (1970) chronology was based on the conventional (but unreliable) dating of just two Greek imports; a *kylix* and a *lekythos*.

Additionally, five kraters bore inscriptions that have been dated on epigraphic data (Bordreuil 1982; 2004), though three of these are unprovenienced. The remaining two kraters were both recovered from Tomb IV, but the epigraphic dates are not in accord with each other, or with Doumet's dates (Bordreuil 1982). Moreover, most Tell Rachidieh vessels were removed with little attention given to context. In the end, the tombs' assemblages contribute little to reconstructions of the Iron Age.

3.36 Ras al Bassit (Syria)

Summary of Excavations

Excavation at Ras al Bassit revealed a virtually uninterrupted occupation from the Late Bronze Age to the Arab conquest of the seventh century AD, and features a complete Iron Age sequence (Courbin 1986; **Table 3.42**). In the Late Bronze Age, Bassit was little more than an outpost of the great mercantile capital Ugarit. Around the end of the Late Bronze Age the site was partly abandoned and burnt; an event associated by the excavator with the arrival of the "Sea Peoples" (Courbin 1990b, 503). But unlike Ugarit, which remained abandoned, Bassit was immediately reoccupied (Courbin 1983, 119). The modest structures of the earliest Iron Age level (Phase 1) were again destroyed and replaced, but by larger, more permanent houses (Courbin 1986, *passim*). Building Phases 3 and 4 were characterised by a straight fortification wall two metres wide. By the eighth century BC (Phase 5), Greek imports were appearing alongside the local repertory (Perreault 1986, 149-150). Only in BY Phase 9, Greek imports equaled that of the local products, which Courbin (1986, 196*ff*; 1974) considers indicative of Greek presence at Bassit, living alongside the local population. The pottery from the tell included a plentiful collection of burnished Red-Slip vessels (140 vessels of complete and incomplete profile), which Braemer (1986) used to define the development of this surface treatment in the Northern Levant (Courbin 1986, 190; **Table 3.43**).

Table 3.42: Stratigraphy of Ras al Bassit Tell - Iron Age

Phase	Date	Period	Comments
1	11th cent.	Early Iron I	Poor architecture
2	Cypro-Geometric I (11th – 10th cent.)	Iron IB	Permanent structures
3	Late 9th cent.	Iron IIA	Silos
4			Incoherent domestic plans
4b	Mid 8th cent.	Iron IIB	Large walls
5	2nd half 8th cent.		Thick fortification wall
6	1st half of 7th cent.	Iron III	Dense population
7	2nd half of 7th cent.		Large 'weaving' building
8	1st half of 6th cent.		Increase in Greek imports
9	2nd half of 6th cent.		Strong Greek 'presence'

(After Courbin 1986)

Mortuary contexts were also attested at Bassit in two different areas: in the necropolis south-west of the tell; and on the tell itself (Courbin 1993a, 7). The necropolis, where 53 rock-cut chamber-tombs were excavated, was characterised by the exclusive use of cremation (Courbin 1990b, 507; 1993a, 115). The cremation burials, comparatively richer than the few *intramural* tombs, contained local and Cypriot pottery. Greek pottery, while abundant within the settlement, was completely absent from the necropolis (Courbin 1983, 120). The *intramural* burials, wherein the bodies were inserted into torpedo-amphorae, were relatively poor. The cremation burials were associated with bag-shaped amphorae (Courbin 1986 192). There appears to be a strong link between the necropolis at Ras al Bassit and the settlement that is not reflected in similar mixed-use cemeteries (e.g. Khalde, Tell Sukas; Courbin 1993a, 7).

Review

The French began excavating Ras al Bassit with the hope that it would illuminate the nature of local material culture for the north-eastern Mediterranean region (Courbin 1983, 119). Despite best intentions, Courbin (1976, 63) focused on the evidence for Greek influence and interaction at the site (Lagarce and Lagarce 2000, 140). Furthermore, he frequently employed Greco-centric terms (e.g. *l'epoque 'archaique'*) when discussing the history of the site, which ensured that the local culture was viewed from an Aegean perspective (Courbin 1990b). Moreover, the local elements of material culture were rarely discussed, as the literature instead focused on Greek and Cypriot pottery (Courbin 1982a; 1990a; 1990b; 1993b). Indeed, Courbin discusses the same imported vessel again and again (e.g. Courbin 1973, Fig. 15; 1986, Fig. 24; 1993a, Pl. 15.2a). In contrast, the tell's local ceramic sequence has not been published in detail, with only Braemer's (1986) study of the Red-Slip pottery appearing (**Table 3.43**).

Courbin's preoccupation with imported pottery resulted in a reconstruction dependent upon the unreliable chronologies of Cyprus and Greece. Moreover, Courbin (1986, 190; 1993a, 115) has not stated which Cypriot chronology, Gjerstad's or Birmingham's, (**§2.5**) he followed. In a separate, but related point, the absence of Greek pottery in the necropolis argues against Courbin's (1983, 120) permanent Greek settlement at Bassit.

Table 3.43: Development of Red Slip at Bassit

Period	Date	Assemb.
Iron IIA	800	A
Iron IIB		B
	700	C
		D
	650	E
Iron III	600	F
		G
	550	
	500	H

(After Braemer 1986, 246)

Braemer's (1986) study of the Red-Slip pottery is the most exhaustive presentation of non-imported material from the tell (**Table 3.43**). But while he isolated eight phases in the ware's development, Braemer's dates are without firm grounding. His comparison with other eastern Mediterranean sites was not a reliable chronological tool, and the Red Slip sequence cannot be considered in relation to the tell's stratigraphy, which has not been published. While Braemer's Red-Slip chronology is of particular importance for the Northern Levant (e.g. Cecchini 1998, 277; Mazzoni 1990a, 79), the dates for is sequence are without firm foundation.

3.37 Ras Ibn Hani (Syria)

Summary of Excavations

Prior to excavation, there was general consensus that Ras Ibn Hani was founded in the Late Bronze Age by the king of Ugarit (e.g. Bounni *et al.* 1998, 7-8). Hence, the aim of the Franco-Syrian project was twofold: to explore the reasons for the king of Ugarit building a city on the Ras Ibn Hani peninsula (*ibid*); and to determine why Ras Ibn Hani was re-occupied in the Iron Age following destruction, when Ugarit was not (Courbin 1990b, 503). Apparently, Ras Ibn Hani was excavated not because the site itself was interesting, but only for what it could reveal regarding the Late Bronze Age historical trajectory of nearby Ugarit. Excavation identified four main periods – Byzantine, Hellenistic (Bounni *et al.* 1981, 229-254), Iron Age, and the highly-anticipated Late Bronze Age (Bounni *et al.* 1976a, 237; 1998, 7-8).

Excavation began on the tell's south-east corner, the highest point of the site, with expectation of uncovering elite architectural remains: indeed, a large Late Bronze Age palace was exposed (Bounni *et al.* 1998, 3-4). Excavation of this "South Palace", however, was stopped: the overburden of later periods hindered progress; exploration of the Late Bronze Age deposits would have destroyed the rare Iron I deposits (*ibid*); and the walls of the palace were significantly disturbed (Bounni *et al.* 1978, 241-242). As a result, excavation shifted to the more readily accessible ruins of the "North Palace" (Lagarce & Lagarce 1992). The "North Palace" was covered by a thick ash deposit, attributed by the excavators to a destruction of the site by the "Sea Peoples" (based on the presence of Mycenaean IIIC pottery above the ash layer) (Bounni *et al.* 1978, 246, Fig. 28; 1979, 245; 1981, 254-271).

A small archive of 130 tablets was found within the Ras Ibn Hani palaces, many of which were written in alphabetic Ugaritic (Bordreuil and Caquot 1979; Bordreuil and Pardee 1995, 29; Bounni *et al.* 1998, 91).

Review

The published chronology for the excavated strata at Ras Ibn Hani was largely derived from the conventional historical narrative. In particular, the Late Bronze-Iron Age transition was dated to 1200 BC based on the conventional dates for the "Sea Peoples" invasion (Bounni *et al.* 1998, 88). This interpretation assumed that the burning of the "North Palace" could be attributed to the "Sea Peoples", though the historicity of this "event" is not secure (**§2.3**). Furthermore, the destruction of the "North Palace" was not representative for the whole site; the "South Palace" witnessed only a partial fire following a period of abandonment (Bounni *et al.* 1998, 86). There was also remarkable continuity of material culture (and population?) across the Late Bronze and early Iron Age strata, suggesting that the Myc. IIIC pottery was not the result of a new Aegean population, but a local adaption of Late Bronze Age imported pottery (Badre 1983, 206). Chronological confirmation for the Ras Ibn Hani sequence was sought through the dating of Cypriot imports; e.g. Cypriot White-Slip III was used to support the 1200 BC date of the "North Palace" destruction (Bounni *et al.* 1998, 85). A reliance on Cypriot chronology, however, is not a reliable chronological tool (**§2.5**). In comparison, the local ceramic evidence for the last phase of the "North Palace" was extremely fragmentary. Little Iron Age pottery has been published as yet (Bounni *et al.* 1976a, 239, 242-245; 1978, 233-311; 1979, 245; 1981, 254-271), except for Cypriot imports of supposed chronological value (Bounni *et al.* 1978, 243). To further complicate matters, most of the Iron Age material comes from ill-defined contexts from a stratigraphy that has not been published in full (Buhl 1983, 118; Lehmann 1996, 199). In the end, the Iron Age sequence from Ras Ibn Hani is selective and fragmentary, historically derivative, and cannot be relied upon to gain chronological definition across the region.

3.38 Rehov, Tel (Israel)

Summary of Excavations

When in 1996 Mazar (1999a, *7ff*) shifted his Jordan Valley Expedition from Beth Shan to Tel Rehov, he opened excavation areas across the lower (**Table 3.44**) and upper mounds of the site (**Table 3.45**). At least five Iron Age phases were isolated across the upper mound. Area B, on the northern slope of the upper mound, revealed the Iron Age fortifications of the acropolis. Preliminary results indicate that Tel Rehov was at its greatest extent during the Late Bronze and early Iron Age, before significantly reducing in size by the Iron IIB period; indeed Mazar (2001, 292) has concluded that Tel Rehov was the main Canaanite city-state in the Beth Shan Valley during the second millennium BC. According to the ceramic horizon at Tel Rehov, Canaanite traditions continue into the Iron I, but change dramatically in the Iron II period when the painted tradition gives way to Red-Slip (Mazar 1999a, 38).

Table 3.44: Correlation of Tel Rehov Lower Mound Strata

Period	BC	Area D	Area C	Area E	Area F
Iron II	$10^{th}/9^{th}$ cent.		1a	1a	1
Iron II	$10^{th}/9^{th}$ cent.		1b	1b	2
Iron II	10th cent.	1	2		3
Iron I/II	$11^{th}/10^{th}$ cent.	2	3		4

(After Mazar 1999a, 9)

Table 3.45: Stratigraphy of Tel Rehov Acropolis

Area A	Area B	Main Features
1	1	Islamic remains, burials
2a		Assyrian period burial
2b	2	Post-732 BC destruction
3	3	Construction of city wall
4	4	Pre-city wall settlement
5	5?	Late 9th – early 8th centuries BC
	6?	Late 9th – early 8th centuries BC

(After Mazar 1999a, 30)

Review

The excavation and interpretation of Tel Rehov has become a key point of contention in the on-going chronology debate in the Southern Levant (*cf.* Bruins *et al.* 2003a; 2003b; Finkelstein 2004; 2005; Finkelstein and Piasetzky 2003a; 2003b, 2003c; 2006; Coldstream and Mazar 2003; Mazar 2000, 2004, 2005; Mazar and Carmi 2001). In summarising the chronology debate, Mazar suggests that only the tenth century BC is in question; "At the two ends of the debated period stand incontrovertible well-dated assemblages" (Coldstream and Mazar 2003, 41). Nevertheless, these "well-dated assemblages" are based on correlations between the conventional historical narrative and trends in the archaeological record: i.e. Egyptian presence in the Southern Levant in the eleventh century BC; ninth century BC pottery from biblical Jezreel; Assyrian military campaigns in the late eighth century BC.

According to the publications of Mazar and various co-authors (see above paragraph), Tel Rehov has made two significant contributions to the chronology debate: precise scientific dating; and relative dating with Greek chronology (*ibid*, 43*ff*). Despite Mazar's claims, the radiocarbon sampling program from Tel Rehov has not proved decisive. Advocates of the "High Chronology" (Bruins *et al.* 2003a; 2003b; 2005; Mazar and Carmi 2001, 1337-1339) and of the "Low Chronology" (Finkelstein and Piasetzky 2003a; 2003b; 2003c; 2006) have both interpreted the same results as supporting each of their views. Part of the problem lies in the way the radiometric dates are being used to locate Rehov within the already established historical narrative, rather than allow the site's cultural record, aided by precisely dated contexts, to construct an alternative, more-meaningful reconstruction of Iron Age society. For instance, the destruction of Stratum V, dated to the tenth century BC, has been associated with the campaign of Shishak, which has then been used to support the historicity of Shishak's (Shoshenq) campaign, which in turn confirms a tenth century BC date for all other Shishak "destruction" levels across the region (Mazar 2005). The date for the Stratum V destruction event, however, does not offer conclusive support for the historicity of Shishak's campaign, only for the date of that specific context (see Sherratt 2005b, 115). Other examples of the archaeological evidence being shoe-horned into a historical "straight-jacket" at Rehov include associating the destruction of Stratum IV with events following the end of the Omride Dynasty between 840-830 BC (Jehu's revolt, Shalmeneser III's invasion, Aramaean wars), and the connection of Stratum III with the 732 BC campaign of Tiglath-Pileser III (Coldstream and Mazar 2003, 31, 42). In the end, Tel Rehov's meticulously constructed "Ladder of Time" has the pre-established historical narrative at its heart (Mazar *et al.* 2005).

Excavations at Tel Rehov have produced 11 sherds of imported Greek pottery (Coldstream and Mazar 2003, 32-36; Mazar 2004, 24-25). While the importance of this pottery is implied by the singular treatment given to so few sherds, exactly why it is important has not been discussed. One might expect that the radiometric dates from Tel Rehov might have been used to refine Greek chronology, but Mazar preferred to use the Greek pottery to confirm and refine the "already established" dates from Tel Rehov (Coldstream and Mazar 2003, 44-45). Greek chronology, however, remains perilously reliant on the presence of Greek pottery in the Levant – its use to refine Levantine dates represents circular argument (**§2.5**). "The significance of the Greek sherds at Tel Rehov" appears to be that it confirms the site's overall importance (*ibid*, 45-46). In other words, Tel Rehov is important because it has evidence of Greek imports, and by implication participated in far-reaching networks. Contrary to Mazar's acceptance of Greek chronology, Coldstream (2003, 251-252) has recently called on the Tel Rehov sequence to help refine Greek chronology (*ibid*, 255).

3.39 Rifa'at, Tell (Syria)

Summary of Excavations

With only two short accounts in an obscure eastern European newspaper, the Czechoslovakian expedition to Tell Rifa'at is not well-known (Seton-Williams 1961, 68). In these articles, Hrozny, the excavator, discusses a large (23 x 30 m) Iron Age palace that incorporated "Syro-Hittite" architectural elements; the palace was only broadly dated to the first millennium BC. No pottery from this palace was recorded.

British interest was drawn to Tell Rifa'at during the 1953 River Qoueiq survey (Seton-Williams 1961, 68). When excavation commenced in 1956, the British expedition exposed an interrupted occupational sequence stretching from the Roman period back to the Chalcolithic (**Table 3.46**). Iron Age occupation was associated with Phase II, which contained three sub-phases (Seton-Williams 1961, 80-82; 1967, 19-21).

Review

The Iron Age levels from the British expedition were assigned absolute dates on two considerations: historical data and relative stratigraphic position. Once the phases of the occupational sequence were defined, Seton-Williams (1961; 1967, 16-17) then attributed each assemblage to an ethnic group: e.g. Assyrians were associated with the presence of "Palace Ware". By associating the historically-attested people-groups with assemblages, Seton-Williams (1961, *passim*) then applied historical dates to the corresponding levels. The history of the site as portrayed in the conventional historical narrative was considered unambiguous.

Table 3.46: Tell Rifa'at Stratigraphic Phasing

Phase	Description	Date
I	a) Roman	1st–4th cent. CE
	b) Hellenistic	4th–1st cent. BC
	c) Persian	6th–5th cent. BC
II	a) Neo-Babylonian	7th–6th cent. BC
	b) Assyrian/Aramaean	9th–7th cent. BC
	c) Aramaean	10th–9th cent. BC
III	a) Aramaean Settlement	14th–12th cent. BC
	b) Pre-Aramaean	?
IV	a) Early – Middle Bronze	2300–2000 BC
	b) Early Bronze	3rd mill. BC
V	Chalcolithic	5th–4th mill. BC

(After Seton-Williams 1967, 16-17)

The Rifa'at sequence was refined through an analysis of ceramic parallels, most of which are also based on the conventional historical narrative. For instance, the pottery from two Iron Age ash layers (G1-6; M6-9b) was considered comparable to that from Hama E; hence Matthers (1981b, 416) and Seton-Williams (1961, 82) interpreted the two Rifa'at ash layers as late-eighth century BC Assyrian destructions. Seton-Williams (1967, 20) was, however, open to an alternate historical destruction – that associated with the Neo-Babylonian conquest of Carchemish in 605 BC. Cypriot and Greek imports were considered important chronological tools (Seton-Williams 1967, 19), despite problems with both these foreign chronological sequences (**§2.5**).

Additionally, there are indications that archaeological practice and stratigraphic control were not adequately maintained by the British project. This is particularly evident in the earliest Iron Age level; Phase IIc was labelled "Aramaean" by Seton-Williams (1967, 19) because it contained pottery comparable to Hama E. However, the "Aramaean" pottery was accompanied by a small collection of Mycenaean sherds; in other words, Late Bronze Age pottery was found alongside Iron II pottery. While the Mycenaean sherds may well be examples of misclassified Iron I sub-Mycenaean pottery, this does not solve the problem. Could Red-Slip, usually associated with the Iron II period, appear at Tell Rifa'at significantly earlier than at other sites (*cf.* Braemer 1986)? Maybe sub-Mycenaean forms were used at Tell Rifa'at significantly longer than at other sites? Most likely, the stratigraphy was mixed and the "Mycenaean" (or sub-Mycenaean) sherds were intrusive. Alternatively, Red-Slip pottery may well be an earlier phenomenon than previously thought (Dornemann, *pers. comm*; Harrison, *pers. comm*) To complicate matters further, the closest parallel for the "Aramaean" East Gate was the Late Bronze Age "Palace Gate" at Alalakh (Woolley 1939a, 238-239, Pl. XLIV). Results from the British excavation of Tell Rifa'at have appeared in only two preliminary reports, with no definitive detailed phasing of the site available for assessment.

3.40 Sarepta (Lebanon)

Summary of Excavations

In the 1920s, Late Bronze Age pottery from a tomb near the village of Sarafand was taken to the AUB Museum. The museum director later visited the site and produced a plan of the tomb (Baramki 1958). In 1968, a further 40 rock-cut tombs were found. Although all but three had been completely robbed, the contents of those not plundered indicated a late Iron Age date (sixth-fifth century BC; Culican 1970, 15-16; Saidah 1969, 134-137).

In 1969, looking for an Iron Age settlement within "Phoenicia", Pritchard (1978, 3-14) began the excavation of the small tell near Sarafand. Two trenches were opened on the acropolis: Soundings X (875 m^2) and Y (100 m^2) (Anderson 1988, 33). Sounding X exposed substantial remains of a major pottery workshop and textile-dyeing industry, though the stratigraphy was greatly disturbed (**Table 3.47**; Pritchard 1978, 74).

Table 3.47: Sarepta Area X Chronology

Period	Absolute Dates
V	c. 1275-1150
VI	c. 1150-1025
VII	c. 1025-800
VIII	c. 800-350

(After Khalifeh 1988, 11-58)

Sounding Y produced a less disturbed sequence from the Late Bronze and Iron Ages (**Table 3.48**). Stratum G was dated to the end of the Late Bronze Age on close parallels with Tyre XV (*cf.* Anderson 1988, 385; Bikai 1978b, Pls 42-43). Anderson (1988, 380, 390) noted remarkable cultural continuity between Stratum G and the early Iron Age, Stratum F, which was dated to the early Iron Age on comparisons with Tyre XIV. Significantly, no ash layer separated Stratum G from Stratum F.

Table 3.48: Sarepta Area Y Chronology

Stratum	Dates BC	Comments
G	ca. 1320/1290-1200/1190	Ceramic continuity
F	ca. 1200/1190-1150/1125	
E	ca. 1150/1125-1050/1025	
	distinct break in material culture	
D	ca. 1025/1000-850-825	Bichrome and Red-Slip
C	ca. 850/825-650	
B	6th-5th cent.	Badly disturbed

(After Anderson 1988, 423)

The transition from Stratum F into Stratum E was again gradual, with the dates for the latter being based on Megiddo VI parallels (Anderson 1988, 395). A complete break from the Late Bronze Age cultural horizon finally came with the close of Stratum E; Stratum D was considered markedly different in both architectural and ceramic data, though the stratigraphy shows no evidence of destruction or abandonment. Stratum D (considered equivalent to Tyre XIII-VIII) marked a new era, one that Anderson (1988, 396) and Pritchard (1978) associated with the "Phoenicians". Distinctive characteristics include the use of ashlar masonry, Trefoil-Lip jugs, Mushroom-Lip jugs, and Bichrome and Red-Slip pottery. Neckless storage jars and a particular type of Persian lamp were used to date Stratum B to the sixth or fifth century BC.

Review

The primary means for dating the Iron Age strata at Sarepta were the sequences from Tyre and Megiddo, neither of which is based on independent chronological data (**§3.28**; **§3.46**). Since the tools used in the Sarepta publications are open to revision, the dates for the Iron Age sequence are tentative. Indeed, Anderson (1988, 423) concedes that the dating of the Late Bronze and Iron Age strata "represent a hypothetical framework which is subject to clarification, supplementation, and even modification". While a number of C^{14} samples were collected from Sounding X, only those that derived from pre-Iron Age levels were analyzed (Khalifeh 1988, 102, 113; Pritchard 1978, 123). Even if Iron Age samples were tested, the mixed nature of Sounding X stratigraphy would not aid interpretation.

3.41 Sheikh Hassan, Tell (Syria)

Summary of Excavations

Excavation at Tell Sheikh Hassan began as part of the Tabqa Dam salvage project. Work started under Syrian direction in the early 1970s, when a Byzantine Basilica was exposed beside the tell (Schneider 1999a, 325). In 1976, Cauvin uncovered the extensive remains of an aceramic Neolithic settlement on the eroded western edge of the tell (*ibid*). In 1981, Orthmann sunk a sondage through the tell and identified several occupational levels belonging to the Uruk Period (*ibid*). Very little from these projects was published. A French-German co-operative project returned to the site in 1984 for the purpose of excavating the entire mound (*ibid*). By this time, the rising waters of the dam were causing irreparable damage to the edges of the tell.

Despite the thick accumulation of material at the site, few periods were attested: Islamic, Byzantine, Roman, Hellenistic, Iron Age and the Uruk Period (**Table 3.49**; Schneider 1999a, 325). The Iron Age settlement was characterised by a large building (*Bau A*) on the acropolis, originally interpreted as a temple (Boese 1995: 37), but later a *bit-hilani* palace (Boese 1986-1987, 71; 1995: 221). Due to the lack of finds within *Bau A*, Boese (1986-1987, 71) refrained from assigning absolute dates. The layout of this building has close parallels at Zincirli (Margueron 1979, Figs 9-11). At Sheikh Hassan, as at Zincirli, the massive mud brick walls were founded on solid stones, but unlike Zincirli, no orthostats or sculpture were found, much to the dismay of the site's excavators (Boese 1988-1989, 164).

Table 3.49: Tell Sheikh Hassan Levels

Level	Period
Ib	Islamic cemetery
1	Late Roman-Byzantine
2	Hellenistic
3	Iron Age
4-22	Late Uruk
	Ubaid?

(After Boese 1986-1987)

The Iron Age pottery from Tell Sheikh Hassan was characterised by "local" and "Assyrianising" forms, but no Red-Slip. The so-called "local" forms were broadly dated to the eight to fifth centuries BC, based on parallels with Tell Jurn Kabir and Tille Höyük (Schneider 1999a, 330). The "Assyrianising" forms were also dated to a similar period, with the forms paralleled from across the Assyrian Empire (Schneider 1999b, 351-361).

Review

Few Iron Age sites have been systematically excavated in the Middle Euphrates region, hence the excavation of Tell Sheikh Hassan was an important opportunity to fill a conspicuous hole in the Iron Age landscape. Unfortunately, the excavation of the Iron Age levels focused upon a single building. No domestic structures were excavated, and little pottery was found in a secure stratigraphic context. Furthermore, no independent scientific method was employed to try and refine the date of this building, and the settlement within which it no doubt stood.

3.42 Shiyukh Fawqani, Tell (Syria)

Summary of Excavations

The site of Tell Shiyukh Fawqani was excavated by a joint French-Italian mission as part of the Tishrin Dam salvage project. The primary objective was to investigate principal periods of occupation (Bachelot 1999, 143; Luciani 2005, 719). Five different areas were excavated, uncovering a stratigraphic sequence running from the Late Uruk to the Islamic period (**Table 3.50**; Bachelot and Fales 2005; Luciani 2000). Area D isolated late Uruk and Early Bronze Age material, while Area E exposed Late Bronze Age levels. The Iron Age was encountered in the remaining three areas: F, G and H (Bachelot 1999, Fig. 1).

Area F was opened to investigate the area where an Aramaic ostracon had been discovered in the 1994 season (Makinson 2005). Excavation worked through Islamic graves, Byzantine stone-walls, and classical levels before exposing Iron Age II deposits, and a small textual archive (Fales *et al.* 2005; Makinson 2005). By 1997, 139 textual fragments had been recovered.

Area G was located on the tell's east flank where archaeological deposits lay exposed from the agricultural plundering of soil (**Table 3.51**; Luciani 2005, 719). Amongst the pre-classical deposits was a group of inhumation burials (11 in total) dated to the Iron III and Persian periods (sixth and fifth centuries BC). The investigation of these graves was considered particularly important because no other local late Iron Age inhumation cemeteries had been systematically excavated (e.g. Deve Höyük, Nayrab – *ibid*). The simple-pit burials and specific grave goods were similar to those of Deve Höyük (Luciani 2000, 803). Underneath the Persian graves, a sizeable architectural complex was exposed, dated to the Iron II and Iron III periods. Within this complex a collection of six rooms bore evidence of domestic and industrial (metallurgical) activities (Luciani 2005, 722-759).

Area H, which was situated 50 metres from the tell, also contained Iron Age burials, though these were human cremations held within tall, narrow cinerary jars. Like with the Yunus cemetery at Carchemish, this cremation cemetery was situated on the outer fringe of the Iron Age settlement (Bachelot 1999, 151; al Bahloul *et al.* 2005, 997).

Table 3.50 Tell Shiyukh Fawqani Phases

Phase	Period
I	Late Uruk
II	Early Bronze Age I
III	Early Bronze Age IV
IV	Middle Bronze Age I
V	Middle Bronze Age II
VI	Late Bronze Age I
VII	Late Bronze Age II
VIII	Iron Age I
IX	Iron Age II-III
X	Achaemenid
XI	Hellenistic
XII	Roman
XIII	Early Medieval
XIV	Late Medieval
XV	Modern

(After Bachelot and Fales 2005, XLII)

Review

The results of the excavations at Tell Shiyukh Fawqani are contained in two large volumes, totalling over 1000 pages. In spite of the apparent enormity of the results, a significant percentage of the publication is accounted for by ceramic catalogues. Furthermore, the presentation of the material is largely descriptive. For instance, Makinson (2005) cites comparative material for most of the illustrations, but offers no analysis of what all the cultural connections mean (nor is there any critical appraisal of sequences he uses for comparison). Instead, the ceramic sequence is tied into a history of the site which is then placed within the broader context of the conventional regional historical narrative. This is evident in the Area F periodisation, where Makinson (2005, 455-456, Tab. 2) suggests that the Iron II-III transition coincided with the arrival of Tiglath-Pileser III on the Euphrates in 744 BC. This date is not based on clear archaeological evidence for Tiglath-Pileser's destruction of, or arrival at, the site. It is instead based on the expected effect a significant historical event would have on the cultural record. But why should we expect Tiglath-Pileser's presence at Tell Shiyukh Fawqani to greatly affect the cultural record? Makinson's chronology is also based on negative evidence: the absence of Middle Assyrian pottery marked the beginning of Period IX, while the absence of Neo-Assyrian pottery marked its end (*ibid*, 457).

Table 3.51: Tell Shiyukh Fawqani Area G Iron Age Strata

Level	Phase	BC	Comments
1	X	6th–5th cent	Inhumation cemetery
2	IX	Late 7th–early 6th cent	Fragmentary architecture
3		8th & 7th cent	Non-domestic buildings

(After Bachelot 1999)

In contrast to Area F, Areas G and H are dated only generally. While typological comparisons for the Area H cremation cemetery were closest at Deve Höyük I and Yunus, there were also clear parallels at Hama, in the Amuq (Oa-b) and on Cyprus (CG I-III) (Al Bahloul *et al.* 2005, 1013). The absence of seventh century BC Assyrian pottery was deemed a *terminus post quem*, which ensured the cremation cemetery was assigned a pre-seventh century BC date (tenth to eighth centuries BC). Al Bahloul *et al.* (2005, 1015) believed the broad dates were "due to the limited numbers of [artefacts]". A similar problem was encountered amongst the relatively poor inhumation burials of Area G, where grave goods were neither abundant nor considered chronologically distinct (Luciani 2005, 807). While some objects suggested a date within the Persian period, Luciani (*ibid*) admits that many were not characteristic of this period. Furthermore, there was a distinct lack of imported material in Area G graves (e.g. Greek imports, which were present in the inhumation cemeteries of Deve Höyük and Kamid el Loz, were absent at Tell Shiyukh Fawqani), though the absence might depend on factors other than chronology, such as geography, economy, status, networks, identity (Luciani 2005, 982). While a radiocarbon sampling program was undertaken at Tell Shiyukh Fawqani, the results were not integrated into the final conclusions and have not contributed significantly to the final sequence (Saliège and Pessin 2005).

3.43 Sidon (Lebanon)

Summary of Excavations

Sidon has witnessed extensive archaeological excavation and exploration over the course of the past century. The earliest discovery of significance was the royal necropolis of Ayaa in 1887, where numerous marble sarcophagi were uncovered (Meurdac and Albanèse 1938; 1939). Another important find was the large Temple to Eshmun, the Phoenician god of healing (Contenau 1920, 1923, 1924; Dunand 1926; 1965; 1983; Macridi-Bey 1904b; Stucky 2000; 2004). "Systematic" excavation, however, did not begin until 1963 when the Department of Antiquities invited Dunand to begin a large scale project at Sidon (Dunand 1965, 1966b, 1967b).

In the late 1960s an extensive necropolis was discovered during building operations south of Sidon where a few hundred tomb groups were recovered from the sand-dunes belonging to one of three distinct periods: classical burials within clay sarcophagi; Iron Age stone tombs; and Late Bronze Age graves laid in the sand; only one of the Iron Age tombs ("Tomb 26") has been published (Culican 1975, 145; Saidah 1969, 122).

A glance at the voluminous publication of work at Sidon reveals a intensive focus on "Greater Sidon", with little work on the oldest occupational area – the tell of "Little Sidon" – where dense occupation has until recently prevented exploration. In 1998, however, the Lebanese government purchased land within the heart of the city and granted permission for the British Museum to undertake the first truly systematic, in-depth excavation of the tell (www.sidonexcavation.org). After seven seasons the British Museum Project (BMP) has identified a continuous ceramic sequence from the early-third millennium to the late-second millennium BC; i.e. the entire Bronze Age (Doumet-Serhal 1998-1999; 2004d, 107). A multitude of Middle Bronze Age burials were uncovered, which contained strong Cretan influences (Doumet-Serhal 2004d, 112-119; 2004e; 2004f; 2004g; MacGillivray 2004), while Late Bronze Age pottery was characterised by Mycenaean imports. The main architectural discovery for the Late Bronze Age was the "sunken" basement, which was destroyed by a fierce conflagration and provided good charcoal samples for radiocarbon testing. One specific sample returned a calibrated date of ± 1390-1120 BC (Doumet-Serhal 2004d, 119). The long uninterrupted Bronze Age sequence and volume of material, especially imports, reflects the prominence of Sidon within Bronze and Iron Age documentary sources. Iron Age Sidon has also been identified in recent years, but is yet to be published in detail (Doumet-Serhal 2004d, 108-121; 2006). While late Iron Age architecture has been isolated, no coherent features were discerned from the early Iron Age (Doumet-Serhal 2006, 5).

Review

The earliest excavations at Sidon concentrated on the classical settlement, especially its temples, monuments, and artistic treasures. The methodology and publication of these early missions lacked an understanding of stratigraphy and rarely made any attempt to relate published material to contexts.

The renewed excavations at Sidon have brought with them an opportunity for the application of modern archaeological techniques to an important, but poorly understood site. A number of scientific methods have been employed at Sidon; e.g. core samples were taken from the harbour and analysed (biological, granulometric and radiocarbon) (www.sidonexcavation.org). Despite the application of scientific methods, a less-secure means has been used to date the Iron Age strata. Architecture in Trench 28 was dated by the presence of Attic wares to the fifth and fourth centuries BC, despite the presence of earlier Iron Age forms (Doumet-Serhal 2006, 2-4, 10). These dates ignored the mixed nature of the material, and the debate surrounding Greek chronology. Furthermore, the Greek pottery appeared to "outrank" the local forms that indicated an earlier date. Cypriot and Aegean imports were used to date the earlier Iron Age assemblage to the Cypro-Geometric III period (Doumet-Serhal 2006, 25). Scientific methods, which proved so useful for analyses of the harbour and burnt Late Bronze Age building, were passed over for the older, more familiar (but less reliable) practice of ceramic comparison. Doumet-Serhal (2006, 25) admits that there is still much work to be done on the Iron Age of Sidon.

3.44 Sukas, Tell (Syria)

Summary of Excavations

In 1934, Forrer made two soundings at Tell Sukas, which revealed a long occupational history with close links to Classical Greece (Riis 1970, 7) Following World War II, the Syrian authorities invited a Danish team to choose a site for excavation; Riis chose Tell Sukas because it offered an opportunity to investigate the Levant and Aegean interaction. Riis (1970, 10) also hoped that Sukas would link two earlier Danish expeditions; Rhodes (1901-1914) and Hama (1931-1938). Riis' (*ibid*) three aims at Sukas were: (a) to contribute towards a safer chronology of the Iron Age culture in Phoenicia, i.e. c. 1200-500 BC; (b) to elucidate the relations between the Near East and Greece during the same period, thus assessing Greek chronology; and (c) to supplement Danish archaeological collections.

Riis central trench was aimed at securing an occupational history of the site, extended from the sixth millennium BC through to the medieval period (**Table 3.52**). The Iron Age was attributed to Phases H and G. The early Iron Age (H2) bears remarkable cultural continuity with the Late Bronze Age (J), despite an ash layer separating the two (Lund 1986, 41, 188), while Phase H1 was assigned to the Iron II period. A significant break in material culture came with Phase G, characterised by an increase in Greek imports. An open air sanctuary was exposed near the Southern Harbour and dated to the Late Bronze Age on ceramic parallels with Cyprus (*cf.* Jensen 1996; Riis 1979, 6-7, 33; 1996, 5-6).

Table 3.52: Summary of Tell Sukas Stratigraphy

Phase	Description	BC
A	Later Middle Ages to Modern Times	
B	Crusaders' Period	
C	Byzantine Period	
D	Roman Period	
hiatus		
E	Late Hellenistic Period	140- 69
F	Neo-Phoenician Period	c.380-140
hiatus		
G	**Period of Greek Community**	
	G1 Phase 3	552-498
	G2 Phase 2	588-552
	G3 Phase 1	677/1-588
H	**Early Iron Age**	
	H1 Phoenician (Iron Age II) (Greek Settlement)	850-677/1
Assyrian destruction 850 BC		
	H2 Phoenician (Iron Age I)	1170-850
Sea Peoples destruction		
J	Late Bronze Age	1600-1170
K	Middle Bronze Age	2000-1600
L	Early Bronze Age	
M	Chalcolithic Period	
N	Neolithic Period	

(After Riis 1970, 12, 127; Buhl 1983, 110; emphasis mine)

Review

The Late Bronze Age (Phase J) settlement at Sukas held evidence of significant burning, which persuaded the excavators to link this "event" with an invasion by the "Sea Peoples" and date the end of Phase J to c. 1170 BC (Riis 1970, 40, 126). This interpretation ignored the architectural and ceramic record, which testified to continuity of culture (Lund 1986, 41, 188). Instead, the "destruction" was seen as representing a new "Sea Peoples" population that was identified with the Iron Age Phoenicians. Clearly, the Phase J date was based on the presumption that the "Sea Peoples" were the agents of destruction and the conventional historical date for this event (Riis 1970, 21, 24-26) – the conventional historical narrative was imposed onto the stratigraphic sequence.

The material from Phases J and H appears to have been mixed, as Buhl (1983, 110-126) has suggested. Phase H1 was dated to the Iron Age despite finds of a Mycenaean female figurine, Ugaritic (fourteenth century BC) cylinder seal, and large number of Late Bronze Age pottery sherds (Riis 1970, 36). The end of Phase H was dated by Riis (1970, 58) to the seventh century BC on historical considerations; Esarhaddon campaigned in the region in 677 and 671 BC. Riis (*ibid*) assumed that this campaign would manifest itself archaeologically and identified an appropriate "destruction" layer.

The "Greek" sanctuary exposed in Phase G (Riis 1970, 52, Figs 18, 23, 31, 33), led some scholars to suggest an actual Greek presence within the community at Sukas. For example, Riis (1970, 129) saw Sukas G as "a Phoenician town with a strong, at times very strong Greek element". Phase G was divided into three sub-phases, with the divisions determined according to the historical narrative: the Phase G3-G2 transition is dated to the Egyptian Pharaoh Apries' attack on Phoenicia in 588 BC (Riis 1970, 58-59); Phase G2-G1 is associated with the campaign of Nabonidus in 552 BC (mentioned in Herodotus, *History III* 19.3); the end of Phase G is equated with Greek defeat in the eastern Mediterranean in 498 BC (also mentioned in Herodotus, *History V* 104, 108-116). No evidence was presented that justified these interpretations, or that indicated how these events might manifest themselves archaeologically. These references to classical history instead highlight Riis' emphasis on the link between Tell Sukas and the Aegean. Furthermore, Riis could not perceive of a history of the site outside of the conventional historical narrative.

While the interpretation of the Iron Age strata at Tell Sukas was dependent on historical data, the presentation of the results was confusing. For instance, Buhl's typology (1983) is difficult to reconcile with the catalogue in the same volume. Certain types are introduced in the catalogue as deriving from the Late Bronze and early Iron Age, yet the listed examples often extend from Persian or post-Iron Age contexts; indeed rarely does a form's parallel match the chronological context proposed by the excavators (e.g. the rounded base jug – Buhl 1983, 33; ring base jug – Buhl 1983, 31; splayed foot jug – Buhl 1983, 35; and ring base bowl – Buhl 1983, 37-39). One must conclude that the stratigraphy at Sukas was either mixed or not accurately recorded. Indeed, Buhl (1983, 110) confirms that "most of the pottery consisted of the remains of storage jars originating from elsewhere and used as filling material", leading us to conclude that during excavation no distinction was made between occupational (primary) contexts and fill (secondary) deposits. As a result, residual pottery has apparently tainted the Sukas sequence. This problem is further accentuated by a typology based largely on base forms, due to a lack of rim sherds (Buhl 1983, 6). There was also a heavy reliance on Cypro-Aegean imports for dating the various occupational phases (Buhl 1983, 124); while this in itself means the Sukas chronology is unreliable (**§2.5**), it is also at odds with the original objective of the project "to confirm current Greek dating" (Riis 1970, 10).

Ta'yinat, Tell (Turkey) - see 3.8 Amuq

3.45 Tille Höyük (Turkey)

Summary of Excavations

Tille Höyük was excavated by the British Institute of Archaeology at Ankara from 1979 until its flooding by the Ataturk Dam in 1990 (Blaylock 1999, 263). Surface pottery suggested that occupation started as early as the fourth millennium BC, though excavation focused on the medieval, classical and Iron Age levels (*ibid*). A sondage excavated in later seasons exposed the Late Bronze-Iron Age transition (Summers 1993, 3). The expedition strategy was to examine architecture over as great an area as possible and recover sufficient *in situ* ceramic material to construct a reliable sequence (Blaylock 1999, 263).

Public architecture characterised Late Bronze Age Tille (*ibid*, 265), which was destroyed by fire (dendrochronology provided a the late-twelfth century BC date; Kuniholm *et al.* 1993). There was demonstrable continuity in ceramics between the Late Bronze and the earliest Iron Age levels (Blaylock 1999, 265; Summers 1993, 3). The Iron Age revealed ten distinct architectural phases (**Table 3.53**), with the remains of Level I cutting the burnt remains of the Late Bronze Age (Blaylock 1999, 266). The technique and motifs of early Iron Age painted pottery were reminiscent of the material from Hama (Riis and Buhl 1990, Figs 64, 67, 81), Tell Afis (Mazzoni 1992b, Figs 10, 11) and Tell Rifa'at (Seton-Williams 1961, Pl. 39).

Table 3.53: Stratigraphy of Tille Höyük Iron Age

Level	Period	Date
X	'Persian'	6th-5th centuries BC
IX	?	7th or 6th century BC
VIII	'Assyrian'	8th century BC
VII VI	? (Poorly preserved)	9th-8th centuries BC
V IV	Iron II ('Neo Hittite')	9th century BC
III II I	Iron I	11th-10th centuries BC

(After Blaylock 1999)

Levels IV and V contained extensive plans of buildings that were considered rural in character (*ibid*, 267). Levels IV and V contained hand-made burnished cooking-pots and the distinctive "Ribbed Ware" known from salvage work in the Keban region, a rare connection to the north (*ibid*). Otherwise, when parallels existed, the complexion was overwhelmingly "North Syrian" and "North Mesopotamian". Cypro-Aegean pottery was rare (*ibid*, 265). Levels VI and VII were largely destroyed by later terracing. Level VIII was characterised by a well-planned complex of structures within a defensive perimeter wall, and built around a large courtyard of pebble-paving in chequerboard design, comparable to the Tell Ahmar example (Roobaert and Bunnens 1999, Fig. 13). The Level VIII pottery was characterised by "Assyrian" carinated-cups and dimpled Palace Ware, with parallels in northern Iraq (Blaylock 1999, 269). The final Iron Age Level (X) contained well-preserved architecture, and pottery reminiscent of Levantine Persian period pottery.

Review

It is not easy to assess the conclusions of the Tille Höyük excavations, primarily because it is difficult to determine what exactly was concluded. The stratigraphy is presented as a very general, relative sequence, with only occasional attempts to date the levels more precisely; e.g. dendrochronological sampling from the burnt Late Bronze Age gateway (Blaylock 1999, 253-264; Summers 1993, 13, 55). While the continuity of culture across the Late Bronze-Iron Age transition discouraged theories of absolute conquest, the "destruction" layer was still attributed to a foreign entity, in this case Tiglath-Pileser I (Summers 1993, 11). While dendrochronology was used to date the Late Bronze Age destruction, the dating for the Iron Age sequence relied on arbitrary estimates of building life and correlations with historical data (Blaylock 1999, 263-264).

3.46 Tyre (Lebanon)

Summary of Excavations

Archaeological work at Tyre has too long a history to fully recount here (see Pierre Bikai 1992b). Of the many excavations, the largest was Chéhab's "city excavation", though the classical overburden prevented access to earlier material (Coldstream and Bikai 1988, 36). While some Iron Age ceramics were recovered (*ibid*), the mixed nature of the material prevented Patricia Bikai from constructing a meaningful typological sequence. It was for this reason that Bikai, in 1973-1974, opened a small trench into the pre-classical levels and excavated a Bronze and Iron Age sequence for ancient Tyre (*ibid*).

In 1990, during the Lebanese Civil War, some stones with Phoenician inscriptions and painted pots were looted in the al Bass district of eastern Tyre, an area originally on the mainland when Tyre was still an island (Seeden 1991). Upon closer inspection, the stelae and cinerary urns appeared to be the remains of an Iron Age cremation cemetery, or *tophet* (Sader 1991; Seeden 1991). Until this discovery, the Iron Age necropolis of Tyre was believed to have existed 5 km south of the site, at Tell Rachidich, where a number of Iron Age tombs had been excavated (Bikai and Bikai 1987, 76). In 1997, agricultural digging in the al Bass area brought more pottery and stelae to light. The ensuing salvage expedition excavated over 80 Iron Age cremation burials, which suggested to Aubet (2004a, 19), the most recent project's director, that this was Phoenician Tyre's main cemetery.

Review

Bikai's (1978b, 3) results presented an almost continuous record of ceramic development from the Roman-Byzantine period back to the third millennium BC. According to Bikai (1978b, 64, 75), the high percentage of imported wares and comparative material made it possible to arrive at an absolute chronology for the strata. For the Late Bronze and Iron Ages (**Table 3.54**), Bikai relied on the Cypriot chronologies of Åström (1972) and Birmingham (1963), respectively.

Table 3.54: Stratigraphy of Bikai's Sounding at Tyre

Stratum	BC	Description
I	700	Iron II-III transition
III-II	740-700	
V-IV	760-740	
VII-VI	800-760	
IX-VIII	850-800	Iron II
X	850	Iron I-II transition
XI	925-850	
XII	1000-925	
XIII	1070-1000	Iron I
XIV	1200-1070	LB-Iron transition
XV	1375-1200	LBII
XVI	1415-1375	LB

(After Bikai 1978b, 68).

Tyre XIV was dated to before the mid-eleventh century BC on the absence of Cypriot White-Painted pottery. While some Levantine Iron Age parallels for Tyre XIV were identified (Hazor XII; Megiddo VI), the lack of Cypro-Geometric I convinced Bikai of a Cypriot Bronze Age date. The paradox of Tyre XIV falling within both the Late Bronze Age (on Cyprus) and the Iron Age (in the Levant) is difficult to reconcile with conventional histories of the eastern Mediterranean, and symptomatic of the cultural continuity that exists between these two periods (**§2.3**).

With the first appearance of "White-Painted" pottery, Tyre XIII was dated to the beginning of the Cypriot Iron Age; i.e. mid-eleventh century BC. At the other end of the spectrum, Tyre I was dated to the late-eighth century BC based on the presence of Cypriot Period V wares and parallels with Hazor IV. With these two dates supposedly confirmed, Bikai (1978b, 66) was left a period of only three-and-a-half centuries to fit thirteen Iron Age strata, which implied a very dynamic and unsettled history of occupation at Tyre; something not immediately evident in the conventional historical narrative. The high number of strata may be due to the small size of the test trench, which would have made the identification of architectural phasing difficult (*ibid*, 66-68). Further chronological definition is offered by Bikai through comparison with Birmingham's (1963) Cypriot chronology, and the Tyre IX appearance of Greek pendant semi-circle *skyphoi* (Åström 1972; Boardman 1959, 163; Coldstream 1968, 152-154). The closest Levantine parallels for Tyre were recorded at Tell Abu Hawam, Megiddo and Hazor; three sites with notoriously complicated and ambiguous chronologies (Bikai 1978b, 66-68). Bikai's sequence is dependent on the unreliable witness of Birmingham's Cypriot chronology, which undermines her absolute dates.

According to Aubet (2004b, 465), five distinct periods of development are discernible within the Tyre-al Bass cremation cemetery, with the bulk of the burials dated to the Iron II period, or Periods IV-II (*ibid*, 458 – **Table 3.55**). No Iron I in-tact burials were discovered, but isolated finds of early Iron Age pottery bear witness to limited use of the cemetery in this period (Núñez 2004b, 352). While the means for dating the Tyre-al Bass sequence were not made explicit, Aubet (2004b, Fig. 312) appears to base her findings on Bikai's (1978b) Tyre sequence, and Bikai's (1987) sequence of "Phoenician" pottery on Cyprus (**Table 3.56**).

Table 3.55: Tyre-al Bass Phases

Phase	Date
I	1100-850
II	850-775
III	775-750
IV	750-700
V	700-600

(After Aubet 2004b, 465)

Aubet (2004b, Appendix A) also undertook a small program of radiocarbon dating, but included only four results in her publication as confirmation for the sequence dates. Moreover, only one of the resulting two-sigma values was accepted as accurate because it alone agreed with Aubet's (2004b, 469) chronology. In other words, rather than refine her chronological sequencing in the light of conflicting dates, Aubet simply dismissed the problematic data. In the end, the ceramic typology appears to have been the main means for dating the Tyre-al Bass sequence, which was then confirmed through the application of radiometric samples.

Table 3.56: Tyre-al Bass Comparative Chronology

BC	Tyre Bikai 1978b	Phoenician Cyprus Bikai 1987	Tyre al Bass Aubet 2004b
1200			
	XIV		
1100			
	XIII		Period I
1000	XII	Kouklia Horizon	
900	XI X		
	IX		
800	VIII	Salamis Horizon	Period II
	VII-VI		
	V-IV		Period III
700	III-II	Kition Horizon	Period IV
	I	Amathus Horizon	Period V
600			

(After Aubet 2004b, Fig. 312)

3.47 Zincirli (Turkey)

Summary of Excavations

Zincirli was excavated by the German Oriental Society in the late nineteenth century, under the direction of Felix von Luschan (Lehmann 1996, 272-273). Five campaigns revealed a heavily fortified acropolis surrounded by an extensive Lower City and an enormous double fortification wall with three gates and 100 evenly spaced bastion towers (Wartke 2005). A number of building phases were discernible in the architecture, though the sequence and dates are still debated (Pucci 2008a, 545). Nevertheless, Zincirli is a celebrated example of "Syro-Hittite" architecture, with its collection of richly decorated orthostats and statuary (Akurgal 1962, 130*ff*).

Review

The Zincirli excavations exemplify the "*deutsche Bauschule*" method, which emphasised the broad exposure and careful documentation of elite architectural units (Lehmann 1996, 272-273), which explains the absence of domestic structures in the site plan (Humann 1898, Tfn 28-29). This approach ignored debris layers, which prevented a true stratigraphic record of the site (Andrae 1943, 38; Naumann 1971, 418-419). The excavators also paid little attention to pottery; none-what-so-ever for stratigraphic purposes. The result is a fairly complete picture of the layout of the city that has been difficult to date accurately (*cf.* Busink 1970, 540; Koldewey 1898; Landsberger 1948, 8-82; Lehmann 1996, 272-273; Naumann 1971, 418-425; Oelmann 1921; Orthmann 1971, 60*ff*; Pucci, 2008a; 2008b, 15-41; Ussishkin 1968; Wachtsmuth 1923-1924; Winter 1973). Method aside, the final publications display a clear pre-occupation with the elite art and architecture (Koldewey 1898; von Luschan 1902). The volume devoted to small finds (Andrae 1943) inconsistently presents only those ceramics of high artistic appeal.

Table 3.57: Summary of Zincirli Building Phases

Phase	
V	Befunde, alter als das Ende des 8 Jahrhunderts
destruction of eighth century buildings – 676 BC	
IV	Hilani 3, Hilani 4, Gebaude J, Gebaude K, Gebaude L, Hof M, Gebaude P, Hof R
III	Hilani 1, Hilani 2, Burgmauer mit Balkenrost, Inneres Burgtor, Ausseres Burgtor, Kasematten
II	Oberer Palast, Burgmauer über der Mauer mit Balkenrost
I	Spätachamenidische und hellenistische Befunde

(After Lehmann 1996, 274)

3.48 Concluding summary

A number of recurring themes are evident in the above survey of Iron Age archaeology in the Northern Levant. These themes can be loosely grouped according to one of two categories: archaeological practice or archaeological interpretation.

The above survey has demonstrated that excavation practice in the Northern Levant in the early-twentieth century was directed toward the broad exposure of elite monuments and architecture (e.g. Arslan Tash, Carchemish, Zincirli). As understanding of tell stratigraphy developed, excavation technique became more systematic in its approach. However, not all "systematic" approaches were sensitive to tell stratigraphy (e.g. Dunand imposed a uniform but artificial stratigraphy onto Byblos) or were accompanied by adequate methodology (e.g. the Hama excavations have been widely criticised for poor stratigraphic control). Furthermore, the excavation of Iron Age sites remained largely concerned with the monumental and significant (e.g. Tell Ta'yinat, Ain Dara, Aleppo citadel) as Iron Age archaeology sought the illumination of the historical narrative (e.g. Tell Nebi Mend and the Battle of Kadesh; the Greek "Orientalising" period at al Mina; the Assyrian destruction of Hama). Nevertheless, focus has shifted from broad horizontal exposures to an understanding of tell stratigraphy and the diachronic development of settlement. And while the stratigraphic excavation of a site is being undertaken with higher levels of care and precision, the underlying interest in an illumination of the historical record has persisted (e.g. Greek presence at Tell Sukas; Assyrian destruction of Tell Shiyukh Fawqani).

This chapter has also shown that there are persistent chronological and ethnic undercurrents in the interpretation of most Iron Age sites, both of which primarily derive from a reliance on a conventional and literal reading of documentary sources. Due to the early focus on elite architecture, other elements of North Levantine Iron Age material culture have lacked adequate understanding by modern archaeologists. This has resulted in an appeal to, and subsequent reliance upon other regions for chronological definition; i.e. imported material culture (usually pottery from Cyprus of Greece) was given prominence in site interpretations (e.g. al Mina; Ras al Bassit, Tyre). Furthermore, interpretations focused on the ethnic identity of a site's population as derived from the historical narrative (e.g. al Mina, Tell Ahmar, Tell Sukas). The means for explaining changes in material culture, therefore, was associated with the contact, interaction, or movement of different populations (e.g. Aegean style pottery at early Iron Age Tell Kazel; the practice of cremation in Hama F). Regardless of how carefully and meticulously some sites had been excavated, interpretation has often been characterised by a reliance on the conventional historical narrative. This echoes Sherratt's (2005b, 122) wonderfully worded appeal for archaeologists to

> *"critically re-examine the whole mythology surrounding the beginning of the Iron IA, and with it the tight chronological constraints of this mythology [which] has merely acted as an agent of distortion, as we manipulate the archaeological record – and the chronology – to fit a straight-jacket created by the need to hang the archaeology around supposed major events of a sudden nature for which pottery styles are seen as surrogate proof."*

What is common to the majority of the excavations surveyed above is an under appreciation of how ceramic material can inform reconstructions of Iron Age society outside of the conventional historical narrative, something that is explored in **Chapter 4**.

CHAPTER FOUR
Approaches to Iron Age Pottery from the Northern Levant

4.1 Introduction

The previous two chapters have demonstrated the widespread application of text/s in the interpretation of the archaeology of the IA-NL, which has resulted in reconstructions of past societies that lack a perspective on human behaviour and archaeological residues. Part of the problem lies with the view of material culture as a reflection of "history" as depicted in the historical narrative; as reflecting peoples and events.

This chapter explores past analyses of Iron Age pottery in the Levant with the aim of highlighting the key assumptions behind these approaches. Emphasis will be given to previous interpretations and why these might be questioned on theoretical grounds. Following this, it will be shown that there are alternative ways of understanding the relationship between people, social constructs, and material culture. While the typo-chronological approach has its place, and has been instrumental in contributing to our initial knowledge of Iron Age material culture, there are many fruitful avenues of analysis that have not been explored due to the reliance on the conventional historical narrative. This chapter will conclude with a brief look at developments in archaeological ceramic studies in other regions and the implications these might have for excavation techniques and methodologies across the Levant.

4.2 Analysis of Iron Age Ceramics in the Levant

Introduction

As was shown in **Chapter 2**, the conventional historical narrative has overdetermined reconstructions of the IA-NL. Texts provide a specific view of a specific past; their use as an interpretative framework for all material culture across the entire Levant is problematic. Material culture will have been used by many different people, at different times, in different places, for different reasons; we cannot assume that all material culture held the same meaning in all these different contexts. Material culture, therefore, represents different pasts, not just the one imparted by a literal reading of the historical narrative (Whiting 2007a, 77). To impose a particular historical perspective of "what happened" onto material culture is to conflate the world in the text, and the world as lived by human beings in all its diversity. As a result, the lives of the people in the IA-NL have been largely subsumed by the historical narrative. Through the identification of "historical events" in material culture, the "whole story" of that narrative has been imposed onto material culture. Since the "total" history of the IA-NL is already (perceived to be) known from the historical narrative, material culture is rarely analysed to its full potential. Instead, analysis tends to focus on classification (which historical "culture"?) and chronology (when in "history"?).

This chapter will outline the study of pottery from the IA-NL by discussing the research of two influential scholars. In the author's view, it is the work of Stefania Mazzoni and Gunnar Lehmann that has made the most significant contribution to our current understanding of ceramic development for the IA-NL (see Akkermans and Schwartz 2003, 361-366). Each scholar's research on Iron Age pottery will be briefly summarised, followed by a discussion of the underlying principles in each approach.

Case Studies

Stefania Mazzoni

Early excavations of Iron Age sites in the Northern Levant focused on elite art and architecture, much to the detriment of pottery (e.g. Arslan Tash; Carchemish). As a result, when scholarly interest in the Iron Age was rekindled in the 1980s, little was known regarding local ceramic style, distribution, and development. Rather than lament the lack of a coherent and reliable ceramic typology for the first millennium BC, Mazzoni (1988a; 1990a; 1991-1992; 2000a) set about constructing an Iron Age periodisation based on her meticulously-constructed stratigraphic sequence from Tell Afis. But while her ceramic sequence is a good example of consistent practice and thorough method, Mazzoni's interpretation of the Tell Afis sequence is less innovative.

As already discussed, the correlation of the conventional historical narrative with the archaeological record is a difficult, and misguided, undertaking. Nevertheless, specific political events and broad regional narratives are key features of Mazzoni's Iron Age periodisation. For instance, the Iron I period is considered distinct from the Late Bronze Age not on consideration of material culture (Mazzoni is aware there is strong continuity – 2000a, 31; 2000d, 1043-1044), but on the historically-derived political crisis and the expected effect this would have on material culture. Indeed, general historical trends have helped Mazzoni characterise each of the main divisions in her scheme (**Table 4.1**).

Table 4.1: Mazzoni's Ethno-Historical Periodisation of the Iron Age

Period	c. BC	Cultural Trends	Historical Correlation
Iron IA	1200–1000	Recovery	Aftermath of political crisis
Iron IB	1000–950	Re-urbanisation	Levant-wide political stability
Iron IC	950–early 9th	Monumentality	Syro-Hittite political prosperity
Iron IIA	early–late 9th	New Trends	Phoenician and Aramaean expansion
Iron IIB	late 9th–720	Regionalisation	Coalition facing Assyrian ascendancy

(After Mazzoni 2000a, *passim*)

Mazzoni's interpretation of the Iron I period is based on the conventional historical narrative. For instance, in her discussions on the Iron I ceramic material, Mazzoni (2000a, 31-33; 2000c, 147) often emphasises its Aegean character, noted as the result of either trade or cultural exchange. However, in comparison to the Late Bronze Age, when imports were common, the Iron I period bears minimal Aegean affinities (e.g. Bikai 1978b, 66). So why does the strong Aegean influence of the Late Bronze Age not encourage an Aegean interpretation of the mid-second millennium BC, while the Iron I period, with its relatively diluted Aegean style, is considered Aegean in nature? It seems that the "Sea People" migration theory contributes to this interpretation; Mazzoni is not alone here (Badre 1983, 204, n.3-6; Bell 2006, 15; Bonatz 1993, 138; Dikaios 1969, 272*ff*; Dothan 1982, 98-106; Karageorghis 1985, 271*ff*; A. Mazar 1988; Muhly 1984, 52*ff*). The interpretation of the archaeological record has been "over-determined" by a particular, traditional interpretation of history. Invoking the Aegean explanation is inadequate, and is really just a coverall to avoid dissecting the complexity of the Iron I period in the Northern Levant.

According to Mazzoni (2000a), her periodisation of the IA-NL is not based on only one element of material culture, but is derived from "A Cross-Cultural Perspective". In other words, Mazzoni's (2000a, Table 2, p. 58; 2000d, *passim*) scheme derived information from not just the pottery, but from many aspects of material culture; i.e. monumental art and architecture, settlement patterns, urban-planning. Despite best intentions, Mazzoni's (see 1997a; 2000d, Tab.1, Figs 1-20) discussion of Iron Age material culture has tended to emphasise the monuments and inscriptions of the "Syro-Hittite" tradition (e.g. lion statues). These are essentially political monuments that correlate well with the historical narrative (**§2.3**). While political monuments constitute an important category of evidence, they do tend to over-emphasise the political history of the region at the expense of more localised cultural traits. The result was a "political" periodisation of the Levant, as defined by "historically-sensitive" elements of culture, that Mazzoni then applied to the Tell Afis stratigraphic and ceramic sequence.

Figure 4.1: Mazzoni's Iron Age Ceramic Provinces

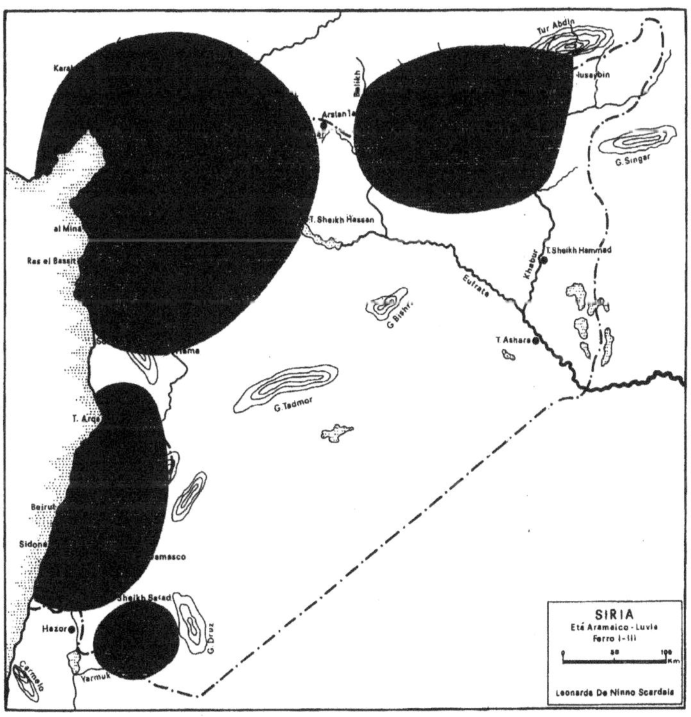

(After Mazzoni 1991-1992, Fig. 3)

Tell Afis remains one of the most-extensively published Iron Age ceramic sequences from the Northern Levant, yet it has been published in only a general catalogue form. No typology has been published for the site and comparative parallels are only infrequently and inconsistently cited (see references by Cecchini, D'Amore, Mazzoni, Oggiano, and Venturi amongst others). Moreover, the Tell Afis ceramic sequence has relied on the presence of imported pottery for absolute dates. For example, Late Helladic IIIC1b and Cypriot White-Painted sherds were used to date the early Iron Age levels (Bonatz 1998; Mazzoni 2000a, 33; 2000d, 1050). Furthermore, Mazzoni (2000c, 147) also accepts the presence of imported pottery as an indicator of trade contact and economic prosperity; how these vessels were viewed, accepted, and used outside of their "homeland" is not explored. Indeed, the "meaning" and use of material culture does not feature strongly in Mazzoni's interpretation. Material culture is either "local" or foreign, and therefore indicative of cultural interaction; trade being the principal means for cultural diffusion (Mazzoni 2000a, 36, 45). For example, the diffusion of Red-Slip surface treatment is associated with "Phoenician" trade (*ibid*, 41). In this way, Mazzoni (2000c, 147-148) is arguing for the correlation between the development of regional ceramic provinces and the formation of political and economic systems (**Figure 4.1**). While we cannot disagree entirely – large-scale socio-political transformations would likely have some resonance within the distribution of ceramic styles – there are problems in suggesting that this is the only explanation (A.T. Smith 2001, 370). There is no consideration of other contexts or mechanisms by which the appropriation of material culture might have taken place, or even that ceramic provinces might be the product of modern frameworks. In the end, Mazzoni's proposed chronology (**Table 2.7**), as opposed to the relative stratigraphic sequence, is based on political history and draws only marginally on ceramic data.

Gunnar Lehmann

The presentation of an extensive, regional ceramic typology for the Northern Levant is credited to the doctoral research of Gunnar Lehmann (1996; 1998), who explored the development and distribution of local pottery production in the later Iron Age and Persian period (eighth to fourth centuries BC). Lehmann's (1998, 7) aim was to present an "archaeological periodisation system... based only on an analysis of the material record", which could then be related to the historical record once the internal consistency of the archaeological data was assured. From the incomplete and somewhat selective corpus of published ceramic data, Lehmann was able to identify eight discrete ceramic assemblages, the distribution of which he then plotted geographically. In particular, Lehmann identified two broad ceramic regions within his first four assemblages; there was a clear distinction between pottery assemblages from coastal sites and those from inland sites (**Figure 4.2**).

Lehmann's coastal/inland dichotomy is comparable to the large ceramic regions evident in Mazzoni's earlier work (**Figure 4.1**), and broadly echoed in the present study (*cf.* **Chart 8.1** and **Chart 8.4**). But while Lehmann's work represents a watershed for the study of Iron Age pottery in the Northern Levant, it afforded little effort in trying to understand the reasons for the large ceramic regions. Lehmann (1996, Abb. 4.4-4.7; 1998, Fig. 14) simply presents the ceramic regions as highly-bounded, geographically-discrete entities, and describes them as products of socio-political and economic developments (1998, 31).

> *"Their [Assemblages 3 and 4] distribution reflects major developments in the political geography of Syria and Lebanon. The Phoenician seaports were the main economic centres during the seventh century BC. In inland Syria there were very few provincial cities with Greek imports. The main road from the coast into inland Syria seems to have led through Al Mina and up the Orontes into northern Syria. The distribution of local pottery still follows older Iron Age traditions. There was a large field of interaction along the coast, reflecting the long-distance sea trade" (Lehmann 1998, 29).*

Ceramic regionalisation is not an explanation of distribution, only a description. Moreover, this diffusionist perspective ignores the effect that non-political and non-economic cultural factors might have had on material culture patterning. Instead, ceramic regions are associated with socio-political histories, which are then equated with socio-political polities depicted within the documentary sources; i.e. "Aramaeans" inland, "Syro-Hittites" in the north, "Phoenicians" on the coast. But Lehmann seems aware of this pitfall:

> *"The late Iron Age and Persian period in Syria and Lebanon have never been dealt with in a comprehensive archaeological study. As a result, the periodization system, even of the archaeological record, reflects historical events rather than internal developments of the material record itself. Often, ethnic or political terms like "Neo-Assyrian," "Neo-Babylonian," "Persian," or "Achaemenid" are used to date the changes and developments of archaeological finds. Too often the material record is related in a somewhat* ***naive*** *way to known historical events" (Lehmann 1998, 7,* ***emphasis mine****)*

Figure 4.2: Distribution of Lehmann's Assemblages 1-2 (left) and 3-4 (right)

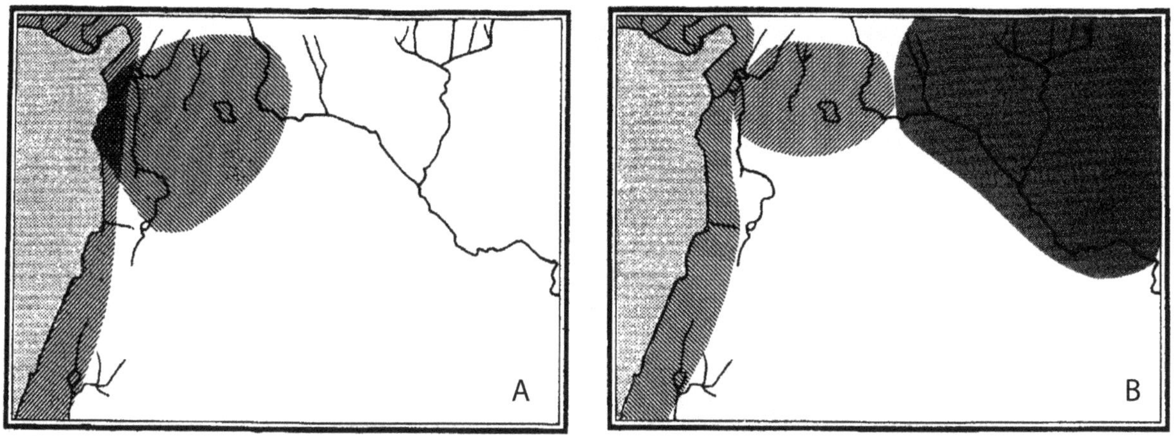

(After Lehmann 1998, Fig. 14A-B)

Despite an awareness of archaeology's reliance on the conventional historical narrative, Lehmann's study perpetuated a number of assumptions within the existing interpretative framework. Lehmann (1998, 9) states that the "pottery of Assemblage 1 comprises the vessels of the Aramaean and Phoenician kingdoms before their destruction by the Assyrians" and supports this statement by an appeal to Gjerstad's (1948) *and* Birmingham's (1963) Cypriot chronologies, as well as the Middle- and Late-Geometric pottery from Greece (Lehmann 1998, 13). He then makes "a connection between the destruction levels of Tell Rifa'at and Hama with the campaigns of Tiglath-Pileser III against Arpad in 740 BC and of Sargon II in 720 BC" (*ibid*) and calls on comparative pottery from the Southern Levant to "point to a date in the eighth century BC" (*ibid*). In the end, Lehmann relies upon the same conventional historical data to date and interpret changes in the ceramic record that he labelled "naive" (*ibid*, 7).

While Lehmann's (1996, 86-92) eight assemblages were derived from the internal structure of his ceramic data, the absolute dates were not. By using the existing chronological framework, with all its inherent circularity (**§2.5**), Lehmann was undermining his own results and, therefore, negating its value for dating material from other sites. Furthermore, Lehmann was authenticating the existing interpretative framework, wherein the archaeological record is overdetermined by an historical interpretative framework (**§2.4**).

Trends in Iron Age Ceramic Studies

Despite the recency of Mazzoni's and Lehmann's publications, the means for studying pottery from the IA-NL is largely based on the culture-historical approach that was standard practice in the early-twentieth century (Faust 2006, 11; Whiting 2007a, 26). The use of this paradigm in the Northern Levant is due to the adoption of the chronology, and accompanying methodology, used in the Southern Levant, where the culture-history paradigm characterised the first pottery studies (**§2.5**). Albright (1933; 1938; 1941-1943) and Wright (1937) were amongst the first to use pottery to disentangle South Levantine tell strata. These two scholars produced the first comprehensive pottery studies that could be used by other scholars to date their sites (Whiting 2007a, 26). As a result, these typologies were relied upon to provide structure and coherency for the archaeology of the Levant.

An underlying principle behind the culture-history paradigm was that pottery held an inherent chronological value (Whiting 2007a, 78). While this chronological emphasis could help order a site's stratigraphy through time, the historical focus of early-twentieth century archaeology meant that changes in material culture were then equated with specific historical episodes as depicted in texts. Furthermore, these archaeologically-attested events, which were also linked to an ethnic group, became the cultural manifestations of ethnicity (Jones 1997, 19, 25). In other words, the pottery from a specific stratum within a site's stratigraphy represented a specific time and a specific ethnic group as gleaned from the historical texts (Shennan 1997, 217).

In addition, the European evolutionary view of progress and civilisation emphasised the historical link between the ancient Aegean and the Near East (**§2.6**). By implication, the linking of the material culture of these two regions was also assumed (Whiting 2007a, 78). But because culture had come to represent ethnicity, the cultural link between the Near East and the Aegean (i.e. evolutionary progress from one culture to the other) could only be established through a model that emphasised cultural diffusion by the movement of population. Hence, the archaeological appearance of new elements in material culture was reduced to the appearance of a new ethnic group (e.g. the Philistines were associated with Aegean-style pottery of the early Iron Age – Killebrew 2005, 14). Moreover, this link drew the eastern Mediterranean into a single cultural sphere, one where material culture was seen as a universal phenomenon and could be represented by a master ceramic sequence

(Orton *et al.* 1993, 34). There was no apparent reason why pottery from spatially-distant regions could not be compared.

The typological study of pottery, as understood and practiced by Albright and Wright, was quickly institutionalised within the archaeology of the Southern Levant. It was, however, a significantly later practice in the Northern Levant. Scholars working in the Northern Levant, who did not have access to local ceramic typologies, drew chronological inferences for local ceramic assemblages through direct comparison with South Levantine typologies (**§2.5**). The creation of local typologies in the Northern Levant was not common practice until the 1980s (e.g. Bikai 1978b; Riis 1948; Woolley 1939b). Pottery in the Northern Levant was dated by comparison with the Southern Levant and given a local "flavour" by appeal to the conventional historical narrative (e.g. "Phoenician" – Culican 1970; 1982; "Aramaean" – Seton-Williams 1961; 1967). By comparing ceramic assemblages with those of the Southern Levant, the culture-history paradigm as developed by Albright and Wright was introduced into the Northern Levant. Archaeologists working in the Northern Levant came to accept, consciously or not, that precise pottery analysis allowed a site to be dated and assigned to a particular ethnic group, and thus be made to reveal and therefore support history (e.g. Culican 1970; Fugmann 1958, 264).

While the culture-historical approach has its place, and has become instrumental in contributing to our knowledge of Iron Age ceramics, it is problematic for a number of reasons. Pots can tell us little about people in any real genetic or ethnic sense. According to ethnographic research, identity is negotiated in such a subjective and dynamic manner that it is difficult to define an individual's ethnicity according to a set of universal parameters (Barth 1969, 9-11; Goodby 1998, 161; S. Jones 1997, 51-55; Sherratt 1998, 294; 2005a, *passim*). Instead, ethnic identity tends to be entangled in the definer's own political preoccupations (Sherratt 2005a, 27). It is because ethnicity is such a vague and value-laden concept, that it cannot be used to define the physical or textual record. We should instead read textually-derived ethnonyms through the eyes of the people who created the texts (*ibid*, 36). Furthermore, archaeologists should not expect to find cultural indicators for clearly defined ethnic groups in the IA-NL because well-defined homogenous ethnic identities simply did not exist. Nevertheless, the modern conceptualisation of "ethnic identity" continues to influence archaeological method in the Northern Levant, where ethnonyms such as "Phoenician", "Aramaean" and "Syro-Hittite" are used to describe and define ceramic cultures: e.g. Aubet (2004b) published the Iron Age cemetery at Tyre al Bass as a "Phoenician Cemetery"; al Maqdissi (2003, Abb. 13) referred to the large pithoi from Tell Mishrife as "Aramaean" vessels; Doumet-Serhal (1992-1993) plotted "Phoenician" expansion across the Mediterranean through the distribution of certain ceramic forms.

Within the culture-historical paradigm the idea of "ethnicity" has been equated with the idea of national identity (Sherratt 2005a, 27). That ancient cultural-groups are seen as highly-bounded, geographically-discrete entities is problematic. Scholars have widely renounced the idea of a past world invariably made up of distinct, relatively homogenous nation-states (e.g. Jones 1997, 16-17; Sherratt 2005a, 27), yet there remains a tendency to present distributive regions of archaeological cultures as past abstractions of modern nation-states (e.g. United Monarchy – *cf.* Liverani 2005b, 308-323; Whitelam 1996, 122-175; "Syro-Hittite/Aramaean" states – Hawkins 1982, Map 14, p.374; Lipinski 2000a, 77*ff*). As a result, traditional cultural-historians seek to understand ceramic distribution across space in terms of *regionalisation*, where the boundaries between ceramic *regions* and cultural (ethnic/nation-state) groups were seen as coterminous (**Figures 4.1-2**).

There is another concern with this focus; in defining ceramic regions, archaeologists have tended to categorise pottery as either "local" or "foreign", whereas any "foreign" elements are viewed as evidence for economic or political interaction (e.g. the so-called "Aegean" element in early Iron Age pottery of the Northern Levant – *cf.* Mazzoni 2000a, 33-34; Sherratt 1998, 292*ff*; 2005a, 33-35). Hence, conventional histories of the ancient Near East have emphasised ethnic geography; plotting the ebb-and-flow, interaction and integration, of different cultural groups across space as evident in the diffusion of different ceramic elements (e.g. Bunnens 1995; 1999; Lipinski 2000b; Na'aman 1995). For instance, the presence of Greek imports has been widely used to plot mercantile interaction between the Aegean and Near East, with arguments often focusing on the actual presence of Greeks in the Near East (Boardman 1959; 2002; Courbin 1990a; 1990b; 1993b; Riis 1970, 129; Waldbaum 1994; 1997). The defining of ceramic regions has prevented the full development of an *explanation* of social processes and behaviours behind material culture patterning. *Regionalisation* ignores diversity and variation in local cultural behaviour and material culture, and instead focuses on broad, regional patterns. What is missing is an account of the social production of certain spatial distributions; i.e. what is it that generates regional ceramic patterns? Critical to such an undertaking is an account of the monitoring of social boundaries, across which certain material culture items move while others do not (see Bowser 2000; Dietler and Herbich 1998; Goodby 1998; Stark 1998; Stark *et al.* 2000).

As in the Southern Levant, the study of Iron Age pottery in the Northern Levant has placed a disproportionate emphasis on the chronological value of pottery (e.g. Lebeau 1983, 24). The means for determining the chronological value of specific pottery styles, however, relies on an archaeologist's ability to place that ceramic horizon, as it was stratigraphically-defined, within history. In other words, a ceramic vessel's chronological value was dependent upon the conventional historical narrative; in practice this was usually achieved through the correlation of archaeological strata with specific historical events. For example, Seton-Williams (1961, 82) interpreted the appearance of Red-Slip pedestal platters (characteristic of Hama E) in Tell Rifa'at Level IIc as evidence for a date in the late-eighth century BC. The end result is a ceramic chronology that is presented as independent and absolute, but which is intricately connected to, and completely dependent upon, the conventional historical narrative.

The emphasis on ceramics as chronological and cultural indictors has also meant that individual components of an assemblage have been emphasised over an understanding of the entire assemblage as a whole. For example, the ceramic sequence from the tell of Ras al Bassit has not been published in detail, yet the Greek imported pottery has appeared in numerous publications; Courbin (1986, 196*ff*) has suggested that these are indicative of Greek presence at the site (i.e. an ethnic indicator). The only local pottery from Ras al Bassit that has been published is the assemblage of Red-Slip Ware that Braemer (1986) studied for the development of this surface treatment in the Northern Levant (i.e. a chronological indicator). In the end, the Ras al Bassit pottery was not studied as a functional assemblage, only for its chronological and/or cultural value. A focus on individual components results in a de-emphasis of variability within and between assemblages.

It is clear, from the above, that current interpretations of the ceramic record for the IA-NL are based on the approaches used in culture-historical archaeology, with a superficial grafting-on of processual/diffusionist theory. This approach is matched by an apparent unwillingness to engage with the archaeological record in all its messy diversity, and investigate the complex relationship between people, social structures, and material culture. The superficial unity of prescribed cultural units, as well as the focus on ethnic groups such as "Phoenicians" and "Aramaeans", has distracted scholarly attention from the existence of significant diversity and complex patterning within the material culture of the IA-NL.

4.3 The Study of Pottery Beyond the Eastern Mediterranean

While cultural-history remains embedded in archaeological practice for the IA-NL, other branches of archaeology have made significant progress in appreciating the complex relationship between people and material culture. The following discussion will explore the manner by which pottery is used, outside the eastern Mediterranean, to provide a more dynamic interpretation of the archaeological record. By outlining what is being done in other areas of the globe, the possibilities of what can be done in the Northern Levant will be emphasised. In particular, the study of pottery in British and North American archaeology has been significantly influenced by interpretative methodologies based on an awareness of diverse social practices (e.g. Barrett 1994; Bowser 2000; Dobres and Hoffman 1999a; Dobres and Robb 2000).

Developments in the Approach to Material Culture

The culture-historical concern with chronology and ethnic groups was largely abandoned in European and North American archaeology in the 1960's, when the "New Archaeology" shifted emphasis from static description to explanation of social processes and change (Binford 1962; 1972). The study of pottery, therefore, sought the explanation of economic and social mechanisms, such as trade and exchange, ceramic technology, and vessel function (Orton *et al.* 1993, 23-35; Whiting 2007a, 78-79). But in prioritising a search for social process, processual archaeology had imposed a rigid functionalist conceptualisation of culture as a universal adaptive mechanism (Shanks and Tilley 1987b, 94). As a result, human agency was subordinated to environmental determinism, within which people were depicted as culturally-determined automatons (Jones 1997, 117). In the end, functionalist approaches were unable to account for cultural diversity or social change.

In *The Constitution of Society: Outline of the Theory of Structuration* (1984) Giddens argued that society is created and maintained through the actions of knowledgeable human agents, whose actions are in turn constrained by patterns of behaviour learnt and deemed appropriate within that society. Giddens (1984, xxiii) was emphasising the human capacity to understand their actions as social agents, instead of merely as cultural puppets. He argued that people create the conditions and structures in which they live and that their actions are meaningful within that given context. Furthermore, the building of social structures is an ongoing and recursive process that is never really complete – always "in process" (Dobres and Hoffman 1999b, 3; Hodder 1987, 6). In other words, social agents are socially-embedded people interacting between the structures in which they exist and, paradoxically, which they create (see Dobres 2000, 4; Dobres and Robb 2000). Gidden's theory of *Structuration* therefore appreciates the way in which social structures are both the medium and outcome of their production through human action. Or as Geertz (1973, 93) has phrased this point, "all these categories of evidence are the remains of models *for* reality as well as models *of* reality".

Like Giddens, Bourdieu (1977) proposed a theory of "practice" in which social agents were both structuring and structured. Unlike Giddens, however, who did not identify specific arenas of social discourse, Bourdieu (1977, 89) emphasised that an individual's awareness of their own social context is both socially and *materially* defined. For Bourdieu, inhabited space is the locale where understanding is generated – a concept he calls

habitus – which is dependent on an individual's understanding of social context, as experienced through the same individual's senses. As such, the physicality of the human body and the world is a primary reference point; material culture therefore has no single, objective meaning. Instead material culture is imbued with many meanings, dependent upon the many discourses into which it is drawn, since meaning only exists in the moment of human agency (Bourdieu 1984, xiii, 1, 170).
In 1997, Ian Morris (1997, 3) sought a return of archaeology to culture-history; not the culture-history as archaeologists had perceived it (Morris 2000, 19), but the New Culture History that had come to recognise material culture and text as human responses to social events, usually in an effort to reshape those events for individual benefit (Brumfiel 2000, 249; Hunt 1989, 7-9; S. Jones 1997, 125-126; Morris 1997, 8). The ancient Greeks themselves make it clear in their writings that they saw material culture as being just as potent a medium in the construction of identity as the spoken word (Morris 1997, 11). Hence, Morris (*ibid*) concluded that all aspects of material culture can be viewed as symbolically constructing and contesting social categories. But taking cultural history seriously means thinking on all three temporal levels described by Braudel (1972, 21; 1980, 25-54): geographical time (*long dureé*), social time, and human time. Archaeologists, until recently, had virtually ignored human time, which Barrett (1994, 47) suggested was due to the way archaeologists thought of "individuals… as given, pre-existing the material consequences of their actions." Instead, we should "move away from asking what kinds of people made these conditions?", to an understanding of what the possibilities were of being human within those material and historical conditions (Barrett 1994, 4-5). In other words, objects may have had different meanings for different people (Morris 1992, 17-18).

Aggregate approaches to material culture interpretation are too simplistic because they fail to recognise the importance of subjective, knowledgeable agency. However, individuals who made, used or witnessed material culture did not necessarily understand it in the same way. Hence, the relationship between people, material culture and social structures is not static and cannot be deduced easily. A single historical reality cannot be determined, primarily because *it* never existed (Barrett 1994, 169, 171). As a result, context is vital to the interpretation of material culture as archaeologists seek to understand the fluid meaning ascribed to objects within a specific social context, as represented by the material realities within which objects are found. This approach has obvious implications for excavation strategy, recording and publication. Furthermore, it means each assemblage should be considered as a whole, and not just key components of these assemblages studied in isolation.

Implications for the Study of North Levantine Pottery

A socially-embedded agency approach to the study of pottery has been successfully applied to archaeological ceramics from various sites and regions across Europe and the Americas in recent years (e.g. Bowser 2000; Dobres 1999; Goodby 1998). Although pottery is considered an important and informative element of material culture in Levantine archaeology, the study of Iron Age pottery from the Northern Levant remains directed towards chronology, ethnicity, and cultural diffusion. The developments in ceramic studies in European and American archaeology since the 1960s highlight a number of methodological and analytical issues that have not been addressed in ceramic studies of the IA-NL. In particular, there are a number of practicalities preventing a more dynamic interpretation of pottery.

One of the more crucial issues that has not been addressed in the study of pottery from the IA-NL is the manner by which ceramic typologies are created. Typologies are fundamental in creating order out of ceramic data, but the nature of this "order" is rarely considered (Whiting 2007a, 78-79). The typology is seen as a "natural" component of material culture rather than the result of both ancient and modern behaviours (Sørensen 1997, 182). From where do the ceramic categories derive? Archaeologists appear to be unaware of the relationship between a modern typology and the social reality that the ceramic vessels emerged from in the past. This is especially the case for the archaeology of the IA-NL, where the lack of understanding of archaeological processes due to the reliance on textual sources means that different pottery types are simply regarded as type fossils for particular chronological periods and ethnic groups (**§4.2**). Furthermore, the categorisation of pottery appears to assess the similarity between objects as if the similarity itself was *the* meaning behind material culture patterning. However, the aim of a typology is not to understand similarity or dissimilarity, but to measure where similarity stops and dissimilarity begins.

Another important area in need of address is the description of ceramic fabrics. The fabric of pottery is rarely presented in detail within published reports of IA-NL excavations. To take a recent example, Núnez's (2004b) typology for the Iron Age cremation cemetery at Tyre-al Bass includes detailed discussions of surface treatment, decoration, shape, and parallels, but makes no mention of the vessel fabrics (see also Schreiber 2003). The few reports that do discuss fabric do so in a summary manner; they simply state a general colour (e.g. orange – see Cecchini 1998 sherd descriptions opposite her figures) or generic term (e.g. common-ware, cooking-pot-ware – see Mazzoni 1998a, 166-169) without compiling a ware series with detailed description (*cf.* Whincop 2007, Tab. 1). This tends to emphasise general similarities and negate any subtle differences in fabric within and between sites. Instead, a discussion of fabric should include a systematic investigation and recording of *all* its constituents; i.e. paste (colour, density, particle size), inclusions (type, colour, size, roundedness, concentration, sorting), firing technology, and surface treatment. Fabric analysis is crucial to the study of trade-patterns, raw material sources, production methods, technological constraints, and vessel function (Orton *et al.* 1993, 132-135). The lack of fabric analyses means the study of pottery can rarely move beyond description. While a small number of petrographic investigations of IA-NL pottery have been undertaken in recent years, the results are usually centred on vessel origins for the purpose of plotting cultural and population diffusion (Lagarce and Lagarce 2000), rather than for the identification of socio-technological behaviours. Hence, even when more progressive and scientific methods are adopted, they are made to fit existing methodologies, and are not used to challenge traditional frameworks.

There also seems to be an unawareness that the social and cultural context of pottery is essential to understanding a vessel's significance. The meaning of material culture is neither coincidental nor inherently objective, but constantly changing according to context. It is therefore only possible to understand the meaning of pottery in conjunction with an awareness of immediate context. The publication of IA-NL pottery rarely includes a systematic description of contexts; usually only the "significant" finds (vessels or architecture) warrant a detailed reconstruction of loci. For example, the majority of pottery from the Hama E "royal quarter" was simply assigned to a building, while the find-spot of the two Greek *skyphoi* sherds were recounted in detail (Riis 1965, 80. n.5-6). A fresh concern with context is likely to highlight the dynamic meanings behind ceramic production, function, use, consumption, and discard. Furthermore, since social boundaries are abstractions and ideological constructs, recognised differently and for different reasons by different people, an understanding of the social and cultural context of pottery might begin to highlight elements of group identity (Goodby 1998, 161; Stark 1998).

Though fabric, context, and typologies might be considered elementary components of archaeological ceramic analysis, work on the IA-NL has continued to rely on static description. The publication of Iron Age ceramics is often unsystematic, and cursory, with only an arbitrary selection of "important" or "significant" finds sometimes making it into publication, usually without explanation of selection procedure (e.g. Courbin 1990a; 1993b). Quantification of the complete assemblage is rarely included (e.g. Anderson 1988; Bikai 1978b), and comprehensive typologies are only a recent phenomenon (e.g. Núñez 2004b). The excavation and publication practices in the Northern Levant have meant that the spatial and functional analysis of ceramic assemblages is virtually impossible.

4.4 Conclusions

Traditional interpretations of pottery from the IA-NL owe less to close consideration of the data than to the tenacity of a disciplinary tradition that has continued to employ rather simplistic concepts regarding the meaning of material culture. The direct correlation of peoples and events in the historical narrative with material culture has resulted in a paradigm for pottery analysis that emphasises only the ethnic and chronological value of pottery. Hence, typo-chronological analyses of Iron Age pottery dominate the discipline.

Establishing meaningful reconstructions of the Iron Age requires careful stratigraphic and typological study of the archaeological record. Iron Age reconstructions based on the conventional historical narrative provide only one particular perspective on "history", a perspective that cannot be affirmed until the archaeological data is sufficiently interrogated so as to extract a framework that can be used with confidence. The ubiquity and dynamic nature of pottery implies that this is one of the most important resources for telling us about the lives of the people who used it, but to a large extent the potential is untapped. The solution is for archaeologists to actually engage with material culture in all its messy, contradictory reality, and seek explanations for cultural traits outside of the historical narrative.

While pottery was indeed chronologically sensitive and likely to reflect elements of cultural identity, ceramics were also an active agent in the negotiation of social structures. To this end, new models of cultural change should emphasise the active role of pottery as social agent. Such an approach warrants a close engagement with *all* elements of material culture, rather than a superficial analysis of isolated categories of "significant" artefacts. Practically speaking, this means that archaeologists need to investigate the cultural context of all ceramic categories, and understand the complexities within a ceramic assemblage as a whole.

The overall aim of the present study is to undertake an analysis of the current flawed ceramic dataset to highlight the potential of alternative approaches to the study of Iron Age pottery for the Northern Levant. The intention was to demonstrate the value and potential of interpretations not based on the historical narrative. While no analytical method can be considered "perfect", Chapters 2, 3, and 4 have shown that traditional interpretations of the archaeology are inadequate and add very little to the conventional history. So although imperfect, it is intended that the methods employed for the analysis of pottery throughout this study will demonstrate that new insights are possible even using the current unsatisfactory data.

SECTION TWO

Presentation of Iron Age Ceramic Data

CHAPTER FIVE
Form and Nature of the Ceramic Data

5.1 Introduction

In recent years, pottery has come to represent the key artefactual material for the archaeological investigation of the IA-NL. Considering the importance attributed to this artefact class, and expenditure of resources in its study and publication, it is surprising that pottery is often presented in such a cursory manner. Pottery is one of the most important resources for telling us about the lives of the people who used it, but to a large extent the potential is ignored. The means for studying pottery has changed little since the first ceramic typologies were created for the Southern Levant early last century. Consequently, the study of Iron Age pottery has tended to focus on its chronological and cultural value. Closely intertwined with this approach is a view of pottery as the passive product of society; how these vessels were used to define these societies has largely been ignored. In other words, it has been the material itself, not the manner by which it was produced, consumed, and discarded, that has been the focus of archaeological enquiry.

The purpose of this chapter is to outline the nature of the ceramic data that was investigated in the present study. It will provide an account of how the ceramic data was conceived, collected, classified, and stored, and is designed to provide the "background" for the later analyses. A brief overview of the electronic database is also presented, so as to help the reader envisage various weights and biases in the data. The charts presented throughout this chapter provide only broad summaries of a few categories; the author feels the need to stress that these are not intended to present an analysis of the data, which is undertaken in more detail in **Chapters** 7 and **8**. A central theme in this chapter is that the level of detail originally desired for this study was not available, either because published data was insufficient or the collection of new data was limited, for whatever reason. Hence, the current study is an exercise in testing the limits of the current corpora of published data in all their insufficiencies and contradictions.

5.2 Collection of Data

We have seen from **Chapter 2** that the Northern Levant was not a single, well-defined cultural entity during the Iron Age (**§2.2**). From the outset, the present study sought to avoid imposing rigid borders onto the data. For this reason, a study-area was proposed that was significantly larger than the modern borders of Lebanon and Syria. It was hoped that this approach would allow the ceramic data to reveal their own cultural borders and limitations (**Maps 1** and **2**). The original aim of the study was to visit ceramic archives from between 10 and 15 sites that had produced significant assemblages of Iron Age pottery within the study area. These key sites were to provide detailed information that could then be extrapolated for a broader perspective of the whole Northern Levant. Archive visits were initially envisaged as the primary means for data collection for this study. It was hoped that an analysis of fabric, surface treatment, manufacturing technique, colour, and shape could be undertaken firsthand and the detailed data be collated into a single database. Access to ceramic archives, however, proved to be a complicated and inconsistent process. Determining where each site's ceramic assemblage was stored was rarely straightforward. The pottery from a number of sites had been lost (e.g. Aleppo citadel; the tell at Akhziv) or broadly dispersed (e.g. Byblos; Deve Höyük; al Mina); while at other sites much of the pottery was discarded (e.g. Arslan Tash). Moreover, identified ceramic archives were not always available for firsthand study – usually due to time constraints, limited access, or lack of an appropriate contact. As a result, the number of archives available for study was significantly smaller then originally envisaged. The first few archive visits in 2004 seemed to confirm the need for a change in strategy; during these visits access to material varied greatly. It was clear that the data to be obtained from archive visits and first-hand study of pottery was likely to be greatly varied in quantity and quality. As a result, it was decided that the current study would instead have to rely largely on published data; rendering a meaningful study of fabric and technology impossible.

The revised strategy was to identify Iron Age excavations in the study area that had published ceramic assemblages; 54 sites in total. Data collection, therefore, took the form of obtaining all publications of ceramic material from those excavations. Nevertheless, site and archive visits remained an important means for supplementing the published data (the author is grateful to many project directors and excavation staff for their time and access to published and unpublished material). Consequently, the resulting ceramic database is reliant on the minimal level of information present in pottery publications. While some publications include comments on fabric, and manufacture, the majority is usually limited to a description, or illustration, of decoration and shape Furthermore, data analysis for this study had to be based on presence/absence data alone, as published ceramics rarely included sufficient statistical information to allow a quantified investigation.

5.3 The Ordering of Data

From the published and unpublished ceramic catalogues, photocopies of over 12,000 Iron Age vessels were collected from the study area, with each site, context and publication reference noted on the back. Before the ceramic material could be collated and analysed, however, it was important to categorise and assess the data.

The first task of categorising the data involved the construction of a ceramic typology for the entire dataset, the practicalities of which are discussed in Chapter 6. Suffice to say here that no existing typologies were imposed on to the data. Rather, the ceramic material was categorised according to overall profile (e.g. hole-mouth cooking-pot), followed by more-subtle variations in form (e.g. bevelled lip, thickened lip). While the end result has parallels with Lehmann's (1996) typology, it is not directly influenced by either his methods or conclusions. The second task of assessing the data was concerned with the reliability of "contexts". For this purpose, a critical review was undertaken for each site in the study area (**Chapter 3**). It was important that unstratified (e.g. Byblos, Nayrab) and temporally inconclusive (e.g. Khirbet Silm; Tell Rachidieh Tomb II) contexts had minimal influence in determining diachronic patterns. Once the typology was constructed and the stratigraphic survey complete, the 12,000 ceramic vessels were ready for entry into an electronic database that would aid spatial and temporal analysis. For this purpose, an MS Access database was constructed, the structure of which is outlined below.

MS Access Database

In constructing an electronic database for this type of data there were a number of variables that had to be considered. In particular, the database needed to incorporate different levels of contextual, temporal, and typological information, whilst also allowing for each individual vessel's distinctiveness. To this end an MS Access database was created by linking four primary tables; SITE TABLE, CONTEXT TABLE, RIM TABLE, and MAIN TABLE (included on appended CD). **Figure 5.1** visually depicts the relationships of the four primary tables and as well as the many "look-up" tables.

Figure 5.1: Screenshot of Relationships in MS Access Database

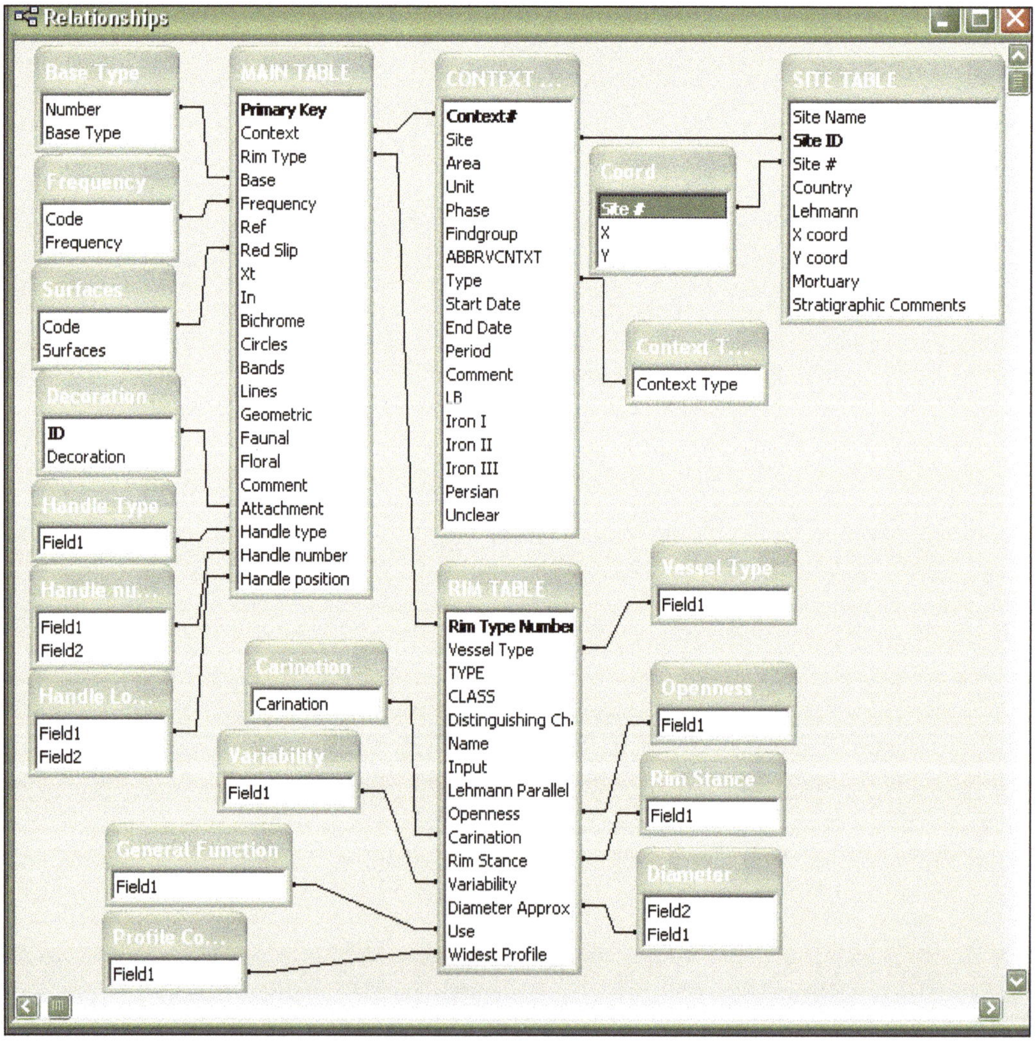

Figure 5.2: Screenshot of MAIN TABLE (design view)

MAIN TABLE : Table

Field Name	Data Type	Description
Primary Key	AutoNumber	
Context	Number	Linked to CONTEXT TABLE
Rim Type	Number	Linked to RIM TABLE
Base	Text	Linked to Base Type table
Frequency	Number	1=Present; 2=common
Ref	Text	Publication reference
Red Slip	Number	Which surface, if any?
Xt	Yes/No	Exterior Paint
In	Yes/No	Interior Paint
Bichrome	Yes/No	If No then paint is monchrome
Circles	Yes/No	Type of painted motif
Bands	Yes/No	Type of painted motif
Lines	Yes/No	Type of painted motif
Geometric	Yes/No	Type of painted motif
Faunal	Yes/No	Type of painted motif
Floral	Yes/No	Type of painted motif
Comment	Text	
Attachment	Number	Functional/ decorative alterations
Handle type	Text	
Handle number	Number	None, one, two, or more
Handle position	Number	Where on the vessel?

The SITE TABLE contains base level information for each site; name and geographic co-ordinates being the most important. The co-ordinates, which were important for the production of distribution maps in ARCGIS, were not derived from any "official" source (as many sites do not appear on "official" maps), but were approximated from maps in excavation reports. The SITE TABLE contains 54 different sites, linked to over 600 contexts in the CONTEXT TABLE.

The CONTEXT TABLE contains information relevant to each individual context within the database. Each entry represents either a whole site (e.g. Joya), a broad occupational phase (e.g. Megiddo VIA), or a specific Area and Level (e.g. Tell Afis Area E_1 Level 9c); the level of detail was determined by publication. This table is linked to the MAIN TABLE via the unique "Context#", as depicted in **Figure 5.1**. Another key field was the context "Type" field, which recorded the nature of each context; i.e. mortuary or settlement. The Yes/No "check-boxes" for the different periods were the means for recording a context's chronological value.

The RIM TABLE contains details of the ceramic typology. Rim and general profile characteristics are described according to terms chosen from linked "look-up" tables (**Figure 5.1**). The RIM TABLE contains 264 entries representing 193 final CLASSES. The base form and overall decoration were not included in the RIM TABLE as these were not always considered characteristic of a vessel's CLASS.

The MAIN TABLE is where other tables come together to represent the presence of a ceramic incident. Each entry represents a combination of CLASS, decoration, base, and attachment. Hence, each entry could represent a single vessel or group of like-vessels, as long as all constituent parts were the same; the "Frequency" field represents only presence or common presence, rather than actual numbers. In the end, 12,000 pots (not all could be assigned to a CLASS) were represented by c. 8000 incidents. Each of the MAIN TABLE fields is briefly described in **Figure 5.2**.

WinBASP

To aid analysis of the data (**Chapters** 7 and **8**), the dataset was also entered into the Bonn Archaeological Software Package, version 5.43 for Windows (WinBASP v. 5.43). The Bonn Archaeological Software Package (BASP) is a non-profit software project for and by archaeologists which has been developed cooperatively since 1973. It includes more than 70 functions for seriation, clustering, correspondence analysis, and mapping tools for archaeologists and is available to download from http://www.uni-koeln.de/~al001/. WinBASP was used to undertake Seriation (**§7.2**), Correspondence Analysis (**§8.2**) and Cluster Analysis (**§8.3**) of the ceramic dataset. Without going into detail here, **Figure 5.3** depicts WinBASP data-entry. The "window" on the left contains the contexts (or "Units"), while the "window" on the right contains each ceramic CLASS (or "Type"). The middle "window" lists the database incidents; either of Types in Units, or Units in Types. While WinBASP contains a facility for importing data, version 5.43 was unable to read MS Access files. Consequently, the entire database was entered manually into WinBASP. The function of this program is generally intuitive and easy to use, and was a valuable resource for the present study.

Figure 5.3: Screenshot of WinBASP data-entry screen

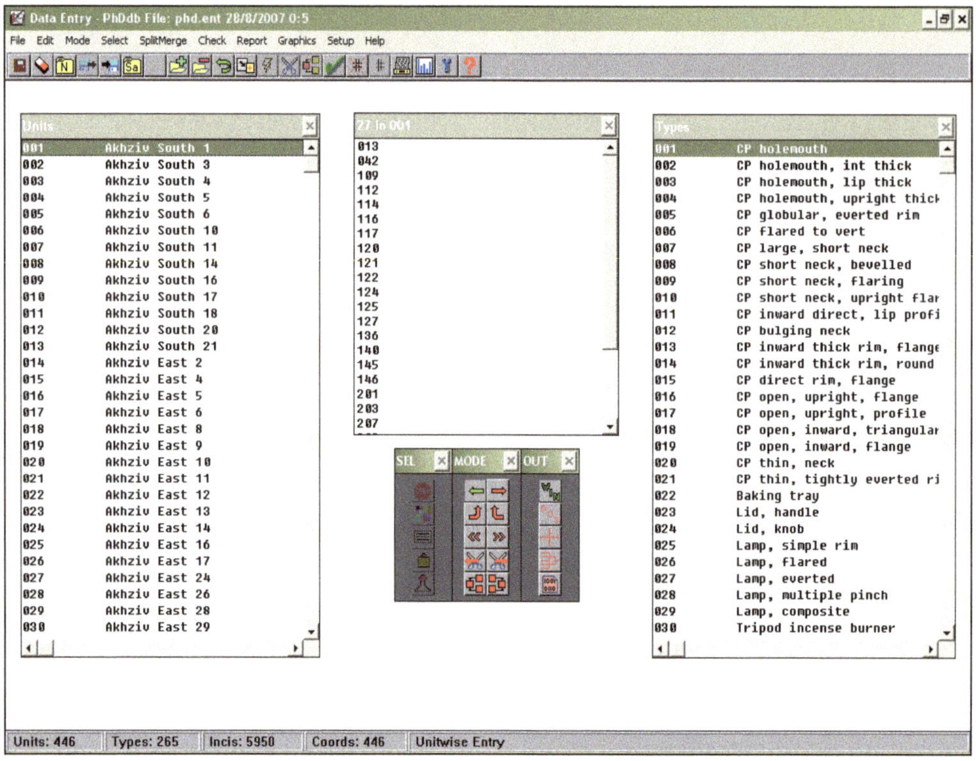

5.4 The Data

The collection, classification, and collation of the dataset was a significant undertaking within the scope of this thesis. The complete dataset is included on the appended CD (please view the readme file before opening the database files). The dataset consists of c. 8000 different incidents representing c. 12,000 vessels across c. 600 different contexts from 54 different sites. The database represents vessels from mortuary and non-mortuary contexts alike, though mortuary contexts represent only one fifth of all ceramic incidents (**Chart 5.1**).

Chart 5.1: Proportion of database incidents - context type

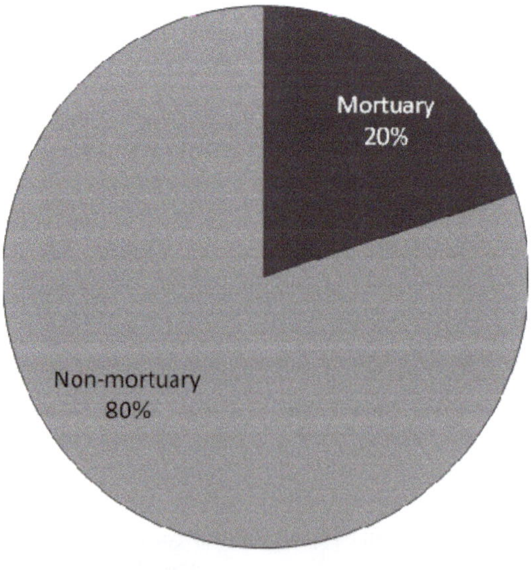

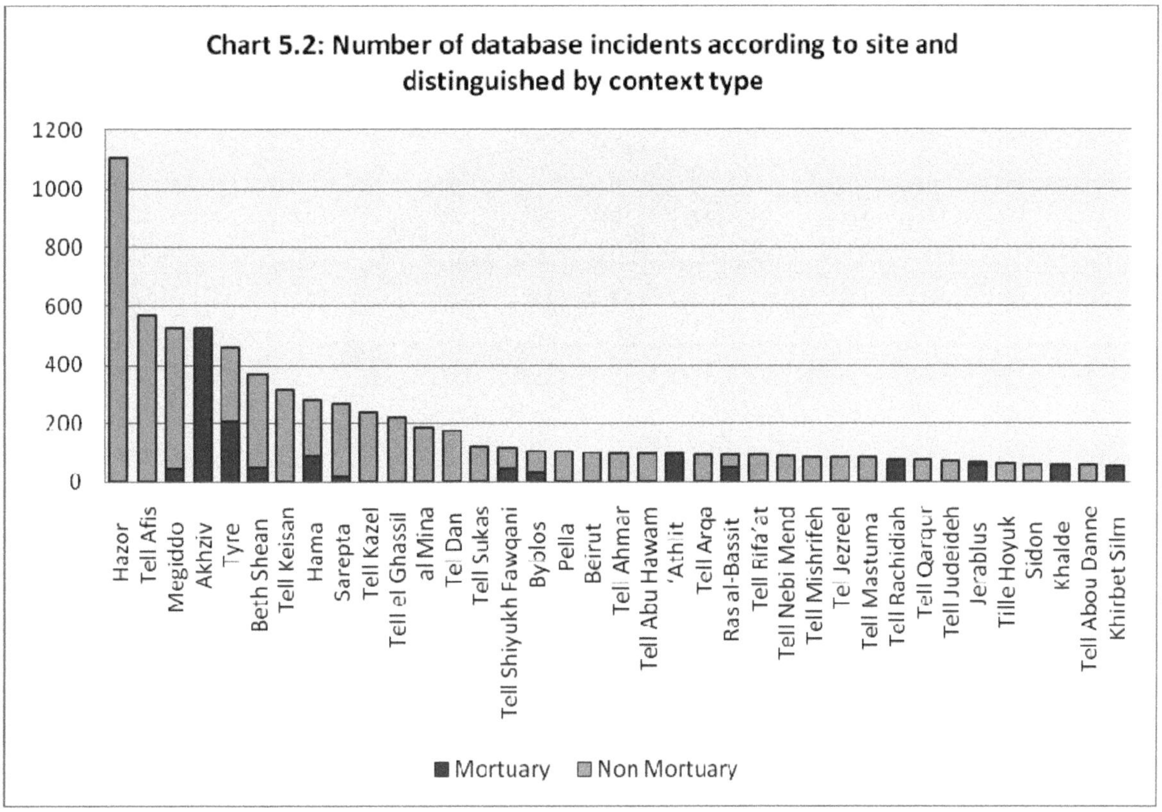

Chart 5.2: Number of database incidents according to site and distinguished by context type

Though the majority of Iron Age contexts in Lebanon were mortuary contexts, many of these sites produced comparatively small amounts of pottery. This is clearly shown in **Chart 5.2**, which depicts the number of incidents per site. This chart also highlights the large contribution to the database made by sites in the Southern Levant, with five of the seven largest assemblages coming from sites in this region. In sharp contrast, the inland Northern Levant has only two large assemblages amongst the ten biggest (i.e. Tell Afis and Hama). This bias in the database emphasises the general lack of knowledge regarding Iron Age pottery from across the Northern Levant. Unfortunately, a chart that includes 54 sites on one axis is difficult to present clearly, therefore sites with less than 50 incidents were not included in **Chart 5.2**; these are, in decreasing order, Tell Ta'yinat, Joya, Kamid el Loz, Tell Sheikh Hassan, Tell Jurn Kabir, Ain Dara, Zincirli, Ras Ibn Hani, Deve Höyük, Chatal Höyük, Qraye, Nayrab, Tel Rehov, Kefrik, Qasmich, and Tambourit. Together these sites accounted for only 133 mortuary incidents, and 264 non-mortuary incidents; or less than 0.5% of the entire database.

To further illustrate the nature of the database, each of the 54 sites were attributed to one of four broad regions; inland Southern Levant, inland Northern Levant, the Mediterranean coast, and the Beqa' Valley. These somewhat arbitrary regions were designed only to loosely group the data to identify any broad spatial bias in the data. The two Beqa' Valley sites were grouped separately because it was not clear to which "region" they belong. Two pie-charts were produced; the first (**Chart 5.3**) represents the percentage of sites in the database according to region, while the second (**Chart 5.4**) depicts the percentage of database incidents according to region.

Chart 5.3 illustrates a clear predominance of North Levantine and coastal sites within the database; as one might expect from a study of inland and coastal Northern Levant. However, when compared to **Chart 5.4**, the small amount of Southern Levantine sites contributed a disproportionately high percentage of incidence data, which reflects much more extensive publication of Iron Age pottery from this region. The coastal and Beqa' regions contributed a number of incidents roughly equivalent to their site representation, while inland Northern Levant was poorly represented. These charts demonstrate the presence of a geographic bias in the data, one that was largely unavoidable because of the infrequent and only partial publication of ceramics from the IA-NL.

While, **Charts 5.1-5.4** present an overview of the data according to context; **Charts 5.5-5.6** present a brief overview of the data according to ceramic form. **Chart 5.5** depicts the percentage of incidents within the database according to vessel function, using the broad functional categories outlined in the typology (**Chapter 6**). What is immediately obvious from this chart is that bowls constitute a large percentage (c. 40%) of all incident data, more than twice as many incidents as the second most-common category, jugs. To make **Chart 5.5** easier to read, the four functional categories with the lowest representation were grouped together to form the "Other" category (i.e. "Assyrian" forms; Bottles, Unguent containers, and Spouted vessels).

Chart 5.3: Percentage of sites in database – broad region

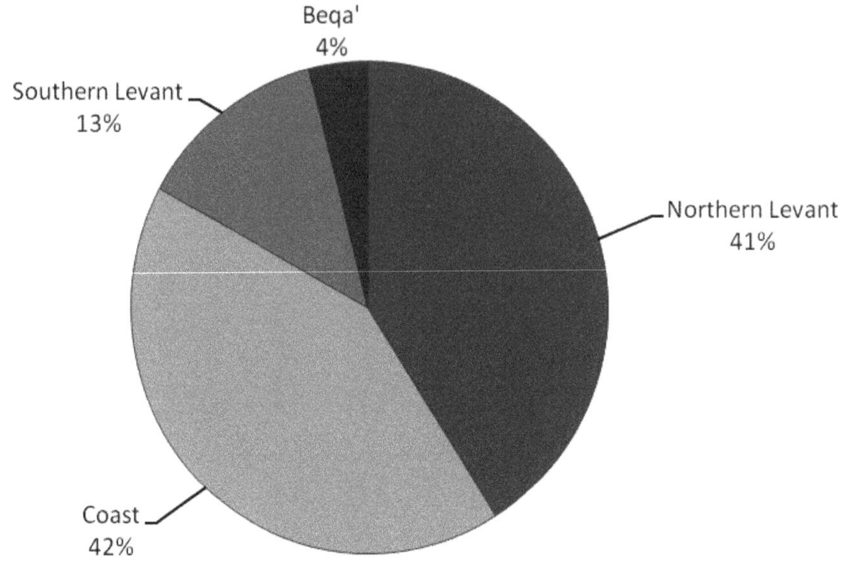

Chart 5.4: Percentage of incidents in database – broad region

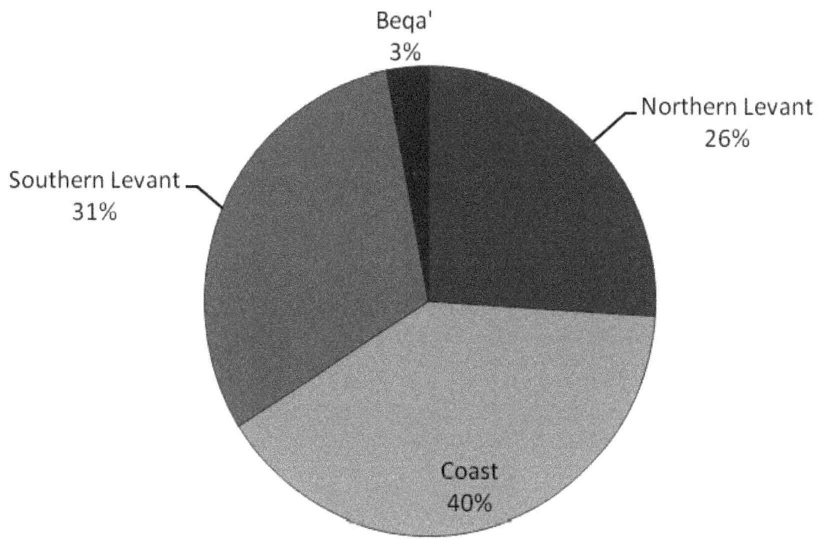

The trends in **Chart 5.5** are generally self-explanatory. However, the same data becomes even more interesting when regional divisions are introduced. Consequently, **Chart 5.6** depicts the percentage of each functional category according to the four broad regions outlined above. To take one example, it shows that less than 1% of all bowls derived from Beqa' Valley contexts, 36% from coastal contexts, 34% from the Northern Levant, and 29% from the Southern Levant. There are a few points worthy of brief comment here, but the majority of patterns are discussed at greater length in **Chapters** 7 and **8**. The fact that the vast majority of jugs were found along the Mediterranean coast is all the more significant considering jugs are the second-most common functional category in the database. The pouring of liquids with ceramic jugs appears to have been an important element in coastal communities, a point emphasised by the predominance of bottles, juglets and flasks in this same region. Transport amphorae were also concentrated on the coast. Pithoi, on the other hand, were predominantly found across the inland Northern Levant. Kraters, cooking-pots, and urns have a presence in each region roughly equivalent to the proportion of incident data for each region (*cf.* **Chart 5.4** and **Chart 5.6**).

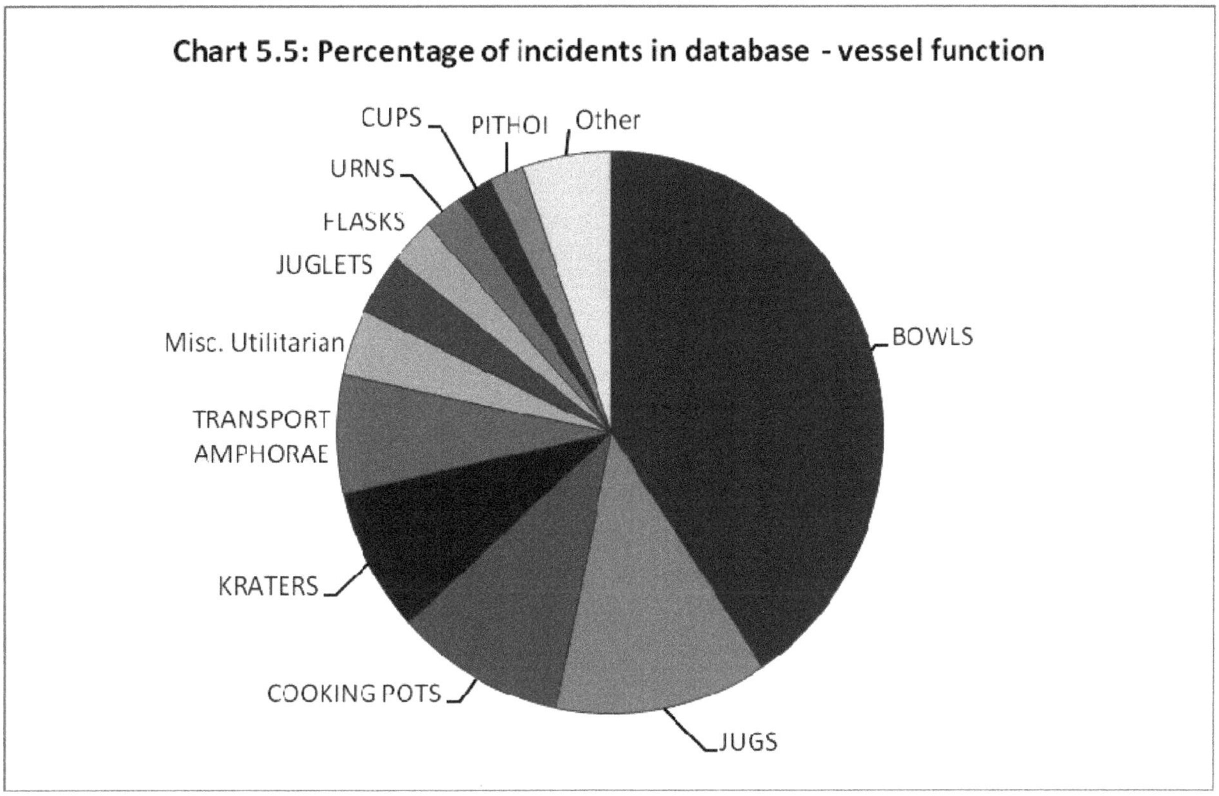
Chart 5.5: Percentage of incidents in database - vessel function
CUPS
PITHOI
Other
URNS
FLASKS
JUGLETS
Misc. Utilitarian
BOWLS
TRANSPORT AMPHORAE
KRATERS
COOKING POTS
JUGS

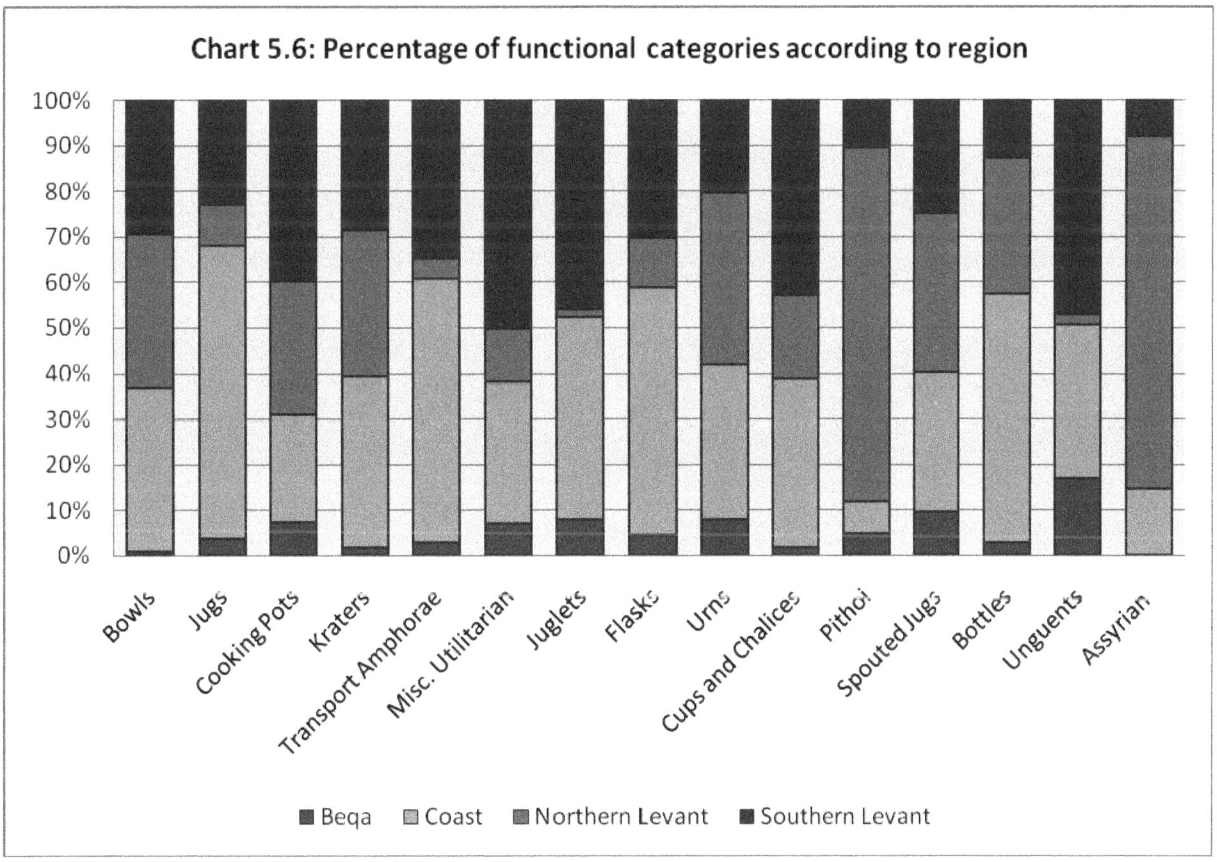
Chart 5.6: Percentage of functional categories according to region
100%
90%
80%
70%
60%
50%
40%
30%
20%
10%
0%
Bowls
Jugs
Cooking Pots
Kraters
Transport Amphorae
Misc. Utilitarian
Juglets
Flasks
Urns
Cups and Chalices
Pithoi
Spouted Jugs
Bottles
Unguents
Assyrian
Beqa
Coast
Northern Levant
Southern Levant

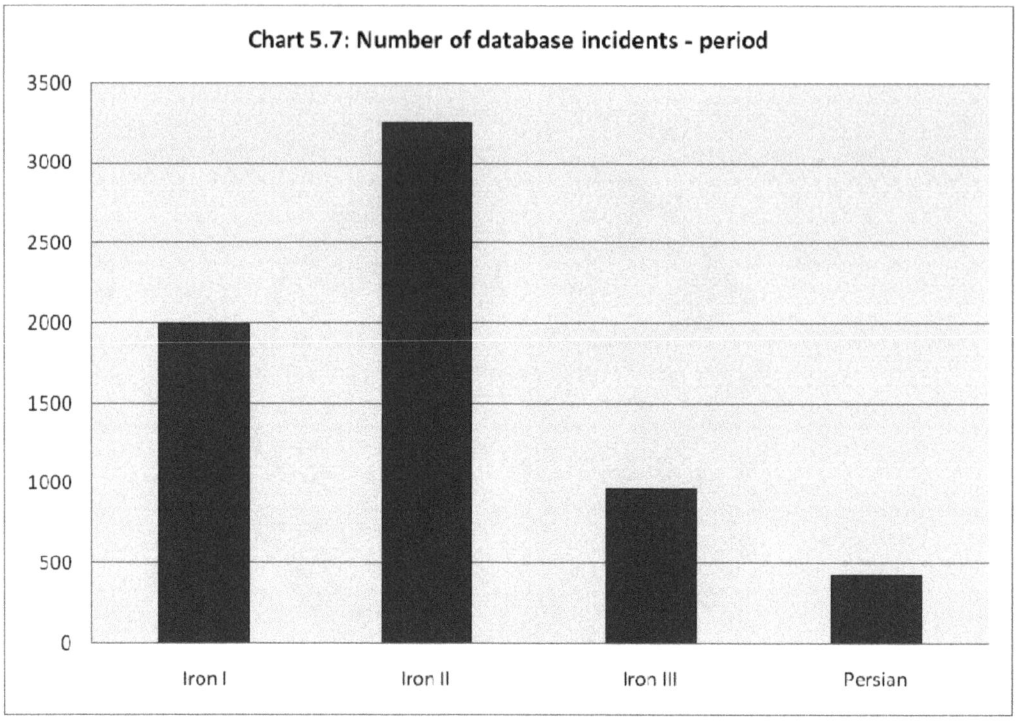

In addition to the regional bias in the data, there is an apparent temporal one. To demonstrate this point, all well-defined contexts associated with each period were isolated. **Chart 5.7** depicts the number of incidents according to period as derived from period-specific contexts, while **Chart 5.8** displays the number of sites represented by "secure" contexts within each period. What should be immediately evident from **Chart 5.7** is that the vast majority of "temporally-distinct" incidents in the database derive from the Iron II period. This is due to the fact that Iron II contexts were identified at more sites in the study area (**Chart 5.8**; **Maps 3-6**). The number of Iron I incidents is also directly proportional to the number of sites with "secure" Iron I contexts. The Iron III and Persian periods, however, have significantly less representation in the database, despite a reasonable number of sites with related contexts. To view this from a slightly different perspective, Iron I contexts were identified at 25 sites in the study area, accounting for 1995 database incidents; i.e. an average of 80 incidents per site. The Iron II period was similarly represented with 79 incidents per site. In comparison, the Iron III and Persian periods were under-represented, with 45 and 20 incidents per site, respectively.

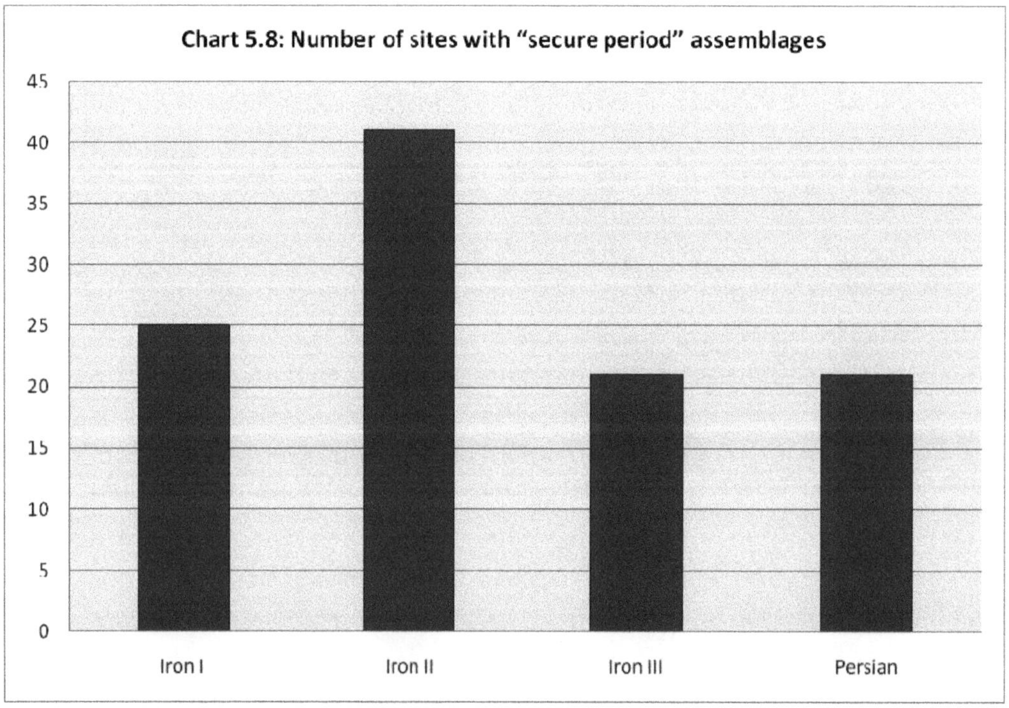

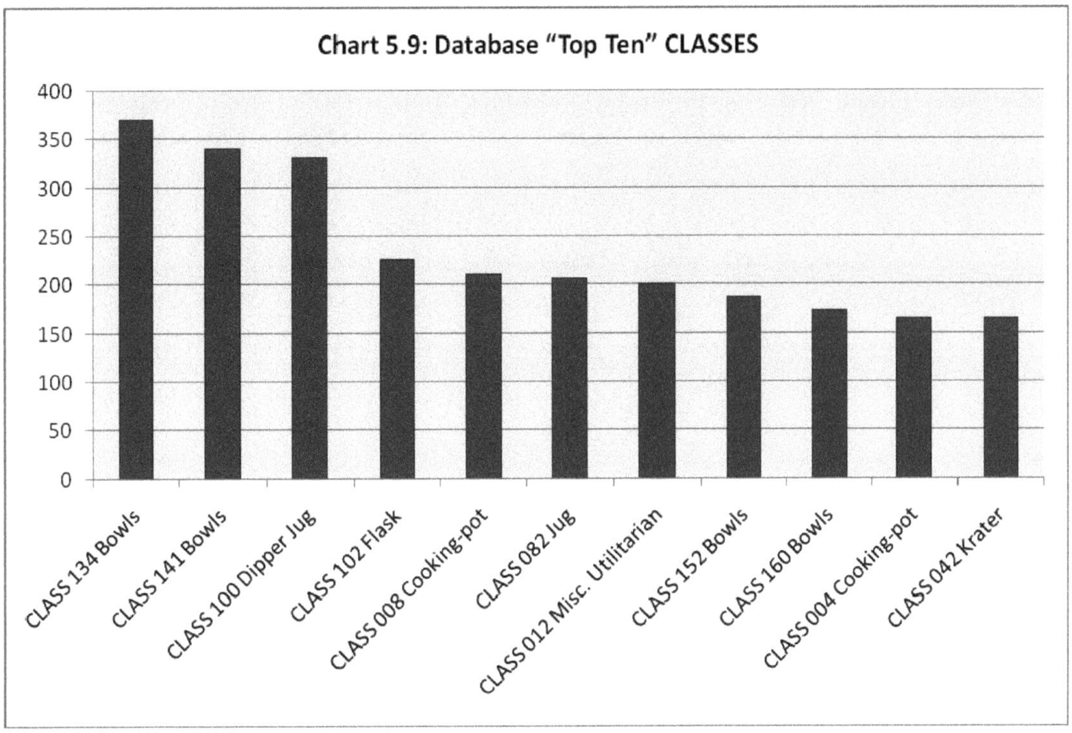

Chart 5.9 presents an overview of the most frequently encountered ceramic CLASSES within the database. Most of the bowl CLASSES within this chart are well-distributed across the study area. On the contrary, the most abundant pouring and cooking forms were generally restricted to the coastal and Southern Levant zones (e.g. CLASS 082 Red-Slip trefoil lip jugs; CLASS 042 kraters; CLASS 102 Pilgrim Flasks; CLASS 100 Dipper juglets).

The above collection of charts was intended as a visual introduction to the nature and weight of the data. The evident patterns raised some interesting questions and helped direct later investigation within the database, as well as going some way towards explaining apparent trends in the data. These charts also highlight the potential insight that unconventional approaches might bring to the study of Iron Age ceramics.

5.5 Limitations in the Data

While the excavation of Iron Age strata has intensified in the Northern Levant, publication remains sporadic and inconsistent. Hence, the dataset used for the present study had a number of unavoidable limitations. Firstly, for both the published and unpublished pottery, quantitative information was insufficient. The criterion by which archaeologists collected, discarded and published the pottery was rarely known, though one may assume that these criteria were not the same. As a result, the typology recorded only the presence or absence of a particular CLASS within a particular context, however that was defined. Secondly, contextual data was inconsistent. While this study was originally intended to provide quantified contextual data, this could not be satisfactorily completed. Such an approach required pottery to have been excavated and published to a high standard, which was clearly not the case. Moreover, the misapplication, or misunderstanding, of the stratigraphic method resulted in a number of ill-defined contexts; these contexts were included in the dataset and noted as "Unstratified". Finally, independent/absolute chronological data is generally unavailable/unreliable. While this research was never intended as a chronological study, reliable chronological information would have made have magnified its results. The conventional Iron Age sequence is used to serve as a relative chronological framework only. So as to avoid circular reasoning, the cross-dating of local assemblages with extra-regional sequences (via Greek and Cypriot imports) was not attempted.

5.6 Concluding Remarks

This chapter has presented an overview of the ceramic data that is investigated in the following chapters. Though originally conceived as an investigation of fabric and surface treatment technologies in the IA-NL, the limited nature of available data shifted the focus of this study to the distribution of ceramic form. The above discussion recounted the means by which the revised data was collected and entered into MS Access and WinBASP databases; the primary tools for later analyses. This account has revealed the structure of the data and the different types of parameters imposed onto the data. It also provided a cursory glimpse of the data so as to reveal the general form of the final dataset; which categories dominate, which sites are under-represented, which regions were over-represented. In addition, the different regional weights in the data emphasise the need for a study such as this; the pottery of the IA-NL is generally under-represented and poorly-understood.

CHAPTER SIX
Typological Patterns in the Data

6.1 Introduction

A necessary part of any regional ceramic study is to establish meaningful categories of data that can produce patterns whilst allowing for significant diversity to exist within and across categories. It is not just similarity and dissimilarity that are interesting but also where the similarity ends and dissimilarity begins. Consequently, the ceramic typology that was to be used in the current study needed to be flexible and allow for variations in the data. If typological criteria were too rigid, then diversity would be lost as vessels were "shoe-horned" into static categories. As a result, the typology presented below, and illustrated in **Figures 1-54**, includes significant variation within some of the 193 categories. This can be contrasted with Lehmann's (1996) typology, which consisted of over 500 Forms for only the later part of the Iron Age.

Within the construction of the below Iron Age typology, the initial exercise was to obtain a ceramic drawing for every known Iron Age vessel from the study area. This included the photocopying of numerous excavation reports, the drawing of ceramics first-hand, and the receipt of unpublished ceramic catalogues and typologies from a number of project directors. Whenever possible the Iron Age site was visited by the author, so that ceramic categories could be studied and discussed with excavation staff (**§5.2**). The final result was a collection of ceramic drawings and notes for over 12,000 individual vessels. Each vessel had its site and context details (as available) written on the back. This large corpus of illustrations was then divided according to sixteen broad categories, primarily derived from functional considerations: cooking-pots; miscellaneous utilitarian vessels; transport-amphorae; pithoi; kraters; urns/storage-amphorae; spouted-amphorae; jugs; juglets; flasks; unguents; spouted-jugs; bottles; Assyrian bottles and cups; cups and chalices; and bowls. These broad categories were then simply laid out before the author and general similarities in shape were slowly grouped together into a CLASS (e.g. CLASS 001; 002; 003). Variations in lip-shape, base or decoration were not always considered significant enough to warrant category distinction. When there was persistent variation within a CLASS, additional sub-classes were created (e.g. CLASS 001a; 001b; 001c). The final typology contains 193 CLASSES, which consists of 264 different forms. Only once the typology was created were vessels, and their accompanying contextual information, entered into the MS Access database, where note was taken of base type, surface treatment, decoration and any other functional attachments.

This chapter presents the typology and any apparent trends within each CLASS or sub-CLASS by discussing, as systematically as possible, a number of specific points. The presence/absence nature of the data prevented a true statistical treatment of these trends; instead terms such as "rare", "known", "common" or "typical" are used to relay degrees of "quantification". If no base, surface or attachment information is available, then these categories are omitted. Reference is also made to figures and key distribution maps found in the appendices; the complete collection of distribution maps can be found on the appended CD. Each entry follows a standard layout, with some fields omitted if information is incomplete or deemed irrelevant.

Description: Characteristic features for each CLASS are described briefly.

Distinction: Distinguishing factors between any Sub-CLASSES are presented.

Bases: Base forms are listed, with the most-abundant or typical presented in **bold**.

Surfaces: Decorative techniques are outlined, with the absence of this field representing undecorated vessels as characteristic.

Attachments: The nature and position of functional attachments is summarily noted; e.g. spouts, handles, lugs.

Distribution: When and where the CLASS appeared in the archaeological record is briefly outlined, with the most-abundant period/s presented in **bold**. Undated contexts are not included.

Parallels: When possible, the current typology links CLASSES to Lehmann's (1996) ceramic typology for the Late Iron Age of Syria and Lebanon. Modern geographic terms are used here to describe smaller areas of the Northern Levant.

Comments: Any additional comments not covered by the above categories are outlined, as well as possible non-Iron Age parallels for the CLASS. Following analysis of the data, it was evident that some ceramic CLASSES used within this study have poor chronological or spatial definition; this is noted in bold type.

6.2 Cooking Pots

CLASS 001 (Maps 07; 08)

The distinctive feature of the hole-mouth cooking-pot is its lack of a neck; the mouth is a simple opening, formed by an in-turning rim. The widest point of the spherical body is the waist. The four sub-classes included here bear different lips. Despite a few late Iron I examples, hole-mouth cooking-pots are most commonly encountered in Iron II and Iron III contexts of the inland Northern Levant, as indicated below.

CLASS 001a (Figure 1)

Distinction: Unthickened lip with bevelled edge

Bases: **Round**

Surfaces: Occasional pressed band under rim

Attachments: Some flat-strap handles under rim

Distribution: Iron I; **Iron II/Iron III** (inland Northern Levant); Persian

Parallels: Lehmann 1996, Forms 438; 449

Comments: Earliest examples extend from late Iron I contexts. Atypical examples (more upright) are known from Hazor (Ben Tor *et al.* 1997, Photo III.35) and Tyre (Bikai 1978b, 50, CP 2).

CLASS 001b (Figure 1)

Distinction: Internally thickened lip

Attachments: Flat-strap handles below rim are common

Distribution: **Iron II/Iron III** (inland Northern Levant)

Parallels: Lehmann 1996, Form 440

Comments: One coastal example known from Tell Arqa.

CLASS 001c (Figure 1)

Distinction: Internally and externally thickened lip

Distribution: **Iron II/Iron III** (inland Syria); Persian

Parallels: Lehmann 1996, Form 439

Comments: One coastal example known from Tell Arqa.

CLASS 001d (Figure 1)

Distinction: Almost upright rim, deeper and less-hemispherical form

Distribution: Iron II/Iron III (Tell Afis and Tell Mastuma)

Parallels: Lehmann 1996, Form 441

Comments: Possible local variant

CLASS 002 (Figure 2)

Description: Hemispherical with short-everted rim.

Distribution: **Iron II** (No pattern discernible); Iron III

Comments: The lack of discernible pattern in distribution implies poor definition. Vessels of similar form with painted decoration or Red-Slip are known, and are not included within the database (e.g. Badre 1997b, Fig. 34a.8, 11; Blaylock 1999, Fig. 11.16; Cecchini 1998, Fig. 35.5; Yadin *et al.* 1958, Pl. 70.13). The Nayrab example was the receptacle for an infant burial. **POORLY DEFINED CLASS**

CLASS 003 (Not Illustrated)

Description: The CLASS 003 cooking pot is characterised by an inward rim with a simple lip that flares up toward the vertical. This CLASS is primarily represented by small rim sherds.

Distribution: Iron I; Iron II; Iron III (not well-defined)

Comments: CLASS 003 appears to be poorly-defined. **POORLY DEFINED CLASS**

CLASS 004 (Maps 09; 10)

The distinctive feature of CLASS 004 is the short flaring neck. These vessels are widest at the lower waist and subtly rounded base, which gives a sagging appearance. The four sub-classes are differentiated by the treatment of the lip.

CLASS 004a (Figure 2)

Distinction: Large cooking-pot with flaring rim and externally thickened, rounded lip
Bases: **Round**

Attachments: Two handles between rim and shoulder, oval in section

Distribution: **Iron II** (Lebanon, north of Southern Levant); Iron III; Persian

Parallels: Lehmann 1996, Form 445

Comments: Four examples derive from mortuary contexts on the Lebanese coast.

CLASS 004b (Figure 2)

Distinction: Short-neck with triangular or bevelled lip exterior

Attachments: Two handles oval or round in section connected to rim

Distribution: Iron I (Inland Syria; Lebanon, Palestine coast); **Iron II** (Lebanon and Palestine coast); Iron III (Palestine coast)

Parallels: Lehmann 1996, Form 450

Comments: Some examples derive from mortuary contexts.

CLASS 004c (Figure 3)

Distinction: Short flaring neck with rounded lip profile

Attachments: Two handles commonly connect rim and shoulder, though single-handled examples are known.

Distribution: **Iron I** (well-spread); **Iron II** (Lebanon); Iron III

Parallels: Lehmann 1996, Forms 454; 455; 457

Comments: The few examples from the inland Northern Levant bear an uncharacteristically tight flare; otherwise distribution is reasonably well-defined.

CLASS 004d (Figure 3)

Distinction: Short flaring neck, upright rim, and thin lip with external depression

Attachments: Handles round in section

Distribution: **Iron I/Iron II** (Lebanon, northern Palestine); Iron III

CLASS 005 (Figure 3)

Description: Characterised by an inwardly direct rim with depressed lip exterior.

Attachments: Two handles connecting the rim and shoulder are common
Distribution: **Iron I**; **Iron II**; Iron III (poorly defined distribution)

Parallels: Lehmann 1996, Form 444

CLASS 006 (Figure 3; Map 011)

Description: These cooking-pots are characterised by short, bulging necks

Attachments: Handles connect the shoulder and rim, but vary in number: two are common, though one-handled and unhandled examples are known.
Distribution: **Iron I** (southern Lebanon, northern Palestine); **Iron II** (Iron I pattern plus Homs Basin area); Iron III; Persian

Parallels: Lehmann 1996, Forms 448; 453

CLASS 007 (Map 12)

The CLASS 007 cooking-pot is characterised by an inwardly direct rim with heavily thickened and rounded lip. The two sub-classes are differentiated by the presence or absence of a lip depression.

CLASS 007a (Figure 4)

Distinction: Inwardly direct rim with externally depressed lip, occasionally protruding to form an external flange

Attachments: Two ovoid handles connect the rim and shoulder

Distribution: **Iron I/Iron II** (southern Lebanon, northern Palestine, Beqa' and Orontes Valleys); Iron III (northern Palestine)

CLASS 007b (Figure 4)

Distinction: Inwardly direct rim with thickened lip exterior, round in profile.

Bases: Ring

Surfaces: Some examples of decorated or slipped surfaces

Attachments: Handles are common.

Distribution: No patterns discernible

Parallels: Lehmann 1996, Forms 147; 446; 451

Comments: The mixed data do not characterise a well-defined CLASS and, therefore, may include other vessel categories. **POORLY DEFINED CLASS**

CLASS 008 (Maps 13;14)

The CLASS 008 cooking-pot is characterised by an open, shallow profile with round base and near vertical rim that commonly bears a flange of varying size. The pots are generally wider than they are high, leaving the interior open and unrestricted. While handles are a common feature on cooking pots, only 20% of CLASS 008 cooking-pots bear direct evidence for the presence of handles. Five sub-classes are included within CLASS 008 and are differentiated on flange length and rim stance. CLASS 008 is considered by Ben Ami (2001) to be characteristic of the Palestinian Iron I and early Iron II period.

CLASS 008a (Figure 4)

Distinction: Flanged cooking-pot rims with open-stance; most examples are too fragmentary to be otherwise classified as CLASS 008b-e.

Distribution: **Iron** I/Iron II (southern Lebanon; northern Palestine); Iron III

Comments: Not a cohesive category.

CLASS 008b (Figure 4)

Distinction: Inwardly oblique rim with external flange

Attachments: Two oval handles occasionally connect rim and shoulder
Distribution: **Iron I/Iron II** (southern Beqa' Valley; northern Palestine)

CLASS 008c (Figure 5)

Distinction: Upright rim with external flange

Bases: **Round**; Flat

Attachments: Handles are rare

Distribution: **Iron I/Iron II** (southern Beqa' Valley; northern Palestine)

Comments: Almost identical distribution between CLASSES 008b and 008c suggests that the distinction between an upright and inward rim is arbitrary.

CLASS 008d (Figure 5)

Distinction: Upright rim with triangular lip exterior

Distribution: **Iron I** (southern Lebanon; northern Palestine); **Iron II** (northern Palestine; Tell Kazel; Tell el Ghassil); Iron III (Tell Keisan)

Comments: Cooking-pot rims from Iron I contexts at Ain Dara resemble the CLASS 008d rims (Stone & Zimansky 1999, Fig. 70.200, 203), but the fragments are too small to confirm the identification.

CLASS 008e (Figure 5)

Distinction: Inwardly oblique rim with triangular lip exterior

Bases: **Round**; Flat

Attachments: Handles are rarely evident

Distribution: **Iron I/Iron II** (southern Lebanon; northern Palestine)

CLASS 009

The CLASS 009 cooking pot with everted rim is characterised by the uniformly thin vessel walls that suggest wheel manufacture; this has also meant few full profiles are extant. Strap handles are commonly attached to the rim. The two sub-classes are distinguished by the relative tightness of the neck and lip stance.

CLASS 009a (Figure 6)
Distinction: Upright short neck, often externally thickened with triangular edge

Attachments: Two strap handles connect rim and shoulder

Distribution: Iron III; **Persian** (North Levantine coast)

Parallels: Lehmann 1996, Forms 458; 459

Comments: Three Iron II examples are known from Abou Danne, though they are probably residual; Lebeau (1983, 350) cannot offer any further eighth century BC parallels (the Tyre parallel cited by Lebeau is a CLASS 004b rim).

CLASS 009b (Figure 6)

Distinction: Restricted neck with tightly-everted rim

Distribution: Late Iron Age

Parallels: Lehmann 1996, Forms 442; 443

Comments: No complete profile is evident. Small dataset.

CLASS 010 (Figure 6)

Description: General form is a very wide and shallow plate with rounded base.

Surfaces: Many baking trays have a heavily scored underside. When turned upside down over a fire, the rough surface would help keep food in place during cooking, and, more importantly, aid its removal – similar baking trays are still used by Bedouin to cook pancake shaped bread (Buhl 1983, 117).

Attachments: Unhandled examples are more common than handled. When present, handles are horizontally aligned along the rim edge.

Distribution: **Iron I/Iron II** (northern Palestine); Iron III

Parallels: Lehmann 1996, Form 437

6.3 Miscellaneous Utilitarian Vessels

CLASS 011

While many different bowl-like vessels were used as lids in the ancient world (e.g. Núñez 2004a, Figs 100, 105, 106), the lids included within CLASS 011 are those that have no other apparent purpose. The two sub-classes are not extensively represented.

CLASS 011a (Figure 6)

Distinction: Characterised by upside-down shallow-bowl topped by a handle

Surfaces: Red-Slip and fenestration are rare (Tyre)

Attachments: Megiddo example has three handles.

Distribution: Only evident at Tyre and Megiddo

CLASS 011b (Figure 6)

Distinction: Characterised by deep dome with knob-like peak.

Attachments: A few examples bear additional small knobs and pierced lugs

Distribution: Iron II (found at only three sites in the study area)

CLASS 012 (Maps 15; 16)

CLASS 012 is characterised by lamps with a pinched lip, within which a wick would be placed and lit. The form is essentially that of a small, shallow bowl with one or more sides pinched together. Four sub-classes are distinguished according to the treatment of the lip, though bases also vary accordingly. CLASS 012 lamps are well represented throughout the Iron Age, but are particularly abundant during the Iron I and Iron II periods.

CLASS 012a (Figure 6)

Distinction: Outwardly oblique rim and unthickened lip.

Bases: **Round**; Flat; Disc

Distribution: **Iron I** (northern Palestine; Beqa' Valley; coastal); Iron II (primarily northern Palestine)

Comments: CLASS 012a is known from sites during the Late Bronze Age (e.g. Ben Dov 2002, Fig. 2.61.70-76; Yadin *et al.* 1960, Pl. 135.1-6, 9-11; 1961, Pl. 267.1-8).

CLASS 012b (Figure 6)

Distinction: Slightly flaring, unthickened lip

Bases: Disc (Iron I); **Round**; Flat (Iron II)

Surfaces: Red-Slip is rare

Distribution: **Iron I/Iron II** (northern Palestine; southern Lebanon); Iron III

Parallels: Lehmann 1996, Form 423

CLASS 012c (Figure 6)

Distinction: Short everted lip

Bases: **Flat**; **Round**; Disc

Distribution: **Iron I** (southern Lebanon; northern Palestine); **Iron II** (same as Iron I plus inland Syria); Iron III/Persian (primarily coastal)

Parallels: Lehmann 1996, Forms 424; 425; 427; 429

Comments: Flat bases are common in the Iron III and Persian periods, while disc and rounded bases appear in the early Iron Age. CLASS 012c is known from some Late Bronze Age contexts (e.g. Yadin *et al.* 1960, Pl. 135.7-8, 12).

CLASS 012d (Figure 7)

Distinction: Multiple-pinched lip

Bases: Round; **Pedestal**

Surfaces: Two examples of Red-Slip are known

Distribution: Never abundant; distribution not well defined

Parallels: Lehmann 1996, Form 422

CLASS 013 (Figure 7)

Description: The composite lamp is characterised by a small conical bowl fixed within a shallow bowl. Two separate rims are discernible, though neither is pinched.

Bases: **Round**; Flat

Surfaces: Three examples of Red-Slip are known from Megiddo

Attachments: An example from Tel Dan, which has a handle reaching from one side of the dish to the other (Biran 1994, Fig. 212), appears atypical.

Distribution: **Iron I/Iron II** (northern Palestine; southern Lebanon); Iron III

Comments: CLASS 013 is also known from Late Bronze Age contexts (e.g. Yadin *et al.* 1960, Pl. 146.8-13; 1961, Pl. 275.20).

CLASS 014 (Figure 7)

Description: The tripod incense burner is a small, perforated vessel with tripod base.

Bases: **Tripod**; Ring (some ring bases are known)

Surfaces: Red-Slip and painted decoration are known.

Attachments: A large percentage bears a single vertical handle attached to the rim.

Distribution: **Iron I/Iron II** (inland sites of northern Palestine and Lebanon)

Parallels: Lehmann 1996, Form 466

Comments: Tripod incense burners extend from contexts that yielded a significant quantity of cooking pots.

CLASS 015 (Figure 7)

Description: Pot-stands are a open-ended cylinders

Surfaces: Painted decoration is known; fenestration and/or plastic decoration is rare.

Attachments: Two examples of multiple vertical handles are known.

Distribution: Difficult to assess; Iron I; **Iron II** (inland regions, not Beqa' Valley); Iron III

Comments: The cylindrical pot-stand has a long history in the Bronze and Iron Ages, but is difficult to classify.

6.4 Transport Amphorae

CLASS 016 (Maps 17; 18)

The CLASS 016 amphora is characterised by an ovoid form, short neck and two handles on the curving shoulders. Three sub-classes are distinguished by differences in shape. These amphorae recall elements of the Late Bronze Age ceramic horizon (e.g. Badre & Gubel 1999-2000, Fig. 32; Ben Dov 2002, Fig. 2.30.6; Capet 2003, Fig. 6.a-b).

Parallels: Lehmann 1996, Form 380

CLASS 016a (Figure 8)

Distinction: Symmetrical oval form with short neck and simple upright lip

Bases: Round; Round-thickened

Surfaces: Monochrome is known; Red-Slip is rare

Attachments: Two vertical handles on the shoulder are characteristic

Distribution: **Iron I/Iron II** (northern Palestine; coastal regions)

CLASS 016b (Figure 8)

Distinction: Slightly top-heavy shape, widest at the shoulder.

Bases: Round; Round-thickened

Surfaces: Four painted examples date from Iron I contexts.

Attachments: Two vertical handles on the shoulder are characteristic

Distribution: **Iron I/Iron II** (northern Palestine; coastal regions)

CLASS 016c (Figure 8)

Distinction: Slightly bottom-heavy form, widest below the two handles.

Bases: Round; Round-thickened

Surfaces: Two examples of painted decoration.

Attachments: Two vertical handles on the shoulder are characteristic

Distribution: **Iron I/Iron II** (northern Palestine; coastal regions)

CLASS 017 (Figure 8)

Description: Squat, barrel-shaped amphora with carinated shoulders and distinct neck with externally thickened lip.

Bases: **Round**; Round-thickened

Attachments: Two tightly-curved handles attached at point of carination.

Distribution: Iron I; **Iron II** (inland northern Palestine)

CLASS 018 (Figure 8; Map 19)

Description: Bag-shaped amphora (low waist) with softly carinated shoulder.

Bases: **Round**

Surfaces: Bichrome is rare; Monochrome examples from Tell Arqa bear a *lmlk* inscription.

Attachments: Two tightly-curved handles on shoulder

Distribution: Iron I; **Iron II** (northern Palestine; Beqa' Valley; coastal regions); Iron III; Persian

Parallels: Lehmann 1996, Forms 318; 382; 402; 403; 409; 411

CLASS 019 (Figure 8; Map 20)

Description: Bag-shaped amphora with carinated shoulder and long everted rim.

Bases: **Round**

Surfaces: Painted bands and lines are common; two Red-Slip examples

Attachments: Two tightly-curved handles on carinated shoulder

Distribution: Iron I; **Iron II** (northern Palestine; Beqa' Valley; coastal regions); Iron III; Persian

Parallels: Lehmann 1996, Form 379

CLASS 020 (Figure 8)

Description: Medium-sized amphora with carinated shoulders and pointed base.

Bases: **Point**

Surfaces: One painted example from Tell Keisan

Attachments: Two small handles on carinated shoulder

Distribution: Never particularly abundant (northern Palestine; southern Lebanon)

Parallels: Lehmann 1996, Forms 377; 406

CLASS 021 (Figure 8)

Description: Small amphora with carinated shoulder, long, narrow neck and roundly pointed base.

Bases: **Point**

Surfaces: Painted decoration is not common

Attachments: Two relatively large handles on carinated shoulder

Distribution: **Iron I/Iron II** (mainly coastal regions)

Parallels: Lehmann 1996, Form 376

Comments: Derives from some mortuary contexts in Iron II period.

CLASS 022 (Figure 8)

Description: Small round amphora.

Bases: **Round**

Surfaces: Monochrome bands and geometric patterns are common

Attachments: Two handles on the rounded waist

Distribution: **Iron I/Iron II** (northern Palestine; Beqa' Valley; coastal regions)

Comments: CLASS 022 recalls Late Bronze Age forms (e.g. Ben Dov 2002, Figs 2.30.4, 10; 2.56.24; Yadin *et al.* 1958, Pl. 86.1, 8, 9; 1960, Pl. 143.12).

CLASS 023 (Figure 9)

Description: "Heavy bag"-shaped amphorae with carinated shoulder and pointed base.

Bases: **Point**; Point-thickened

Attachments: Two handles on carinated shoulders

Distribution: Iron II; Iron III; **Persian** (northern Palestine; al Mina)

Parallels: Lehmann 1996, Forms 393; 399; 404

CLASS 024 (Maps 21; 22)

The distinctive feature of CLASS 024 amphorae is the elongated form with pointed base and the thin waist narrower than the shoulders and lower body. Two sub-classes are distinguished on shape and relative length.

CLASS 024a (Figure 9)

Distinction: Accentuated hourglass profile with wide "hips" and well-pointed base.

Bases: **Point**

Attachments: Two tightly-curved handles just under the carinated shoulders

Distribution: Iron II; **Iron III; Persian** (northern Palestinc; coastal Lebanon)

Parallels: Lehmann 1996, Forms 383; 384

Comments: Some mortuary (inhumation?) association in Iron III and Persian periods.

CLASS 024b (Figure 9)

Distinction: Elongated hourglass profile with subtle "hips" and pointed base.

Bases: **Point**

Attachments: Two tightly-curved handles just under the carinated shoulders

Distribution: **Iron II** (northern Palestine; Lebanon); **Iron III** (coastal Levant); **Persian**

Parallels: Lehmann 1996, Forms 385; 386; 387; 397; 398

Comments: Some mortuary (inhumation?) association in Iron III and Persian periods.

CLASS 025 (Figure 9; Map 23)

Description: Angular torpedo amphora with wide "hips" and long pointed base.

Bases: **Point**

Attachments: Two tightly-curved handles on or below the carinated shoulders

Distribution: **Iron III**; **Persian** (Levantine coast)

Parallels: Lehmann 1996, Forms 394; 395; 396

Comments: Strong presence in inhumation contexts of northern Palestine

CLASS 026 (Figure 9; Map 24)

Description: Top-heavy amphorae with tapering body and wide, carinated shoulders.

Bases: Point; Point-thickened

Attachments: Two handles immediately under carinated shoulder

Distribution: Iron I; **Iron II** (northern Palestine; coastal Lebanon); Iron III; Persian

Parallels: Lehmann 1996, Forms 388; 389; 390; 391; 392

Comments: Some Iron III examples known from Syrian coast and Amuq.

CLASS 027 (Figure 9; Maps 25; 26)

Description: Basket-handled amphora

Bases: Flat-thickened; Point-thickened

Attachments: Two basket-handles on sloping shoulders

Distribution: Iron III; **Persian** (Levant coast)

Parallels: Lehmann 1996, Form 421

CLASS 028 (Figure 9; Maps 27; 28)

Description: Long and thin body with low, diagonal shoulders

Bases: Round; **Point**
Attachments: One or two handles below shoulder.

Distribution: Iron II (inland Syria; northern Palestine); **Iron III** (Northern Levant); Persian (Levant coast)

Parallels: Lehmann 1996, Forms 414; 417; 418; 420

CLASS 029 (Figure 9)

Description: Small torpedo-shaped *amphoriskos* with pointed base

Bases: **Point**; Point-thickened

Surfaces: Monochrome and Bichrome bands are common; Red-Slip is rare

Attachments: Two handles on or below shoulder

Distribution: Never abundant (limited to northern Palestine; coastal Lebanon)

CLASS 030 (Figure 9)

Description: Small flask-like amphora with "nipple" base

Bases: Nipple; Round

Surfaces: Byblos example of Monochrome bands

Attachments: Two handles (vertical or horizontal) on top of shoulder

Distribution: Poorly-represented (at only three sites)

6.5 Pithoi

CLASS 031 (Figure 10)

Description: Large teardrop-shaped pithos with diagonal shoulders and flared rim.

Bases: Round; Point

Surfaces: Examples of rope impressions and pressed-ridge decoration.

Attachments: Two handles on shoulder

Distribution: **Iron I** (northern Palestine; Beqa' Valley); Iron II

Comments: CLASS 031 is known from the Late Bronze Age (e.g. Yadin *et al.* 1958, Pl. 88.11). Termed "Galilean Pithos" by Biran (1994, 130) because it was first identified at Tuleil in the Upper Galilee (Amiran 1970, Pl. 77.1).

CLASS 032 (Figure 10)

Description: Large teardrop-shaped pithos with upright neck and heavy rolled rim.

Bases: Point-solid

Distribution: Only two examples

Comments: The limited numbers suggest CLASS 032 is not indicative of Phoenician trade, as suggested by Biran (1994, 137)

CLASS 033 (Figure 11)

Description: Wide pithos with distinct upright neck and long solid pointed base.

Bases: Point-solid

Surfaces: Pressed-ridge and plastic bands are common.

Attachments: Jerablus examples with two vertical handles

Distribution: Iron II; Iron III (inland Syria)

Parallels: Lehmann 1996, Form 364

Comments: Broadly similar to the "Phoenician" pithos (CLASS 032), this CLASS incorporates a shorter height-to-width ratio, and more open form.

CLASS 034 (Figure 11)

Description: Top heavy pithos with rounded base and collared rim.

Bases: **Round**

Surfaces: Rope impressions are known.

Attachments: Two handles under shoulders

Distribution: Iron I (northern Palestine)

Comments: Within the archaeology of Palestine, the "collared-rim" pithos has been associated with the settlement of the biblical Israelites in the early Iron Age, a correlation not supported by the archaeological record (Biran 1989a; Dever 1995b; Esse 1992; Finkelstein 1988). Recent finds of collared-rim pithoi outside of traditional "Israelite" contexts has challenged this interpretation (Artzy 1994; London 2003, 148-149).

CLASS 035 (Figure 12)

Description: Short pithos with wide waist and terminating in a heavy rolled rim.

Bases: Point-thickened; Disc-thick; Round; Ring

Surfaces: Incised lines, plastic bands, and pressed-ridge

Distribution: Iron I (not well-defined)

CLASS 036 (Figure 12)

Description: Large, open pithos with upright, heavy rolled rim

Bases: Flat; Flat-thickened; Point; Point-thickened; Round; Ring

Surfaces: Plastic bands and pressed ridge common

Attachments: Two examples bear handles

Distribution: **Iron I** (Northern Levant); Iron II; Iron III

Parallels: Lehmann 1996, Forms 368; 369

CLASS 037 (Maps 29; 30)

The "Aramaean" pithos is characterised by a relatively narrow and deep form; what Mazzoni has termed "cigar-shaped". Bases vary from flat, thickened flat, point, solid point, and thickened point. Handles are rare, suggesting these pithoi were not meant to be transported; their sheer weight also argues against any form of movement. The two sub-classes are distinguished by the stance and roundness of the lip. Associated with inland Syria, al Maqdissi (2003, Fig. 13) has called these "Aramaean" pithoi.

Parallels: Lehmann 1996, Forms 363; 371

CLASS 037a (Figure 13; Map 29)

Distinction: Inwardly direct rim with angular lip interior.

Surfaces: Finger impressions, rope impressions, and pressed ridge are common.

Distribution: **Iron I** (inland Syria); Iron II

Comments: Few full profiles are known; possible early Iron Age form.

CLASS 037b (Figure 13; Map 30)

Distinction: Close to upright rim with rounded lip profile.

Bases: Point; Point-solid; Point-thickened; Flat

Surfaces: Finger impressions, rope impressions, and pressed ridge are common.

Distribution: Iron I; **Iron II** (inland Northern Levant)

Comments: Possible Iron II form.

6.6 Kraters

CLASS 038 (Map 31)

The CLASS 038 krater is characterised by an open form, general lack of handles, and sinuous s-curved rim. The two sub-classes are distinguished by size and depth. This CLASS is related to the s-curve bowls (CLASS 134) characteristic of the early Iron Age in the Southern Levant.

CLASS 038a (Figure 14)

Distinction: Large handle-less krater with upright s-curve rim.

Bases: **Ring**; Disc

Surfaces: Six examples of Red-Slip on interior surface.

Distribution: **Iron I** (northern Palestine; southern Lebanon); Iron II

Parallels: Lehmann 1996, Forms 354; 355

CLASS 038b (Figure 14)

Distinction: Small handle-less krater with upright s-curve rim

Bases: **Ring**

Surfaces: Red-Slip and Bichrome are rare

Distribution: **Iron I** (northern Palestine); Iron II

Parallels: Lehmann 1996, Form 356

CLASS 039 (Figure 14)

Description: Krater with inward, direct rim and softly carinated shoulder.

Bases: **Ring**; Looped

Surfaces: Only Iron II examples are decorated.

Attachments: Two handles connecting rim and shoulder

Distribution: Iron I (northern Palestine); Iron II

Comments: Similarly profiled vessels are known at Jerablus, but are atypical: looped base, low handles, large size (Woolley 1939b, Pls 14.d; 22.K1, K18).

CLASS 040 (Map 32)

The distinctive feature of the CLASS 040 krater is a bulging shoulder and long, upright rim. The three sub-classes are distinguished on the curve and depth of profile, and character of the handles.

CLASS 040a (Figure 15)

Distinction: Medium-sized krater with carinated waist and upright rim.

Bases: **Ring**

Surfaces: Exterior surface usually painted with geometric designs. Two Red-Slip examples are known from the Iron II period.

Attachments: Two or more handles connect the rim and shoulder.

Distribution: **Iron I** (inland Northern Levant); Iron II

Parallels: Lehmann 1996, Form 194

CLASS 040b (Figure 15)

Distinction: Medium-sized krater with rounded waist and upright rim.

Bases: **Ring**

Surfaces: Monochrome decoration known in Iron I. Bichrome example from Iron II Hazor.

Attachments: Two handles on shoulder

Distribution: Not well-defined

Parallels: Lehmann 1996, Form 131

CLASS 040c (Figure 15)

Distinction: Medium-sized krater with carinated waist and horizontal handles.

Bases: **Ring**

Surfaces: Monochrome is rare.

Attachments: Two horizontal handles on shoulder.

Distribution: Iron I; Iron II; (poorly-defined).

Parallels: Lehmann 1996, Form 207

Comments: The profile resembles Cypriot-White-Painted kraters of the Iron I (e.g. Anderson 1988, Pl. 32.2).

CLASS 041

The CLASS 041 krater is characterised by an open form, lack of handles, and relatively large size. The two sub-classes are distinguished by lip profile.

CLASS 041a (Figure 16)

Distinction: Large krater with heavy rolled rim.

Bases: **Flat**

Distribution: Iron I (poorly-defined)

Comments: **POORLY DEFINED CLASS**

CLASS 041b (Figure 16)

Distinction: Large krater with short everted rim.

Bases: **Ring**

Distribution: Iron I (only two sites)

Table 6.1: Frequency of CLASS 042 Components

Period	Base	Mono	Bichr.	Band	Circ.	Slip	Comment
Iron I	Ring (18)	8	5	13	0	0	Coastal
28	Pinched ring (2)	2	0	2	0	0	-
	Unknown (8)	1	2	3	0	1	-
Iron II/III	Ring (45)	23	19	38	7	1	N Pal. / S. Leb.
103	Pinched ring (40)	17	14	30	0	1	Leb. mortuary
	Unknown (18)	13	6	15	4	1	Coastal

CLASS 042 (Figure 17; Maps 43; 44)

Description: Deep krater with bulging waist, wide neck, upright rim, and square lip.

Bases: Iron I (**Ring**); Iron II (**Ring**; **Pinched-ring**)

Surfaces: Monochrome or Bichrome is common; Red-Slip is rare.

Attachments: Two handles connect the rim and shoulder.

Parallels: Lehmann 1996, Forms 202; 206

Distribution: Iron I (northern coastal regions); **Iron II** (western half of the Northern Levant); Iron III (coastal regions); Persian

Comments: CLASS 042 is one of the most abundant forms within the study area. CLASS 042 kraters are known from the Late Bronze Age (e.g. Ben Dov 2002, Fig. 2.54.14, 16). IN the Iron Age the majority of these kraters derive from mortuary contexts, usually as containers for cremated human remains in southern Lebanon. The different base and decoration combinations are summarized in **Table 6.1**.

CLASS 043 (Figure 18)

Description: Deep krater with bulging waist, very square lip and horizontally-looped handles.

Bases: **Ring**

Surfaces: Monochrome decoration is rare.

Attachments: Two horizontal handles loop up from shoulder to ledge on square lip.

Distribution: Persian (Syrian coast)

Parallels: Lehmann 1996, Form 208

CLASS 044 (Figure 19)

Description: Small-sized krater with relatively closed form of CLASS 042 profile.

Bases: Ring; Pinched-ring

Surfaces: Monochrome bands are common; Red-Slip is known.

Attachments: Two handles connect bulging shoulder and lower neck.

Distribution: Iron II (north Levantine coast; northern Palestine); Iron III

Parallels: Lehmann 1996, Form 361

CLASS 045 (Figure 19)

Description: Straight neck krater with two small handles on bulging shoulder.

Bases: Ring; Pinched-ring

Surfaces: Monochrome or Bichrome bands are characteristic; more complex designs are known.

Attachments: Two tightly-curved handles, either horizontal or vertical.

Distribution: Iron II/Iron III (mainly cremation contexts of northern Palestine/southern Lebanon coast)

CLASS 046 (Figure 20)

Description: Short, painted krater with spherical body and short everted rim.

Bases: **Ring**; Pinched-ring

Surfaces: Monochrome or Bichrome bands are characteristic, but geometric patterns and circles are also known. Four Black-on-Red examples are known.

Attachments: Two handles are common, but vary from horizontal to vertical, and are positioned on either shoulder or rounded waist. All vertical handled examples derive from the Tyre Al-Bass cemetery. One example bears more than two handles.

Distribution: Iron II/Iron III (only four coastal sites)

CLASS 047 (Figure 21; Map 35)

Description: Oval krater with incurving rim and painted decoration.

Bases: Ring; Pinched-ring; Looped

Surfaces: Monochrome or Bichrome bands and lines are common.

Attachments: Some horizontal handles are known.

Distribution: Iron I; **Iron II** (mainly mortuary contexts of northern Palestine and southern Lebanese coast)

Parallels: Lehmann 1996, Form Z14

CLASS 048

The distinctive feature of this CLASS is its deep, barrel-like form and slightly closing rim. The two sib-classes are distinguished by the lip profile and stance. While the majority of this CLASS is found in northern Palestine during the Iron I and Iron II periods, a few examples from the inland Northern Levant are also known.

CLASS 048a (Figure 21)

Distinction: Deep, barrel-shaped krater with incurving rim and thickened lip.

Bases: **Ring**; Looped (Megiddo)

Surfaces: Decoration is not common; Hama produced one Red-Slip example and one painted with geometric and faunal scenes.

Attachments: Two or more handles are usually found on or just below the rim.

Distribution: Iron I; Iron II; Iron III (not well-defined)

Parallels: Lehmann 1996, Form 374

CLASS 048b (Figure 21)

Distinction: Deep, barrel-shaped krater with inwardly direct rim.

Bases: **Ring**

Surfaces: Monochrome or Bichrome geometric patterns are not common. Red-Slip is known only from Hama and Hazor.

Attachments: Two or more handles connect rim and shoulder

Distribution: Iron I; **Iron II** (non-coastal sites from Hazor to Hama); Iron III

Parallels: Lehmann 1996, Forms 195; 199

CLASS 049 (Figure 22)

Description: Short, rounded krater with in-turning rim

Bases: **Ring**; Looped

Surfaces: Red-Slip is the most common decorative technique.

Attachments: Two or more vertical handles connect rim and shoulder. Horizontal handles are rare.

Distribution: Iron I; Iron II (Lebanon; northern Palestine); Iron III

Parallels: Lehmann 1996, Form 193

CLASS 050 (Figure 22)

Description: Heavily-decorated krater with deep, rounded form and inwardly direct, thickened rim.

Bases: High-ring

Surfaces: Monochrome geometric designs with some floral and faunal motifs.

Attachments: Two handles are located on either the shoulder or rim.

Distribution: Iron II/III (Euphrates region mortuary contexts)

Parallels: Lehmann 1996, Form 188

CLASS 051 (Figure 23)

Description: Deep krater with upright rim and handles below the rim.

Bases: Round

Surfaces: Red-Slip appears characteristic

Attachments: Two handles below rim

Distribution: Only two examples known

Parallels: Lehmann 1996, Forms 198; 200

CLASS 052 (Figure 23)

Description: Deep, straight-sided krater with looped base.

Bases: **Looped**

Surfaces: Monochrome geometric patterns only at Jerablus

Attachments: Two handles are common, but positioning varies.

Distribution: Iron II (Euphrates)

Parallels: Lehmann 1996, Form 189

CLASS 053 (Map 36)

This CLASS is characterised by its small size and relatively open, but deep form. The three sub-classes are distinguished according to the depth of the form, and/or base type.

CLASS 053a (Figure 24)

Distinction: Small krater with looped base.

Bases: Looped

Attachments: Two handles on waist

Distribution: Iron II/Iron III (one example from Deve Höyük)

Parallels: Lehmann 1996, Form 190

CLASS 053b (Figure 24; Map 36)

Distinction: Small krater with relatively deep form.

Bases: Disc; Ring; Round

Surfaces: Monochrome and Bichrome geometric designs are common. Bichrome decoration is used in conjunction with one handle.

Attachments: Two, one or no handles are common; usually positioned below the rim.

Distribution: **Iron I**; Iron II; Iron III (mainly inland Northern Levant)

Parallels: Lehmann 1996, Forms 138; 191; 192; 362

Comments: Ring bases are typical for single-handled examples (usually decorated). Round bases are common amongst unhandled examples (usually undecorated). A few similarly-profiled kraters are known from northern Palestine and southern Lebanon, but are atypically angular in profile (e.g. Briend & Humbert 1980, Pl. 45.1; Lamon and Shipton 1939, Fig. 10.43) or extremely thick in section (e.g. S.V. Chapman 1972, Fig. 22.82; Lamon & Shipton 1939, Fig. 11.53).

CLASS 053c (Figure 24)

Distinction: Short krater with relatively open form.

Bases: **Ring**; **Disc**

Attachments: Two handles attached to rim

Distribution: Small dataset

Parallels: Lehmann 1996, Form 197

CLASS 054 (Figure 24)

Description: Open krater with s-shaped, flaring rim.

Bases: **Ring**; **Flat**; Looped

Surfaces: Monochrome is prevalent in Iron I period; Red-Slip and Bichrome evident in post-Iron I contexts;.

Attachments: Two handles are known

Distribution: **Iron I/Iron II** (coastal regions; Orontes, Beqa' and Jordan Valleys); Iron III (inland Syria)

Parallels: Lehmann 1996, Forms 61; 203; 209; 447

Comments: Late Bronze Age parallels (e.g. Bounni *et al.* 1976a, Fig. 27.3; Yadin *et al.* 1960, Pl. 124.12). **POORLY DEFINED CLASS.**

CLASS 055 (Figure 24)

Description: Open, handless krater with bulging waist and thickened rim.

Bases: **Ring**

Surfaces: Red-Slip only evident at Tel Jezreel

Distribution: **Iron I/Iron II** (northern Palestine)

Parallels: Lehmann 1996, Forms 201; 204

CLASS 056 (Not Illustrated)

Description: Krater rims with thickened exterior

Distribution: **Iron I** (Northern Levant); **Iron II** (northern Palestine; inland Syria); Iron III (inland Syria)

6.7 Storage Amphorae/Urns

CLASS 057 (Map 37)

CLASS 057 is characterised by a closed vessel with low, rounded waist and long, wide neck. The five sub-classes are distinguished by the number and position of any handles. Bases are also variable (ring, flat, disc), though the ring base is the most common.

CLASS 057a (Figure 25)

Distinction: Handless urn with rounded waist and long neck.

Bases: **Ring**; Flat; Disc

Surfaces: Monochrome geometric decoration is typical

Distribution: **Iron I/Iron II** (primarily inland Northern Levant)

Parallels: Lehmann 1996, Form 313

Comments: Large percentage of these vessels extends from mortuary contexts.

CLASS 057b (Figure 25; Map 37)

Distinction: High-handled urn with rounded waist and long neck

Bases: **Ring**

Surfaces: Monochrome bands and geometric designs are common in Iron I; Red-Slip and Bichrome is common in Iron II.

Attachments: Two handles connecting rim and shoulder

Distribution: **Iron I/Iron II** (northern Syria; northern Palestine); Iron III (coastal regions)

Parallels: Lehmann 1996, Form 321

Comments: During the Iron I period, north Syrian examples are all painted with Monochrome bands and simple geometric decorations, while examples from northern Palestine are undecorated. In the Iron II period, the "southern" examples are usually decorated with Bichrome or Red-Slip.

CLASS 057c (Figure 26)

Distinction: Low-handled urn with rounded waist and long neck

Bases: Ring; Pinched-ring

Surfaces: Monochrome geometric designs are common; some faunal and floral motifs

Attachments: Two handles connect neck and shoulder.

Distribution: Iron I (poorly-defined); **Iron II** (northern Levant); Iron III (north Syria)

Parallels: Lehmann 1996, Forms 317; 360; 378

Comments: Strong mortuary association during the Iron II period.

CLASS 057d (Figure 26)

Distinction: Many-handled urn with rounded waist and long neck

Bases: **Ring**

Surfaces: Painted decoration is standard

Attachments: More than two handles on shoulder

Distribution: Iron I (not well-defined)

Parallels: Lehmann 1996, Form 319

CLASS 057e (Figure 26)

Distinction: Horizontal-handled urn with rounded waist and long neck

Bases: **Ring**

Surfaces: Monochrome decoration is typical

Attachments: Two horizontal handles on shoulder

Distribution: Iron I (Hama only); Iron II

Parallels: Lehmann 1996, Form 314

CLASS 058 (Figure 27; Map 38)

Description: Long-handled amphora with heavily painted, rounded body and very long and relatively narrow neck

Bases: **Ring**

Surfaces: Monochrome geometric decoration, with thick bands and floral motifs are common.

Attachments: Two long handles connect the neck and low shoulder

Distribution: Iron III; (Syrian coast; Amuq); **Persian** (Syrian coast)

Parallels: Lehmann 1996, Form 359

Comments: These vessels are regarded by Lehmann (1996, Pls 60-61) as being Greek in nature.

CLASS 059 (Figure 27)

Description: Double-handled amphora with round base and simple flaring rim.

Bases: **Round**

Surfaces: Monochrome decoration is rare

Attachments: Two handles connect rim and shoulder.

Distribution: Iron I; **Iron II** (west Syria); Iron III

Comments: CLASS 059 is similar in form to the CLASS 068 jug, which bears only one handle: the two forms are often found together.

CLASS 060 (Figure 28)

Description: Amphora with long, diagonal shoulders and carinated waist.

Bases: Ring

Surfaces: Monochrome decoration is common on Iron I examples; Red-Slip is known from one Iron II example.

Attachments: Two or more handles on shoulder

Distribution: Iron I; Iron II (not well-defined)

Parallels: Lehmann 1996, Form 320

Comments: The few examples identified display significant variability; in the height of the carination, size of the vessel, and decoration. **POORLY DEFINED CLASS**

CLASS 061 (Figure 28)

Description: Small amphora with two handles and very long neck

Bases: **Ring**

Surfaces: Two Iron II Red-Slip examples; one Iron I Monochrome

Attachments: Two handles on shoulder or waist

Distribution: Iron I; **Iron II** (Jordan, Beqa' and Orontes Valleys)

Parallels: Lehmann 1996, Form 318

Comments: Small dataset.

CLASS 062 (Figure 29; Map 39)

Description: Tall, painted amphora with long upright neck and square lip.

Bases: **Ring**

Surfaces: Monochrome or Bichrome geometric pattern is standard

Attachments: Two horizontal handles on shoulder are common; some vertical handles are known.

Distribution: Iron II; Iron III (coastal mortuary contexts)

Parallels: Lehmann 1996, Form 315

CLASS 063 (Figure 29)

Description: Tall, slender urn with rounded base and flaring neck.

Bases: Round

Distribution: Evident at only one site

Comments: While there are some similarities with Late Bronze Age jars at Hazor (e.g. Yadin *et al.* 1960, Pl. 122.1-6), and a possible Iron I period amphora from Megiddo (Harrison 2004b, Pl. 13.9), no further Iron Age parallels have been identified. Both examples derive from mortuary contexts at Tell Shiyukh Fawqani.

CLASS 064 (Figure 29)

Description: Long, cylindrical storage jar with simple flaring lip.

Bases: **Round**

Distribution: **Iron I/Iron II** (northern Palestine); Iron III

Parallels: Lehmann 1996, Forms 352; 370

CLASS 065 (Figure 30)

Description: Small urn with bulging, angular body, narrow neck and short flaring lip.

Bases: Ring

Surfaces: Red-Slip

Parallels: Lehmann 1996, Form 256

6.8 Spouted Amphorae

CLASS 066 (Figure 30)

Description: Medium-sized amphora with spout (CLASS 060 profile)

Bases: **Ring**

Surfaces: Monochrome bands and lines are typical; Red-Slip and Bichrome are rare.

Attachments: Two handles on opposite shoulders; spout positioned between handles

Distribution: Iron I (northern Palestine)

CLASS 067 (Figure 30)

Description: Amphora with rounded body, short neck and wheel-turned spout

Bases: **Ring**

Surfaces: Red-Slip and Monochrome decoration are rare
Attachments: Three vertical handles connect neck and shoulder; wheel-turned spout in position of fourth handle

Distribution: Iron I; **Iron II** (northern Palestine); Iron III; Persian

6.9 Jugs

CLASS 068 (Figure 31)

Description: Undecorated jug with flaring neck and rounded base.

Bases: **Round**

Surfaces: Monochrome is rare

Attachments: One handle connecting shoulder with either rim or neck.

Distribution: Iron I; **Iron II** (inland regions); Iron III/Persian (Jezreel Valley)

Parallels: Lehmann 1996, Forms 333; 334; 335

Comments: Similar to CLASS 059 double-handled amphora; the two often appear together. Three handle-less examples from Hama are atypically decorated with Monochrome bands.

CLASS 069 (Figure 31)

Description: Large, wide-shouldered jug with flat base.

Bases: **Flat**
Attachments: One handle connects the shoulder and rim

Distribution: Iron I (Tell Kazel)

CLASS 070 (Figure 32)

Description: Large, thin-walled jug with everted neck and flattened base

Bases: Flat

Surfaces: Bichrome bands are common

Attachments: One handle connects rim and shoulder

Distribution: Iron II/Iron III (northern Palestine; southern Lebanon)

Parallels: Lehmann 1996, Form 344

Comments: Two handle less vessels with similar profile are not included in dataset (Courbin 1993a, Fig. 12.1(1076); Lamon & Shipton 1939, Fig. 10.47).

CLASS 071 (Map 40)

The CLASS 071 jug is characterised by a low, rounded waist and relatively wide neck, a one-handled form of the CLASS 057 urn/amphora. The two sub-classes are distinguished on the position of the handle.

CLASS 071a (Figure 32)

Distinction: High-handled jug with rounded waist and wide neck.

Bases: **Ring**; Flat; Disc

Surfaces: Undecorated surfaces are typical, Monochrome and Bichrome are well-attested. Red-Slip is rare.

Attachments: One handle connects rim and shoulder
Distribution: **Iron I** (primarily northern Palestine); Iron II; Iron III; Persian

Parallels: Lehmann 1996, Forms 337; 338; 339

Comments: The profile is known from Late Bronze Age contexts (e.g. Ben Dov 2002, Fig. 2.57.34, 38; Yadin *et al*. 1960, Pl. 133.1-3)

CLASS 071b (Figure 32)

Distinction: Low-handled jug with rounded waist and wide neck.

Bases: **Ring**; Flat

Surfaces: Monochrome and Red-Slip are known, but not typical.

Attachments: One handle connects shoulder and neck
Distribution: Iron I/Iron II (mainly northern Palestine); Persian

Parallels: Lehmann 1996, Form 282

Comments: The profile is known from Late Bronze Age contexts (e.g. Ben Dov 2002, Fig. 2.57.37; Yadin *et al.* 1960, Pl. 133.4-6).

CLASS 072 (Figure 32)

Description: Human-faced jug.

Bases: Ring

Surfaces: Human face applied in plastic decoration

Attachments: One handle on shoulder

Distribution: small dataset

Parallels: Lehmann 1996, Form 283

Comments: The handle is located on the opposite side of the jug from the face; i.e. the face is viewed by the person being served.

CLASS 073 (Figure 33)

Description: Unslipped trefoil-lip jug with rounded waist

Bases: **Ring**; **Pinched-ring**; Disc; Flat; Round

Surfaces: Monochrome and Bichrome are known

Attachments: One handle connects rim and shoulder

Distribution: **Iron I/Iron II** (northern Palestine; Lebanon); Iron III; Persian

Parallels: Lehmann 1996, Form 298

Comments: There is some association of CLASS 073 jugs with mortuary contexts.

CLASS 074 (Figure 33)

Description: Medium-sized jug with rounded waist and relatively wide, short neck.

Bases: Low-ring

Surfaces: Geometric Monochrome/Bichrome are rare

Attachments: One handle connects rim and shoulder

Distribution: Iron I; **Iron II** (mainly Lebanon); Persian
Parallels: Lehmann 1996, Forms 336; 343; 347

Comments: More than half of these jugs derive from mortuary contexts.

CLASS 075 (Figure 33)

Description: Medium-sized jug with rounded waist and relatively wide, long neck.

Bases: Disc; Flat

Surfaces: Megiddo examples are Red-Slip; Beth Shan examples are mostly undecorated.

Attachments: One handle connects rim and shoulder

Distribution: **Iron I** (mainly Jezreel Valley)

CLASS 076 (Figure 33)

Description: Medium-sized jug with ovoid body and long thin neck.

Bases: **Ring**

Surfaces: Monochrome bands are common; Red-Slip and Bichrome are rare.

Attachments: One handle connects the lower neck and shoulder

Distribution: **Iron I/Iron II** (northern Palestine; Lebanon)

Parallels: Lehmann 1996, Form 248

CLASS 077 (Figure 33; Map 41)

Description: Painted spherical jug with long neck and distinct base.

Bases: **Ring**; **Disc**

Surfaces: Handle and body decorated with Monochrome or Bichrome circles, lines and horizontal bands. Five Black-on-Red examples are known.

Attachments: One tightly-curved handle connects lower neck and shoulder.

Distribution: Iron I (northern Palestine); **Iron II** (northern Palestine; south Lebanese coast); Iron III (coast)

Parallels: Lehmann 1996, Forms 260; 263

Comments: The majority of CLASS 077 jugs derive from mortuary contexts.

CLASS 078 (Figure 34; Map 42)

Description: Painted spherical jug with flaring neck and rounded base.

Bases: **Round**

Surfaces: Handle and sphere are decorated with Bichrome circles, bands and simple geometric patterns; Monochrome, Red-Slip and Black-on-Red are rare.

Attachments: One handle connects shoulder and lower neck.

Distribution: **Iron I** (northern Palestine; southern Lebanon); **Iron II** (northern Palestine; Lebanon; Orontes Syria)

Parallels: Lehmann 1996, Form 261

Comments: "Inland" sites are more commonly decorated in Monochrome, while coastal sites used Bichrome. Around one quarter of CLASS 078 jugs derive from mortuary contexts.

CLASS 079 (Figure 34)

Description: Small decorated jug with round base and long flaring neck.

Bases: **Round**

Surfaces: Decoration is common, but the technique varies (Bichrome; Monochrome; Red-Slip).

Attachments: One handle connects the shoulder with either rim or neck.
Distribution: Iron II (coastal region of northern Palestine and southern Lebanon)
Comments: Primarily found in mortuary contexts.

CLASS 080 (Figure 34; Map 43)

Description: Barrel-shaped, painted jug with thin, flaring neck.

Bases: **Round**

Surfaces: Monochrome, Bichrome, Red-Slip, Black-on-Red, Cypriot-White-Slip are all attested

Attachments: One handle connects the barrel shoulder and lower neck.

Distribution: **Iron I** (northern Palestine; southern Lebanon); **Iron II** (western half of Northern Levant); Iron III (north Syrian coast)

Parallels: Lehmann 1996, Form 261

CLASS 081 (Figure 34; Map 44)

Description: Decorated spherical jug with long thin neck and trefoil rim.

Bases: Ring; Pinched-ring; Disc; Flat

Surfaces: Black-on-Red is the most common decorative scheme, though Monochrome and Bichrome are also known. Decorative motifs incorporate concentric circles painted on the spherical body and horizontal bands across the upper body and neck.

Attachments: One double- or single-strap handle connects the shoulder and rim

Distribution: Iron I; **Iron II** (Jezreel and Litani Valleys); Iron III/Persian (coastal regions)

Parallels: Lehmann 1996, Form 296

CLASS 082 (Map 45)

The CLASS 082 jug is characterised by Red-Slip and a pinched trefoil lip. Considered a hallmark of Phoenician culture, these distinct jug types are reminiscent of Late Bronze Age *oinochoai* well-attested at Ras Shamra (*Ugarit Recent* 2-3/1450-1200 BC) and Tell Sukas (Period J), both located on the Syrian coast (see Riis *et al.* 1996, 35-37; Fig. 24 Class VI). The five sub-classes are distinguished by neck profile.

CLASS 082a (Figure 35)

Distinction: Red-Slip trefoil jug with inwardly tapering neck.

Bases: Pinched-ring

Surfaces: Red-Slip is typical

Attachments: Double- or single-strap handle connects shoulder and lip

Distribution: Iron I; **Iron II/Iron III** (mainly coastal regions of Lebanon and northern Palestine; a few Amuq examples)

Parallels: Lehmann 1996, Forms 302; 307

Comments: The imprecise dating/long period of use at the Akhziv and Tyre Al Bass cemeteries prevents accurate dating. Most CLASS 082a jugs derive from mortuary contexts.

CLASS 082b (Figure 35)

Distinction: Red-Slip trefoil jug with piriform body and long flaring neck

Bases: Ring; Pinched-ring

Surfaces: Red-Slip is typical; Monochrome and Bichrome jugs of same profile are known from Khirbet Silm

Attachments: Double- or single-strap handle connects shoulder and lip

Distribution: Iron I; **Iron II** (northern Palestine; southern Lebanon); Iron III (coastal)

Parallels: Lehmann 1996, Forms 299; 300; 305

Comments: The majority derives from mortuary contexts

CLASS 082c (Figure 35)

Distinction: Small Red-Slip trefoil jug with piriform body and short flaring neck

Bases: Ring; Pinched-ring

Surfaces: Red-Slip is typical; Monochrome and Black-on-Red are rare.

Attachments: One handle connects rim and shoulder

Distribution: Iron I; **Iron II** (northern Palestine; southern Lebanon)

Parallels: Lehmann 1996, Forms 297; 303; 304

Comments: Around half of these jugs were found in mortuary contexts of two sites.

CLASS 082d (Figure 35)

Distinction: Red-Slip trefoil jug with globular body and short, everted neck

Bases: Flat; Disc; Round; Ring

Surfaces: Red-Slip is typical

Attachments: One handle connects the shoulder and lip

Distribution: **Iron II** (northern Palestine; southern Lebanon; Hama); Iron III

Parallels: Lehmann 1996, Form 342

Comments: CLASS 082d is primarily found in mortuary contexts.

CLASS 082e (Figure 35)

Distinction: Red-Slip trefoil jug with sinuous profile and flaring neck.

Bases: Ring; Pinched-ring

Surfaces: Red-Slip is typical; Monochrome and Bichrome are known in Iron I period.

Attachments: One handle connects shoulder and rim.

Distribution: Iron I (Jezreel Valley); **Iron II** (southern Lebanon); Iron III

Parallels: Lehmann 1996, Forms 262; 280; 301; 306

Comments: Undecorated jugs with similar profiles are known from the Late Bronze Age (e.g. Ben Dov 2002, Fig. 2.57.29-30). Less than half of the dataset is associated with mortuary contexts.

CLASS 083 (Figure 36)

Description: Thin-walled, spherical jug with thin, flaring neck.

Bases: Flat

Attachments: Handle connects shoulder with rim or neck

Distribution: Small dataset

CLASS 084 (Figure 36; Map 46)

Description: Small decorated jugs with spherical body and neck-ridge

Bases: **Disc**; **Flat**; Ring

Surfaces: Monochrome, Bichrome, Black-on-Red and Red-Slip are all common. Motifs include bands and concentric circles.

Attachments: The tightly-curved handle connects shoulder and lower neck

Distribution: Iron I (northern Palestine; southern Lebanon); **Iron II** (Northern Levant); Iron III

Comments: This jug form is considered by Doumet-Serhal (1993-1994) to be a fossil-type for plotting Phoenician expansion throughout the Mediterranean. CLASS 084 is associated with only a few mortuary contexts.

CLASS 085 (Map 47)

The CLASS 085 jug is characterised by a well-rounded body, and narrow neck that ends in a lip that flares tightly out, which in profile resembles a "mushroom-lip". The two sub-classes are distinguished by general body shape.

CLASS 085a (Figure 36)

Distinction: Mushroom-lip jug with oval body.

Bases: Pinched-ring

Surfaces: Upper neck area is usually decorated with Monochrome or Bichrome bands

Attachments: The tightly-curved handle connects shoulder and lower neck.

Distribution: **Iron II/Iron III** (coasts of Lebanon and northern Palestine)

Parallels: Lehmann 1996, Forms 233; 236; 237; 240

Comments: Over two thirds of these jugs derive from mortuary contexts.

CLASS 085b (Figure 36)

Distinction: Mushroom-lip jug with round body.

Bases: Pinched-ring

Surfaces: Upper neck area is usually decorated with Bichrome bands

Attachments: The tightly-curved handle connects shoulder and lower neck.

Distribution: Iron I (northern Palestine); **Iron II** (northern Palestine; Lebanon; Syrian coast); **Iron III** (Levant coast)

Parallels: Lehmann 1996, Forms 234; 235; 239

Comments: A large percentage of CLASS 085b derive from the Tyre Al Bass and Akhziv cemeteries.

CLASS 086 (Figure 36; Map 48)

Description: Square jug with carinated shoulders and flaring 'mushroom lip'.

Bases: **Ring**

Surfaces: Painted bands are rare.

Attachments: The tightly-curved handle connects shoulder and lower neck

Distribution: **Iron II** (northern Palestine; southern Lebanon); Iron III

CLASS 087 (Figure 36; Map 49)

Description: Square Red-Slip mushroom-lip jug.

Bases: Ring; Pinched-ring

Surfaces: Red-Slip; Black-on-Red is rare

Attachments: The tightly-curved handle connects shoulder and lower neck

Distribution: **Iron II** (northern Palestine; southern Lebanon); **Iron III** (coastal Northern Levant)

Parallels: Lehmann 1996, Forms 238; 241

Comments: CLASS 087 holds some mortuary association, though a large percentage derives from one cemetery.

CLASS 088 (Figure 37; Map 50)

Description: Globular jug with sinuous profile and flaring lip.

Bases: **Ring**; Disc

Surfaces: Decorative schemes vary; Black-on-Red, Red-Slip, Monochrome and Bichrome. Motifs are rarely more complicated than horizontal bands.

Attachments: The tightly-curved handle connects shoulder and lower neck

Parallels: Lehmann 1996, Forms 243; 244

Distribution: Iron I; **Iron II** (Lebanon; northern Palestine); Iron III

Comments: Some mortuary association within this class.

CLASS 089 (Figure 37)

Description: Small oval jug with flaring ridge-neck.

Bases: Pinched-ring

Surfaces: Monochrome or Bichrome bands are known.

Attachments: The tightly-curved handle connects shoulder and lower neck

Distribution: Iron I; Iron II; Iron III (Lebanese coast)

Parallels: Lehmann 1996, Form 252

Comments: Small dataset, but with some possible mortuary association.

CLASS 090 (Map 51)

The CLASS 090 jug is characterised by a globular body, and short, narrowing neck that ends in small rolled-lip. The two sub-classes are distinguished by the length of the neck.

CLASS 090a (Figure 37)

Distinction: Short-globular jug with straight neck with small ridge.

Bases: Ring; Pinched-ring

Surfaces: Monochrome or Bichrome bands on upper neck are typical. Red-Slip and Black-on-Red are rare.

Attachments: The tightly-curved handle connects shoulder and neck-ridge

Distribution: **Iron II** (southern Lebanon; northern Palestine); Iron III

Parallels: Lehmann 1996, Forms 247; 250

Comments: Over half of these jugs are associated with mortuary contexts.

CLASS 090b (Figure 37)

Distinction: Globular jug with very short, straight neck and rolled lip.

Bases: **Pinched-ring**; Round

Surfaces: Monochrome or Bichrome bands are common; Red-Slip is rare.

Attachments: The tightly-curved handle connects shoulder and under rim

Distribution: Iron II; Iron III (northern Palestine; southern Lebanese coast)

Parallels: Lehmann 1996, Forms 250; 255

CLASS 091 (Figure 37)

Description: Square-shouldered jug with straight sides and short neck.

Bases: Ring

Surfaces: Decoration is rare.

Attachments: The tightly-curved handle connects shoulder and neck

Distribution: **Iron II** (Jordan and Beqa' Valleys); Iron III (northern Palestine; North Levantine coast)

Parallels: Lehmann 1996, Forms 257; 258

CLASS 092 (Figure 37; Map 52)

Description: Small globular jug with narrow neck and thick flaring lip.

Bases: Flat; Ring; Pinched-ring

Attachments: The tightly-curved handle connects shoulder and neck

Distribution: Iron II; Iron III (coast of northern Palestine and southern Lebanon); Persian

Parallels: Lehmann 1996, Forms 251; 253; 281

CLASS 093

The CLASS 093 jug is characterised by a bulging form, slanting shoulders and narrow, sinuous neck. The two sub-classes are distinguished by the relative width and size of the jug.

CLASS 093a (Figure 38)

Distinction: Thin-walled, bulging jug with curves slowly into thin, flaring neck.

Bases: Flat; Ring

Surfaces: Monochrome bands and geometric patterns occasional.

Attachments: The handle connects shoulder and neck

Distribution: Iron III; Persian (north Syria)

Parallels: Lehmann 1996, Forms 267; 268; 270; 271

CLASS 093b (Figure 38)

Distinction: Very wide, bulging jug with narrow, short neck.

Bases: **Ring**

Attachments: The handle connects lower shoulder and rim

Distribution: Persian (Deve Höyük)

Parallels: Lehmann 1996, Form 279

CLASS 094 (Figure 38)

Description: Tall and narrow jug with narrow rim.

Bases: Flat; Disc; Ring

Surfaces: Red-Slip is common

Attachments: The handle connects shoulder and neck

Distribution: **Persian** (North Levantine coast; Deve Höyük)

Parallels: Lehmann 1996, Form 229

CLASS 095 (Figure 38; Map 53)

Description: Jug with rounded body curving into long, thin neck and upright lip.

Bases: **Ring**

Surfaces: Monochrome geometric designs are known.

Attachments: The handle loops between shoulder and lower neck

Distribution: Iron III; **Persian** (Northern Levant)

Parallels: Lehmann 1996, Forms 246; 269; 274

CLASS 096 (Figure 38)

Description: Piriform jug with distinct painted design.

Bases: Ring

Surfaces: Monochrome decoration on handles and upper half – concentric circles, bands, wavy lines

Attachments: The handle loops between lower neck and waist; two additional horizontal handles on the waist

Distribution: Iron II/III (Deve Höyük)

Parallels: Lehmann 1996, Form 282

CLASS 097 (Figure 38)

Description: Thin-walled jug with rounded body, straight neck and stepped-out lip.

Bases: Ring; Pinched-ring

Surfaces: Monochrome, Bichrome, and Red-Slip are known

Attachments: Thin handle connects lower neck and shoulder

Distribution: **Iron II** (southern Lebanon; northern Palestine); Iron III

Parallels: Lehmann 1996, Form 245

CLASS 098 (Figure 38; Map 54)

Description: Small jug with disc base and narrow neck ending in thickened lip.

Bases: **Disc**

Surfaces: Generally undecorated; Monochrome bands are known

Attachments: A tightly-curved handle connects neck and shoulder

Distribution: **Persian** (northern Palestine)

CLASS 099 (Figure 38)

Description: Squat decorated jug with round base and short neck.

Surfaces: Monochrome decoration – geometric, floral

Attachments: The tightly-curved handle connects rim and shoulder

Distribution: Persian (Deve Höyük)

6.10 Juglets

CLASS 100 (Maps 55; 56)

The distinctive feature of the CLASS 100 juglet is its small size, round base, single handle and teardrop shape. These vessels are known as dipper juglets, presumably because they were used in the same fashion as a ladle. The five sub-classes are distinguished by general variations in form.

CLASS 100a (Figure 39)

Distinction: Elongated dipper juglet.

Bases: Round; Flat; Ring

Surfaces: Red-Slip is rare

Attachments: Handle connects rim and shoulder

Distribution: **Iron I/Iron II** (Jezreel Valley); Iron III; Persian (Syrian coast)

Parallels: Lehmann 1996, Forms 218; 219

CLASS 100b (Figure 39)

Distinction: Dipper juglet with open-mouth

Bases: Round; Nipple

Surfaces: Decoration is not characteristic, but Red-Slip is known.

Attachments: Handle loops between lip and shoulder

Distribution: **Iron I/Iron II** (northern Palestine; southern Lebanon; Syrian coast); Iron III (North Levantine coast); Persian (coastal regions)

Parallels: Lehmann 1996, Forms 210; 211; 212; 213; 217

Comments: Trefoil-pinched lips are common in the Iron I. Little mortuary association.

CLASS 100c (Figure 39)

Distinction: Dipper juglet with tight flaring neck

Bases: **Round**

Surfaces: Decoration is not characteristic, but Red-Slip is known.

Attachments: Small, tightly-curved handle attached under rim

Distribution: Iron I/**Iron II** (northern Palestine; southern Lebanon); Iron III; Persian

Parallels: Lehmann 1996, Forms 214; 215

CLASS 100d (Figure 39)

Distinction: Dipper juglet with long narrow straight neck

Bases: Disc; Flat; Round; Nipple

Surfaces: Decoration is not characteristic, but Red-Slip is known.

Attachments: Small, tightly-curved handle attached under rim

Distribution: **Iron I**/Iron II (northern Palestine)

Parallels: Lehmann 1996, Form 216

CLASS 100e (Figure 39)

Distinction: Dipper juglet with ridge-neck and slightly flaring rim

Bases: **Round**; Nipple

Surfaces: Decoration not characteristic; Red-Slip known.

Attachments: Small handle connects lower neck and shoulder

Distribution: Iron I; Iron II (only three sites)

6.11 Flasks

CLASS 101 (Figure 39)

Description: Round flask with distinct base and short flaring neck.

Bases: Flat; Ring

Attachments: Two tightly-curved handles connect shoulder and lower neck
Distribution: Iron I; Iron II (inland Syria)

Parallels: Lehmann 1996, Form 472

Comments: Possibly associated with mortuary practices.

CLASS 102 (Maps 57; 58)

Pilgrim flasks are characterised by a lentoid body shape with round base and flaring neck. Painted decoration usually emphasises the lentoid shape. The more complex flask forms (e.g. CLASS 102e) incorporate more complex decorative schemes. The five sub-classes are distinguished by variations in decoration, handles, and rim.

CLASS 102a (Figure 39)

Distinction: Painted standard pilgrim flask

Surfaces: Monochrome/Bichrome circles/bands common.

Attachments: Two tight handles connecting shoulder and neck, are set parallel to flask's thinnest plane.

Parallels: Lehmann 1996, Form 309

Distribution: **Iron I/Iron II** (northern Palestine; south Lebanon); Iron III

Comments: Painted pilgrim flasks are known from Late Bronze Age contexts (e.g. Ben Dov 2002, Figs 2.59.54-55; 2.60.56-59, 61-67; 2.85.100-101, 103; Yadin *et al.* 1960, Pl. 130.10-13; 1961, Pl. 293.1). Around 40% of the Iron Age CLASS 102a flasks were recovered from contexts with clear mortuary associations.

CLASS 102b (Figure 39)

Distinction: Unpainted standard pilgrim flask

Surfaces: Decoration is not characteristic, but Red-Slip is known.

Attachments: Two tightly-curved handles connect shoulder and neck; set parallel to flask's thinnest plane.

Parallels: Lehmann 1996, Forms 309; 311

Distribution: **Iron I** (northern Palestine; Lebanon); **Iron II** (northern Palestine; North Levantine coast); Iron III (coastal regions); Persian

Comments: Unpainted pilgrim flasks are known from Late Bronze Age contexts (e.g. Ben Dov 2002, Figs 2.59.53; 2.60.60; Yadin *et al.* 1960, Pl. 130.8-9). Around one third of the Iron Age CLASS 102b flasks were recovered from mortuary contexts.

CLASS 102c (Figure 39)

Distinction: Single-handled pilgrim flask.

Surfaces: Monochrome or Bichrome circles are common.

Attachments: Tightly-curved handle connects shoulder and neck; set perpendicular to flask's thinnest plane.

Distribution: **Iron I**; Iron II (not well-defined)

CLASS 102d (Figure 39)

Distinction: Wide-mouthed pilgrim flask

Surfaces: Monochrome and Red-Slip are known.

Attachments: Two tightly-curved handles connect shoulder and neck.

Distribution: Iron I (two sites only)

Parallels: Lehmann 1996, Form 310

Comments: All decorated examples derive from the site of Megiddo.

CLASS 102e (Figure 40)

Distinction: Spoon-mouthed pilgrim flask

Surfaces: Monochrome or Bichrome bands, lines, geometric patterns, and circles.

Attachments: Two pierced lugs on shoulders

Distribution: Iron I (northern Palestine); Iron II

6.12 Unguent Containers

CLASS 103 (Map 59)

Pyxides are characterised by squat forms with low carinated waist, short neck and diagonal shoulder. The two sub-classes are distinguished by the form of the handles on the shoulder. Iron Age alabaster vessels with a similar profile are also known (e.g. James 1966, Fig. 66.13), but are not included here. Iron Age pyxides are rarely associated with mortuary contexts.

Parallels: Lehmann 1996, Form 348

CLASS 103a (Figure 40)

Distinction: Small pyxis with two pierced lugs.

Bases: Flat; Disc; Round

Surfaces: Decoration is not common, but Monochrome bands and lines are known.

Attachments: Two pierced lugs on shoulder

Distribution: **Iron I** (northern Palestine; Lebanon; Syrian coast); Iron II (northern Palestine; southern Lebanon)

Comments: The CLASS 103a form is closely related to vessels from the Late Bronze Age (e.g. Yadin *et al.* 1958, Pl. 86.3).

CLASS 103b (Figure 40)

Distinction: Small pyxis with two horizontal strap-handles

Bases: **Flat**

Surfaces: Monochrome bands and lines are common; Bichrome is also known.

Attachments: Two horizontal strap handles on shoulder

Distribution: Iron I (northern Palestine; Lebanon; Syrian coast); Iron II (southern Lebanese coast; northern Palestine)

Comments: The one Red-Slip example bears an atypical ring base. CLASS 103b is closely related to vessels from the Late Bronze Age (e.g. Ben Dov 2002, Fig. 2.83.88-92).

CLASS 104 (Figure 40; Map 60)

Description: Small spouted "Stirrup Jar"

Bases: Flat; Disc; Ring

Surfaces: Monochrome bands and lines are common; Bichrome is known.

Attachments: Small spout on shoulder; two "stirrups" connect shoulder and lip

Distribution: **Iron I** (northern Palestine; Lebanon; southern Syrian coast); Iron II

Comments: Stirrup jars, both large and small, are known from a number of Late Bronze Age contexts (e.g. Ben Dov 2002, Fig. 2.84.93-99; Bounni *et al.* 1998, Fig. 152.5; Yadin *et al.* 1960, Pl. 137.6-12).

6.13 Spouted Jugs

CLASS 105 (Figure 40)

Description: Bulging juglet with small round spout.

Bases: Disc; **Flat**; **Round**; Ring

Surfaces: Monochrome is common; Bichrome is rare.

Attachments: Handle connects rim and shoulder; spout on shoulder or waist.

Distribution: **Iron I/Iron II** (inland Northern Levant); Iron III; Persian

Parallels: Lehmann 1996, Forms 284; 285; 308

CLASS 106 (Figure 40)

Description: Bulging juglet with strainer.

Bases: **Ring**; **Pinched-ring**; Flat; Disc

Surfaces: Red-Slip common; Bichrome/Monochrome known.

Attachments: Handle connects rim and shoulder; small strainer on shoulder.

Distribution: **Iron I** (northern Palestine; Beqa' Valley); Iron II (extends to inland Syria)

Parallels: Lehmann 1996, Forms 291; 292; 293

CLASS 107 (Map 61)

The CLASS 107 "beer-jug" is characterised by a tall, narrow neck set atop a globular body that boasts a long strainer or spout. The two sub-classes are differentiated by the form of the spout.

CLASS 107a (Figure 41)

Distinction: "Beer-jug" with pipe-like spout

Bases: Ring; Flat

Surfaces: Red-Slip, Monochrome, and Bichrome evident

Attachments: Handle connects shoulder and neck; spout extends diagonally from shoulder

Distribution: Iron I; **Iron II** (southern Lebanon)

Parallels: Lehmann 1996, Forms 287; 288

Comments: Red-Slip only derives from southern Lebanese mortuary contexts.

CLASS 107b (Figure 41; Map 61)

Distinction: "Beer-jug" with long half-pipe strainer

Bases: **Ring**

Surfaces: Bichrome bands, lines, and geometric evident.

Attachments: Handle connects shoulder and neck; strainer extends diagonally from shoulder

Distribution: **Iron I**/Iron II (northern Palestine; Lebanon)

Parallels: Lehmann 1996, Forms 289; 290

Comments: Associated with southern Lebanese mortuary.

CLASS 108 (Figure 41)

Description: Small strainer bowl with rounded base

Bases: Round

Surfaces: Decoration is rare

Attachments: Handle on rim common at Megiddo.

Distribution: Iron I/Iron II (northern Palestine; inland Syria)

Comments: The form has close parallels with metal artefacts recovered from the region (e.g. Harrison 2004, Pl. 33.6-7).

CLASS 109 (Figure 41)

Description: Small basket-handled jar with spout

Bases: **Ring**; Flat

Surfaces: Bichrome bands and lines are rare.

Attachments: Single basket-handle on rim; small spout or strainer on shoulder

Distribution: Iron I (Jezreel Valley; Lebanon); Iron II; Iron III

Comments: One basket-handled jar from Tell el-Ghassil is not spouted, while on other examples it is unclear whether a spout was present (e.g. Briend & Humbert 1980, Pl. 61.18; Harrison 2004b, Pl. 17.3; Lebeau 1983, Pl. 144.1).

CLASS 110 (Figure 41)

Description: Askos

Bases: Round; Pinched-ring

Surfaces: Red-Slip is known.

Attachments: Handle connects spout and shoulder; spout on shoulder; spout occasionally bears trefoil-pinched lip
Distribution: Iron I/Iron II (widespread)

Parallels: Lehmann 1996, Form 462

Comments: CLASS 110 possibly holds some mortuary association.

CLASS 111 (Figure 42; Map 62)

Description: Spouted zoomorphic vessel

Bases: **Foot**; Looped

Surfaces: Monochrome or Bichrome geometric patterns common

Attachments: Wheel-turned spout on animal's back; handle linking back and spout; zoomorphic spout

Distribution: Iron I (southern Lebanon; northern Palestine); Iron II (widespread)

Comments: The animal represented varies and is often ambiguous; a ram and bull are two possible representations. Spouted zoomorphic vessels are also known from Late Bronze Age contexts (e.g. Yadin *et al.* 1960, Pl. 152.12; 1961, Pl. 277.3).

CLASS 112 (Figure 42)

Description: Large urn-shaped jug with spout

Bases: Ring

Surfaces: Bichrome bands, lines, circles, and geometric patterns

Attachments: Handle connects rim and shoulder; spout on shoulder by handle

Distribution: Not well-represented

Comments: Small dataset

6.14 Bottles

CLASS 113 (Figure 42)

Description: Large decorated bottle with short, flaring rim.

Bases: Pinched-ring; Flat

Surfaces: Monochrome bands

Distribution: Iron III; Persian

Parallels: Lehmann 1996, Form 322

Comments: Small dataset

CLASS 114 (Figure 42)

Description: Small bottle with narrow base and piriform body – no rims known

Bases: Disc; Ring

Surfaces: Monochrome bands

Distribution: Lebanese coast

Parallels: Lehmann 1996, Form 272

Comments: Small dataset

CLASS 115 (Figure 42)

Description: Elongated bottle with solid pedestal base.

Bases: Pedestal-solid

Distribution: Iron II; Iron III (Tell Kazel); **Persian** (northern Palestine)

CLASS 116 (Figure 42)

Description: Long, thin bottle with pointed base

Bases: **Point**

Attachments: One handle on shoulder; trefoil lips are known

Distribution: **Iron I** (northern Palestine; Lebanon); Iron II (southern Lebanese cemeteries)

Comments: CLASS 116 examples are also known from a number of Late Bronze Age contexts (e.g. Bounni *et al.* 1998, Fig. 152.6; Yadin *et al.* 1960, Pls 120.1-9; 131.1-23; 1961, Pl. 281.4-11).

CLASS 117 (Figure 42)

Description: Ceramic "alabastron"; handle-less, bottom-heavy bottle

Bases: **Round**

Attachments: Two small lugs on shoulder are common

Distribution: Persian (mortuary Northern Levant)

Parallels: Lehmann 1996, Forms 220; 506

Comments: CLASS 117 is similar to a number of Persian *alabastra* made from alabaster-stone (e.g. Johns 1932, Fig. 19; Poppa 1978, Gr. 2.17; 76.38; Woolley 1938b, Fig. 19.1), especially the small "lugs". Much of the literature does not describe vessels of CLASS 117 profile; hence, some may well be alabaster vessels.

CLASS 118 (Figure 43)

Description: Torpedo-shaped bottle with pointed base and long curving neck.

Bases: **Point**

Distribution: Iron I; Iron II; Iron III; Persian (inland Northern Levant)

Parallels: Lehmann 1996, Form 332

CLASS 119 (Figure 43)

Description: Small torpedo-shaped bottle with bulging waist and short neck

Bases: **Point**; Round

Surfaces: Monochrome or Bichrome bands are known; Glaze (al Mina) and Red-Slip (Tyre) are rare.

Distribution: Iron II; **Iron III** (northern Palestine; north Syria); Persian (northern Palestine)

Parallels: Lehmann 1996, Forms 222; 223; 224; 225

Comments: CLASS 119 holds little association with mortuary contexts.

CLASS 120 (Figure 43)

Description: Small, decorated bottle with wide base and narrow neck.

Bases: Disc; Flat

Surfaces: Monochrome bands are characteristic; Black-on-Red is rare

Attachments: On handle connecting neck and shoulder is characteristic.

Distribution: Iron II (northern Palestine; southern Lebanon); Iron III (coastal regions)

Parallels: Lehmann 1996, Form 221

6.15 Assyrian Bottles and Cups

CLASS 121 (Maps 63; 64)

CLASS 121 incorporates vessels of varying form, united by a common origin in Assyria. While these vessels do appear outside of northern Iraq, the concentration of these vessels increases close to northern Iraq. Within the study area, CLASS 121 vessels are primarily found across inland regions, except the Beqa' Valley. The six sub-classes are distinguished according to form.

CLASS 121a (Figure 43)

Distinction: Undecorated spherical bottle with very short, rolled-out rim.

Bases: **Round**

Distribution: Iron III (only three sites in study area)

Parallels: Lehmann 1996, Forms 264; 265

Comments: CLASS 121a is found in abundance at sites east of the Euphrates; such as Tell Halaf (von Oppenheim 1962, Pl. 66.93) and Nimrud (Oates 1959, Pl. XXXVIII.81, 83-85).

CLASS 121b (Figure 43)

Distinction: Round bodied bottle with carinated neck and flaring neck

Bases: **Round**

Surfaces: Etched "collar"

Distribution: Iron III (north Syria)

Parallels: Lehmann 1996, Form 266

Comments: CLASS 121b is found at sites east of the Euphrates; such as Tell Halaf (von Oppenheim 1962, Pl. 66.91) and Nimrud (Oates 1959, Pl. XXXVIII.97).

CLASS 121c (Figure 43)

Distinction: Pointed bottle with short flaring neck and significantly dimpled surfaces

Bases: **Point**

Surfaces: Characteristic dimpled surface of Palace-Ware

Distribution: Iron II; Iron III (northern Syrian)

Parallels: Lehmann 1996, Form 329

Comments: The origin of the Palace-Ware bottles and cups is conventionally located in northern Iraq (Oates 1959, Pl. XXXVII.60-67).

CLASS 121d (Figure 43)

Distinction: Long bottle with pointed base, slightly bulging shoulders, and a wide upright neck.

Bases: **Point**; Point-thickened

Distribution: Iron II; **Iron III** (northern Syria)

Parallels: Lehmann 1996, Forms 323; 328; 330; 351

Comments: CLASS 121d is also found east of the Euphrates at Tell Halaf (von Oppenheim 1962, Pl. 66.96).

CLASS 121e (Figure 43)

Distinction: Thin carinated cup with flaring rim.

Bases: **Round**

Surfaces: Red-Slip is rare

Distribution: Iron II (inland Syria); **Iron III** (widespread); Persian

Parallels: Lehmann 1996, Forms 97; 98; 119; 326; 327

Comments: CLASS 121e is also found east of the Euphrates, at Nimrud (Oates 1959, Pl. XXXVII.59) and at Tell Halaf (von Oppenheim 1962, Pl. 66.84).

CLASS 121f (Figure 43)

Distinction: Cup with sinuous profile and flaring lip.

Bases: Point

Distribution: Iron II; Iron III (northern Palestine coast; north Syria)
Parallels: Lehmann 1996, Forms 324; 325

Comments: CLASS 121f is also found east of the Euphrates at Nimrud (Oates 1959, Pl. XXXVII.78-79) and at Tell Halaf (von Oppenheim 1962, Pl. 66.83).

6.16 Cups and Chalices

CLASS 122 (Figure 44)

Description: Cup with large looping handles

Attachments: Two handles loop from belly to rim

Comments: The validity of CLASS 122 is uncertain considering both examples bear different bases, and decoration. **POORLY DEFINED CLASS**

CLASS 123 (Figure 44)

Description: One-handled cup with carinated belly and upright rim.

Bases: Ring; Flat; Disc; Round

Surfaces: Monochrome is known; Bichrome is rare

Attachments: Handle connects rim and shoulder

Distribution: Iron I (Jezreel Valley; West Syria); Iron II; Iron III

Comments: The form is known from Late Bronze Age contexts of northern Palestine (e.g. Ben Dov 2002, Fig. 2.58.39-42; Yadin *et al.* 1960, Pl. 134.9-11).

CLASS 124 (Figure 44)

Description: One-handled cup with well-rounded belly and upright rim.

Bases: **Round**

Attachments: Handle connects rim and shoulder

Distribution: Iron I; Iron II; Iron III; Persian (not well-defined)

Comments: One Black-on-Red example with trefoil lip is known from Byblos (Homsy 2003, Pl. 4d).

CLASS 125 (Figure 44)

Description: Short one-handled cup with a strainer/spout

Bases: Ring

Attachments: Handle attached to rim; strainer positioned on, or under belly

Distribution: Iron II

Parallels: Lehmann 1996, Form 294

Comments: Small dataset

CLASS 126 (Figure 44)

Description: Small one-handled cup with tripod base

Bases: **Tripod**

Surfaces: Monochrome bands are rare

Attachments: Handle attached to rim

Distribution: Iron I; **Iron II**; Iron III (mainly northern Palestine)

Comments: Small dataset. The shape of this cup is often very similar to the “tripod incense burner” (CLASS 014), and is only distinguished by a lack of holes.

CLASS 127 (Figure 44)

Description: Wishbone-handled cup with ring base.

Bases: **Ring**

Surfaces: Red-Slip and Black-Slip are rare

Attachments: Wishbone-handle attached under rim

Distribution: **Iron I** (northern Palestine; southern Lebanon); Iron II

Comments: The majority have mortuary associations. CLASS 127 is well-attested in the Late Bronze Age (e.g. Ben Dov 2002, Fig. 2.66.79; Bikai 1978b, Pl. 42.3).

CLASS 128 (Figure 44)

Description: Angular "thistle-shaped" vase

Bases: Ring; Disc

Surfaces: Monochrome bands and Bichrome geometric patterns are known

Attachments: Handle attached to rim is rare (on Bichrome examples); shoulder lugs (Jerablus) are rare

Distribution: **Iron I**; Iron II (not well defined)

Parallels: Lehmann 1996, Forms 139; 140; 349

CLASS 129 (Figure 44)

Description: Tall, footed "thistle-shaped" vase

Bases: Pedestal

Surfaces: Monochrome bands/lines, and Red-Slip known.

Distribution: **Iron I** (mainly Megiddo); Iron II; Iron III; Persian

Comments: CLASS 129 examples are known from a few Late Bronze Age contexts (e.g. Yadin *et al.* 1960, Pl. 118.28-29; 1961, Pl. 273.7-10).

CLASS 130 (Figure 44)

Description: Pedestal goblet.

Bases: Pedestal

Distribution: Iron I; Iron II; Iron III; Persian (sporadic)

Parallels: Lehmann 1996, Form 125

CLASS 131 (Map 65)

The distinctive feature of the CLASS 131 chalice is its shallow bowl mounted on a tall, pedestal base. The three variants are distinguished on the form of the pedestal. Similar chalices are known from the Late Bronze Age (e.g. Schaeffer 1949, 269, Fig. 115; Yadin *et al.* 1958, Pls 90.14; 91.18, 21; 1960, Pls 118.21-22; 129.18-19; 1961, Pls 273.1, 4-6; 280.3-4).

Parallels: Lehmann 1996, Form 178

CLASS 131a (Figure 44)

Distinction: Shallow-bowl chalice with high flaring pedestal base.

Bases: Pedestal

Surfaces: Red-Slip is known; Painted interiors are known.

Distribution: **Iron I** (northern Palestine; southern Lebanon; coastal Syria); **Iron II** (Lebanon)

CLASS 131b (Figure 44)

Distinction: Shallow-bowl chalice with stepped-pedestal base

Bases: Pedestal

Surfaces: Red-Slip is rare (only in Iron I southern Levant)

Distribution: **Iron I**/Iron II (northern Palestine; southern Lebanon)

CLASS 131c (Figure 45)

Distinction: Shallow-bowl chalice with fenestrated pedestal base.

Bases: Pedestal

Surfaces: Fenestrated triangles and Monochrome bands

Distribution: Iron I Megiddo

Comments: Small dataset

CLASS 132 (Figure 45)

Description: "Horned" chalice

Bases: Pedestal

Surfaces: Monochrome bands or Red-Slip are common; Black-on-Red is rare

Attachments: Individual horn-like pendants or disk-like ridge positioned under rim

Distribution: Iron I; **Iron II/Iron III** (north Syria; southern Lebanon)

Parallels: Lehmann 1996, Forms 179; 180

CLASS 133 (Figure 45)

Description: Shallow-bowl for use on pot-stand

Surfaces: Red-Slip is known

Attachments: Horn-like pendants are known

Distribution: Iron I/Iron II (northern Palestine)

6.17 Bowls

CLASS 134 (Map 66)

CLASS 134 is characterised by a distinctive s-curve rim. The five sub-classes are distinguished by vessel-depth, lip-angle, and rim stance. CLASS 134 examples are known from Late Bronze Age contexts (e.g. Bounni *et al.* 1998, Fig. 159.5, 8), and is considered a hallmark of the Iron I in northern Palestine (Ben Ami 2001, 160). Red-Slip examples do not appear in inland Syria until the Iron III period, despite appearing along the coast and in northern Palestine during the Iron I and Iron II periods.

CLASS 134a (Figure 46)

Distinction: Deep s-curve bowl

Bases: **Ring**; Flat; Disc

Surfaces: Red-Slip is known

Attachments: Handle attached to rim is rare

Distribution: **Iron I** (well spread); **Iron II** (northern Palestine; inland Syria); Iron III (north Syria; northern Palestine coast)

Comments: Despite "deep" being used here, the form is still wider than it is deep.

CLASS 134b (Figure 46; Map 66)

Distinction: Shallow s-curve bowl

Bases: **Disc**; Ring; Flat

Surfaces: Internal painted bands are rare; Red-Slip is known

Distribution: **Iron I/Iron II** (northern Palestine; Syria); Iron III; Persian

Parallels: Lehmann 1996, Forms 20; 21; 40; 134

Comments: Generally missing from Lebanon. CLASS 134b is not associated with mortuary practices.

CLASS 134c (Figure 46)

Distinction: Shallow-bowl with direct rim and short bent lip

Bases: Disc; Ring

Surfaces: Red-Slip is known; Monochrome is rare

Distribution: **Iron II** (northern Palestine; inland Syria); Iron III (two sites)

Parallels: Lehmann 1996, Form 41

CLASS 134d (Figure 46)

Distinction: Small-bowl with short, everted lip (**Primarily rims**)

Bases: Disc; Flat; Ring

Surfaces: Red-Slip is common; Monochrome or Bichrome bands are rare.

Distribution: Iron I (northern Palestine); **Iron II** (northern Palestine; inland Syria); Iron III; Persian

Parallels: Lehmann 1996, Forms 17; 83; 157; 158

Comments: This sub-class includes a number of bowls that have a large hole bored through the base, indicative of a funnel; known from Iron II Hama and Qatna.

CLASS 134e (Figure 46)

Distinction: Medium-sized s-curve bowl with bevelled lip-interior.

Surfaces: Red-Slip is common

Distribution: Iron I (northern Palestine; west Syria); **Iron II** (northern Palestine; Syria); Iron III (inland Syria); Persian

Parallels: Lehmann 1996, Forms 19; 64; 70

CLASS 135 (Figure 46)

Description: Carinated wide-bowl with upright rim and long-everted lip.

Bases: Ring; Disc

Surfaces: Red-Slip is rare

Attachments: Two handles connect rim and carination

Distribution: Iron III; Persian (not well-defined)

Parallels: Lehmann 1996, Form 205

Comments: Small dataset. While a bowl from Iron II Hazor (Yadin *et al*. 1960, Pl. 98.43) is generally similar in profile, it lacks the long everted lip.

CLASS 136 (Figure 46)

Description: Large-bowl with heavy, everted lip

Bases: Ring

Surfaces: Red-Slip is rare

Distribution: Iron III (north Syria)

Parallels: Lehmann 1996, Form 53

Comments: Small dataset.

CLASS 137 (Figure 47)

Description: Everted bowl rim with grooved upper-lip.

Bases: Pedestal

Surfaces: Red-Slip is known

Distribution: Iron II/Iron III (north Syria)

Parallels: Lehmann 1996, Forms 181; 182; 183

CLASS 138 (Figure 47)

Description: Large-bowl with high-ring base and external flange under rim.

Bases: High-ring

Surfaces: One example of etched design

Distribution: Iron III/Persian (Euphrates)

Parallels: Lehmann 1996, Form 44

CLASS 139 (Figure 47)

Description: Medium-sized bowl with small external flange under rim.

Bases: **Ring**; Disc

Surfaces: Red-Slip is typical

Attachments: Short bone-shaped lugs on the rim are common with Red-Slip

Distribution: **Iron II/Iron III** (northern Palestine/inland Syria); Persian

Parallels: Lehmann 1996, Forms 55; 99; 100; 101

Comments: Red-Slip is particularly well-represented at Hazor during the Iron II period and at Tell Afis during the Iron III period.

CLASS 140 (Figure 47)

Description: Shallow-bowl with thick incurving rim.

Bases: Disc; Low-ring

Surfaces: Red-Slip is common

Attachments: Bone-shaped lugs are known (usually at Hazor)

Distribution: Iron II (only three sites)

Parallels: Lehmann 1996, Form 33

CLASS 141 (Maps 67; 68)

CLASS 141 is characterised by a medium-sized bowl with concave sides and an upright, slightly thickened rim. The four sub-classes are distinguished by the shape of the thickened-lip.

CLASS 141a (Figure 47)

Distinction: Externally thickened lip with tooling underneath.

Bases: **Disc**; Flat; Ring

Surfaces: Red-Slip is common; Painted decoration is rare (all from coast)

Distribution: Iron I (northern Palestine); **Iron II** (Northern Levant); **Iron III** (inland Syria; northern Palestine coast); Persian

Parallels: Lehmann 1996, Forms 46; 49; 56; 103; 146

CLASS 141b (Figure 47)

Distinction: Externally thickened lip with depression underneath

Bases: **Ring**

Surfaces: Red-Slip is known; Monochrome is rare

Distribution: Iron I; **Iron II/Iron III** (inland Syria; northern Palestine); Persian

Parallels: Lehmann 1996, Form 124

CLASS 141c (Figure 47)

Distinction: Externally rounded lip

Bases: **Disc**

Surfaces: Red-Slip is common; Painted decoration is rare

Distribution: Iron I; **Iron II** (Northern Levant); **Iron III** (north Syria; northern Palestine coast); Persian

Parallels: Lehmann 1996, Forms 48; 50; 52; 59; 62; 66; 133

Comments: Red-Slip appears in northern Palestine during the Iron I and Iron II periods, and common in Syria during the Iron III period.

CLASS 141d (Figure 47)

Distinction: Externally and internally rounded lip

Bases: Flat

Surfaces: Red-Slip is known

Distribution: Iron I (northern Palestine); **Iron II** (inland northern Palestine; inland Syria); Iron III

Parallels: Lehmann 1996, Forms 67; 68

Comments: There is a distinct lack of coastal examples.

CLASS 142 (Figure 47)

Description: Medium bowl with hanging everted lip.

Bases: **Flat**

Surfaces: Red-Slip is known

Distribution: Iron II; Iron III/Persian (coastal sites)

Parallels: Lehmann 1996, Forms 45; 65

CLASS 143 (Figure 48)

Description: Bowl with externally-thickened lip that is flattened on top.

Bases: **Flat**

Surfaces: Bichrome bands and lines on interior are rare; Red-Slip is common

Distribution: **Iron I** (mainly inland regions); **Iron II** (Northern Levant); Iron III (mainly coastal regions); Persian

Parallels: Lehmann 1996, Form 73

CLASS 144 (Maps 69; 70)

The distinctive feature of a CLASS 144 bowl is the upright rim with short, triangular lip. Handles are rare and when present are in the form of a bone-shaped lug attached to the rim. The two sub-classes are distinguished by the presence or absence of a sharp carination below the rim.

CLASS 144a (Figure 48; Map 69)

Distinction: Deep sub-class with outwardly direct rim

Bases: **Disc**; Flat

Surfaces: Red-Slip is typical

Distribution: **Iron II/Iron III** (Levantine coast)

Parallels: Lehmann 1996, Forms 120; 121

Comments: Some mortuary association.

CLASS 144b (Figure 48; Map 70)

Distinction: Shallow sub-class with carination under rim

Bases: **Disc**; Flat; Ring

Surfaces: Red-Slip is typical; Painted decoration is rare

Distribution: **Iron II** (northern Palestine; North Levantine coast); **Iron III** (North Levantine coast)

Parallels: Lehmann 1996, Forms 84; 86; 88

Comments: CLASS 144b holds little association with mortuary contexts.

CLASS 145 (Maps 71; 72)

CLASS 145 is characterised by a shallow bowl with outwardly direct rim and an everted, thickened rim. The two sub-classes are differentiated by the angle of the thickened lip. Many of the CLASS 145 bowls were recovered from mortuary contexts, where they were used either as lids for CLASS 042 cinerary urns or funerary deposits within inhumation contexts.

CLASS 145a (Figure 48; Map 71)

Distinction: Shallow-bowl with downwardly-everted thickened lip.

Bases: **Flat**; **Disc**

Surfaces: Red-Slip is rare

Distribution: **Iron II/Iron III** (coastal regions); Persian (northern Palestine coast)

Parallels: Lehmann 1996, Forms 28-30; 32; 36; 37; 84d; 148

Comments: Few inland examples known. Strong mortuary association.

CLASS 145b (Figure 48; Map 72)

Distinction: Shallow-bowl with upwardly-everted thickened lip.

Bases: **Flat**; Disc

Surfaces: Red-Slip is rare

Distribution: **Iron II/Iron III** (Levantine coast)

Parallels: Lehmann 1996, Forms 26; 27; 31

Comments: Strong mortuary association.

CLASS 146 (Figure 48)

Description: Small-bowl with inverted and everted lip

Bases: Disc

Parallels: Lehmann 1996, Form 43

Comments: Only one example identified; possibly unique CLASS

CLASS 147 (Figure 48)

Description: Bowl with long everted lip and depressed "gutter".

Bases: Ring

Distribution: North Syria

Parallels: Lehmann 1996, Form 177

Comments: Small dataset. CLASS 147 derives only from unstratified contexts and cannot be dated, though Lehmann (1966, Pl. 29) places it in the Persian period.

CLASS 148 (Figure 48)

Description: Shallow-bowl with long, sinuous flaring rim

Bases: Disc; Low-ring

Surfaces: Red-Slip is common at Iron II Hazor and Iron III al Mina; Painted decoration is rare.

Parallels: Lehmann 1996, Forms 149; 150; 151

Distribution: Iron II (Hazor); Iron III (north Syria); Persian

Comments: Examples from northern Palestine/southern Lebanon tend to bear shorter lips. A bowl from the earlier Deve Höyük cemetery is decorated with a fiance glaze, a technique also present at Al Mina (Peltenberg 1969).

CLASS 149 (Figure 48)

Description: Bowl with sinuous, flaring rim and thin lip.

Bases: Disc; Ring

Surfaces: Red-Slip and Bichrome are rare.

Distribution: Iron I; Iron II/**Iron III** (inland northern Levant); Persian

Parallels: Lehmann 1996, Forms 116; 117; 118

CLASS 150 (Figure 48)

Description: Shallow-bowl with direct, very thin lip.

Surfaces: Red-Slip is typical

Distribution: Iron II/Iron III (widespread)

Parallels: Lehmann 1996, Forms 57; 75a

Comments: Small dataset

CLASS 151 (Figure 49)

Description: Shallow-bowl with outwardly-oblique, but everted sides.

Bases: **Flat**; **Disc**

Surfaces: Red-Slip is known; Monochrome and Bichrome decoration is rare

Distribution: Iron I (northern Palestine; southern Lebanon); Iron II/Iron III (Levantine coast); Persian

Parallels: Lehmann 1996, Forms 75b; 76

Comments: Around one third of these bowls were recovered from mortuary contexts.

CLASS 152 (Figure 49)

Description: Medium-sized bowl with carinated sides and direct rim.

Bases: **Ring**; Disc

Surfaces: Red-Slip is common; Painted decoration is rare

Parallels: Lehmann 1996, Forms 58; 109-111; 113; 114

Distribution: **Iron I** (northern Palestine; southern Lebanon); **Iron II** (widespread); Iron III (northern Palestine; north Syria); Persian

Comments: During the Iron I period, Red-Slip examples are restricted to northern Palestine and the southern coast of Lebanon, while in the Iron II period they are found at sites along the Syrian coast. Only in the Iron III period are Red-Slip examples well distributed across the inland Northern Levant.

CLASS 153 (Figurc 49)

Description: Shallow-bowl with horizontal sides and vertically tapering lip.

Bases: **Disc**; Flat; Ring

Surfaces: Red-Slip is common

Distribution: Iron I/**Iron II** (northern Palestine; southern Lebanon; Hama); Iron III

Parallels: Lehmann 1996, Forms 81; 82; 130

CLASS 154 (Figure 49)

Description: Carinated small-bowl with flat base, diagonal sides and upright lip.

Bases: **Flat**

Surfaces: Monochrome interior bands are known; Red-Slip is rare

Distribution: **Iron I**/Iron II (northern Palestine); Iron III; Persian

Parallels: Lehmann 1996, Form 107

Comments: Around one third of CLASS 154 bowls came from mortuary contexts.

CLASS 155 (Figure 49; Map 73)

Description: Thin-walled shallow-bowl with carinated sides and direct lip.

Bases: **Round**; Flat; Ring

Surfaces: Red-Slip is typical (reserve-slip on base is common); Black-on-Red is rare

Parallels: Lehmann 1996, Forms 78; 80

Distribution: Iron I (northern Palestine; southern Lebanon; Hama); **Iron II/Iron III** (northern Palestine; north Syria)Persian (northern Palestine coast)

Comments: The extra attention to decorating what would be the rounded base suggests that these bowls doubled as lids, possibly in conjunction with burial urns; around one third of CLASS 155 bowls were recovered from mortuary contexts.

CLASS 156 (Figure 49)

Description: Thin-walled small-bowl with carinated sides and near-upright lip

Bases: **Flat**

Surfaces: Red-Slip (with reserve-slip base) is common

Distribution: **Iron II** (northern Palestine; southern Lebanese coast); Iron III; Persian

Parallels: Lehmann 1996, Form 79

Comments: The majority of CLASS 156 bowls derives from mortuary contexts.

CLASS 157 (Figure 49; Map 74)

Description: Small hemispherical bowl with upright rim

Bases: **Round**; Disc; Flat; Low-ring

Surfaces: Red-Slip is common; Painted bands are known

Distribution: **Iron I** (northern Palestine; Lebanon); **Iron II** (northern Palestine; Lebanon; north Syria); Iron III (north Syria; northern Palestine coast); Persian

Parallels: Lehmann 1996, Forms 144; 145a

CLASS 158

CLASS 158 is characterised by a small, conical bowl. The two sub-classes are distinguished by the presence/absence of handles inside the bowl. The CLASS 158 profile appears to have its origins in the Late Bronze Age (e.g. Bounni *et al*. 1998, Fig. 159.3-4).

CLASS 158a (Figure 49)

Distinction: Small conical bowl

Bases: **Flat**; **Disc**; Low-ring

Surfaces: Red-Slip is common

Distribution: **Iron I/Iron II** (northern Palestine; Lebanon); Iron III

Comments: One quarter of these bowls were recovered from mortuary contexts.

CLASS 158b (Figure 49)

Distinction: Small conical bowl with internal handle

Bases: Flat

Attachments: Internal handles

Distribution: Iron I (two sites)

CLASS 159 (Figure 50)

Description: Bowl with tri-looped base

Bases: Looped

Surfaces: Bichrome geometric design is rare; Red-Slip is rare

Distribution: Iron I/Iron II (widespread)

CLASS 160 (Maps 75; 76)

CLASS 160 is characterised by a small hemispherical bowl with incurving rim. The two sub-classes are distinguished by the severity of curve in the rim.

CLASS 160a (Figure 50; Map 75)

Description: Small hemispherical bowl with gently incurving rim

Bases: Flat; Disc; Round

Surfaces: Red-Slip is common; Painted decoration is known

Distribution: **Iron I** (northern Palestine; Lebanon; west Syria) **Iron II** (most of study area); Iron III (North Levantine coast); Persian (northern Palestine coast)

Parallels: Lehmann 1996, Form 145b

Comments: Amongst the Red-Slip examples, a few Iron II bowls also employ the reserve-slip technique. One quarter of CLASS 160a were recovered from mortuary contexts.

CLASS 160b (Figure 50; Map 76)

Description: Small hemispherical bowl with tightly incurving rim

Bases: Flat; Disc

Surfaces: Red-Slip is common; Monochrome or Bichrome bands generally limited to Tyre

Distribution: Iron I/**Iron II** (northern Palestine; southern Lebanon; Orontes Syria)

Parallels: Lehmann 1996, Forms 127; 128; 129

Comments: Around one fifth of CLASS 160b was recovered from mortuary contexts.

CLASS 161 (Figure 50)

Description: Medium-bowl with incurved rim and internally bevelled lip

Bases: **Flat**; **Disc**

Surfaces: Red-Slip is known; Painted decoration is rare

Attachments: Horizontal bone-shaped lugs are rare

Distribution: Iron I/Iron II (northern Palestine; Lebanon; Orontes Syria); Iron III (northern Palestine; inland Syria); Persian

Parallels: Lehmann 1996, Forms 47; 60

Comments: Horizontal bone-shaped lugs are generally limited to Lebanese Iron I contexts. CLASS 161 holds no association with mortuary practice.

CLASS 162 (Figure 50; Maps 77; 78)

Description: Decorated shallow-bowl with sides that curve upright

Bases: **Ring**; **Disc**; Flat

Surfaces: Internal Monochrome and Bichrome (post-Iron I) bands are common; Painted decoration is generally limited to coastal examples; Red-Slip is known, as too is Black-on-Red.

Attachments: Short horizontal handles under rim are common

Distribution: **Iron I/Iron II** (northern Palestine; coastal Lebanon; west inland Syria); Iron III (north Syria; coast of northern Palestine and southern Lebanon); Persian

Parallels: Lehmann 1996, Forms 51; 106; 153; 156

Comments: The CLASS 162 form is known from Late Bronze Age contexts (e.g. Ben Dov 2002, Fig. 2.85.105). Less than 20% of CLASS 162 derives from mortuary contexts.

CLASS 163 (Figure 50)

Description: Deep hemispherical decorated bowl

Bases: **Ring**

Surfaces: Monochrome or Bichrome designs are typical (bands, lines, geometric patterns, circles, fauna, flora); Black-on-Red and Cypriot-White-Slip are known

Attachments: Horizontal handles under rim

Distribution: Iron I; **Iron II** (northern Palestine; south Lebanese coast); Iron III (coastal regions)

Parallels: Lehmann 1996, Form 154

Comments: Similar bowl forms are known from the Late Bronze Age (e.g. Yadin *et al*. 1958, Pl. 91.26).

CLASS 164 (Figure 51)

Description: Shallow-bowl with up-turned sides and flat base.

Bases: **Flat**; **Disc**

Surfaces: Monochrome or Bichrome bands are known; Red-Slip is rare

Distribution: **Iron I** (coast of southern Lebanon and northern Palestine); **Iron II** (primarily northern Palestine; southern Lebanon); Iron III

Parallels: Lehmann 1996, Forms 7; 23; 175

Comments: CLASS 164 is similar to Late Bronze Age forms (e.g. Bounni *et al*. 1998, Fig. 152.7). Less than one quarter of CLASS 164 was recovered from mortuary contexts.

CLASS 165 (Maps 79; 80)

CLASS 165 is characterised by a shallow-bowl with an upturned rim, unthickened lip and ring base. The two sub-classes are distinguished according to height of the ring base.

CLASS 165a (Figure 51; Map 79)

Description: Low ring base shallow-bowl

Bases: **Ring**

Surfaces: Red-Slip is rare; Monochrome bands and lines are known

Distribution: Iron I/**Iron II** (most of study area); Iron III (north Syria)

Parallels: Lehmann 1996, Form 2

Comments: Minor mortuary association.

CLASS 165b (Figure 51; Map 80)

Description: High-ring base shallow-bowl

Bases: High-ring

Surfaces: Monochrome designs are common on bowl interior; Red-Slip is rare.

Distribution: Iron I; **Iron II** (only three sites)

Parallels: Lehmann 1996, Forms 3; 8; 9

Comments: Common at Hama

CLASS 166 (Figure 51; Map 81)

Description: Pedestal platter ("Hama fruit-stand")

Bases: **Pedestal**

Surfaces: Red-Slip is typical; undecorated are known

Distribution: **Iron II**/Iron III (inland west Syria)

Parallels: Lehmann 1996, Form 4; 5

Comments: The ratio between platter and base diameters is c. 3:1. Two coastal vessels recall the general form, but are deeper and thicker with flaring base (Buhl 1996, Fig. 29.XIII A 3 3400/1; S.V. Chapman 1972, Fig. 28.154).

CLASS 167 (Figure 51)

Description: Shallow-bowl with tripod base

Bases: **Tripod**

Surfaces: Red-Slip is known

Distribution: Iron I; **Iron II** (Orontes Syria; inland northern Palestine); Iron III (north Syria; northern Palestine coast)

Parallels: Lehmann 1996, Forms 185; 186; 187

Comments: CLASS 167 is the ceramic replication of the ubiquitous basalt mortar (Culican 1970, 14-16; Lehmann 1996, Pl. 92(Form 500); Yadin *et al.* 1958, Pl. 73.11).

CLASS 168 (Figure 51; Map 82)

Description: Shallow-bowl with outwardly direct rim and flattened lip

Bases: **Ring**; Disc; Flat

Surfaces: Monochrome bands rare; Red-Slip is known.

Distribution: **Iron I/Iron II** (most of study area); Iron III

Parallels: Lehmann 1996, Forms 1; 2

CLASS 169 (Figure 51)

Description: Shallow-bowl with rolled-out lip.

Bases: Disc; Flat

Surfaces: Red-Slip is common

Distribution: **Iron I** (inland Syria); **Iron II** (inland Syria; northern Palestine); Iron III (north Syria)

Comments: This simple profile is similar to Late Bronze Age forms (e.g. Bounni *et al.* 1998, Fig. 152.4).

CLASS 170 (Figure 52)

Description: Shallow-bowl rims with thickened lip inner

Bases: Disc

Surfaces: Monochrome and Bichrome bands and Red-Slip are all known.

Distribution: **Iron I** (northern Palestine; southern Lebanon; west Syria); **Iron II** (most of the study area); Iron III (north Syria; coast of northern Palestine and southern Lebanon)

Parallels: Lehmann 1996, Forms 16; 22

Comments: **POORLY DEFINED CLASS**

CLASS 171 (Figure 52)

Description: Shallow-bowl with ribbed surface

Bases: Ring

Surfaces: Etched bands

Distribution: **Iron II** (widespread); Iron III/Persian (northern Palestine coast)

CLASS 172

CLASS 172 is characterised by a shallow-bowl with sharply-bent profile and upright rim. The two sub-classes are distinguished by the sharpness of the bend.

Parallels: Lehmann 1996, Forms 6; 39; 85; 104; 105; 115; 155

CLASS 172a (Figure 52)

Distinction: Carinated shallow-bowl with upright rim

Bases: Disc

Surfaces: Red-Slip is known; Monochrome and Bichrome bands are rare

Distribution: **Iron I/Iron II**/Iron III (northern Palestine; southern Lebanon; west Syria)

Comments: CLASS 172a is known from the Late Bronze Age (e.g. Bounni *et al.* 1998, Fig. 159.14; Johns 1938, Fig. 13.2). **POORLY DEFINED CLASS**

CLASS 172b (Figure 52)

Distinction: Shallow-bowl with tightly-bent rim (**Primarily Rims**)

Bases: Disc; Flat

Surfaces: Red-Slip is common; Monochrome and Bichrome bands are known

Attachments: Horizontal lugs and handles are occasionally evident.

Distribution: **Iron I** (primarily northern Palestine and southern Lebanon); **Iron II** (most of study area); **Iron III** (north Syria; northern Palestine); Persian

Comments: **POORLY DEFINED CLASS**

CLASS 173 (Figure 52)

Description: Very shallow platter with up-turned and tapering lip (**Primarily Rims**)

Surfaces: Red-Slip is typical

Distribution: **Iron II/Iron III** (inland Syria)

Parallels: Lehmann 1996, Form 4

Comments: Likely pedestal platter. A rim from the Tyre Al Bass cemetery is the only example of a possible pedestal platter beyond inland Syria, and is the only example from a mortuary context.

CLASS 174 (Figure 52)

Description: Shallow-bowl with up-turned rim (**Primarily Rims**)

Surfaces: Monochrome and Bichrome bands are common; Red-Slip is common

Distribution: **Iron I** (northern Palestine; Beqa' and Orontes Valleys); **Iron II** (northern Palestine; southern Lebanon; inland Syria); Iron III; Persian

Parallels: Lehmann 1996, Form 34

Comments: CLASS 174 includes a few shallow-bowls with slipped interior and pendant semi-circles painted on the exterior; some of these bowls have been published as East Cycladic, or Euboean, imports (e.g. Bikai 1978b, Pls. 11.20; 22.5-6; 24.5, pp. 53; Courbin 1982a, Fig 4; see Desborough 1952, 118, Pl. 12). **POORLY DEFINED CLASS**

CLASS 175 (Figure 52)

Description: Bowl with direct rim and flattened lip

Surfaces: Red-Slip is common

Distribution: **Iron I/Iron II** (northern Palestine; inland Syria); Iron III; Persian

CLASS 176 (Figure 52)

Description: Shallow-bowl with direct flattened rim

Surfaces: Red-Slip or Monochrome bands are known

Parallels: Lehmann 1996, Form 1?

Distribution: same as CLASS 175

Comments: Reminiscent of Late Bronze Age forms (e.g. Bounni *et al.* 1976a, Fig. 27.4).

CLASS 177 (Figure 52)

Description: Large, heavy plate with rounded rim

Surfaces: Red-Slip is known

Distribution: Iron II (inland Syria)

CLASS 178 (Figure 52)

Description: Small but thick plate

Bases: Flat

Surfaces: Black-on-Red is rare

Distribution: Iron I; Iron II; Iron III (widespread)

Parallels: Lehmann 1996, Forms 132; 137

CLASS 179 (Figure 52; Map 83)

Description: Small bowl with direct bevelled rim.

Bases: **Flat**; **Disc**

Surfaces: Interior Monochrome or Bichrome bands are common; Red-Slip is known

Distribution: Iron I ("Phoenician coast"); **Iron II** (includes Inland Syria); Iron III (see Iron I)

Parallels: Lehmann 1996, Forms 10; 11; 12; 13; 132

Comments: Strong mortuary association.

CLASS 180 (Figure 52; Map 84)

Description: Shallow-bowl with tapering lip and flat base

Bases: **Flat**

Surfaces: Red-Slip is common; Black-on-Red is rare

Distribution: Iron I; **Iron II** (northern Palestine; Homs Basin); Iron III

Parallels: Lehmann 1996, Form 78b?

CLASS 181 (Figure 53)

Description: Thin-walled shallow bowl

Bases: Round

Surfaces: Red-Slip is typical

Distribution: Iron I; **Iron II** (northern Palestine; inland Syria); Iron III

Parallels: Lehmann 1996, Form 78b?

CLASS 182

CLASS 182 is characterised by a fine-bowl with thin lip. The three sub-classes are distinguished by the curve and stance of the rim.

CLASS 182a (Figure 53)

Distinction: Fine-bowl with outwardly direct rim

Surfaces: Red-Slip is typical; painted decoration is rare

Attachments: One handle is attached to rim on very few inland Syrian bowls

Distribution: Iron I/**Iron II** (most of study area but Lebanon); Iron III; Persian

Parallels: Lehmann 1996, Forms 143; 184

CLASS 182b (Figure 53)

Distinction: Fine-bowl with curving sides (**Primarily Rims**)

Surfaces: Red-Slip is typical; painted decoration is rare

Distribution: Iron I (northern Palestine; west Syria); **Iron II** (Euphrates; Orontes; northern Palestine); Iron III (north Syria); Persian

Parallels: Lehmann 1996, Forms 18; 42

CLASS 182c (Figure 53)

Distinction: Fine-bowl with upright rim (**Primarily Rims**)

Surfaces: Monochrome and Bichrome bands are common; Red-Slip is common

Distribution: Iron I/**Iron II** (inland Syria; northern Palestine); Iron III; Persian

CLASS 183 (Figure 53)

Description: Fine-bowl with flaring rim (**Primarily Rims**)

Surfaces: Monochrome and Bichrome bands are known; Red-Slip is common

Distribution: Iron I (west Syria); **Iron II** (west Syria; northern Palestine); Iron III

CLASS 184 (Figure 53; Map 85)

Description: Wide, carinated fine-bowl

Bases: Round; Ring

Surfaces: Red-Slip is common; painted decoration is rare

Attachments: Rim handle known; horizontal handle rare

Distribution: **Iron I** (northern Palestine; Lebanon, north Syria); **Iron II** (most of study area); Iron III; Persian

Parallels: Lehmann 1996, Forms 89c; 92; 94

Comments: The CLASS 184 form has its roots in the Late Bronze Age (e.g. Ben Dov 2002, Figs 2.29.7; 54.5-7; Yadin *et al.* 1961, Pl. 272.1-16). The CLASS 184 profile has been connected with Assyrian bowls (e.g. Adachi 1997), but are widespread before the rise of Assyrian influence west of the Euphrates.

CLASS 185 (Figure 53)

Description: Fine carinated bowl with pattern burnishing

Bases: Low-ring

Surfaces: Burnished lines

Attachments: Horizontal handles under rim

Distribution: Iron I/Iron II (two sites)

CLASS 186 (Figure 53)

Description: Narrow fine-bowl with sharp carination

Bases: Disc

Surfaces: Monochrome painted bands

Distribution: Iron I (Tell Afis and Tell Rifa'at)

Comments: Small dataset

CLASS 187 (Figure 53; Map 86)

Description: Fine-bowl with long-flaring rim

Bases: **Round**
Surfaces: Monochrome decoration is rare; Red-Slip is known

Attachments: Horizontal handles under rim are known on Iron I examples

Distribution: Iron I (west Syria); **Iron II** (inland Syria); Iron III

Parallels: Lehmann 1996, Forms 96; 123

Comments: CLASS 187 holds little, if any, association with mortuary contexts.

CLASS 188 (Figure 53; Maps 87; 88)

Description: "Sub-Mycenaean" s-curve small-bowl

Bases: **Ring**; Flat; Disc; Round

Surfaces: Monochrome bands and lines are typical; painted circles are known; Bichrome decoration is known; Red-Slip is known; Black-on-Red is known

Attachments: Horizontal handles on shoulder are common

Distribution: **Iron I** (most of the study area); **Iron II** (not Euphrates); Iron III

Parallels: Lehmann 1996, Form 96

Comments: A number of similarly profiled and decorated bowls are known from Late Bronze Age contexts (e.g. Bounni *et al.* 1978, Fig. 28.1; Yadin *et al.* 1958, Pl. 87.7; 1961, Pl. 273.11-12), when Mycenaean imports were frequently encountered along the Levantine coast. CLASS 188 bowls are often referred to as Late Helladic IIIC in style (Koehl 1985, No. 193), but are likely local in manufacture. The odd Mycenaean import might be included here (e.g. Anderson 1988, Pl. 30.10).

CLASS 189 (Figure 54)

Description: Fine-bowl with bent shoulder and inwardly direct rim

Bases: Round; Ring; Disc

Surfaces: Red-Slip is common in Iron II period

Distribution: Iron I/Iron II (inland Syria; northern Palestine coast)

Parallels: Lehmann 1996, Forms 91; 95

CLASS 190 (Figure 54)

Description: Fine-bowl with bulging, carinated profile and upright rim

Bases: Round

Surfaces: Red-Slip is typical
Distribution: Iron II (Orontes Syria)

Parallels: Lehmann 1996, Forms 89a; 89b

Comments: Considering the nature of the sites from which examples have been recovered, it seems likely CLASS 190 is associated with elite contexts.

CLASS 191 (Figure 54; Map 89)

Description: Wide *skyphos* with carinated, flaring rim

Bases: **Ring**; **Flat**

Surfaces: Monochrome decoration is typical

Attachments: Two horizontal handles under shoulder carination

Distribution: Iron I; **Iron II/Iron III** (coastal Levant; Orontes Syria); Persian

Parallels: Lehmann 1996, Form 152

Comments: The "pendant semi-circles" motif is widely considered indicative of imported Euboean *skyphoi* (Descœudres and Kearsley 1983, 44-46); a provenience that has been scientifically tested (Popham *et al.* 1983). Other less-distinct decorative schemes are also evident, and again are evidence for importation from the Greek world during the Proto-Geometric period, or Iron I period (Descœudres and Kearsley 1983).

CLASS 192 (Figure 54)

Description: Medium-sized, deep-bowl with flaring rim

Bases: Ring

Surfaces: Red-Slip is rare

Attachments: Two handles under rim are known

Distribution: Iron I; Iron II; Iron III (west inland Syria)

Parallels: Lehmann 1996, Form 122

CLASS 193 (Figure 54; Map 90)

Description: Shallow-bowl (mortarium?) with flat base and heavily thickened lip

Bases: **Flat**; Disc; Low-ring

Surfaces: Red-Slip and Monochrome are rare

Distribution: Iron II; Iron III; **Persian** (north Syria; Levantine coast)

Parallels: Lehmann 1996, Forms 159-163; 165; 167-173

SECTION THREE

Analysis of Iron Age Ceramic Data

CHAPTER SEVEN
Exploratory Data Analysis

7.1 Introduction

The purpose of this chapter is to explore some of the variables within the ceramic database presented in **Chapters 5** and **6**; such as context type, geographic location, decoration, and vessel function and form. More specifically, these elements are investigated in order to reveal their change geographically and diachronically. However, listing the different ceramic variables for each of the 54 different sites seemed an arduous task. Hence, the definition of more manageable categories of time and space was an important part of this process.

The exploratory data analysis within this chapter was undertaken using MS Access and Excel, and WinBASP v. 5.43. The data was originally entered into the main MS Access database, as described in **Chapter 5**, and then imported into MS Excel for ease of analysis. Excel pivot-tables and pivot-table charts were used to filter the data and explore any perceivable or conceivable patterns. Many avenues were explored, and not all were rewarded with interesting patterns; nevertheless, important insight into the structure and patterning of the data was gained from positive and negative outcomes. This chapter only presents lines of enquiry that were seen as providing appropriate background for, or directly contributed toward, the interpretation contained within **Chapter 9**.

It was important that the following analyses never lost sight of the fact that the database is comprised of incidence data. The different charts and tables discuss percentages only with regard to incidents in the database, rather than actual numbers of vessels. This is why the exploratory data analysis concentrates on multivariate analyses; by which different categories are compared with each other according to a common measurement. Absolute numbers and percentages are only used to illustrate different general trends.

7.2 Temporal Analysis

In order to understand many of the distribution patterns evident in this chapter, it was important to identify meaningful temporal categories in the data. Hence, when diachronic change was investigated, chronologically imprecise contexts were removed from the analyses. For example, to plot the development of Red-Slip (**§7.6**) only single-period contexts could be used; contexts that could fall into more than one sub-period would "blur" the results. Unfortunately, many assemblages have derived from contexts of poor or vague provenience, rendering the "period specific" database smaller than expected. "Period-specific" data was used to determine which CLASSES could be considered "abundant" within each period. While the number of incidents of each CLASS was an important consideration, only those that appeared at four or more sites during a single period were considered "abundant"; a CLASS that was well-represented but present at only one or two sites was not included in the results. In other words, only those CLASSES that appeared repeatedly in Iron I contexts were considered representative of the Iron I period. The results are presented in **Tables 7.1** and **7.2**.

Table 7.1: "Abundant" CLASSES in Iron I and Iron II periods

Iron I CLASSES		Iron II CLASSES
4, 6, 8	Cooking-pots	1, 2, 3, 4, 5, 6, 7, 8
12, 13	Lamps/burners	12, 13, 14
16, 21, 22, 26	Transport Amphorae	16, 17, 18, 19, 22, 24, 26, 28
31, 34, 35, 37	Pithoi	37
38, 40, 42, 53	Kraters	38, 40, 42, 44, 45, 46, 47, 48, 49, 53
57	Urns/Amphorae	57, 59, 62, 64
71, 73, 78, 80	Jugs	68, 71, 73, 74, 77, 78, 80, 81, 82, 84, 85, 86, 87, 88, 90, 97
100	Dipper Juglets	100
102	Flasks	102
103, 104, 105, 107	Spouted	103, 105, 106, 107, 111
116	Bottles	119
123, 131	Cups/Chalices	131, 132
134, 157, 158, 160, 161, 164, 165a, 168, 184, 187, 188	Bowls	134, 139, 141, 143, 144, 145, 151, 152, 153, 154, 155, 157, 158, 160, 161, 162, 163, 164, 165, 166, 167, 168, 179, 180, 181, 184, 187, 188, 189, 191, 193

Table 7.2: "Abundant" CLASSES in Iron III and Persian periods

Iron III CLASSES		Persian CLASSES
1, 4, 6, 7, 9, 10	Cooking-pots	1, 6, 9
12	Lamps/burners	12
18, 24, 26, 27, 28	Trans. Amph.	18, 23, 24, 25, 27, 28
37	Pithoi	
42, 49, 53	Kraters	
57, 62	Urns/Amphorae	58
73, 77, 80, 82, 84, 85, 87, 88, 90, 91	Jugs	73, 93, 95, 98
100	Dipper Juglets	100
102	Flasks	
119, 120	Bottles	117
134, 139, 141, 143, 144, 145, 148, 151, 152, 153, 155, 157, 160, 161, 162, 163, 164, 167, 168, 179, 183, 184, 187, 188, 191, 193	Bowls	134, 141, 160, 162, 193

While **Tables 7.1** and **7.2** list the CLASSES considered "abundant" within each period, they do not easily demonstrate *change* over time. For this reason, **Tables 7.1** and **7.2** are summarised in **Chart 7.1**, which presents the persistence of ceramic traditions and rate of ceramic change across conventional sub-divisions of the Iron Age. In addition, the ceramic CLASSES considered typical of the Iron I period were compared with Late Bronze Age assemblages from Tell Bazi, Ras Ibn Hani, Tell Afis, Tell Kazel, Tell Arqa, Sarepta, Tyre, Dan, and Hazor (Anderson 1988; Badre & Gubel 1999-2000; Ben Dov 2002; Bikai 1978b; Bounni *et al.* 1998; Mazzoni 2002e; Otto 2006; Thalmann 2006b; Yadin *et al.* 1958; 1960). This exercise was undertaken in order to assess the apparent similarity between Late Bronze Age pottery and Iron I pottery. Indeed, **Chart 7.1** confirms that Iron I pottery relied heavily on vessel forms that were already present in the Late Bronze Age. Of the 42 typical Iron I CLASSES, 36 (or 86%) were identified in Late Bronze Age contexts (**Appendix B**). In other words, much of the Iron I ceramic repertoire drew upon Late Bronze Age ceramic traditions. This is in sharp contrast to the Iron II period, when less than 40% of typical Iron II CLASSES were consistently found in the preceding Iron I contexts. While many of the typical Iron I forms were present in the Iron II period, there is a marked increase in new ceramic forms. **Chart 7.1** demonstrates the strong ceramic link between the Late Bronze Age and the Iron I period, with only minimal ceramic change evident in the Iron I period. In contrast, the Iron II period is depicted as a time of rapid ceramic change, yet it still retained a large percentage of earlier ceramic traditions. The Iron III and Persian periods witnessed little change among existing ceramic traditions. **Chart 7.1** also clearly depicts the diminishing influence of earlier ceramic traditions over time.

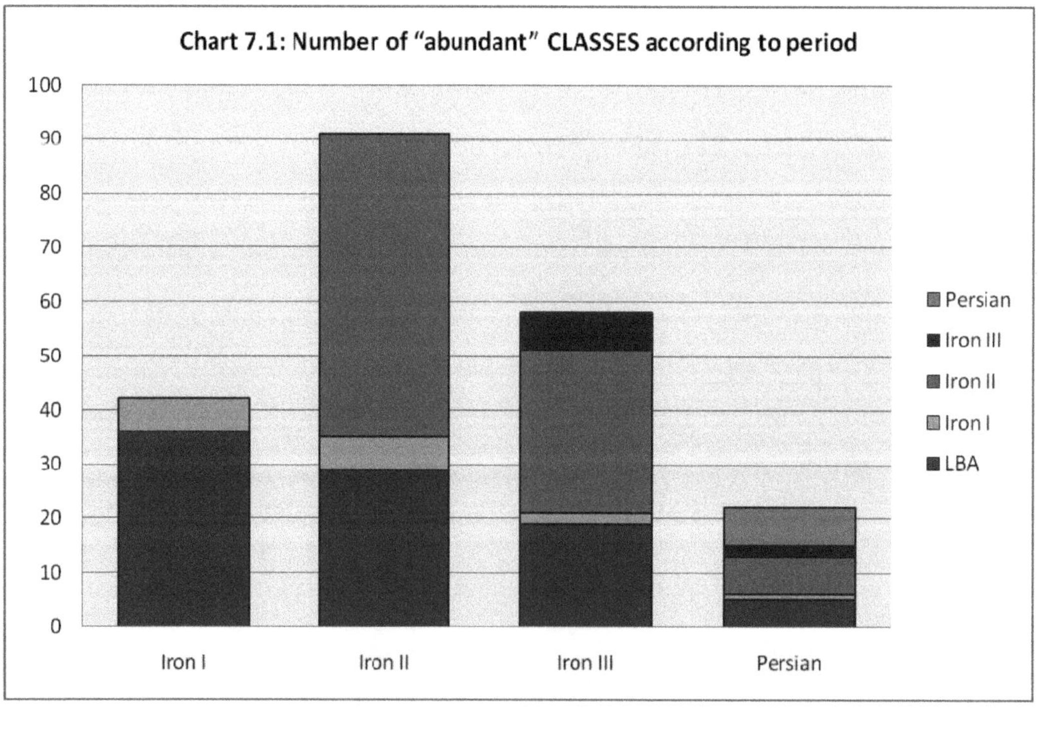

The strong continuity of ceramic culture across the conventional Bronze/Iron Age transition has important implications for the way we might view early Iron Age societies. Since early Iron Age pottery is similar to that of the Late Bronze Age, it attests to continuity of manufacturing traditions and ceramic technology, and thus probably also population, and therefore to an early Iron Age society and culture based on that of the Late Bronze Age (Mazzoni 2000d; 2001; Caubet 1992; Peckham 2001, 20-21). Continuity of population also implies that no significant (limited in scale or impact) migration of "Sea Peoples" or other new entities took place (**§2.3**). The term Iron I, therefore, appears to be misleading; this period appears to have been more closely aligned with Late Bronze Age culture than with that of the later Iron Age. Might we consider the early Iron Age of the Northern Levant to be, in essence, a sub-Late Bronze Age? A conventional and literal reading of the historical narrative does not adequately explain this continuity.

While **Chart 7.1** is interesting, not every "abundant" CLASS in **Tables 7.1** and **7.2** can be considered temporally specific. For example, CLASS 134 bowls are "abundant" throughout the entire Iron Age and are, therefore, a poor representation of any single sub-period. To help isolate which CLASSES could be considered temporally-sensitive (or characteristic of a single period) a seriation was undertaken using WinBASP version 5.43 (the Seriation matrices presented below are difficult to read in detail, and are included on the appended CD for closer inspection). The first seriation was performed using only stratified sites with good-sized assemblages, preferably with a long Iron Age sequence. The key site for the inland Northern Levant was Tell Afis, with its well-published ceramic assemblage covering the length of the Iron Age; other sites used were Tell Keisan, Hazor, Tyre, Megiddo, Sarepta, Hama, and Tell Jurn Kabir. The first seriation included the "abundant" ceramic CLASSES from **Table 7.1** and **Table 7.2**. The resulting seriation matrix is crowded, though a diagonal "curve" is discernible (**Seriation 1**).

Despite limitations in the data, some temporal ordering is present in **Seriation 1**. The progression of time can be "read" from left to right, and top to bottom (i.e. the most recent contexts and forms are in the bottom right of the matrix, and the older examples are found in the top left). While there is much confusion within the centre of the matrix, the two extremities show good chronological distinction. The sequential ordering of the horizontal axis does not perfectly reflect stratigraphic succession from left to right, but this can be expected considering the degree of material cultural residuality within tell sites. Nevertheless, there is a general grouping of "early" contexts to the right, and "later" contexts to the left. The shades/colours represent different ceramic functional categories, as presented on the next page.

Seriation 1 (CD/SERIATION/SERIATION1)

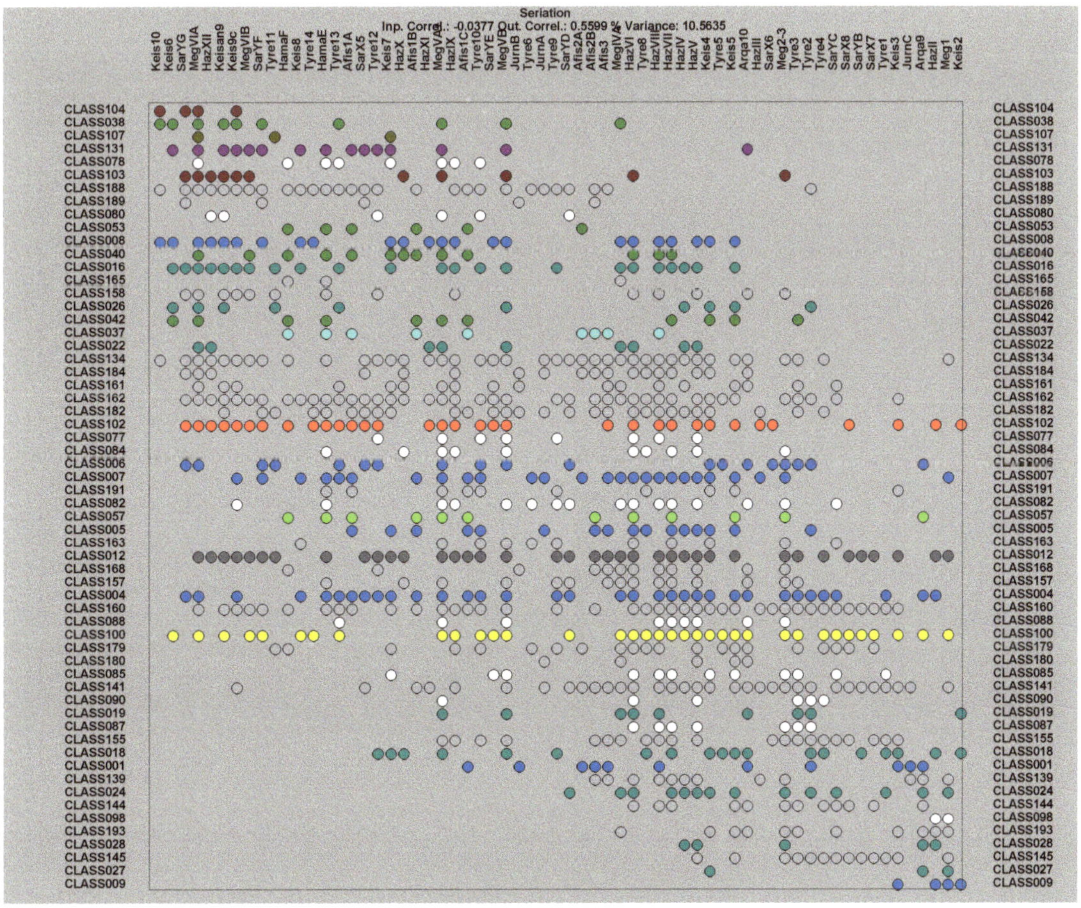

Seriation Legend

Blue • Cooking-pots		Teal •	Transport Amphorae
Turquoise •	Pithoi	Dark Green •	Kraters
Green •	Urns	White •	Jugs
Yellow •	Dipper Juglets	Red •	Flasks
Burgundy •	Unguents	Grey •	Lamps
Purple •	Chalice/Cups	Olive Green •	Spouted Vessels
No fill °	Bowls		

A second seriation was performed on the data using the same sites but with the removal of obvious long-lived CLASSES (e.g. CLASS 134). It was hoped that the removal of these persistent forms would give the seriation curve some definition, which would help emphasise the chronological value of some CLASSES. Indeed, **Seriation 2** shows better definition in the "curve", especially at the two extremities (note the matrix order was inexplicably reversed – progression of time is now from right to left, and bottom to top). The shades/colours in the matrix imply that some cooking-pots, transport amphorae, jugs, pithoi and bowls were chronologically sensitive for much of the Iron Age (see above legend). Kraters, unguent containers, and chalices/cups were generally indicative of the earlier Iron Age periods.

While there is some general chronological order evident amongst the contexts on the horizontal axis, there remains some confusion regarding specific sequencing. For instance the Megiddo sequence adheres to its published order (VIB, VIA, VB, etc...), while the Keisan and Tyre phasing are somewhat mixed. For this reason, some additional seriations were performed on further restricted datasets.

Seriation 2 also revealed an interesting relationship between the Hama and Tell Afis assemblages; i.e. Tell Afis followed good stratigraphic order, but the two main Hama strata were reversed and were *both* considered comparable with the Tell Afis Iron I assemblages. This trend was worth further investigation.

Seriation 2 (CD/SERIATION/SERIATION2)

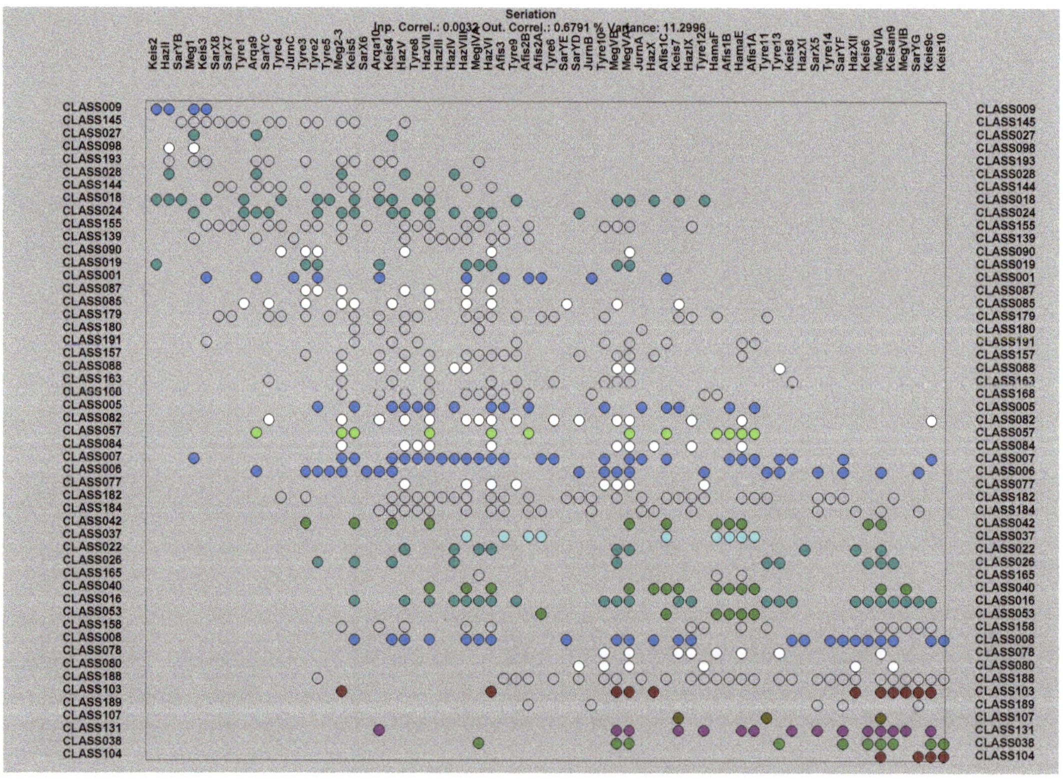

Seriation 3 confirms the general order of the Tell Afis sequence, though slightly altered in detail. The order of the Hama E and F assemblages, however, remain reversed and, more importantly, both were positioned before the Tell Afis Iron I assemblages. This persistent pattern (also **Seriation 2**) seems to imply that the current interpretation of the Hama E pottery as a clear Iron IIB assemblage is problematic. Either the material is mixed, and contains a high level of earlier Iron Age pottery, or Hama E has been incorrectly dated to the Iron IIB period. The ceramic data confirms the need for caution regarding the assumed Assyrian destruction of Hama E (**§2.4**). Of course, not all patterns evident in a seriation are strictly chronological; the difference between the Hama and Tell Afis assemblages may instead be due to regional variation (there is 70 km between the sites). Hence, the possibility of variation due to geographic positioning needed further exploration. For this reason, **Seriation 4** included four geographically-distant sites from the Northern Levant; Tell Afis, Tyre, Tell Jurn Kabir, and Tell Arqa. A good "curve" is again present, but the two inland sites are grouped separately to the two coastal sites. Nevertheless, there is some order within the two groups. For instance, the position of the Tell Jurn Kabir assemblages amidst those from Tell Afis confirms the published chronology for this site: a short life-span in the late-Iron II and Iron III periods. This matrix seems to indicate that geographic position had a significant effect on the make-up of a ceramic assemblage. More importantly, the results appear to confirm the coastal-inland distinction evident in Lehmann's (1998) and Mazzoni's (1991-1992) work.

Seriation3 (CD/SERIATION/SERIATION3)

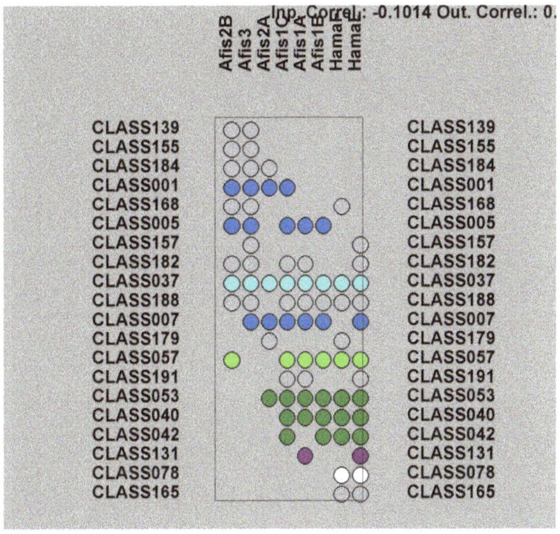

Seriation 4 (CD/SERIATION/SERIATION4)

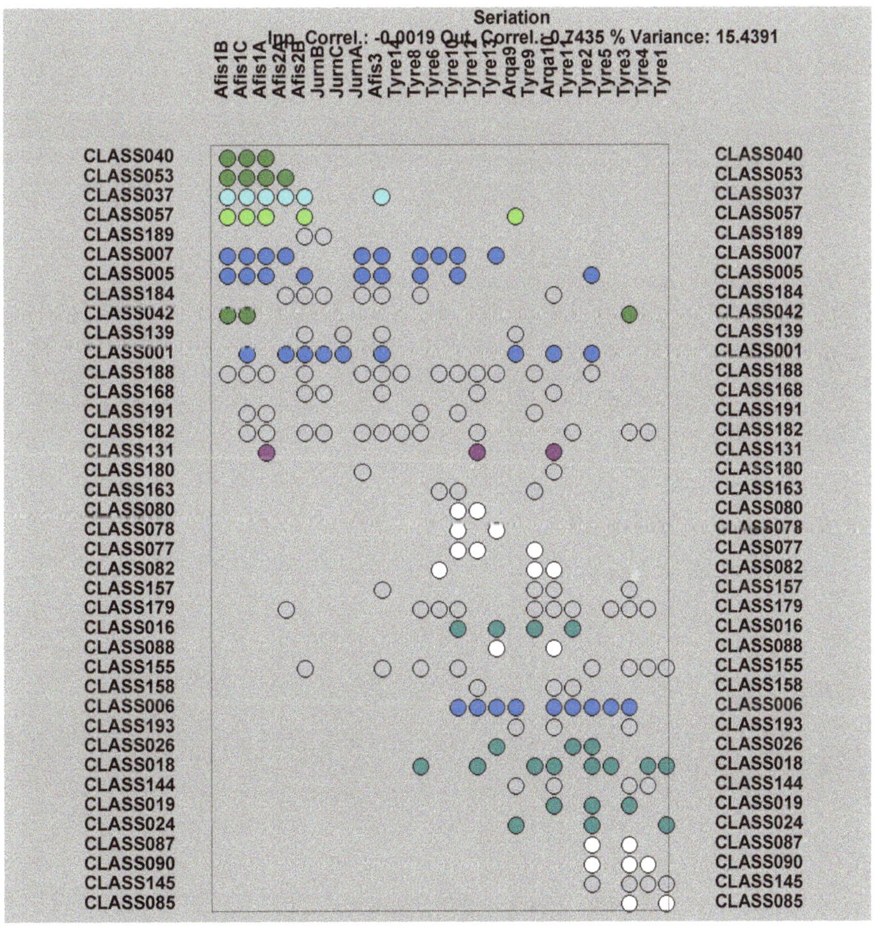

Seriation 5 (CD/SERIATION/SERIATION5)

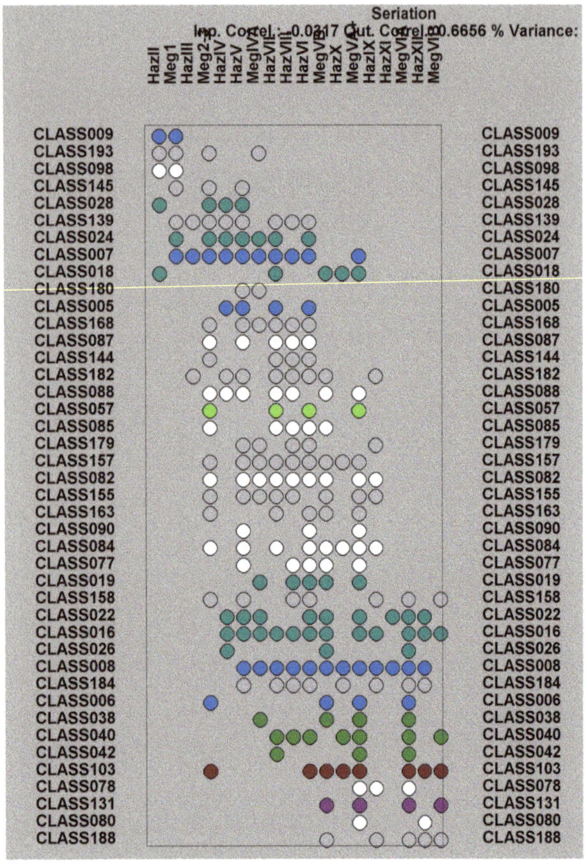

A fifth and final seriation was undertaken on the Hazor and Megiddo assemblages for comparative purposes. **Seriation 5** again shows good order in the data, with each site's sequence following a general order. Furthermore, there are a few vessel categories that appear to be good chronological indicators; kraters, chalices/cups, and unguent containers are primarily early Iron Age phenomena. **Seriation 5** also shows a large cluster of Iron II forms in the middle of the matrix that had no Iron I precedents, confirming that the Iron II period experienced rapid ceramic change (**Chart 7.1**).

The five seriation analyses have demonstrated that there is some general chronological ordering in the database (at least relatively if not absolutely), despite the many limitations in using presence/absence data. In addition, there appears to be some vessel forms that might be considered good chronological indicators. Finally, it is worth re-iterating that not all the patterns in the matrices reflect chronological order, and might instead be the result of different cultural and/or geographic factors. Nevertheless, **Seriation 4** appears to confirm that the Northern Levant is split into at least two different regions; coastal and inland. Geographic distribution and ceramic regionalisation will be explored further throughout this and the following chapters.

7.3 Spatial Analysis

The geographical areas used in **Chapter 5** (**Charts 5.3** and **5.4**) were considered too broad for an analysis of spatial distribution. Hence, nine smaller geographic zones were created for use throughout the remainder of the study. Though essentially arbitrary categories, these nine zones were based on geographic considerations. They are defined in **Table 7.3** and depicted in **Map 91**. Before exploring each of these zones in more detail, it was deemed important to understand any geographical weighting within the data. **Chart 7.2** depicts the percentage of database incidents according to these nine zones. Apart from confirming the bias towards the Southern Levant evident in **Chapter 5**, it shows the limited nature of the data from the inland Northern Levant; only Orontes Syria is well represented. Furthermore, despite only two sites representing the Northern Palestine zone, it is the most abundantly attested. The Beqa' Valley, on the other hand, is also represented by only two sites, but accounts for a much smaller percentage of incidents. North-west Syria and the Euphrates are represented by a good sample of sites, but only a relatively small percentage of database incidents. The full publication of the OI Amuq sequence would help correct this imbalance.

Table 7.3: Local ceramic zones

Zone	Modern Equiv.	Sites
1	Syrian Coast	Al Mina; Ras al Bassit; Ras Ibn Hani; Tel Sukas; Tell Kazel; Tell Arqa
2	Lebanese Coast	Byblos; Beirut; Khalde; Sidon; Tambourit; Qraye; Qasmieh; Joya; Khirbet Silm; Tyre; Tell Rachidieh; Sarepta
3	Beqa' Valley	Tell el Ghassil; Kamid el Loz
4	Jezreel Valley	Pella; Tel Rehov; Beth Shan; Tel Jezreel; Megiddo
5	Northern Palestine	Tel Dan; Hazor
6	Palestinian Coast	Akhziv; Tell Keisan; Tell Abu Hawam; Athlit
7	Orontes Syria	Tell Nebi Mend; Tell Mishrife; Hama; Tell Qarqur; Tell Mastuma; Tell Afis
8	North-west Syria (incl. Amuq)	Deve Höyük; Tell Rifa'at; Zincirli; 'Ain Dara; Nayrab; Tell Abou Danne; Tell Judeideh; Tell Ta'yinat; Chatal Höyük
9	Euphrates Syria	Tille Höyük; Jerablus; Kefrik; Tell Shiyukh Fawqani; Tell Ahmar; Tell Jurn Kabir; Tell Sheikh Hassan

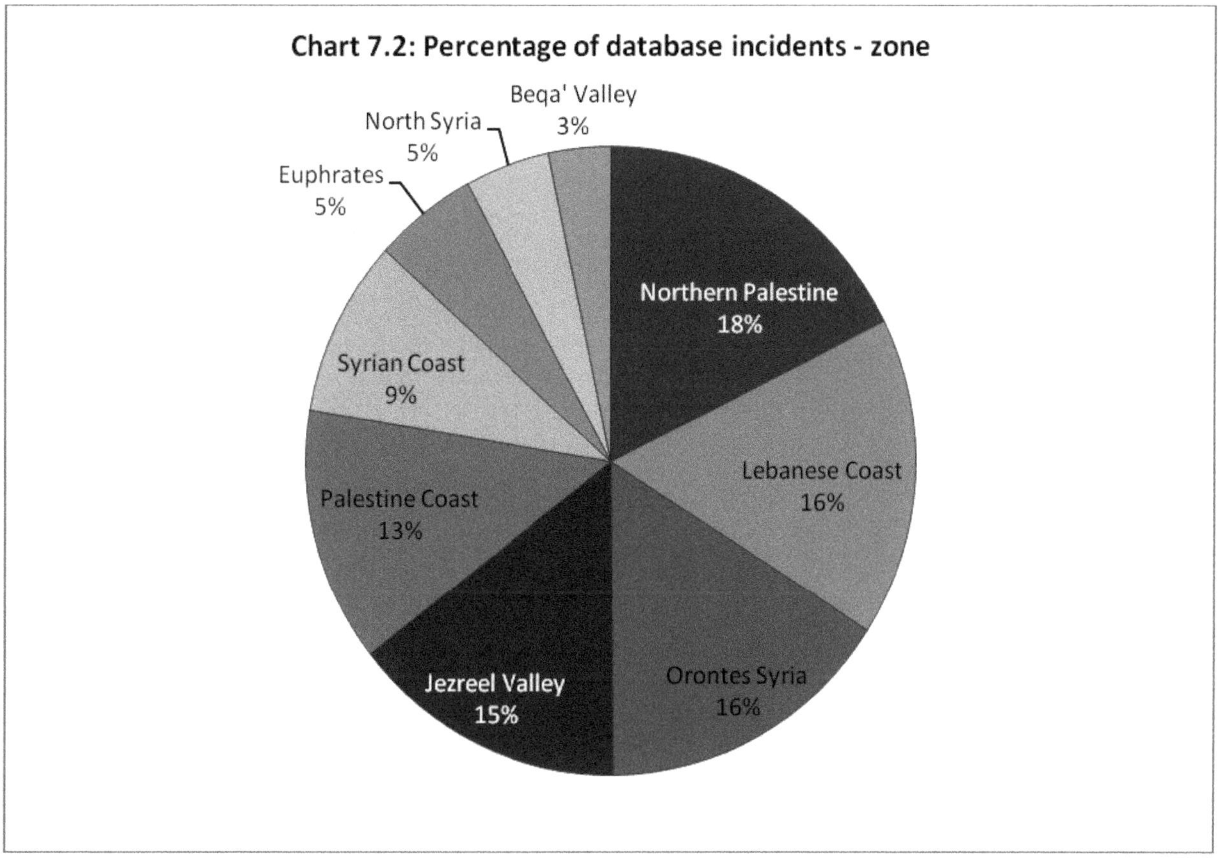

Chart 7.3 shows that despite an overall bias in the database toward settlement assemblages, the Palestine and Lebanese Coasts show a strong mortuary component; there is a distinct lack of reliable, stratified settlement contexts from these two zones. The mortuary bias within coastal zones may account for the two-region model presented in the literature (**§4.2**).

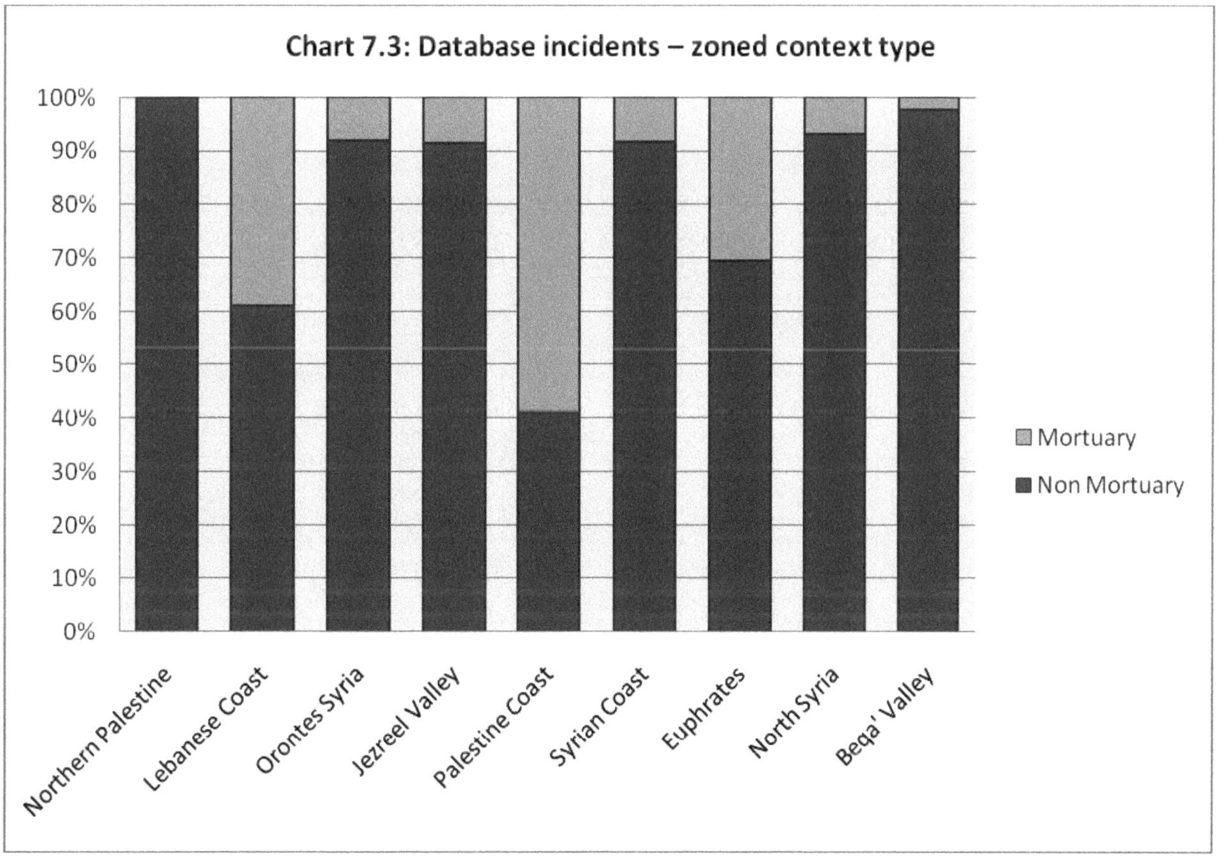

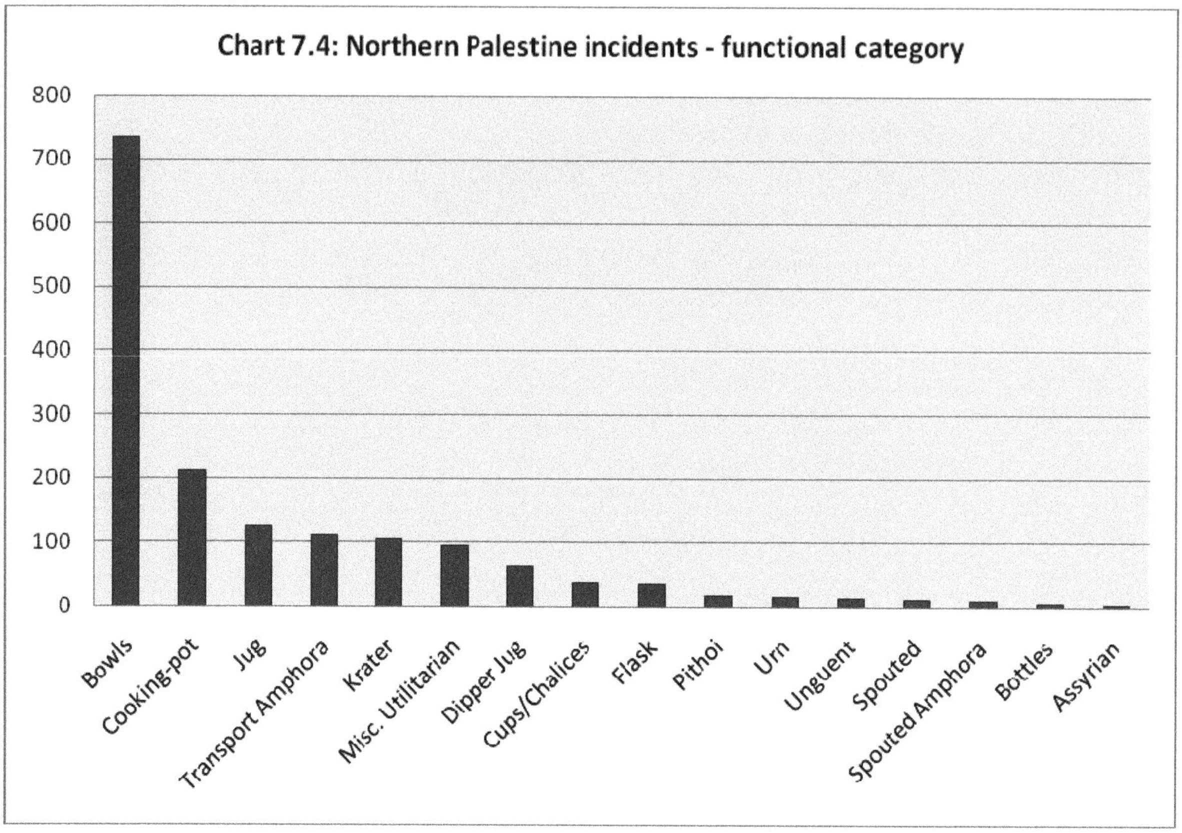

Ceramic trends in each of the nine zones are discussed in order of overall size (**Chart 7.2**), from largest to smallest. **Chart 7.4** shows bowls as the most common functional category in Northern Palestine, followed by cooking-pots. The most abundant forms are CLASS 134, 141, and 152 bowls (all amongst the most abundant within the database – **Chart 5.9**), and CLASS 007 and 008 cooking-pots (**Chart 7.5**).

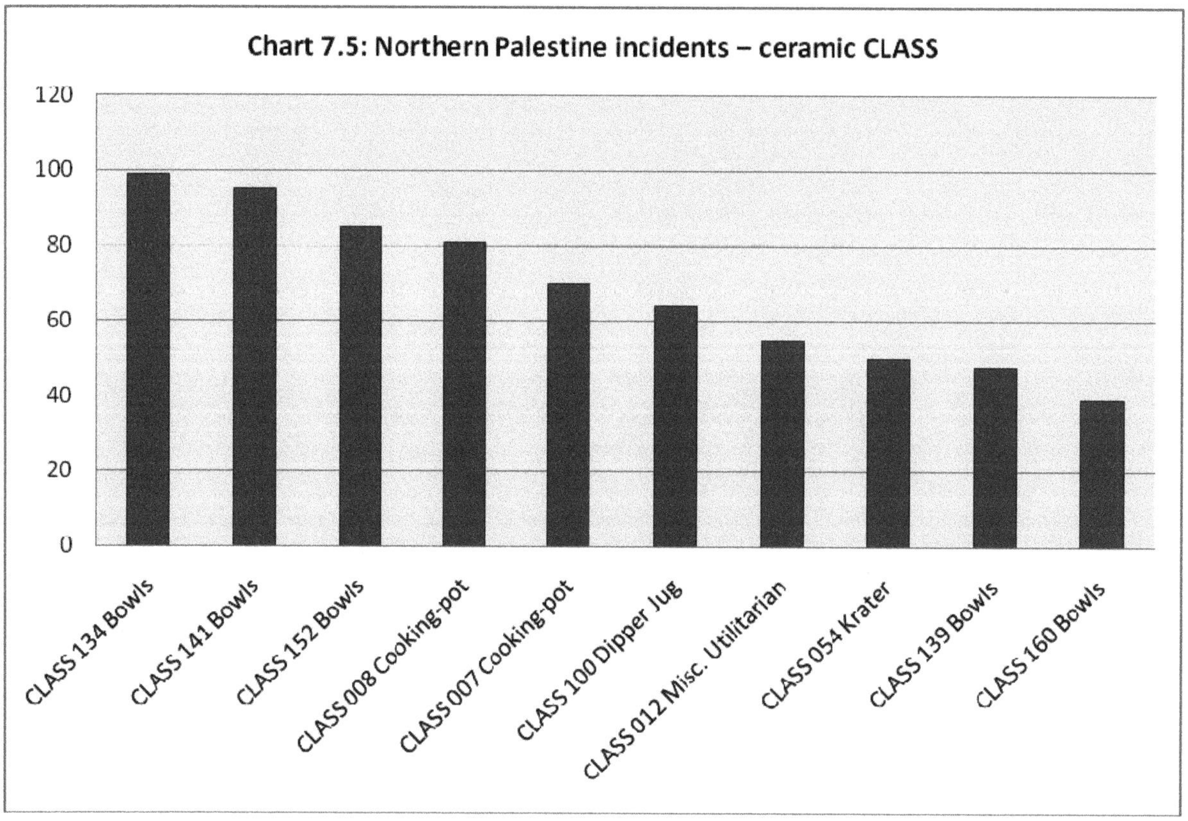

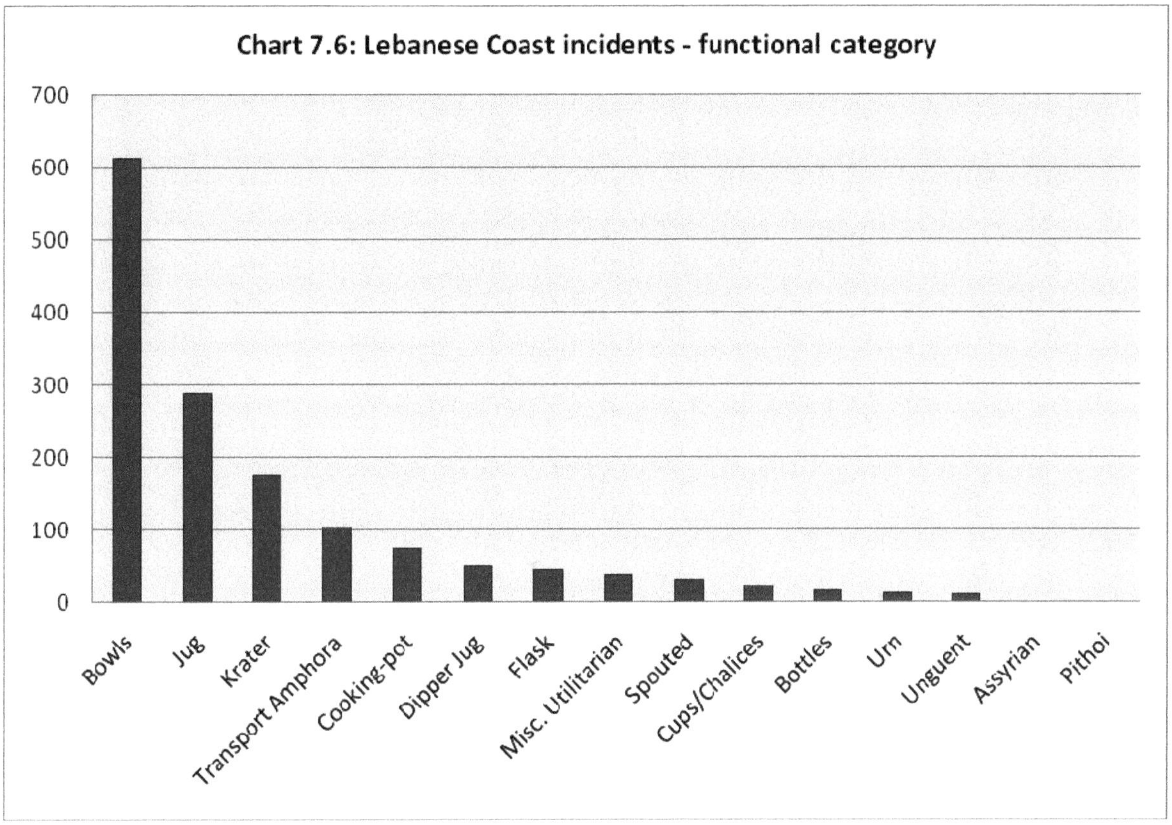

Chart 7.6 shows that the most abundant categories on the Lebanese Coast are bowls, followed by an abundance of jugs and kraters. The two most abundant individual forms, however, are not bowls, but CLASS 042 kraters and CLASS 082 jugs. Nevertheless, five of the ten most abundant CLASSES are bowls (**Chart 7.7**); the other five are associated with the storing or pouring of liquids.

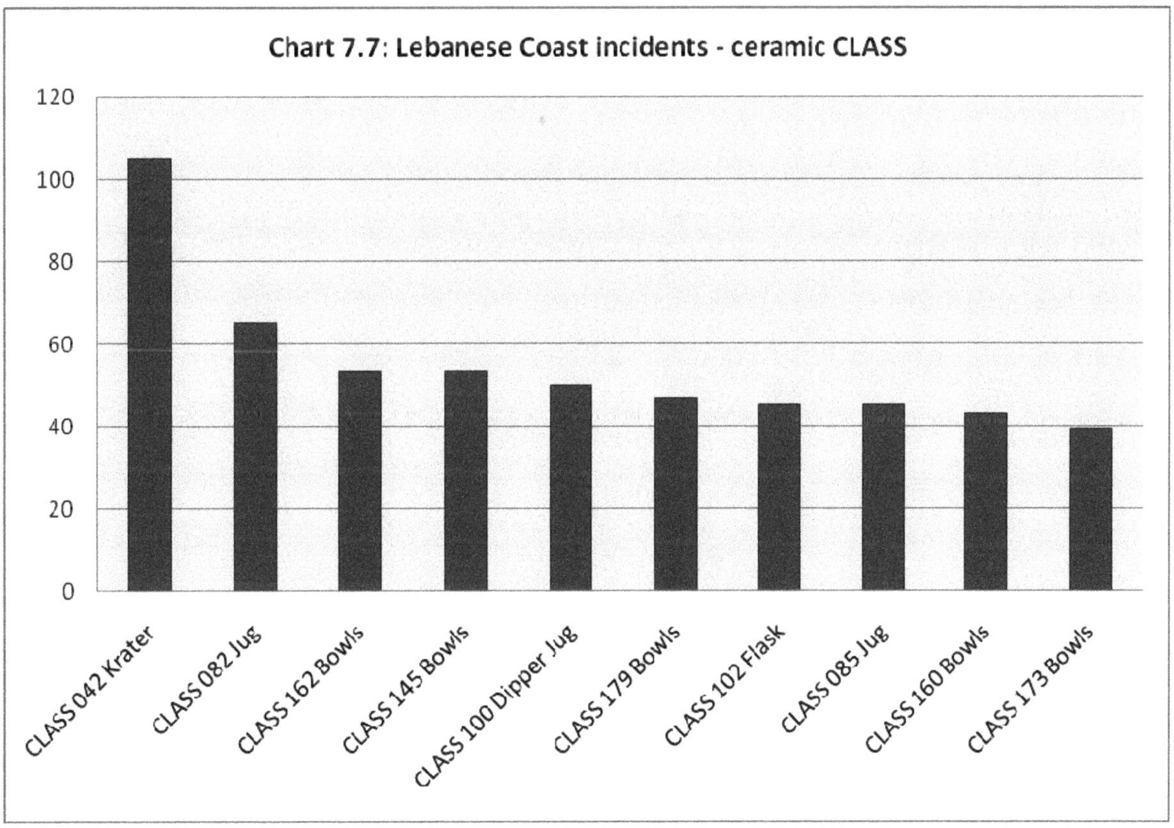

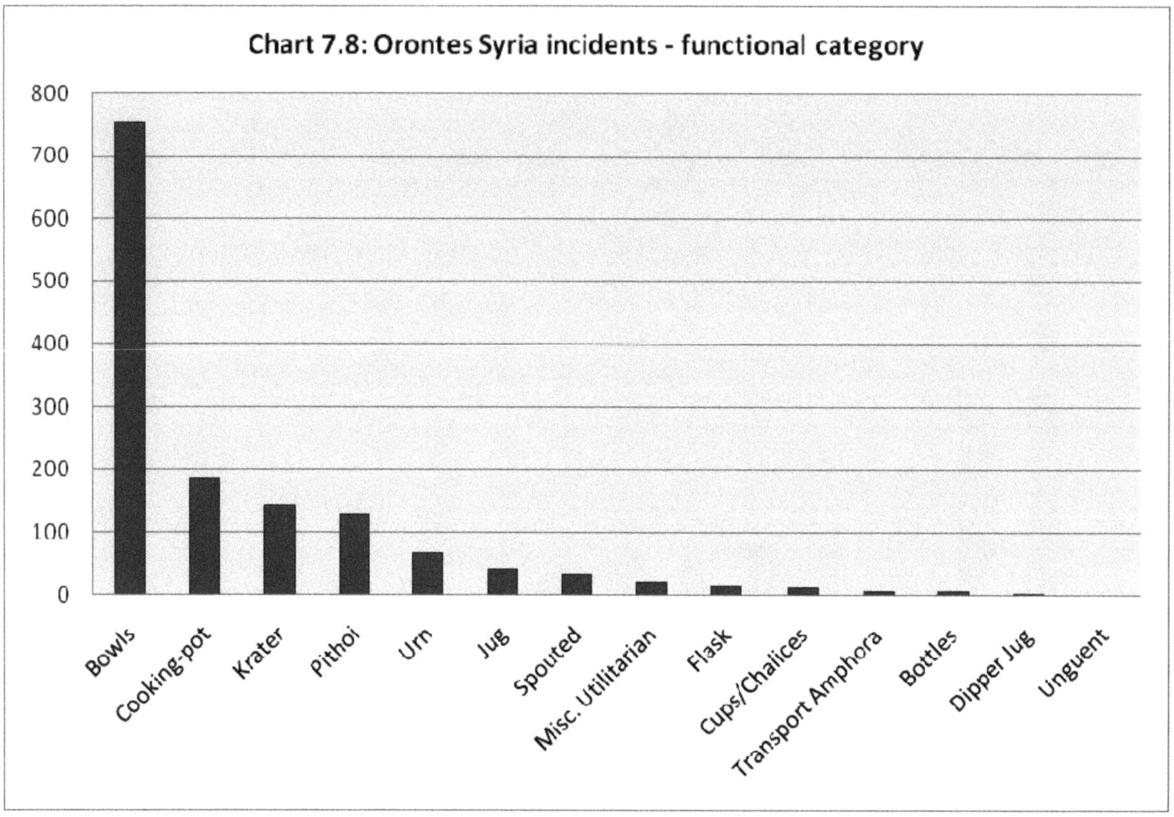

Chart 7.8 shows that bowls are the most abundant category for Orontes Syria, followed by cooking-pots, kraters, and pithoi. Indeed, CLASS 037 pithoi and CLASS 001 cooking-pots are the two most abundant forms; CLASS 057 is also well-attested (**Chart 7.9**). Bowls account for the majority of other abundant forms (i.e. CLASSES 134 and 141); small carinated bowls are also well-attested (CLASSES 182 and 188).

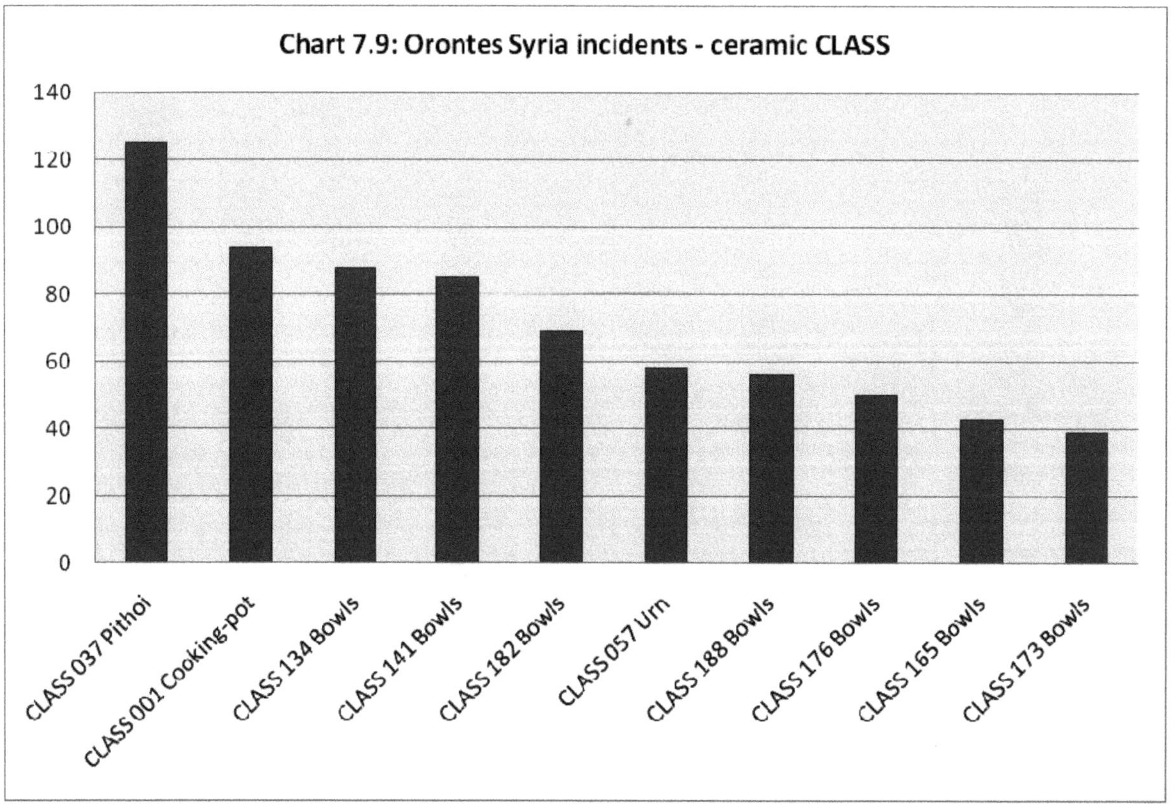

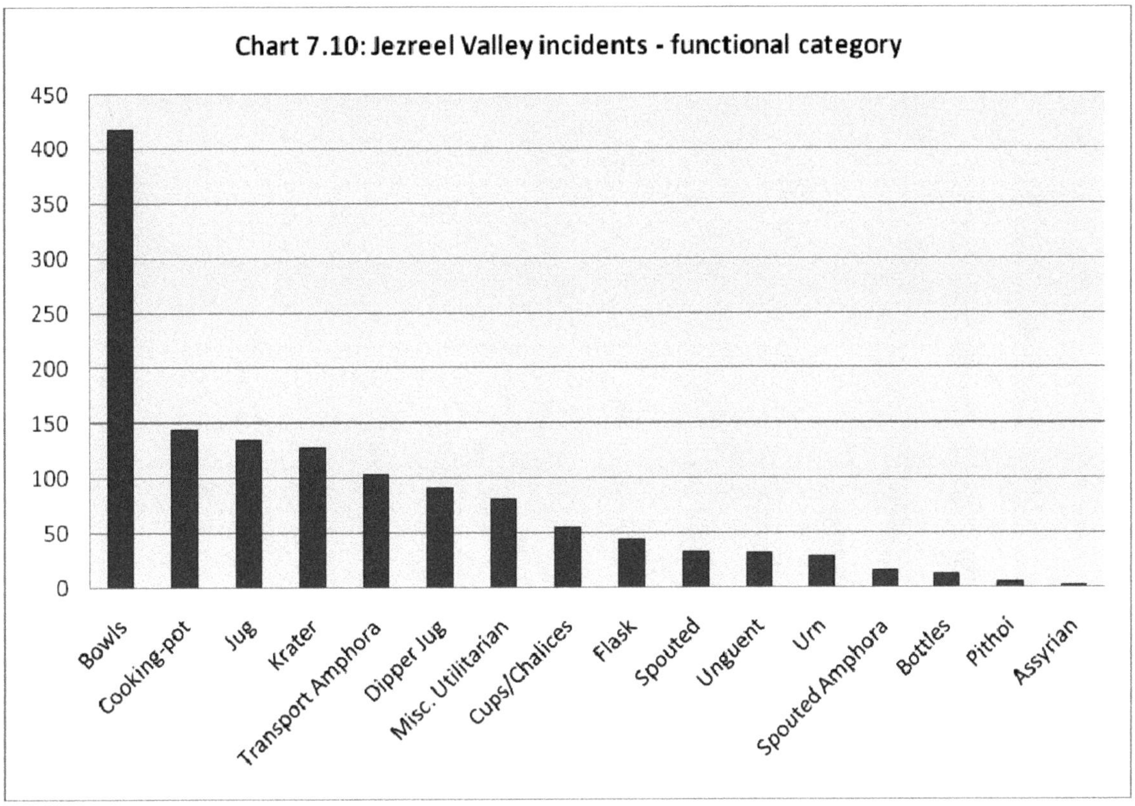

Chart 7.10 shows that bowls are the most abundant category for the Jezreel Valley, followed by cooking-pots, jugs and kraters. The most abundant forms are the CLASS 100 dipper juglet, CLASS 134 and 152 bowls, CLASS 008 cooking-pot, and CLASS 102 flasks (**Chart 7.11**).

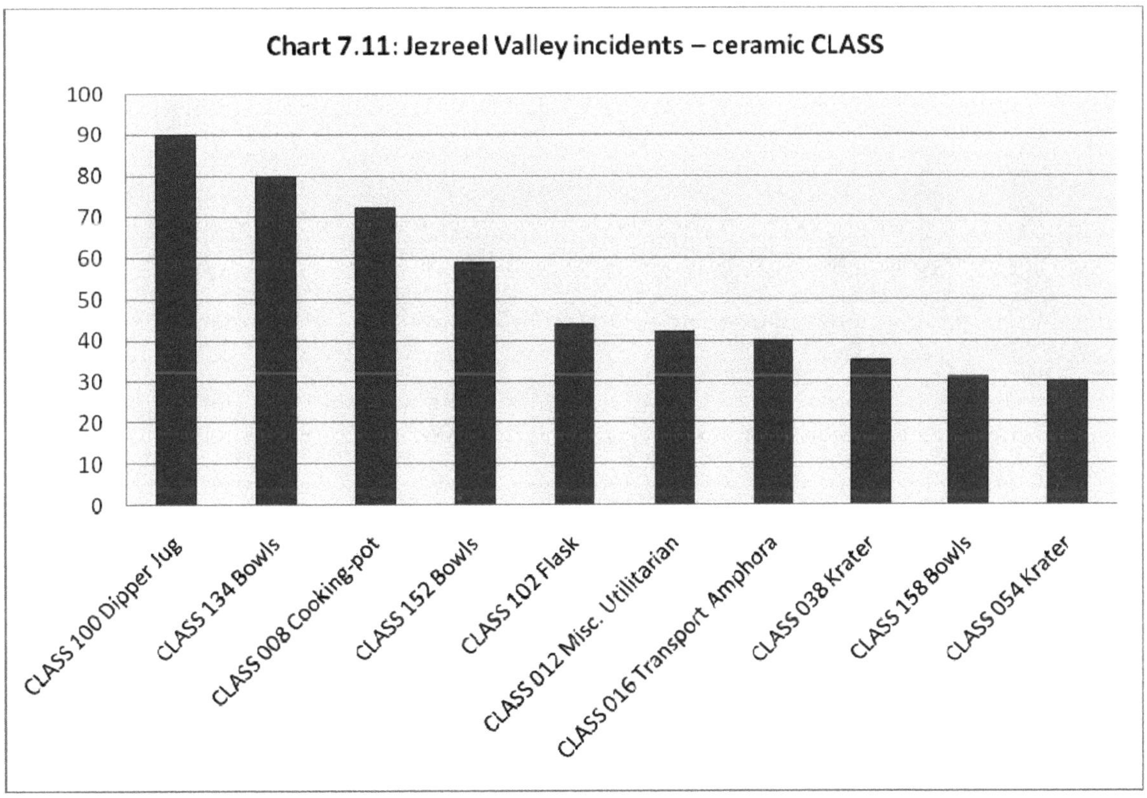

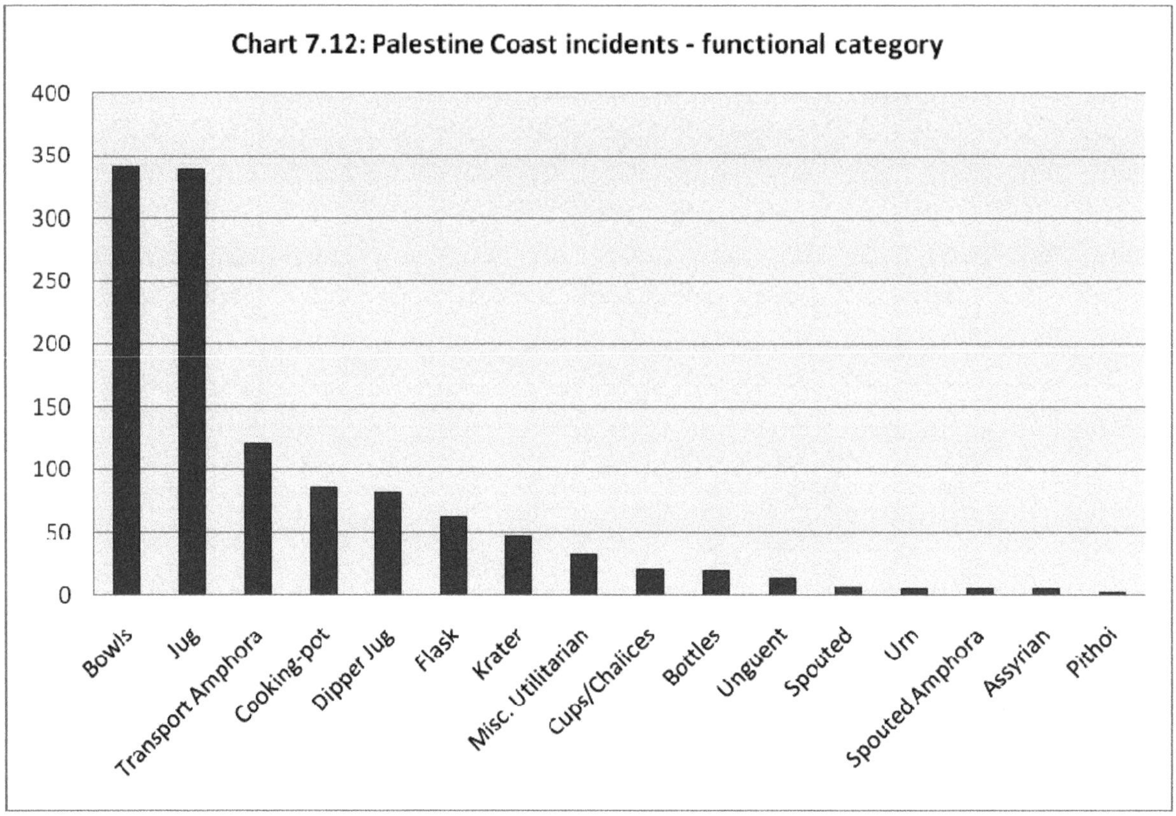

Chart 7.12 shows that bowls and jugs are both roughly equal as the most abundant on the Palestine coast, followed by transport amphorae. The four most abundant forms are all associated with the pouring of liquids; CLASS 082 and 085 jugs, CLASS 100 dipper jugs, and CLASS 102 flasks (**Chart 7.13**). CLASS 024 and 025 transport amphorae are also well attested.

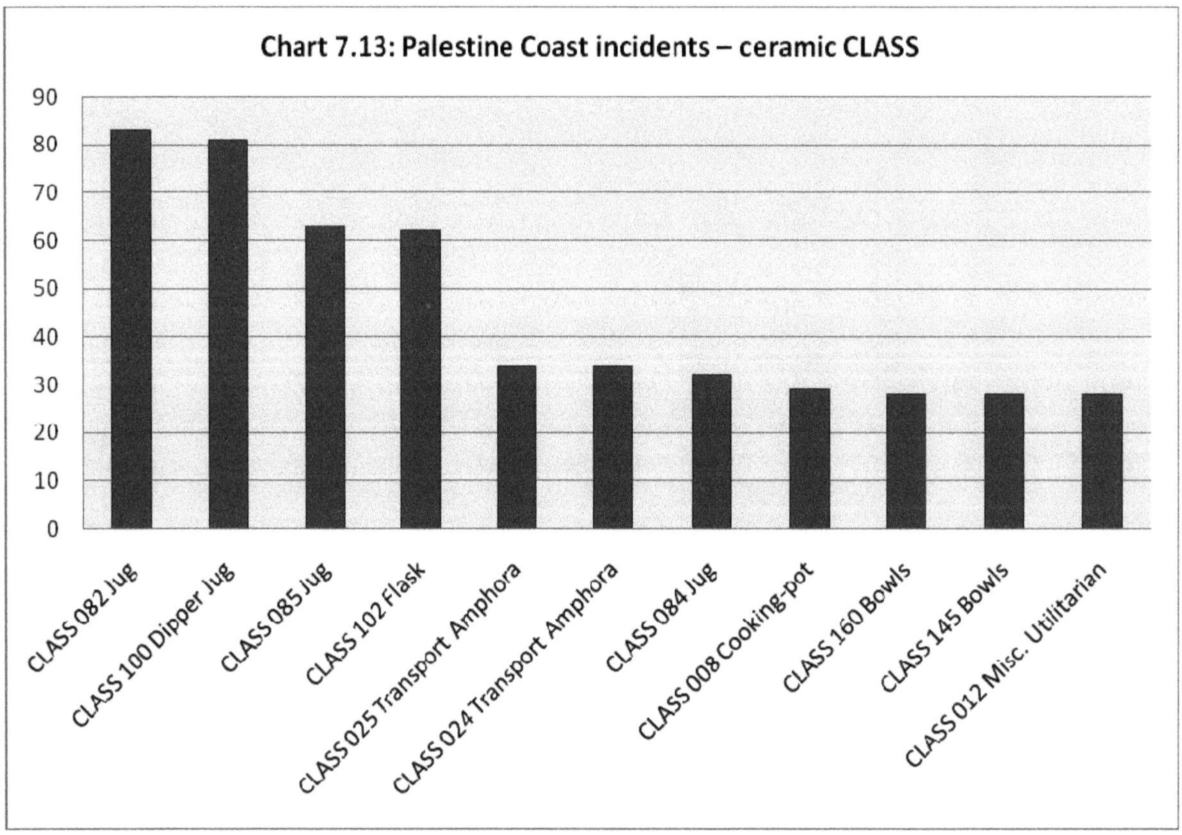

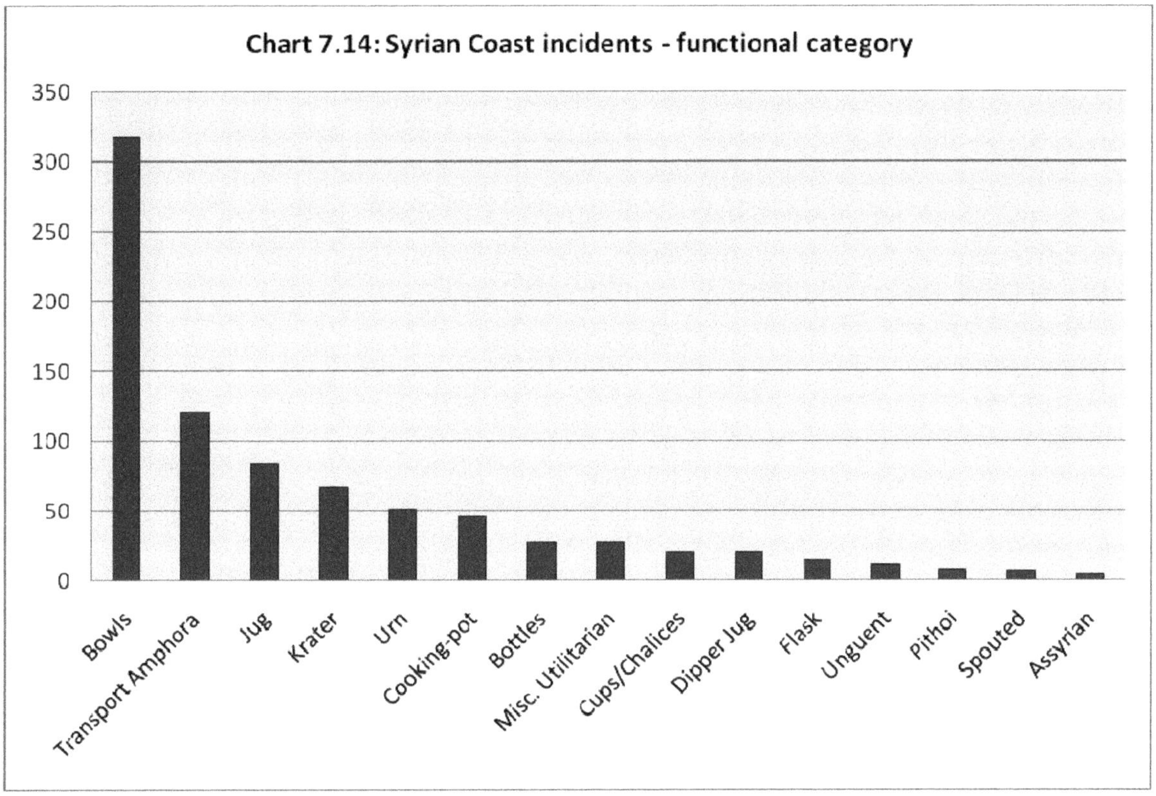

Chart 7.14 shows the abundance of bowls on the Syrian coast, with transport amphorae next, followed by jugs and kraters. CLASS 042 kraters and CLASS 188 bowls are the most abundant forms (**Chart 7.15**). Also well-attested are CLASS 016 and 018 transport amphorae and CLASS 144, 134, and 191 bowls.

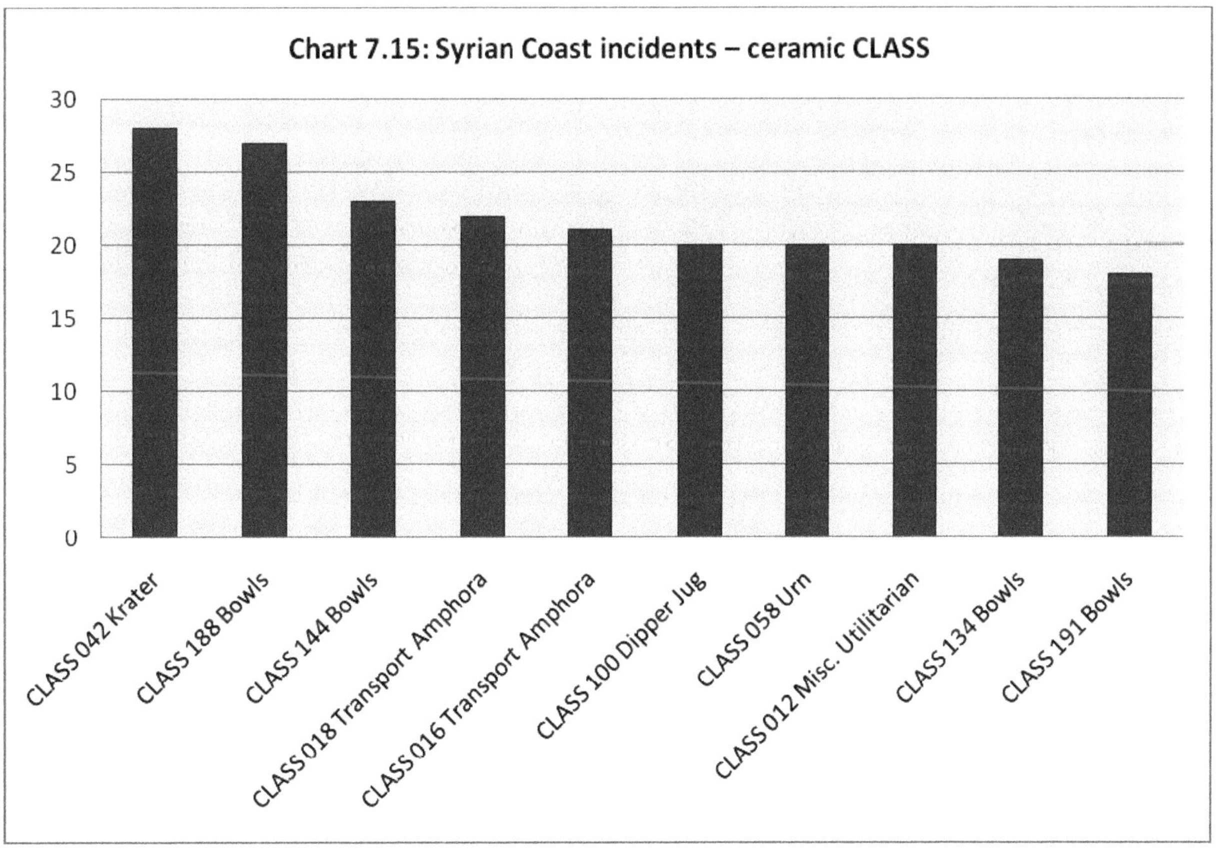

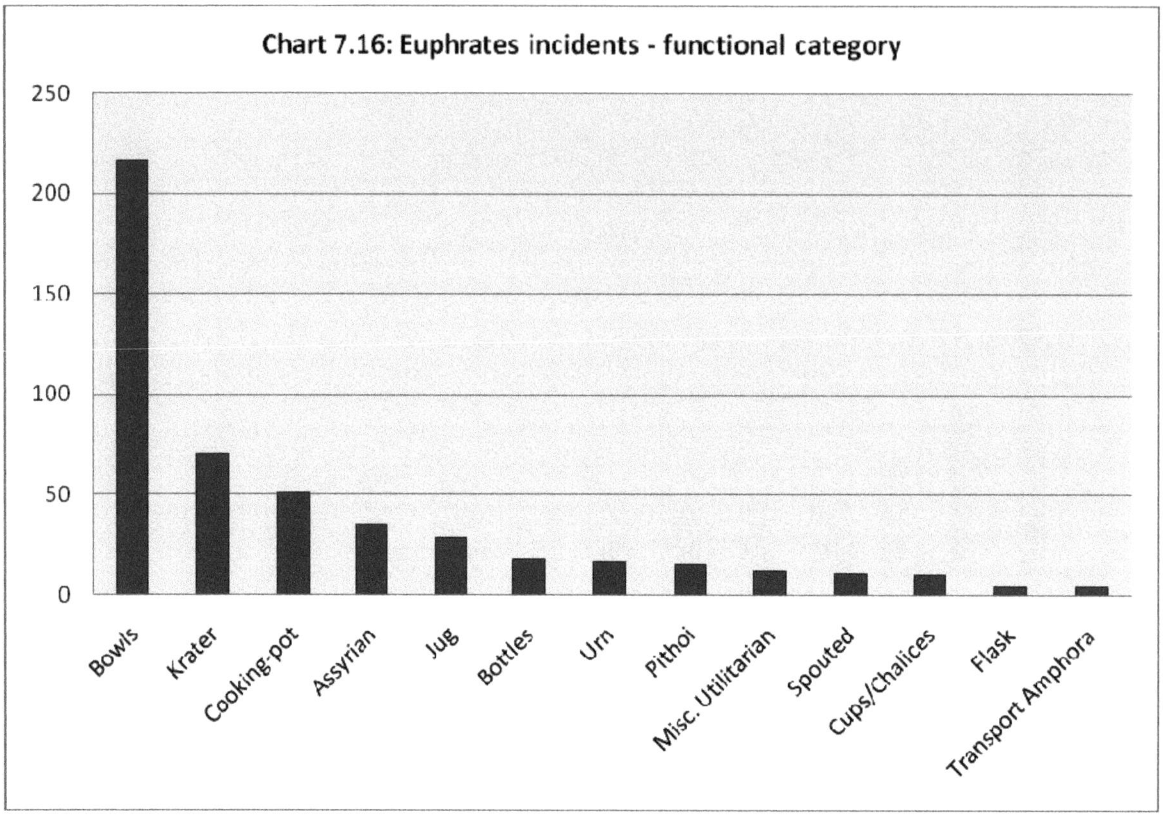

Chart 7.16 shows the by-now familiar pattern of inland Northern Levant regions, with bowls, kraters, and cooking-pots the most abundant categories. It is also significant that the relatively rare Assyrian forms are well-represented in this zone. The most abundant forms are CLASS 134 and 141 bowls, CLASS 121 "Assyrian" cups and bottles, and CLASS 001 cooking-pots (**Chart 7.17**).

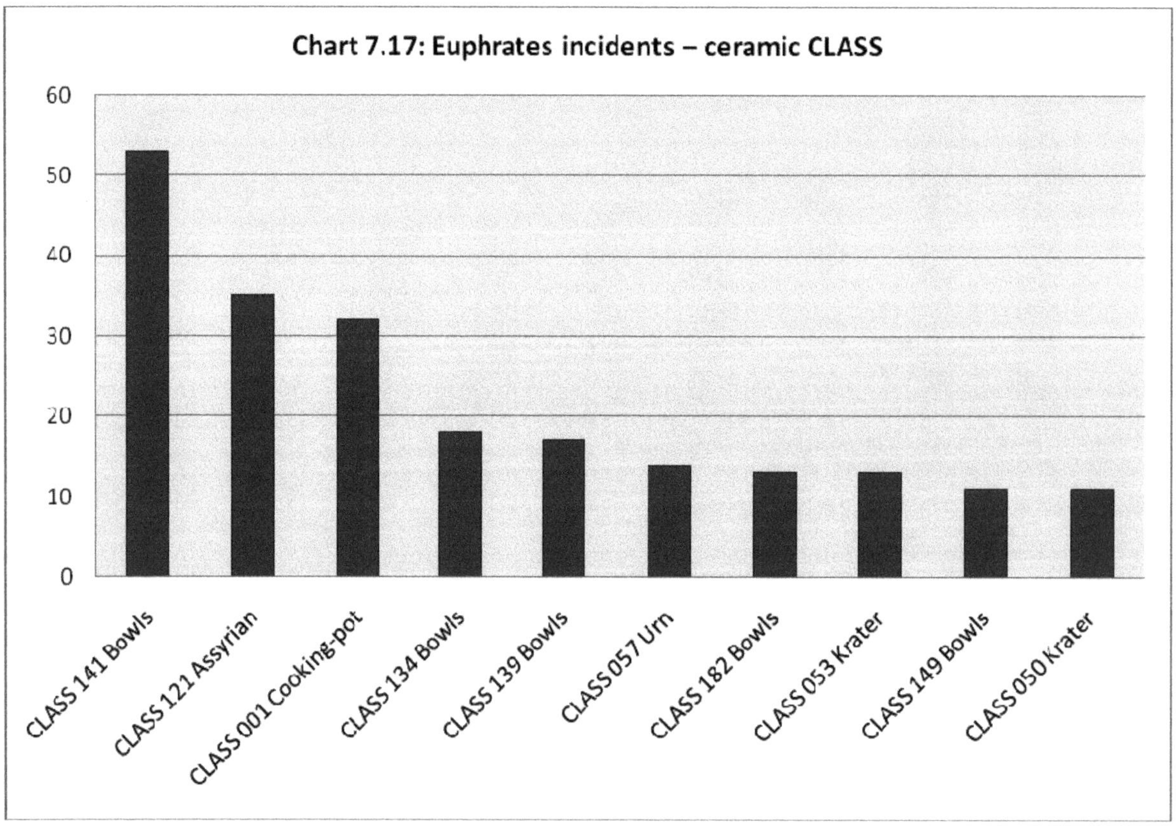

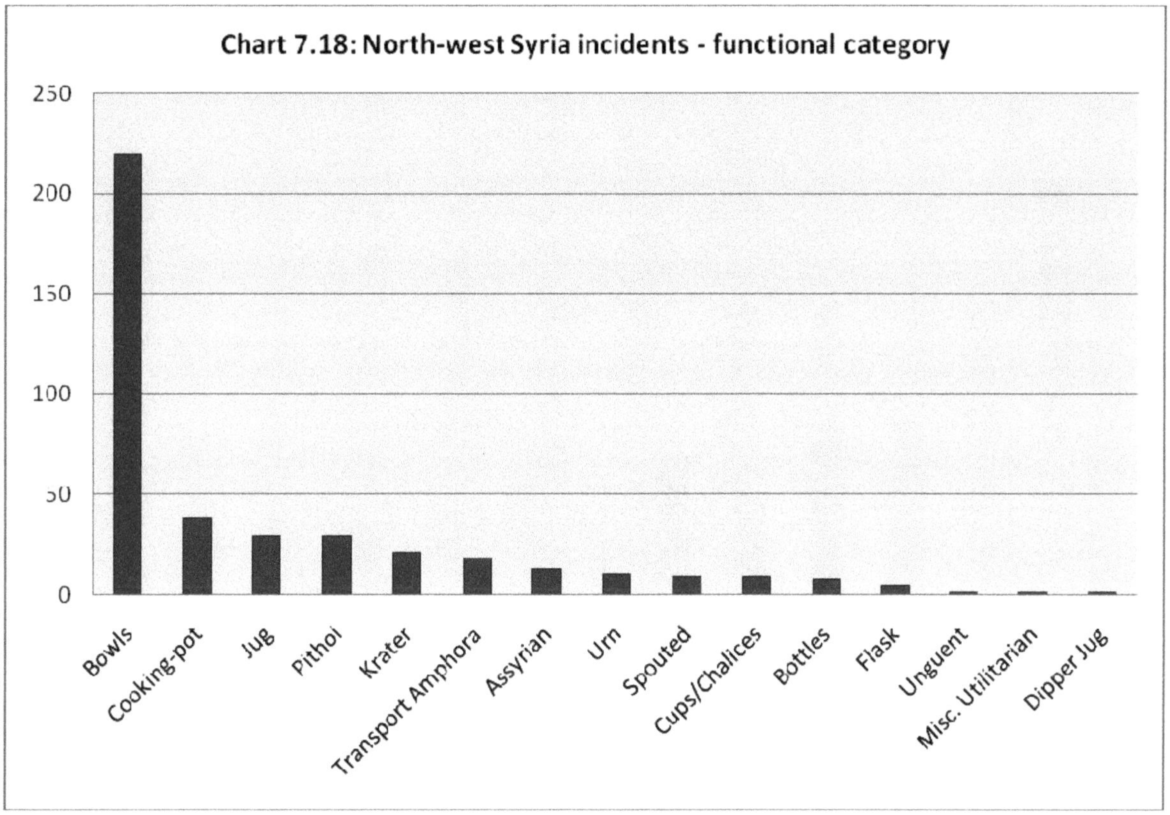

Chart 7.18 shows the standard abundance of bowls in the North-west Syria zone. There is a significant differentiation between the number of bowl incidents and the next most common category, cooking-pots. Once again the CLASS 134 and 141 bowls are amongst the most abundant forms, along with CLASS 001 cooking-pots and CLASS 037 pithoi (**Chart 7.19**). However, no single form is dominant; possibly due to the limited dataset for this zone.

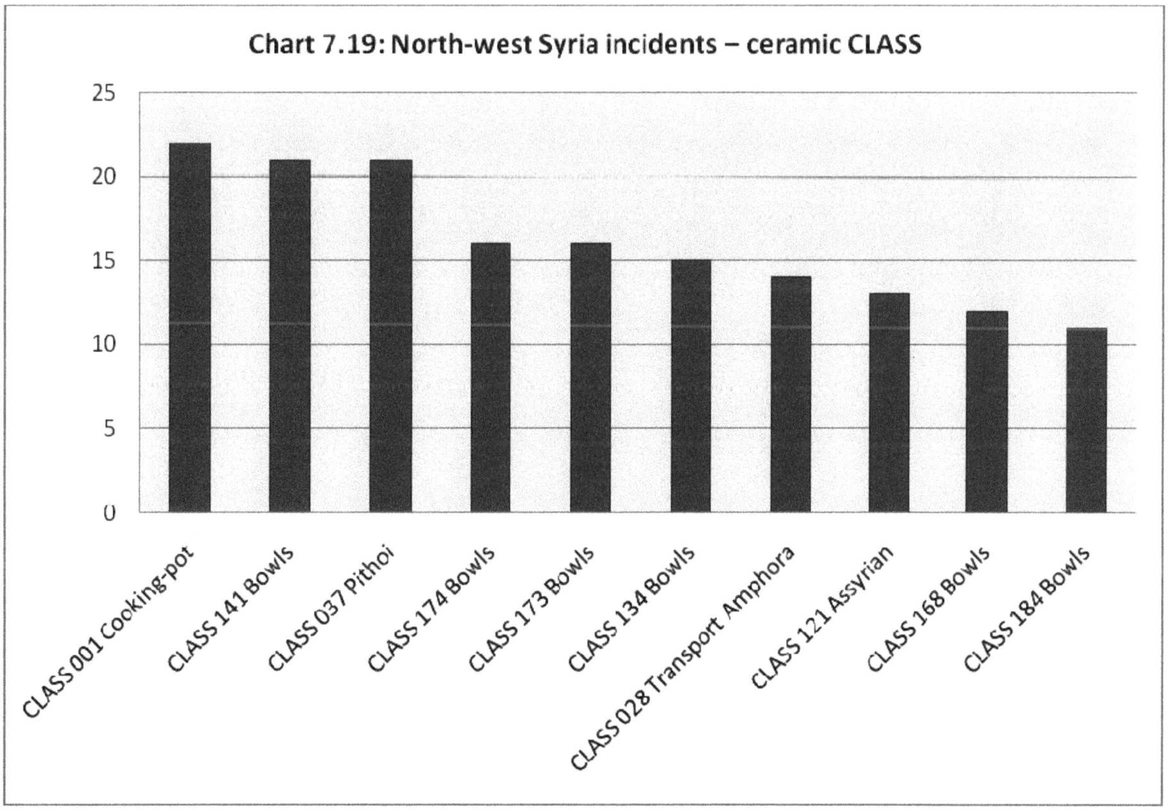

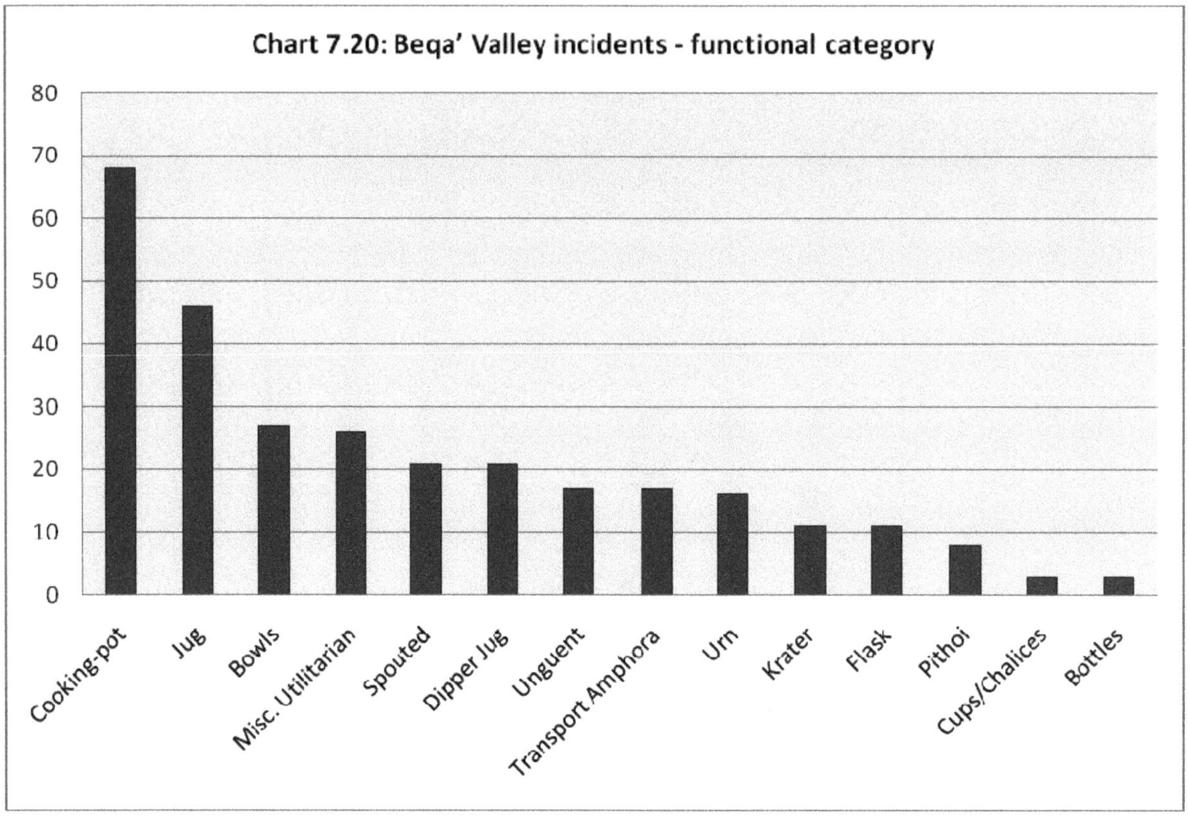

Chart 7.20 shows that the Beqa' Valley is the only regional zone that is not dominated by bowls. Instead, bowls are the third most abundant category, behind cooking-pots and jugs. The CLASS 004 cooking-pot is clearly the most dominant form, but the results from this zone warrant caution. The pottery from Kamid el Loz has been published sporadically, leaving only the assemblage of one site, Tell el Ghassil, as contained within an unpublished Master's thesis (Joukowsky 1972) to characterise the data.

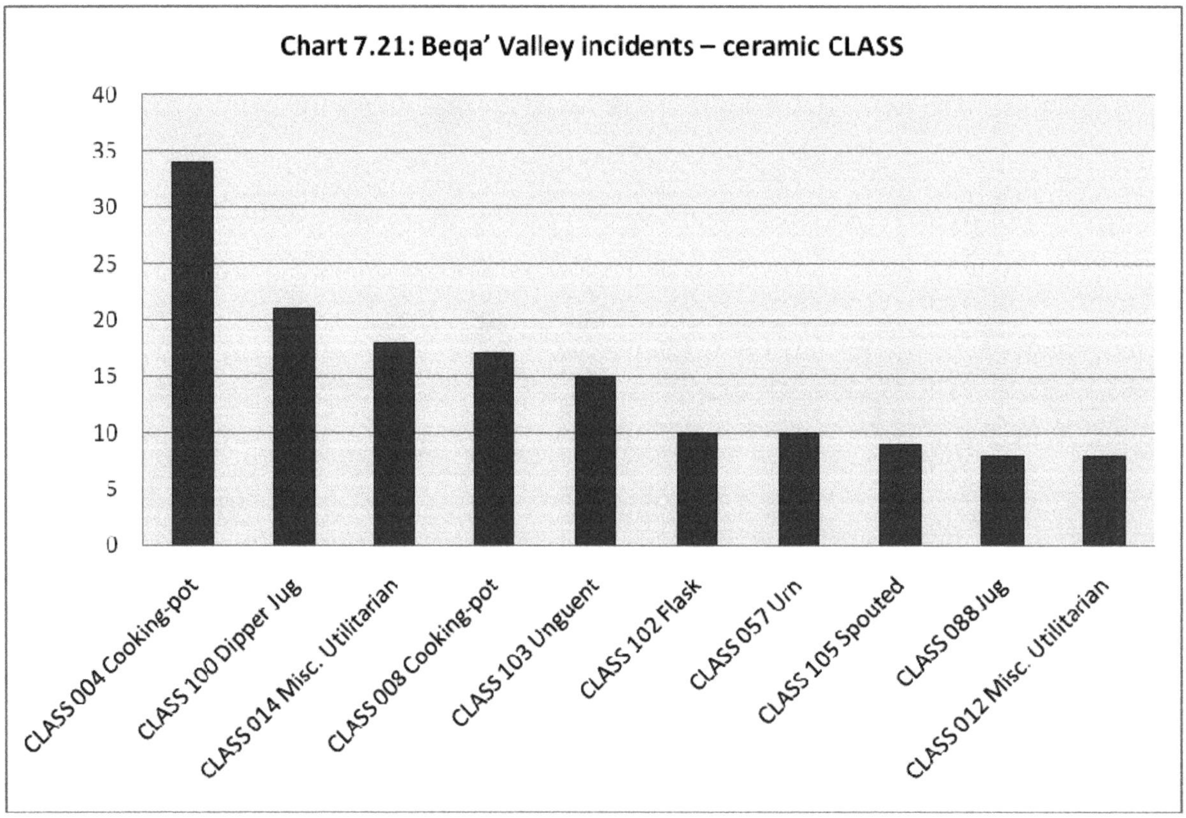

7.4 Functional Analysis

As discussed in **Chapter 5**, there has been an emphasis placed on broad ceramic regions in research on the Iron Age Northern Levant, wherein the coast is generally considered culturally distinct from the interior (**Figures 5.1-2**). While this general distinction is also apparent in this study, researchers have not sought to understand why the two regions might be different, apart from the ethnicity of the regional populations. Regionalisation appears to have been considered explanation enough, when it is really only a descriptive tool. For this reason, this study is interested in possible differences in the functional requirements of the two "regional" assemblages. To this end, the functional categories of **Chapter 6** were applied to the data. These functional categories are:

- Assyrian vessels
- Bottles
- Bowls
- Cooking pots
- Cups and chalices
- Dipper juglets
- Flasks
- Jugs
- Kraters
- Pithoi
- Spouted vessels
- Transport amphorae
- Unguent containers
- Urns

Each functional category is presented in a pie chart and in a column chart. Each pie chart, which incorporates some chronological distinction (but does not divide the Iron II and Iron III periods due to the chronological ambiguity of some contexts), presents the diachronic trends in the different functional categories. While most categories are best represented by Iron II/III examples (due to the high concentration of Iron II/III incidents in the database), a few appear to be largely Iron I phenomena. Cups and chalices (**Chart 7.30**), unguent containers (**Chart 7.46**), and flasks (**Chart 7.34**), all have a distinctively strong Iron I presence. In addition to temporal data, the second chart for each category, a column chart, includes geographic information. Hence, these charts are designed to plot functional differences according to region. The most compelling regional differences are evident in jugs (**Chart 7.37**) and transport amphorae (**Chart 7.45**) (both are primarily restricted to inland northern Palestine and the Levantine coast), and pithoi (**Chart 7.41**) (limited to inland contexts of the Northern Levant). While these trends alone are able to confirm the general differences between in the inland and coastal regions of the Northern Levant, they offer a rather simple explanation for these differences. In other words, these two "regions" imposed different demands on ceramic their assemblages. The people on the coast needed ceramic transport amphorae and used ceramic jugs, whilst the inland population required large ceramic vessels for storage purposes. How or why these differences manifested themselves is discussed in more detail in **Chapter 9**.

Chart 7.22: "Assyrian" vessels - period

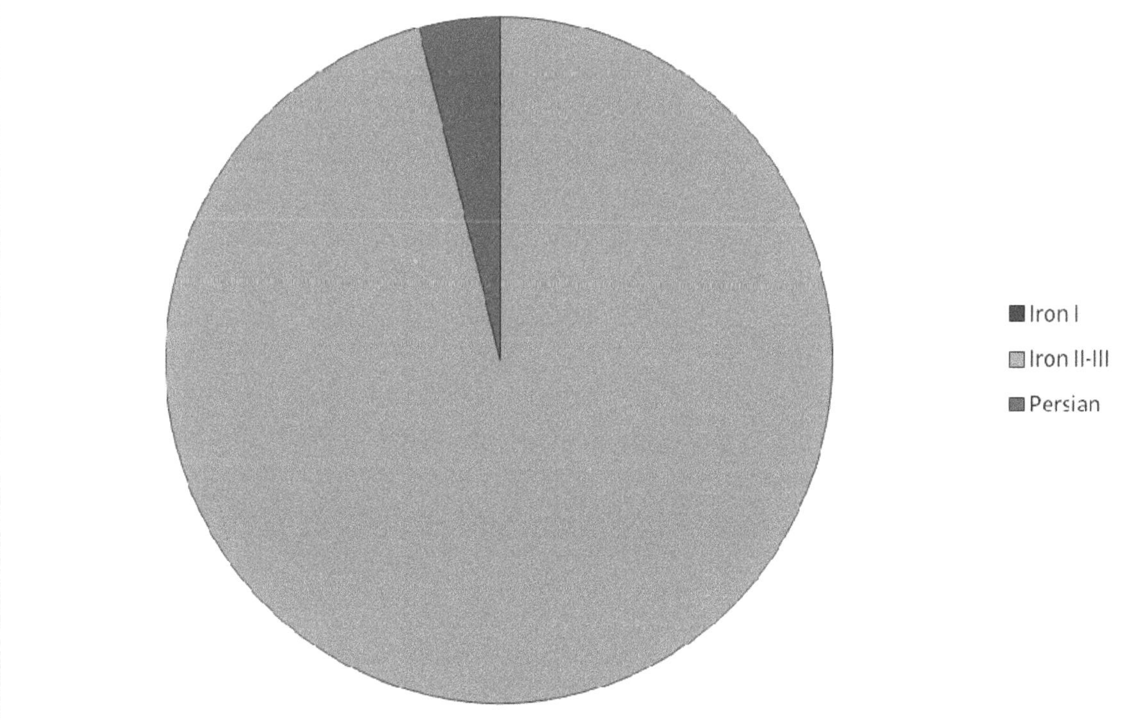

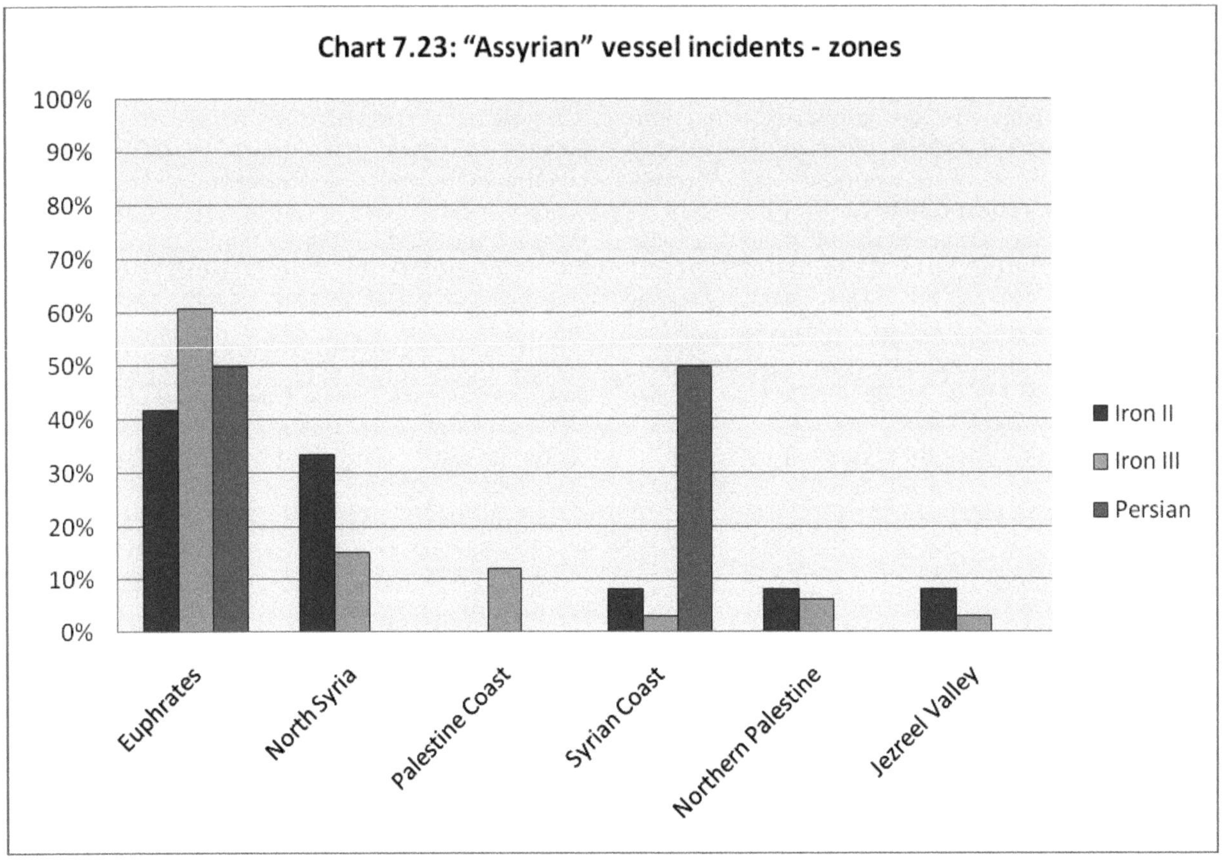

"Assyrian" vessels were mostly encountered in the Iron II-III period (**Chart 7.22**) of the inland Northern Levant, though conspicuously absent from the Orontes zone (**Chart 7.23**). Both of these trends are as we would expect from a functional category that is historically associated with an Iron II-III eastern phenomenon.

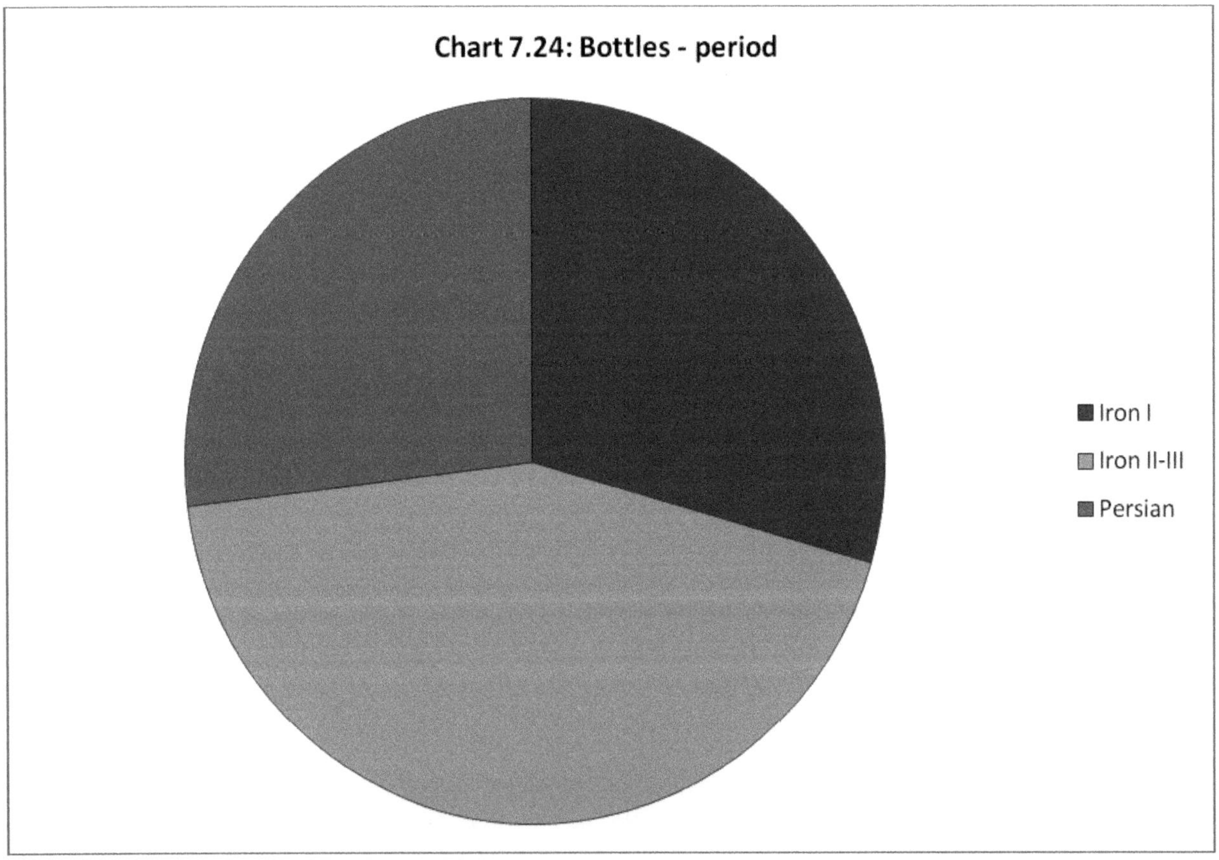

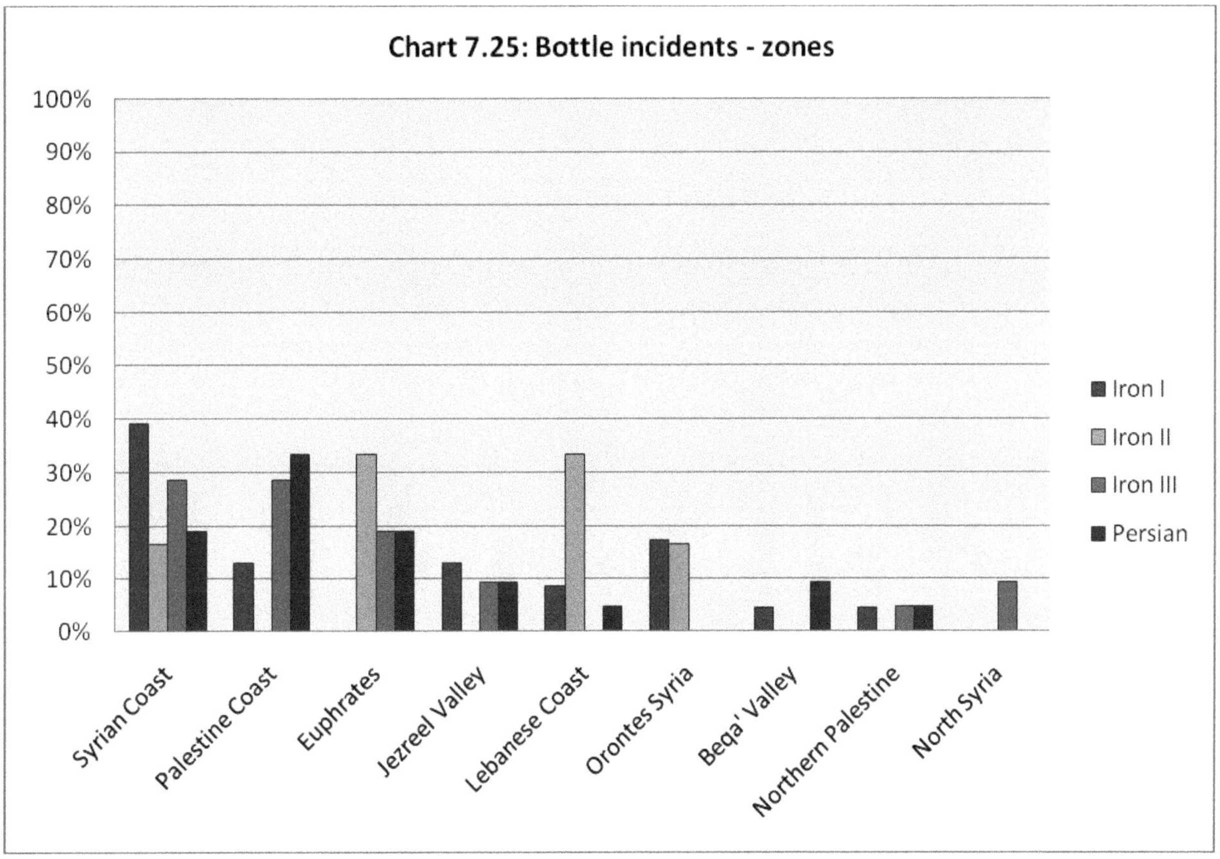

Chart 7.24 shows that the presence of bottles in the database was evenly spread throughout the Iron Age and Persian period. There also appeared to be a lack of any distinct patterns in the regional distribution of bottles (**Chart 7.25**).

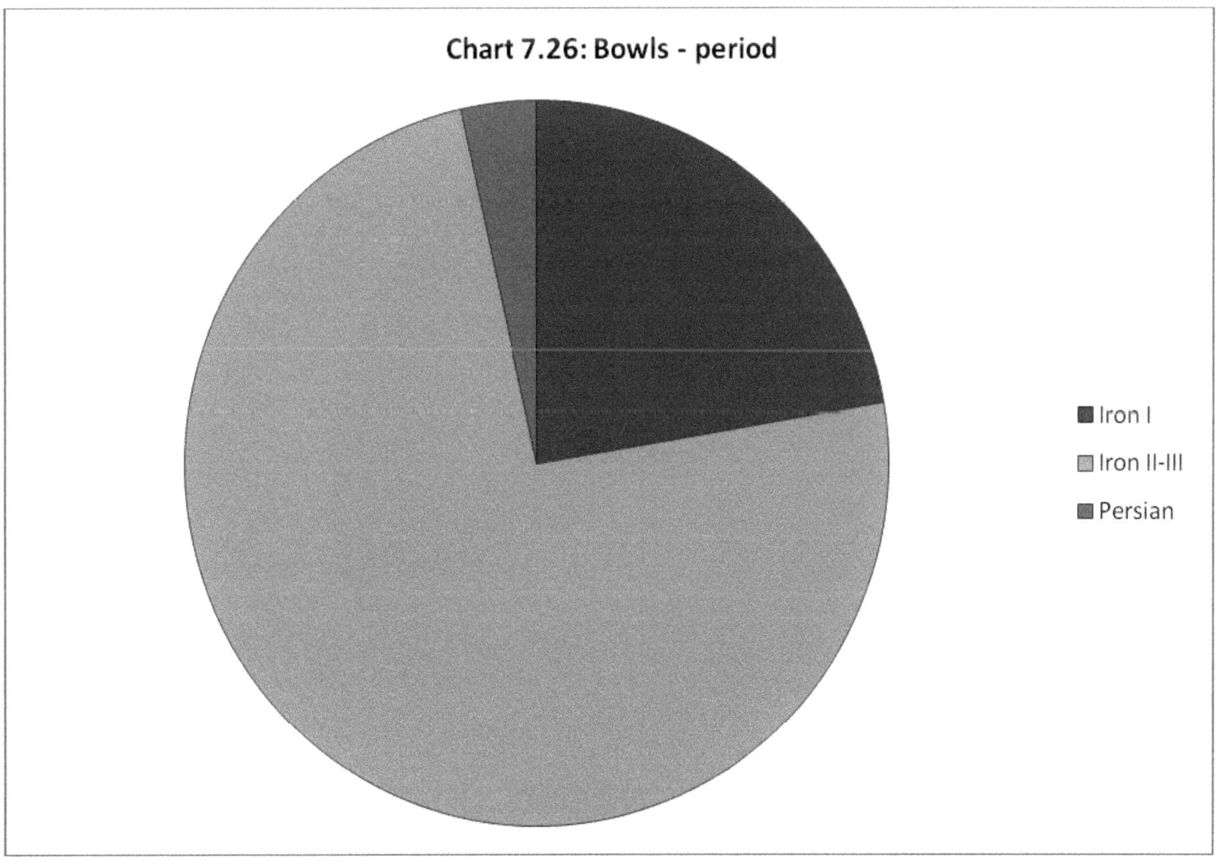

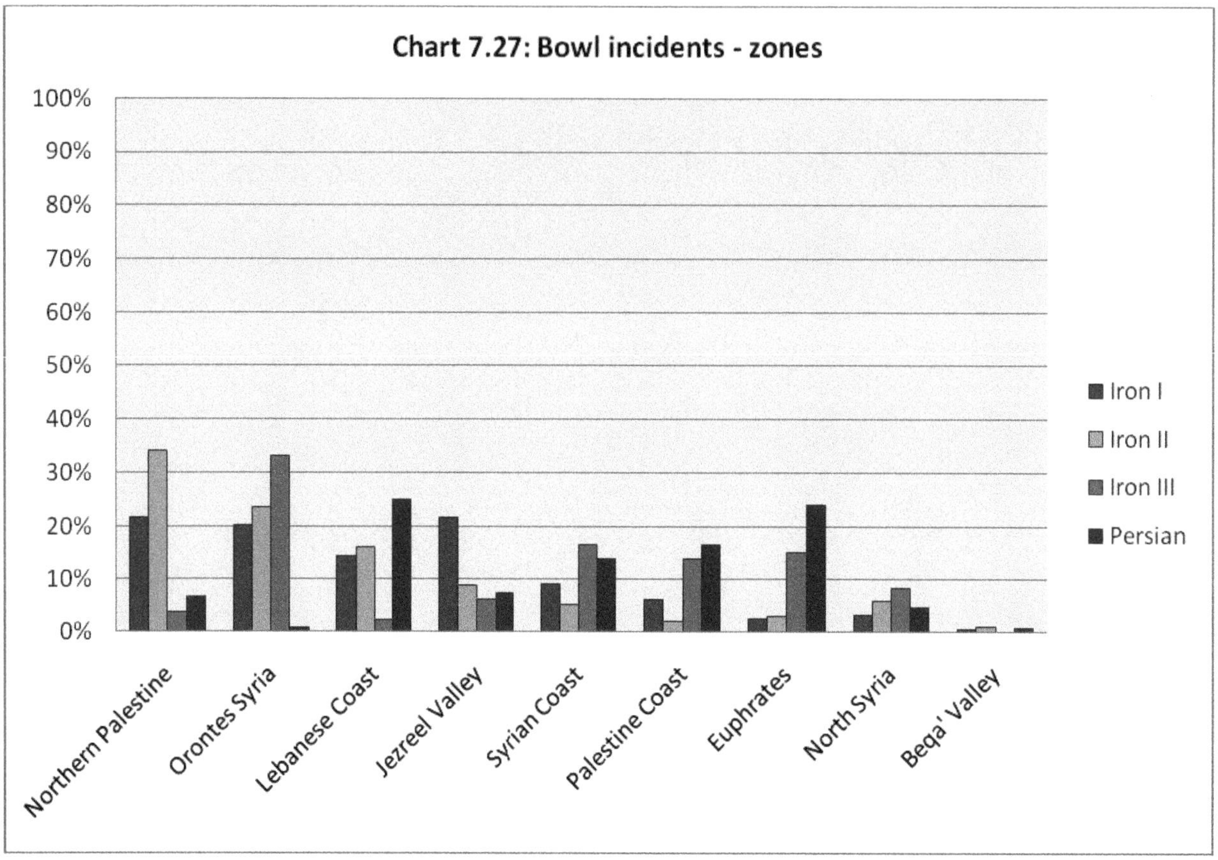

Bowls are the most common category within the database (**Chart 5.5**). **Chart 7.26** shows that bowls were most frequently encountered during the Iron II-III period; less so in the Iron I period. Bowls were well-spread across the study area (except the Beqa' Valley), with no particular concentration discernible (**Chart 7.27**).

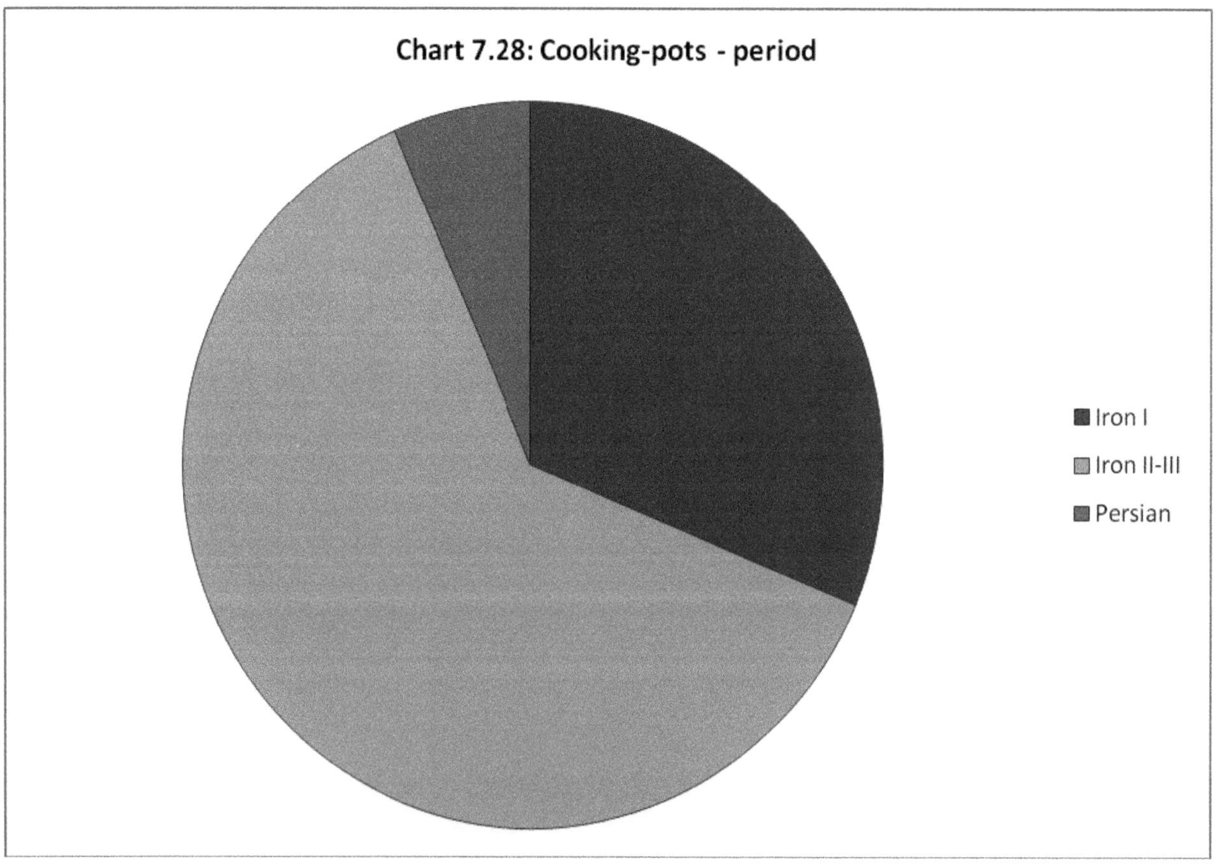

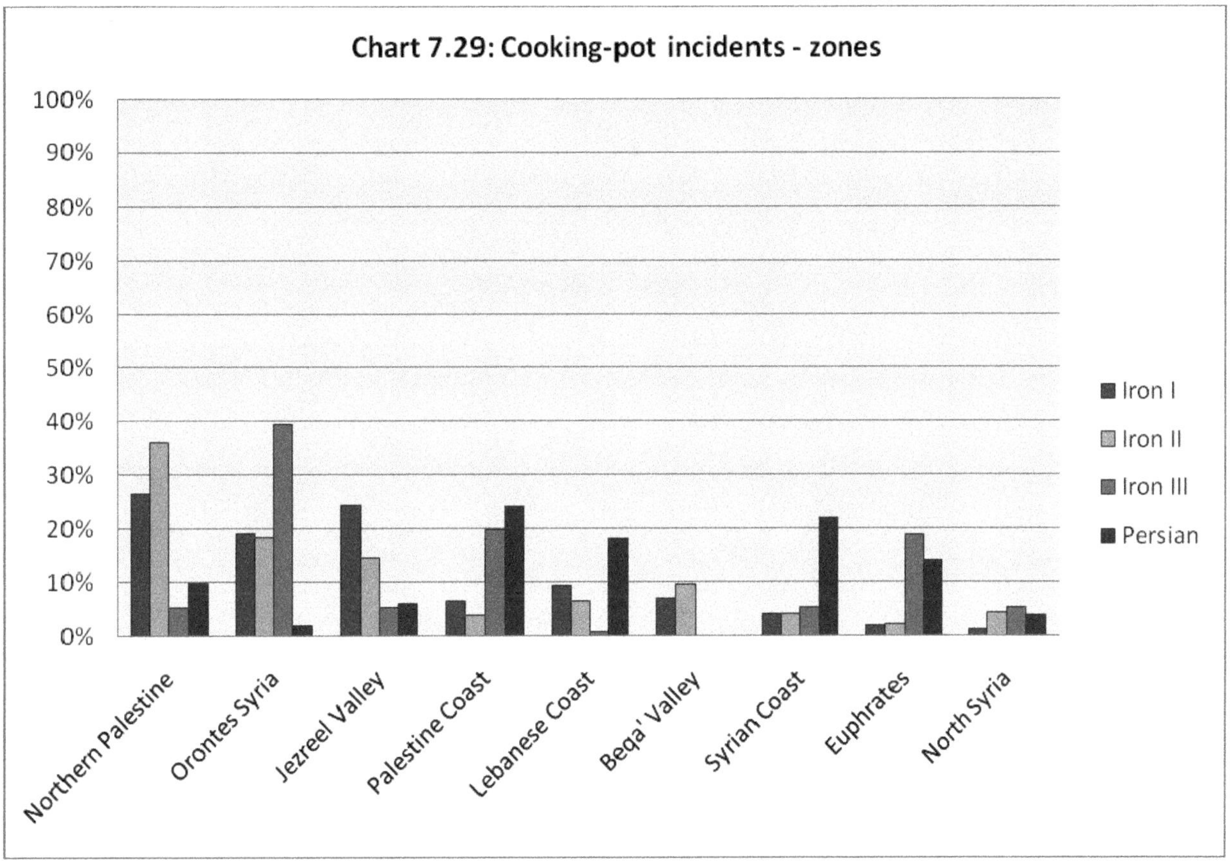

Charts 7.28 and **7.29** show cooking-pots were well-spread across the study area in most periods, with no particular concentration discernible. Patterns within this category might be evident in the distribution of individual ceramic CLASSES.

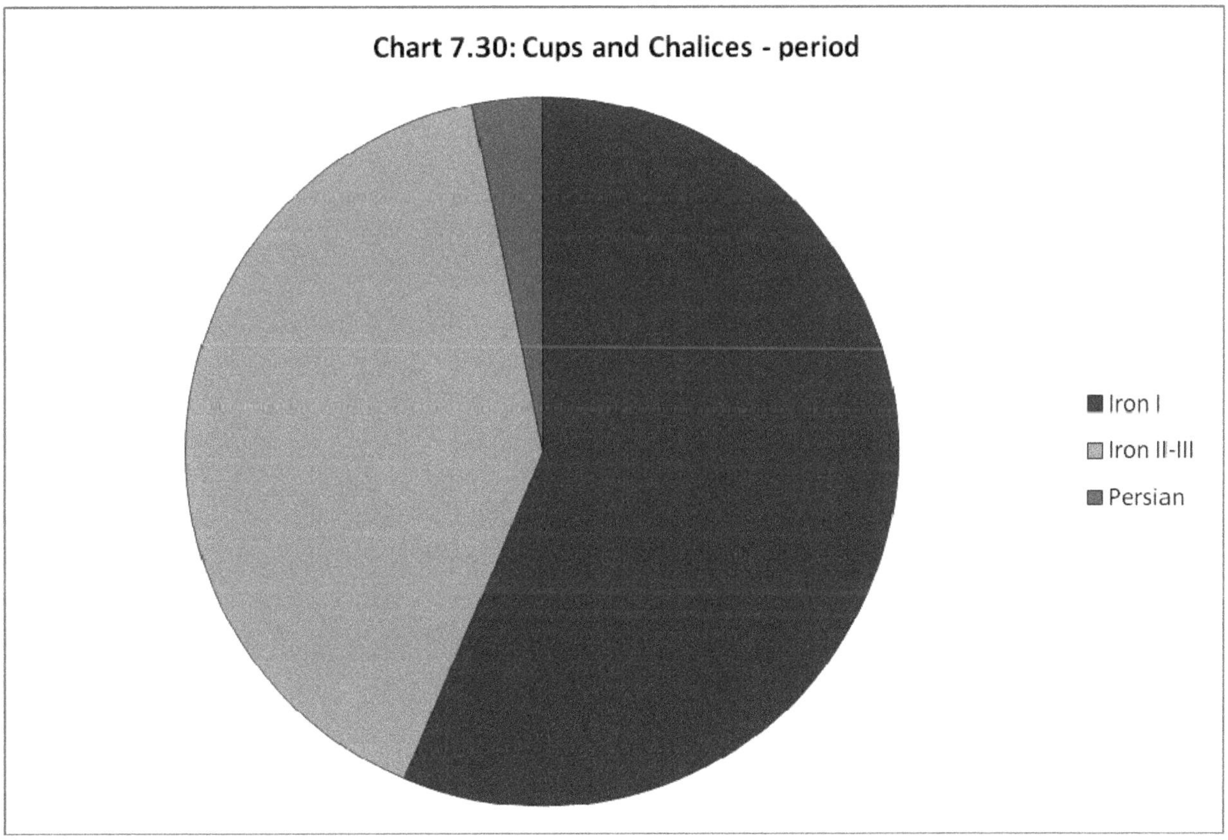

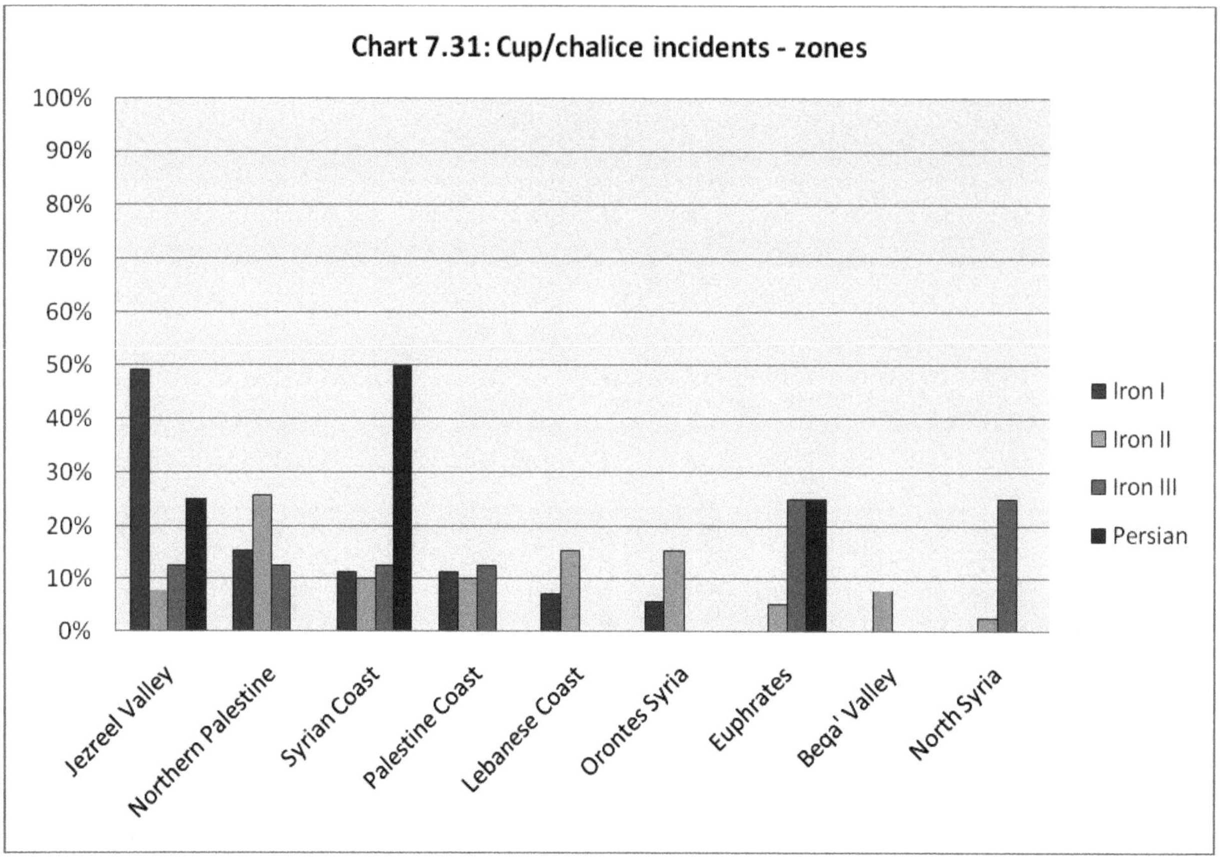

Cups and chalices were well-represented in the Iron I period (**Chart 7.30**), but by the Persian period had all but disappeared. This category was rarely encountered in the zones of the inland Northern Levant, and is instead more frequently encountered along the coast and in the Southern Levant (**Chart 7.31**).

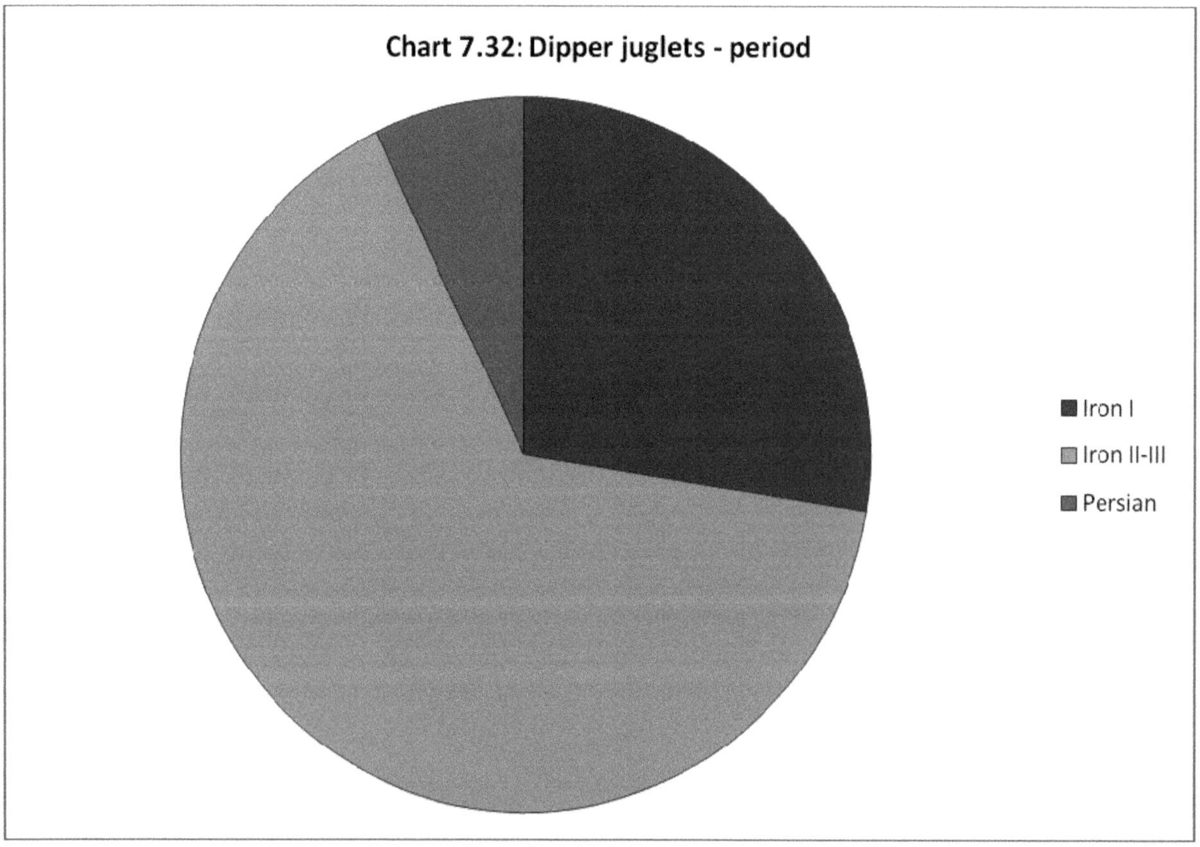

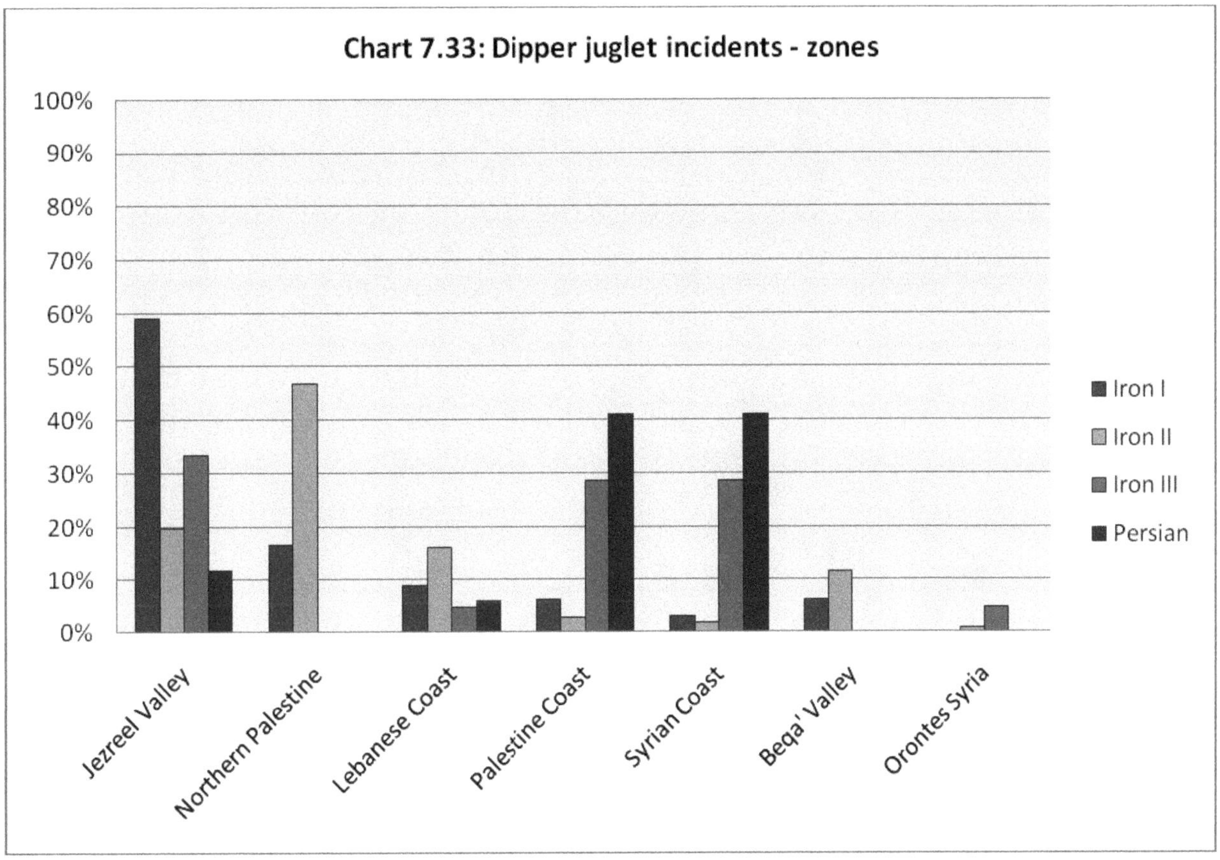

Dipper juglets were well-represented during the Iron I and Iron II/IIII periods (**Chart 7.32**), and were primarily encountered in the zones of the coast and Southern Levant (**Chart 7.33**). Dipper juglets were not encountered in the Euphrates and North Syria zones, and only occasionally along the Orontes.

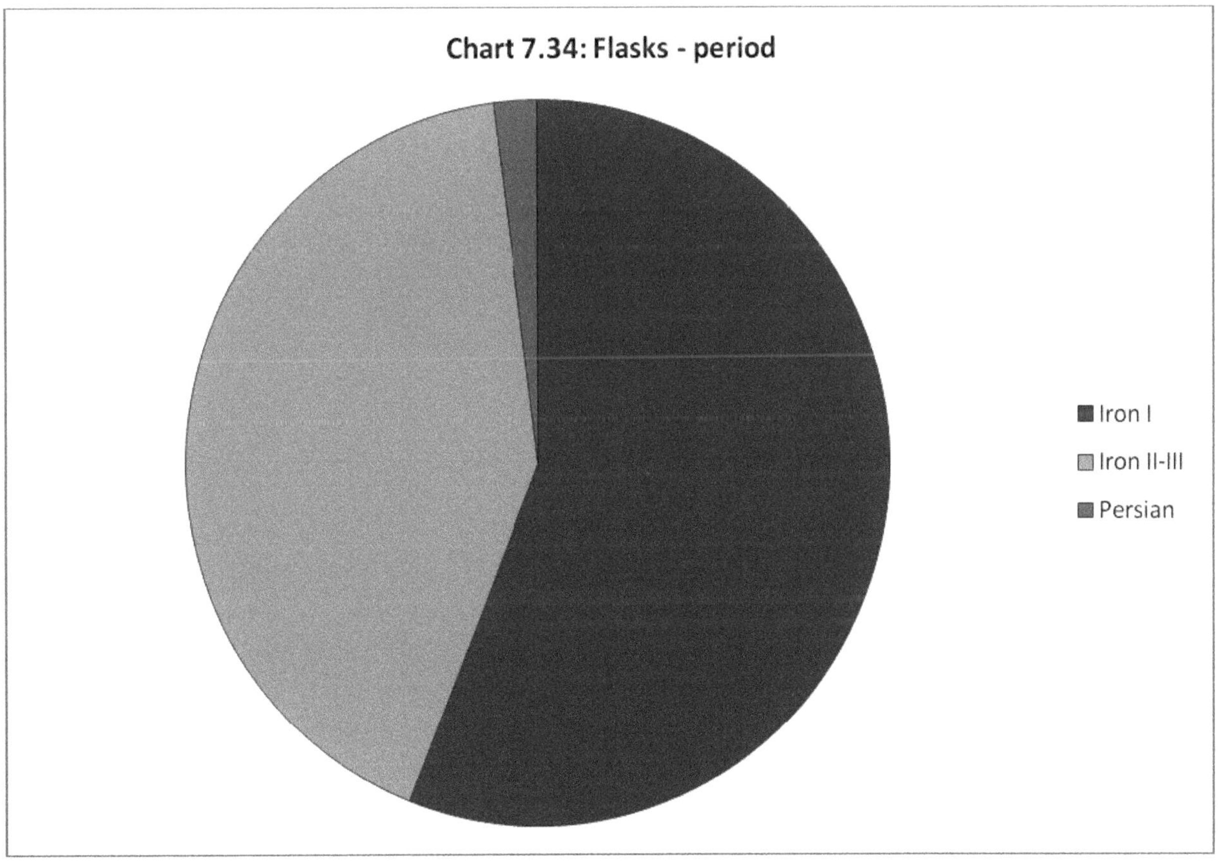

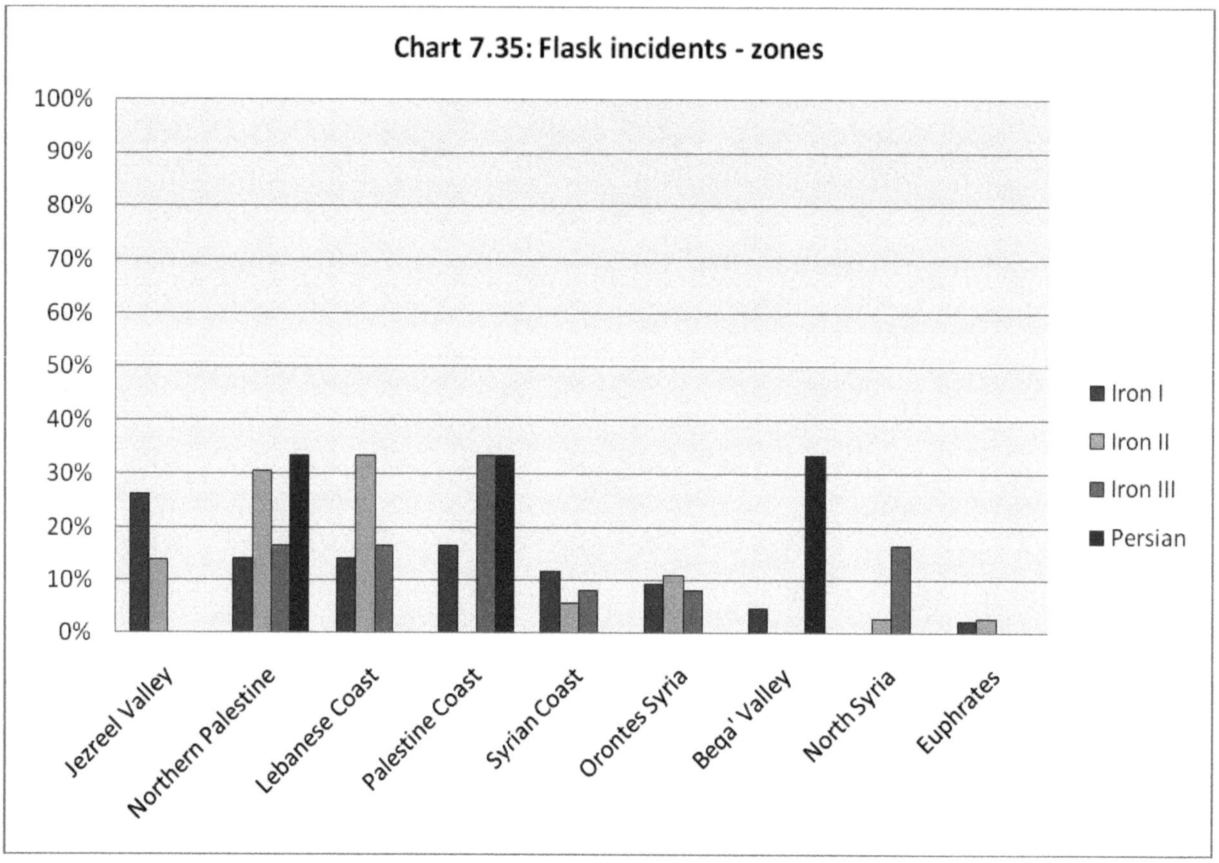

Flasks were commonly encountered in Iron I contexts (**Chart 7.34**), when they were well spread across the coastal and Southern Levant zones (**Chart 7.35**). The absence of flasks from the Palestine coast during the Iron II period might be due to the general lack of reliable Iron II contexts from this region.

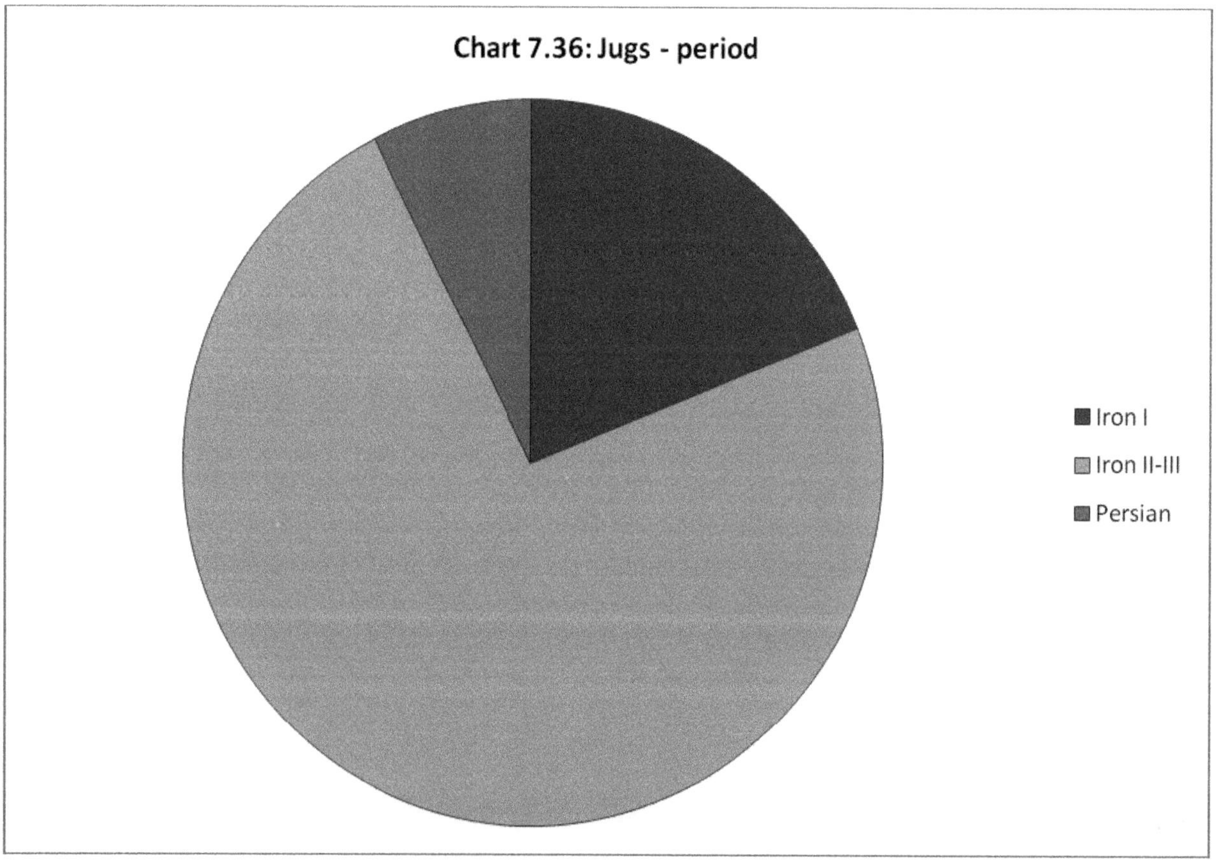

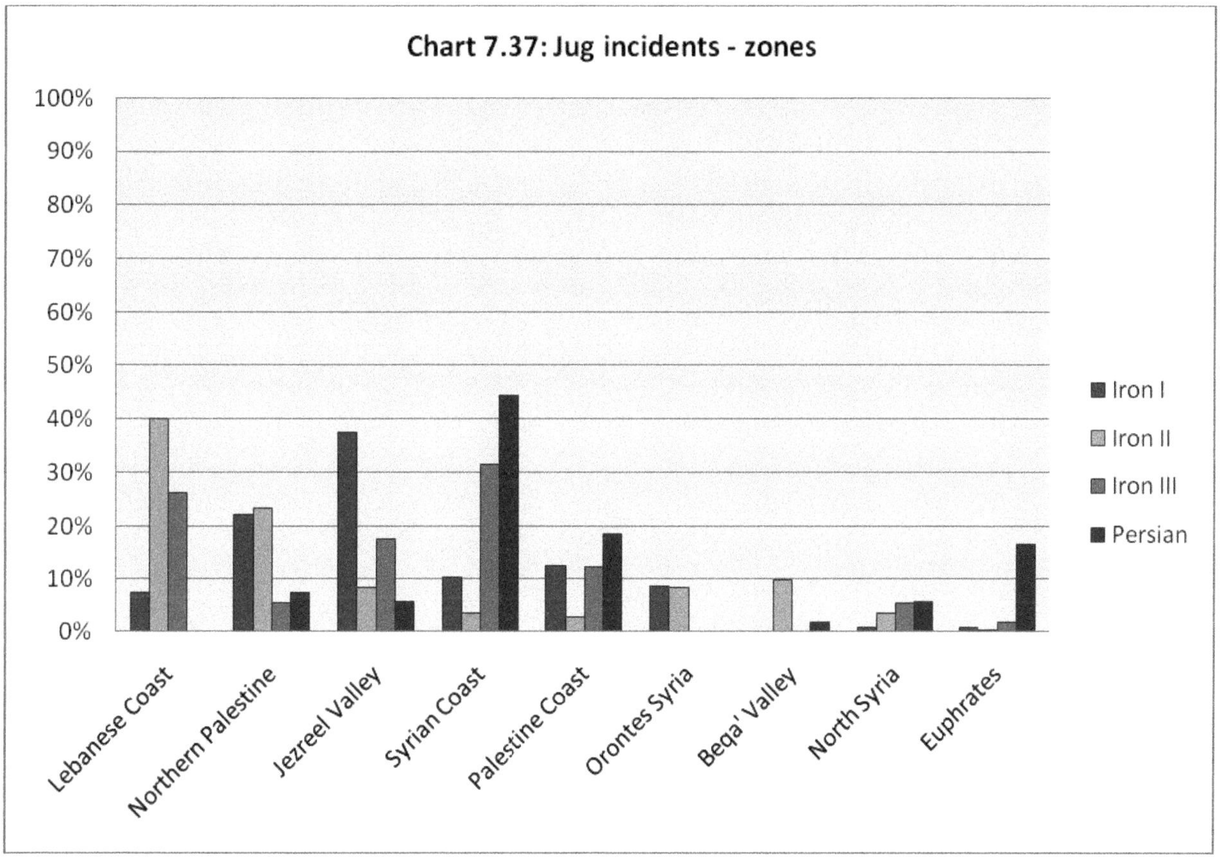

Jugs are the second most common category within the database (**Chart 5.5**), particularly in the Iron II-III periods (**Chart 7.36**), when they were concentrated on the Lebanese coast, Northern Palestine, and Syrian coast (**Chart 7.37**). Jugs were relatively rare across the inland Northern Levant.

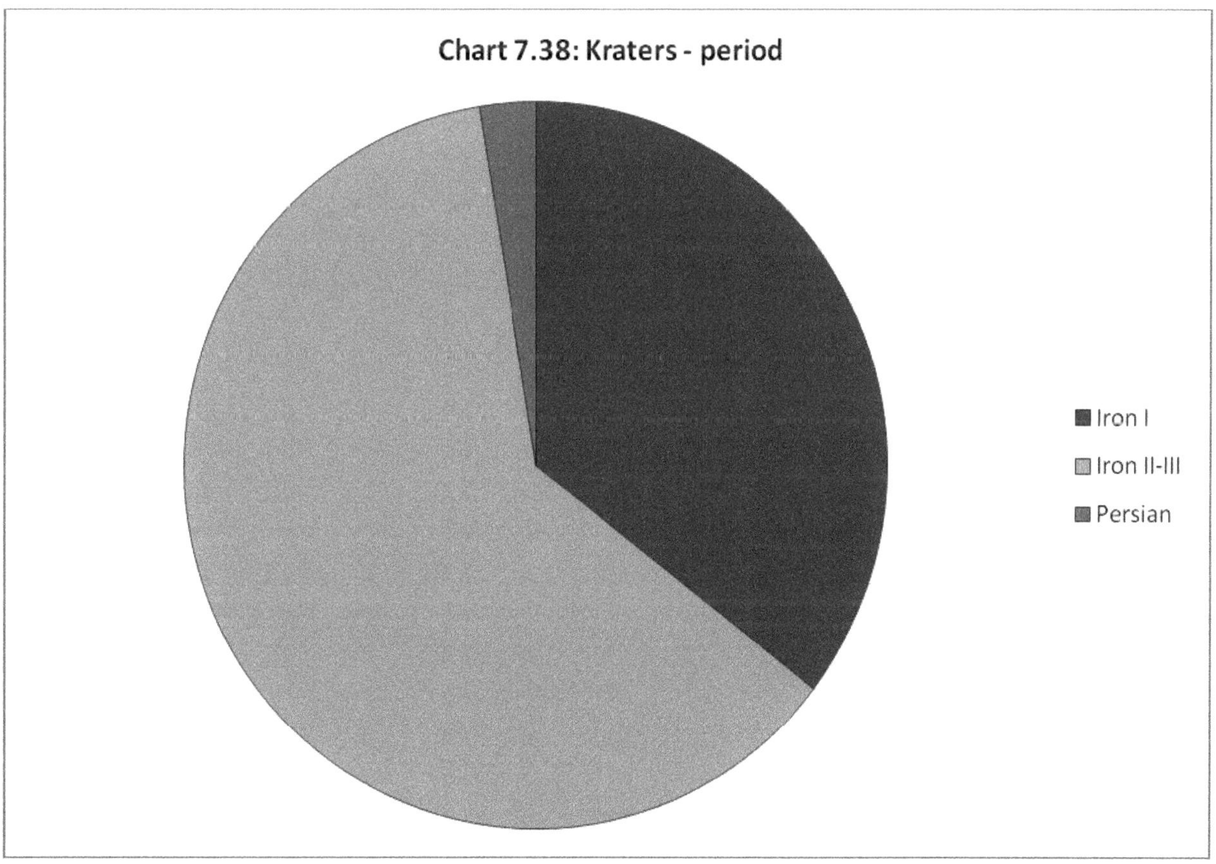

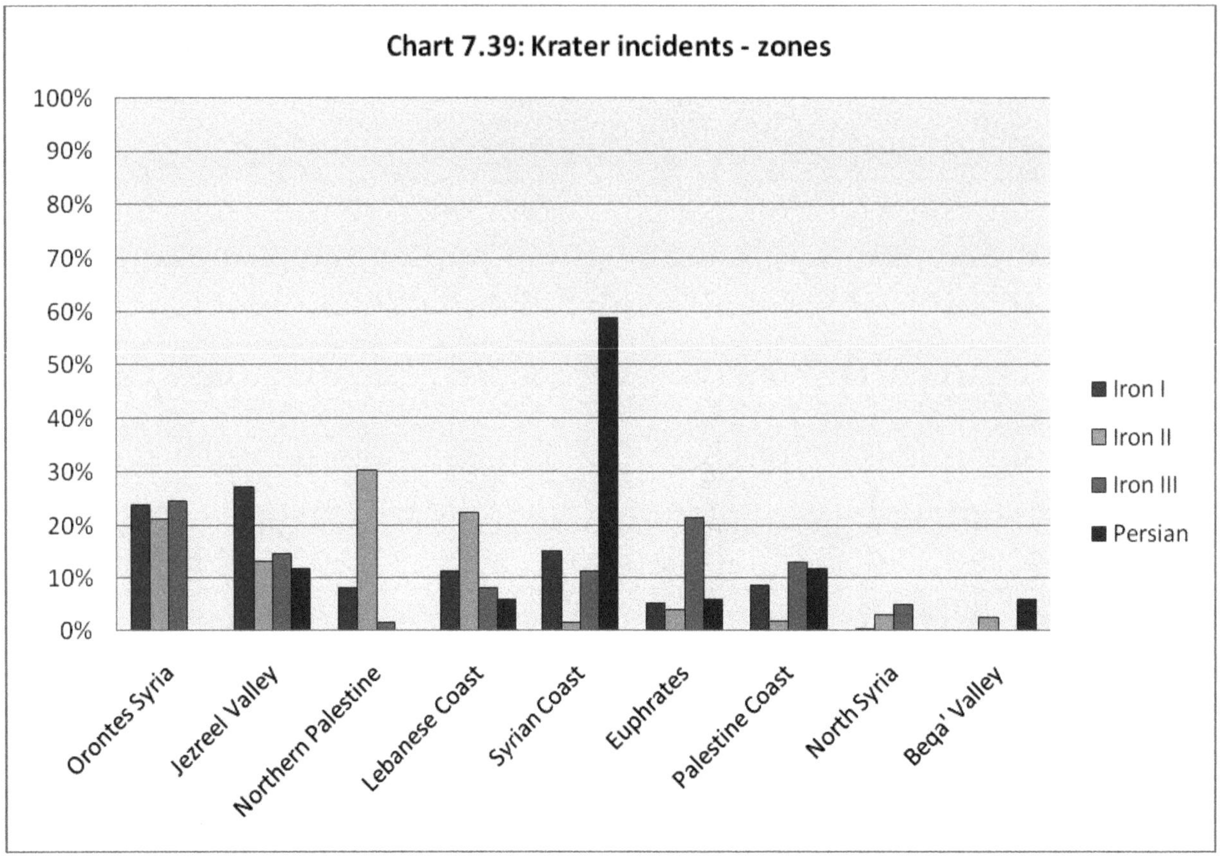

Kraters were encountered throughout the Iron I and Iron II/III periods, but were relatively rare in the Persian period (**Chart 7.38**). Kraters were well-spread across the different zones, though the low incidents in North Syria and Beqa' Valley may be as much to do with the limited published data from these areas (**Chart 7.39**).

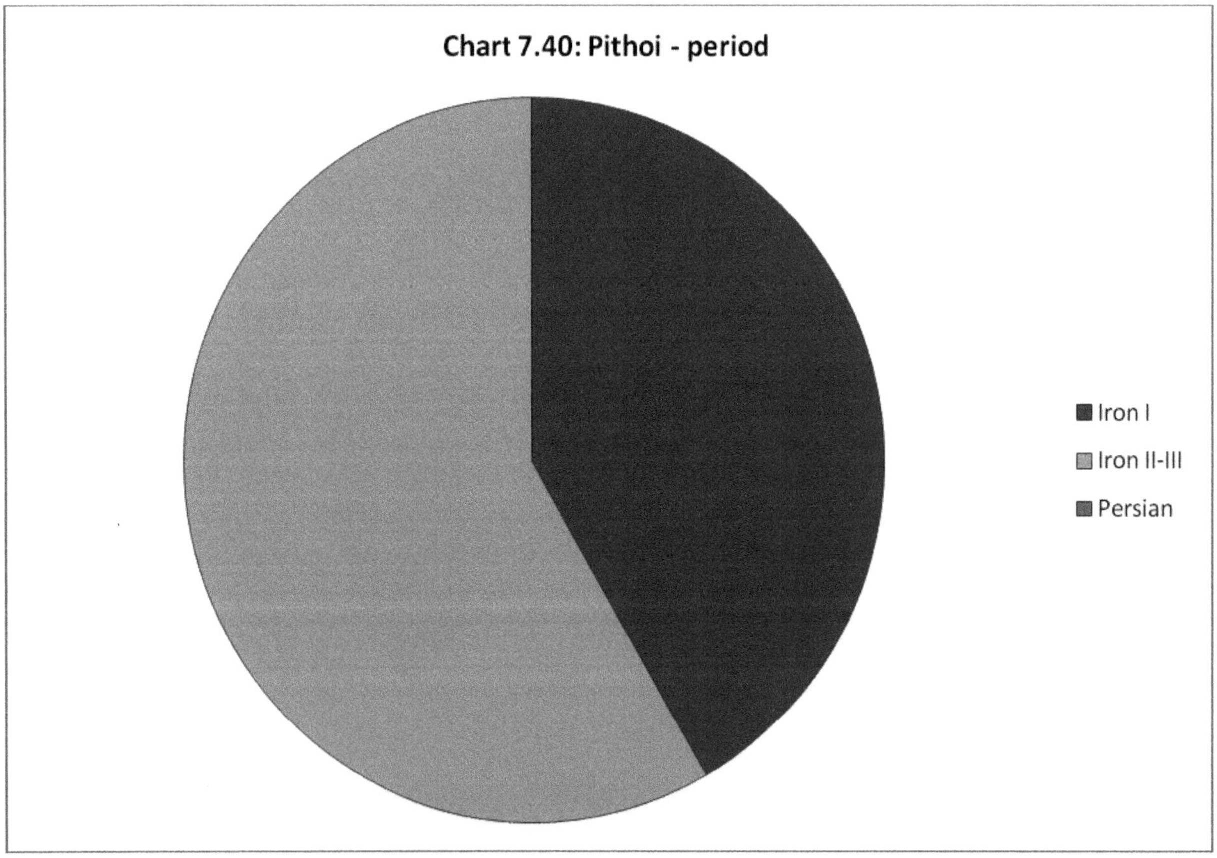

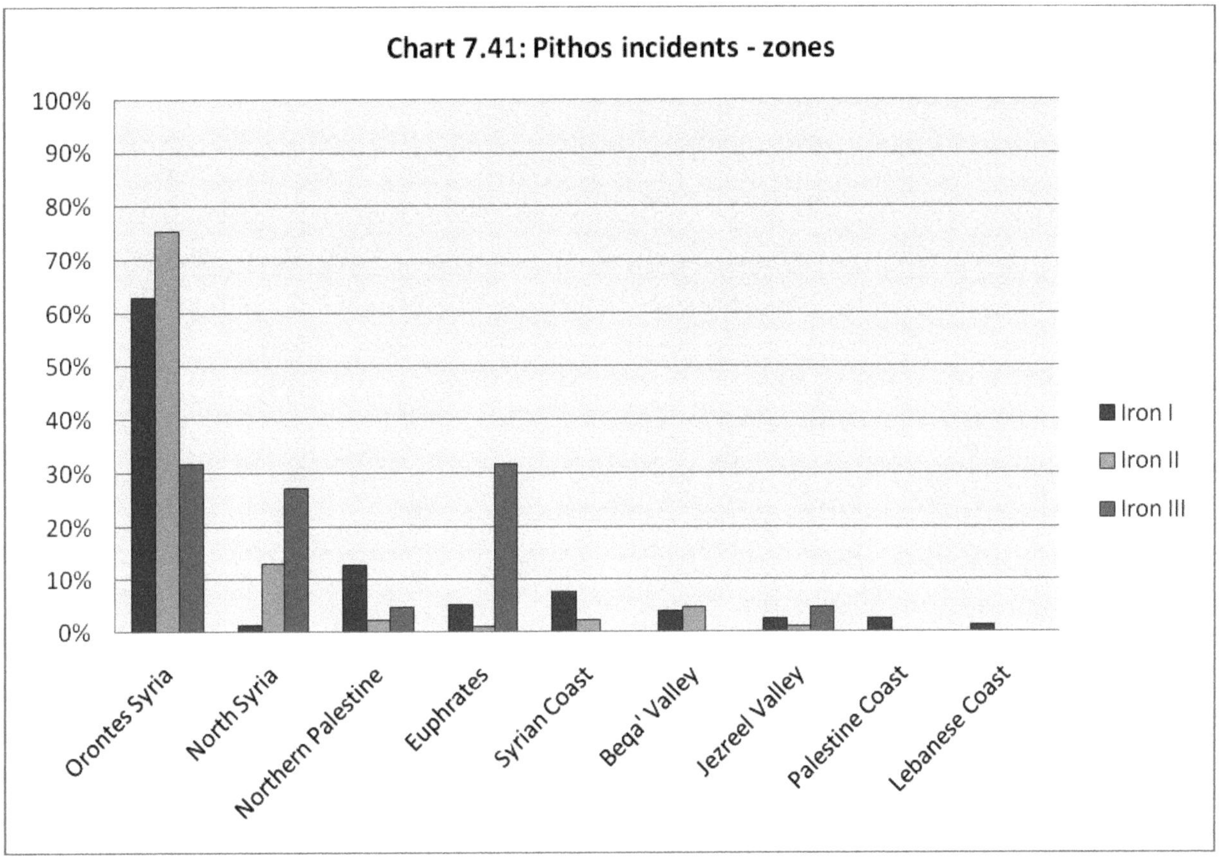

No pithoi were identified from Persian period contexts (**Chart 7.40**). One of the more regionally-restricted categories, pithoi were rare in coastal sites, especially after the Iron I period (**Chart 7.41**). The majority of pithoi were found across the inland Northern Levant (**Maps 29; 30**).

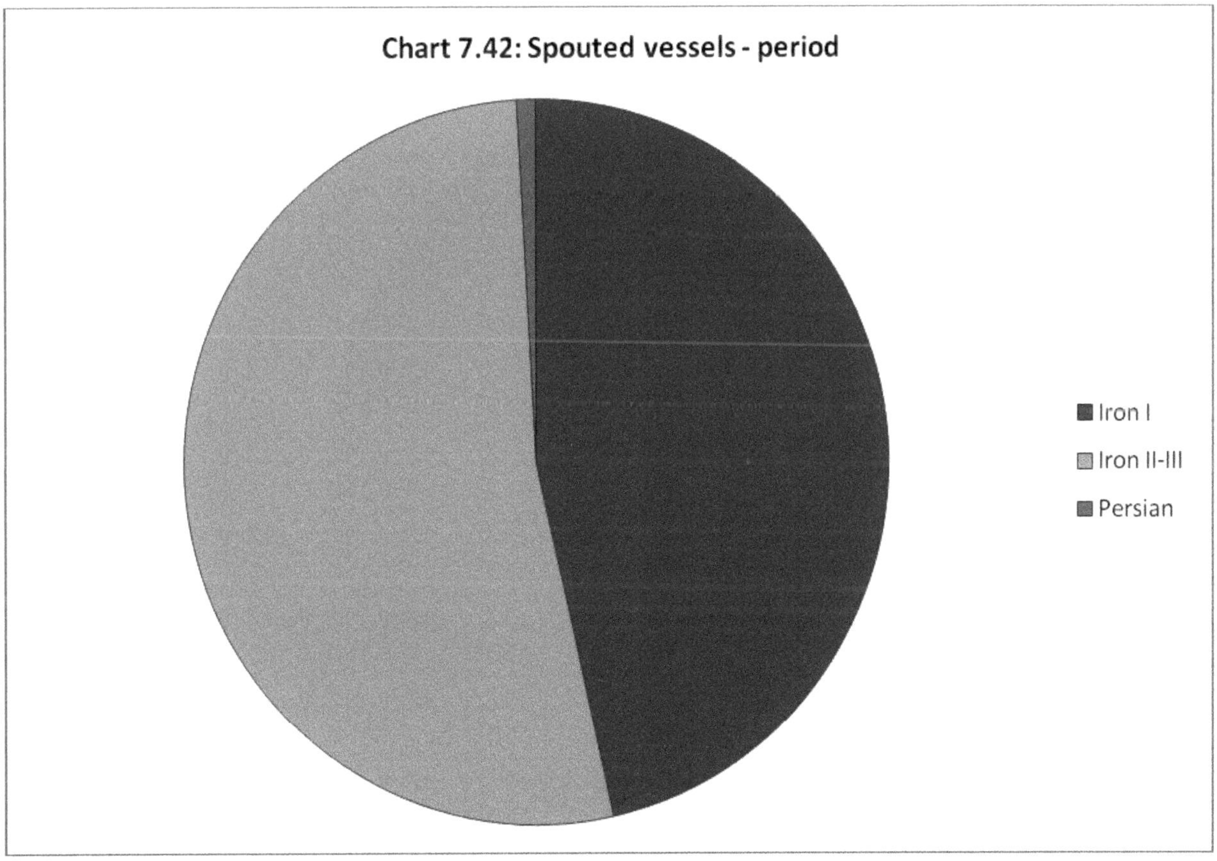

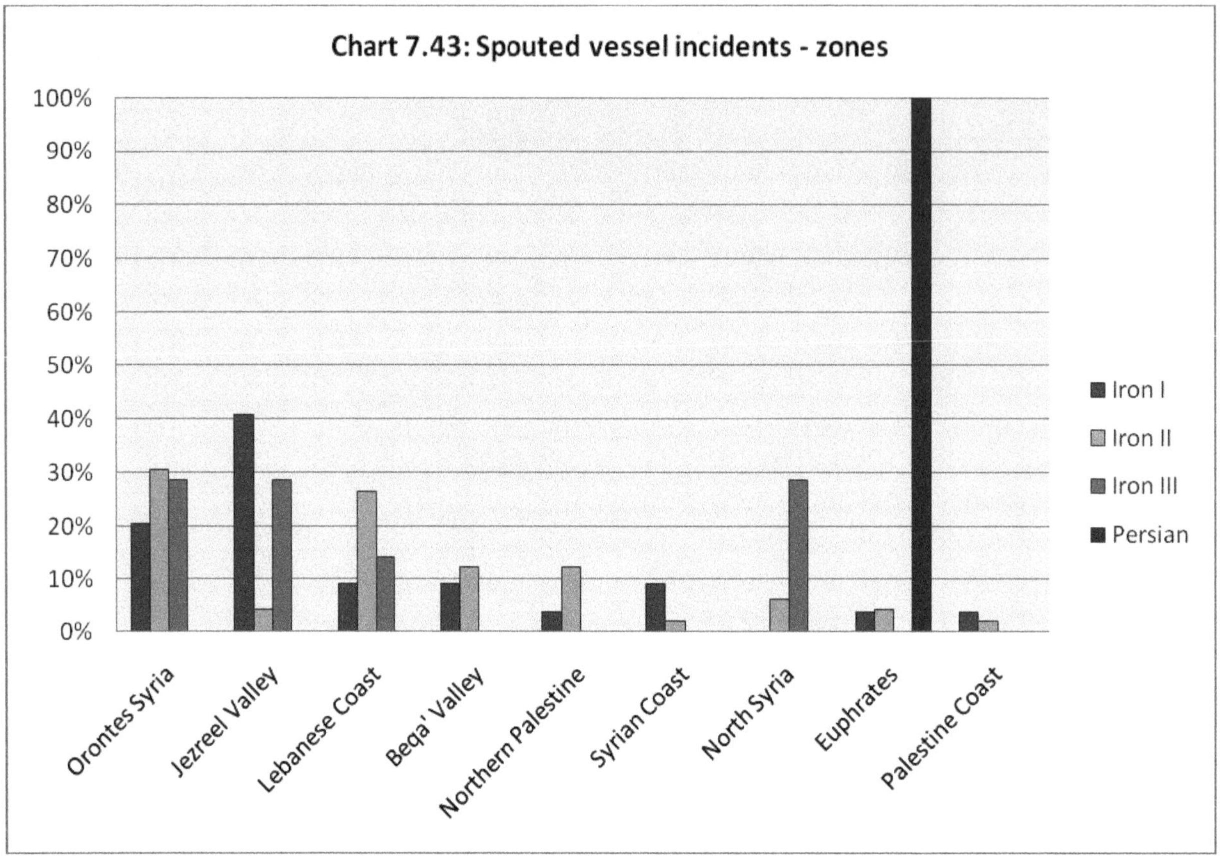

Chart 7.42 shows a large amount of spouted vessels were found in Iron I and Iron II-III period contexts, while they were rare in the Persian period. The distribution of spouted vessels is sporadic and no pattern discernible in the category as a whole (**Chart 7.43**).

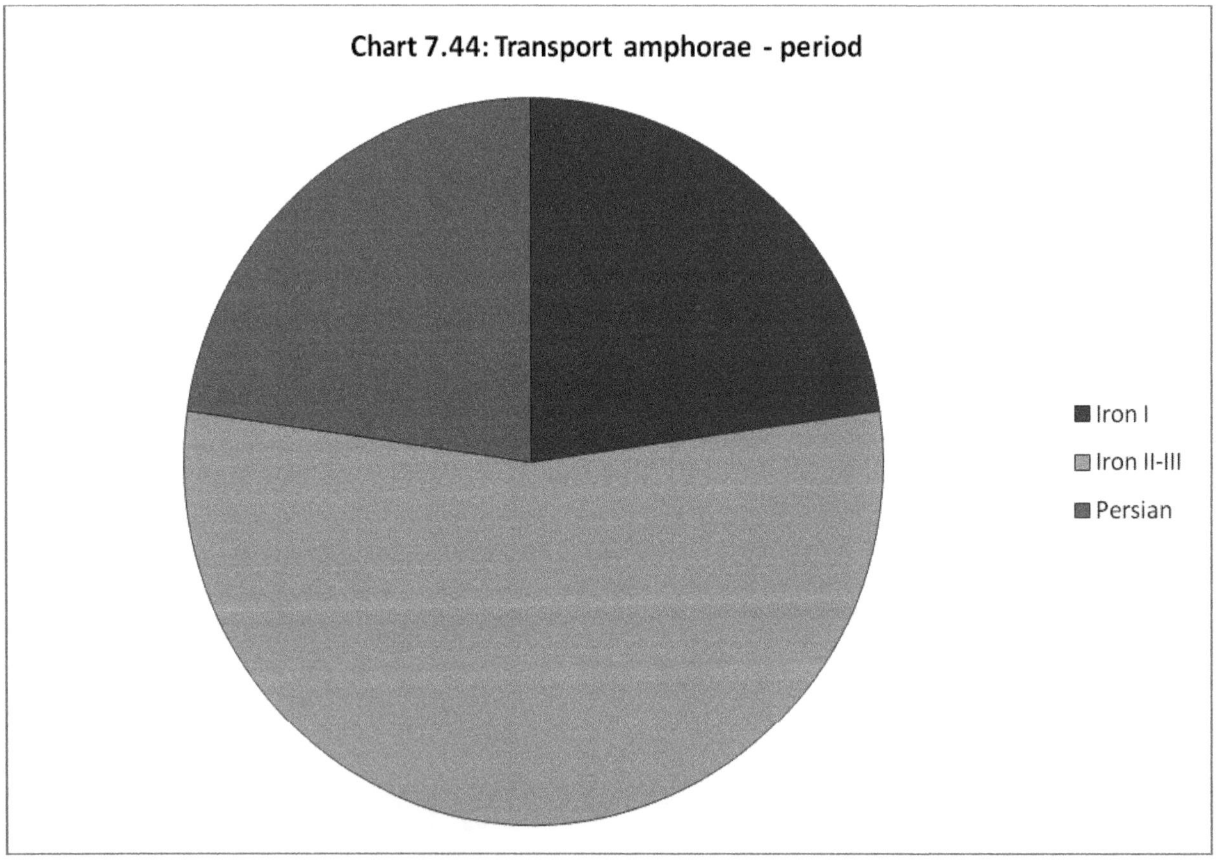

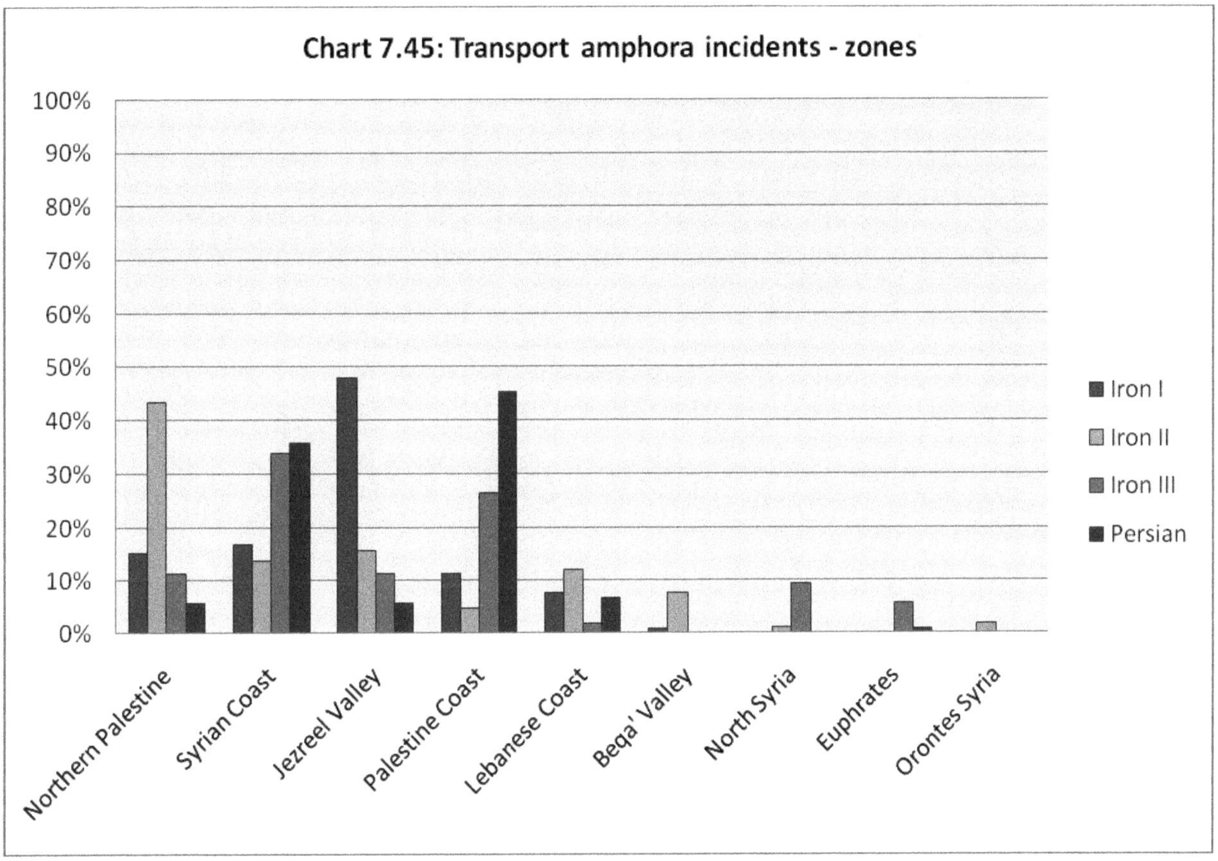

Transport amphorae were present throughout the Iron Age (**Chart 7.44**) Transport amphorae were primarily encountered in the coastal and Southern Levant zones (**Chart 7.45; Maps 17-28**). The inland Northern Levant is poorly represented in all periods, but especially so in the early Iron Age.

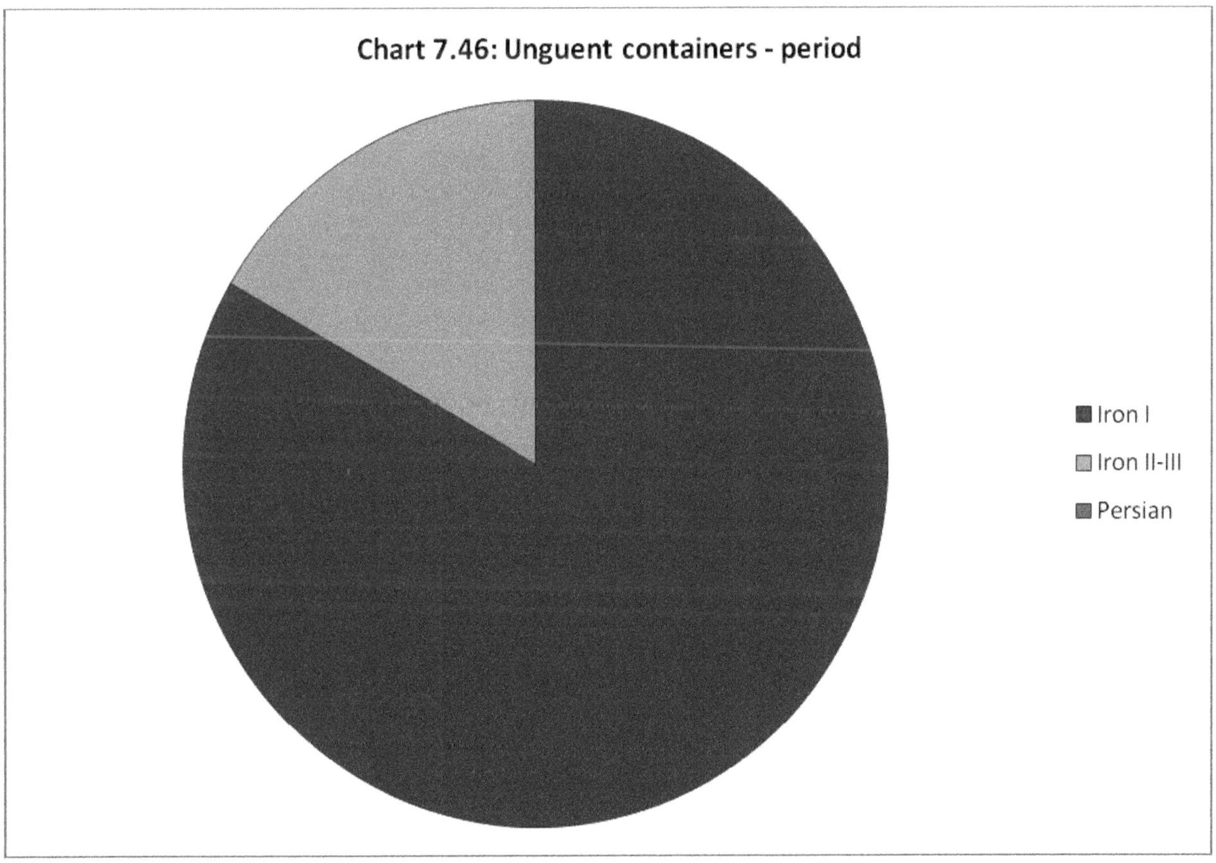

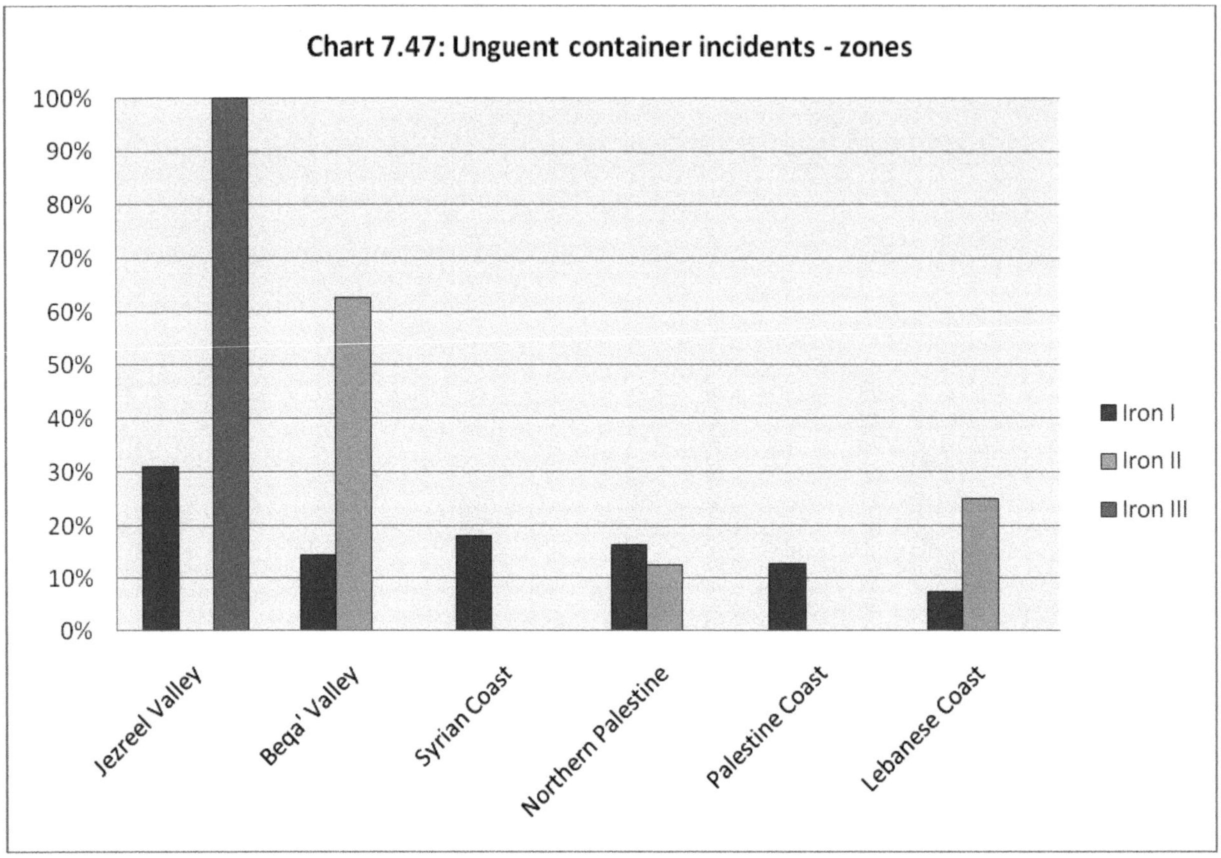

Unguent containers were generally encountered in the Iron I period (**Chart 7.46**), and were not evident in inland zones of the Northern Levant (**Chart 7.47; Maps 59; 60**). These vessels were restricted to coastal zones, and inland Lebanon and Palestine. Most of the Iron II and Iron III examples are from inland zones.

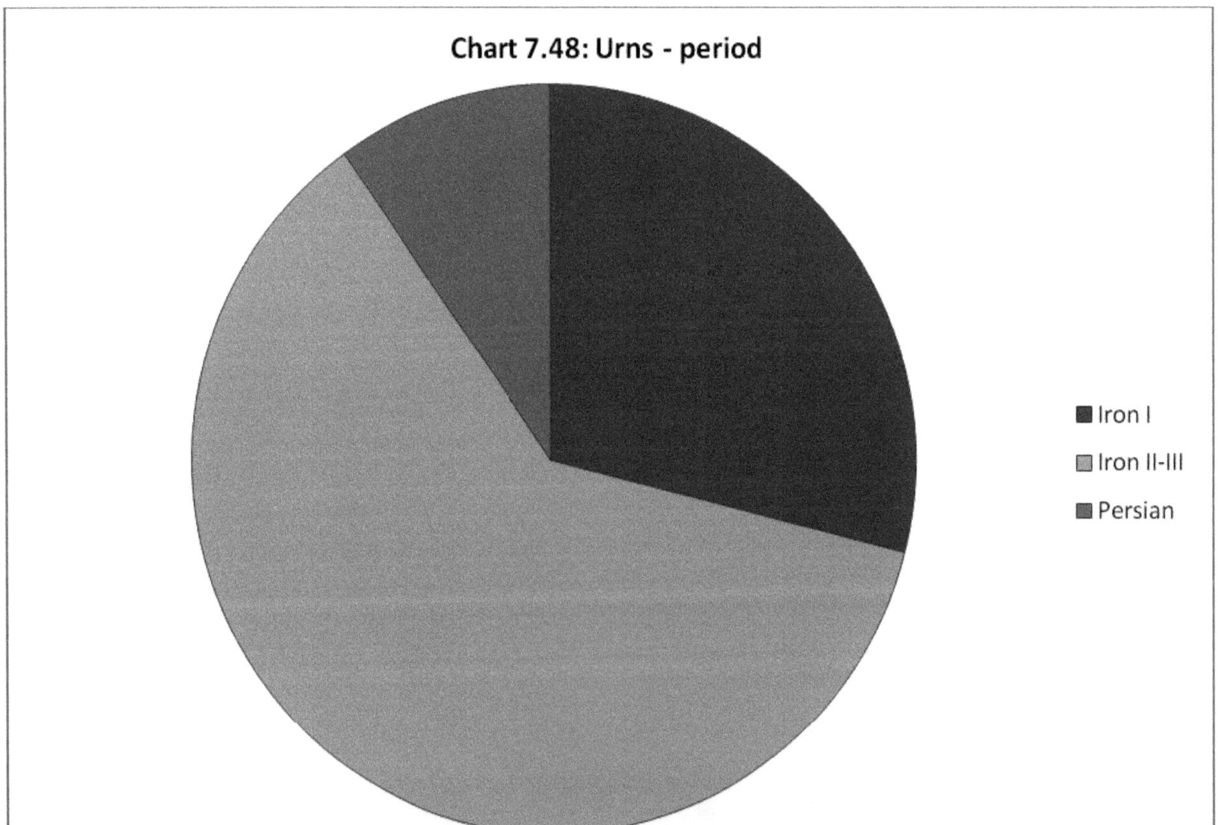

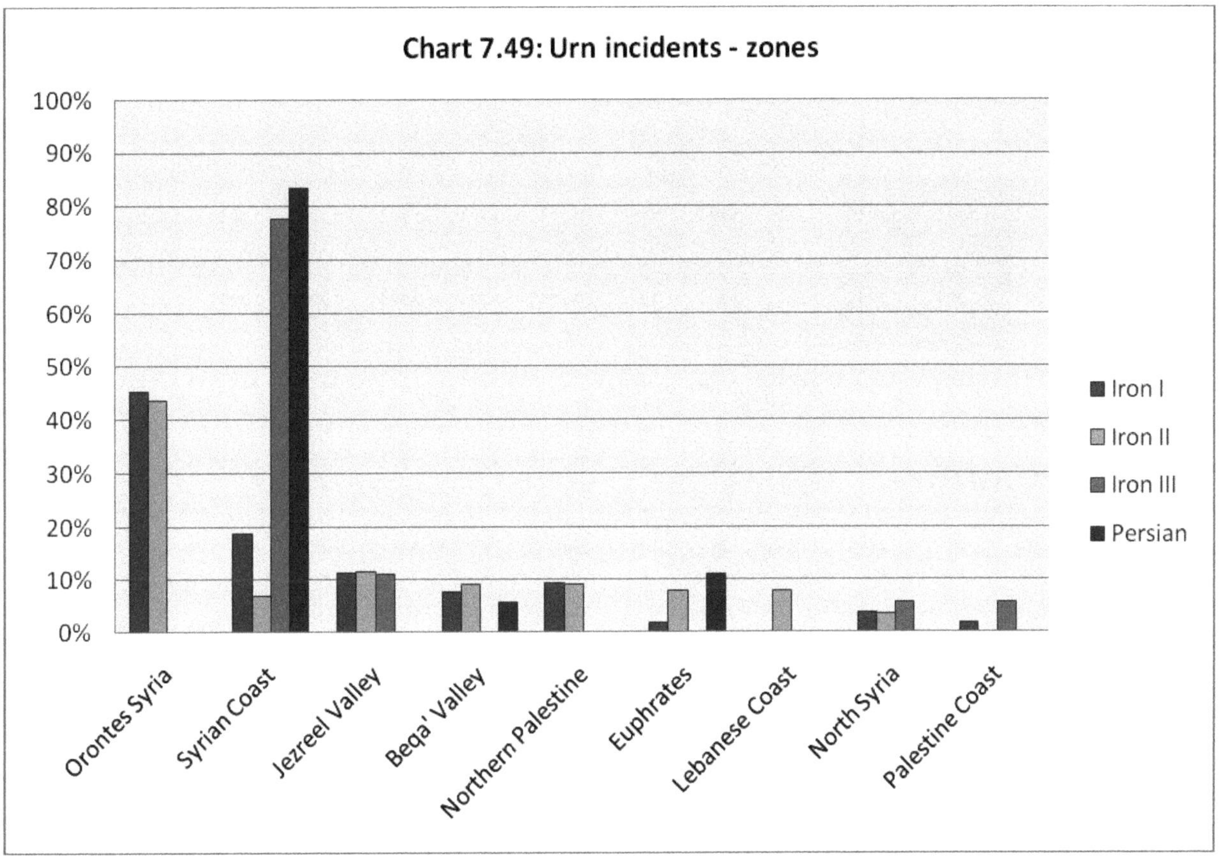

Urns were most abundant during the Iron II/III period, but were still well-attested in the Iron I and Persian periods (**Chart 7.48**). The majority of urns during the Iron I and Iron II periods derive from the inland Northern Levant, while during the Iron III and Persian periods were concentrated on the northern coast (**Chart 7.49**).

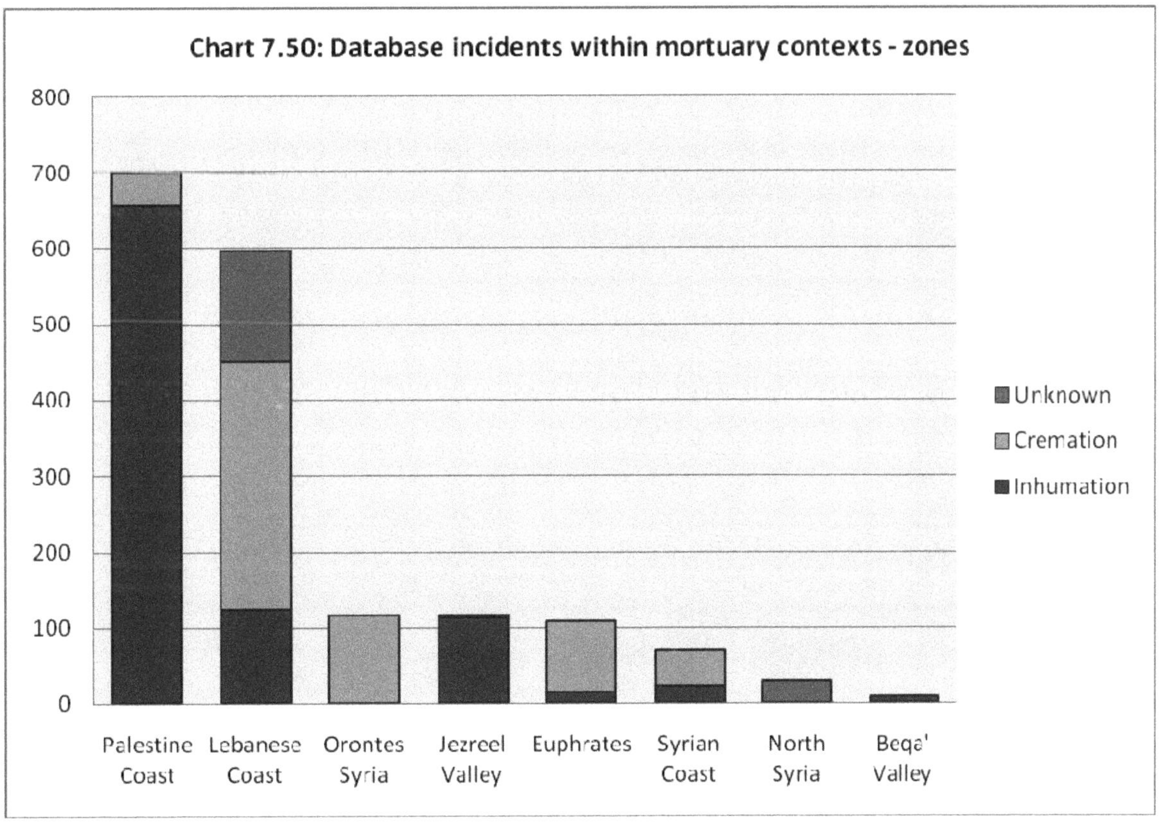

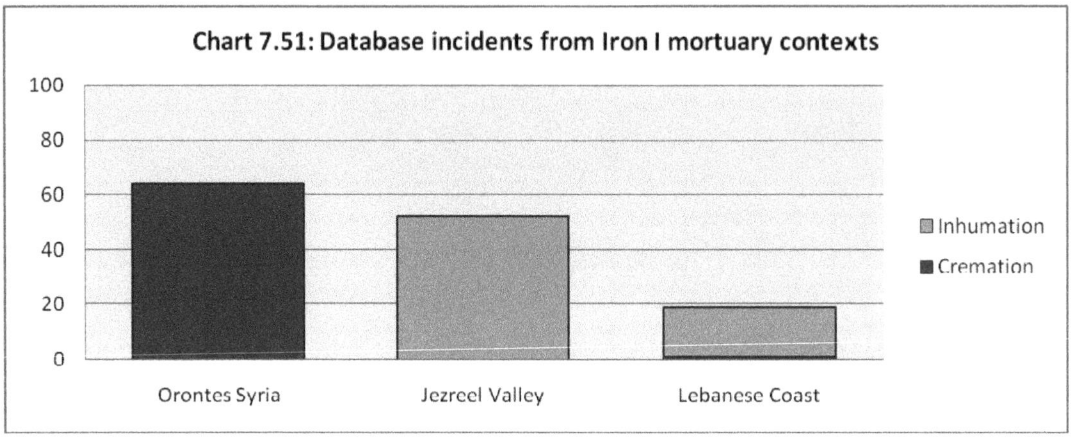

7.5 Mortuary Analysis

Chart 5.1 presented the overall ratio of mortuary versus non-mortuary contexts within the database as 1:4; in other words, there are four non-mortuary incidents in the database for every one mortuary incident. **Chart 7.50** depicts the general breakdown of all mortuary incidents according to geographic zones and type of mortuary contexts. The Palestine and Lebanese coasts represent the vast majority of all mortuary incidents, with inhumation the general rule for the former, and cremation for the latter. Cremation was the sole form of mortuary incident amongst the Orontes sites, and represents a large proportion of the Syrian Coast and Euphrates incidents. Inhumation was the general rule amongst the Jezreel Valley and Palestine coast sites. The high number of incidents from the Palestine coast may be related to the nature of inhumation burials and the fact that they are easy to identify (tombs are hard to miss). In contrast, the nature of cremation burials might make these contexts more difficult to identify. Many of the cremation burials on the Lebanese coast were placed inside tombs (e.g. Tell Rachidieh), hence the high representation for this zone.

Charts 7.51 to **Chart 7.53** plot the development of different mortuary types (inhumation and cremation) throughout the Iron Age. **Chart 7.51** presents a distinction between the inland Northern Levant and coastal/Southern Levant regions in Iron I mortuary behaviour. In this period, cremation was only identified in the inland Northern Levant, while inhumation was known from the Southern Levant and Lebanese Coast.

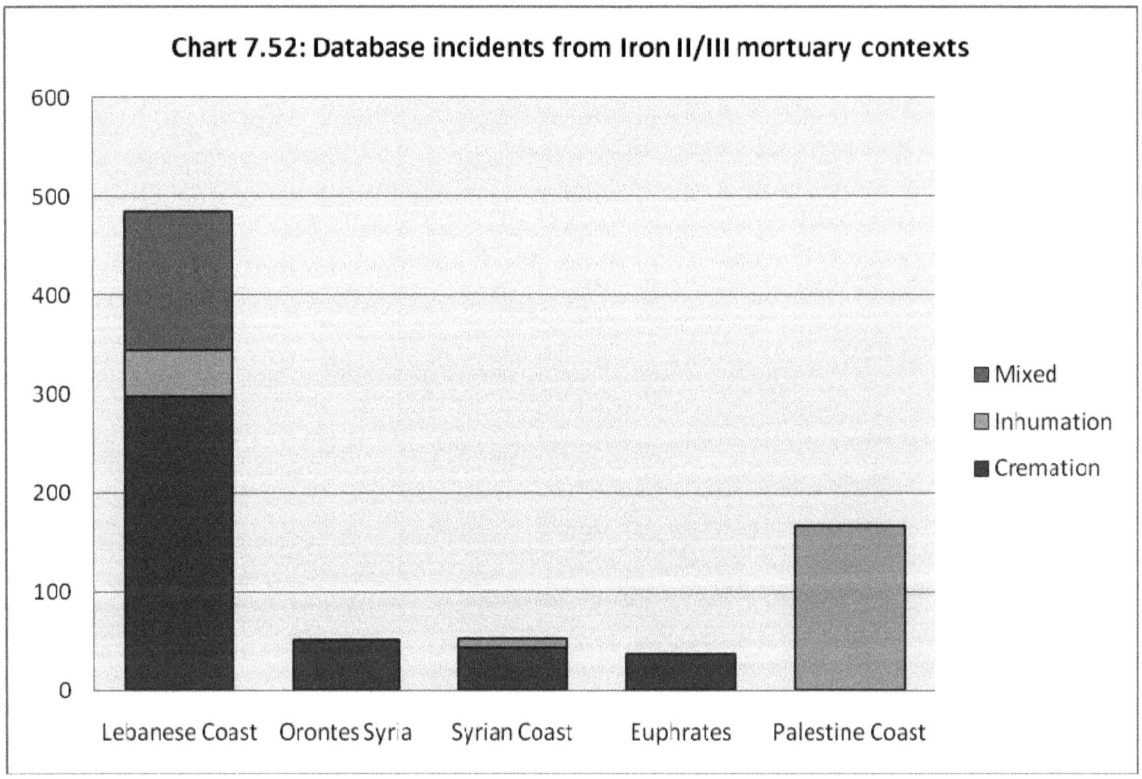

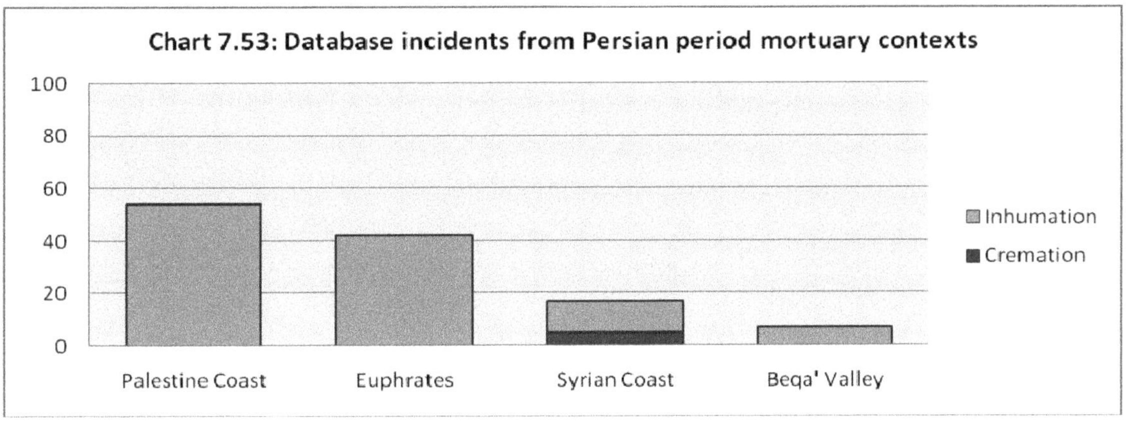

Chart 7.52 presents the Iron II and Iron III periods together because many mortuary contexts could not be conclusively dated. This chart indicates that for the Iron II-III period cremation remained the rule for the inland Northern Levant, while only inhumation was evident for the Southern Levant. In contrast, cremation and inhumation were both attested along the coast. In particular, the Lebanese coast experienced both a sharp increase in identified mortuary incidents and a change in mortuary behaviour. During the Iron I period, this region was characterised by inhumation, but cremation became the predominant burial type in the subsequent Iron II/III period ("mixed" mortuary assemblages are those either where the context was unclear, or the mixed use of cremation and inhumation were evident – e.g. the deposition of cremated remains within a tomb). **Chart 7.53** clearly shows that inhumation was the main mortuary rite during the Persian period.

Since 20% of the database represented mortuary incidents, it seemed worth exploring any possible differences between mortuary and non-mortuary ceramic assemblages. To this end, **Chart 7.54** presents the percentage of different functional categories for each of the two context types. As already indicated (**Chart 5.5**), bowls accounted for a large percentage of the database; but while bowls represented the most frequent non-mortuary incident (c. 44%), they were the second-most common mortuary incident. Instead, jugs (c. 30%) accounted for more mortuary incidents than any other category, followed by bowls (c. 26%) and kraters (12%). Moreover, flasks, dipper juglets, and urns, were also associated with mortuary contexts. In contrast, cooking-pots, pithoi, and miscellaneous utilitarian vessels were all associated with non-mortuary contexts. Interestingly, transport amphorae were equally represented by the two different context types.

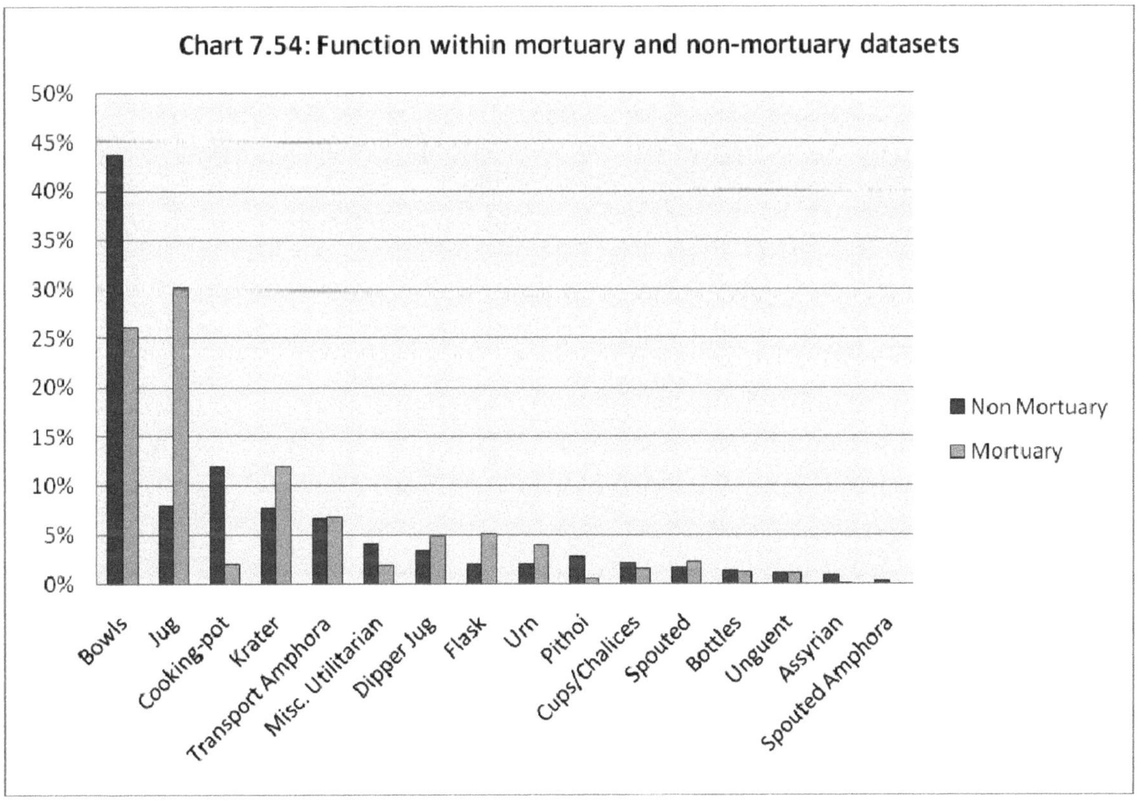

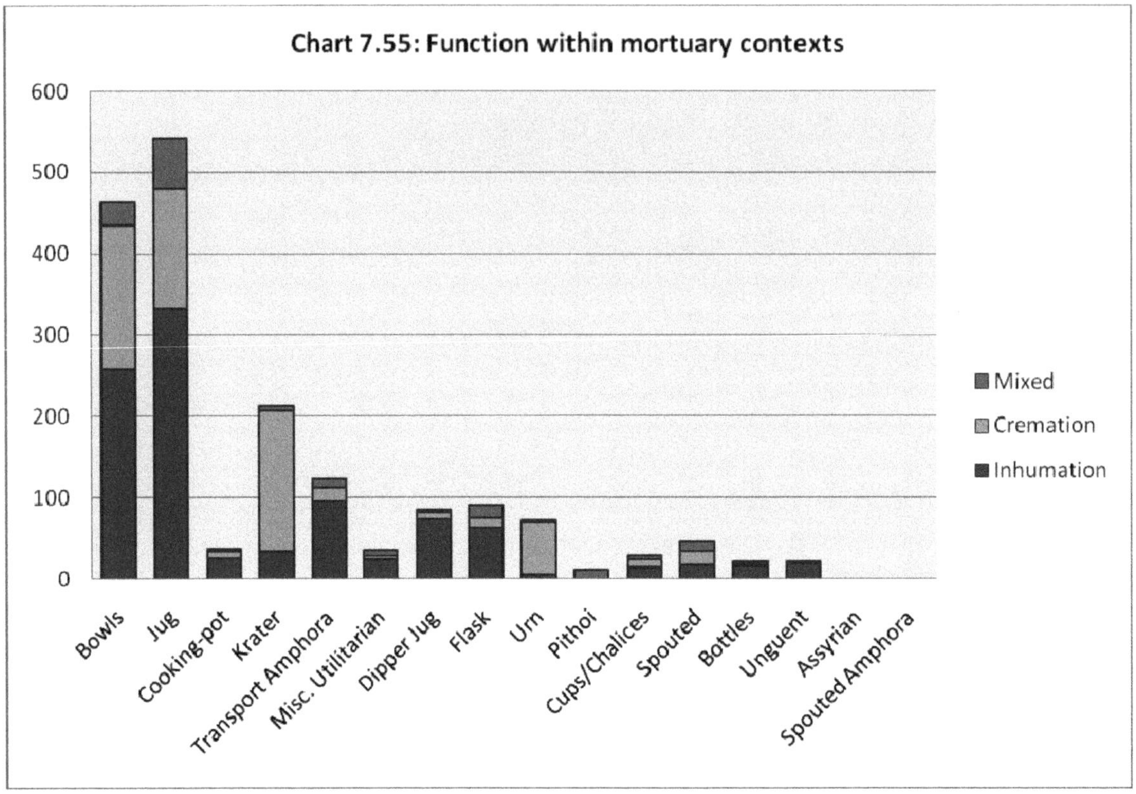

To concentrate on the nature of mortuary assemblages, the non-mortuary incidents were removed from **Chart 7.54** and differences in mortuary contexts were included. The vertical (y) axis now represents incident counts. The results are presented in **Chart 7.55**, which confirms that bowls, jugs, and kraters were the most abundant components within mortuary contexts, followed by transport amphorae, flasks, dipper juglets, and urns. Cremation and inhumation contexts appear to be different in nature. For instance, cremation contexts show a marked preference for jugs, bowls, kraters, and urns, while inhumation is better represented by jugs, bowls, transport amphorae, dipper juglets, and flasks.

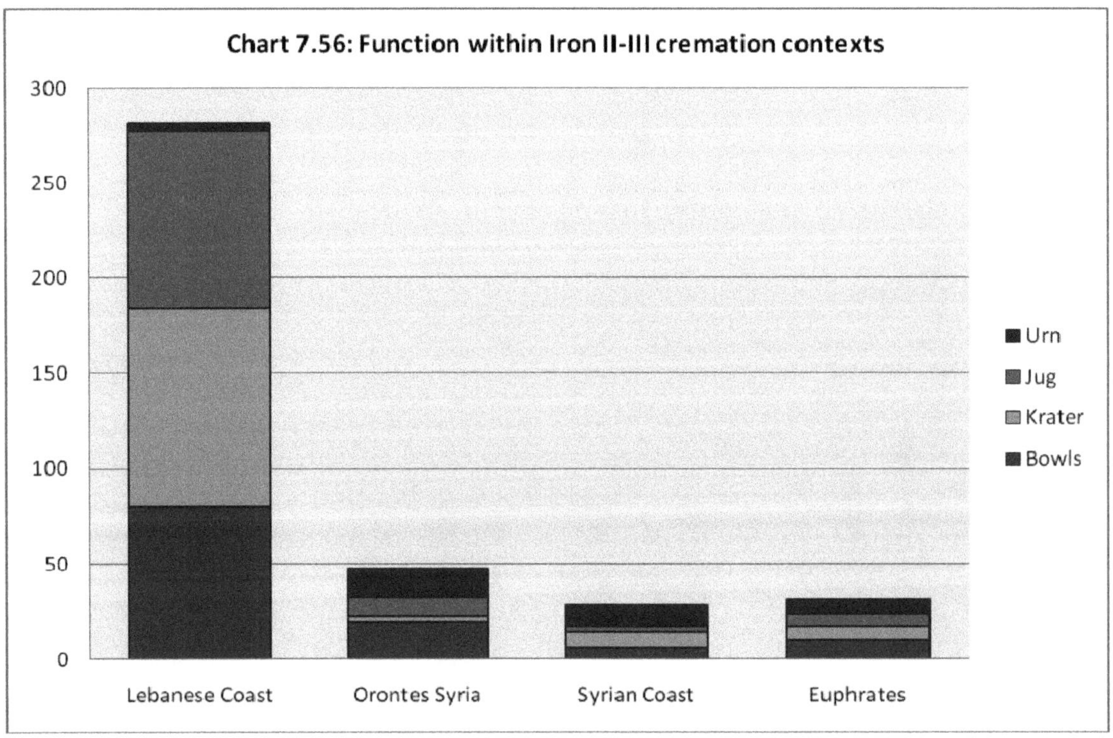

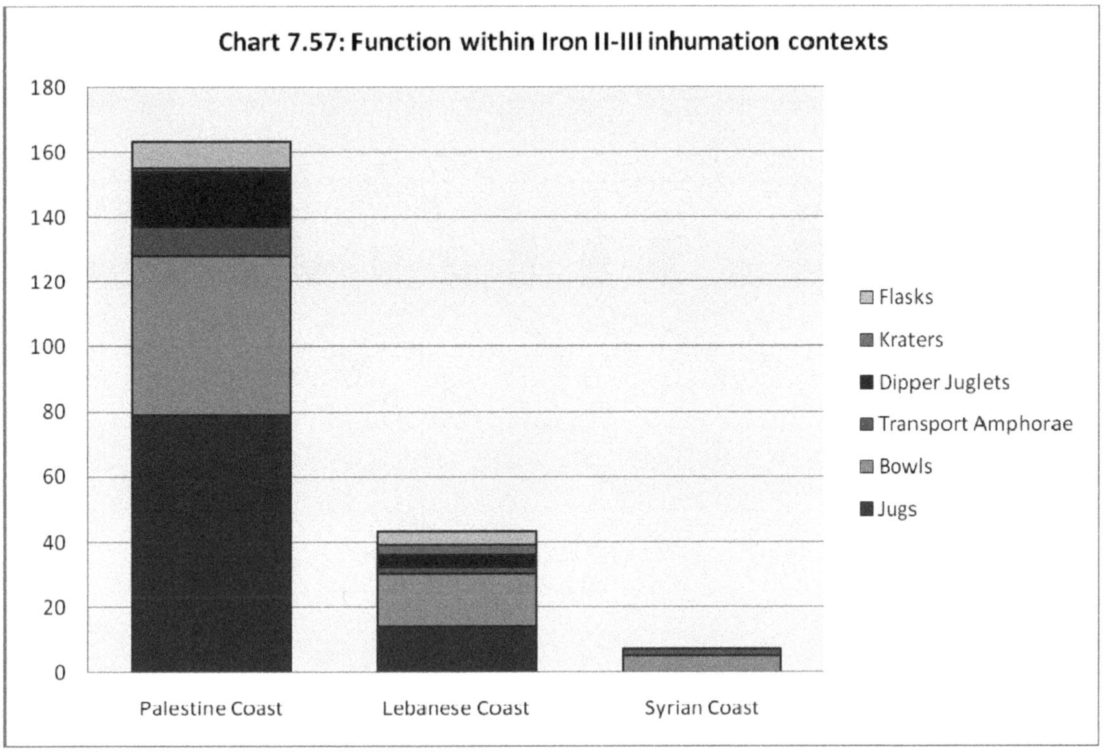

The four functional categories most commonly evident in cremation contexts were bowls, jugs, kraters, and urns. **Chart 7.56** presents the distribution of these four functional categories within cremation contexts of the Iron II-III periods, when cremation was at its most abundant. Probably the most striking trend in this chart is the preference for urns in the inland Northern Levant, while the Lebanese coast is dominated by kraters. This trend seems to suggest that different vessel forms were used as cinerary containers according to regional traditions/customs.

Chart 7.57 presents the inhumation equivalent of **Chart 7.56**; it displays the six most common functional categories in Iron II-III inhumation contexts, but shows little regional variation – inhumation was apparently restricted to the coast in this period. There is more variety in inhumation assemblages than in cremation.

Charts 7.58-60 explore three different regional patterns in mortuary behaviour for the Iron II-III period. **Chart 7.58** presents the most common CLASSES within mortuary contexts of the Lebanese coast. These "top ten" CLASSES correspond with the distinctive cluster in **Dendrogram 8.8** and confirms the presence of a rigid repertoire of pottery within coastal cremation contexts. The results from the Palestine coastal inhumation contexts (**Chart 7.59**) and inland Northern Levant cremation contexts (**Chart 7.60**) present a less-distinctive pattern. The coastal inhumation contexts were well-represented by some of the same vessels as in coastal cremation (e.g. CLASS 190 jug; CLASS 145 bowl), but with much greater variety (**Chart 7.59**). In contrast with the coastal cremation, the inhumation contexts did not contain many kraters. The inland cremation contexts show a clear preference for urns over kraters; no other vessel forms were well represented (**Chart 7.60**).

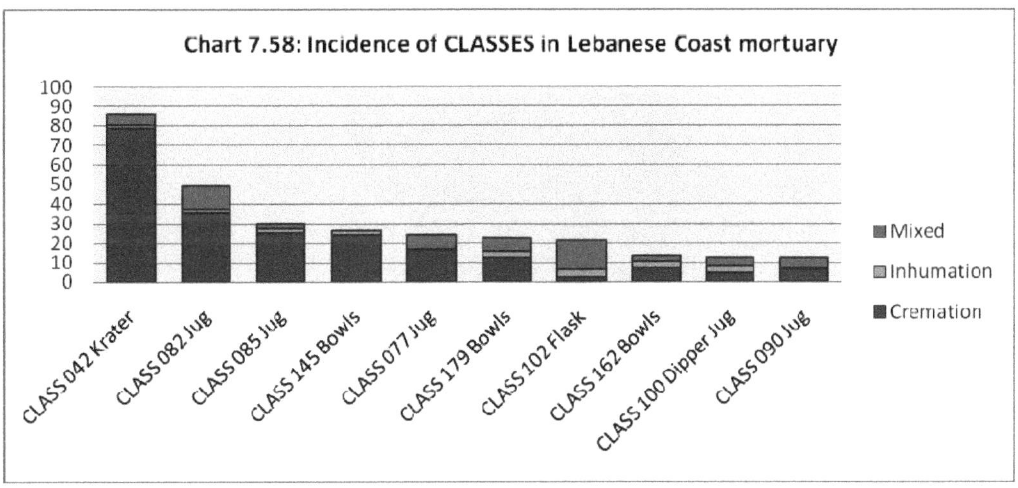

Chart 7.59: Incidence of CLASSES in Palestine coast inhumation

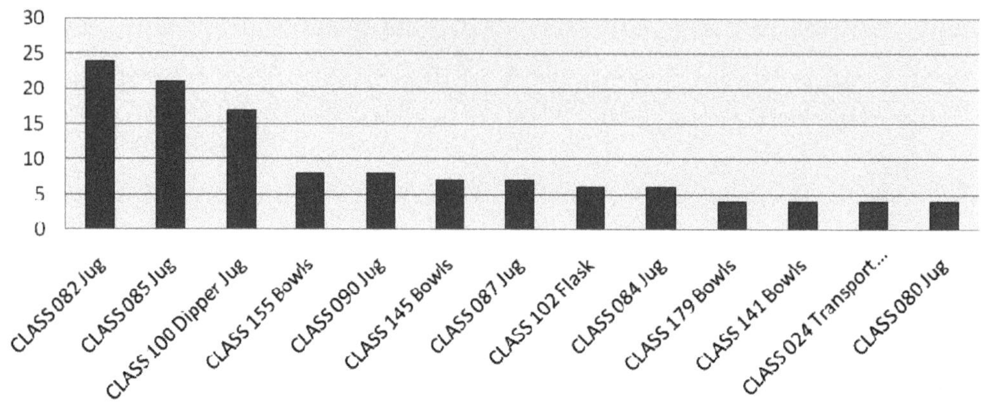

Chart 7.60: Incidence of CLASSES from inland Northern Levant mortuary contexts

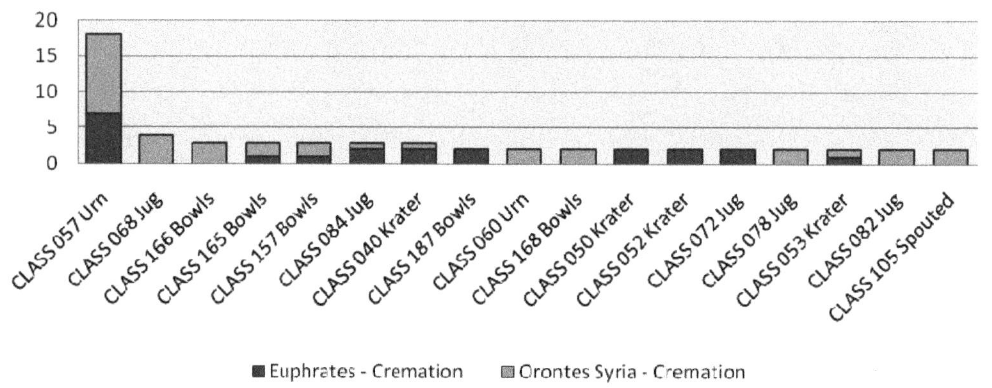

Chart 7.61: Relative % of decoration in database

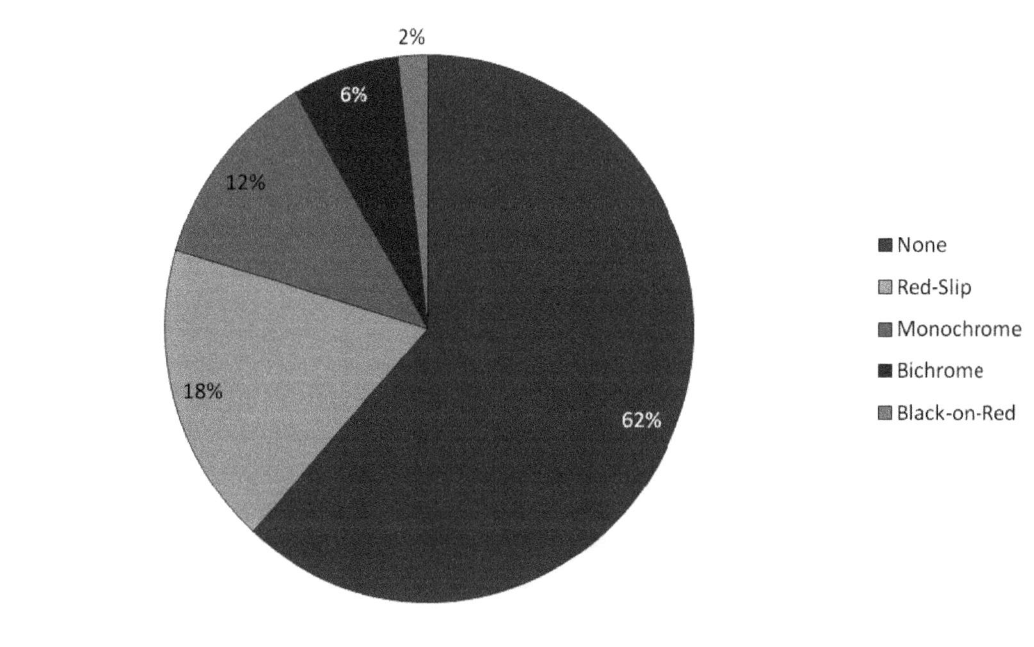

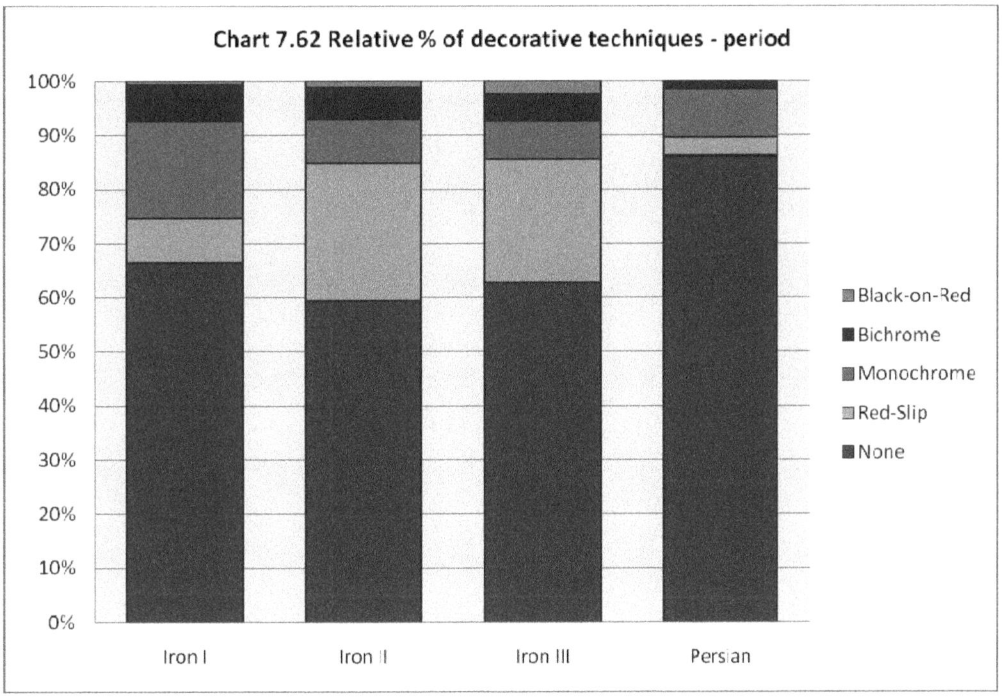

7.6 Decoration Analysis

An important element of the database was the recording of different decorative techniques; the most common four are discussed in this section – Red-Slip, Bichrome (also known as "Phoenician Bichrome" and distinct from so-called "Philistine Bichrome"), Monochrome, and Black-on-Red. **Chart 7.61** shows the percentage of different decorative techniques within the ceramic database. It demonstrates that the vast majority of pottery within this study had no recorded evidence of paint or slip decoration, or none was noted in publication. Red-Slip is the most common decorative scheme, followed by Monochrome, Bichrome, and Black-on-Red. The predominance of Red-Slip might be due to the predominance of Iron II-III period incidents in the database, when Red-Slip characterised pottery (**Chart 5.7**). To confirm this supposition, the dataset was categorised according to periods, the results of which are presented in **Chart 7.62**. This exercise was also a convenient means of plotting the development of the different decorative techniques over time.

Chart 7.62 again confirms that a large percentage of pottery within the database was not decorated. However, there is also evident a marked shift from Monochrome decoration in the Iron I period, to Red-Slip decoration in the Iron II and Iron III periods. Very little decoration is evident in the Persian period, though there appears to be some tendency toward Monochrome decoration. The percentage of Bichrome in the Iron I period is surprising, as this particular decoration technique is usually associated with the shift away from Monochrome.

While the **Chart 7.62** plots a somewhat simplistic development of decoration over time, there was also likely to be regional variation within the trends. By categorising the decoration data according to the nine geographic zones (**§7.3**), different regional models for the use of Red-Slip, Monochrome, Bichrome, and Black-on-Red were developed – **Charts 7.63-66** present the regional trends per period. The four charts are pivot-table charts, hence the different geographic zones and decorative techniques are ordered according to their representation in each period, from greatest to least. There is a change in legend between the Iron I and Iron II charts.

Chart 7.63 presents the number of incidents of Iron I decoration according to region. Monochrome is the predominant decorative technique overall. Red-Slip is primarily restricted to inland regions of the Southern Levant, with only a few examples known from the coast; Red-Slip was virtually absent from the inland Northern Levant. Bichrome decoration is consistently represented in most areas, but not in the inland Northern Levant.

Chart 7.64 presents the number of incidents of Iron II decoration according to region. Red-Slip is the most abundant decorative technique. While it remained well-attested in the Southern Levant, Red-Slip rose in use along the Northern Levant coast and the Orontes. The Beqa' and Euphrates Valleys, and the Southern Levant coast are poorly represented. It is interesting to note that Monochrome remained an important feature of the Lebanese coast and the Orontes during this period, despite the rise in Red-Slip. In contrast, the use of Monochrome in the inland Southern Levant was affected by Red-Slip. As common as Monochrome and Red-Slip were on the Lebanese Coast, Bichrome was the most abundant (**Chart 5.2**).

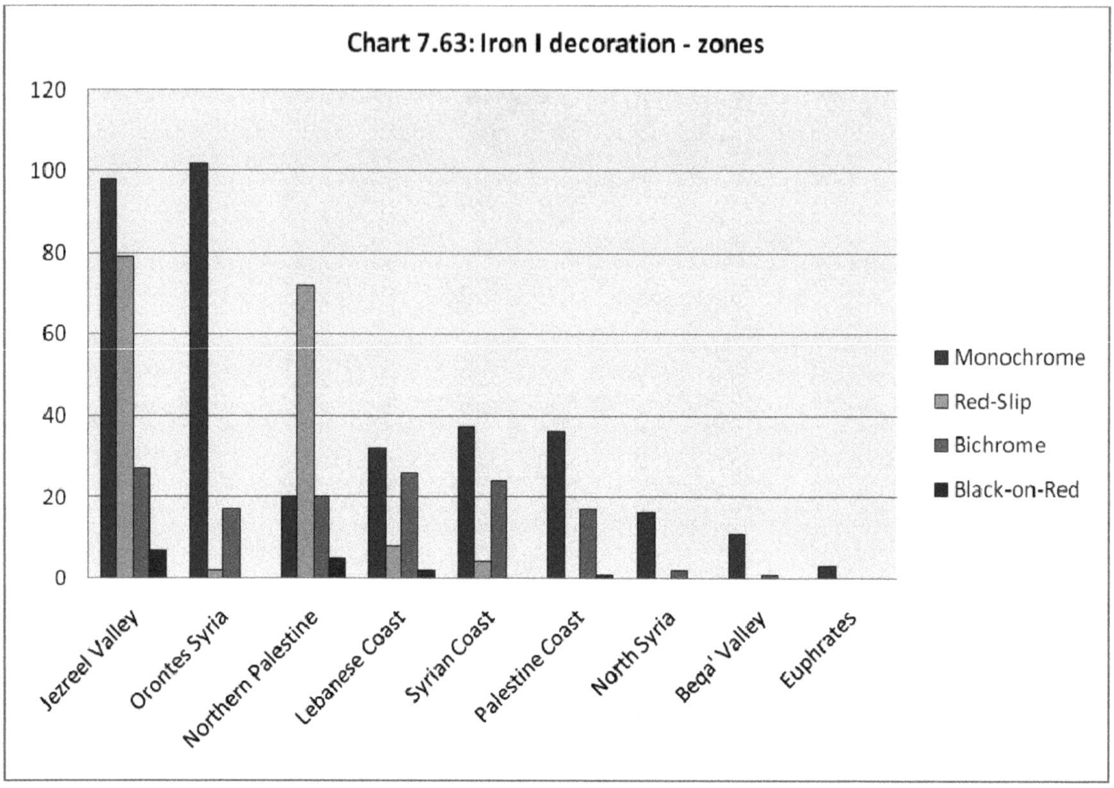

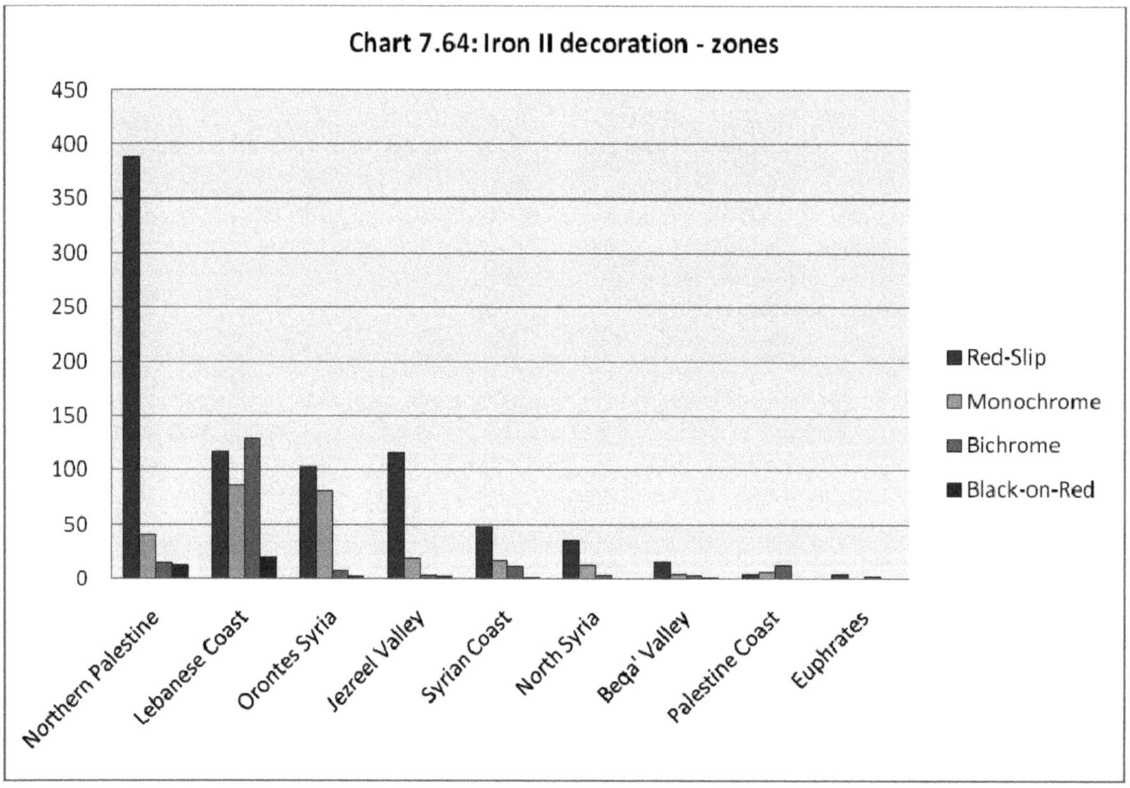

Chart 7.65 presents the number of incidents of Iron III decoration. In this period, Red-Slip was the primary decorative technique for the Northern Levant. In Orontes Syria, for instance, Red-Slip was the main decorative technique. While Red-Slip was the most common on the Syrian and Palestinian coasts, Monochrome, Bichrome, and Black-on-Red were all well-attested. Despite the marked preference for Red-Slip in all regions, the actual number of incidents is reduced; hence we are unable to draw.

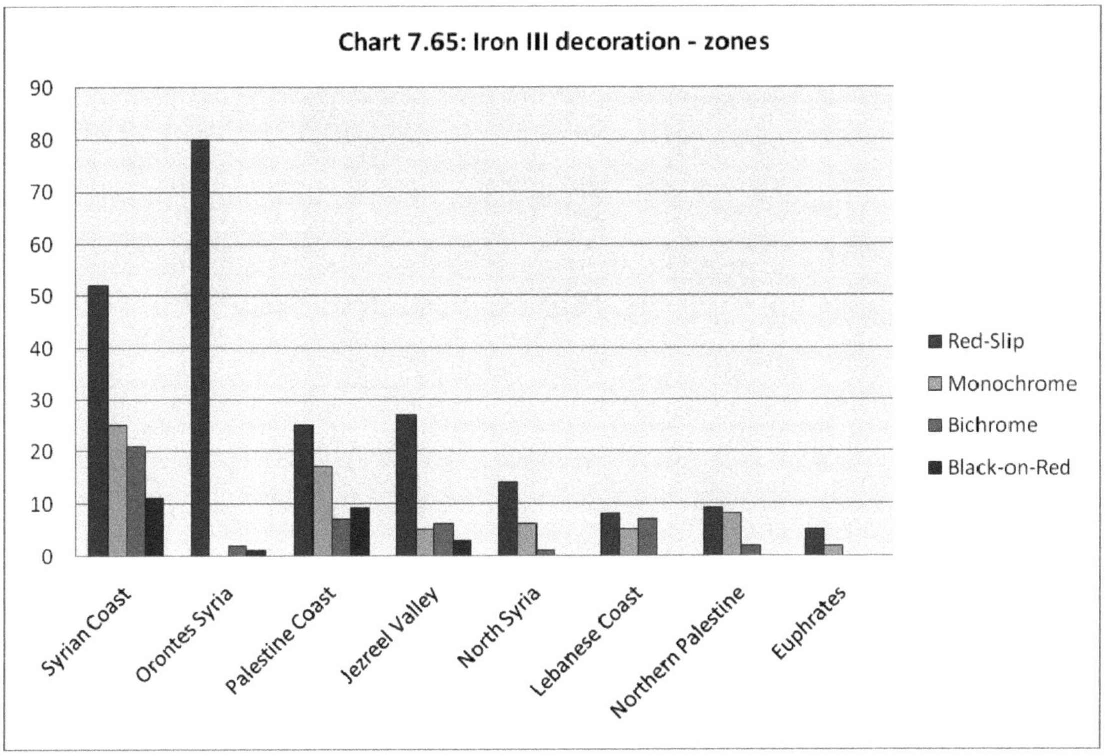

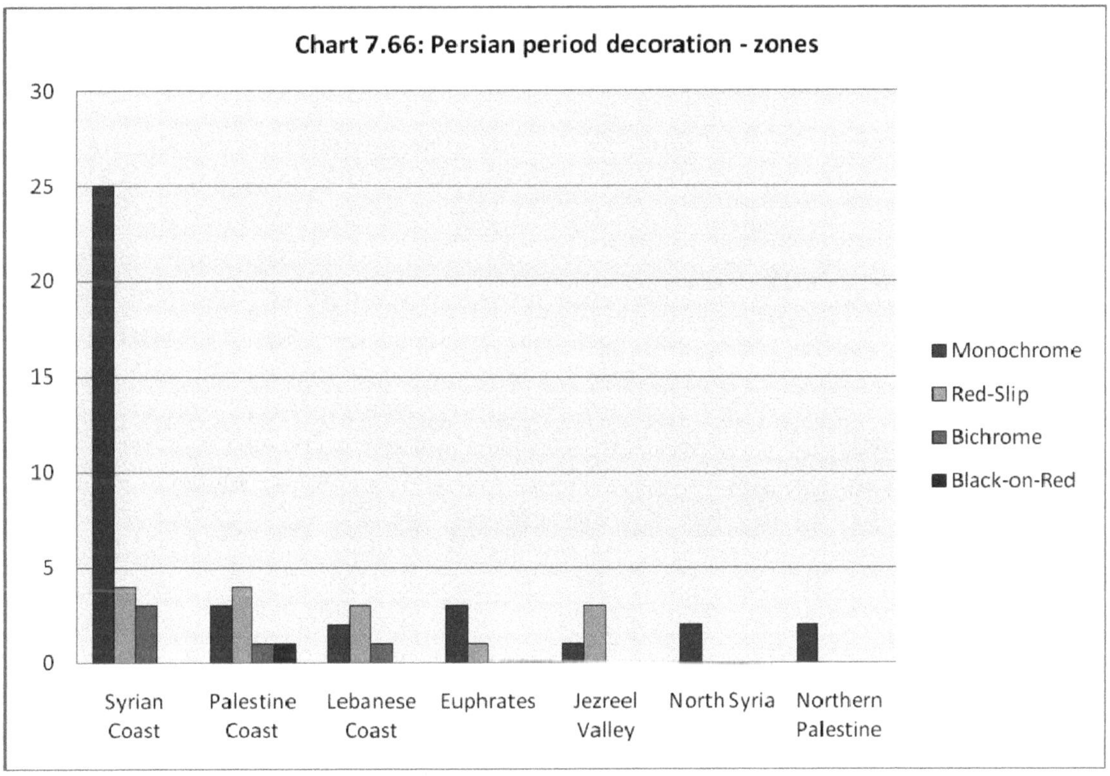

Chart 7.66 presents the number of regional incidents for the Persian period, when decoration was generally limited to the coast. Monochrome was the main decorative scheme on the Syrian Coast, but over 85% of all Persian period vessels were undecorated (**Chart 7.62**). The number of incidents for each period was relatively low.

In addition to regional and diachronic variations, decoration was related to vessel function. Hence, the relationship between the four key decorative techniques and ceramic function was investigated in **Charts 7.67-7.71**. **Chart 7.67** presents the relative percentage of ceramic function attested by each decorative technique. While too "busy" to understand fully (especially grey-scale), it does demonstrate that both Red-Slip and Black-on-Red were primarily represented by bowls and jugs. In contrast, the two painted traditions were represented by greater diversity of ceramic function. Monochrome and Bichrome were evident on bowls, jugs, kraters, flasks, urns, spouted vessels, and amphorae.

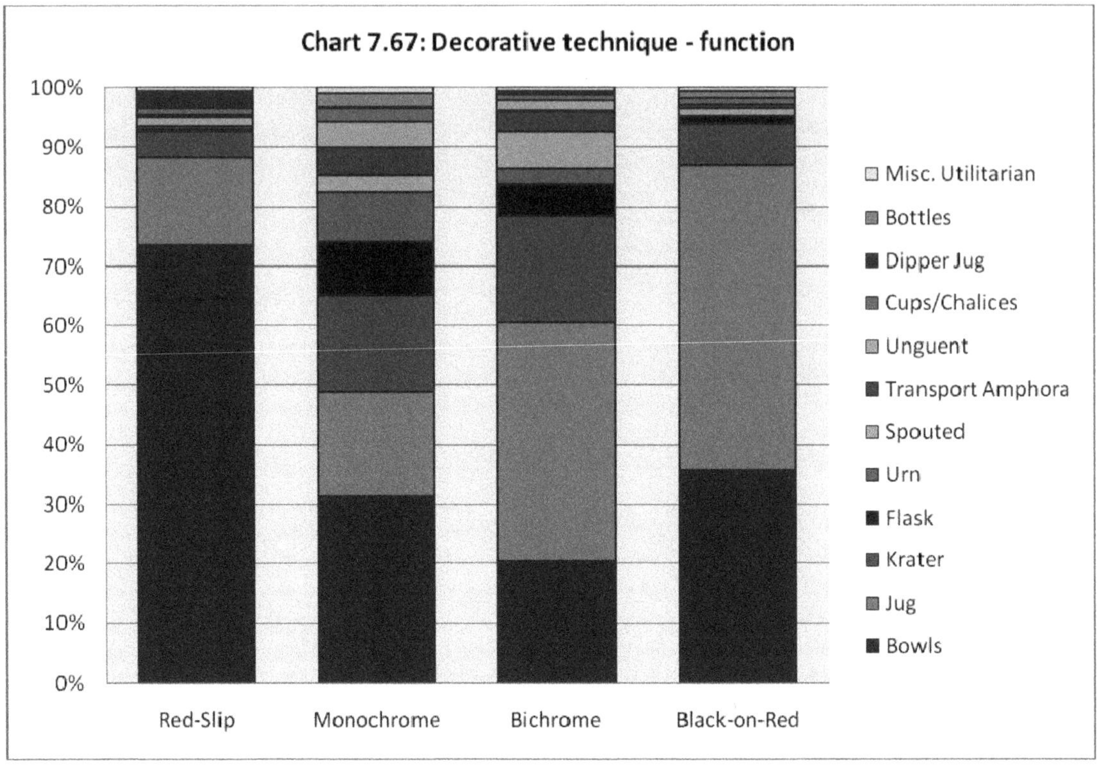

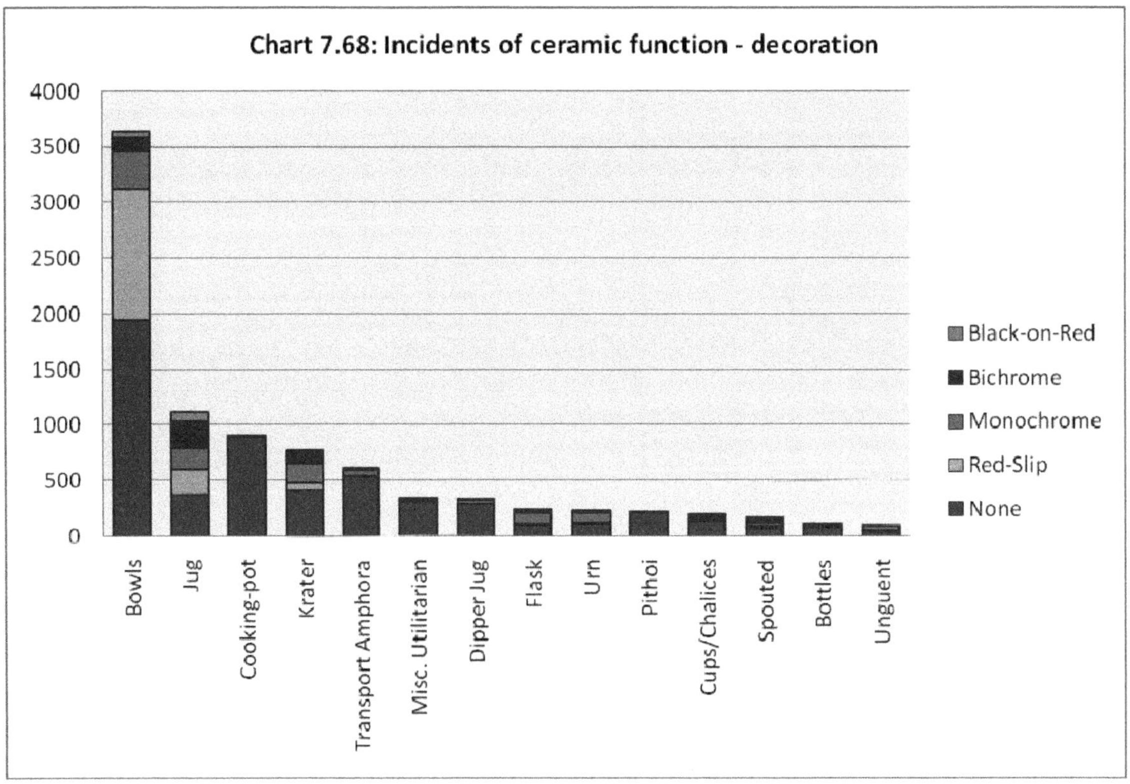

Chart 7.68 presents the same data as **Chart 7.67** but in the opposite relationship; i.e. it depicts ceramic function according to decoration. The vertical axis represents incidents. The different functional categories are arranged from most to least on the horizontal axis. **Chart 7.68** emphasises the large proportion of decorated vessels represented by bowls. Indeed, a significant component of all bowls, jugs, and kraters are decorated (and to a lesser extent flasks and urns). In contrast, some categories have little decorative representation; e.g. cooking-pots, pithoi. **Charts 7.69-7.71** present the different decorative schemes diachronically for bowls, jugs, and kraters. **Chart 7.69** indicates that during the Iron I and Persian periods the majority of bowls were undecorated. During the Iron II and Iron III periods, however, it seems that Red-Slip bowls were almost as common as undecorated

bowls. Monochrome bowls were generally restricted to the Iron I period. **Chart 7.70** indicates that the decoration of jugs largely follows a different pattern to that of bowls. In particular, Bichrome was the most common decorative technique overall. Red-Slip was roughly equivalent to Bichrome and Monochrome during the Iron II and Iron III periods, but was poorly represented in other times. Apparently, painted decoration was a significant factor in jug presentation. Furthermore, jugs were more often decorated than not, except in the Persian period. **Chart 7.71** indicates that the decoration of kraters was primarily restricted to painted decoration, though Red-Slip was relatively common during the Iron II period. Interestingly, Monochrome and Bichrome show a reverse pattern; with Monochrome slowly becoming less common, and Bichrome intensifying diachronically.

A large percentage of bowls were Red-Slip (**Charts 7.67-7.69**). For this reason, **Chart 7.72** explores the regional development of Red-Slip bowls. Geographic zones are arranged on the horizontal axis following geographic progression from the inland Southern Levant to the coast, northwards, and inland again. Red-Slip bowls were present in the Southern Levant during the Iron I (*cf.* A. Mazar 1998). A similar pattern is evident in the Iron II period, though the Northern Levant was better represented. The Iron III period, however, marked a reversal of the earlier pattern, with the concentration of Red-Slip bowls centring on the Northern Levant, and tapering off southwards. Red-Slip bowls during the Persian period were centred on coastal regions. While the patterns in **Chart 7.72** are clear, explaining them is more difficult; something that is attempted in **Chapter 9**

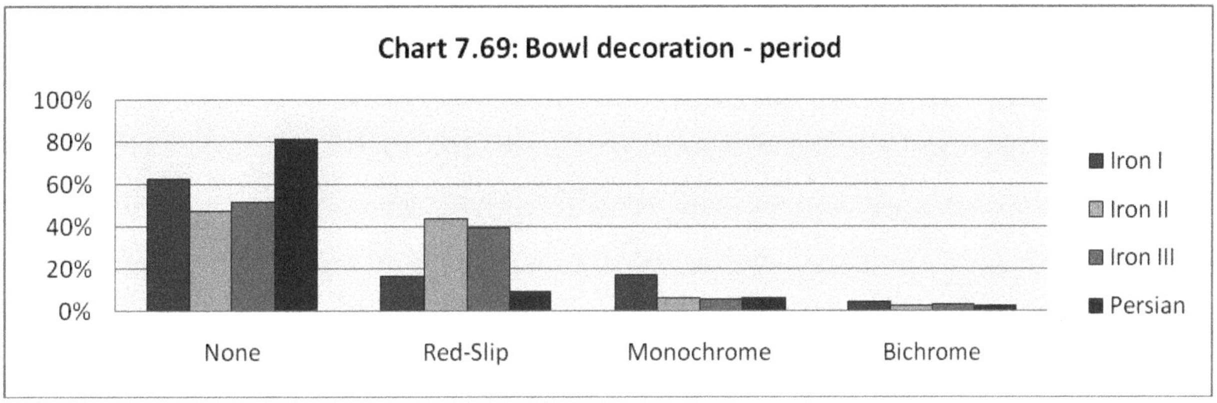

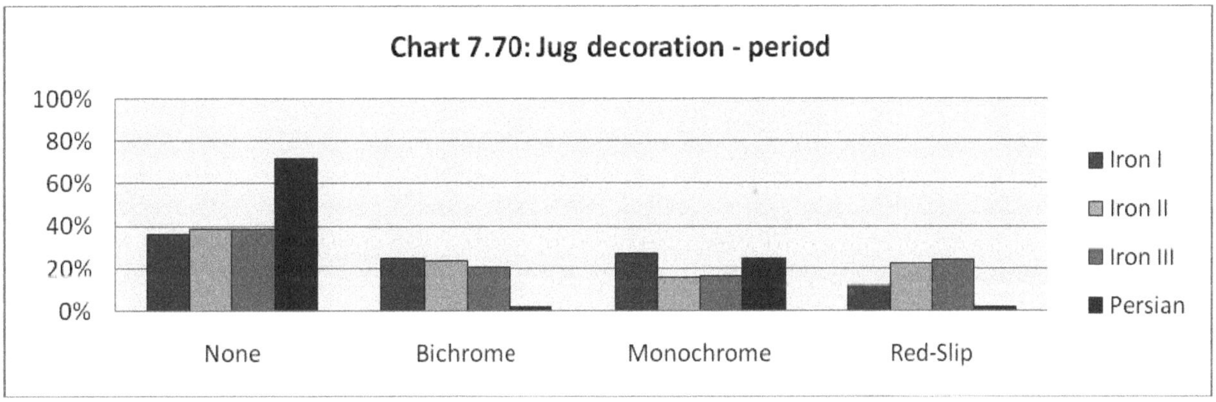

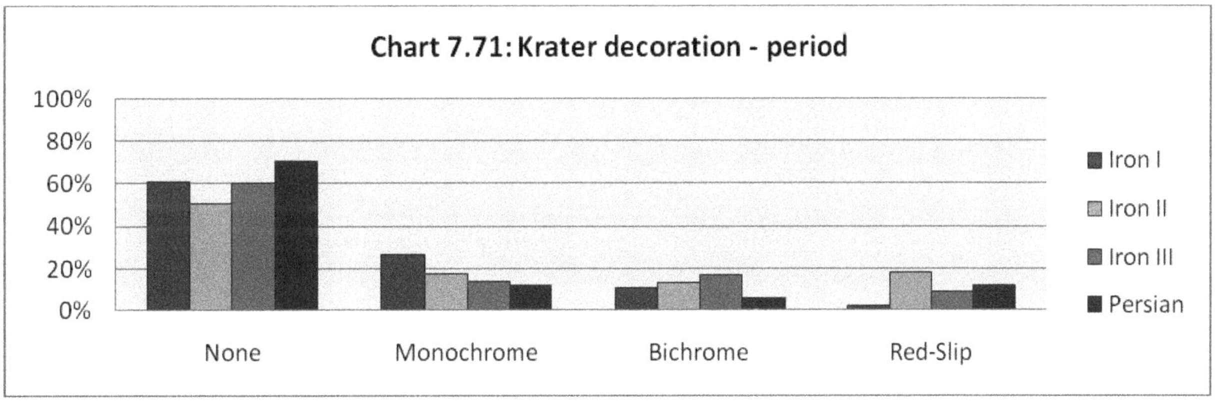

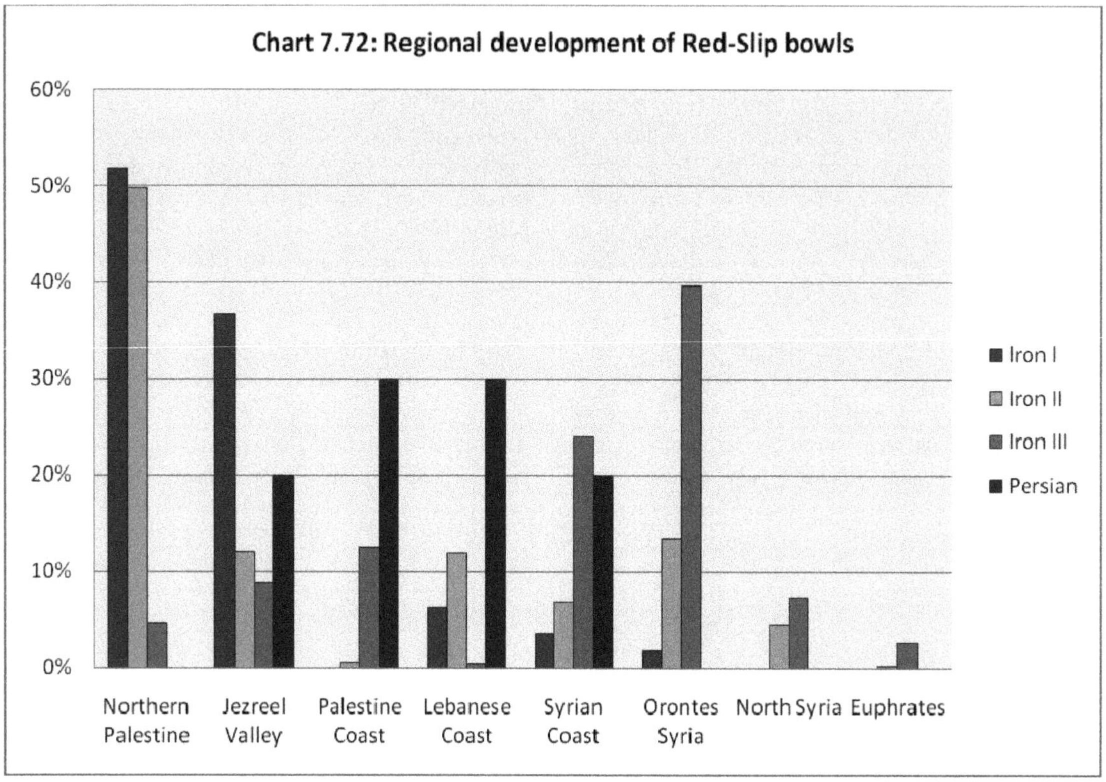

7.7 Summary

The exploratory data analysis undertaken within this chapter has demonstrated that even with the present flawed data broad patterns are evident. Nevertheless, there are also more subtle patterns in the data, patterns that appear to hold no single organising principle and which cross-cut the conventional broad political regions; patterns within and between CLASS, function, period, decoration, and mortuary behaviour. This evident diversity cannot be easily linked to ethnic or political units. Instead, the ceramic data depicts a high level of complexity within cultural behaviour for the Iron Age Northern Levant; complexity that is analysed further in **Chapter 8**.

CHAPTER EIGHT
Multivariate Data Analysis

8.1 Introduction

This chapter presents the investigation of this study's ceramic database using two multivariate analysis techniques, both of which are widely accepted as appropriate for incidence (presence/absence) data (Baxter 2003, 140). The first technique is that of Correspondence Analysis which, broadly speaking, compares whole assemblages based on the association of ceramic types within each assemblage. The second technique, Cluster Analysis, focuses on similarities between ceramic categories within assemblages, which in this study constitute CLASSES. The results of the two techniques are presented visually. In the case of Correspondence Analysis, the output is a chart (or plot) that positions assemblages in multi-dimensional space. Cluster Analysis, on the other hand, produces a dendrogram that emphasises clustering of data. Each technique and its output are described in the relevant sections. It is intended that these two techniques will reveal some of the more subtle patterns in the data that **Chapter** 7 might have passed over.

8.2 Correspondence Analysis

Introduction

Despite the great effort and resource expended on its study, Iron Age pottery is rarely discussed beyond its supposed chronological and cultural value. However, pottery is likely to be the product of different influences: chronology and geography are obvious factors, but others may be present, e.g.. status, function, cultural boundaries, consumption, distribution mechanisms, the ways in which the material enters the archaeological record, and any number of unknown factors. Current method and practice, and publication standards conspire to "rob" the ceramic record of its true value. Correspondence Analysis (hereafter CA) is an analytical technique available for the systematic comparison of archaeological assemblages, based on the premise that regularly occurring patterns of associations are likely to reflect patterns of social activity and behaviour. It is an exploratory data-analytic technique which produces a graphic view of the relationship between ceramic assemblages (Shennan 1997, 308-327). Through the comparison of ceramic assemblages from sites of different type, date, and region, recurrent patterns emerge in the data; patterns that may have very little to do with chronological or geographical factors. Despite the potential, CA has rarely been used in Near Eastern archaeology.

The Technique

For the purpose of aiding comparison, assemblages can be expressed as tables where the rows represent contexts (or units), the columns represent vessel types, and their interconnected cells represent the occurrence (or incidence) of each type per context (Baxter 2003, 137). CA is a descriptive technique designed to reveal patterns within such a table and identify associations between rows and columns (Shennan 1997, 308-341). The technique works by converting the values of each row or column to percentages of the total value of that row or column, which can then be plotted within multi-dimensional space. The CA results are presented as two graphs; the row plot and the column plot. The intersection of the axes in each plot represents the average profile within the data (Shennan 1997, 321). In the row plot, contexts with relatively similar assemblages will appear in the same area, while ceramic categories that are closely associated will appear in the same area of the column plot. In other words, the row plot displays relationships between contexts, while the column plot displays relationships between ceramic categories. A ceramic category is deemed to characterise certain contexts when they appear in similar areas of superimposed column and row plots (Baxter 2003, 138-139). CA also weights the different categories according to the number of incidents; contexts with more incidents are given more weight than those with very few incidents, and vice versa for the different ceramic categories. Single incident categories, however, are excluded from the analysis. In addition to the graphic output, CA also produces statistical tables that indicate how well the plots represent the data. Inertia is the term used for describing how much variation in the data is explained by each axis in the plot; this value is noted under each chart. A plot is considered a good representation of the data if combined inertia of the axes (or components) forms a high percentage of total inertia.

Analysis

All CA calculations and plot productions were undertaken with the Bonn Archaeological Statistics Package (WinBasp) version 5.43 for Windows; as a result, all table and matrix data are not included here, as they can only be viewed using the WinBasp program (a screen-shot is presented in **Figure 5.3**). Each CA plot is presented below as well as being included on the appended CD; the file name is noted beneath each plot.

Broad Patterns

In emphasising the historical narrative, Near Eastern archaeologists have largely ignored the more localised patterns of cultural behaviour. With this in mind, the approach adopted below is one that moves from the general to the more particular, successively focussing the analyses around more narrowly-defined data categories. The aim is for the broad dominating influences, which may mask other more specific patterns, to be isolated and their effects noted, helping to reveal those patterns produced by less dominant factors. An understanding of the broader patterns will also direct the more specifically targeted analyses.

The first CA examines broad patterns in the data. To this end, each site's assemblage, regardless of stratigraphy, was treated as a single unit; in other words, no chronological or spatial divisions were maintained. It was hoped this would negate any chronological patterns, and reveal other influences. This would also allow more meaningful comparison with "unstratified" assemblages (e.g. Joya). The CA results are presented in **Chart 8.1**, which depicts two distinct groups of assemblages (as indicated). Under closer scrutiny, these two groups appear to represent a broad geographical distinction in the data: the right-side cluster derives from the inland Northern Levant while the left-side cluster represents the rest of the study area (coastal regions and the Southern Levant). These two broad geographic categories mirror a similar distinction in Lehmann's (1996, Fig. 4.4) research on the late Iron Age pottery of Syria and Lebanon, though Lehmann makes no effort to either explain these "regions" or isolate more localised patterns (**§4.2**). Two assemblages to the top of the plot (i.e. Deve Höyük and Kefrik) are affecting the scale of the plot. Hence, the same CA results are presented in **Chart 8.2** with the two outlying assemblages removed. This second chart maintains the geographical distinction whilst highlighting some smaller clusters.

While **Charts 8.1** and **8.2** highlight the very general effect that a site's geographical location has upon the make-up of a site's assemblage, they may mask other variables within the data. For instance, it is reasonable to assume that mortuary and non-mortuary assemblages would contain different functional elements. To this end, the next CA takes context type (mortuary versus non-mortuary) into consideration: in other words, the pottery from mortuary contexts at a particular site is treated as a separate assemblage to the pottery from settlement contexts at the same site. The two context types are represented in **Chart 8.3** by two different colours: the lighter colour denoting mortuary assemblages; blue denoting non-mortuary assemblages. The resulting CA plot maintains the broad dichotomy (right side = inland Syria; left side = coastal and southern areas) present in the first two charts while highlighting another trend in the data; mortuary assemblages are largely distinct from non-mortuary assemblages.

Chart 8.1: CA row plot – site assemblages

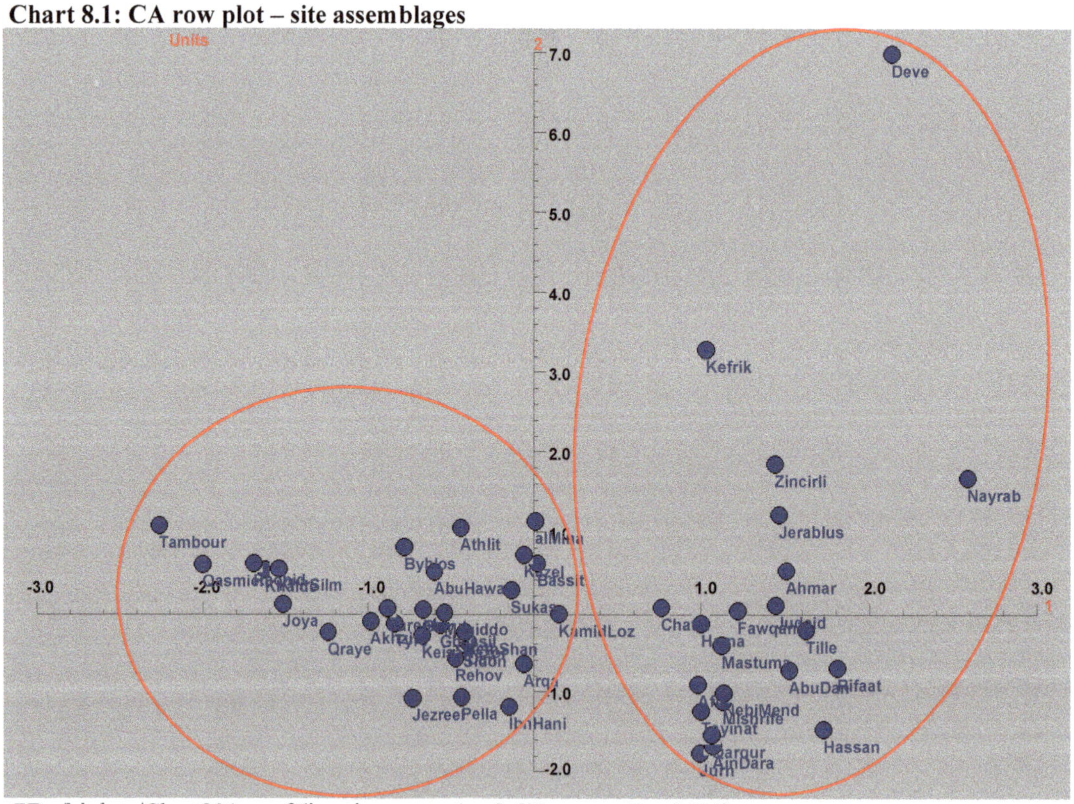

CD: /biplots/Chart801.wmf (inertia: comp. 1 – 9.6%; comp. 2 – 6.0%)

Chart 8.2: CA row plot – site assemblages (outlying points removed)

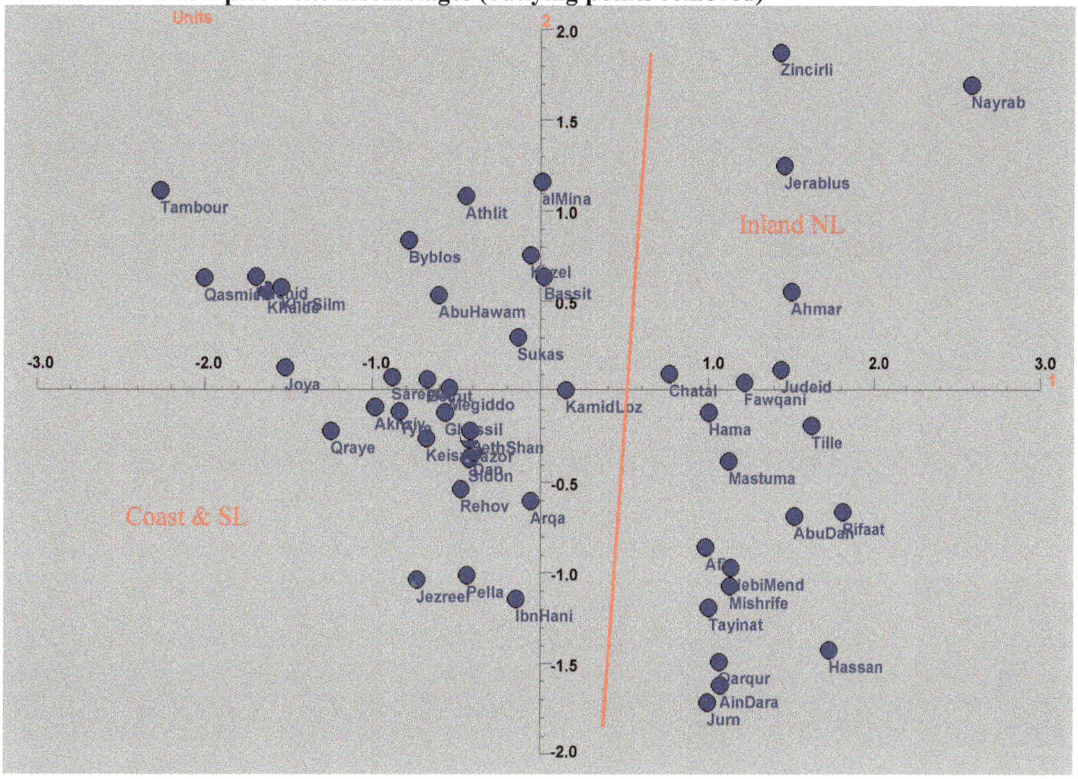

CD: /biplots/Chart802.wmf (inertia: comp. 1 – 9.6%; comp. 2 – 6.0%)

Chart 8.3: CA row plot – mortuary and non-mortuary assemblages

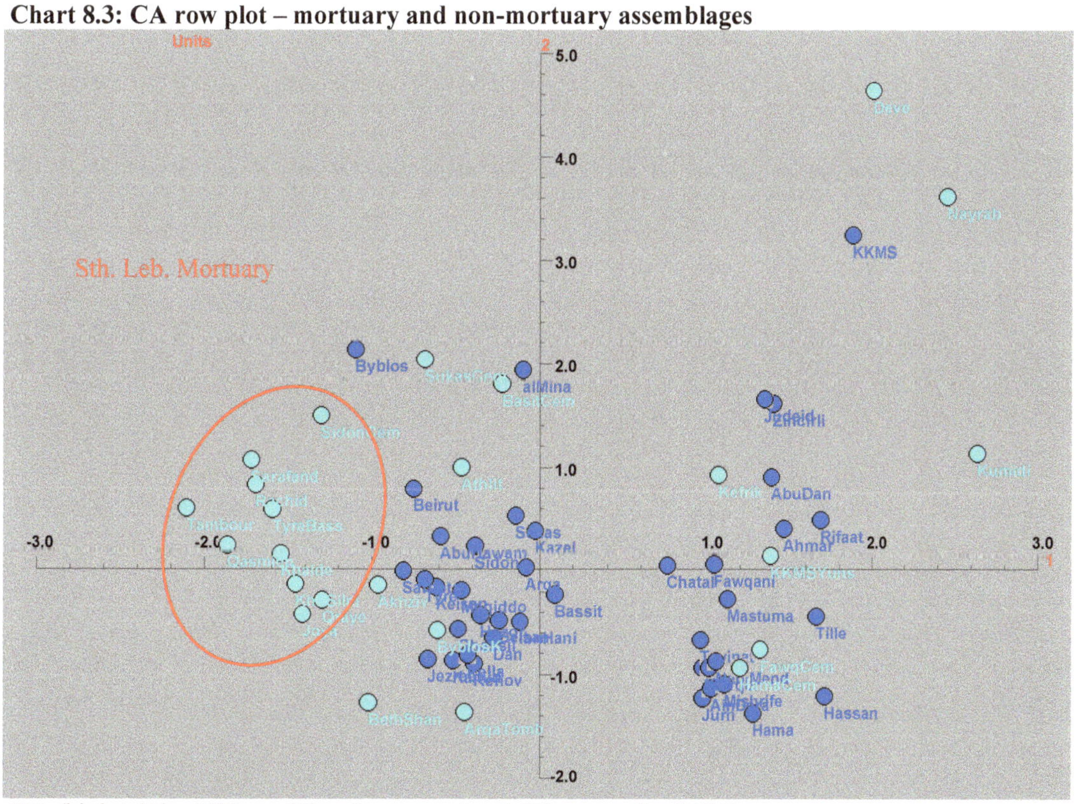

CD: /biplots/Chart803.wmf (inertia: comp. 1 – 8.1%; comp. 2 – 5.2%)

Blue •	Non-mortuary	Turquoise •	Mortuary

Chart 8.3 confirms that the nature of the archaeological context has a significant influence on the composition of an assemblage. This is particularly evident in the cluster of southern Lebanese mortuary assemblages; .the distinct cluster in the far left of the plot is well removed from non-mortuary contexts of the same and nearby regions. The mortuary/non-mortuary distinction is not as pronounced amongst the mortuary assemblages of the inland Northern Levant on the right side of the plot. Nevertheless, **Chart 8.3** confirms that the mortuary assemblages of inland Northern Levant are clearly separate from the coastal mortuary assemblages. Evidently the collection of ceramics used in mortuary practices of the inland Northern Levant is markedly different to that from coastal mortuary contexts, though the difference may be partly explained by chronology and/or mortuary technique (cremation/inhumation). This is particularly important regarding the development of cremation in the Northern Levant, which conventionally is attributed to Aegean immigrants. The above chart suggests that the mortuary assemblages of the Northern Levant may have developed along a separate trajectory from that of the coastal regions. The effect that mortuary ritual has on the nature of a ceramic assemblage is explored further in **Charts 8.6-8.10**.

Localised Geography

Before mortuary and non-mortuary assemblages were explored in more detail, the effect of geographic location on an assemblage was explored further. To this end, sites across the study area were divided according to the nine local geographic zones presented in **Table 7.3** and depicted in **Map 91**: i.e. Syrian coast (blue), Lebanese coast (teal), Beqa' Valley (turquoise), Jezreel Valley (dark green), Northern Palestine (green), Palestinian coast (white), Orontes Syria (yellow), North-west Syria (red), and the Euphrates Valley (burgundy). While these zones were based on geographic factors, their boundaries were assigned in an arbitrary (but apparently meaningful) fashion. These zones covered smaller areas than the two broad regions already evident in the data with the aim of identifying any localised geographic influences in the data. The data categories used for this CA were the same as those in **Chart 8.3**, but with each zone represented by the different colours used in **Map 91**. While the distinction between mortuary and settlement assemblages was maintained, they both follow the colour coding of the regional zone to which they belong. For instance, the mortuary and settlement contexts of Hama, in Orontes Syria, appear as two yellow points. While **Chart 8.4** maintains the broad regional trends witnessed in **Charts 8.1-8.3**, there are also smaller clusters that roughly coincide with the nine local zones. The grouping of assemblages according to colour within each broad region is significant; apparently the two broad regions are not homogenous but incorporate a patchwork of smaller, localised patterns. Because conventional interpretations of the IA-NL have emphasised the broad politico-historical regions, the more dynamic and subtle variations in local cultural behaviour have not been considered. Clearly, interpretations based on the historical narrative provide only a superficial understanding of Iron Age society. The fact that the colours in **Chart 8.4** largely cluster together suggests that the nine zones are, generally speaking, valid categories. This does not, however, suggest that all vessel forms strictly adhere to these local zones; rather it is the specific correspondence of many forms that characterise these assemblages. Some pottery forms will adhere to the general patterns, while others will crosscut them. A few specific trends evident in the chart are worthy of note.

Despite its geographic proximity to Orontes Syria (yellow), the Beqa' Valley (turquoise) is more closely aligned in the plot with Northern Palestine (green) and the Jezreel Valley (dark green). Also, Tell Nebi Mend's assemblage falls firmly within Orontes Syria (yellow) on the chart, despite its geographical proximity to the Beqa' Valley. This is possibly due to the limited occupation of the Beqa' Valley throughout much of the Iron Age (Marfoe 1998, 225). The Palestinian Coast (white) is also more closely aligned with Lebanese (teal) and Syrian (blue) coastal settlement assemblages, than those from Northern Palestine (green), and the Jezreel Valley (dark green). The clustering of Lebanese mortuary contexts away from the non-mortuary Lebanese assemblages is significant; mortuary and non-mortuary assemblages in this zone are distinct. Al Mina's assemblage, which is exclusively non-mortuary, is comparable to mortuary assemblages from the Syrian Coast. This is possibly due to the high level of imported pottery recovered at al Mina. The common wares from al Mina have not been published and are not, therefore included in this study, but their absence may have "created" an assemblage artificially high in prestige items, something akin to a selective burial assemblage. The cluster of Orontes Syria sites is particularly well-defined, while the North-west Syrian and Euphrates zones are much wider spread. On the one hand, one might wonder whether the variability associated with the wider spread is the result of early excavation techniques and poor publication, while on the other hand, the tighter Orontes Syria cluster might highlight the disproportionate influence that the Hama assemblage has held over inland assemblages, being the primary comparative assemblage for much of the Northern Levant.

There are four outlying points in the right of **Chart 8.4** (indicated); three of these represent mortuary assemblages, two of which have been dated to the Persian period. The position of the Kamid el Loz Persian cemetery assemblage (Kumidi) is a conspicuous anomaly, as it represents the only Lebanese assemblage in the right-hand group. The position of the Carchemish assemblage (KKMS) within the vicinity of these two points is possibly due to the limited publication of the Carchemish non-mortuary ceramics; indeed none of the four outlying assemblages are represented by a good-sized, well-published ceramic corpus.

Chart 8.4: CA row plot – geographic zones

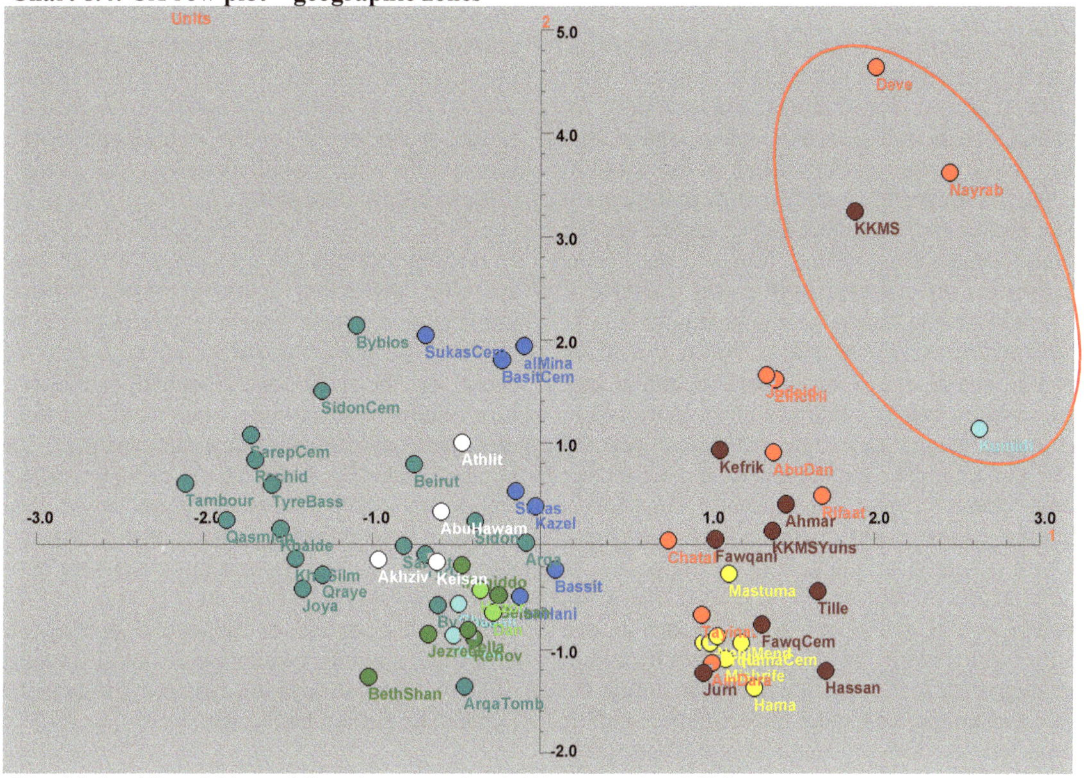

CD: /biplots/Chart804.wmf (inertia: comp. 1 – 8.1%; comp. 2 – 5.2%)

Chart 8.5: CA row plot – period assemblages (mortuary/non-mortuary)

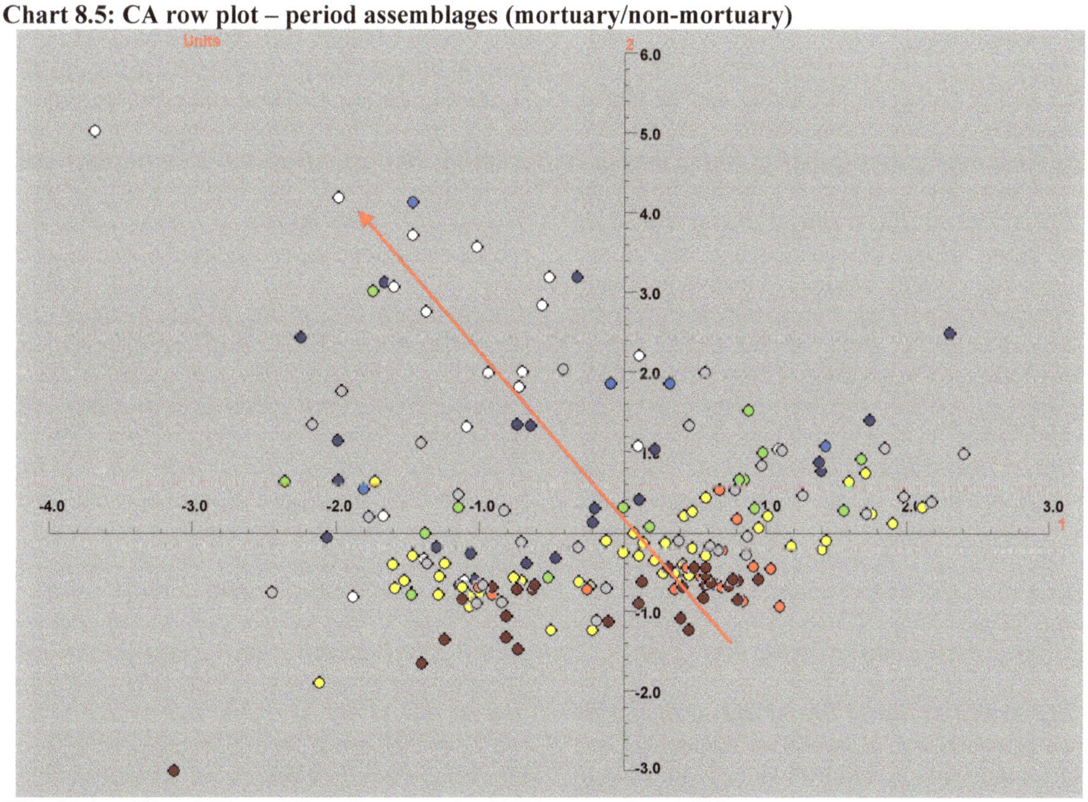

CD: /biplots/Chart805.wmf (inertia: comp. 1 – 7.8%; comp. 2 – 6.2%)

• Iron I • Iron I-II • Iron II • Iron II-III • Iron III • Iron III-Persian • Persian ° Undated

General Chronological Factors

Geography is a significant factor in the makeup of an assemblage, as is chronology. Indeed these two variables, time and space, are conventional explanations for material-culture patterning. Hence, the **Chart 8.5** investigates broad chronological patterns in the data. To this end, site assemblages were divided according to period, with each period represented by a specific colour. As with most chronological studies, however, the results of the analysis are directly dependent on the standard of the original publications. While the chronology of the IA-NL is unsecured, errors in absolute chronology will not greatly alter the CA results as it is the relative chronology of each assemblage that is important. Poor stratigraphic control, though, would affect the results. **Chart 8.5** presents the CA row plot for comparison of assemblages across time. While the plot is particularly crowded in places (adding labels would compound this problem), there is a clear progression in time from the bottom of the plot in the direction of the arrow. The progression is not always a sharp change, but nor is the transition from one period to the next. During the Iron I and Iron II periods the cluster of assemblages is much tighter than during the Iron III and Persian periods. What this pattern means is difficult to determine with any certainty, but it appears to indicate greater variation in assemblages in the later Iron Age. This might indicate changes in production methods, or poorer quality data for this period.

Mortuary Assemblages

Charts 8.3 and **8.4** suggest distinction can be drawn between mortuary and settlement assemblages. For this reason, this section explores associations within mortuary contexts (non-mortuary contexts were removed from the dataset). When available, the nature of mortuary practice was noted; either inhumation, cremation, or the mixed use of the two. Unfortunately, many mortuary assemblages were not recovered in a systematic fashion, leaving their context a point of conjecture. These assemblages have been presented as they are described in the publications. The following analyses also incorporated a basic level of chronological information, with the hope of isolating change across time. This was achieved with distinction being made, when possible, between the sub-divisions of the Iron Age; viz. Iron I, Iron II, Iron III, and Persian period. Because the chronological conclusions for many sites are without firm foundation (**Chapter 3**), the CA results were expected to isolate only general chronological trends in the data.

Chart 8.6 presents the row plot for mortuary assemblages with distinction between inhumation, cremation, and mixed-use assemblages established through the use of colour (blue, red and yellow, respectively). Chronological data is also evident in the label for each assemblage; the suffix A, B, C, and D represents the Iron I, II, III and Persian periods, respectively. Five outlying points on the far right of the plot were removed so as to enhance the clarity of the main group of data: two of the outliers were Persian period inhumation assemblages of North-west Syria, while the other three were cremation assemblages from the Euphrates (none of which derive from well-published excavations). With outliers removed, a number of patterns are evident. Probably the most obvious pattern is the by-now familiar grouping of the inland Northern Levant assemblages (to the right) away from all other areas (to the left). Indeed, the geographic interpretation of this plot seems compelling.

Cremation assemblages are located in different areas of the plot dependent upon geographic location. The implication is that cremation practices of the inland Northern Levant (to the right of the y-axis) and of the coast (to the left) used distinctly different ceramic assemblages. Furthermore, the inland Northern Levant cremation assemblages have an "internal" geographic ordering. Starting at the Yunus cemetery at the top and following a half horseshoe shape down through the mortuary assemblages of Hama, Ras al Bassit and Tell Arqa, there is a geographical procession from inland toward the coast. The fluctuating position of Ras al Bassit on the plot appears to indicate a shifting of cultural influences upon the population of Ras al Bassit as a result of its geographic location between the inland Northern Levant and the southern coast.

While geographical factors influence the results, the nature of mortuary practice is also important, as is evident on the left side of the plot. For instance, there is a clustering of inhumation assemblages conventionally dated to the LB/early Iron Age (i.e. Megiddo tombs, Beth Shan Northern Cemetery, Byblos Necropolis K; Khalde A; the earliest use of Akhziv dated to the very beginning of the Iron II period). This Iron I inhumation group (marked by the short line) is distinct from later inhumations. During the Iron II and Iron III periods, however, the inhumation assemblages of the Phoenician coast are positioned close to mixed use cemeteries.

Coastal cremation contexts are located together in the far left of **Chart 8.6**. Though only three sites are represented, they are geographically limited to the Phoenician coast. The assemblages used within these cemeteries differ most significantly from those present in cremation cemeteries of the inland Northern Levant (right side of plot), but also from the early Iron Age inhumation group. Tyre-al Bass is the main example of cremation on the coast for much of the Iron Age. Despite the extended period of use at Tyre-al Bass there is little change evident in the assemblage. It may be that the form of cremation ritual used rigidly dictated the use of specific ceramic vessels. The four phases of the Hama cemetery display a similar (but slightly looser) grouping over time, which suggests that cremation ritual there prescribed a specific ceramic assemblage. So, despite obvious differences in the make-up of cremation assemblages at Tyre-al Bass and Hama, the two sites seem to agree on one point; that cremation ritual required the use of an assemblage of ceramic vessels that was culturally prescribed by the fact that cremation was being undertaken. The two sites simply differ in the specifics of vessel and, likely, ritualised behaviour.

Chart 8.6: CA row plot – mortuary assemblages (outlying points removed)

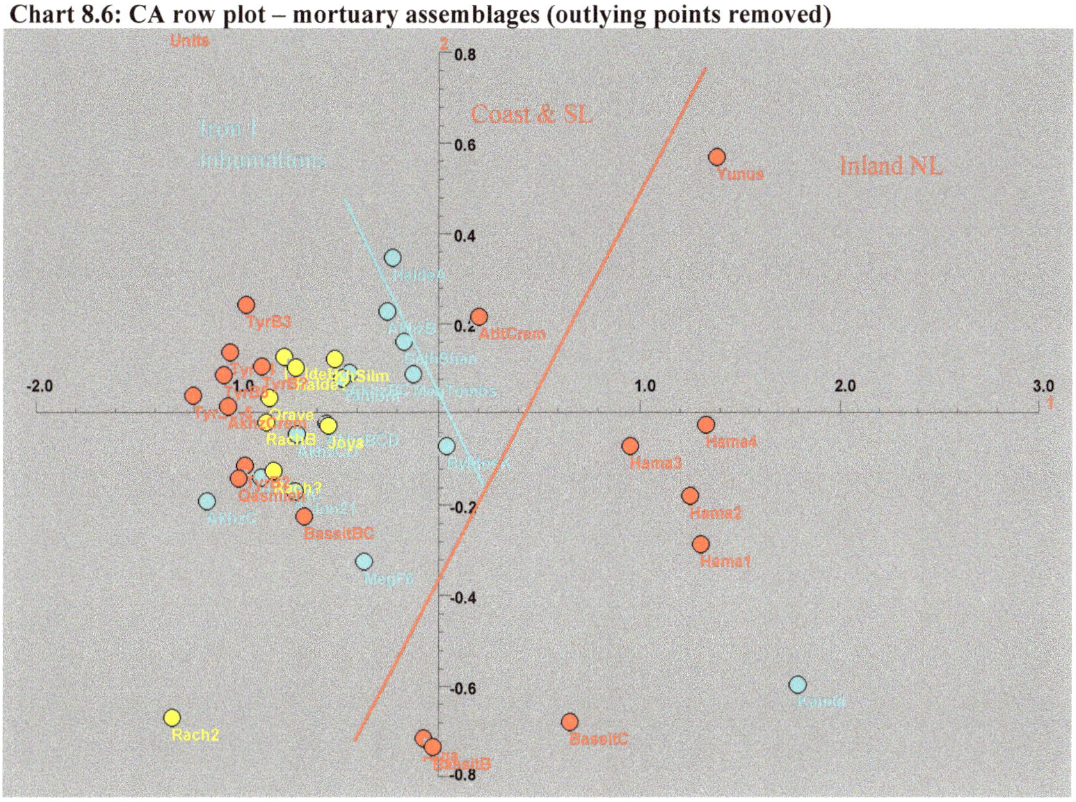

CD: /biplots/Chart806.wmf (comp. 1 – 7.7%; comp. 2 – 6.2%)

• (Cremation) • (Inhumation) • (Mixed or unclear mortuary practice)

Mixed-use mortuary contexts can take the form of cremation and inhumation being attested alongside each other, or the use of cremation within a chamber tomb, as is attested at Tell Rachidieh. Not surprisingly, mixed-use assemblages in **Chart 8.6** (yellow) are plotted between the inhumation (blue) and cremation (red) groups. The assemblages of Joya and Khirbet Silm are plotted close to inhumation assemblages of the Iron II and Iron III periods, suggesting that cremation played only a minor role in the use of these cemeteries; though neither of these assemblages was systematically excavated.

Geography and chronology had an influence on mortuary assemblages, as did the nature of the mortuary practice. It is also possible to explore the general types of activity that mortuary ritual might entail. By grouping ceramic types into broad functional categories, simplistic models of mortuary behaviour will be evident. **Charts 8.7** and **8.8** show the row and column plots for mortuary assemblage functions (outlying points removed). **Chart 8.7** depicts similarity in function between different mortuary assemblages (e.g. the Hama I and Yunus cemeteries). **Chart 8.8** presents the relationship between functional categories within mortuary assemblages. **Chart 8.7** maintains the inland/coastal dichotomy, though with a comparatively less emphatic division. In other words, material culture may be clearly different between these regions (**Chart 8.6**), but the functional differences are less distinct. In **Chart 8.7** the majority of cremation assemblages are positioned to the left of the y-axis, with inland Northern Levant sites in general located above the x-axis and coastal/southern sites located below. A comparison of the row and column plots reveals Hama's preference for urns and fine bowls, while Tyre-al Bass displays a closer link with kraters, bowls, shallow bowls and jugs. The Yunus cemetery is not associated with any specific functional categories, despite a published typology characterised by kraters (Woolley 1939b). Apart from highlighting the limitations of presence/absence data, these results emphasise the variability of functional categories present in the Yunus cemetery (or the limited publication of this assemblage). It is particularly interesting that kraters in **Chart 8.8** are plotted near bowls and jugs: kraters have long been associated with the mixing of wine with water at banquets (Buhl 1983, 126), and their association with bowls and jugs suggest that drinking and/or pouring wine (or the implication of wine drinking equipment) was an important component of cremation symbolism, especially within the coastal cremation sites. Patterns are less obvious amongst the inhumation assemblages: of interest is the association of the Akhziv shaft tombs with transport amphorae and dipper juglets. Once again, the early Iron Age inhumation cemeteries of Beth Shan, Megiddo, Byblos, and Khalde are loosely grouped together (indicated), and are joined by the enigmatic Joya and Khirbet Silm assemblages. The plotted position of the Qasmieh assemblage implies a cremation context.

Chart 8.7: CA row plot – mortuary assemblage function

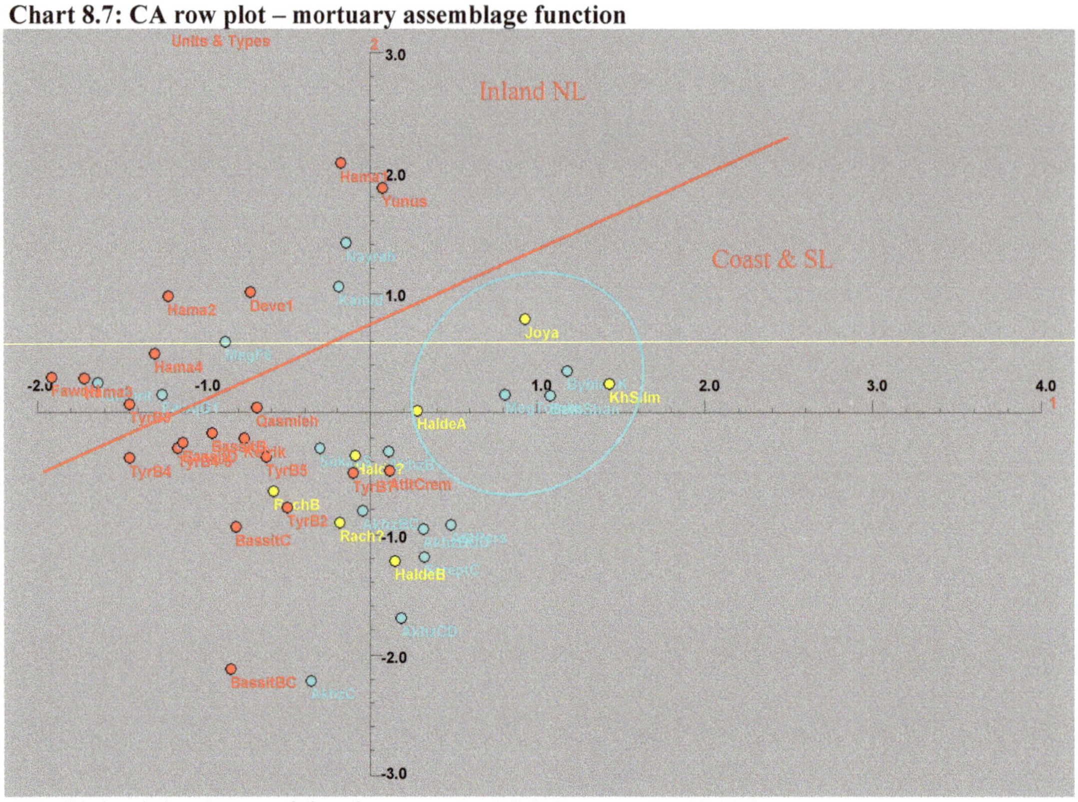

CD: /biplots/Chart807.wmf (inertia: comp. 1 – 15.4%; comp. 2 – 12.3%)

• (Cremation) • (Inhumation) • (Mixed or unclear mortuary practice)

Chart 8.8: CA column plot – mortuary assemblage function

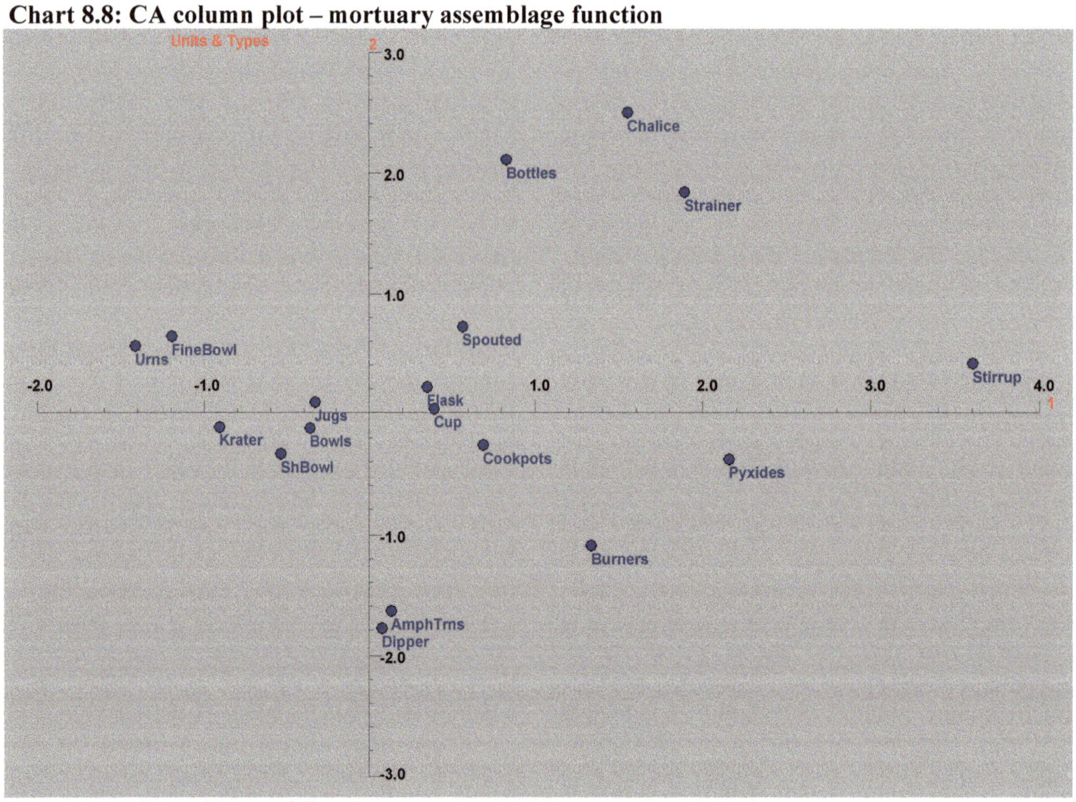

CD: /biplots/Chart808.wmf (inertia: comp. 1 – 15.4%; comp. 2 – 12.3%)

Chart 8.9: CA row plot – Hama and Tyre assemblage function

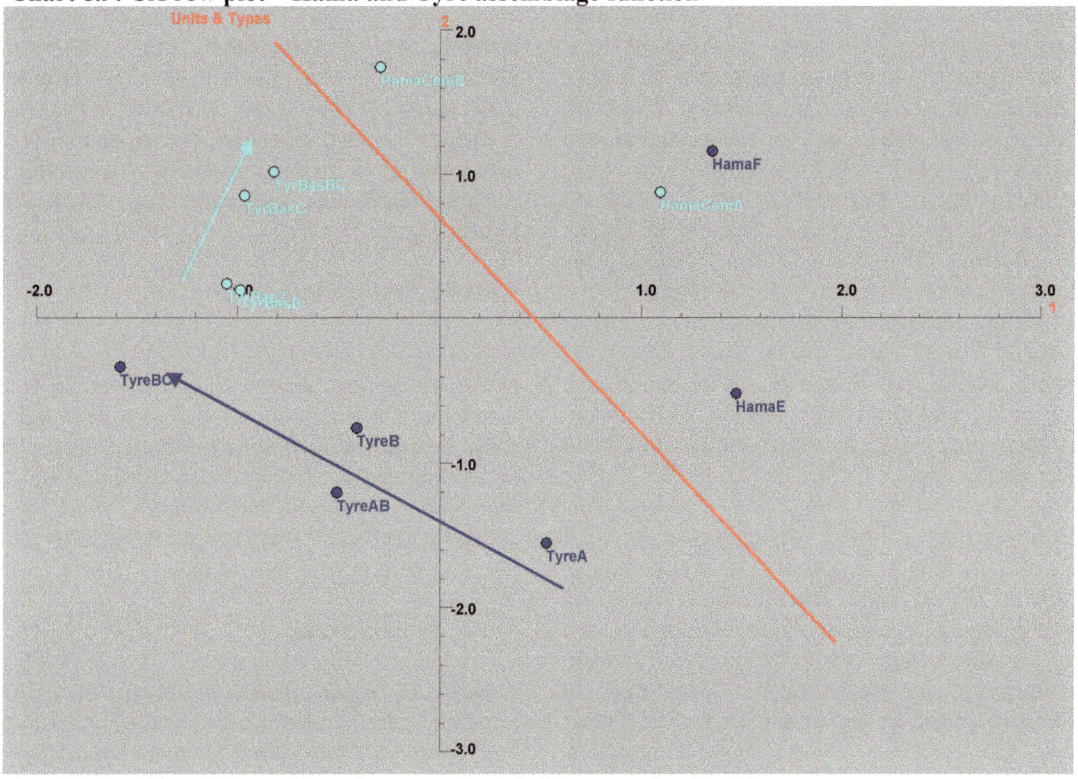

CD: /biplots/Chart809.wmf (inertia: comp. 1 – 33.5%; comp. 2 – 23.6%)

•(Non-mortuary) • (Mortuary)

Chart 8.10: CA column plot – Hama and Tyre assemblage function

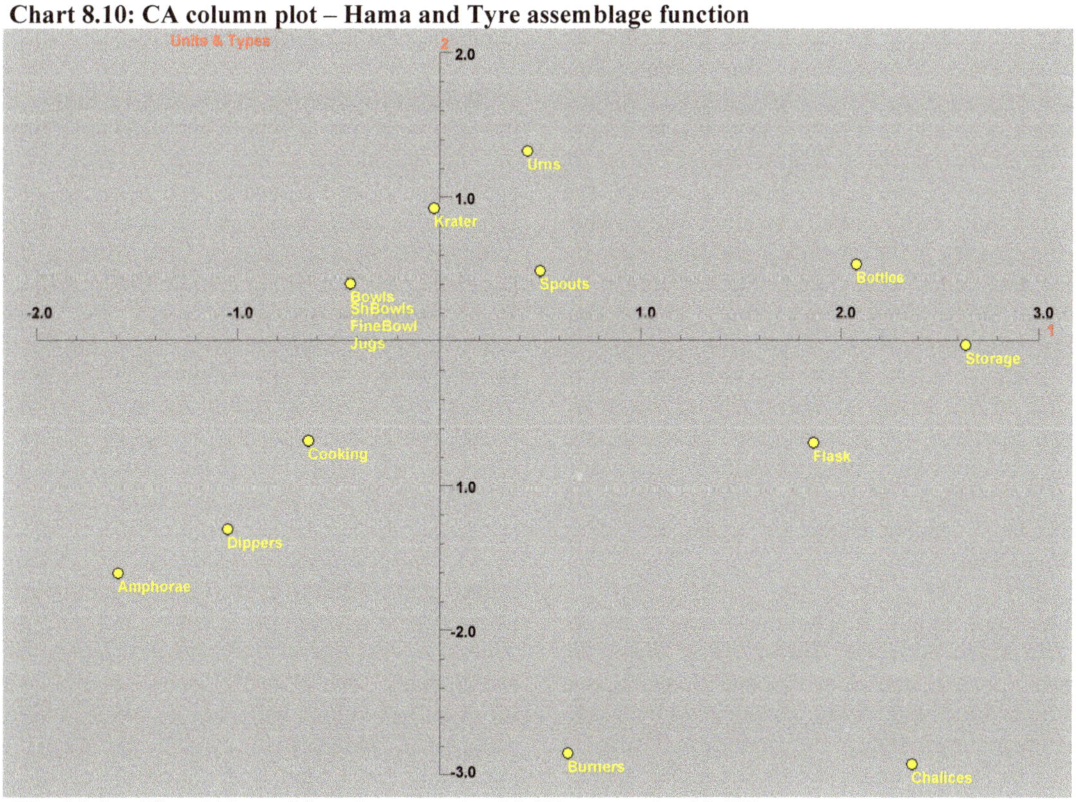

CD: /biplots/Chart810.wmf (inertia: comp. 1 – 33.5%; comp. 2 – 23.6%)

Chart 8.9 explores the effects of time on cremation assemblages. Unfortunately, the majority of cremation assemblages are without chronological data: only at Tyre-al Bass and Hama were archaeologists able to isolate specific periods of use within a cremation cemetery. While the absolute dates for these two cremation cemeteries are not based on reliable evidence, it is the relative phasing that is important for this CA (though there also some doubts over the reliability of the Hama sequence – **§3.18**). The Hama and Tyre non-mortuary assemblages are also included, in order to emphasize the difference, if any, between mortuary and non-mortuary contexts. Hama F and E represent the Iron I and Iron II periods, while Tyre A, AB, B and BC denote the Iron I, Iron I-II, Iron II, and Iron II-III periods, respectively. **Chart 8.9** represents the CA row data, while **Chart 8.10** presents the column data.

Chart 8.9 implies that a variety of factors have a direct influence on the functional components of an assemblage, and possibly social behaviour. Three of these factors are particularly prominent: context type, geography, and chronology. *Context type*: there is a clear division between mortuary and non-mortuary assemblages in **Chart 8.9** as characterised by vessel function in **Chart 8.10**. With the exception of the Iron I Hama F assemblage, non-mortuary assemblages lie below the x-axis, while mortuary assemblages are found above it. *Geography*: there is a clear distinction between the functions of the Hama assemblages (right of long line), and those of Tyre (left), a general division once again mirroring the coastal and inland distinction already encountered. *Chronology*: the ordering of the Tyre non-mortuary assemblages follows a clear chronological progression moving from Tyre A leftwards and slightly upwards (dark blue arrow). Tyre's mortuary assemblages are also affected by time, moving upwards away from the site's non-mortuary assemblages (turquoise arrow). The implication is that, on the coast, cremation assemblages became functionally more distinct from non-mortuary assemblages over time. At the inland site of Hama, there is a similar distinction between the function of mortuary and non-mortuary assemblages in the Iron II period. In the Iron I period, however, there is a close proximity in the position of the Hama mortuary and non-mortuary assemblages in **Chart 8.9**. It is possible that, during the early Iron Age, cremation ritual used ceramic vessels that were functionally similar to everyday vessels, though the limited Hama F sample (Fugmann 1958, Fig. 165) warrants caution here.

A few functional trends in **Charts 8.9** and **8.10** are worthy of further note: the Tyre-al Bass cremation assemblages are characterised by a distinct combination of tightly associated jugs, bowls and kraters, while the Hama cremation assemblages are more closely aligned with urns. The Hama cremation cemeteries also contain some jugs and kraters, but they are represented only as a small proportion of the ceramic material, unlike at Tyre-al Bass where jugs and kraters were clearly an important component. If quantified (abundance) data, rather than presence/absence (incidence) data, was analysed, the resulting CA plot would possibly emphasise the differences between these two cemeteries even further; the Hama cemetery contained numerous urns and kraters, while the Tyre al-Bass assemblage was dominated by red-slip trefoil jugs, mushroom jugs, small bowls, and kraters.

Non-Mortuary Assemblages

Following the investigation of mortuary assemblages, the non-mortuary data underwent a series of Correspondence Analyses. For this purpose, mortuary data was removed from the database. The non-mortuary data was much more abundant than mortuary and, to avoid getting lost in the detail, was split into smaller, more manageable categories; namely assemblages defined by geography, period, function, and context type.

General Trends

Chart 8.11 below presents the results of a CA on non-mortuary site assemblages with no chronological, functional, or contextual divisions imposed upon the data: geographic location is indicated through the same use of colour as employed in **Chart 8.4**. While the resulting pattern resembles that of **Chart 8.4**, the nine geographic zones are better defined here due to the removal of the mortuary data. The main division between the inland Northern Levant and the coast and Southern Levant is maintained (as indicated).

Well-defined clustering in **Chart 8.11** further highlights the significant impact that local geographic ordering has on the assemblages. Not only is each of the nine geographic zones reasonably clear but there appears to be some level of geographical progression across these zones. Starting with Orontes Syria (bottom right), it is possible to trace an arc in the plot northwards toward the Syrian Euphrates, across North-west Syria to the Syrian coast, down the Lebanese then Palestinian coasts, to the Jezreel Valley, Northern Palestine and finally the Beqa' Valley. This pattern is significant, as it confirms that the inland Northern Levant was culturally a very different entity to the inland Southern Levant and the Beqa' Valley, despite possible overland links.

In an effort to further illuminate this broad regional progression, **Chart 8.12** groups non-mortuary site assemblages together according to the nine geographic zones used throughout this study, producing nine regional assemblages. The plotted results clearly confirm the localised geographic influence on material culture. The geographic progression evident in **Chart 8.11** is maintained in **Chart 8.12** (arrows), as is the main division between the inland Northern Levant and the coastal/southern zones (line).

Chart 8.11: CA row plot – non-mortuary site assemblages (outliers removed)

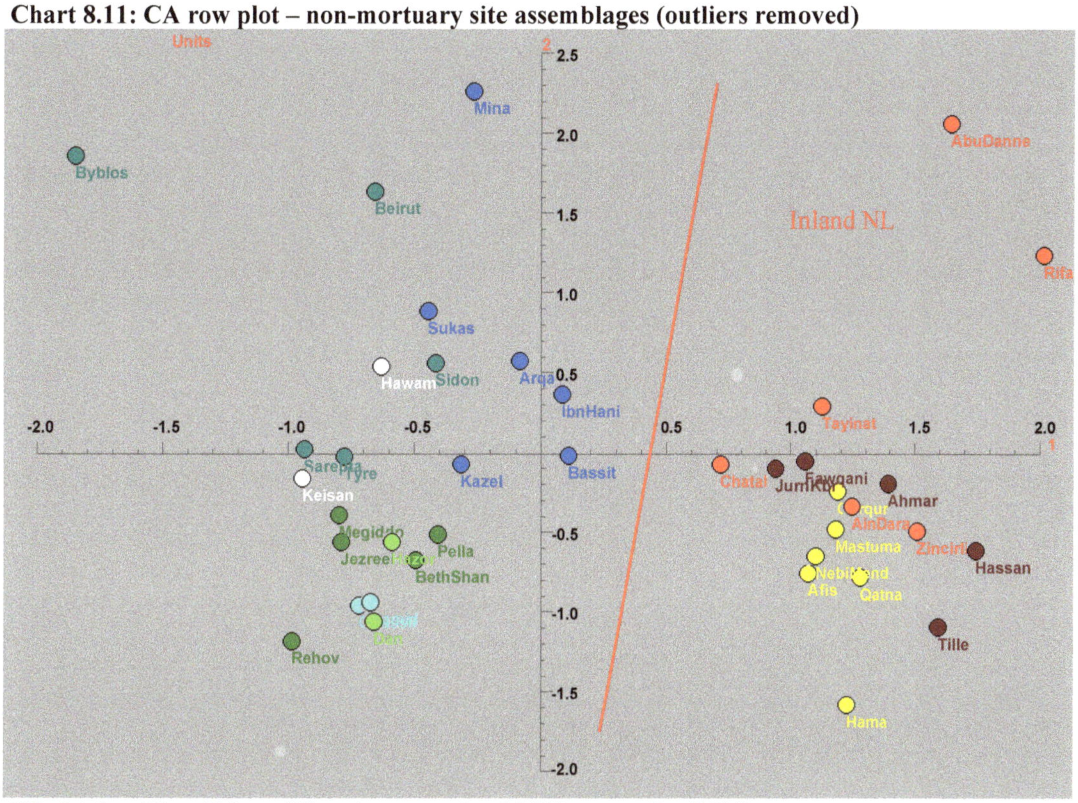

CD: /biplots/Chart811.wmf (inertia: comp. 1 – 10.2%; comp. 2 – 7.1%)

• Syrian Coast • Lebanese Coast • Beqa' Valley • Jezreel Valley
• Northern Palestine • Palestinian Coast • Orontes Syria • North Syria
• Euphrates Syria

Chart 8.12: CA row plot – non-mortuary regional assemblages

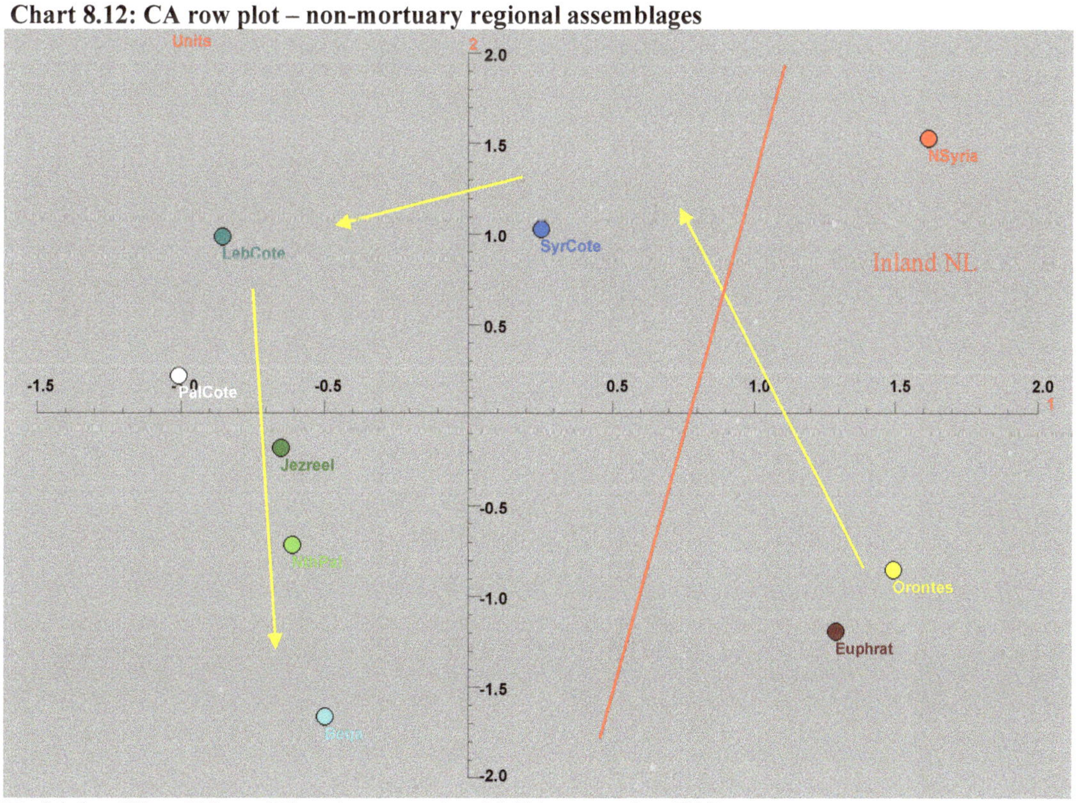

CD: /biplots/Chart812.wmf (inertia: comp. 1 – 35.3%; comp. 2 – 22%)

Period Specific Assemblages

In an effort to reduce the data set into smaller, more manageable portions, the following four Correspondence Analyses compared non-mortuary assemblages according to the four main sub-periods of the Iron Age. The aim was to identify how geographical patterns in the data might have changed over time. No functional distinction was maintained in this section.

Chart 8.13 plots the CA results of Iron I non-mortuary assemblages. The results maintain the broad distinction between the inland Northern Levant and the rest of the study area (line). The Ras Ibn Hani assemblage is the sole exception (circled), which might be indicative of this site's connections with the interior. While the fragmentary and unreliable nature of most of the Iron I data warrants caution, geographical factors appear to be influencing the plotted assemblages.

Chart 8.14 plots the CA results of Iron II non-mortuary assemblages. Once again, the results are characterised by the clustering of the nine local geographic zones (indicated by colour), though the coastal assemblages are wider spread than inland zones. The inland Northern Levant zones are tightly clustered on the left side of the plot, while the inland Southern Levant appear on the right: coastal assemblages are again the link between the two, though the inland Northern Levant assemblages are more "removed" from the coastal assemblages than are the inland Southern Levant group. Hence, the main division in the data separates the inland Northern Levant assemblages from all others. **Chart 8.14** confirms Lehmann's (1996, Abb. 4.4) dual distribution (coastal and inland) of Iron II pottery (Assemblages 1 and 2) in the Northern Levant. However, there are more localised clusters in the data suggesting that the two-region model might mask some of the more localised cultural patterns. For instance, the Orontes Syria (yellow) assemblages are clustered, implying a well-defined localised cultural profile. **Chart 8.14** contains clear evidence for geographic ordering of the data. Generally speaking, the right side of the plot, if read right to left (arrow), progresses geographically through the Beqa' Valley, Northern Palestine, Jezreel Valley, and the Palestinian and Lebanese coasts. While it is impossible to explain all patterns in the data, it is clear that **Chart 8.14** is influenced by local geographic factors.

Chart 8.15 displays the CA results for Iron III non-mortuary assemblages; it presents a distinct change from earlier periods. Despite fewer assemblages representing the coast and Southern Levant, the most obvious change is the position of the Southern Levant assemblages between those from the inland Northern Levant and the coast: for the first time in the Iron Age, the pottery of the inland Northern Levant was more closely comparable to the Southern Levant than the coast. This change is significant, as it appears to signal changes in pottery production and distribution systems.

Chart 8.13: CA row plot – non-mortuary Iron I assemblages

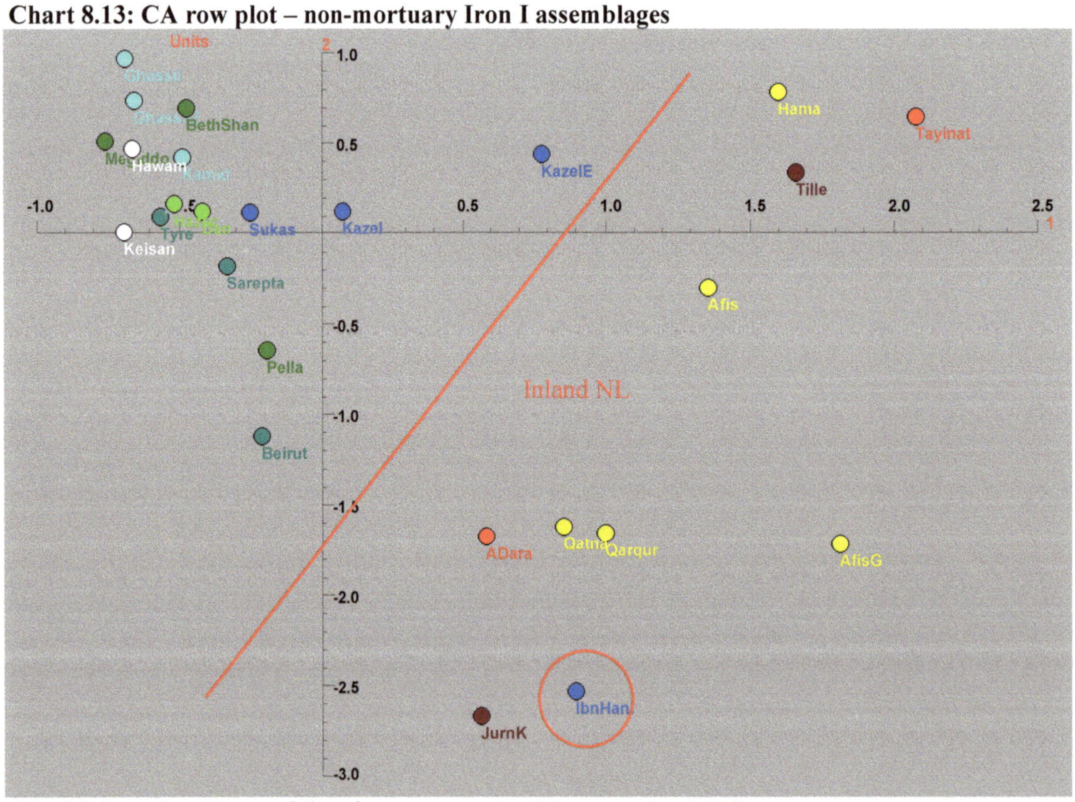

CD: /biplots/Chart813.wmf (inertia: comp. 1 – 11.4%; comp. 2 – 8.1%)

Chart 8.14: CA row plot – non-mortuary Iron II assemblages

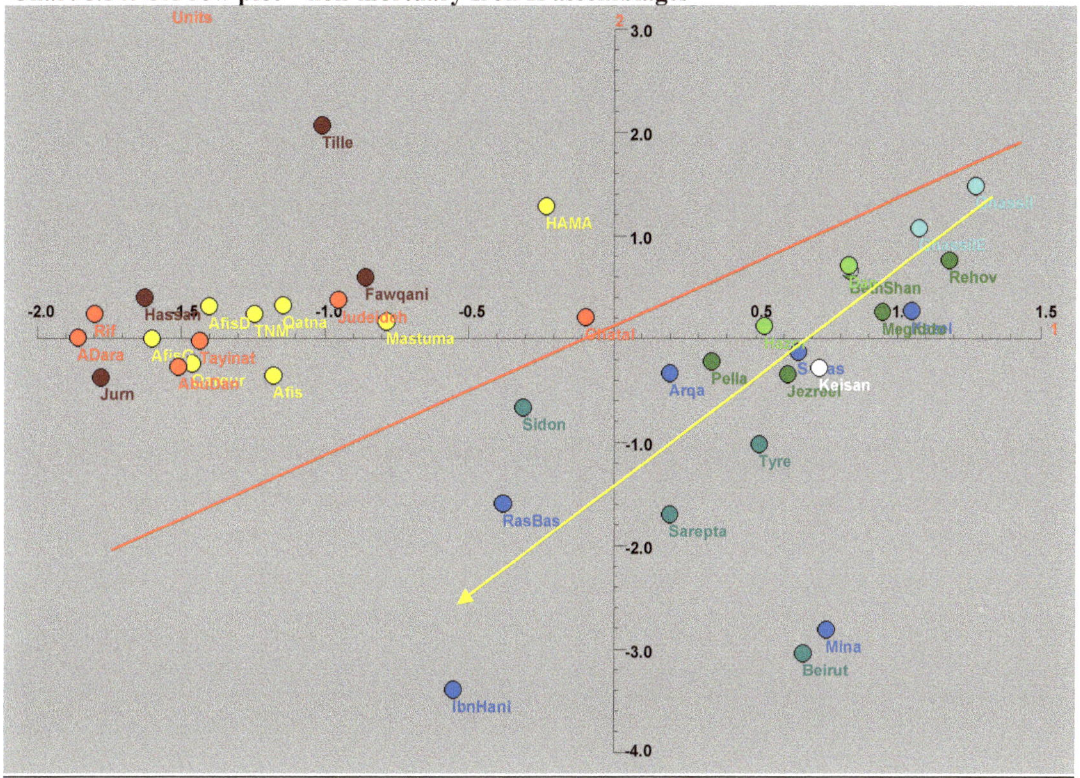

CD: /biplots/Chart814.wmf (inertia: comp. 1 – 10.7%; comp. 2 – 7.7%)

Chart 8.15: CA row plot – non-mortuary Iron III assemblages

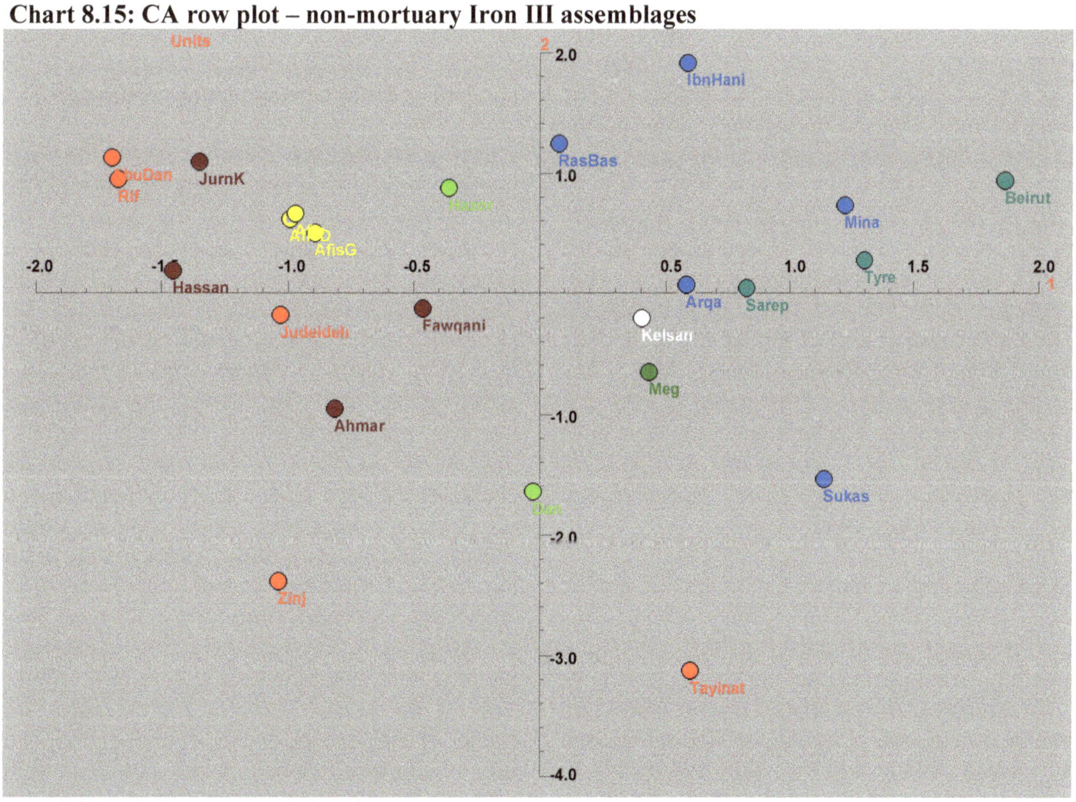

CD: /biplots/Chart815.wmf (inertia: comp. 1 – 11.6%; comp. 2 – 8.5%)

Chart 8.16: CA row plot – non-mortuary Persian period assemblages

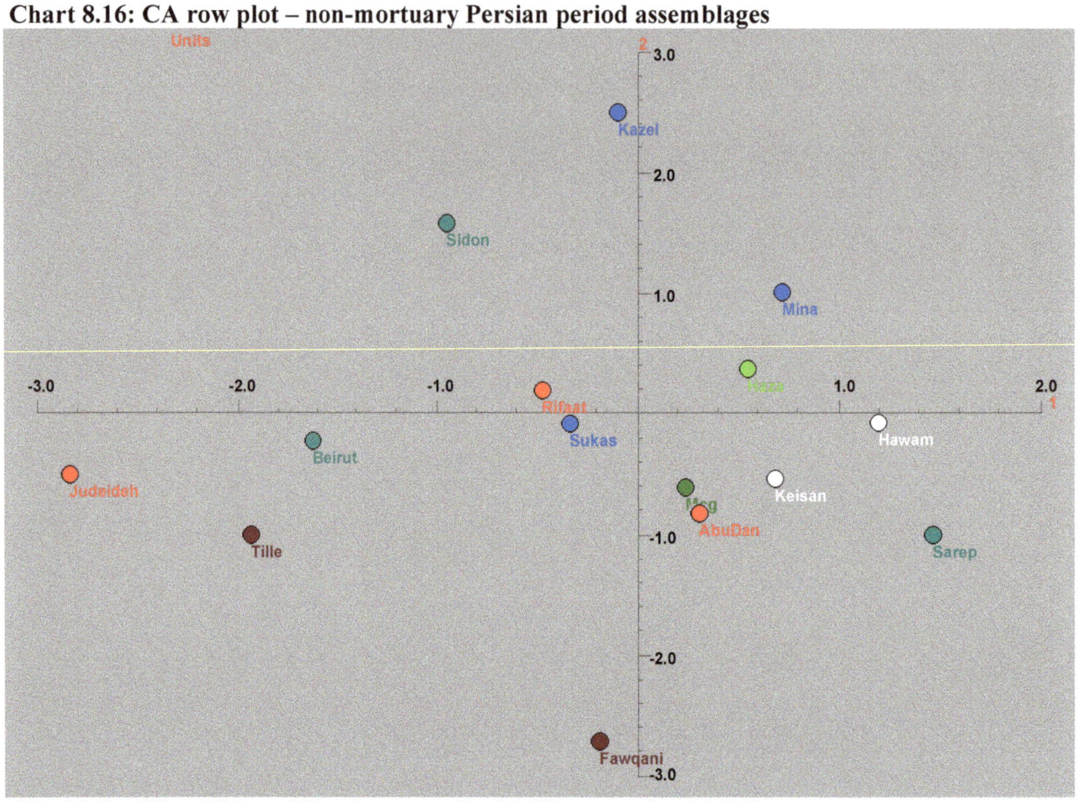

CD: /biplots/Chart816.wmf (inertia: comp. 1 – 16.4%; comp. 2 – 14.7%)

Chart 8.16 displays the CA results for all Persian period non-mortuary assemblages, though these are conspicuously few. As a result, the CA plot is characterised by a lack of discernible patterns. The geographical location of a site during the Persian period appears to have minimal effect on ceramic culture.

The Persian period as defined in this study is roughly equivalent to Lehmann's (1996, 87) Assemblages 5-8 (his *Eisenzeit III*). Within his distribution analysis for this period, Lehmann (1996, Abb. 4.6-7) identified a complex series of five overlapping areas of diffusion, yet the data for this period is both unreliable and incomplete. For any patterns in this data to be isolated, a number of generous extrapolations are needed. Furthermore, Lehmann does not appear to make any concession for possible differences between mortuary and non-mortuary assemblages in the Persian period.

Period Specific Function

The above discussion has demonstrated that geography and chronology were significant influences on non-mortuary assemblages. By grouping ceramic types into broad functional categories, this section will investigate patterns in function that change over time and space. To this end, the next four analyses (each Iron Age sub-period) introduced functional categories into the CA data. The resulting row and column plots are superimposed for each period. When a site assemblage is positioned near a functional category, a correlation between the two is evident; i.e. the functional category is a significant component of the corresponding assemblage.

Chart 8.17 displays the CA results for Iron I non-mortuary assemblages according to function. Apart from the by-now familiar distinction between the inland Northern Levant (line - Ras Ibn Hani is again the exception) and other zones, **Chart 8.17** is difficult to interpret. Nevertheless, there appears to be a tight grouping of functional categories to the left of the y-axis; this group consists of shallow bowls, kraters, cooking-pots, bowls, and fine bowls. No site assemblages, however, are particularly close to this group. In contrast, the assemblages from the coast and Southern Levant are characterised by a wide variety of functional categories (i.e. transport amphorae, lamps, chalices, dipper juglets, pyxides, bottles, urns, flasks, and spouted vessels), while the inland Northern Levant assemblages are only marginally associated with pithoi. Jugs are peripheral to the Beqa' and Northern Palestine assemblages. These patterns raise some interesting questions. What were the sites of the inland Northern Levant using for pouring liquids if not ceramic jugs? Were the possession and/or use of valuable liquids/oils not important amongst the inland communities of the IA-NL? What did coastal sites use for storage if not pithoi?

Chart 8.17: CA row and column plots superimposed – Iron I non-mortuary function

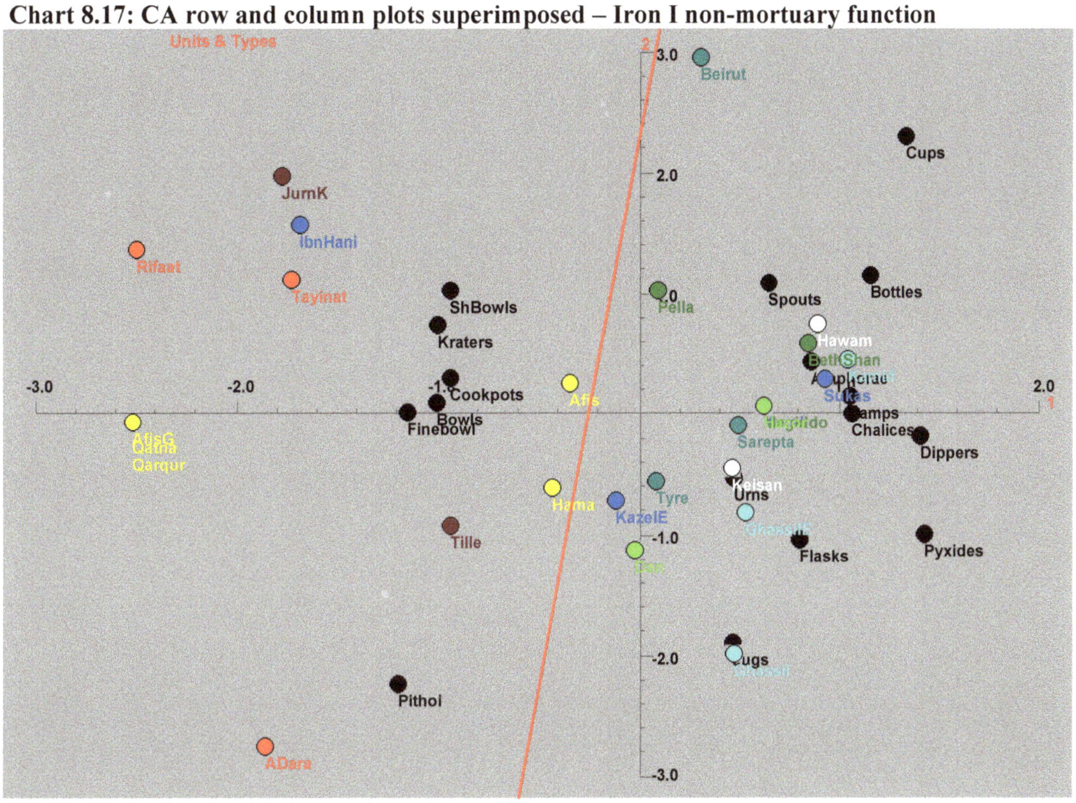

CD: /biplots/Chart817.wmf (inertia: comp. 1 – 34.4%; comp. 2 – 14.8%)

• Syrian Coast • Lebanese Coast • Beqa' Valley • Jezreel Valley
• Northern Palestine • Palestinian Coast • Orontes Syria • North Syria
• Euphrates Syria Black (Functional Category)

Chart 8.18: CA row and column plots superimposed – Iron II non-mortuary function

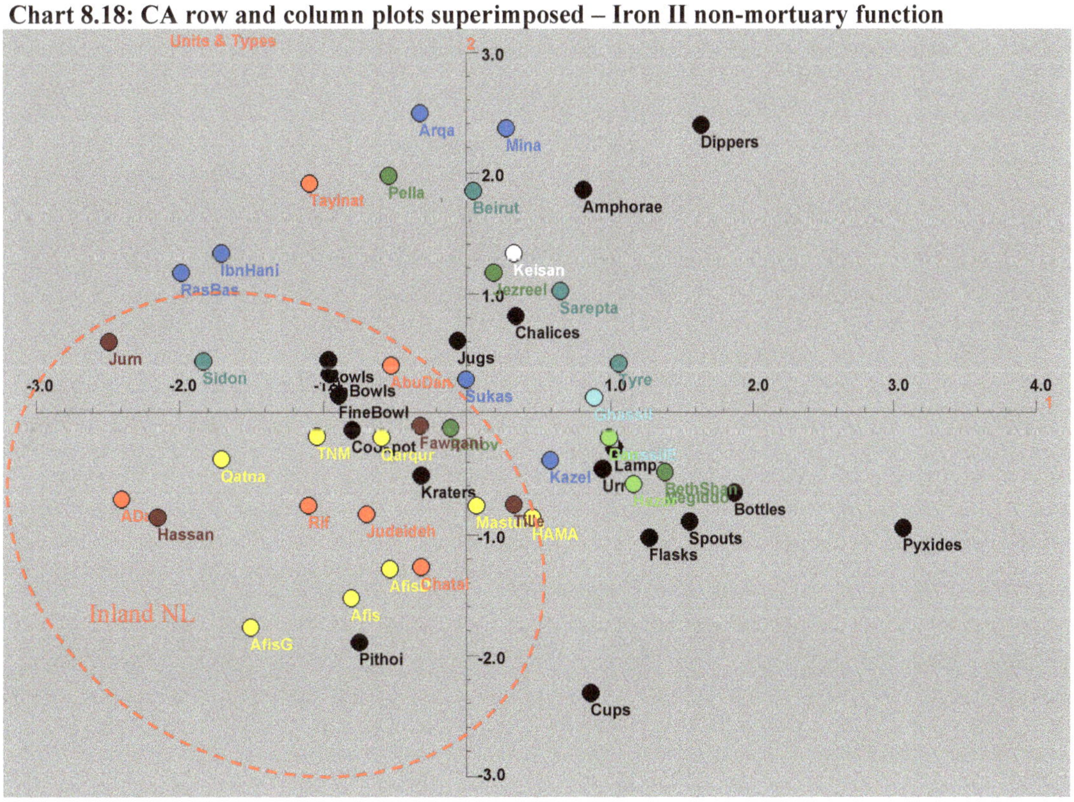

CD: /biplots/Chart818.wmf (inertia: comp. 1 – 20.9%; comp. 2 – 16.1%)

Chart 8.19: CA row and column plots superimposed – Iron III non-mortuary function

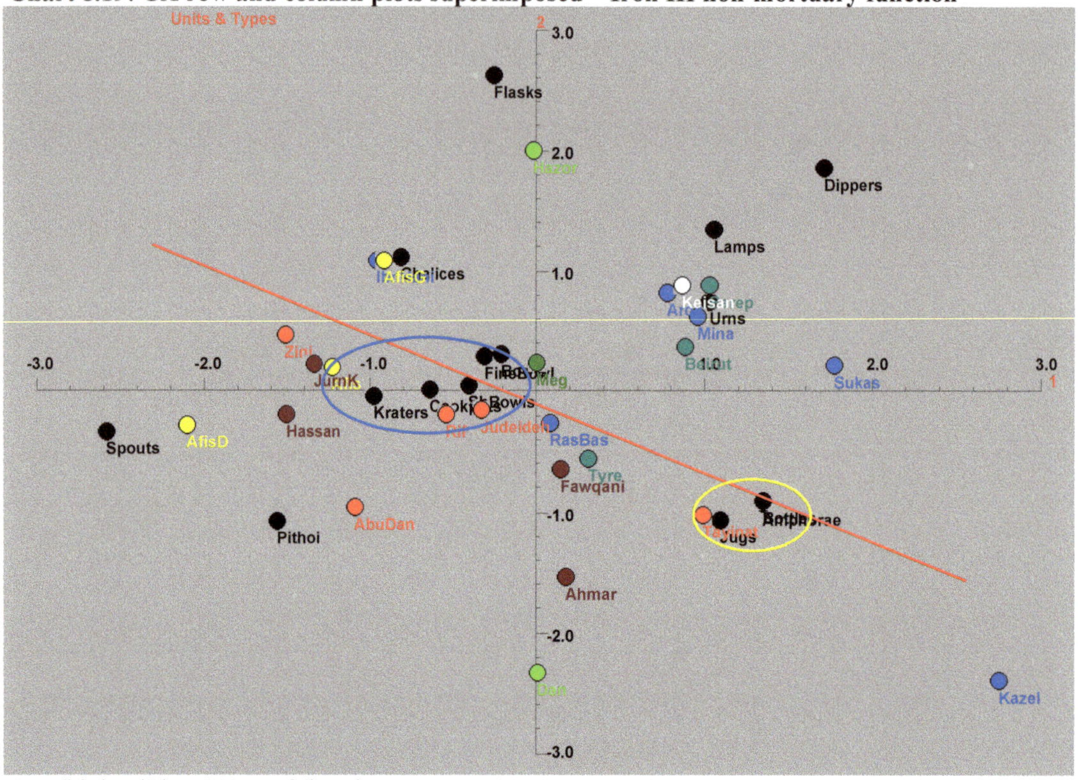

CD: /biplots/Chart819.wmf (inertia: comp. 1 – 21.1%; comp. 2 – 16.5%)

Chart 8.18 displays the CA results for Iron II non-mortuary assemblages characterised by function. Once again the most obvious pattern is the distinction between the inland Northern Levant and all other zones (dashed line). The majority of assemblages are again positioned around a core functional group, the nature of which has changed little from the core Iron I functional group; bowls, shallow bowls, fine bowls, cooking pots, and craters. The craters, however, are slightly further removed than previously. In addition to the core functions, the Southern Levant and coastal assemblages are characterised by several functional categories (bottles, urns, lamps, cups, flasks, spouted vessels, pyxides, dipper juglets, amphorae). Some assemblages from the inland Northern Levant are positioned so close to the core group that they are probably only characterised by these functions (e.g. Tell Qarqur), while other inland Northern Levant assemblages are associated with pithoi (e.g. Tell Afis). The conspicuous shift in the position of jugs from the Iron I periphery to near the Iron II core implies that jugs became a more important functional component of Iron II society, especially within coastal sites.

Chart 8.19 displays the CA results for Iron III non-mortuary assemblages according to function. While the ever-present distinction between the inland Northern Levant and other areas (red line) is roughly maintained, for the first time in the Iron Age the distinction is not a marked one (**Chart 8.15**). For instance, the Tyre assemblage is plotted close to that from Tell Shiyukh Fawqani on the Euphrates. The Dan assemblage is positioned amongst the inland Northern Levant group, suggesting a link between Northern Palestine and the inland Northern Levant: unfortunately no Iron III assemblages are known from the Beqa' Valley or Damascus region. Syrian Coastal assemblages are, as usual, well spread across the plot, suggesting this area was not a cohesive cultural unit but one affected by localised traditions. In other words, pottery was used in a broadly different manner between coastal sites; though there is a general lack of non-mortuary data from much of the coast. When the row and column plot are compared, the majority of assemblages are positioned around a core functional group, the nature of which has changed little since the Iron I period: bowls, shallow bowls, fine bowls, cooking pots, and kraters (larger oval). This group is positioned closer to the intersection of the two axes than previously, suggesting it represents the average functional profile. In other words, a large percentage of Iron III assemblages are characterised, at least initially, by these functional elements. Pithoi continue to be associated with inland assemblages, but transport amphorae no longer characterise only coastal assemblages (smaller oval).

Chart 8.20 displays the CA results for Persian period non-mortuary assemblages characterised by function. No geographical pattern is discernible, despite most regions being represented. Nevertheless, **Chart 8.20** depicts a marked change in the relationships between the functional categories – no core functional group is evident. Positioned close to the average profile, amphorae alone appear characteristic for this period. The lack of discernible meaning within these superimposed plots appears to indicate an insufficient dataset for this period.

Chart 8.20: CA row and column plots superimposed – Persian period non-mortuary function

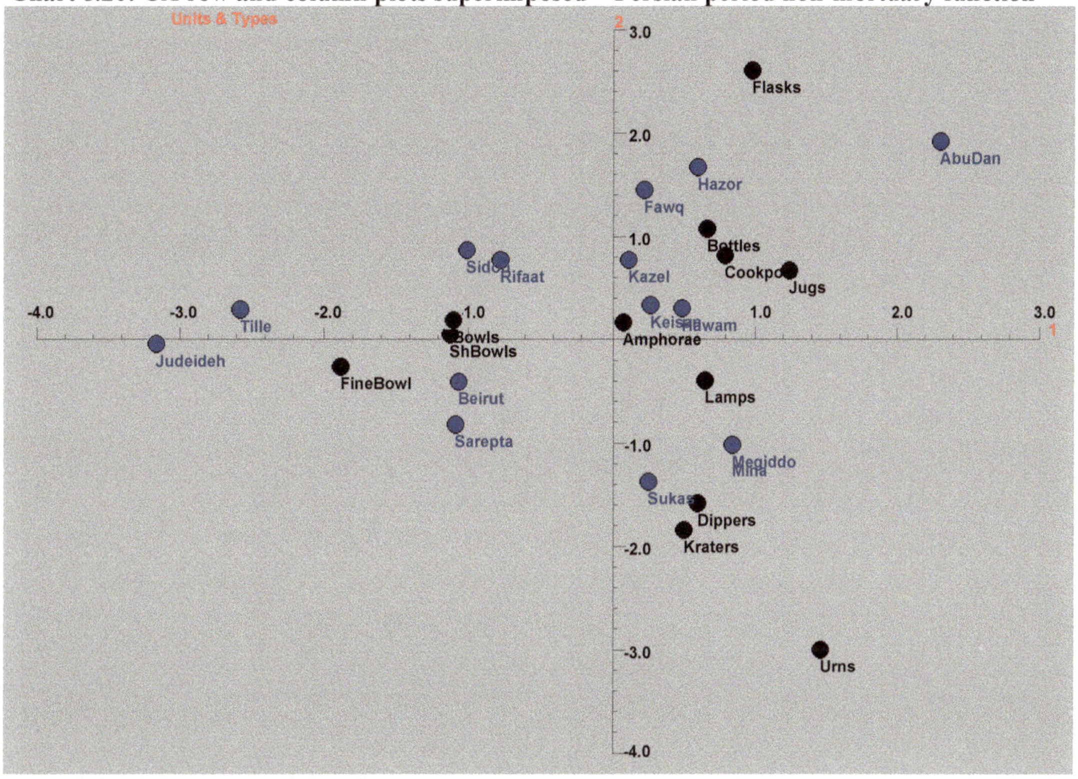

CD: /biplots/Chart820.wmf (inertia: comp. 1 – 26%; comp. 2 – 20.1%)

• Site Assemblages • Function Category

Type of Non-Mortuary Context: Tell Afis

Throughout the above Correspondence Analyses, non-mortuary assemblages have occasionally been split according to differences in context type: cremation, inhumation, or mixed/unspecified. There are, however, conceivable differences in non-mortuary context types too; elite buildings (e.g. Hama), domestic quarters (e.g. Tell Mastuma). Hence, it was deemed worthwhile to explore the possible difference in the nature of assemblages from elite and domestic non-mortuary contexts, as well as those unspecified. To avoid any possible regional variation influencing the data, a single site was needed with both domestic and elite assemblages available. Unfortunately, only one site has yielded significant evidence for these two context types. As a result, the following CA will focus on the non-mortuary assemblages of Tell Afis, located within the Orontes Syria zone. Comparison of the three main excavation areas of Tell Afis hoped to determine what, if any, differences existed between what the excavators had interpreted as domestic (Area D), elite (Area G), and non-specific (Areas E^1 and E^2) settlement contexts.

Chart 8.21 distinguishes context type on basis of colour, with each label denoting the excavation area and stratigraphic layer: note E17 represents Area E^1 level 7, not Area E level 17, and E24 represents Area E^2 level 4, not Area E level 24. At first glance **Chart 8.21** suggests that there is a general distinction between the three settlement types exposed at Tell Afis. The left side of the plot is dominated by non-specific contexts, while the right side is mainly occupied by the elite and domestic contexts. Despite the general trends, a few elite assemblages appear on the left of the plot (short oval), and a few non-specific assemblages are located to the right (longer oval). It is interesting that elite and domestic contexts are largely comparable.

The few "anomalies" aside, **Chart 8.21's** grouping suggests that non-mortuary context type might be an important factor on the nature of assemblage; but this is illusory, the pattern is best explained by chronological factors, as is clearly shown in **Chart 8.22**. The second Tell Afis CA plot ignores context type and instead codes the assemblages by period. The left hand side is clearly associated with the Iron I period, while the right hand side includes assemblages from the Iron II and Iron III periods; in addition to this clear trend, the four assemblages that were previously considered problematic are now explained. The plot largely depicts chronological progression from left to right (arrow); the non-mortuary assemblages at Tell Afis are determined more by chronological factors than context type. We might conclude from Charts 8.21 and 8.22 then, that while the type mortuary context has an impact on the nature of a ceramic assemblage, the type of settlement context does not greatly affect the ceramic assemblage. Whether this is true for all non-mortuary site data can not been demonstrated, as suitable data remains unavailable.

Chart 8.21: CA row plot – Tell Afis non-mortuary contexts

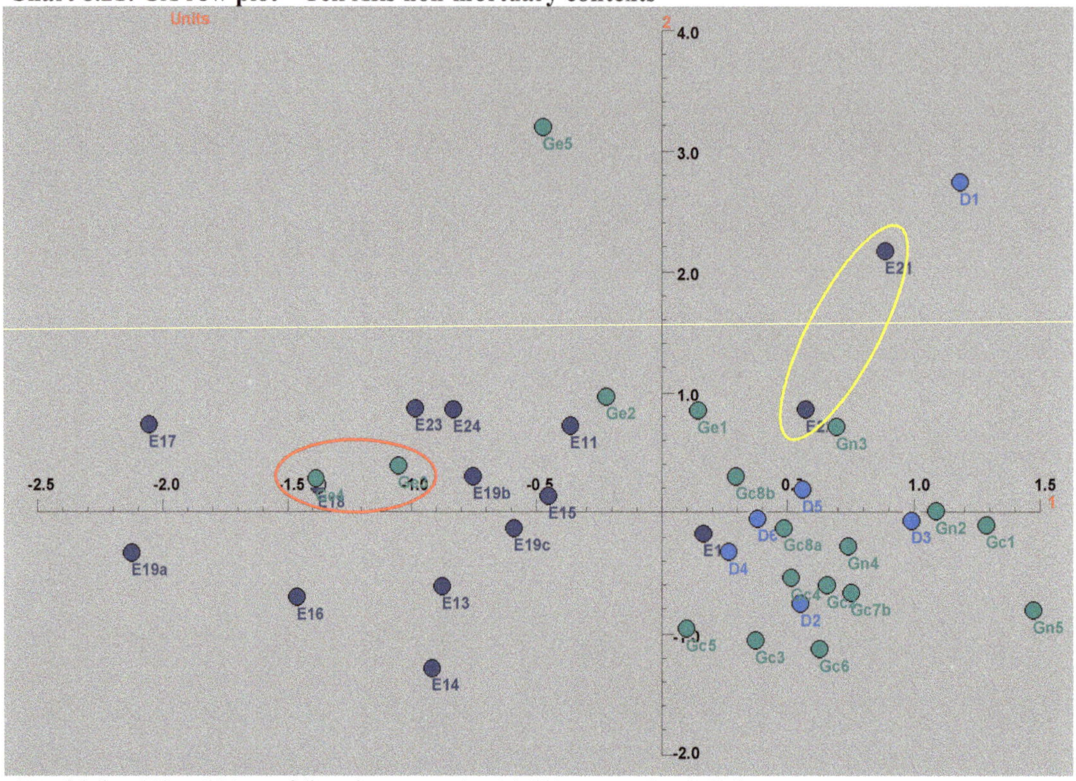

CD: /biplots/Chart821.wmf (inertia: comp. 1 – 14.8%; comp. 2 – 9.5%)

• Domestic • Elite • Non-specific

Chart 8.22: CA row plot – Tell Afis non-mortuary periods

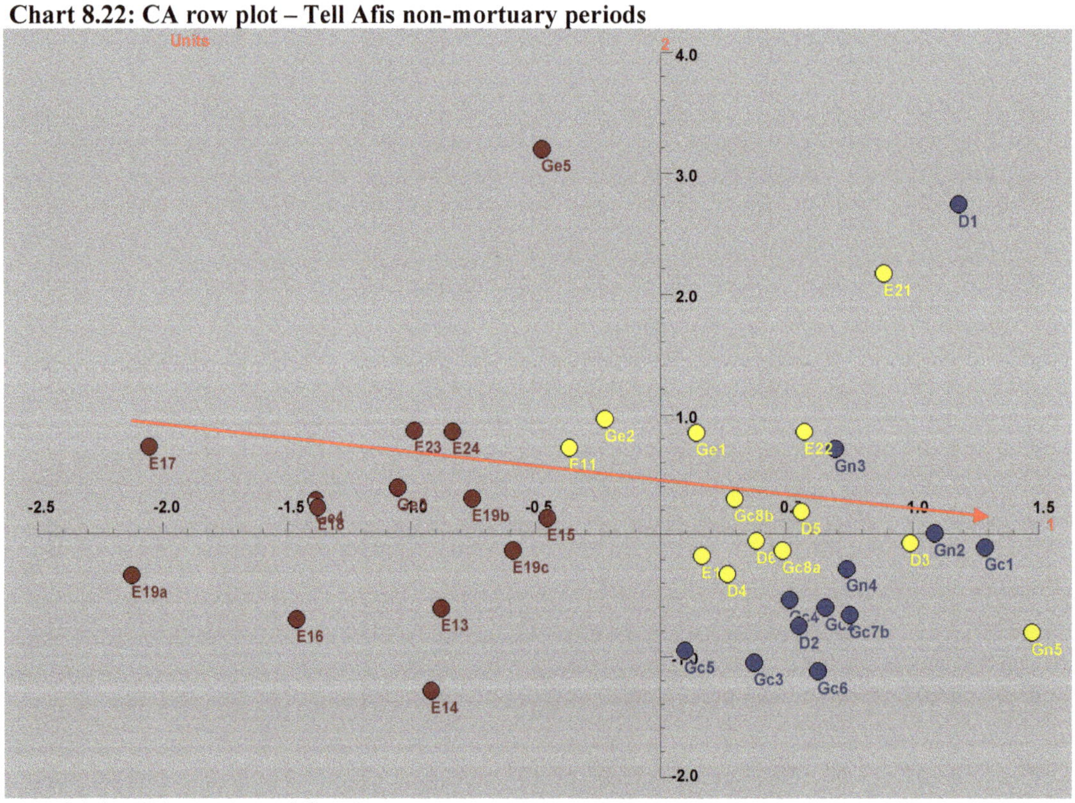

CD: /biplots/Chart822.wmf (inertia: comp. 1 – 14.8%; comp. 2 – 9.5%)

• Iron I • Iron II • Iron III

Summary of CA Trends

The above Correspondence Analyses have explored the relationship between different assemblages and have highlighted a number of influences on ceramic assemblages; geography, chronology, function, context-type (e.g. mortuary vs. non-mortuary), and mortuary type (cremation vs. inhumation), are all important considerations. They have confirmed the presence of a clear division between the coastal and inland regions of the IA-NL, but have also highlighted significant local patterns in the data. Clearly the identification of broad ceramic regions is valid, but it does not offer a detailed understanding of diverse local cultural behaviours. While CA does not explain the localised patterns, it does identify changes in the patterns over time which can inform interpretation. Clearly, time and space have an impact on material culture, or more accurately, a number of ceramic patterns can be defined according to geographic and chronological constraints. A vessel's function also has an effect; it seems there were geographic differences in the functional "make-up" of ceramic assemblages. For instance, the pouring and mixing of wine were important amongst coastal mortuary assemblages, but less so in the inland Northern Levant where large storage pithoi were a key functional element.

A final pattern evident in the CA results is associated with the nature of context. In particular, there is a clear distinction between mortuary and non-mortuary assemblages. In addition, there are different patterns within the mortuary contexts according to type of burial; inhumation, cremation, and mixed-use all follow slightly different trajectories. A distinction within non-mortuary contexts, however, is less marked, with no apparent difference between elite and domestic contexts, though this observation is based on the evidence from only one site.

8.3 Cluster Analysis

Cluster Analysis is a generic term for a wide range of methods used to discover homogenous groups or clusters in data. The method aims to express the relationships between individual components by measuring the level of similarity and dissimilarity between the components; in this case, ceramic CLASS. Depending upon the nature of the data, the analyst specifies an appropriate algorithm that will group components on the basis of their similarity, or more importantly their dissimilarity (Baxter 2003, 90-104). There are many ways of specifying the means for measuring dissimilarity, hence many different ways in which a Cluster Analysis might be carried out. The following Cluster Analyses utilise Jaccard's Similarity Coefficient, which is a hierarchical clustering method and is appropriate for binary (presence/absence) data. Without reviewing the entire mathematical structure of this calculation principle, or algorithm, it is sufficient to note that Jaccard's Similarity Coefficient disregards negative matches; i.e. mutual absence is not indicative of similarity (for a more detailed discussion of this principle see Baxter 2003, 94; Shennan 1997, 228-234).

The Technique

The aim of Cluster Analysis is to derive a partition, or a sequence of partitions, of a set of objects based on their similarities to one another, so that objects clustered into the same group can be considered similar in presence. The resulting clustering procedure can be conveniently represented in the form of a tree diagram or *dendrogram*. Interpreting the resulting *dendrogram*, however, can be confusing. The interpretation of the following *dendrograms* presented below is in no way considered exhaustive, and is generally restricted to patterns of obvious significance. The socio-cultural implications of the different clusters within the *dendrograms* are discussed in **Chapter 9**. All Cluster Analysis calculations and dendrogram creation were undertaken with the Bonn Archaeological Statistics Package version 5.43 for Windows (WinBasp) using Jaccard's Similarity Coefficient. Each dendrogram is also included on the appended CD for closer inspection – the file name/path is listed below each dendrogram.

Analysis

Cluster Analysis was considered an attractive analytical tool within this study because its focus on single ceramic categories complements that of Correspondence Analysis, which explores the relationship between collections of units or types (whole assemblages). Unfortunately, the multivariate nature of the dataset resulted in a single *dendrogram* that was large and unmanageable. For this purpose, the dataset was split into more manageable components, and a *dendrogram* produced for each sub-set. As a result, the following discussion begins with an analysis of the main ceramic functional categories (i.e. cooking pots; amphorae; pithoi; urns; kraters; jugs; bottles; bowls) before investigating the relationships across these categories within mortuary and non-mortuary contexts.

Cooking-pots

The cooking-pot Cluster Analysis (**Dendrogram 8.1**) indicates that there is not much clustering of cooking-pot CLASSES. There are two outliers in the data, CLASSES 010 (baking trays) and 002, and only one small group: CLASSES 004, 006, and 008 derive from the Southern Levant and coastal Lebanon (**Maps 10-11; 13-14**). This single group, however, is only one neighbour removed from most other cooking-pot classes, and is, therefore, not overly distinct. The general lack of clustering implies that within the majority of contexts one cooking-pot class was used, rather than a collection of classes. This does not, however, infer any geographic limits on each class.

Transport Amphorae

The transport amphora Cluster Analysis (**Dendrogram 8.2**) suggests that there are three main groups within this dataset, with two outliers (CLASSES 017 and 029). The top group is also the broadest, incorporating CLASSES 016, 018, 019, 020, 021, 022, and 026. This group occurs across the study area in the Iron I and Iron II periods, but is generally absent from the inland Northern Levant (**Maps 17-20**). Within this group is a small sub-group of CLASSES 021 and 022, which implies there is little reason to distinguish between these two similar forms in the dataset. The middle group consists of CLASSES 024, 025, and 030, which are also generally restricted to coastal areas but appear mainly in the Iron II-III periods (**Maps 21; 22**). The few amphorae known from the inland Northern Levant are associated with the bottom group, which is characterised by CLASSES 023, 027 and 028 (**Maps 27; 28**). The majority of transport amphora CLASSES occur in the Southern Levant and coastal regions, but not the inland Northern Levant. It is interesting to note that of the few amphorae known from inland Northern Levant, the most commonly encountered are elongated in form and undecorated (e.g. CLASS 028). The significance of this is unclear, though it might indicate that elongated, narrow vessels were possibly easier to carry overland by animal. In the end, the three main clusters of transport amphora appear to reflect both chronological and geographic relationships.

Pithoi

The pithos Cluster Analysis (**Dendrogram 8.3**) suggests little clustering of pithos CLASSES. Apart from two small groups (CLASSES 032 and 034 – from the early Iron Age Southern Levant, and CLASSES 035 and 036 from the early Iron Age Northern Levant), each CLASS is as similar to the other pithoi classes as they are dissimilar. Nevertheless, the CLASSES in the top of the dendrogram are generally associated with the inland Southern Levant, while the CLASSES in the lower half are associated with the Northern Levant (circled).

Dendrogram 8.1: Cooking-pot Cluster Analysis

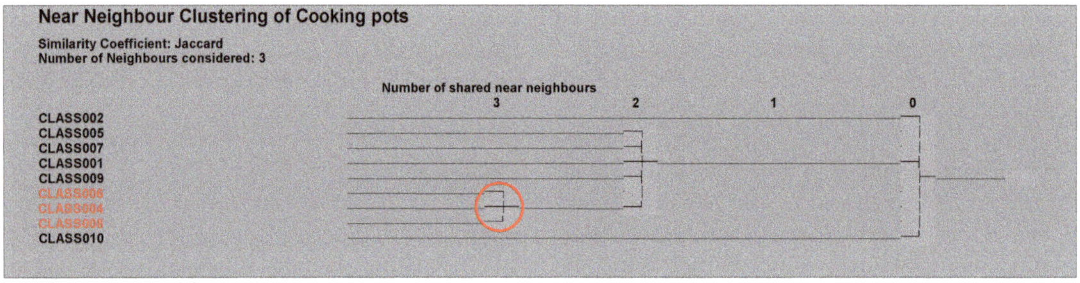

CD/clusters/cooking

Dendrogram 8.2: Transport Amphora Cluster Analysis

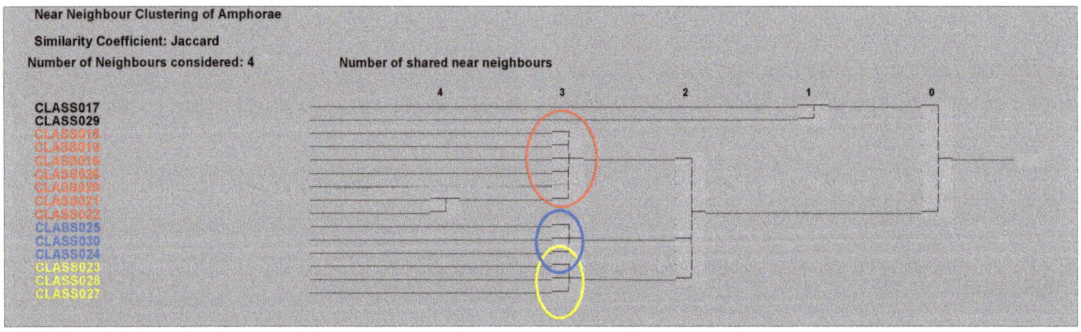

CD/clusters/amphorae

Dendrogram 8.3: Pithos Cluster Analysis

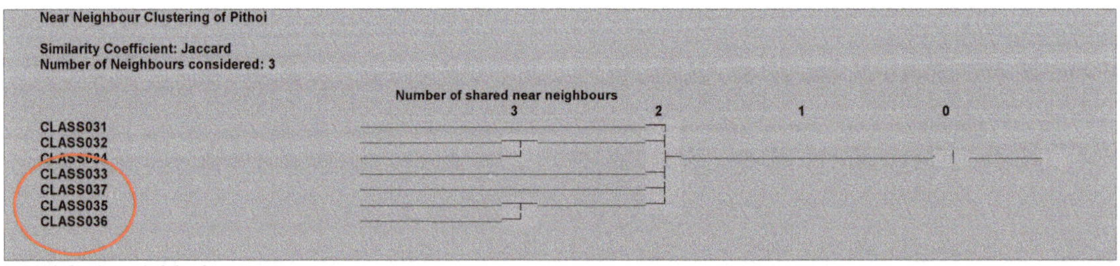

CD/clusters/pithoi

Kraters

The krater Cluster Analysis (**Dendrogram 8.4**) depicts minimal clustering of kraters. There are three outliers in the data; CLASS 043 is a late Iron Age krater, while CLASSES 050 and 052 are unique to the Jerablus region. Nevertheless there are three small groups in the data. The top group consists of CLASSES 038, 039, and 055, generally associated with the Southern Levant (**Map 31**). The middle group consists of three krater CLASSES that mainly derive from Lebanese mortuary contexts (**Map 35**). CLASSES 044 and 049 form the third group.

Urns

The urn Cluster Analysis (**Dendrogram 8.5**) has two outliers; CLASSES 063 and 065 occur at only one mortuary site each. The remaining urns appear to be characterised by two small clusters (CLASSES 058 and 062; CLASSES 060 and 061). No pattern is immediately discernible in the urn category, possibly due to their limited number throughout the study area (**Chart 5.5**).

Jugs, Flasks, and Spouted Vessels

The Cluster Analysis for jugs, flasks, and spouted vessels (**Dendrogram 8.6**) suggests a number of divisions in the data. The outlying group at the bottom of the diagram (CLASSES 093, 094, 095 and 098) are all common in the Persian period. The top group consists of CLASSES 068, 071, 101 and 105, all of which are generally undecorated vessels found across inland areas. The group second-from-top consists of a variety of juglets (CLASS 100 – **Maps 55; 56**), flasks (CLASS 102 – **Maps 57; 58**), unguent containers (CLASS 103 – **Map 59**), beer jugs (CLASS 107 – **Map 61**), and decorated jugs (CLASSES 077, 078, 080, 082, 084, 085, 087, 088, and 090 – **Maps 41-43; 45-47; 49-51**), the majority of which are associated with coastal mortuary contexts. It is interesting to note the very close association of CLASSES 082 (Red-Slip trefoil jugs) and 085 (Mushroom-lip jugs), as they continually appear along side each other in Lebanese cremation contexts. The third (from top) cluster is that of CLASSES 075, 076, 104, and 109, all of which are slightly more oriented toward the inland Southern Levant (**Map 60**). The fourth group consists of CLASSES 074, 081, 086, and 097.

Bowls

The Cluster Analysis for bowls (**Dendrogram 8.7**) is both large and confusing and is reproduced here only to demonstrate the main groupings; the reader is referred to the appended CD for a more detailed inspection. While the large dendrogram makes bowl analysis difficult, a few clusters are apparent. The top group consists of CLASSES 143, 161, 170, 174, 175, 176, 182, 183, and 184, primarily shallow bowl rims and fine bowl rims. The second group consists of CLASSES 134, 141, 160, 162, and 172, the most abundant bowls within the study (**Chart 5.9**); it seems that their over-abundance has created an artificial cluster. Finally, the third group consists of CLASSES 144, 145, 151, 155, 163, 179, and 191, which largely derive from coastal mortuary assemblages (**Maps 69-73; 83; 89**).

Dendrogram 8.4: Krater Cluster Analysis

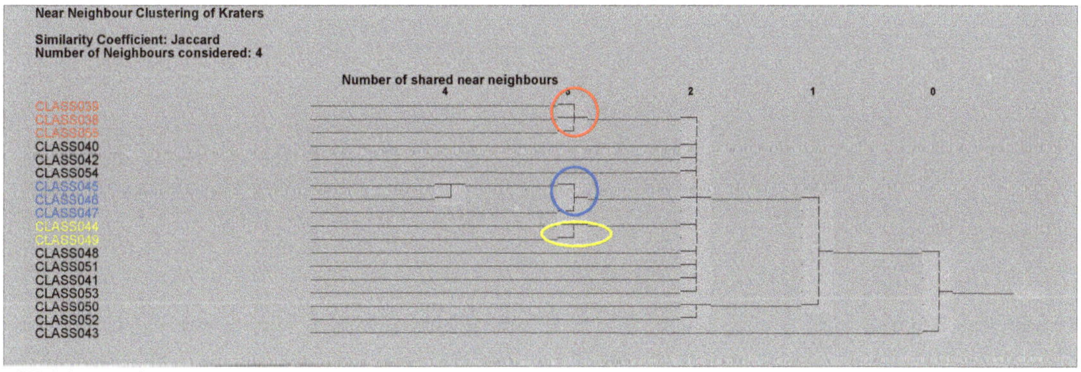

CD/clusters/craters

Dendrogram 8.5: Urn Cluster Analysis

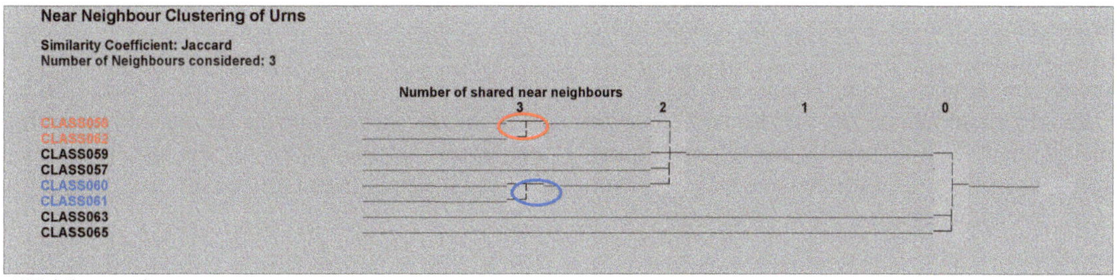

CD/clusters/urns

Dendrogram 8.6: Jug, Flask and Spouted Vessels Cluster Analysis

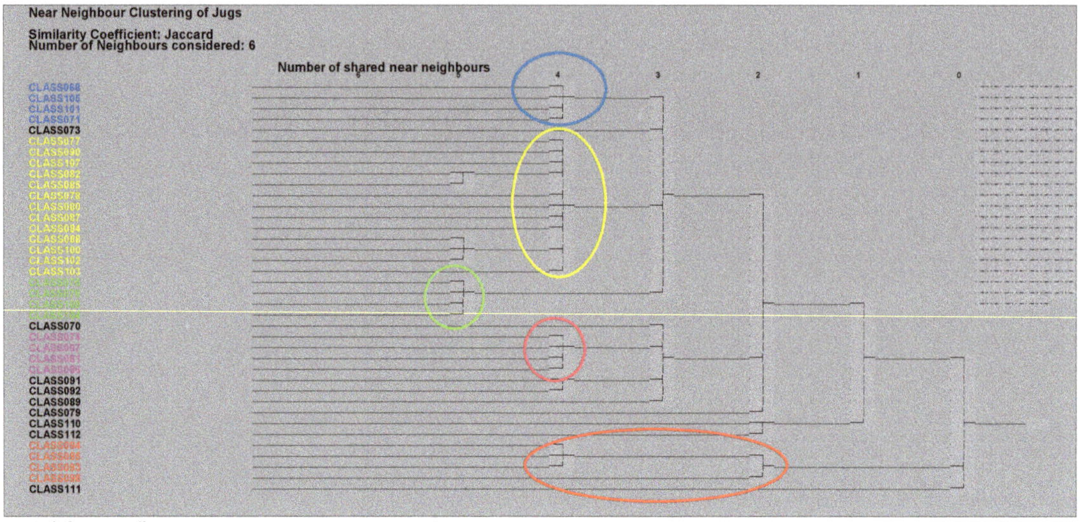

CD/clusters/jugs

Dendrogram 8.7: Bowl Cluster Analysis

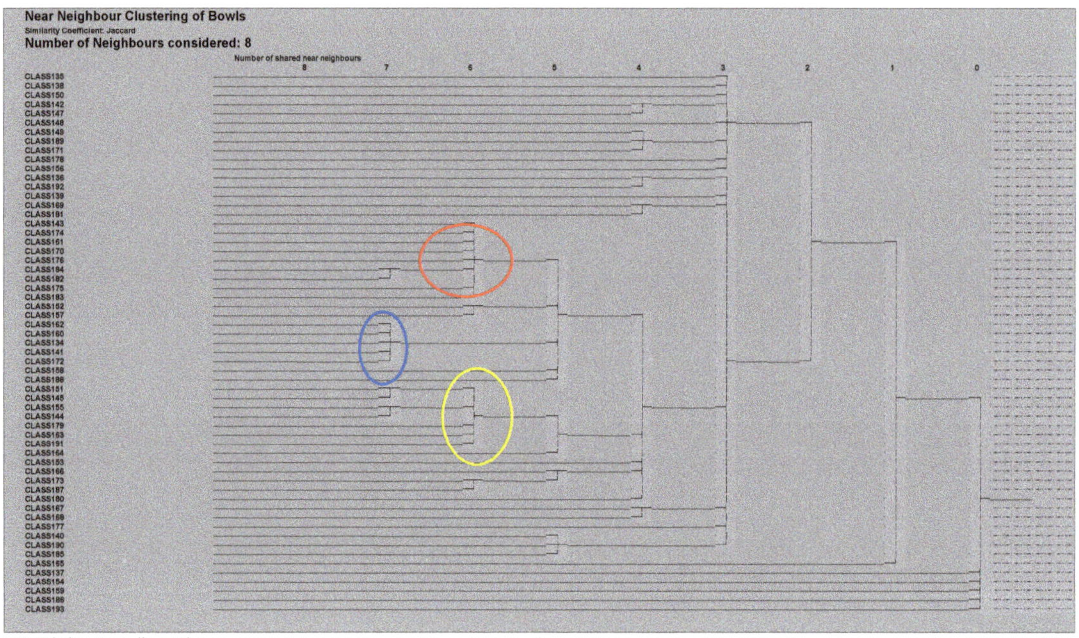

CD/clusters/bowls

Mortuary and Non-Mortuary

The segmentation of the data above greatly eased dendrogram analysis, but it also prevented an understanding of clusters across functional classes. In other words, by only comparing pithoi with pithoi, the relationship of pithoi with other categories, such as bowls or amphorae, was not explored. For this purpose, Cluster Analyses were undertaken for mortuary and settlement contexts. The size and complexity of the two resulting dendrograms argued against their inclusion here, though the mortuary dendrogram is included over the page (both dendrograms are included on the appended CD - /clusters/mortuary; */settlement). The complexity of these dendrograms also meant that the relationships were difficult to decipher. Nevertheless, the Cluster Analysis of mortuary contexts (**Dendrogram 8.8**) presents one well-defined cluster of ceramic vessels, all of which are characteristic of coastal mortuary contexts (**Chart 7.58**). This group consists of the CLASS 042 krater, CLASS 077, 082, 085, 088, 090 jugs, CLASS 102 flasks, and CLASS 144, 145, 155, and 179 bowls. An equivalent assemblage is not evident amongst inland mortuary or settlement contexts. This implies that mortuary ritual on the coast rigidly prescribed a "set" of ceramic vessels.

Dendrogram 8.8: Mortuary Assemblage Cluster Analysis

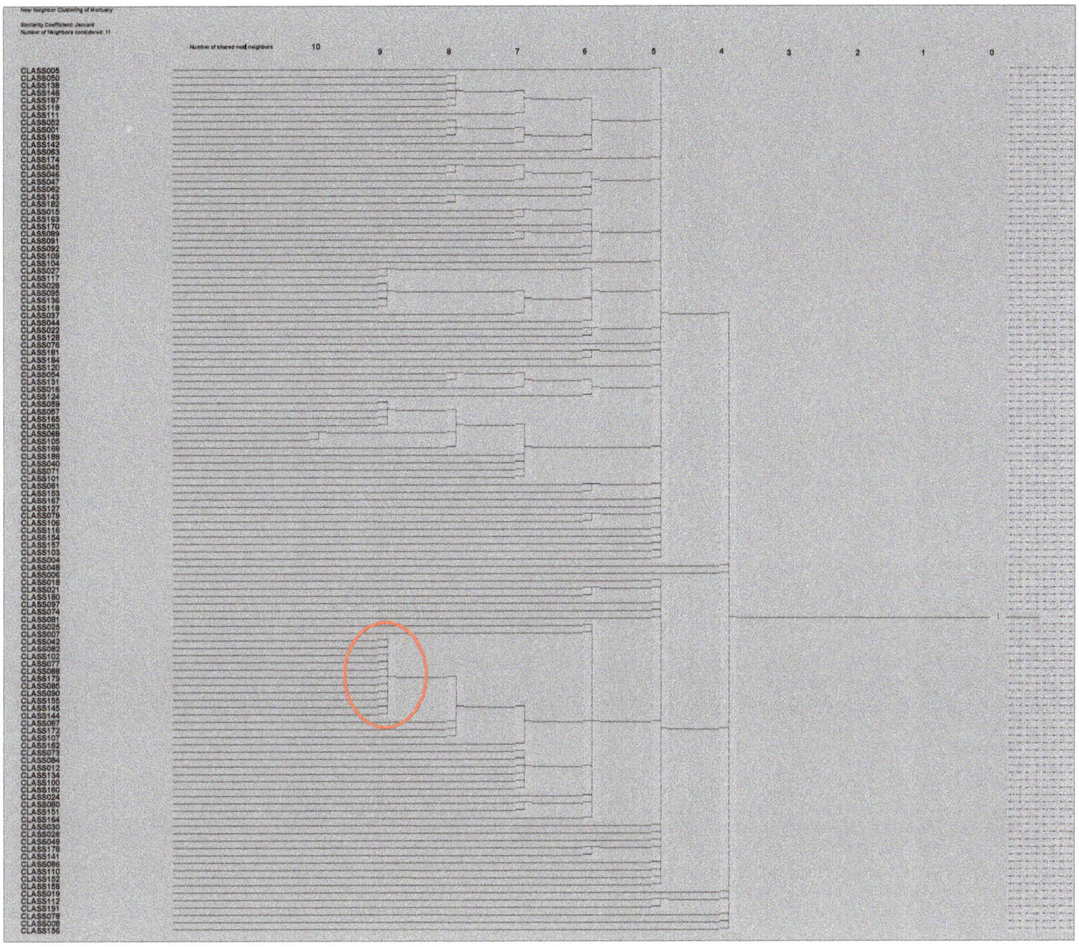

CD/clusters/mortuary

8.4 Summary

This chapter has employed two statistical techniques designed to visually depict the multivariate ceramic relationships in the large database created for the present study. The first technique, Correspondence Analysis, explored the relationship between different assemblages, different contexts, and between assemblages and contexts (The author would like to express his gratitude to Prof. Mike Baxter and Dr Andrew Millard for their comments and guidance with regard to the above CA results and their interpretation). The second, Cluster Analysis, presented the relationship between specific ceramic forms. While Correspondence Analysis and Cluster Analysis are complementary statistical techniques for the investigation of incidence data, they do have limitations. On the one hand, the CA plots are appropriate for isolating only general trends in the data, and cannot be used to draw specific inferences: there is a danger of over-interpretation. Cluster Analysis, on the other hand, can produce more specific relationships in the data, but with such a large dataset as is being negotiated here, the high level of detail tends to obscure the more general patterns. In other words, there is a danger of not being able to 'see the forest for the trees'. Consequently, the results outlined in this chapter were used in conjunction with the exploratory data analysis in **Chapter 7** to inform and direct the interpretation presented in **Chapter 9**.

SECTION FOUR

An Alternative Reconstruction for the Iron Age Northern Levant

CHAPTER NINE

Ceramics and Identity in the Iron Age of Syria

Material culture and pottery provide not an exclusive but certainly an important clue for understanding economic trends and social behaviour...Ecological, economic and social frontiers were an essential feature of this scenery; their changes and alterations over the long term resulted in a fluctuation of regional patterning which had an effect on pottery production and distribution (Mazzoni 2000c, 139).

9.1 Introduction

SECTION ONE of the current study (**Chapters 2, 3, and 4**) highlighted a number of problems associated with the idea that the distribution of Iron Age pottery was, to a large extent, determined by "history", as conventionally derived from a literal reading of the documentary sources. Hence, ceramic distribution, and rapid change in ceramic style, has often been closely linked with either the movement of people groups (ethnicity) or sudden "historical" events (chronology). Furthermore, an almost exclusive focus on artefact typology in many ceramic studies has meant that intra-regional variability has not been considered an important research question. Instead, "regionalisation" was invoked as explanation for variation. The aim of this chapter is to offer alternative explanations for material culture patterning other than that extracted from a literal reading of the historical narrative. These explanations are based on similarities and dissimilarities in the data that were observed and explored in SECTIONS TWO and THREE of the present study (**Chapters 5, 6, 7** and **8**).

As discussed in **Chapter 4**, material culture is a dynamic medium by which social agents are potentially able to negotiate, uphold and challenge a multitude of different social identities, both individual and collective. Material culture can be viewed and used differently according to context, social convention, and individual choice. **Chapters 2** and **3** demonstrated that current interpretations of the IA-NL do not adequately explore the multi-faceted dynamics of social behaviour that run much deeper than politico-ethnic concerns. The following discussion aims to transcend description by considering elements of social life evident in the production, function, use, and discard of ceramic culture. These potential explanations are presented without an appeal to historical events, but do occasionally draw upon iconography upon political monuments to illustrate how certain vessels might have been used.

9.2 Categories of Use

Introduction

A pot's appearance and structure is not dependent on isolated choices, but on precise technological responses to its functional or stylistic requirements. Indeed the overall form can be said to be determined by its designed purpose, whether considered practical or symbolic (Orton *et al.* 1993, 28, 217). Hence, elements of shape, fabric, and surface treatment are indicative of certain functions: closed vessels restrict exposure; open vessels display the contents; cooking-pot fabrics need to withstand thermal shock; polished surfaces reduce permeability; and decorative surfaces imply some level of symbolic importance (Faust 2002, 56-60). For this reason, the following discussion is structured according to function and the implied use of the different functional categories.

Transport and Trade

Within the data for this study, there is a distinct concentration of transport amphorae along the Mediterranean coast and inland Southern Levant (**Charts 7.4; 7.6; 7.10; 7.12 7.14**), a pattern that changes little throughout the Iron Age (**Chart 7.45**). This concentration is in stark contrast to the general absence of these vessels across the inland Northern Levant (**Charts 7.8; 7.16; 7.18; 7.20**). But rather than explain this pattern as an indicator of ethnicity or ceramic regionalisation, it appears to be a direct result of the nature of use of these vessels. In other words, it is their use within maritime trade networks, for which they were intended, that determines the coastal concentration. This is to some degree also affected by geographical factors. The inland Northern Levant was not directly involved in maritime trade because it was not directly linked to the sea, and consequently had little need for transport amphorae; the imposing Jebel Anasariya was an effective barrier between the interior and the coast. The northern areas of the Southern Levant, on the other hand, had reasonably direct access to the Palestine coast via the Jezreel Valley, as indicated by the presence of Cypriot and Greek imported pottery inland throughout the Iron Age (Clairmont 1955; 1956-1957; Coldstream and Mazar 2003; Fantalkin 2001; Mazar 2004; Waldbaum 1994; Wriedt-Sorenson 1997). Geography, however, is not the only factor.

Despite the natural barriers between the inland Northern Levant and the coast, few archaeologists would suggest that no trade was undertaken between these two regions. It is, therefore, worth considering what type of products were transported inland, and why these goods were apparently not transported in amphorae. Was it simply that the shape and size of most transport amphorae were not conducive to overland transportation? Or were these amphorae used to transport specific products and goods that were not transported inland? The few transport amphorae evident across the inland Northern Levant are usually of a narrow, elongated form (e.g. CLASS 028 – see **Chart 7.19; Maps 27-28**), which was presumably easier to secure upon a pack animal than the wider, larger and heavier amphorae associated with coastal trade. Indeed, when animals are depicted in Assyrian reliefs carrying goods from the Northern Levant (usually in the form of tribute or plunder), they take the form of bales or packaging, rather than ceramic vessels (King 1915, Pl. XXIII; Parpola 1987, Fig. 32). Similarly, wine or oil, which was commonly transported in amphorae within maritime contexts, is depicted as being transported by people carrying what appear to be animal-skin bladders (Parpola 1987, Fig. 2, 12). The dearth of transport amphorae inland does not exclude the possibility for the trade of the same products normally shipped in these vessels. The same products were conceivably being transported inland, but within a different container; one that might be archaeologically less-visible (e.g. organic material). Nevertheless, different containers might indicate a different origin for transported goods, such as a local primary producer of these goods, or inland distribution centre. Alternatively, these products may have originated from the coastal centres, but the size and nature of most transport amphorae prohibited their long distance carriage across the Jebel Anasariya by pack animals. Whichever interpretation seems more likely, the distribution of transport amphorae may be as much the result of geography or economy, as the indicator of political identity.

While geography might offer an explanation for the distribution of transport amphorae, it does not answer all questions regarding distribution. For instance, explaining the presence of transport amphorae in the inland Southern Levant only as the result of geography misses the strong cultural link that this region had with the coast in other ceramic categories (**Charts 8.1; 8.2; 8.3; 8.4**). So the question remains: why was the inland Southern Levant more closely connected with the mid-Levant coast (traditionally referred to as "Phoenicia"), while the inland Northern Levant was not? Is it just a question of geography or topography? Or was there a social "condition" that meant the communities of the inland Southern Levant were more closely aligned with a Mediterranean lifestyle and its trappings; a way of life that might have been formed through geographic proximity but which had come to represent something more meaningful? This may be reflected in the particularly strong presence of Cypriot and Mycenaean imported pottery in the Jezreel Valley during the Late Bronze Age (Amiran 1969, 172-186).

The use of transport amphorae was not restricted to actual transportation; a number of inhumation burials from the Levant coast also produced these vessels.

Storage

Pithoi were generally restricted to inland contexts of the Northern Levant (**Charts 7.41; 8.18**). In particular, the "cigar-shaped" "Aramaean" pithos (CLASS 037) was characteristic of the Iron II and Iron III periods for this area (**Maps 29-30**). The distribution of this functional category is in sharp contrast with the concentration of transport amphorae along the coast. Broadly speaking, these two categories appear to confirm the inland/coastal dichotomy that frequently appears in literature on the Iron Age (the two-region model of Lehmann's 1998; the "Port Power" model of Stager 2001). The few coastal pithoi evident in the database primarily dated to the early Iron Age (**Chart 7.41**). This seems to indicate that the division between inland and coast was less rigid in the Iron I period than in the Iron II period, when the division was stark (*cf.* **Charts 8.13; 8.18**). Pithoi are so large and heavy that they were probably intended only for storage, and not transportation (*contra* Artzy 1994). Indeed, the Balawat gates carry a depiction of Assyrian soldiers transporting a CLASS 037 pithos upon a flat-bed cart and requiring the attention of at least 13 men – clearly not intended for ease of transportation (Dion 1997, Fig. 19). The fact that many pithoi have been found fixed within the remains of large buildings, as permanent features, also argues against their use for transport (Fugmann 1958, Fig. 299). Nevertheless, Late Bronze Age Cypriot pithoi were found on the Ulu Burun shipwreck suggesting that, at least in that period, similarly sized pithoi were involved in maritime trade (Webb and Frankel 1994, 18). Archaeological investigations throughout the eastern Mediterranean have suggested that pithoi were used for the storage of dry goods (pulses, grains), as well as fruits, oils, wine, water, and preserved (salted) goods (Christakis 1999, 2). Their permanence might also explain the conservative nature of CLASS 037, existing relatively unchanged throughout much of the Iron Age. Their vast size and permanent placement prevented them from being handled, implying that they rarely needed replacing (*contra* King 1915, Pl. VII). There is also the possibility that such large vessels would have been left *in situ* in abandoned buildings, and later reused.

The fact that these vessels were "hidden" within storage magazines and not handled openly suggests that they were unlikely to embody significant symbolic messages. Nevertheless, the manipulation and storage of foodstuffs is a social process, and one within which identities and social structures are negotiated (Webb and Frankel 1994, 18-20). The abundance of CLASS 037 pithoi during the Iron II and Iron III periods is significant, not least because it coincided with the use of food and foodways for the negotiation of social identity, as evident in the nature of serving wares. Considering their large size, it is interesting to note that pithoi were rarely encountered in mortuary contexts of the Northern Levant; we might expect them to be ample coffins/cinerary urns (Iron I pithos burials have been excavated at Tel Nami – Killebrew 2005, Fig. 3.12). This lack of mortuary association may be due to their significance in food control strategies, or simply because inhumation was not used in the inland Northern Levant where these vessels were most abundant.

The distribution of the most abundant pithos form within the study area (CLASS 037) is intriguing; these vessels were found only in inland regions of the Northern Levant, stretching from Ain Dara in the north to Tel Dan in the south (**Maps 29-30**). But why were pithoi virtually absent from the coast? Or, to rephrase the question; why did the coastal communities not use pithoi to store large quantities of agricultural products? Is it that they had no need for the storage of agricultural produce? Part of the reason may again be attributable to geographic factors. The Orontes region, for instance, is famous for its abundantly fertile soil, and is today still called the "bread-basket" of Syria. In marked contrast, the coastal regions are desperately short of arable land and are, therefore, less likely to produce significant agricultural surpluses requiring storage. But this does not explain why people stored produce? Could the difference between coast and interior be because of cultural differences; one society produced foodstuffs that needed storage in large ceramic vessels, while the other stored produce, possibly of a different nature requiring different storage methods, in non-ceramic containers. Alternatively, the coastal communities might have simply imported what was needed. While the coast had a constant influx of different goods, the interior may have been less "connected" to trade networks and instead was required to maintain a considerable store of local products. This would also explain the presence of pithoi in areas that were poorly represented by transport amphorae. Of course, the lack of pithoi along the coast does not preclude the use of other storage vessels there; kraters or transport amphorae might have been used for storage, or instead they may have used vessels that are less "visible" in the archaeological record (e.g. wooden barrels, cloth sacks, basketry). If pithoi represent surplus produce, then they may also reflect a certain level of centralised authority that may not be evident amongst coastal societies (i.e. private control of goods – see Webb and Frankel 1994).

Regardless of how we interpret the concentration of transport amphorae on the coast and pithoi across the interior, it is clear that there are a number of different factors possibly influencing their distribution, few of which could be said to reflect ethnic or political identity (**§9.4**).

Domestic Appliances

Cooking-pots were identified across all zones within the study area (**Chart 7.29**). Many of the different cooking-pot CLASSES experienced a distinct spatial and temporal distribution (the more informative patterns are those for CLASSES 001, 005, 006, 007, 008, and 009). Following the patterning evident in **Seriation 2**, the earliest of these forms is the CLASS 008 open cooking-pot, which was distributed across the inland and coastal regions of the Southern Levant during the Iron I period (**Maps 13-14**). CLASS 006 and 007 were also found in the Iron I period but, according to the seriation matrix, appear to be slightly later phenomena. CLASS 006 was primarily restricted to the Southern Levant and Lebanon during the Iron I and Iron II periods (**Map 11**). CLASS 007 was also present in both periods, but experienced a broad distribution across much of the study area (**Map 12**). The same seriation matrix indicates that CLASS 005 has few early occurrences, but was mainly present in the Iron I and Iron II periods, when it was mostly distributed along the interior of the study area. Probably one of the most distinctive cooking-pot forms in the present study is the CLASS 001 hole-mouth cooking-pot, which is an Iron II-III component of the inland Northern Levant (**Maps 7-8**). **Seriation 2** clearly confirms that CLASS 009 is a late Iron Age cooking-pot. The above review of cooking-pot distributions illustrates the high level of complexity and diversity within the ceramic record of the Iron Age. These many different patterns rarely map onto the broad regions derived from a literal reading of the documentary sources. Instead, these patterns cut across each other and imply that conventional historical interpretations provide only a superficial explanation of material culture patterning.

Cooking-pots are often considered a mundane and unimportant component of most ceramic assemblages, yet they too reflect a variety of social behaviours: culinary practice and cuisine; unostentatious declarations of identity; and socio-technological choice of domestic potters, to name just a few. In contrast to pithoi, cooking-pots experience harsh treatment (fire, hot liquids, hurried handling, suspension) and break often; therefore, they need replacing regularly, making them highly sensitive to change. This temporal sensitivity appears to be evident in **Chapter 7**'s seriation analysis, and may help explain why there are a number of small but distinct cooking-pot distributions across the study area. One possible interpretation is that these smaller distributions reflect regional differences in diet and/or cuisine. For instance, the more open cooking-pots of the Southern Levant would allow for greater evaporation of liquids during cooking than their hole-mouth counterparts from inland Northern Levant, while short-necked, flanged cooking-pots were possibly used in conjunction with lids, restricting evaporation even further. As a result, the different cooking-pots may represent different degrees of "wetness" of cooked food, or different types of food requiring different cooking techniques. Access to specific produce, or local tastes and convention may have played an important role in cooking-pot design.

While visible vessels might be seen as holding deliberate symbolic meaning for the definition of social identity (see below), the same may be true of the less conspicuous cooking-pots. As much as conspicuous vessels embody self-ascription, domestic production of pottery and/or use of inconspicuous vessels may well embody unintentional, or even intentional, statements of affiliation (Bowser 2000, 231*ff*). Hence, cooking-pots can embody a wide range of social identities, from the subversive to the blatant. For many domestic potters, the only available means for self-expression may have been the form of cooking-pots they produced and/or used. Furthermore, the division of household tasks, such as cooking and food preparation, is a potent means for enforcing domestic, or familial, social structures.

In addition to cooking-pots, lamps are another oft-ignored category of domestic vessel. The predominant lamp category within this study is the pinched-lip lamp (CLASS 012 – **Maps 15; 16**), which is associated with the burning of oil. Pinched-lip lamps were common throughout the Southern Levant and coastal regions during the Iron I and Iron II periods. They were, however, rare amongst sites in the North Syrian and Euphrates zones, but present in Iron II contexts of Orontes Syria. The "northern" absence may reflect different traditions regarding the equipment used for lighting internal spaces, or different substances being burnt; e.g. sheep fat would not require the same vessels that the burning of olive oil demands. Additionally, the distribution of CLASS 012 lamps might represent areas with sufficient olive production to make the burning of olive oil a viable activity and, therefore, may be related to climate and farming methods. Alternatively, the use of pinched-lip lamps may be of long-standing tradition with its origins in earlier oil-burning practices of the Bronze Age, when these vessels were abundant across the eastern Mediterranean (Amiran 1969, Pl. 59).

Conspicuous Consumption

Transport amphorae, pithoi, and cooking-pots are all generally considered to be utilitarian in purpose (i.e. they are designed for specific practical functions). Consequently, painted decoration and Red-Slips were virtually absent within these categories; surfaces were only minimally treated, and fabrics tended to be more coarse than fine (**Charts 7.67-68**). Vessel function encompasses much more than the practical, however, and may include various forms of intentional symbolism associated with specific vessels (MacClancy 1992, 5). From sociological research into the symbolic power of food, it is clear that the manipulation of food and drink (procurement, distribution, consumption) is a moral process through which ideologies and social relations are articulated, upheld or transformed (*cf.* Levi-Strauss 1970, Douglas 1975). Food is, therefore, both a *product* of society and very much its *agent*, as certain cuisines and culinary behaviours *actively* define individual and social identities (Sørensen 2000, 8). Hence, the consumption of food and drink is closely intertwined with the creation and negotiation of social identity, as too are the accoutrements for their preparation, service, presentation, and consumption. Eating and drinking are also acts of consumption, wherein the significance of the symbolism is incorporated into the body (embodied) of the individual or group (Hamilakis 1999).

The nature of table-wares is that they are visible and, consequently, a potential tool for impressing meaning upon guests within a banqueting and/or feasting context, or even day-to-day consumption. The potency of dining and drinking symbolism is evident in the frequency by which kings and rulers are depicted in banqueting scenes upon political monuments (see Adachi 1997). Ceramic serving vessels hold deliberate symbolic meaning that communicates particular statements about the host or owner. Such a statement is essentially a declaration of identity, whether economic (wealth), cultural (belonging), ethnic (kin-based belonging), or something completely different. But the symbolic meaning is not only in the appearance of the vessel, it is also in the association this vessel has within the culinary context; with which foods it is associated, how and when it is used, and what it represents, all contribute to its symbolic message. For instance, a number of scholars have suggested that Red-Slip is a substitute for bronze, with its burnished sheen and bright colour mimicking that of bronze vessels (e.g. Braemer 1986; Holladay 1990; Mazzoni 2000a; Steel 2002; 2004). This is known as *skeumorphism*, where the intention of these vessels is to invoke the appearance of vessels regularly made from another material (Vickers and Gill 1994, 105-153). In other words, the form and surface finish of the jugs and bowls were emulating bronze vessels (**Table 9.1**). While the ceramic counterparts would be less expensive and remain unaffected by recycling strategies, they would have held a symbolic significance similar to the original bronze vessels. Red-Slip vessels appear to be indicative of emulation strategies, wherein a sub-elite element sought to acquire social significance through the emulation of distinct elite behaviours (Sherratt 1999, 185-186).

Since pottery can be used as an agent of social change, ceramic style is affected by tensions between all forms of social categories being negotiated: i.e. male/female, culture/nature, public/private, sacred/profane, belonging/separateness, young/old, single/married, wealthy/poor, free/enslaved, dependent/independent. As the symbolism associated with these high-visibility vessels changes, so too do the vessels themselves. Hence, the decorated pottery associated with the ostentatious display and conspicuous consumption of food and drink is particularly sensitive to change.

Conspicuous consumption encompasses a variety of formal and informal behaviours, many of which are unlikely to manifest themselves archaeologically. Two particular forms of display that are attested, and particularly pertinent to this study of Iron Age ceramics from the Northern Levant, are the serving and consumption of food, and the pouring and drinking of liquids (usually wine).

Serving of Food

The present study has defined and investigated a wide variety of bowl forms from the IA-NL. The large number of different bowl forms has made it particularly difficult to identify trends; for this reason, only the obvious, general, and most significant trends are presented here. Nevertheless, the very diversity of bowl forms is interesting in itself; especially considering their abundance across inland regions, and somewhat restricted repertoire on the coast.

Within the **Iron I** period, the majority of serving bowls in the Northern Levant are undecorated (**Charts 7.69; 7.72**); forms are generally quite simple and sinuous in profile, with carinations and attachments kept to a minimum (e.g. CLASS 134). In the Southern Levant, however, Iron I bowls could be either undecorated or bear a Red-Slip. If the consumption of food is indeed an important medium for conspicuous symbolism, which it can be, the communities of the Northern Levant in the Iron I period put little effort into "eye-catching" designs of serving vessels. Hence, they seem to have made little use of food consumption and culinary behaviour as an arena for the negotiation of social identity; i.e. there was little emphasis on the presentation and consumption of food within the ceramic record. In contrast, the use of Red-Slip in the Iron I Southern Levant (**Chart 7.72**) suggests that these communities understood, at least to some degree, the symbolic value of food presentation.

During the **Iron II** period, the Southern Levant continued to use Red-Slip serving bowls (**Chart 7.72**). Throughout the inland Northern Levant (except the Euphrates zone) there was a distinct shift in serving wares in this period – bowls became generally much shallower, with larger diameters and higher bases, and were more frequently decorated: Monochrome shallow bowls (CLASS 165 – **Maps 79-80**) and Red-Slip platters (CLASS 166 – **Map 81**) were characteristic. The Iron II communities of the inland Northern Levant west of the Euphrates appear to have placed emphasis on food consumption as an arena for the display or affirmation of social identities and or relationships. It is for this reason that the majority of open vessels (bowls) are decorated in some manner (**Chart 7.72**). The presence of pedestal platters (CLASS 166), used to elevate some foods above others in certain situations, suggests that some foods held symbolic priority (Sørensen 2000, 112). Likewise, the internal decoration of Monochrome shallow bowls implies that the decoration was intended to draw attention to the inside of these pots; i.e. their contents. Moreover, shallow, open vessels would not have held great quantities of food; instead they appear to be used for the service of dry foods in a manner that emphasises display, reflecting a high-value foodstuff – however that might be measured. Furthermore, Red-Slip bowls appear to be a substitute for bronze vessels; the colour, burnish, some shapes, and occasional attachments all suggest a mimicry. For instance, the bone-shaped lug primarily appears on Red-Slip vessels and is remarkably similar to bronze clasps on bronze vessels (*cf.* Matthäus 1985, Tf. 25; Yadin *et al.* 1958, Pl. 71.15-16). This notion of substitution (*skeumorphism*) might explain the general lack of Red-Slip bowls along the Euphrates; either the symbolic use of food manifested itself differently there or conspicuous consumption was restricted to non-ceramic elements of material culture, such as bronze bowls or basketry.

On the Northern Levant coast, however, serving bowls were few during the Iron II period. They were generally smaller than those from the inland Northern Levant, with minimal decoration. The majority of Red-Slip bowls on the coast were small, thin-walled, shallow bowls that might have been part of drinking sets (**Charts 7.64; 7.67; 7.69; 7.72**) – generally speaking, the smaller the bowl, the more likely it was decorated. The scarcity of serving wares from coastal non-mortuary contexts, however, is significantly affected by archaeological sampling, with very few non-mortuary contexts excavated and sufficiently published.

During the **Iron III** period, the general trends discussed above changed little, though Red-Slip becomes even more abundant throughout all areas of the Northern Levant except the Euphrates (**Chart 7.64-65; 7.72**). The lack of significant change is in contradiction to most periodisations of the Iron Age, which emphasise change in material culture due to Assyrian annexation of the west. This periodisation model is primarily based on the historical narrative (**§2.7**). There is a general lack of serving wares during the **Persian period**, which could suggest that either our archaeological sampling is incomplete, or the communities of the Northern Levant served food in vessels that are not archaeologically attestable; e.g. basketry, wooden or metal bowls. Nevertheless, there is also the possibility that Persian period communities no longer considered food presentation and banqueting an important means for the negotiation of social categories. The exploratory data analysis suggested a percentage increase in undecorated bowls during the Persian period, though the dataset is small (**Chart 7.62**). On the other hand, this change in ceramics may reflect a change in political structures. If there were new ways of marking "eliteness", the ceramic emulation of elite behaviours might also have altered.

Drinking

In addition to the conspicuous manipulation of food, the drinking of alcohol is a potent means by which social structures (e.g. identity, power) can be negotiated and enforced (Douglas 1987, 8-12). Alcohol is particularly effective because of its intoxicating and, therefore, liminal qualities (Hamilakis 1999, 40). But drinking is not exclusively a functional adaptation that serves the community by holding it together; it can also incorporate elements of exploitation, competition, coercion, and resistance (Parker-Pearson 2000, 10-11, 27). The power of drinking rituals, acquired through the elements of communal consumption, generosity, embodied pleasure, and intoxication, can transform, mask, or legitimise other less pleasant and more serious forms of power (Bourdieu 1977, 411). In other words, while the serving and consumption of food may embody significant symbolic messages, drinking alcohol might be considered a more powerful medium because of its "extra dimension".

Iron I – Aegean Influence

Much has been written regarding Mycenaean-style (sub-Mycenaean) ceramics in the early Iron Age Levant. Indeed, research has emphasized its Aegean affiliations (e.g. Mazzoni 2000a; 2000c), presenting these influences as the product of either economic exchange, or cultural diffusion through commercial contacts or migration (*cf.* Badre 1983; Knapp 1998; Larsen 1987b; Sherratt 1998b; Sherratt and Sherratt 1991; 1998). Such approaches are grounded in concepts of production and exchange (see Steel 1998; 2002; van Wijngaarden 1999; 2002). An alternative approach (to those of ethnic movement and trade models) prioritises the transmission of esoteric knowledge surrounding the consumption of intrusive pottery styles. These elements allow for the element of human agency and choice, and explore changing styles in the context of internal social transformations. The extent to which one culture adopts elements from another and the mode of adoption is dependent on the role that the adopted elements play within the recipient society (Steel 2002, 27). In ceramic studies, little attention has been paid to the role of the consumer, the various contexts in which the consumption of imported pottery took place and the ways in which such pottery was incorporated into the material culture of the recipient societies.

The so-called "Aegean" repertory from the Iron I period in the Northern Levant does not include all forms of ceramic vessels. Particularly noticeable is the general dearth of forms associated with the preparation of food. Instead, "Aegean" influence is evident in a repertory of vessel forms that, during the Late Bronze Age, were associated with drinking (cups – CLASS 188, jugs, juglets, flasks – CLASS 102, kraters for mixing wine – CLASS 052, chalices) and/or funerary activities (pyxides – CLASS 103, and stirrup jars – CLASS 104, for the storage of precious oils and scented unguents – see below) (van Wijngaarden 2002, 13-15).

Imported Late Bronze Age Aegean pottery has been found at over 100 sites in the Levant (Hankey 1967; 1993, 105-107) and appears to have played an important role in drinking and funerary rituals (van Wijngaarden 2002). During the Iron I period, however, the Aegean-styled vessels were primarily of local manufacture (Badre 1983; Dornemann *pers. comm.*; *contra* Bonatz 1998) and are found in significantly lower quantities. Nevertheless, the style and form of the Iron I vessels resemble imports from the Late Bronze Age, when Aegean patterns of wine consumption were prominent. Hence, the early Iron Age "Aegean" influence might be best understood as a continuation of local Late Bronze Age culture, expressed differently (local manufacture) so as to adapt to a new reality (lack of access to Mycenaean production).

That the Late Bronze Age Cypro-Aegean imports were associated with drinking or funerary activities is important as it implies that local populations exercised considerable discrimination in the choices they made as to the adoption of foreign material culture. This has led some scholars (e.g. Knapp 1998, 202-204; Sherratt 1994b; Steel 2002) to propose a syncretic model of acculturation, where a cross-cultural drinking behaviour was adopted widely by Late Bronze Age elites of the eastern Mediterranean. Did Aegean drinking sets accompany shipments of wine? Once incorporated into the local setting, imported vessels would then have acquired new uses and/or meanings according to the values of the recipient culture (Howes 1996, 5; Steel 2002, 26). For instance, Mycenaean pottery was closely integrated into local patterns of consumption at Ugarit, where Aegean patterns of wine consumption formed an integral component within the centuries-old *marzeah* ritual, a kind of mourning ritual involving drinking and music (Carter 1995, 300*ff*; Yon 1987; Yon *et al.* 2000). Indeed, ritual drinking appears to have become an important component of mortuary display in many areas of the Late Bronze Age eastern Mediterranean (Carter 1995; Steel 1998, 290). Hence, the Iron I production is thought to represent a continuation of Late Bronze Age traditions, based on an Aegean model, and *not* new Aegean links (*contra* van Wijngaarden 1999, 22).

Considering the cultural continuity evident in the eastern Mediterranean (**§2.3**), there is little reason to suggest that the meaning of drinking paraphernalia and their use changed between the Late Bronze Age and Iron I period. Generally speaking, the Aegean-style pottery of the Iron I period is a continuation of Late Bronze Age drinking behaviours. The general dearth of Iron I mortuary contexts from Lebanon, where drinking sets were so prominent during the Iron II and Iron III periods, prevents a fuller investigation into the meaning ascribed to Aegean-style drinking paraphernalia within funerary customs. A number of large kraters came from Iron I cremation cemeteries from inland Northern Levant, but these burials generally lacked the accompanying jugs and small cups. While this might suggest that the krater had taken on a different significance within Iron I mortuary contexts, it is possible that mortuary drinking rituals continued, but were undertaken using bronze drinking vessels (Steel 2002; 2004).

Iron II-III – Red-Slip

The beginning of the Iron II period in the Northern Levant is conventionally associated with the disappearance of Aegean-style pottery and the advent of Red-Slip (**§2.7; Chart 7.62**). Red-Slip was not present within all categories of the database, but was generally limited to vessel forms associated with the drinking of wine or serving of food. Apart from the many small, Red-Slip fine-bowls with round bases that are known across the inland Northern Levant (most likely cups), the majority of Red-Slip drinking forms derive from mortuary contexts (**Chart 7.54**). Jugs and kraters were rare in Iron II and Iron III settlement (non-mortuary) contexts of the Northern Levant (**Chart 7.37**). Once again this introduces the question of what category of artefact might instead have performed the function of pouring and/or mixing of wine inland - bronze jugs and bronze kraters possibly?

Table 9.1: Possible Bronze/Ceramic Substitutes

CLASS	Form	Bronze Ref	Context
012	Pinched lip lamps	Matthäus 1985, Tf. 81-83 Gershuny 1985, Pl. 13.133-136	LB-Iron Cyprus LB-Iron Israel/Jordan
042	Krater	Matthäus 1985, Tf. 68	LB Cyprus
082	Trefoil jug	Matthäus 1985, Tf. 71	Iron Age Cyprus
085	Mushroom lip jug	Matthäus 1985, Tf. 70.533	Iron Age Cyprus
107	Beer jug	Matthäus 1985, Tf. 73.552	Iron Age Cyprus
108	Strainers	Matthäus 1985, Tf. 78-80	LB-Iron Cyprus
127	Cup	Matthäus 1985, Tf. 47.454-455	LB Cyprus
144	Small Bowl	Matthäus 1980, Tf. 50.430	Bronze Age Aegean
155	Cup	Matthäus 1985, Tf. 17.298-302	Iron Age Cyprus
157/182	Cup	Matthäus 1980,Tf. 49.414-417 Gershuny Pl. 1-2 Matthäus 1985, Tf. 1-16	Bronze Age Aegean LB-Iron Israel/Jordan LB-Iron Cyprus
160	Bowl	Gershuny 1985, Pl. 2.28 Matthäus1985, Tf. 18.308-311	LB-Iron Jordan LB-Iron Cyprus
184	Carinated cup	Matthäus 1985, Tf. 31.421-422	Iron Age Cyprus
185	Carinated cup	Matthäus 1980, Tf. 52.445	Bronze Age Aegean
191	Skyphos	Matthäus 1985, Tf. 49.460-464	Iron Age Cyprus

Within mortuary contexts Red-Slip drinking-sets are much more common and complete (**§7.5**). Hence, it might be argued that drinking activity within a settlement context could have been associated with metal artefacts, but that within a funerary context the society sought a ceramic substitute as opposed to removing large quantities of metal from circulation (Steel 2002; 2004). **Table 9.1** lists a number of ceramic forms that have clear bronze parallels. The close similarity of these forms with known bronze vessels from the eastern Mediterranean confirms their *skeumorphic* nature.

This apparent mimicking of metal vessels may indicate that metal artefacts were too valuable to be taken out of circulation by deposition within a mortuary context. Nevertheless, epigraphic and iconographic evidence detail the rich collection of metal vessels used by North Levantine communities of the Iron II period: texts and reliefs depict booty taken by victorious Assyrian kings that include large numbers of bronze and other metal vessels (e.g. Budge 1914, Pl. XX.b; King 1915, Pls XIV, XXVI-XVIII, XXXI-XXXIV; Smith 1938, Pl. XLVII; Yamada 2000, 225-272).

Iron II-Persian Period – Greek Imported Pottery

Greek pottery was an important commodity of the later Iron Age in the Northern Levant (Bonatz 1993; Collombier 1987; Crielaard 1999a; 1999b). Like with the earlier Aegean-styled pottery, these imports were dominated by drinking wares: e.g. two handled *skyphoi* (CLASS 191); amphorae decorated with vine-leaves (CLASS 058); square-handled kraters (CLASS 043) (Crielaard 1999b, 281). Were these vessels replacing the old Aegean styles? The importance of these vessels has been over-emphasised in a number of publications (e.g. Riis 1970), usually with regard to their chronological value (**§2.5**). Their functional significance, however, is often over-looked. These Greek wine-drinking sets might represent the importation of ceramic vessels as products, or the importation of the Greek-style of wine-drinking (*symposia*). Regardless, these vessels are indicative of continued importance given to the drinking of wine. The low numbers of such imports (except at al Mina) argue for their specialised and symbolic use by elites, who might have had privileged access to shipments of wine or wine-drinking equipment from Greece. Indeed, these vessels may have been more than just an attractive drinking cup symbolising a Greek style of drinking, but may have come to imply the drinking of actual Greek wine, and the status and symbolism that accompanied it.

Unguent Containers

As discussed above, the so-called "Aegean" ceramic repertory of the Iron I period of the Northern Levant was restricted to wine drinking sets and to small, spouted vessels (CLASS 103 - pyxides, CLASS 104 - "stirrup jars") which, during the Late Bronze Age, were associated with the storage of precious oils and scented unguents (Hamilakis 1996; Steel 2002, 43). The Iron Age examples appear to have been remnants of the Late Bronze Age fascination with Aegean styles and products (Steel 2002, 32). While these small vessels were primarily found in non-mortuary contexts dating from the Iron I period (**Seriation 1; Charts 7.46-47**), possibly due to a lack of Iron I mortuary contexts on the Mediterranean coast, in the Late Bronze Age unguent containers held clear association with funerary ritual. At Late Bronze Age Tell el-Ajjul, for instance, the Mycenaean "stirrup jar" was found foremost in mortuary contexts (Steel 1998, 294-295; 2002, Figs 5-6), which reflects its original use as a common funerary gift in indigenous Late Bronze Age Aegean contexts (Mountjoy 1993, 127). During the Bronze Age, perfumed oils were used in the Aegean to anoint the deceased and, therefore, played a central role in funerary ritual (Hamilakis 1999, 47, 49). Epigraphic and archaeological evidence suggests that the treatment of the body with oil or libations was also an integral element of funerary ceremonies at Late Bronze Age Ugarit (Ginsberg 1969b, 154; Kinet 1981; Salles 1995, 176; van Wijngaarden 2002, 71) and possibly Cyprus (Keswani 1989, 59-60). In the IA-NL, however, the ceramic record indicates that "stirrup jars" and pyxides were only rarely being used within funerary ritual, signalling a change in the symbolism and/or functional use of these vessels (**Charts 7.54-7.55**). When these vessels did appear in the archaeological record, they were associated with non-mortuary contexts (e.g. the numerous *pyxides* from Kamid el-Loz). It is commonly accepted that the Late Bronze Age circulation of "stirrup jars" reflects the trade of Mycenaean perfumed oil (Baker 2006, 1; Steel 2002, 39), which has led to the conclusion that these vessels were closely associated with their contents. The fact that these Mycenaean-styled vessels were being made locally in the Iron Age might be indicative of limited access to the associated oils and unguents. Nevertheless, these vessels would have still symbolised the exotic products without having to actually hold them. Their importance as *signifier* had probably equalled the importance of the *signified*, so much so that the contents were no longer necessary for the desired message to be conveyed; i.e. the Aegean style was symbolic of wealth and luxury, even in the early Iron Age derivative style (van Wijngaarden 2002, 71-72). As a result, the original use of unguent containers within mortuary ritual appears to have been largely forgotten in the IA-NL. In the IA-SL, however, where pyxides and "stirrup jars" *are* found in inhumation contexts (**Charts 7.50; 7.51**), the anointing of the body with unguents and perfumed oils continued to be an important element in funerary ritual (Dayagi Mendels 1993, 130). The change in use of these vessels across the Northern Levant may be indicative of the form of burial being used. For instance, cremation was the dominant mortuary ritual in the IA-NL, which would have left no body available for anointing (**Chart 7.50 7.51; 7.52**). The northern regions of the Southern Levant instead used inhumation wherein the body could still be anointed.

During the Iron II and Iron III periods in the Northern Levant, perfumed oils and unguents appear to have been associated with small jugs (e.g. CLASS 084) and bottles (e.g. CLASS 119), though once again they were only occasionally associated with mortuary contexts (**Charts 7.54-7.55**).

Mortuary Assemblages

One fifth of all incidents in the database within this study derive from mortuary contexts (**Chart 5.1**), with a particularly large portion of these being located along the Palestinian and Lebanese coasts (**Chart 7.50**). Despite the wide variety of mortuary assemblages within the present study, archaeologists have consistently interpreted ceramic "grave goods" as passively reflecting different aspects of essentially two phenomena; first, as the treasured possessions of the dead intended for their use in the afterlife; second, as faithful reflections of the status of the dead person according to a regular system of rules (e.g. Aubet 2004b, 449*ff*; Baker 2006, 1; Doumet 1982; Doumet-Serhal 2003a; Keswani 2004, 6-7; Saidah 1966; 1977; see Campbell 1995, 29-30). While these approaches might attempt to isolate the symbolic meaning of material culture, they ignore the active role that mortuary assemblages play within the realm of the *living*, incorporating statements of those who use the burial of a group member consciously or subconsciously to proclaim their own social identity and place within society (Morris 1987, 38-42; Parker Pearson 2000, 3). From this perspective, the goods are viewed as carrying current social significance and not only as products for the deceased or reflections of social organisation. Mortuary assemblages, therefore, are likely to embody a whole complex of ideas about social structures (Morris 1987, 38-39), often not coherently formulated and even potentially contradictory.

When working with mortuary data, it should be remembered that mortuary assemblages may only represent a portion of the population, and that it is possible a section of the community was socially excluded from using archaeologically detectable types of burial (Dickinson 2006, 174-175; Morris 2000, xxviii; *contra* Snodgrass 1980b, 21). Furthermore, the "grave" itself preserves only the material residue of burials rather than the totality of rituals associated with the funeral (Hall 1997, 130; Parker Pearson 2000, 5). Rituals of mourning, funerary procession, actual burial, secondary rituals around the deceased, and the observances at the grave on later occasions, could all well have presented a potent arena for the enactment of social distinctions than the grave's layout and contents. The location of the burial in relation to the inhabited landscape and to other graves may also have held considerable significance (Dickinson 2006, 178).

Drinking within Mortuary Contexts

Within mortuary contexts of the IA-NL, there is a clear emphasis on ceramic forms associated with wine drinking (e.g. kraters, jugs, and small bowls). During the Iron I period, few mortuary assemblages have been identified (**Chart 7.51**). Nevertheless, there is a clear difference in the early Iron Age between the cremation cemeteries of inland Northern Levant (e.g. Jerablus, Hama) and the inhumation cemeteries of the coast (e.g. Khalde, Akhziv).

At the inland sites of Jerablus and Hama the emphasis was on kraters and urns, both of which were used to contain the cremated remains of the deceased. Both forms have held associations with wine drinking; the krater for mixing, the urn for storing. These cinerary containers were usually decorated in a mix of Monochrome motifs that find parallels within Aegean (e.g. wavy line – *cf.* Bonatz 1993, 134-135; Buhl 1983, Fig. 16.281; Dikaios 1969, Pls 69.40; 76.11-13, 15, 17) and/or Anatolian contexts (e.g. full-pointed stag – *cf.* Akurgal 1955, Figs 1-9; Fugmann 1958, Fig. 188(5A842); Woolley 1939b,

Pl. 13.11). What is most interesting, however, is the general lack of other wine-drinking equipment within the ceramic repertory of inland cremation contexts; jugs, juglets, flasks, and cups are rare within the mortuary assemblages from both sites. Did these communities use drinking sets made from materials other than pottery? Was wine drinking not an important component of funerary ritual at these sites? Wine-drinking sets simply may have not been deposited in cremation burials because of cultural or economic reasons. Regardless, this is in sharp contrast with the Iron II and Iron III cremation contexts of the coast, where wine-drinking sets were an important consideration. Had the kraters and urns, like the "stirrup jars" and pyxides on the coast, taken on a symbolic importance different to their original purpose?

No Iron I cremation cemeteries are known from the coastal regions (**Chart 7.51**; though Tell Sukas has published evidence of Late Bronze Age cremation – Riis 1961, 140-141). Instead, inhumation was the rule at Khalde and Akhziv, where small liquid containers/pourers ("pilgrim" flasks and dipper juglets) and serving bowls were found in abundance during the Iron I period. Hence, the emphasis was apparently on both wine *and* food. Funerary meals are known in the Levant from epigraphic and pictorial evidence from throughout the Bronze Age (Pinnock 1994, 21-24; Pl. VI-IX). Ugarit's archives make frequent mention of the *marzeah*, a kind of ancestral mourning ritual involving drinking and music, as well as food (Carter 1995, 300*ff*; Ginsberg 1969a; 1969b; Healey 1995, 189-190). The *marzeah* was a long-established tradition in the Levant, evident from the Late Bronze Age to the late Iron Age and into the Roman period (Carter 1995, 303-304). These funerary meals were not only associated with the burial event, but were periodically "re-enacted" as a means for remembering and connecting with the deceased. The locale of the ritual is not clear, but the high concentration of serving-bowls and dipper juglets in single burial contexts at Akhziv suggests it was undertaken near the burial/in the tomb, with the accoutrements of the meal being deposited with the deceased. The symbolism of these rituals is also not clear. Was the funerary meal considered a form of nourishment for the living and departed, wherein the link between the two could be emphasised? Funerary stele from Nayrab and Zincirli depict the deceased partaking of the funerary meal (Dion 1997, Figs 11, 16). Or was it a mnemonic device for refreshing memories of the dead within the minds of the living? (Chesson 2007, 122). Nevertheless, the communal nature of these secondary rituals, and the frequency of their repetition would have provided community members with ample opportunities to assert, negotiate, or undermine different social structures, some of which may have had little real relevance to the deceased. In other words, the deceased continued to act as social agent within the community of the *living*, long after their departure from that community.

During the Iron II and Iron III periods, there is little discernible change within the cremation assemblages of the inland Northern Levant. Within the inhumation assemblages of the North Levantine coast, however, the Iron II period witnessed an increase in jugs, kraters, cups, flasks, and bowls, all indicating a continued emphasis on the role of wine *and* food in inhumation ritual (**Chart 7.57**), and the occasional appearance of small unguent jugs and bottles. In addition to this continuity of earlier patterns, mortuary ritual of the North Levantine coast took on a new character. During the Iron II period, cremation and mixed-use cemeteries appeared on the Mediterranean coast (**Chart 7.52**). With the advent of cremation in this area came a distinct ceramic assemblage indicative of drinking activities (**Charts 7.56; 7.58**). This "funerary kit", as Baker (2006) would term it, is greatly standardised and rigidly adhered to within cremation contexts from the Lebanese coast. The "funerary kit" is clearly visible in the Cluster Analysis dendrogram for mortuary assemblages (**Dendrogram 8.8; CD/clusters/mortuary**). The distinct "funerary kit" was comprised of Red-Slip and undecorated vessels that appear to be mimicking bronze drinking sets (Dayagi-Mendels 1999, 59); essentially kraters, trefoil-lip jugs, mushroom-lip jugs, and cups (**Chart 7.58**). Moreover, few of the "funerary kit" forms commonly appear in non-mortuary contexts; it seems that they were almost exclusively for the purpose of funerary ritual. This is an important point. It implies that the bronze drinking-sets were too important or expensive to remove from circulation and use, and instead each burial was provided with a ceramic equivalent. Hence, cremation ritual was replicating the symbolic importance of drinking wine with bronze vessels, but through the use of a much more readily available medium, pottery. As a result, a greater portion of the community was able to be associated with this elite symbolism. The rigid standardisation of the "funerary kit" suggests that each individual received equal funerary treatment, as though something other than persona, rank, or status was being conveyed (Baker 2006). But drinking within cremation ritual of the Lebanese coast was not just an important consideration, it was essential. This also implies that cremation ritual was predictable and, therefore, a potent arena for the maintenance and negotiation of social structures (Baker 2006, 5). In contrast to inhumation burials, cremation ritual was likely a single event; the ceramic assemblage bears little evidence for secondary rituals, as associated with funerary meals (*ibid*, 7). Nevertheless, cremation ritual may have been a long, drawn-out affair entailing the preparation of the body, actual cremation, deposit of cremated remains, drinking rituals, deposit of "funerary kit", and sealing of the grave. Despite the lack of secondary rituals, there was clearly enough opportunity within cremation ritual to transfer the "life" of the deceased (as it was deemed it should be selectively remembered, or forgotten) into the communal consciousness of the living (Chesson 2007, 109-110). The presence of funerary stelae by the burial would have served as a mnemonic aid for this communal memory (Gras *et al*. 1991; Sader 1991). The cremation of the individual meant that the body was discarded and thereafter inaccessible. Hence, cremation funerary ritual was the sole means for institutionalising memories of the deceased.

"Funerary kits" were also known from the Bronze Age Levant (though different in detail), when food, drink and, unguents were important components of mortuary ritual (Baker 2006, 1). In fact, elite drinking associated with a cult of the dead appears in various forms amongst a number of Bronze Age cultures; Anatolia, the Levant, Egypt, the Aegean. It seems that the drinking "funerary kit" of the Iron Age is essentially an extension of earlier traditions. The population was apparently employing well-organised and time-honoured funerary traditions that were prescribed by social convention, ensuring it was a medium for the negotiation of existing social structures.

This drinking "funerary kit" is also apparent in inhumation and mixed-use cemeteries of the same period, but in these contexts the "funerary kit" appears amongst a wider variety of vessel forms (**Chart 7.59**). Ceramic assemblages from coastal inhumation contexts and inland Northern Levant cremation cemeteries attest to a more varied ritual, one that was open to individual interpretation.

Storage in Mortuary Contexts

Despite the strong emphasis on drinking in mortuary ritual of the IA-NL, there is another, separate ceramic category that, while not common, appears with some regularity within inhumation and mixed-use mortuary contexts. The presence of transport amphorae in mortuary contexts of the Iron II and Iron III periods is primarily attested on the Mediterranean coast (**Charts 7.57; 7.59**), while in the Persian period they are found across most of the Northern Levant (**Chart 7.53**). There is a clear typological distinction between amphorae included within inhumation burials and those found associated with mixed-use contexts (cremation burials laid inside a tomb). In the case of inhumations, torpedo amphorae (CLASSES 024; 025; 026) tend to accompany the deceased, while at sites like Ras al Bassit (a mixed-use cemetery) bag-shaped amphorae (CLASSES 018 and 019) were used to inter the cremated remains of the deceased. The ideological significance of such a distinction is not clear, though it may simply be that torpedo amphorae were too small to hold adult remains (Courbin 1993a, 13). If this is the case, we might then wonder why torpedo-amphorae were included within inhumations at all. It seems likely that the inclusion of amphorae with the deceased held some significance; provisions for the afterlife, gifts for the gods, representation of the journey to the next world.

"Assyrian" Wares

Much has been written regarding the presence of "Assyrian Palace-Ware" vessels in the Levant (e.g. Gilboa 1996; Hestrin and Stern 1973; Jamieson 1999; Schneider 1999b; Van Beek 1987). The small cups and bottles labelled as "Assyrian" in the present study (**Chapter 6**) may not all be true "Assyrian" imports but might include local imitations of these forms. Indeed, the imitation of "Assyrian Palace-Ware" is well-documented in the Southern Levant, though their date is debated – either eighth or seventh century BC (Na'aman and Thareani-Sussely 2006). These ceramic forms were primarily encountered in the late Iron II to Iron III periods (**Chart 7.22; Maps 63-64**). The earliest examples are concentrated in the north-east of the study area, though a few examples also derive from the northern areas of the Southern Levant (**Chart 7.23**). Dating the appearance of "Assyrian" vessels is difficult, however, since their presence at many sites has been taken as evidence of Assyrian ascendancy, which some scholars have then used to date the context from which they were recovered. Hence, stating that these vessels come from Iron III contexts is a circular argument. Surprisingly, no "Assyrian" vessels were recovered from the Orontes Syria, Beqa' Valley, and Lebanese coast zones (**Chart 7.23**). This is at odds with the historical narrative, which depicts a strong Assyrian presence in the Hamath province during the eighth century BC (**§2.4.3**). Furthermore, the two Persian period examples within this study (**Chart 7.22**) suggest these vessels may have continued to be used well after the historical conquest of Nineveh and the end of the Assyrian Empire (Kuhne 2002; Oates and Oates 2001, 257*ff*).

Within their original context in Assyria, "Palace-Ware" cups and bottles were associated with royal or elite drinking rituals (Adachi 1997; Oates and Oates 2001, Figs 23; 84; 158). Hencc, thc presence of these and similar vessels (imitations) in the Levant might reflect their use for the emulation of elite drinking behaviour.

9.3 Complexity and Diversity in the Iron Age Northern Levant

Current reconstructions of the ceramic record for the IA-NL seem inadequate. Through an emphasis on chronology, ethnic identity and geographic patterning, archaeologists have ignored the significant diversity and complexity within the archaeology. A literal reading of the historical narrative has essentially "overdetermined" the archaeology. As the patterning of ceramics indicates, social identities run much deeper than current political/ethnic interpretations allow. Indeed, cultural connections occur on many different levels; economic, domestic, ritual, communal, geographic, technological, symbolic, ideological, etc. This implies that we must think in terms not of a homogenous Iron Age "culture" but of an Iron Age world that encompassed the coexistence of diverse communities and lifestyles. Rather than resist the complexity within the archaeological record, an approach that views material culture as a multi-faceted, dynamic agent of social construction (ideological view of individual and communal identity), not as a passive reflection of social organisation (hierarchical communal relationship), should reveal the manner by which pottery was drawn upon in the construction of a range of social identities.

Pottery is ultimately a cultural resource that can be actively employed within a range of social strategies for the construction, expression, and negotiation of social identities. Highlighting the role of agency and seeing the activities of daily life as social practice has important implications for our understanding of the IA-NL. The complex and multi-dimensional character of material culture undermines the idea that pottery directly reflects the ethnicity of peoples, historical events, and social processes. Moreover, the dynamic agency of material culture implies that it is not used arbitrarily but is appropriated to mark a range of specific identities. This concept of appropriation is particularly pertinent to understanding Aegean stylistic elements within the Iron I ceramic horizon – by no means a homogenous phenomenon.

Iron I

Much of the Iron I pottery investigated within the present study consists of forms that were largely based on the Late Bronze Age ceramic repertory (**§2.3**). While many of these forms followed their earlier uses, a number took on new meanings (e.g. pyxides) as they were used within different types of contexts. It was also noted that the majority of painted vessels in the Iron I period employed a number of Aegean-stylistic elements within their form and decoration. These Aegean-styled vessels consisted overwhelmingly of vessels that, during the Late Bronze Age, were associated with the preparation and consumption of wine. There appears to be a specific choice in the Late Bronze Age of certain vessels for this social activity. The local communities of the Northern Levant apparently exercised a degree of discernment in the adoption of material culture. The catalyst for change, therefore, was not external, as current reconstructions suggest, but internal: specific ceramic forms were selected for specific socio-cultural reasons. It would seem that the drinking of wine with the aid of Aegean-styled implements was very important in the construction of identity during the late second millennium BC. Hence, any symbolism associated with Aegean-style drinking paraphernalia in the Iron I period was probably derived from the Late Bronze Age perception of imported Mycenaean drinking-sets and the manner in which they were used for elite display. This supports the notion that the Iron I period was essentially a sub-Late Bronze Age (**§2.7**).

In contrast to the decorated drinking wares, serving bowls of the Iron I period of the Northern Levant were inconspicuous. Yet, undecorated serving wares do not imply a passive role in social discourse; the daily meal, its preparation, presentation, and consumption, can be seen as one of the key areas in which familial structures manifest themselves and are consolidated (Douglas 1975). The presence of decorated Iron I bowls from the Southern Levant implies a greater symbolic importance associated with the presentation of food in those communities.

Cooking-pots from the Iron I period experienced wide distributions (relative to the Iron II period) of only a few forms. The CLASS 008 cooking-pot was the most-restricted in distribution, being limited to the northern areas of the Southern Levant. Other cooking–pots, however, covered much of the Northern Levant. Transport amphorae and pithoi of the early Iron Age experienced starkly different geographic concentrations, though slightly less-well defined than in the later Iron Age.

Amongst the mortuary contexts of the Iron I period two patterns are discernible: food and drink were both important within the inhumation assemblages of the coast and Southern Levant; while the cremation assemblages of the inland Northern Levant were loosely associated with wine-drinking equipment. Apparently, the coastal emphasis on funerary meals was a continuation of Late Bronze Age funerary traditions. Interestingly, the small unguent containers, common amongst Late Bronze Age inhumation assemblages, lost their funerary association in the Iron I period. Though still resembling their original Aegean forms, the small "stirrup jars" and *pyxides* had been re-appropriated for non-mortuary purposes. Inland cremation assemblages contained some Aegean elements associated with drinking rituals (e.g. kraters and urns), but jugs and cups were generally missing from most burials. Hence, if drinking was important here, it was undertaken with non-ceramic drinking implements, or the burial of such implements was not important, or infeasible.

Iron II and Iron III Periods

The Iron II and Iron III periods across the inland Northern Levant are associated with the advent of Red-Slip vessels (**§2.7**), the majority of which were either small drinking cups, or were associated with the serving and presentation of food. Hence, meals had become a key arena for the reproduction and negotiation of social structures. The lack of decorated jugs and other wine-drinking paraphernalia in the inland Northern Levant indicates that drinking rituals were either unimportant there, or being enacted with the use of bronze drinking sets, as the Red-Slip cups might imply. The presence of "Assyrian" forms in the Iron III period in the north-eastern region of the study area might further indicate that elite emulation was being undertaken through pottery drinking sets.

It is within the context of this period, when food became an important social agent across the inland Northern Levant, that we might also understand the presence of numerous pithoi in these areas; the accumulation, storage, and control of food was just as potent a symbol as its consumption. Also, cooking-pots took on more regionally distinct/localised forms during this period, with more forms attested and smaller, well-defined distribution patterns evident. Food appears to have become an important social agent at many different levels; acquisition, storage, manipulation, allocation, preparation, cooking, presentation, serving, and consumption.

In contrast to the inland Northern Levant, the coast regions produced numerous complete ceramic drinking-sets of kraters, jugs, and cups. Many of these drinking-sets came from mortuary contexts on the coast, which for the first time in the Iron Age included cremation and mixed-use burials. These vessels made significant use of Red-Slip, which again appears to be mimicking bronze drinking vessels. The implication is that bronze drinking sets were being used within settlement contexts, but within mortuary contexts less-expensive ceramic drinking-sets were employed for the conspicuous disposal associated with the cremation/burial of the deceased. Nevertheless, the frequency by which drinking sets were found within mortuary contexts implies that drinking was an important social tool. Indeed, the ceramic assemblages found within coastal cremations are highly standardised, implying the presence of a standardised funerary ritual within which the living community could negotiate a prescribed range of social structures. In contrast, the inland cremation assemblages were more varied, which implies a less "prescribed" funerary event.

In short, the Iron II and Iron III periods of the inland Northern Levant are associated with a greater variety in local cooking wares, an increase in symbolism within serving wares, and an increase in imported Greek, Cypriot, and Assyrian vessels. All of these changes imply that a greater range of ceramic tools were now being used, which points to important changes in cuisine and how a meal would be cooked, served, presented, eaten, enacted, and "embodied". The greater distinction between methods of cooking and the new emphasis on serving – in other words the new style(s) of eating –represent a distinct shift in social structures and the way these were being negotiated.

Persian Period

While geographic factors of the earlier periods continued to be felt in the Persian period, there was a distinct change in the ceramic record; ceramic vessels became plainer implying that food and drink were less important social tools. While bowls constituted a high percentage of Persian period assemblages, they tended to be undecorated, suggesting a more mundane, domestic arena for the symbolism of food and meals, possibly associated with private/domestic identities (**Charts 7.62; 7.69**). Small juglets and bottles, previously unseen, were also an important element in this period, implying that perfumed oils and unguents once again became important (*cf* **Charts 5.7; 7.24**). In sharp contrast to the Iron II-III periods, a lack of ceramic drinking wares implies that drinking was a less important social activity in the Persian period, or that these activities were being undertaken solely through non-ceramic means. Alternatively, a change in drinking behaviours away from elite emulation might signal a loss of visible elites to emulate.

There is also significant change in mortuary practices in the Persian period, as cremation was largely abandoned in preference for inhumation. Within the inhumation cemeteries, amphorae were becoming increasingly common, as too were small unguent bottles and juglets, as well as undecorated serving bowls (**Chart 7.53**). Nevertheless, there is little distinction between mortuary and non-mortuary assemblages in the Persian period. Emphasis was apparently on food consumption (funerary meal) and the anointing of the deceased. Within this context, funerary ritual was a periodically enacted remembrance of the deceased.

Summary

The ceramic evidence of the Iron I period suggests that communal identity was being negotiated within the arena of wine drinking, while food appears to be reserved for the construction of individual and familial identities. Mortuary practice of the Mediterranean coast was exclusively inhumation, while the inland Northern Levant was characterised by cremation. During the Iron II and Iron III periods there was a distinct shift across the inland Northern Levant toward the importance of food (manipulation, presentation, consumption) in the negotiation of identity, while on the coast cremation burials were intricately connected to wine drinking rituals and inhumation burials were associated with occasional funerary meals. Cremation continued to be the predominant burial rite across the inland Northern Levant, though more-variable than that of the coast. The Persian period saw a return to inhumation practices and an apparent abandonment of ceramic drinking sets; food too becomes less important within a social context. The ceramic record of the Persian period apparently places little emphasis on communal identities.

9.4 Historical Considerations

The above discussion has made occasional reference to political monuments (e.g. Assyrian stelae) and documentary sources (e.g. the *marzeah* from Ugaritic texts) in order to understand the function or use of material culture. It largely tried to avoid referencing conventional political histories of the period precisely because the patterns in the archaeological record were interrogated as evidence for social behaviour rather than historical processes. That is not to say that the historical record cannot aid archaeological investigations. On the contrary, documentary sources provide an important depth to reconstructions of Iron Age society, but the archaeological record and text are *different* types of artefact. Each should be studied separately before comparing the two for important points of reference. So, how does the above interpretation compare with the conventional historical narrative? Are there any connections between the ceramic patterns within this study and the historical narrative?

Some of the ceramic patterns described above roughly follow the broad regional patterns evident in the historical narrative. For instance, during the Iron II period large storage pithoi were generally restricted to the inland regions while transport amphorae were found concentrated along the coast. These two distributions appear to coincide with the conventional placement of Phoenician merchants on the coast and Aramaean tribal-states across the inland regions. This two-region model has been confirmed in the research of Mazzoni and of Lehmann (**§4.2**), and is largely confirmed in the Correspondence Analysis above (e.g. **Chart 8.1**). Nevertheless, this model provides a generalised explanation of material culture patterning, and as a result apparently overlooks the complexity and diversity of Iron Age societies across the Northern Levant.

Despite the apparent homogeneity of the two broad regional units, there is significant complexity in the ceramic record of the Iron Age. A number of subtleties in the data suggest the presence of a diverse range of behaviours and intricate knowledge systems that cut across traditional boundaries. These patterns are evident in the well-defined distribution of Iron Age cooking-pots, as well as the broad distributions of kraters and bowls, amongst other forms. This diversity was also evident in the Correspondence Analyses of the dataset (**Chapter 8**), which confirmed the presence of well-defined, localised patterns within the broader regional patterns (**Chart 8.4**). Correspondence Analysis also revealed different tightly-clustered groups of sites according to different mortuary behaviour, none of which coincide with general historical patterns (**Chart 8.6**). Furthermore, the inland distribution of the "Aramaean" pithos includes the Beqa' Valley and Tel Dan, areas outside the traditional "Aramaean" polities. Under closer scrutiny, the broad socio-political units conventionally derived from a literal reading of documentary sources do not adequately explain the level of complexity and diversity evident in the ceramic record of the IA-NL. It seems that the conventional historical narrative has largely *overdetermined* interpretations of the archaeological record. In other words, material culture patterning appeared to be the result of broad historico-political processes, but may well be the result of a combination of less apparent effects. As a result, other possible factors have been largely overlooked as explanations for material culture patterning. Furthermore, an unconventional "reading" of the historical texts may instead reveal a level of social complexity akin to that depicted in the archaeological record.

9.5 Summary and Implications

The aim of this chapter has been to demonstrate the potential of alternate approaches for the study of pottery from the IA-NL. Approaching pottery from a people-centred perspective has highlighted the role of agency and the social practices that shape the activities of daily life as causal factors in material culture variability. Conventional interpretations sought only to explain material culture variability as the results of geographic, ethnic, and/or chronological factors. Conventional histories have provided only crude descriptions of the data. The implication of this alternative view is that material culture patterning can no longer be viewed as the *result* of poorly-defined external causes: instead, pottery was appropriated by communities for specific purposes. This approach has highlighted the existence of significant complexity and diversity in the ceramic record of the IA-NL and the varying means by which social identity was being negotiated through material culture. Consequently, conventional reconstructions have been shown to be inadequate. The conclusions drawn above have demonstrated the value of approaching existing archaeological data from a less-conventional perspective.

CHAPTER TEN

Conclusions

The aim of this study was to interrogate the ceramic record of the Iron Age Northern Levant without relying upon existing historical preconceptions. It was hoped that this approach would then assess the compatibility of these two records, material culture and text. Foremost, there was a desire to identify patterns in the archaeological record aside from a conventional historical interpretation. For this purpose, the first section undertook a deconstruction of current reconstructions of the Iron Age Northern Levant and an assessment of the applicability of the historical interpretative framework within archaeological research. This exercise revealed the fragile foundations of Iron Age chronology and highlighted a number of key research issues for this period; e.g. the dating of the Hama destruction, the appearance of Red-Slip pottery in the Northern Levant, the linking of early Iron Age Aegean-style pottery with the "Sea Peoples". Part of the problem appears to lie in the lack of a universal definition of the Iron Age Northern Levant as based upon archaeological evidence. Instead, definitions have been derived from the conventional historical narrative, which has resulted in an archaeological method and practice that views material culture largely as the expected product of assumed historical processes. An alternative perspective was needed.

While the first section of this study concluded that a reassessment of the archaeology was needed, the second section set-out the parameters for this exercise. This involved the creation of a comprehensive ceramic database for the Iron Age. Hence, a region-wide typology was constructed, and the dataset categorised accordingly. The compiling of the database, however, highlighted a number of limitations with the available data. Foremost was the lack of quantified data for the majority of published pottery. Consequently, the limited nature of presence/absence information prevented a statistical analysis of the ceramic data.

The third section of this study identified and described the many patterns in the ceramic data. A number of analytical techniques were used for this purpose; some that systematically sought specific relationships between categories of data (e.g. Correspondence Analysis), and others that isolated more general trends (**Chapter 7**). The combined result was an interrogation of the archaeological record on both a micro and macro level to allow the patterning in the data to inform the interpretation presented in the fourth section.

The fourth section of this study discussed the patterns in the ceramic data and their implications for an alternative reconstruction of the Iron Age Northern Levant. More specifically, a number of patterns in the data cross-cut the broader ceramic regions depicted in the conventional historical narrative. This demonstrated that current reconstructions of the Iron Age are largely overdetermined by a literal reading of the documentary sources. While the conventional historical narrative has been considered the ostensible cause behind material culture patterning, it has been shown that ceramic trends are the combined result of a wide range of different causes, few of which can be considered wholly the product of broad socio-political events.

This study has highlighted a number of key issues in need of address for the future study of the Iron Age of the Northern Levant. We might consider the most important of these is a secure chronology. At the moment, the periodisations and site sequences of the Northern Levant are largely reliant on political histories and the unreliable witness of inter-regional pottery comparison. What is needed is closer engagement with material culture in all its messy diversity, as advocated throughout this study, combined with programs of reliable scientific dating programs for all excavations. Only through a more systematic ordering of site data will a reliable regional sequence become available. This should also be accompanied by a revision of the existing structures and the identification of assumptions upon which the current chronology rests. Probably the most important, and therefore urgent, Iron Age sequence in need of thorough revision is that of Hama. The present study has highlighted sufficient reason to doubt the attribution of the Hama E destruction to Sargon II, and has identified some evidence that seems to indicating the "royal quarter" might date considerably earlier than the eighth century BC. If Hama E was revised, this would have a flow on effect for other important issues; the advent of Red-Slip, the length of the Iron I "dark age", and Greek chronology – to name only a few.

The continued use of the term "Iron I" to denote what is essentially a Sub-Late Bronze Age material culture is problematic. I propose that in due course the Iron I period be given a more meaningful terminology: i.e. Late Bronze Age III or Sub-Late Bronze Age. This revision does not ignore material culture, as in James' (*et al.* 1991) scheme, but proposes a chronological framework that better represents it. By extending the Late Bronze Age, the beginning of the "Iron Age" would shift to when iron became the preferred working metal, and substituted rather than supplemented the use of bronze (**Table 10.1**).

Table 10.1: Comparison of Chronologies for Bronze-Iron Age Transition

Ceramic Material	James (*et al.* 1991)	Possible revision
Late Bronze (1600-1200)	Late Bronze (1600-950)	Late Bronze
Iron I (1200-950)	Ignored	Sub-Late Bronze
Iron II (950-720)	Iron I (950-720)	Iron I

This study is presented as a platform from which future research can consider more meaningful interpretations of the archaeological data and the study of the Iron Age Northern Levant can reveal its own local and individual character. Also, archaeologists currently working in the discipline are encouraged to reconsider the role of ceramics versus the conventional historical narrative in reconstructions of the Iron Age Northern Levant.

BIBLIOGRAPHY

Abdi, K. 2001. Nationalism, Politics, and the Development of Archaeology in Iran. *American Journal of Archaeology* 105(1):51-76.

Abel, F. M., and A. Barrois. 1928. Fouilles de l'Ècole archéologique française de Jérusalem, effectuées à Neirab du 12 septembre au 6 novembre 1927. *Syria* 9:187-206, 303-319.

Abou Assaf, 1978-1979. Tall Ain Dara. *Archiv fur Orientforschung* 26:147-148.

Abou Assaf, A. 1985. "'Ain Dara," in *Ebla to Damascus: Art and Archaeology of Ancient Syria: An Exhibition from the Director-General of Antiquities and Museums, Syrian Arab Republic*. Edited by H. Weiss, pp. 347-350. Washington DC: Smithsonian Institution Travelling Exhibition Service.

Abou Assaf, A. 1989/1990. Tall Ain Dara. *Archiv fur Orientforschung* 36/37:203-207.

Abou Assaf, A. 1990. *Der Temple von 'Ain Dara. Damaszener Forschungen Band 3*. Mainz am Rhein: Philipp von Zabern.

Abou Assaf, A. 1996. Die Kleinfunde aus 'Ain Dara. *Damaszener Mitteilungen* 9:47-111.

Abou Assaf, A. 1997 "Ain Dara" in *The Oxford Encyclopedia of Archaeology in the Near East. Vol. 1*. Edited by E. M. Meyers, pp. 33-35. Oxford: Oxford University Press.

Adachi, T. 1997. The Fine Carinated Bowl in the Iron Age. *Bulletin of the Ancient Orient Museum* 18:41-55.

Adams, W. Y. 1968. Invasion, Diffusion, Evolution? *Antiquity* 42:194-215

Adams, W. Y. 1979. On the Arguments from Ceramics to History: A Challenge Based on Evidence from Medieval Nubia. *Current Anthropology* 20:727-734.

Aharoni, Y. 1972. The Stratification of Israelite Megiddo. *Journal of Near Eastern Studies* 31:302-311.

Aharoni, Y. 1978. *The Archaeology of the Land of Israel*. Philadelphia: Westminster Press.

Aharoni, Y., and R. Amiran. 1958. A New Scheme for the Sub-Division of the Iron Age in Palestine. *Israel Exploration Journal* 8:171-184.

Aharoni, Y., Y. Yadin, and Y. Shiloh. 1993. "Megiddo," in *The New Encyclopaedia of Archaeological Excavations in the Holy Land*. Edited by E. Stern, pp. 1003-1023. New York: Simon and Schuster.

Ahituv, S. 1984. *Canaanite Toponyms in Ancient Egyptian Documents*. Jerusalem: Magnus Press, Hebrew University.

Åhlström, G. W. 1993. "Pharaoh Shoshenq's Campaign to Palestine," in *History and Traditions of Early Israel*. Edited by A. Lemaire and B. Otzen, pp. 1-16. Leiden.

Akkermans, P. M. M. G., and G. M. Schwartz. 2003. *The Archaeology of Syria: From Complex Hunter-Gatherers to Early Urban Societies (ca. 16,000-300 BC). Cambridge World Archaeology*. Cambridge: Cambridge University Press.

Akurgal, E. 1955. *Phrygische Kunst*. Ankara: Türk Tarih Kurumu Basimevi.

Akurgal, E. 1962. *The Art of the Hittites*. London: Thames and Hudson.

Albright, W. F. 1922. The Excavations at Ascalon. *Bulletin of the American Schools of Oriental Research* 6:11-18.

Albright, W. F. 1932. The Chalcolithic Age in Palestine. *Bulletin of the American Schools of Oriental Research* 48:10-13

Albright, W. F. 1933. The Excavations at Tell Beit Mirsim IA: The Bronze Age Pottery of the Fourth Campaign. *Annual of the American Schools of Oriental Research* 13.

Albright, W. F. 1934. Excavations in 1933 in Palestine, Transjordan, and Syria. *American Journal of Archaeology* 38(2):191-199.

Albright, W. F. 1935. A Summary of Archaeological Research during 1934 in Palestine, Transjordan, and Syria. *American Journal of Archaeology* 39(1):137-148.

Albright, W. F. 1936. Archaeological Exploration and Excavation in Palestine and Syria, 1935. *American Journal of Archaeology* 40(1):154-167.

Albright, W. F. 1937. Further Light on the History of Israel from Lachish and Megiddo. *Bulletin of the American Schools of Oriental Research* 68:22-26.

Albright, W. F. 1938. The Excavations at Tell Beit Mirsim II: The Bronze Age. *Annual of the American Schools of Oriental Research* 17.

Albright, W. F. 1941-43. The Excavations at Tell Beit Mirsim III: The Iron Age. *Annual of the American Schools of Oriental Research* 21-22.

Albright, W. F. 1949. *The Archaeology of Palestine*: Penguin Books.

Albright, W. F. 1950. Some Oriental Glosses on the Homeric Problem. *American Journal of Archaeology* 54(1):162-176.

Albright, W. F. 1953. Correspondence with Professor Einar Gjerstad on the Chronology of 'Cypriote' Pottery from Early Iron Levels in Palestine. *Bulletin of the American Schools of Oriental Research* 130:22.

Albright, W. F. 1957. *From the Stone Age to Christianity: Monotheism and the Historical Process*. Baltimore.

Albright, W. F. 1964a. *History, Archaeology and Christian Humanism*. New York: McGraw Hill.

Albright, W. F. 1964b. The Beth Shemesh Tablet in Alphabetic Cuneiform. *Bulletin of the American Schools of Oriental Research* 173:51-53.

Albright, W. F. 1975. "Syria, the Philistines, and Phoenicia," in *History of the Middle East and the Aegean Region c. 1380-1000 B.C., Cambridge Ancient History II.2 (First Edition)*. Edited by I. E. S. Edwards, C. J. Gadd, N. G. L. Hammond, and E. Sollberger, pp. 507-536. Cambridge: Cambridge University Press.

Albright, W. F., and N. Glueck. 1937. Archaeological Exploration and Excavation in Palestine, Transjordan, and Syria during 1936. *American Journal of Archaeology* 41(1):146-153.

Alt, A. 1944. "Der Rhythmus des Geschichte Syriens und Palästinas im Altertum," in *Beiträge zur Arabistik, Semitistik und Islamwissenschaft*. Edited by R. Hartmann and H. Scheel, pp. 284-306. Leipzig.

Amadasi Guzzo, M. G. 2001. Une empreinte de sceau de Tell Afis. *Orientalia* 70/3:318-324.

Amadasi Guzzo, M. G. 2005. il frammento di stele in basalto con iscrizione. *Egitto e Vicino Oriente* 28:21-23.

Amiran, R. 1969. *Ancient Pottery of the Holy Land: From its Beginnings in the Neolithic Period to the End of the Iron Age*. Jerusalem: MAsada Press LTD.

Anastasio, S., M. Lebeau, and M. Sauvage. 2004. *Atlas of Preclassical Upper Mesopotamia. Subartu XIII*: Brepols.

Anderson, B. 1991. *Imagined Communities: Reflections on the Origin and Spread of Nationalism*. London: Verso.

Anderson, W. P. 1988. *Sarepta I, The Late Bronze and Iron Age Strata of Area II, Y*. Beruit.

Anderson, W. P. 1990. The Beginnings of Phoenician Pottery: Shape, Style, and Ceramic Technology Early Phases of the Phoenician Iron Age. *Bulletin of the American Schools of Oriental Research* 279:35-54.

Andrae, W. 1943. *Ausgrabungen in Sendschirli V. Die Kleinfunde von Sendschirli. Mittheilungen aus den Orientalischen Sammlungen Heft 15*. Berlin: Verlag von Walter de Gruyter & Co.

Andren, A. 1998. *Between Artifacts and Texts: Historical Archaeology in Global Perspective*. New York.

Arie, 2006. "The Iron Age I Pottery: Levels K-5 and K-4 and an Intra-Site Spatial Analysis of the Pottery from Stratum VIA," in *Megiddo IV. The 1998-2002 Seasons, Tel Aviv University Sonia and Marco Nadler Institute of Archaeology Monograph Series Number 24*. Edited by I. Finkelstein, D. Ussishkin, and B. Halpern, pp. 191-298. Jerusalem: Institute of Archaeology, Tel Aviv University.

Arnaud, P., E. Llopis, and M. Bonifay. 1996. Bey 027 Rapport préliminaire. *Baal* 1:98-134.

Aro, S. 2003. "Art and Architecture," in *The Luwians, Handbuch der Orientalistik 68*. Edited by H. C. Melchert, pp. 281-337. Leiden: Brill.

Artzy, M. 1994. Incense, Camels and Collared Rim Jars: Desert Trade Routes and Maritime Outlets in the Second Millennium. *Oxford Journal of Archaeology* 13(2):121-147.

Artzy, M. 2006. The Carmel Coast during the Second Part of the Late Bronze Age: A Center for Eastern Mediterranean Transshipping. *Bulletin of the American Schools of Oriental Research* 343: 45-64.

Artzy, M. 2007. "Tell Abu Hawam: News from the Late Bronze Age," in *The Synchronisation of Civilisations in the Eastern Mediterranean in the Second Millennium B.C. III*. Edited by M. Bietak and E. Czerny, pp. 357-366. Vienna: Verlag der Österreichischen Akademie der Wissenschaften.

Åström, P. 1972. *The Late Cypriote Bronze Age. The Swedish Cyprus Expedition IV.1C*. Stockholm: The Swedish Cyprus Expedition.

Aubet, M. E. 1993. *Phoenicians and the West. Politics, Colonies and Trade*. Cambridge.

Aubet, C. 1996. Bey 002 Rapport préliminaire. *Baal* 1:60-84.

Aubet, M. E. 2004a. "The Tyre-Al Bass Necropolis," in *Decade: A Decade of Archaeology and History in the Lebanon*. Edited by C. Doumet-Serhal, pp. 16-27. Beirut: The Lebanese British Friends of the National Museum.

Aubet, M. E. 2004b. "Discussion and Conclusions," in *The Phoenician Cemetery at Tyre El-Bass, Excavations 1997-1999, Baal, Hors Serie 1*. Edited by M. E. Aubet, pp. 449-466. Beirut: Ministère de la Culture Direction Générale des Antiquités.

Aubet, M. E., F. J. Nunez, and L. Trellisó. 1998-1999. The Phoenician Cemetery of Tyre-Al Bass. *Baal* 3:267-294.

Aurenche, O. 1986. Chronique Archeologique. *Syria* 63:385-415.

Bachelot, L. 1996. Burmarina, découverte d'une cité araméene. *Archéologia* 323:34-39.

Bachelot, L. 1999. "Tell Shioukh Faouqani (1994-1998)," in *Archaeology of the Upper Syrian Euphrates: The Tishrin Dam Area*. Edited by G. del Olmo Lete and J.-L. Montero Fenollos, pp. 143-162. Barcelona.

Bachelot, L., and F. M. Fales. 2005. "Introduction," in *Tell Shiyukh Fawqani 1994-1998*. Edited by L. Bachelot and F. M. Fales, pp. I-LI. Padova: Sargon Editrice e Libreria.

Bachelot, L., F. M. Fales, M. Luciani, M. Makinson, and D. Morandi. 1995. Tell Shioukh Faouqâni, la deuxième campagne de fouille. *Orient-Express* 95/3:81-84.

Bachelot, L., F. M. Fales, M. Luciani, M. Makinson, and D. Morandi. 1996. Mission Archéologique de Tell Shioukh Faouqâni. La troisième campagne de fouille. *Orient-Express* 96/3:80-84.

Bachelot, L., F. M. Fales, M. Luciani, M. Makinson, and D. Morandi. 1997. La quatrième campagne de fouilles à Tell Shioukh Faouqâni (Syria). *Orient-Express* 97:79-85.

Badre, L. 1982. "Tell el Ghassil: Tomb 1," in *Archeologie au Levant. Recueil R Saidah, Collection de la Maison de l'Orient Mediterraneen No. 12*. Edited by M. Yon, pp. 123-132. Paris: Maison de l'Orient Mediterraneen.

Badre, L. 1983. "Les Peuples de la Mer à Ibn Hani?," in *Atti del I Congresso Internazionale di Studi Fenici e Punici, Roma, 5-10 Novembre 1979, Volume Primo*. Edited by S. Moscati, pp. 203-210. Roma: Consiglio Nazionale Delle Ricerche.

Badre, L. 1990a. Tell Kazel, Syria: Excavations of the AUB Museum, 1985-1987, Preliminary Reports, Introduction. *Berytus* 38:10-15.

Badre, L. 1990b. Tell Kazel, Syria: Excavations of the AUB Museum, 1985-1987, Preliminary Reports: Area II. *Berytus* 38:55-86.

Badre, L. 1991. "Recent Phoenician Discoveries at Tell Kazel," in *Atti del 2 Congresso Internazionale di Studi Fenici e Punici, Roma, 9-14 Novembre 1987*. Edited by E. Acquaro, pp. 627-639. Roma: Consiglio Nazionale Delle Ricerche.

Badre, L. 1992. Tel Kazel 1992. *Chronique Archeologique en Syrie* 1:127-129.

Badre, L. 1997a. "Khaldeh" in *The Oxford Encyclopedia of Archaeology in the Near East. Vol. 3*. Edited by E. M. Meyers, pp. 290-291. Oxford: Oxford University Press.

Badre, L. 1997b. Bey 003 Preliminary Report; Excavations of the American University of Beirut Museum 1993-1996. *Baal* 2:6-94.

Badre, L. 1998. "Late Bronze and Iron Age Imported Pottery from the Archaeological Excavations of Urban Beirut," in *Eastern Mediterranean: Cyprus-Dodecanese-Crete, 16th-6th Cent. B.C.: Proceedings of the International Symposium, Organized by the University of Crete, Rethymnon and the Anastasios G. Leventis Foundation, Nicosia, Rethymnon, 13-16 May 1997*. Edited by V. Karageorghis and N. Stampolidis, pp. 73-83. Athens: University of Crete.

Badre, L. 2000a. "Recently Discovered Bronze Age Temples: Middle Bronze Beirut and Late Bronze Tell Kazel," in *Proceedings of the First International Congress on the Archaeology of the Ancient Near East, Rome, May 18-23, 1998. Volume I*. Edited by P. Matthiae, A. Enea, L. Peyronel, and F. Pinnock, pp. 35-54. Rome: Universita degli Studi di Roma 'La Sapienza'.

Badre, L. 2000b. "Les Premieres decouvertes Pheniciennes a Beyrouth," in *Actas del IV Congreso Internacional de Estudios Fenicios Y Punicos: Cadiz, 2 al 6 de Octubre de 1995 vol. III*. Edited by M. E. Aubet and M. Barthelemy, pp. 941-961. Cadiz: Servico de Publicaciones Universidad de Cadez.

Badre, L. 2001-2002. The Bronze Age of Beirut: Major Results. *Aram* 13-14:1-26.

Badre, L. 2003. "The Handmade Burnished Ware and the Contemporary Local Pottery from Tell Kazel," in *Sea Routes... Interconnections in the Mediterranean 16th-6th c. BC: Proceedings of the International Symposium Held at Rethymnon, Crete, September 29th-October 2nd, 2002*. Edited by N. Stampolidis and V. Karageorghis, pp. 83-99. Athens: University of Crete and Leventis Foundation.

Badre, L. 2006. Tell Kazel-Simyra: A Contribution to a Relative Chronological History in the Eastern Mediterranean during the Late Bronze Age. *Bulletin of the American Schools of Oriental Research* 343:65-95.

Badre, L., and E. Gubel. 1999-2000. Tell Kazel (Syria): Excavations of the AUB Museum, 1993-1998, Third Preliminary Report. *Berytus* 44:123-204.

Badre, L., E. Gubel, E. Capet, and N. Panayot. 1994. Tell Kazel (Syrie): Rapport Preliminaire sur les 4eme-8eme Campagnes de Fouilles (1988-1992). *Syria* 71:259-359.

Badre, L., M.-C. Boileau, R. Jung, and H. Mommsen. 2005. The Provenance of Aegean and Syrian-Type Pottery Found at Tell Kazel (Syria). *Ägypten und Levante* 15:15-47.

al-Bahloul, K., A. Barro, and L. d'Alfonso. 2005. "Area H: The Iron Age Cremation Cemetery," in *Tell Shiyukh Fawqani 1994-1998*. Edited by L. Bachelot and F. M. Fales, pp. 997-1048. Padova: Sargon Editrice e Libreria.

Bahn, P. G. Editor. 1996. *The Cambridge Illustrated History of Archaeology*. Cambridge: Cambridge University Press.

Bahrani, Z. 1998. "Conjuring Mesopotamia: Imaginative Geography and a World Past," in *Archaeology Under Fire: Nationalism, Politics and Heritage in the Eastern Mediterranean and Middle East*. Edited by L. Meskell, pp. 159-174. London.

Bahrani, Z. 2000. "The Extraterrestrial Orient: Colonizing Mesopotamia in Space and Time," in *Landscapes: Territories, Frontiers and Horizons in the Ancient Near East: Papers Presented to the XLIV Recontre Assyriologique Internationale, Venezia, 7-11 July 1997; Part III Landscape in Ideology, Religion, Literature and Art, History of the Ancient Near East/ Monographs - III/3*. Edited by L. Milano, S. de Martino, F. M. Fales, and G. B. Lanfranchi, pp. 5-10. Padova.

Baines, J. 1996. "Contextualizing Egyptian Representations of Society and Ethnicity," in *The Study of the Ancient Near East in the Twenty-First Century*. Edited by J. S. Cooper and G. M. Schwartz, pp. 339-384. Winona Lake, Indiana: Eisenbrauns.

Baker, J. L. 2006. The Funeral Kit: A Newly Defined Canaanite Mortuary Practice Based on the Middle and Late Bronze Age Tomb Complex at Ashkelon. *Levant* 38:1-32.

Balensi, J. 1985. Revising Tell Abu Hawam. *Bulletin of the American Schools of Oriental Research* 257:65-74.

Balensi, J., M. D. Herrera, and M. Artzy. 1993. "Abu Hawam, Tell," in *The New Encyclopedia of Archaeological Excavations in the Holy Land*. Edited by E. Stern, pp. 7-14. New York: Simon & Schuster.

Baram, A. 1994. A Case of Imported Identity: the Modernizing Secular Ruling Elites of Iraq and the Concept of Mesopotamian Inspired Territorial Nationalism. *Poetics Today* 15:279-319.

Baramki, D. C. 1958. A Late Bronze Age Tomb at Sarafend, Ancient Sarepta. *Berytus* 12:129-142.

Baramki, D. C. 1961. Preliminary Report on the Excavations at Tell el Ghassil. *Bulletin du Musée de Beyrouth* 16:87-102.

Baramki, D. C. 1964. Second Preliminary Report on the Excavations at Tell el Ghassil. *Bulletin du Musée de Beyrouth* 17:47-103.

Baramki, D. C. 1966. Third Preliminary Report on the Excavations at Tell el Ghassil. *Bulletin du Musée de Beyrouth* 19:29-49.

Barkay, G. 1992. "The Iron Age II-III," in *The Archaeology of Ancient Israel*. Edited by A. Ben-Tor, pp. 302-373. London.

Barnett, R. D. 1963. Hamath and Nimrud: Shell Fragments from Hamath and the Provenance of the Nimrud Ivories. *Iraq* 25:81-85.

Barnett, R. D. 1975. "The Sea Peoples," in *History of the Middle East and the Aegean Region c. 1380-1000 B.C., Cambridge Ancient History II.2 (First Edition)*. Edited by I. E. S. Edwards, C. J. Gadd, N. G. L.

Hammond, and E. Sollberger, pp. 359-378. Cambridge: Cambridge University Press.

Barnett, R. D. 1987. "Layard's Influence on British Orientalism in the 19th Century," in *Austen Henry Layard Tra l'Oriente e Venezia*. Edited by F. M. Fales and B. J. Hickey, pp. 177-182. Roma.

Barrett, J. C. 1988. Fields of Discourse: Reconstituting a Social Archaeology. *Critique of Anthropology* 7:5-16.

Barrett, J. C. 1994. *Fragments from Antiquity: An Archaeology of Social Life in Britain, 2900-1200 BC*. Oxford: Blackwell.

Barro, A. 2002. "Excavations in the Eastern Part of the Palace (Operation H)," in *Excavating Qatna, I. Preliminary Report on the 1999 and 2000 Campaigns of the Joint Syrian-Italian-German Archaeological Research Project at Tell Mishrifeh, Documents d'Archeologie Syrienne IV*. Edited by M. al-Maqdissi, M. Luciani, D. Morandi Bonacossi, M. Novák, and P. Pfälzner, pp. 111-122. Damascus.

Barth, F. 1969. *Ethnic Groups and Boundaries: The Social Organisation of Culture Difference*. Boston: Little, Brown and Company.

Bartl, K. 1995a. Some Remarks on Early Iron Age in Eastern Anatolia. *Anatolica* 21:205-212.

Bartl, K. 1995b. "Das Ende der Spätbronzezeit und das "dunkle Zeitalter" im westlichen Vorderasien," in *Zwischen Euphrat und Indus. Aktuelle Forschungsprobleme inder Vorderasiatischen Archäologie*. Edited by K. Bartl, R. Bernbeck, and M. Heinz, pp. 193-208. Hildesheim: Georg Olms Verlag.

Bass, G. F. 1987. Oldest Known Shipwreck Reveals Splendors of the Bronze Age. *National Geographic Magazine* 172/6:693-733.

Bass, G. F., C. Pulak, D. Collon, and J. Weinstein. 1989. The Bronze Age Shipwreck at Ulu Burun (Kas): 1986 Campaign. *American Journal of Archaeology* 93(1):1-29.

Batiuk, S., T. P. Harrison, and L. Pavlish. 2005. "The Ta'yinat Survey, 1999-2002," in *The Amuq Valley Regional Projects Volume 1: Surveys in the Plain of Antioch and Orontes Delta, Turkey, 1995-2002, Oriental Institute Publications 131*. Edited by K. A. Yener, pp. 171-192. Chicago: The Oriental Institute of the University of Chicago.

Bauer, A. A. 1998. Cities of the Sea: Maritime Trade and the Origin of Philistine Settlement in the Early Iron Age Southern Levant. *Oxford Journal of Archaeology* 17(2):149-168.

Baurin, M. P. 1923. La Citadelle d'Alep. *Syria* 4:180.

Baxter, M. J. 1992. "Archaeological uses of the biplot - a neglected technique?," in *Computer Applications and Quantitative Methods in Archaeology 1991, BAR International Series S577*. Edited by G. Lock and J. Moffett, pp. 141-148. Oxford: BAR Publishing.

Baxter, M. J. 1994. *Exploratory Multivariate Analysis in Archaeology*. Edinburgh: Edinburgh University Press.

Baxter, M. J. 2003. *Statistics in Archaeology. Arnold Publications in Statistics*. London: Arnold.

Beazley, J. D. 1963. *Attic Red-Figure Vase Painters*. Oxford: Clarendon Press.

Bell, C. 2006. *The Evolution of Long Distance Trading Relationships across the LBA/Iron Age Transition on the Northern Levantine Coast. Crisis, Continuity and Change - a study based on imported ceramics, bronze and its constituent metals. BAR International Series 1574*. Oxford: BAR Publishing.

Ben Ami, D. 2001. The Iron Age I at Tel Hazor in Light of the Renewed Excavations. *Israel Exploration Journal* 51:148-170.

Ben Ami, D. 2006. Early Iron Age Cult Places - New Evidence from Tel Hazor. *Tel Aviv* 33(2):121-133.

Ben Dov, R. 2002. "The Late Bronze Age 'Mycenaean' Tomb," in *Dan II. A Chronicle of the Excavations and the Late Bronze Age 'Mycenaean' Tomb, Annual of the Nelson Glueck School of Biblical Archaeology*. Edited by A. Biran and R. Ben-Dov, pp. 33-208. Jerusalem: Hebrew Union College.

Ben Tor, A. 1993a. "Hazor," in *The New Encyclopaedia of Archaeological Excavations in the Holy Land*. Edited by E. Stern, A. Lewinson-Gilboa, and J. Aviram, pp. 594-606. New York: Simon and Schuster.

Ben Tor, A. 1993b. "Qiri, Tel," in *The New Encyclopaedia of Archaeological Excavations in the Holy Land*. Edited by E. Stern, pp. 1228-1229. New York: Simon and Schuster.

Ben Tor, A. 1997. "The Yigael Yadin Memorial Excavations at Hazor, 1990-1993: Aims and Preliminary Results," in *The Archaeology of Israel: Constructing the Past, Interpreting the Present, Journal for the Study of the Old Testament Supplement Series 237*. Edited by N. A. Silberman and D. Small, pp. 62-81. Sheffied: Sheffield Academic Press.

Ben Tor, A. 1998. "The Fall of Canaanite Hazor – The "Who" and "When" Questions", in *Mediterranean Peoples in Transition: Thirteenth to Early Tenth Centuries BCE-In Honour of Professor Trude Dothan*. Edited by S. Gitin, A. Mazar, and E. Stern, pp. 456-467. Jerusalem: Israel Exploration Society.

Ben Tor, A. 2000. Hazor and Chronology of Northern Israel: A Reply to Israel Finkelstein. *Bulletin of the American Schools of Oriental Research* 317:9-15.

Ben Tor, A. 2003. Notes and News - Excavations and Surveys: Tel Hazor, 2003. *Israel Exploration Journal* 53:218-223.

Ben Tor, A. 2004. Notes and News. Excavations and Surveys. Tel Hazor, 2004. *Israel Exploration Journal* 54:230-235.

Ben Tor, A., R. Bonfil, Y. Garfinkel, R. Greenberg, A. M. Maeir, and A. Mazar. 1997. *Hazor V: An Account of the Fifth Season of Excavation, 1968*. Jerusalem: Hebrew University of Jerusalem: Israel Exploration Society.

Ben Tor, A., and D. Ben-Ami. 1998. Hazor and the Archaeology of the Tenth Century B.C.E. *Israel Exploration Journal* 48:1-37.

Ben Tor, A., and S. Zuckerman. 2008. Hazor at the End of the Late Bronze Age: Back to Basics. *Bulletin of the American Schools of Oriental Research* 350:1-6.

Ben Yehuda, N. 1995. *The Masada Myth: Collective Memory and Mythmaking in Israel*. Madison, WI: University of Wisconsin.

Ben Yehuda, N. 2002. *Sacrificing Truth: Archaeology and the Myth of Masada*. Amherst, NY: Humanity Books.

Bentley, M. 1999. *Modern Historiography. An Introduction*. London: Routledge.

Bernhardsson, M. T. 2005. *Reclaiming a Plundered Past. Archaeology and Nation Building in Modern Iraq*. Austin, TX: University of Texas Press.

Betancourt, P. 2000. "The Aegean and the Origin of the Sea Peoples," in *The Sea Peoples and Their World: A Reassessment*. Edited by E. D. Oren, pp. 297-304. Philadelphia: University of Pennsylvania.

Bettles, E. 2003a. Carinated-Shoulder Amphorae from Sarepta, Lebanon: A Phoenician Commodity and its Intra-Regional Distribution. *Archaeology and History in the Lebanon* 17:60-79.

Bettles, E. 2003b. *Phoenician Amphora Production and Distribution in the Southern Levant: a multi-disciplinary investigation into carinated-shoulder amphorae of the Persian period (539-332 BC). BAR International Series 1183*. Oxford: BAR Publishing.

Bieber, A. M., D. W. Brooks, G. Harbottle, and E. V. Sayre. 1976. Application of Multivariate Techniques to Analytical Data on Aegean Ceramics. *Archaeometry* 18(1):59-74.

Bienkowski, P. 1982. Some Remarks on the Practice of Cremation in the Levant. *Levant* 14:80-89.

Bigazzi, C. 2002. Afis 2000-2001. Area E: le strutture conservative del Ferro I. *Egitto e Vicino Oriente* 25:38-41.

Bikai, Patricia M. 1978a. The Late Phoenician Pottery Complex and Chronology. *Bulletin of the American Schools of Oriental Research* 229:47-56.

Bikai, Patricia M. 1978b. *The Pottery of Tyre*. Warminster: Aris and Phillips.

Bikai, Patricia M. 1983. "The Imports from the East," in *Palaeopahos-Skales: an Iron Age Cemetery in Cyprus*. Edited by V. Karageorghis, pp. 396-406. Konstanz: Universitätsverlag Konstanz.

Bikai, Patricia M. 1987. *The Phoenician Pottery of Cyprus*. Nicosia: Levantis Foundation.

Bikai, Patricia M. 1992. "Phoenician Tyre," in *The Heritage of Tyre. Essays on the History, Archaeology, and Preservation of Tyre*. Edited by M. S. Joukowsky, pp. 45-54. Dubuque, Iowa: Kendall Hunt Publishing Company.

Bikai, Patricia M. 1994. "The Phoenicians and Cyprus," in *Proceedings of the International Symposium, Cyprus in the Eleventh Century B.C.* Edited by V. Karageorghis, pp. 31-37. Nicosia.

Bikai, Patricia M., and Pierre M. Bikai. 1987. Tyre at the End of the Twentieth Century. *Berytus* 35:67-96.

Bikai, Pierre M. 1992a. "The Land of Tyre," in *The Heritage of Tyre. Essays on the History, Archaeology, and Preservation of Tyre*. Edited by M. S. Joukowsky, pp. 13-24. Dubuque, Iowa: Kendall Hunt Publishing Company.

Bikai, Pierre M. 1992b. "History of Excavations," in *The Heritage of Tyre. Essays on the History, Archaeology, and Preservation of Tyre*. Edited by M. S. Joukowsky, pp. 25-36. Dubuque, Iowa: Kendall Hunt Publishing Company.

Binford, L. 1962. Archaeology as Anthropology. *American Antiquity* 28:217-225.

Binford, L. 1965. Archaeological Systematics and the Study of Cultural Process. *American Antiquity* 31(2.1):203-210.

Binford, L. 1972. *An Archaeological Perspective*. New York: Seminar Press.

Biran, A. 1982. The Temenos at Dan. *Eretz Israel* 16:15-43.

Biran, A. 1989a. "The Collared-rim Jars and the Settlement of the Tribe of Dan," in *Recent Excavations in Israel: Studies in Iron Age Archaeology, Annual of the American Schools of Oriental Research 49*. Edited by S. Gitin and W. G. Dever, pp. 71-96. Indiana: American Schools of Oriental Research.

Biran, A. 1989b. The Evidence for Metal Industry at Dan. *Eretz Israel* 20:120-134 (Hebrew).

Biran, A. 1993. "Dan," in *The New Encyclopaedia of Archaeological Excavations in the Holy Land*. Edited by E. Stern, pp. 323 New York: Simon and Schuster.

Biran, A. 1994. *Biblical Dan*. Jerusalem: Israel Exploration Society.

Biran, A. 1996a. "A Chronicle of the Excavations 1966-1992," in *Dan I, Annual of the Nelson Glueck School of Biblical Archaeology*. Edited by A. Biran, D. Ilan, and R. Greenberg, pp. 7-63. Jerusalem: Nelson Glueck School of Biblical Archaeology.

Biran, A. 1996b. High Places at the Gates of Dan? *Eretz Israel* 25:55-58 (Hebrew).

Biran, A. 1996c. "Introduction," in *Dan I, Annual of the Nelson Glueck School of Biblical Archaeology*. Edited by A. Biran, D. Ilan, and R. Greenberg, pp. 1-5. Jerusalem: Nelson Glueck School of Biblical Archaeology.

Biran, A. 1999. Two Bronze Plaques and the Hussot of Dan. *Israel Exploration Journal* 49:43-54.

Biran, A. 2001. "The High Places of Biblical Dan," in *Studies in the Archaeology of the Iron Age in Israel and Jordan, Journal for the Study of the Old Testament Supplement Series 331*. Edited by A. Mazar and V. T. Mathias, pp. 148-155. Sheffield: Sheffield Academic Press.

Biran, A. 2002. "A Chronicle of the Excavations," in *Dan II. A Chronicle of the Excavations and the Late Bronze Age 'Mycenaean' Tomb, Annual of the Nelson Glueck School of Biblical Archaeology*. Edited by A. Biran and R. Ben-Dov, pp. 3-32. Jerusalem: Hebrew Union College.

Birmingham, J. 1963. The Chronology of Some Early and Middle Iron Age Cypriot Sites. *American Journal of Archaeology* 67(1):15-42.

Blaylock, S. R. 1999. "Iron Age Pottery from Tille Hoyuk, South-Eastern Turkey," in *Iron Age Pottery in Northern Mesopotamia, Northern Syria and South-Eastern Anatolia*. Edited by A. Hausleiter and A. Reiche, pp. 263-286. Munster.

Boardman, J. 1959. Greek Potters at al Mina? *Anatolian Studies* 9:163-169.

Boardman, J. 1965. Tarsus, Al Mina and Greek Chronology. *Journal of Hellenic Studies* 85:5-15.

Boardman, J. 1980. *The Greeks Overseas, Their Early Colonies and Trade*. London: Thames and Hudson.
Boardman, J. 1990. Al Mina and History. *Oxford Journal of Archaeology* 9:169-190.
Boardman, J. 2002. "Greeks and Syria; Pots and People," in *Greek Settlements in the Eastern Mediterranean and the Black Sea*. Edited by G. R. Tsetskhladze and A. M. Snodgrass, pp. 1-16. Oxford: BAR International Series 1062.
Boese, J. 1986-1987. Excavations at Tell Sheikh Hassan, Preliminary Report on the Year 1987 Campaign in the Euphrates Valley. *Les Annales Archeologiques Arabes Syriennes* 36/37:67-101.
Boese, J. 1988-1989. Excavations at Tell Sheikh Hassan. A Preliminary Report on the 1988 Campaign in the Euphrates Valley. *Les Annales Archeologiques Arabes Syriennes* 38/39:158-189.
Boese, J. 1992. Ausgrabungen in Tell Sheikh Hassan 1992. *Chronique Archeologique en Syrie* 1:40-42.
Boese, J. 1995. *Ausgrabungen in Tell Sheikh Hassan I. Vorläufige Berichte über die Grabungskampagnen 1984-1990 und 1992-1994. Schriften zur vorderasiatischen Archäologie; Bd. 5.* Saarbrücken: Saarbrücker Druckerei und Verlag.
Bonatz, D. 1993. Some Considerations on the Material Culture of Coastal Syria in the Iron Age. *Egitto e Vicino Oriente* 16:123-157.
Bonatz, D. 1998. "Imported Pottery (Area E1)," in *Tell Afis, Siria : scavi sull'acropoli 1988-1992 = The 1988-1992 excavations on the Acropolis, Ricerche di archeologia del Vicino Oriente; 1*. Edited by S. M. Cecchini and S. Mazzoni, pp. 211-229. Pisa: Edizioni ETS.
Bonatz, D. 2000. "Syro-Hittite Funerary Monuments: A Phenomenon of Tradition or Innovation?" in *Essays on Syria in the Iron Age., Ancient Near Eastern studies. Supplement 7*. Edited by G. Bunnens, pp. 189-210. Louvain: Peeters Press.
Bordreuil, P. 1982. Deux Epigraphes Pheniciennes Provenant des Fouilles de Tell Rachidieh. *Annales d'Histoire et d'Archéologie* 1:137-140.
Bordreuil, P. 1993. *Les Royaumes Araméens de Syrie. Syrie: Mémoire et civilisation.* Paris.
Bordreuil, P. 2003. A Propos des Jarres Inscrites de Tell Rachidieh. *Archaeology and History in the Lebanon* 17:52-57.
Bordreuil, P. 2004. "A Propos des Jarres Inscrites de Tell Rachidieh," in *Decade: A Decade of Archaeology and History in the Lebanon*. Edited by C. Doumet-Serhal, pp. 80-87. Beirut: The Lebanese British Friends of the National Museum.
Bordreuil, P., and A. Caquot. 1979. Les textes en cunéiformes alphabétiques découverts en 1977 à Ibn Hani. *Syria* 56:295-315.
Bordreuil, P., and D. Pardee. 1989. *Ras Shamra-Ougarit V. La Trouvaille Épigraphique de l'Ougarit 1. Concordance.* Paris: Editions Recherche sur les Civilsations.
Bordreuil, P., and D. Pardee. 1995. "L'épigraphie ougaritique: 1973-1993," in *Les Pays d'Ougarit autour de 1200 av. J.-C. Histoire et archéologie. Actes du Colloque International Paris, 28 juin - 1er juillet 1993, Ras Shamra-Ougarit XI*. Edited by M. Yon, M. Sznycer, and P. Bordreuil, pp. 37-32. Paris: Éditions Recherche sur les Civilisations.
Bounni, A. 1991. "La Syrie, Chypre et l'Egée d'après les fouilles de Ras Ibn Hani," in *Proceedings of an International Symposium "The Civlizations of the Aegean and thier Diffusion in Cyprus and the Eastern Mediterranean, 2000-600 B.C." 18-24 September 1989*. Edited by V. Karageorghis, pp. 105-110. Larnaca: Pierides Foundation.
Bounni, A. 1997. "Kazel, Tell," in *The Oxford Encyclopedia of Archaeology in the Near East. Vol. 3*. Edited by E. M. Meyers, pp. 275-276. Oxford: Oxford University Press.
Bounni, A., E. Lagarce, J. Lagarce, and N. Saliby. 1976a. Rapport préliminaire sur la première campagne de fouilles (1975) à Ibn Hani (Syrie). *Syria* 53:233-279.
Bounni, A., E. Lagarce, and N. Saliby. 1976b. La Première campagne de fouille à Ras Ibn Hani. *Les Annales Archeologiques Arabes Syriennes* 26:27-64 (Arabic).
Bounni, A., E. Lagarce, J. Lagarce, and N. Saliby. 1978. Rapport préliminaire sur la deuxième campagne de fouilles (1976) à Ibn Hani (Syrie). *Syria* 55:233-301.
Bounni, A., J. Lagarce, E. Lagarce, N. Saliby, and L. Badre. 1979. Rapport préliminaire sur la troisième campagne de fouilles (1977) à Ibn Hani (Syrie). *Syria* 56:217-291.
Bounni, A., J. Lagarce, E. Lagarce, N. Saliby, L. Badre, P. Leriche, and M. Touma. 1981. Rapport préliminaire sur la quatrième campagne de fouilles (1978) à Ibn Hani (Syrie). *Syria* 58:215-299.
Bounni, A., E. Lagarce, and J. Lagarce. 1998. *Ras Ibn Hani, I. Le palais nord du Bronze récent. Fouilles 1979-1995, synthèse préliminaire. Bibliothèque Archéologique et Historique: 151.* Beirut: Institut Francais d'Archeologie du Proche-Orient.
Bourdieu, P. 1977. *Outline of a Theory of Practice*. Cambridge: Cambridge University Press.
Bourdieu, P. 1984. *Distinction: A Social Critique of the Judgement of Taste* (trans. R Nice). Cambridge, MA.
Bourke, S. J. 1993. The Transition from the Middle to the Late Bronze Age in Syria: The Evidence from Tell Nebi Mend. *Levant* 25:155-195.
Bourke, S. J. 1997. Pre-Classical Pella in Jordan: A Conspectus of Ten Years' Work (1985-1995). *Palestine Exploration Quarterly* 129(2):94-115.
Bourke, S. J., R. T. Sparks, K. N. Sowada, and L. D. Mairs. 1994. Preliminary Report on the University of Sydney's Fourteenth Season of Excavations at Pella (Tabaqat Fahl) in 1992. *Annual of the Department of Antiquities of Jordan* 38:81-126.
Bourke, S. J., R. T. Sparks, K. N. Sowada, P. McLaren, and L. D. Mairs. 1998. Preliminary Report on the University of Sydney's Sixteenth and Seventeenth Seasons of Excavations at Pella (Tabaqat Fahl) in 1994/1995. *Annual of the Department of Antiquities of Jordan* 42:179-211.
Bourke, S. J., R. T. Sparks, P. McLaren, K. N. Sowada, L. D. Mairs, J. Meadows, T. Hikade, and W. Reade. 2003. Preliminary Report on the University of Sydney's Eighteenth and Nineteenth Seasons of

Excavations at Pella (Tabaqat Fahl) in 1996/1997. *Annual of the Department of Antiquities of Jordan* 47:335-388.

Bouzek, J. 1996. Bey 069 Sondage A. *Baal* 1:135-147.

Bowser, B. J. 2000. From Pottery to Politics: An Ethnoarchaeological Study of Political Factionalism, Ethnicity, and Domestic Pottery Style in the Ecuardorian Amazon. *Journal of Archaeological Method and Theory* 7/3:219-248.

Braemer, F. 1982. *L'Architecture Domestique du Levant à l'Âge du Fer*. Paris: editions Recherches sur les Civilisations.

Braemer, F. 1986. La ceramique a engobe rouge de l'âge du Fer à Bassit. *Syria* 63:221-246.

Braemer, F. 1997. "Architecture domestique de l'âge du fer en Syrie du Nord," in *Les Maisons dans la Syrie antique du IIIe millenaire aux debuts de l'Islam*. Edited by C. Castel, pp. 61-71. Beirut: IFAPO.

Braidwood, R. J. 1937. *Mounds in the Plain of Antioch: An Archaeological Survey. University of Chicago. Oriental Institute publications, 48*. Chicago: The University of Chicago Press.

Braidwood, R. J. 1940. Report on Two Sondages on the Coast of Syria, South of Tartous. *Syria* 21:183-226.

Braidwood, R. J., and L. S. Braidwood. 1960. *Excavations in the Plain of Antioch, Vol. 1. University of Chicago Oriental Institute Publications, 61*. Chicago: University of Chicago.

Braudel, F. 1972. *The Mediterranean and the Mediterranean World in the Age of Philip II*. Glasgow.

Braudel, F. 1980. *On History*. Chicago: University of Chicago Press.

Breasted, J. H. 1906a. *Ancient Records of Egypt v. 3 The Nineteenth Dynasty*. Chicago: University of Chicago Press.

Breasted, J. H. 1906b. *Ancient Records of Egypt v. 4 The Twentieth through the Twenty-Sixth Dynasties*. Chicago: University of Chicago Press.

Breasted, J. H. 1933. "The Anatolian Hittite Expedition," in *The Oriental Institute, The University of Chicago Survey 12*. Edited by J. H. Breasted, pp. 265-309. Chicago: University of Chicago Press.

Breisach, E. 1983. *Historiography: Ancient, Medieval, & Modern*. Chicago: University of Chicago Press.

Briend, J. and J.-B. Humbert. 1980. *Tell Keisan, 1971-1976: Une cite phenicienne en Galilee*, Paris: J Gabalda.

Bruins, H. J., J. van der Plicht, and A. Mazar. 2003a. C14 Dates from Tel Rehov: Iron Age Chronology, Pharaohs, and Hebrew Kings. *Science* 300:315-318.

Bruins, H. J., J. van der Plicht, and A. Mazar. 2003b. Response to Comment on "C14 Dates from Tel Rehov: Iron Age Chronology, Pharaohs, and Hebrew Kings". *Science* 302: 568c.

Bruins, H. J., J. van der Plicht, D. Ilan, and E. Werker. 2005. "Iron-Age 14C Dates from Tel Dan: A High Chronology," in *The Bible and Radiocarbon Dating: Archaeology, Text and Science*. Edited by T. E. Levy and T. Higham, pp. 323-336. London: Equinox.

Brumfiel, E. M. 2000. "On the Archaeology of Choice. Agency Studies as a Research Stratagem," in *Agency in Archaeology*. Edited by M.-A. Dobres and J. E. Robb, pp. 249-255. London: Routledge.

Bryce, T. R. 1992. Lukka Revisited. *Journal of Near Eastern Studies* 51:121-130.

Bryce, T. R. 2003. "History," in *The Luwians, Handbuch der Orientalistik 68*. Edited by H. C. Melchert, pp. 27-127. Leiden: Brill.

Buccellati, G. 1967. *Cities and Nations of Ancient Syria: An Essay on Political Institutions with Special Reference to the Israelite Kingdoms*. Rome: Istituto di Studi Del Vicino Oriente Università di Roma.

Budge, E. A. W. 1914. *Assyrian Sculptures in the British Museum: Reign of Ashur-nasir-pal, B.C. 885-860*. London: British Museum.

Buhl, M.-L. 1983. *Sukas VII: The Near Eastern Pottery and Objects of Other Materials from the Upper Strata. Publications of the Carlsberg Expedition to Phoenicia 9*. Copenhagen: Munksgaard.

Buhl, M.-L. 1992. "Hamath," in *The Anchor Bible Dictionary. Vol. III*. Edited by D. N. Freedman, pp. 33-36. New York: Doubleday.

Buhl, M.-L. 1996. "The Objects of Pottery and Other Materials," in *Sukas X: The Bronze and Early Iron Age Remains at the Southern Harbour, Publications of the Carlsberg Expedition to Phoenicia 12. Historisk-filosofiske skrifter (Kongelige Danske Videnskabernes Selskab) 17*. Edited by P. J. Riis, J. Jensen, M.-L. Buhl, and B. Otzen, pp. 25-61. Copenhagen: Munksgaard.

Bulliet, R. W. 1990. *The Camel and the Wheel*. New York: Columbia University Press.

Bunimovitz, S. 1990. Problems in the Ethnic Identification of the Philistine Culture. *Tel Aviv* 17:210-222.

Bunimovitz, S., and A. Faust. 2001. Chronological Separation, Geographical Segregation, or Ethnic Demarcation? Ethnography and the Iron Age Low Chronononology. *Bulletin of the American Schools of Oriental Research* 322:1-10.

Bunnens, G. 1979. *L'expansion phénicienne en Mediterranée*. Bruxelles.

Bunnens, G. 1983. "La Distinction Entre Phéniciens et Puniques Chez Les Auteurs Classiques," in *Atti del I Congresso Internazionale di Studi Fenici e Punici, Roma, 5-10 Novembre 1979, Volume Primo*. Edited by S. Moscati, pp. 233-238. Roma: Consiglio Nazionale Delle Ricerche.

Bunnens, G. 1985. "Le luxe phenicien d'apres les inscriptions royales assyriennes," in *Phoenicia and Its Neighbours, Studia Phoenicia 3*. Edited by E. Gubel and E. Lipinski, pp. 121-133. Leuven: Uitgeverij Peeters.

Bunnens, G. 1990a. "Tell Ahmar After Fifty Years," in *Tell Ahmar: 1988 Season, Abr-Nahrain Supplement Series 2*. Edited by G. Bunnens, pp. 1-10. Leuven: Orientaliste.

Bunnens, G. 1990b. "Area A. Excavations on the Acropolis," in *Tell Ahmar: 1988 Season, Abr-Nahrain Supplement Series 2*. Edited by G. Bunnens, pp. 11-24. Leuven: Orientaliste.

Bunnens, G. 1995. "Hittites and Aramaeans at Til Barsip: A Reappraisal," in *Immigration and Emigration within the Ancient Near East, Orientalia Lovaniensia Analecta 65*. Edited by K. van Lerberghe and A. Schoors, pp. 19-27. Leuven: Peeters.

Bunnens, G. 1997a. Carved Ivories from Til-Barsip. *American Journal of Archaeology* 101(3):435-450.

Bunnens, G. 1997b. "Til Barsip under Assyrian Domination," in *Assyria 1995: Proceedings of the 10th Anniversary Symposium of the Neo-Assyrian Text Corpus Project Helsinki, September 7-11, 1995*. Edited by S. Parpola and R. M. Whiting, pp. 17-28. Helsinki: The Neo-Assyrian Text Corpus Project.

Bunnens, G. 1998. Tell Ahmar/Til Barsip, the ninth and tenth seasons (1996-1997). *Orient-Express* 1998:27-30.

Bunnens, G. 1999. "Aramaeans, Hittites and Assyrians in the Upper Euphrates Valley," in *Archaeology of the Upper Syrian Euphrates, The Tishrin Dam Area: Proceedings of the International Symposium Held at Barcelona, January 28th-30th 1998*. Edited by G. del Olmo Lete and J.-L. Montero Fenollos, pp. 605-624. Barcelona: Editorial AUSA.

Bunnens, G. 2000a. "Syria in the Iron Age: Problems and Definition," in *Essays on Syria in the Iron Age., Ancient Near Eastern studies. Supplement 7*. Edited by G. Bunnens, pp. 3-19. Louvain: Peeters Press.

Bunnens, G. 2000b. "Géographie historique de la région du barrage de Tishrin," in *La Djéziré et l'Euphrate Syriens de la Protohistoire à la fin du IIe Millénaire av. J.-C. Tendances dans l'Interprétation Historique des Données Nouvelles, Subartu VII*. Edited by O. Rouault and M. Wäfer, pp. 299-308: Brepols.

Bunnens, G. 2001. Tell Ahmar/Til Barsip, the eleventh, twelfth and thirteenth seasons (1998-2000). *Orient-Express* 2001:65-68.

Bunnens, G. 2002-2003. Til Barsip before the Assyrians. *Annales Archéologiques Arabes Syriennes* 45-46:163-172.

Bunnens, G. 2006. *A New Luwian Stele and the Cult of the Storm-God at Til Barsib - Masuwari. Publications de la Mission archéologique de l'Université de Liège en Syrie. Tell Ahmar II*. Louvain: Peeters.

Burdajewicz, M. 1990. *The Aegean Sea Peoples and Religious Architecture in the Eastern Mediterranean at the Close of the Late Bronze Age. BAR International Series 558*. Oxford: BAR.

Busink, T. A. 1970. *Der Tempel von Jerusalem*. Leiden.

Butcher, K. 2003. *Roman Syria and the Near East*: The J. Paul Getty Museum.

Campbell, S. 1995. "Death for the Living in the late Neolithic in north Mesopotamia," in *The Archaeology of Death in the Ancient Near East, Oxbow Monographs 51*. Edited by S. Campbell and A. Green, pp. 29-34. Oxford: Oxbow Books.

Canby, J. V. 1985. "Guzana (Tell Halaf)," in *Ebla to Damascus: Art and Archaeology of Ancient Syria: An Exhibition from the Director-General of Antiquities and Museums, Syrian Arab Republic*. Edited by H. Weiss, pp. 332-338. Washington DC: Smithsonian Institution Travelling Exhibition Service.

Capet, E. 2003. Tell Kazel (Syrie), Rapport Preliminaire sur les 9e - 17e Campagnes de Fouilles (1993-2001) du Musée de l'Universite Americaine de Beyrouth, Chantier II. *Berytus* 47:63-121.

Capet, E. 2006-2007. Les Peuples des céramiques 'barbares' à Tell Kazel (Syrie). *Scripta Mediterranea* XXVII-XXVIII: 187-207.

Capet, E., and E. Gubel. 2000. "Tell Kazel: Six Centuries of Iron Age Occupation (c. 1200-612 B.C.)," in *Essays on Syria in the Iron Age., Ancient Near Eastern studies. Supplement 7*. Edited by G. Bunnens, pp. 425-457. Louvain: Peeters Press.

Carrière, B., and A. Barrois. 1927. Fouilles de l'Ècole archéologique française de Jérusalem, effectuées à Neirab du 24 septembre au 5 novembre 1926. *Syria* 8:126-142, 201-212.

Carter, E., and A. Parker. 1995. "Pots, People and The Archaeology of Death in Northern Syria and Southern Anatolia in the latter half of the Third Millennium BC," in *The Archaeology of Death in the Ancient Near East, Oxbow Monographs 51*. Edited by S. Campbell and A. Green, pp. 96-116. Oxford: Oxbow Books.

Carter, J. B. 1995. "Ancestor Cult and the Occasion of Homeric Performance," in *The Age of Homer: A Tribute to Emily Townsend Vermeule*. Edited by J. B. Carter and S. P. Morris, pp. 285-312. Austin: University of Texas Press.

Casana, J., and T. J. Wilkinson. 2005. "Appendix A: Gazetteer of Sites," in *The Amuq Valley Regional Projects Volume 1: Surveys in the Plain of Antioch and Orontes Delta, Turkey, 1995-2002, Oriental Institute Publications 131*. Edited by K. A. Yener, pp. 203-280. Chicago: The Oriental Institute of the University of Chicago.

Caubet, A. 1992. "Reoccupation of the Syrian Coast after the Destruction of the 'Crisis Years'," in *The Crisis Years: The 12th Century B.C. From Beyond the Danube to the Tigris*. Edited by W. A. Ward and M. S. Joukowsky, pp. 123-131. Dubuque: Kendall/Hunt.

Caubet, A. 2002. "Le Travail de l'ivoire au Levant, de la fin de l'Âge du Bronze au début de l'âge du Fer: Continuité et rupture," in *Die Nahöstlichen Kulturen und Griechenland an der Wende vom 2. zum 1. Jarhtausend v. Chr.* Edited by E. A. Braun-Holzinger and H. Matthaus, pp. 109-122. Mohnesee-Wamel: Bibliopolis.

Caubet, A. 2003. "The Case of Ugarit and Carchemish: A Contrast," in *Symbiosis, Symbolism, and the Power of the Past. Canaan, Ancient Israel, and Their Neighbours from the Late Bronze Age through Roman Palaestina*. Edited by W. G. Dever and S. Gitin, pp. 17-22. Winona Lake, IN: Eisenbrauns.

Cauvin, J. 1968. *Les outillages néolithiques de Byblos et du littoral libanais*. Paris.

Cecchini, S. M. 1988. Il sondaggio stratigrafico. *Egitto e Vicino Oriente* 10/2:85-91.

Cecchini, S. M. 1993. Tell Afis (Siria). *Orient-Express* 1993:10-11.

Cecchini, S. M. 1994. Tell Afis (Syrie): Travaux de la mission archeologique italienne en 1994. *Orient-Express* 1994:84-86.

Cecchini, S. M. 1998. "Architecture, Pottery and Finds (Area G)," in *Tell Afis, Siria : scavi sull'acropoli*

1988-1992 = The 1988-1992 excavations on the Acropolis, Ricerche di archeologia del Vicino Oriente ; 1. Edited by S. M. Cecchini and S. Mazzoni, pp. 273-365. Pisa: Edizioni ETS.

Cecchini, S. M. 2000a. "Un batiment mysterieux sur l'acropole de Tell Afis," in *Proceedings of the First International Congress on the Archaeology of the Ancient Near East, Rome, May 18-23, 1998. Volume I*. Edited by P. Matthiae, A. Enea, L. Peyronel, and F. Pinnock, pp. 199-212. Rome: Universita degli Studi di Roma 'La Sapienza'.

Cecchini, S. M. 2000b. "The Textile Industry in Northern Syria During the Iron Age According to the Evidence of the Tell Afis Excavations," in *Essays on Syria in the Iron Age., Ancient Near Eastern studies. Supplement 7*. Edited by G. Bunnens, pp. 211-233. Louvain: Peeters Press.

Cecchini, S. M. 2002. Afis 2000-2001. Area N: presentazione e cronologia. *Egitto e Vicino Oriente* 25:47-53.

Cecchini, S. M. 2005. Tell Afis, Siria (2002-2004) Area N: presentazione e cronologia. *Egitto e Vicino Oriente* 28:77-82.

Cecchini, S. M., and S. Mazzoni. 1998. "Preface," in *Tell Afis, Siria : scavi sull'acropoli 1988-1992 = The 1988-1992 excavations on the Acropolis, Ricerche di archeologia del Vicino Oriente ; 1*. Edited by S. M. Cecchini and S. Mazzoni, pp. 1-9. Pisa: Edizioni ETS.

Chaaya, A. 2000. "L'evolution et le changement cultural a Tell 'Arqa apres l'invasion de Tiglath-Pileser III (Niveaux 10-9)," in *Proceedings of the First International Congress on the Archaeology of the Ancient Near East, Rome, May 18-23, 1998. Volume I*. Edited by P. Matthiae, A. Enea, L. Peyronel, and F. Pinnock, pp. 213-222. Rome: Universita degli Studi di Roma 'La Sapienza'.

Chapman, S. V. 1972. A Catalogue of Iron Age Pottery from the Cemeteries of Khirbet Silm, Joya, Qraye and Qasmieh of South Lebanon. *Berytus* 21:55-194.

Chapman, W. 1989. "Toward an Institutional History of Archaeology: British Archaeologists and Allied Interests in the 1860s," in *Tracing Archaeology's Past: The Historiography of Archaeology, Publications in Archaeology*. Edited by A. L. Christensen, pp. 151-162. Carbondale, IL.

Chartier, R. 1988. *Culture History: Between Practices and Representations*. Ithaca, New York.

Chavalas, M. W. 1992. "Ancient Syria: A Historical Sketch," in *New Horizons in the Study of Ancient Syria*. Edited by M. W. Chavalas and J. L. Hayes, pp. 1-21. Malibu: Undena Publications.

Chesson, M. S. 2007. "Remembering and Forgetting in Early Bronze Age Mortuary Practices on the Southeastern Dead Sea Plain, Jordan," in *Performing Death. Social Analyses of Funerary Traditions in the Ancient Near East and Mediterranean, Oriental Institute Seminars 3*. Edited by N. Laneri, pp. 109-140. Chicago: Chicago University Press.

Chevalier, N. 2002. *La Recherche Archéologique Française au Moyen-Orient 1842-1947*. Paris: Éditions Recherche sur les Civilisations.

Childe, V. G. 1929. *The Danube in Prehistory*. Oxford.

Chiti, B. 2005. Tell Afis, Siria (2002-2004) Area E4b nord: Bronzo Tardo II - Ferro I. *Egitto e Vicino Oriente* 28:62-69.

Christakis, K. S. 1999. Pithoi and Food Storage in Neopalatial Crete: A Domestic Perspective. *World Archaeology* 31(1):1-20.

Christensen, A. L. 1989. "The Past Is Still Alive: The Immediacy Problem and Writing the History of Archaeology," in *Tracing Archaeology's Past: The Historiography of Archaeology, Publications in Archaeology*. Edited by A. L. Christensen, pp. 163-168. Carbondale and Edwardsville, IL.

Ciafardoni, P. 1988. Tell Afis: un insediamento del ferro nella Regione di Idlib. *Egitto e Vicino Oriente* 10/2:5-23.

Cichocki, O. 2000. "Methods and Aims of Dendrochronology; Dating Wooden Objects from the Eastern Mediterranean in the Second Millennium B.C. (Absolute Chronology IV)," in *The Synchronisation of Civilisations in the Eastern Mediterranean in the Second Millennium B.C.* Edited by M. Bietak, pp. 62-67. Vienna: Verlag der Österreichischen Akademie der Wissenschaften.

Clairmont, C. 1955. Greek Pottery from the Near East. *Berytus* 11:85-141.

Clairmont, C. 1956-1957. Greek Pottery from the Near East. *Berytus* 12:1-34.

Clark, E. A. 2004. *History, Theory, Text: Historians and the Linguistics Turn*. Harvard University Press: Cambridge, MA.

Cline, E. H. 2003. "Trade and Exchange in the Levant," in *Near Eastern Archaeology: A Reader*. Edited by S. Richard, pp. 360-366. Winona Lake, Indian: Eisenbrauns.

Coldstream, N. 1968. *Greek Geometric Pottery: A Survey of Ten Local Styles and their Chronology*. London: Methuen & Co Ltd.

Coldstream, N. 1989. "Early Greek Visitors to Cyprus and the Eastern Mediterranean," in *Cyprus and the East Mediterranean in the Iron Age. Proceedings of the Seventh British Museum Classical Colloquium, London, April 1988*. Edited by V. Tatton-Brown, pp. 90-96. London: Trustees of the British Museum.

Coldstream, N. 1998. "The First Exchanges between Euboeans and Phoenicians: Who took the initiative?," in *Mediterranean Peoples in Transition: Thirteenth to Early Tenth Centuries BCE, in Honour of Trude Dothan*. Edited by S. Gitin, A. Mazar, and E. Stern, pp. 353-360. Jerusalem: Israel Exploration Society.

Coldstream, N. 2003. Some Agean Reactions to the Chronological Debate in the Southern Levant. *Tel Aviv* 30:247-258.

Coldstream, N., and P. M. Bikai. 1988. Early Greek Pottery in Tyre and Cyprus: Some preliminary comparisons. *Report of the Department of Antiquities Cyprus* 1988(2):35-43.

Coldstream, N., and A. Mazar. 2003. Greek Pottery from Tel Rehov and Iron Age Chronology. *Israel Exploration Journal* 53:29-48.

Collombier, A.-M. 1987. "Céramique grecque et échanges en Méditerranée orientale: Chypre et la côte Syro-Phénicienne (fin VIIIe - fin IVe siècles av. J.-

C.)," in *Phoenicia and the East Mediterranean in the First Millennium B.C., Studia Phoenicia 5*. Edited by E. Lipinski, pp. 239-248. Leuven: Uitgeverij Peeters.

Conder, C. R. 1881. Kadesh. *Quarterly Statement: Palestine Exploration Fund* 1881:163-175.

Conder, C. R. 1882. Notes: Kadesh on Orontes. *Quarterly Statement: Palestine Exploration Fund* 1882:155-157.

Conder, C. R. 1883. Notes: Kadesh on Orontes. *Quarterly Statement: Palestine Exploration Fund* 1883:100.

Contenau, G. 1920. Mission archéologique à Sidon (1914). *Syria* 1:108-154.

Contenau, G. 1923. Deuxième mission archéologique à Sidon (1920). *Syria* 4:261-281.

Contenau, G. 1924. Deuxième mission archéologique à Sidon (1920). *Syria* 5:123-134.

Cool, H. E. M., and M. J. Baxter. 1999. Peeling the Onion: an approach to comparing vessel glass assemblages. *Journal of Roman Archaeology* 12:72-100.

Cook, R. M. 1972. *Greek Painted Pottery*. London.

Courbin, P. 1972. Rapport sur la Fouille de Ras el Bassit. *Les Annales Archeologiques Arabes Syriennes* 22:45-61.

Courbin, P. 1973. Ras el Bassit Rapport sur la Campagne de 1972. *Les Annales Archeologiques Arabes Syriennes* 23:25-38.

Courbin, P. 1974. Ras el Bassit, Al Mina & Tell Sukas. *Revue Archeologique* 1974:174-178.

Courbin, P. 1974-1975. Rapport sur la 4eme Campagne de Fouille (1974) a Ras el Bassit. *Les Annales Archeologiques Arabes Syriennes* 24-25:59-71.

Courbin, P. 1976. Rapport sur la 5 eme Campagne de Fouille a Ras el Bassit. *Les Annales Archeologiques Arabes Syriennes* 26:63-69.

Courbin, P. 1977-1978. Rapport sur la Sixieme Campagne de Fouille (1976) a Ras el Bassit (Syrie). *Les Annales Archeologiques Arabes Syriennes* 27-28:29-40.

Courbin, P. 1978. Une nouvelle fouille francaise sur la cote syrienne. *Archeologia* 116:48-62.

Courbin, P. 1981-1982. Ras al-Bassit. *Archiv fur Orientforschung* 28:231-232.

Courbin, P. 1982a. "Une Assiette Cycladique a Ras el Bassit," in *Archeologie au Levant. Recueil R Saidah, Collection de la Maison de l'Orient Mediterraneen No. 12*. Edited by M. Yon, pp. 193-204. Paris: Maison de l'Orient Mediterraneen.

Courbin, P. 1982b. Gourdes. *Revue Archeologique* 1982:9-25.

Courbin, P. 1983. Bassit. *Les Annales Archeologiques Arabes Syriennes* 33(2):119-127.

Courbin, P. 1986. Bassit. *Syria* 63:175-220.

Courbin, P. 1986-1987. Rapport sur la Xeme et Derniere Campagne a Ras el Bassit. *Les Annales Archeologiques Arabes Syriennes* 36-37:107-120.

Courbin, P. 1990a. "Fragments d'amphores protogeometriques grecques a Bassit," in *Resurrecting the Past: A Joint Tribute to Adnan Bounni*. Edited by P. Matthiae, M. Van Loon, and H. Weiss, pp. 49-64. Istanbul: Nederlands Historisch-Archaeologisch Instituut te Istanbul.

Courbin, P. 1990b. "Bassit-Posidaion in the Early Iron Age," in *Greek Colonists and Native Populations. Proceedings of the First Australian Congress of Classical Archaeology held in honour of Emeritus Professor A. D. Trendall, Sydney 9-14 July 1985*. Edited by J.-P. Descoeudres, pp. 503-509. Oxford: Clarendon Press.

Courbin, P. 1993a. *Fouilles de Bassit. Tombes du Fer*. Paris.

Courbin, P. 1993b. Fragments d'amphores protogeometriques grecques á Bassit. *Hesperia* 62:95-113.

Crielaard, J. P. 1999a. "Production, Circulation and Consumption of Early Iron Age Greek Pottery (eleventh to seventh centuries BC)," in *The Complex Past of Pottery. Production, Circulation and Consumption of Mycenaean and Greek Pottery (Sixteenth to Early Fifteenth Centuries B.C.)*. Edited by J. P. Crielaard, V. Stissi, and G. J. van Wijngaarden, pp. 49-81. Amsterdam: J. C. Gieben.

Crielaard, J. P. 1999b. "Early Iron Age Greek Pottery in Cyprus and North Syria: a consumption-oriented approach," in *The Complex Past of Pottery. Production, Circulation and Consumption of Mycenaean and Greek Pottery (Sixteenth to Early Fifteenth Centuries B.C.)*. Edited by J. P. Crielaard, V. Stissi, and G. J. van Wijngaarden, pp. 261-290. Amsterdam: J. C. Gieben.

Cross, F. M., and T. O. Lambdin. 1960. A Ugaritic Abecedary and the Origins of the Proto-Canaanite Alphabet. *Bulletin of the American Schools of Oriental Research* 160:21-26.

Culican, W. 1959. Aspects of Phoenician Settlement in the West Mediterranean. *Abr Nahrain* 1:36-55.

Culican, W. 1970. Phoenician Oil Bottles and Tripod Bowls. *Berytus* 19:5-18.

Culican, W. 1973. The Graves at Tell er-Reqeish. *Australian Journal of Biblical Archaeology* 2:66-106.

Culican, W. 1975. Sidonian Bottles. *Levant* 7:145-150.

Culican, W. 1975-1976. Some Phoenician Masks and other Terracottas. *Berytus* 24:47-88.

Culican, W. 1982. "The Repertoire of Phoenician Pottery," in *Phönizier im Westen: die Beiträge des Internationalen Symposiums über "Die Phönizische Expansion im Westlichen Mittelmeerraum" in Köln vom 24. bis 27. April, 1979, Madrider Beiträge ; Bd. 8*. Edited by H. G. Niemeyer, pp. 45-82. Mainz am Rhein: P. von Zabern.

Culican, W. 1991. "Phoenicia and Phoenician Colonization," in *The Assyrian and Babylonian Empires and other States of the Near East, from the Eighth to the Sixth Centuries B.C., Cambridge Ancient History III.2 (Second Edition)*. Edited by J. Boardman, I. E. S. Edwards, N. G. L. Hammond, E. Sollberger, and C. B. F. Walker, pp. 461-546. Cambridge: Cambridge University Press.

Curvers, H. H. 2001-2002. The Lower Town of Beirut (1200-300 BC). A Preliminary Synthesis. *Aram* 13-14:51-72.

Curvers, H. H., and B. Stuart. 1996. Bey 008 The 1994 Results. *Baal* 1:228-234.
Curvers, H. H., and B. Stuart. 1997. The BCD Infrastructure Archaeology Project, 1995. *Baal* 2:167-205.
Curvers, H. H., and B. Stuart. 1998-1999. The BCD Archaeology Project, 1996-1999. *Baal* 3:13-30.
Curvers, H. H., and B. Stuart. 2004. "Beirut Central District Archaeology Project 1994 - 2003," in *Decade: A Decade of Archaeology and History in the Lebanon.* Edited by C. Doumet-Serhal, pp. 248-265. Beirut: The Lebanese British Friends of the National Museum.
Dalley, S. 2000. "Shamshi-ilu, Language and Power in the Western Assyrian Empire," in *Essays on Syria in the Iron Age., Ancient Near Eastern studies. Supplement 7.* Edited by G. Bunnens, pp. 79-88. Louvain: Peeters Press.
D'Amore, P. 1998a. "Architecure, Pottery and Finds (Area L)," in *Tell Afis, Siria: scavi sull'acropoli 1988-1992 = The 1988-1992 excavations on the Acropolis, Ricerche di archeologia del Vicino Oriente; 1.* Edited by S. M. Cecchini and S. Mazzoni, pp. 371-381. Pisa: Edizioni ETS.
D'Amore, P. 1998b. "The Iron Age Clay Figurines," in *Tell Afis, Siria: scavi sull'acropoli 1988-1992 = The 1988-1992 excavations on the Acropolis, Ricerche di archeologia del Vicino Oriente ; 1.* Edited by S. M. Cecchini and S. Mazzoni, pp. 415-425. Pisa: Edizioni ETS.
D'Amore, P. 2002. Afis 2000-2001. Area A. *Egitto e Vicino Oriente* 25:9-12.
D'Amore, P. 2005. Tell Afis, Siria (2002-2004) Area A1: il settore centrale. *Egitto e Vicino Oriente* 28:17-21.
Daniel, G. 1967. *The Origins and Growth of Archaeology*: Penguin Books.
Darnton, R. 1984. The Great Cat Massacre and Other Episodes in French Cultural History. London: Allen Lane.
Daviau, P. M. M. 2001. "Family Religion: Evidence for the Paraphernalia of the Domestic Cult," in *The World of the Aramaeans II: Studies in History and Archaeology in Honour of Paul-Eugene Dion, Journal for the Study of the Old Testament Supplement Series 325.* Edited by P. M. M. Daviau, J. W. Wevers, and M. Weigl, pp. 199-229. Sheffield: Sheffield Academic Press.
Davies, P. 1992. *In Search of Ancient Israel.* Sheffield: Sheffield Academic Press.
Day, P., and D. Wilson. 2004. "Ceramic Change and the Practice of Eating and Drinking in Early Bronze Age Crete," in *Food, Cuisine and Society in Prehistoric Greece, Sheffield Studies in Aegean Archaeology 5.* Edited by P. Halstead and J. Barrett, pp. 45-62. Oxford: Oxbow Books.
Dayagi-Mendels, M. 1993. *Perfumes and Cosmetics in the Ancient World.* Jerusalem: The Israel Museum.
Dayagi-Mendels, M. 1999. *Drink and Be Merry: Wine and Beer in Ancient Times.* Jerusalem: The Israel Museum.
Dayagi-Mendels, M. 2002. *The Akhziv Cemeteries: The Ben Dor Excavations, 1941-1944. Israel Antiquities Authority Reports 15.* Jerusalem: Israel Antiquities Authority.
Degli Esposti, M. 1998. "I livelli del Ferro I-II. Architettura e materiali (Area E2)," in *Tell Afis, Siria : scavi sull'acropoli 1988-1992 = The 1988-1992 excavations on the Acropolis, Ricerche di archeologia del Vicino Oriente ; 1.* Edited by S. M. Cecchini and S. Mazzoni, pp. 231-269. Pisa: Edizioni ETS.
Del Vesco, P. 2002. Afis 2000-2001. Area M. *Egitto e Vicino Oriente* 25:45-47.
Del Vesco, P., V. Melchiorri, and I. Oggiano. 2005. Tell Afis, Siria (2002-2004) Area J. *Egitto e Vicino Oriente* 28:76-77.
Desborough, V. 1952. *Protogeometric Pottery.* Oxford: Clarendon Press.
Desborough, V. 1957. A Group of Vases from Amathus. *Journal of Hellenic Studies* 77(2):212-219.
Descœudres, J.-P., and R. A. Kearsley. 1983. Greek Pottery at Veii: Another Look. *Annual of the British School at Athens* 78:9-54.
Dever, W. G. 1980. The Impact of the 'New Archaeology' on Syro-Palestinian Archaeology. *Bulletin of the American Schools of Oriental Research* 242:15-29.
Dever, W. G. 1992. "The Late Bronze-Early Iron I Horizon in Syria-Palestine: Egyptians, Canaanites, "Sea Peoples" and Proto-Israelites," in *The Crisis Years: The 12th Century B.C. From Beyond the Danube to the Tigris.* Edited by W. A. Ward and M. S. Joukowsky, pp. 99-110. Iowa: Kendall Hunt.
Dever, W. G. 1993. What Remains of the House that Albright Built? *Biblical Archaeologist* 56:25-35.
Dever, W. G. 1995a. The Death of a Discipline. *Biblical Archaeology Review* 21:50-55.
Dever, W. G. 1995b. Ceramics, Ethnicity, and the Question of Israel's Origins. *Biblical Archaeologist* 58:200-213.
Dever, W. G. 1997a. "Keisan, Tell" in *The Oxford Encyclopedia of Archaeology in the Near East. Vol. 3.* Edited by E. M. Meyers, pp. 278-279. Oxford: Oxford University Press.
Dever, W. G. 1997b. "Philology, Theology, and Archaeology: What Kind of History Do We Want, and What is Possible?" in *The Archaeology of Israel: Constructing the Past, Interpreting the Present, Journal for the Study of the Old Testament Supplement Series 237.* Edited by N. A. Silberman and D. Small, pp. 290-310. Sheffied: Sheffield Academic Press.
Dever, W. G. 1997c. "Abu Hawam, Tell," in *The Oxford Encyclopedia of Archaeology in the Near East. Vol. 1.* Edited by E. M. Meyers, pp. 9. Oxford: Oxford University Press.
Dever, W. G. 2000. "Imaging Ancient Israel: Archaeology, Social History, and the Sociology of Knowledge," in *Proceedings of the First International Congress on the Archaeology of the Ancient Near East, Rome, May 18-23, 1998. Volume I.* Edited by P. Matthiae, A. Enea, L. Peyronel, and F. Pinnock, pp. 323-332. Rome: Universita degli Studi di Roma 'La Sapienza'.

Dever, W. G. 2003. Visiting the Real Gezer: A Reply to Israel Finkelstein. *Tel Aviv* 30:259-282.

Dhorme, E. 1938. Les Fouilles Francaises a Till Barsip. *Journal des Savants* 36:109-120, 155-164.

Dhorme, P. 1927. Note sur les tablettes de Neirab. *Syria* 8:213-215.

Diaz-Andreu, M. 1995. "Archaeology and Nationalism in Spain," in *Nationalism, Politics, and the Practice of Archaeology*. Edited by P. L. Kohl and C. Fawcett, pp. 39-56. Cambridge: Cambridge University Press.

Dickinson, O. 2006. *The Aegean from Bronze Age to Iron Age. Continuity and change between the twelfth and eighth centuries BC*. London: Routledge.

Dietler, M., and I. Herbich. 1998. "Habitus, Techniques, Style: An Integrated Approach to the Social Understanding of Material Culture and Boundaries," in *The Archaeology of Social Boundaries*. Edited by M. T. Stark, pp. 232-263. Washington and London: Smithsonian Institution Press.

Dikaios, P. 1969. *Enkomi, Excavations 1948-1958, vol. 1: The Architectural Remains, the Tombs*. Mainz am Rhein: Verlag Philipp von Zabern.

Dion, P.-E. 1997. *Les Arameens de l'Age du Fer: Histoire Politique et Structures Sociales*. Paris: Librairie Lecoffre.

Dobres, M.-A. 1999. "Technology's Links and *Chaines*: The Processual Unfolding of Technique and Technician," in *The Social Dynamics of Technology: Practice, Politics and World Views*. Edited by M.-A. Dobres and C. R. Hoffman, pp. 124-146. Washington and London: Smithsonian Institution Press.

Dobres, M.-A. 2000. *Technology and Social Agency: Outlining a Practice Framework for Archaeology*. Oxford: Blackwell.

Dobres, M.-A., and C. R. Hoffman. Editors. 1999a. *The Social Dynamics of Technology: Practice, Politics and World Views*. Washington and London: Smithsonian Institution Press.

Dobres, M.-A., and C. R. Hoffman. 1999b. "Introduction: A Context for the Present and Future of Technology Studies," in *The Social Dynamics of Technology: Practice, Politics and World Views*. Edited by M.-A. Dobres and C. R. Hoffman, pp. 1-22. Washington and London: Smithsonian Institution Press.

Dobres, M.-A., and C. R. Hoffman. 1999c. "Conclusion: Making Material Culture, Making Culture Material," in *The Social Dynamics of Technology: Practice, Politics and World Views*. Edited by M.-A. Dobres and C. R. Hoffman, pp. 209-222. Washington and London: Smithsonian Institution Press.

Dobres, M.-A., and J. E. Robb. Editors. 2000a. *Agency in Archaeology*. London: Routledge.

Dobres, M.-A., and J. E. Robb. 2000b. "Introduction," in *Agency in Archaeology*. Edited by M.-A. Dobres and J. E. Robb, pp. 3-17. London: Routledge.

Dornemann, R. H. 1997. "Hama," in *The Oxford Encyclopedia of Archaeology in the Near East. Vol. 2*. Edited by E. M. Meyers, pp. 466-468. Oxford: Oxford University Press.

Dornemann, R. H. 1998a. Tell Qarqur 1996. *Chronique Archeologique en Syrie* 2:81-87.

Dornemann, R. H. 1998b. Tell Qarqur 1997. *Chronique Archeologique en Syrie* 2:153-159.

Dornemann, R. H. 1999. Comparisons in the Bronze and Iron Age Inventories between Orontes Valley sites, Ugarit and Ebla, from the point of view from Tell Qarqur. *Les Annales Archeologiques Arabes Syriennes* 43:137-148.

Dornemann, R. H. 2000. "Iron Age Remains at Tell Qarqur in the Orontes Valley," in *Essays on Syria in the Iron Age., Ancient Near Eastern studies. Supplement 7*. Edited by G. Bunnens, pp. 459-492. Louvain: Peeters Press.

Dornemann, R. H. 2003a. "Seven Seasons of American Schools of Oriental Research Excavations at Tell Qarqur, Syria, 1993-1999," in *Preliminary Excavation Reports and Other Archaeological Investigations: Tell Qarqur, Iron I Sites in the North-Central Highlands of Palestine, Annual of the American Schools of Oriental Research; v. 56*. Edited by N. L. Lapp, pp. 1-141. Boston, Mass.: American Schools of Oriental Research.

Dornemann, R. H. 2003b. "State Formation in Syria at the Beginning of the Iron Age," in *One Hundred Years of American Archaeology in the Middle East*. Edited by D. R. Clark and V. H. Matthews, pp. 199-214. Boston: American Schools of Oriental Research.

Dornemann, R. H. 2008. "Evidence from Two Transitional Periods, Early Bronze IV and Iron I, Tell Qarqur, Syria," in *Proceedings of the 4th International Congress of the Archaeology of the Ancient Near East, 29 March-3April 2004, Freie Universität Berlin. Volume 2*. Edited by H. Kuhne, R. M. Czichon, and F. J. Kreppner, pp. 81-96. Wiesbaden: Harrassowitz Verlag.

Dothan, M. 1985. "Terminology for the Archaeology of the Biblical Periods," in *Biblical Archaeology Today*. Edited by J. Amitai, pp. 136-141. Jerusalem: Israel Exploration Society.

Dothan, T. 1982. *The Philistines and their Material Culture*. Jerusalem: Israel Exploration Society.

Dothan, T. 1998. "Initial Philistine Settlement: From Migration to Coexistence," in *Mediterranean Peoples in Transition: Thirteenth to Early Tenth Centuries BCE-In Honour of Professor Trude Dothan*. Edited by S. Gitin, A. Mazar, and E. Stern, pp. 148-161. Jerusalem: Israel Exploration Society.

Dothan, T., and M. Dothan. 1992. *People of the Sea: The Search for the Philistines*. New York: Macmillan Publishing Company.

Douglas, M. 1975. "Deciphering a Meal," in *Implicit Meanings*. Edited by M. Douglas, pp. 249-275. London: Routledge.

Douglas, M. 1987. "A distinctive anthropological perspective," in *Constructive Drinking: Perspectives on Drink from Anthropology, Editions de la Maison des Sciences de l'Homme*. Edited by M. Douglas, pp. 3-15. Cambridge: Cambridge University Press.

Doumet, C. 1982. Les tombes IV et V de Rachidieh. *Annales d'Histoire et d'Archéologie* 1:89-135.

Doumet, C., and I. Kawkabani. 1995. "Les Tombes de Rachidieh: Remarques sur les Contacts Internationaux et le Commerce Phénicien au VIIIe Siècle Av. J.-C.," in *Actes du IIIe Congrès International des Études Phéniciennes et Puniques. Tunis, 11-16 Novembre 1991*, pp. 379-395. Tunis: Institut National du Patrimoine.

Doumet-Serhal, C. 1993-1994. La Cruche a 'Arete sur le Col' un Fossile Directeur de l'Expansion Phenicienne en Mediterranee. *Berytus* 41:99-136.

Doumet-Serhal, C. 1995-1996. Le Bronze Moyen II B/C et le Bronze Recent I au Liban: l'evidence de Tell el Ghassil. *Berytus* 42:37-70.

Doumet-Serhal, C. 1996. *Les Fouilles de Tell el-Ghassil de 1972 a 1974. Etude du Materiel. Bibliotheque Archeologique et Historique T. 146*. Beyrouth: IFAPO.

Doumet-Serhal, C. 1998-1999. First Season of Excavation at Sidon Preliminary Report. *Baal* 3:181-224.

Doumet-Serhal, C. 2000. Second Season of Excavation at Sidon Preliminary Report. *Baal* 4:75-122.

Doumet-Serhal, C. 2001. Third Season of Excavation at Sidon Preliminary Report. *Baal* 5:153-172.

Doumet-Serhal, C. 2002. Fourth Season of Excavation at Sidon Preliminary Report. *Baal* 6:179-210.

Doumet-Serhal, C. 2003a. Jars from the First Millennium BCE at Tell el Rachidieh (south of Tyre): Phoenician Cinerary Urns and Grave Goods. *Archaeology and History in the Lebanon* 17:42-51.

Doumet-Serhal, C. 2003b. Jars from the Second Millennium BC at Tell el Ghassil (Beka'a Valley): Child Burials or Finds in Domestic Installations. *Archaeology and History in the Lebanon* 17:22-24.

Doumet-Serhal, C. 2003c. Fifth Season of Excavation at Sidon Preliminary Report. *Baal* 7:175-208.

Doumet-Serhal, C. 2003d. Jars from the Third Millennium BC at Sidon (south Lebanon): A Standardised Production in the Emergence of an Urban Culture. *Archaeology and History in the Lebanon* 17:2-6.

Doumet-Serhal, C. 2004a. "The Location and Ancient Names of Mainland Tyre and the Role of Rachidieh in this Context," in *Decade: A Decade of Archaeology and History in the Lebanon*. Edited by C. Doumet-Serhal, pp. 60-69. Beirut: The Lebanese British Friends of the National Museum.

Doumet-Serhal, C. 2004b. "Jars from the First Millennium BC at Tell Rachidieh: Phoenician Cineray Urns and Grave Goods," in *Decade: A Decade of Archaeology and History in the Lebanon*. Edited by C. Doumet-Serhal, pp. 70-79. Beirut: The Lebanese British Friends of the National Museum.

Doumet-Serhal, C. 2004c. "Tell Rachidieh: Foreign Relations," in *Decade: A Decade of Archaeology and History in the Lebanon*. Edited by C. Doumet-Serhal, pp. 88-99. Beirut: The Lebanese British Friends of the National Museum.

Doumet-Serhal, C. 2004d. "Sidon British Museum Excavations 1998-2003," in *Decade: A Decade of Archaeology and History in the Lebanon*. Edited by C. Doumet-Serhal, pp. 102-123. Beirut: The Lebanese British Friends of the National Museum.

Doumet-Serhal, C. 2004e. "Jars from the Second Millennium BC at Sidon: Child Burials or Deposited Goods in Graves," in *Decade: A Decade of Archaeology and History in the Lebanon*. Edited by C. Doumet-Serhal, pp. 132-145. Beirut: The Lebanese British Friends of the National Museum.

Doumet-Serhal, C. 2004f. Sidon (Lebanon): Twenty Middle Bronze Age Burials from the 2001 Season of Excavation. *Levant* 36:89-154.

Doumet-Serhal, C. 2004g. Sixth and Seventh Seasons of Excavation at Sidon Preliminary Report. *Baal* 8:47-82.

Doumet-Serhal, C. 2004h. The Palm Tree Motif in the Late Bronze and Early Iron Age: Evidence from Sidon and Tell Rachidieh. *Archaeology and History in the Lebanon* 19:34-43.

Doumet-Serhal, C. 2006. Preliminary Report on the Iron Age at Sidon: British Museum Excavations 2003-2004. *Archaeology and History in the Lebanon* 23:2-29.

Doumet-Serhal, C., and H. Sayegh. 1998. "La Ceramique du Locus 130," in *Un Quartier du Port Phénicien de Beyrouth au Fer III/Perse. Les Objets, Supplement no. 6 a Transeuphraténe*. Edited by J. Elayi and H. Sayegh, pp. 11-24. Paris: Gabalda.

Drews, R. 1993. *The End of the Bronze Age. Changes in Warfare and the Catastrophe ca. 1200 B.C.* New Jersey: Princeton University Press.

Drower, M. S. 1995. *Flinders Petrie: A Life in Archaeology. Wisconsin Studies in Classics*. Madison, WI: University of Wisconsin Press.

Dunand, M. 1926. Sondages archéologiques effectués à Bostan Ech-Cheikh près Saida. *Syria* 7:1-8; Pls 1-VI.

Dunand, M. 1928. La Sixième Campagne de Fouilles de Byblos. *Syria* 9:1-5; 173-186.

Dunand, M. 1929. La Septième Campagne des Fouilles de Byblos. *Syria* 10:206-216.

Dunand, M. 1939a. *Fouilles De Byblos. Tome I 1926-1932 (Texte). Etudes et Documents d'Archeologie I.* Paris: Librairie Orientaliste Paul Geuthner.

Dunand, M. 1939b. *Fouilles De Byblos. Tome I 1926-1932 (Atlas). Etudes et Documents d'Archeologie I.* Paris: Librairie Orientaliste Paul Geuthner.

Dunand, M. 1940. Chronique. III. La Vie du Musée. *Bulletin du Musée de Beyrouth* 4:123-125.

Dunand, M. 1950. *Fouilles De Byblos. Tome II 1933-1938 (Atlas). Etudes et Documents d'Archeologie III.* Paris: Librairie d'Amerique et d'Orient Adrien Maisonneuve.

Dunand, M. 1954. *Fouilles De Byblos. Tome II 1933-1938 (Texte). Etudes et Documents d'Archeologie III.* Paris: Librairie d'Amerique et d'Orient Adrien Maisonneuve.

Dunand, 1955. Rapport Préliminaire sur les Fouilles de Byblos en 1950, 51, 52. *Bulletin du Musée de Beyrouth* 12:7-23.

Dunand, M. 1956. Rapport Préliminaire sur les Fouilles de Byblos en 1954, 55. *Bulletin du Musée de Beyrouth* 13:73-86.

Dunand, M. 1961. Rapport Préliminaire sur les Fouilles de Byblos en 1957, 58, 59. *Bulletin du Musee de Beyrouth* 16:69-85.

Dunand, M. 1964. Rapport Préliminaire sur les Fouilles de Byblos en 1960, 61, 62. *Bulletin du Musee de Beyrouth* 17:21-36.

Dunand, M. 1965. Nouvelles inscriptions phéniciennes du temple d'Echmoun à Bostan ech-Cheikh près de Sidon. *Bulletin du Musée de Beyrouth* 18:105-110.

Dunand, M. 1966a. Rapport Préliminaire sur les Fouilles de Byblos en 1964. *Bulletin du Musée de Beyrouth* 19:95-102.

Dunand, M. 1966b. Rapport Préliminaire sur les Fouilles de Sidon en 1963-1964. *Bulletin du Musée de Beyrouth* 19:103-106.

Dunand, M. 1967a. Rapport Préliminaire sur les Fouilles de Byblos en 1965. *Bulletin du Musée de Beyrouth* 20:21-26.

Dunand, M. 1967b. Rapport Préliminaire sur les Fouilles de Sidon en 1964-1965. *Bulletin du Musée de Beyrouth* 20:27-44.

Dunand, M. 1973. *Byblos: Its History, Ruins and Legends*. Paris.

Dunand, M. 1983. "L'iconographie d'Echmoun dans son temple sidonien," in *Atti del I Congresso Internazionale di Studi Fenici e Punici, Roma, 5-10 Novembre 1979, Volume Secondo*. Edited by S. Moscati, pp. 515-520. Roma: Consiglio Nazionale Delle Ricerche.

Dunand, M., A. Bounni, and N. Saliby. 1964. Fouilles de Tell Kazel: Rapport Préliminaire. *Les Annales Archeologiques Arabes Syriennes* 14:3-14.

Dunand, M., and N. Saliby. 1957. A la recherche de Simyra. *Les Annales Archeologiques Arabes Syriennes* 7:3-16.

Dussaud, R. 1928. Observations sur la ceramique du IIe millenaire avant notre ere. *Syria* 9:131-150.

Dussaud, M. 1931. Fouilles de M. Ploix de Rotrou dans la Citadelle d'Alep. *Syria* 12:95-96.

Dussaud, M. 1934. L'enciente fortifiée d'Alep. *Syria* 15:300-301.

Dussaud, M. 1937. La date des grandes tombes de Til-Barsib. *Syria* 18:410.

Echt, R. 1984. *Kamid el-Loz 5: Die Stratigraphie. Saarbrucker Beitrage zur Alterumskunde 34*. Bonn: Rudolph Habelt Verlag GMBH.

Edgerton, W. F., and J. Wilson. 1936. *Historical Records of Rameses III: the Texts in 'Medinet Habu'*. Chicago: University of Chicago Press.

Edwards, I. E. S. 1992. "Clays and Clay Use at Pella," in *Pella in Jordan 2: The Second Interim Report of the Joint University of Sydney and College of Wooster Excavations at Pella 1982-1985, Mediterranean Archaeology Supplement 2*. Edited by A. W. McNicoll, P. C. Edwards, J. Hanbury-Tenison, J. B. Hennessy, T. F. Potts, R. H. Smith, A. Walmsley, and P. Watson, pp. 281-298. Sydney: MEDITARCH.

Egami, N. 1983. The Archaeological Researches in Idlib Prefecture. *Les Annales Archeologiques Arabes Syriennes* 33:75-82.

Egami, N. 1988. *Tell Mastuma: Excavations in Idlib District, Syria, 1986. Preliminary report of archaeological researches in Syria; v. IV*. Tokyo: Ancient Orient Museum.

Egami, N., and S. Masuda. 1982. *Tell Mastuma: Excavations in Idlib District, Syria, 1980. Preliminary report of archaeological researches in Syria; v. II*. Tokyo: Ancient Orient Museum.

Egami, N., and S. Masuda. 1984. *Tell Mastuma: Excavations in Idlib District, Syria, 1982. Preliminary Report of Archaeological Researches in Syria III*. Tokyo: Ancient Orient Museum.

Eidem, J. 1999a. Across the Euphrates The Region of Qal'at Najm and Reflections on Some Trans-Euphrates Routes. *Les Annales Archeologiques Arabes Syriennes* 43:149-157.

Eidem, J. 1999b. "The 'Tishrin Project' and Salvage Archaeology," in *Archaeology of the Upper Syrian Euphrates, The Tishrin Dam Area: Proceedings of the International Symposium Held at Barcelona, January 28th-30th 1998*. Edited by G. del Olmo Lete and J.-L. Montero Fenollos, pp. 19-23. Barcelona: Editorial AUSA.

Eidem, J., and R. Ackermann. 1999. "The Iron Age Ceramics from Tell Jurn Kabir," in *Iron Age Pottery in Northern Mesopotamia, Northern Syria and South-Eastern Anatolia : papers presented at the meetings of the international "table ronde" at Heidelberg (1995) and Nieborów (1997) and other contributions, Altertumskunde des vorderen Orients ; Bd. 10*. Edited by A. Hausleiter and A. Reiche, pp. 309-324. Münster: Ugarit-Verlag.

Eidem, J., and K. Putt. 1994. An Iron Age Site on the Euphrates: The First Season of Excavations at Tell Jurn Kabir (Syria). *Orient-Express* 1994:8-9.

Eidem, J., and K. Putt. 1995. Tell Jurn Kabir (Syria), 1994. *Orient-Express* 1995:39-40.

Eidem, J., and K. Putt. 1999. "Tell Jurn Kabir and Tell Qadahiye: Danish Excavations in the Tishrin Dam Area," in *Archaeology of the Upper Syrian Euphrates, The Tishrin Dam Area: Proceedings of the International Symposium Held at Barcelona, January 28th-30th 1998*. Edited by G. del Olmo Lete and J.-L. Montero Fenollos, pp. 193-204. Barcelona: Editorial AUSA.

Eidem, J., and K. Putt. 2001. Iron Age Sites on the Upper Euphrates. *Les Annales Archeologiques Arabes Syriennes* 44:83-96.

Einwag, B. 2000. "Die West-Gazira in der Eisenzeit," in *Essays on Syria in the Iron Age., Ancient Near Eastern studies. Supplement 7*. Edited by G. Bunnens, pp. 307-325. Louvain: Peeters Press.

Elayi, J. 1982. Studies in Phoenician Geography during the Persian Period. *Journal of Near Eastern Studies* 41:83-110.

Elayi, J. 1986-1987. Les Importations Grecques a Tell Kazel (Simyra) a l'Epoque Perse. *Les Annales Archeologiques Arabes Syriennes* 36/37:132-135.

Elayi, J. 1987. "Al-Mina sur l'Oronte a l'epoque perse," in *Phoenicia and the East Mediterranean in the First Millennium B.C., Studia Phoenicia 5*. Edited by E. Lipinski, pp. 249-265. Leuven: Uitgeverij Peeters.

Elayi, J., and H. Sayegh. 2000. *Un Quartier du Port Phénicien de Beyrouth au Fer III/Perse. Archeologie et Histoire. Supplement no. 7 à Transeuphraténe*. Paris: Gabalda.

Elon, A. 1971. *The Israelis: Founder and Sons*. New York: Holt, Rinehart and Winston.

Elon, A. 1997. "Politics and Archaeology," in *The Archaeology of Israel: Constructing the Past, Interpreting the Present, Journal for the Study of the Old Testament Supplement Series 237*. Edited by N. A. Silberman and D. Small, pp. 34-48. Sheffied: Sheffield Academic Press.

Emberling, G., and N. Yoffee. 1999. "Thinking About Ethnicity in Mesopotamian Archaeology and History," in *Fluchtpunkt Uruk. Archaologische Einheit aus Methodischer Vielfalt. Schriften fur Hans Jorg Nissen, Internationale Archaologie Studia Honoria 6*. Edited by H. Kuhne, R. Bernbeck, and K. Bartl, pp. 272-281. Rahden.

Engberg, R. M. 1940. Historical Analysis of Archaeological Evidence: Megiddo and the Song of Deborah. *Bulletin of the American Schools of Oriental Research* 78:4-9.

Esse, D. L. 1992. The Collared Pithos at Megiddo: Ceramic Distribution and Ethnicity. *Journal of Near Eastern Studies* 51:81-104.

Esse, D. L., and T. P. Harrison. 2004. "History of Excavations," in *Megiddo 3. Final Report on the Stratum VI Excavations, Oriental Insititute Publications Volume 127*. Edited by T. P. Harrison, pp. 1-6. Chicago: University of Chicago.

Evans, J. 1949. The Royal Archaeological Instituite: A Retrospect. *Archaeological Journal* 106:1-11.

Evans, J. 1956. *The History of the Society of Antiquaries*. Oxford: Oxford University Press.

Fales, F. M. 1990. *The Rural Landscape of the Neo-Assyrian Empire: A Survey. State Archives of Assyria Bulletin IV (2)*. Padova: Sargon.

Fales, F. M. 1995. "Rivers in Neo-Assyrian Geography," in *Neo-Assyrian Geography: Papers Presented at the International Conference Held in the University of Rome 'La Sapienza', November 10th to 12th, 1993., Quaderni di geografia storica ; 5*. Edited by M. Liverani, pp. 203-215. Roma: Università di Roma 'La Sapienza' Dipartimento di Scienze storiche archeologiche e antropologiche dell'Antichità.

Fales, F. M. 1999. "The Tablets from Tell Shioukh Fawqani/Burmarina in the Context of Assyro-Aramaic Studies," in *Archaeology of the Upper Syrian Euphrates, The Tishrin Dam Area: Proceedings of the International Symposium Held at Barcelona, January 28th-30th 1998*. Edited by G. del Olmo Lete and J.-L. Montero Fenollos, pp. 625-636. Barcelona: Editorial AUSA.

Fales, F. M., K. Radner, C. Pappi, and E. Attardo. 2005. "The Assyrian and Aramaic Texts from Tell Shiukh Fawqani," in *Tell Shiyukh Fawqani 1994-1998*. Edited by L. Bachelot and F. M. Fales, pp. 595-701. Padova: Sargon Editrice e Libreria.

Falsone, G. 1988. "Phoenicia as a bronzeworking centre in the Iron Age," in *Bronzeworking centres of Western Asia c. 1000-539 BC*. Edited by J. Curtis, pp. 227-252. London: Keegan Paul International.

Falsone, G. 1995. Salvage excavations at Tell Shiyukh Tahtani. *Akkadica* 92:37-38.

Falsone, G. 1998. Tell Shiyukh Tahtani on the Euphrates. The University of Palermo escue excavations in North Syria (1993-1994). *Akkadica* 109-110:22-64.

Falsone, G. 1999. "Tell Shiyukh Tahtani," in *Archaeology of the Upper Syrian Euphrates: The Tishrin Dam Area*. Edited by G. del Olmo Lete and J.-L. Montero Fenollos, pp. 137-142. Barcelona.

Fantalkin, A. 2001. Low Chronology and Greek Protogeometric and Geometric Pottery in the Southern Levant. *Levant* 33:117-125.

Fantalkin, A., and I. Finkelstein. 2006. The Sheshonq I Campaign and the 8th-Century-BCE Earthquake-More on the Archaeology and History of the South in the Iron I-IIA. *Tel Aviv* 33(1):18-42.

Faust, A. 2000. Ethnic Complexity in Northern Israel During Iron Age II. *Palestine Exploration Quarterly* 132:1ff.

Faust, A. 2002. Burnished Pottery and Gender Hierarchy in Iron Age Israelite Society. *Journal of Mediterranean Archaeology* 15(1):53-73.

Faust, A. 2004. Mortuary Practices, Society and Ideology; The Lack of Iron Age I Burials in the Highlands in Context. *Israel Exploration Journal* 54:174-190.

Faust, A. 2006. *Israel's Ethnogenesis: Settlement, Interaction, Expansion and Resistance*. London: Equinox.

Fay, B. 1998. "Introduction: The Linguistic Turn and Beyond in Contemporary Theory of History," in *History and Theory: Contemporary Readings*. Edited by B. Fay, P. Pomper, and R. T. Vann, pp. 1-12. Malden, MA: Blackwell Publishers.

Felli, C., and E. Merluzzi. 2002. Afis 2000-2001. Area E: l'occupazione del Bronze Medio. *Egitto e Vicino Oriente* 25:24-29.

Ferguson, L. 1977. "Historical archaeology and the importance of material things," in *Historical Archaeology and the Importance of Material Things, Society for Historical Archaeology, Special Publication 2*. Edited by L. Ferguson, pp. 5-8.

Finkbeiner, U. 2001-2002. BEY 020 - The Iron Age Fortification. *Aram* 13-14:27-36.

Finkbeiner, U., and H. Sader. 1997. Bey 020 Preliminary report of the excavations 1995. *Baal*:114-166.

Finkelstein, I. 1988a. *The Archaeology of the Israelite Settlement*. Jerusalem.

Finkelstein, I. 1988b. Arabian Trade and Socio-Political Conditions in the Negev in the Twelfth-Eleventh Centuries BCE. *Journal of Near Eastern Studies* 47(4):241-252.

Finkelstein, I. 1990. On Archaeological Methods and Historical Considerations: Iron Age II Gezer and Samaria. *Bulletin of the American Schools of Oriental Research* 277/278:109-119.

Finkelstein, I. I. 1994. "The Emergence of Israel: A Phase in the Cyclic History of Canaan in the Third and Second Millennia BCE," in *From Nomadism to Monarchy: Archaeological and Historical Aspects of Early Israel*. Edited by I. Finkelstein and N. Na'aman, pp. 150-178. Jerusalem: Israel Exploration Society.

Finkelstein, I. 1995a. The Date of the Settlement of the Philistines in Canaan. *Tel Aviv* 22:213-239.

Finkelstein, I. 1995b. *Living on the Fringe: The Archaeology and History of the Negev, Sinai and Neighbouring Regions in the Bronze and Iron Ages. Monographs in Mediterranean Archaeology.* Sheffield: Sheffield Academic Press.

Finkelstein, I. 1996a. The Archaeology of the United Monarchy: An Alternative View. *Levant* 28:177-188.

Finkelstein, I. 1996b. The Stratigraphy and Chronology of Megiddo and Beth Shean in the 12th-11th Centuries BCE. *Tel Aviv* 23:170-184.

Finkelstein, I. 1996c. "Toward a New Periodization and Nomenclature of the Archaeology of the Southern Levant," in *The Study of the Ancient Near East in the Twenty-First Century.* Edited by J. S. Cooper and G. M. Schwartz, pp. 103-123. Winona Lake, Indiana: Eisenbrauns.

Finkelstein, I. 1997. "Pots and People Revisited: Ethnic Boundaries in the Iron Age I," in *The Archaeology of Israel: Constructing the Past, Interpreting the Present, Journal for the Study of the Old Testament Supplement Series 237.* Edited by N. A. Silberman and D. Small, pp. 216-237. Sheffied: Sheffield Academic Press.

Finkelstein, I. 1998a. Notes on the Stratigraphy and Chronology of Iron Age Ta'anach. *Tel Aviv* 25:208-218.

Finkelstein, I. 1998b. "Philistine Chronology: High, Middle or Low?," in *Mediterranean Peoples in Transition: Thirteenth to Early Tenth Centuries BCE-In Honour of Professor Trude Dothan.* Edited by S. Gitin, A. Mazar, and E. Stern, pp. 140-147. Jerusalem: Israel Exploration Society.

Finkelstein, I. 1998c. Bible Archaeology or Archaeology of Palestine in the Iron Age? A Rejoinder. *Levant* 30:167-174.

Finkelstein, I. 1999a. Hazor and the North in the Iron Age: A Low Chronology Perspective. *Bulletin of the American Schools of Oriental Research* 314:55-70.

Finkelstein, I. 1999b. State Formation in Israel and Judah: A Contrast in Context, A Contrast in Trajectory. *Near Eastern Archaeology* 62:35-52.

Finkelstein, I. 2000. Hazor XII-XI with an Addendum on Ben-Tor's Dating of Hazor X-VII. *Tel Aviv* 27:231-247.

Finkelstein, I. 2002. The Campaign of Shoshenq I to Palestine: A Guide to the 10th Century BCE Polity. *Zeitschrift des Deutschen Palästina-Vereins* 118:109-135.

Finkelstein, I. 2003. "City States and States: Polity Dynamics in the 10th-9th centuries BCE," in *Symbiosis, Symbolism, and the Power of the Past. Canaan, Ancient Israel, and Their Neighbours from the Late Bronze Age through Roman Palaestina.* Edited by W. G. Dever and S. Gitin, pp. 75-83. Winona Lake, IN: Eisenbrauns.

Finkelstein, I. 2004. Tel Rehov and Iron Age Chronology. *Levant* 36:181-188.

Finkelstein, I. 2005. "A Low Chronology Update: Archaeology, history and bible," in *The Bible and Radiocarbon Dating: Archaeology, Text and Science.* Edited by T. E. Levy and T. Higham, pp. 31-42. London: Equinox.

Finkelstein, I. 2006. "The Iron Age Pottery: Levels L-5, L-3, H-5 and H-4," in *Megiddo IV. The 1998-2002 Seasons, Tel Aviv University Sonia and Marco Nadler Institute of Archaeology Monograph Series Number 24.* Edited by I. Finkelstein, D. Ussishkin, and B. Halpern, pp. 303-314. Jerusalem: Institute of Archaeology, Tel Aviv University.

Finkelstein, I. 2007. "Is the Philistine Paradigm Still Viable?," in *The Synchronisation of Civilisations in the Eastern Mediterranean in the Second Millennium B.C. III.* Edited by M. Bietak and E. Czerny, pp. 517-523. Vienna: Verlag der Österreichischen Akademie der Wissenschaften.

Finkelstein, I., D. Ussishkin, and B. Halpern. 2000. "Introduction," in *Megiddo III. The 1992-1996 Seasons.* Edited by I. Finkelstein, D. Ussishkin, and B. Halpern, pp. 1-13. Jerusalem: Institute of Archaeology, Tel Aviv University.

Finkelstein, I., O. Zimhoni, and A. Kafri. 2000. "The Iron Age Pottery Assemblages from Areas F, K and H and their Stratigraphic and Chronological Implications," in *Megiddo III. The 1992-1996 Seasons, Tel Aviv University Sonia and Marco Nadler Institute of Archaeology Monograph Series Number 18.* Edited by I. Finkelstein, D. Ussishkin, and B. Halpern, pp. 244-324. Jerusalem: Institute of Archaeology, Tel Aviv University.

Finkelstein, I., and E. Piasetzky. 2003a. Recent Radiocarbon Results and Biblical History. *Antiquity* 77:771-779.

Finkelstein, I., and E. Piasetzky. 2003b. Wrong and Right; High and Low C14 Dates from Tel Rehov and Iron Age Chronology. *Tel Aviv* 30:283-295.

Finkelstein, I., and E. Piasetzky. 2003c. Comment on "C14 Dates from Tel Rehov: Iron Age Chronology, Pharaohs, and Hebrew Kings". Science 302: 568b.

Finkelstein, I., and E. Piasetzky. 2006. The Iron I-IIA in the Highland and Beyond: 14C Anchors, Pottery Phases and The Shoshenq I Campaign. *Levant* 38:45-62.

Finkelstein, I., D. Ussishkin, and B. Halpern. 2006. "Archaeological and Historical Conclusions," in *Megiddo IV. The 1998-2002 Seasons, Tel Aviv University Sonia and Marco Nadler Institute of Archaeology Monograph Series Number 24.* Edited by I. Finkelstein, D. Ussishkin, and B. Halpern, pp. 843-859. Jerusalem: Institute of Archaeology, Tel Aviv University.

Forsberg, S. 1995. *Near Eastern Destruction Datings as Sources for Greek and Near Eastern Iron Age Chronology. Archaeological and historical studies. The cases of Samaria (722 B.C.) and Tarsus (696 B.C.). Uppsala Studies in Ancient Mediterranean and Near Eastern Civilisations 19.* Uppsala: Acta Universitatis Upsaliensis.

Fortin, M. 1999. *Syria: Land of Civilizations.* Montreal.

Fortin, M. 2006a. "Tell 'Acharneh: rapport préliminaire sur la campagne exploratoire de 1998," in *Tell 'Acharneh 1998-2004. Preliminary Reports on Excavation Campaigns and Study Seaon, Subartu XVIII.* Edited by M. Fortin, pp. 83-106. Turnhout: BREPOLS Publishers.

Fortin, M. 2006b. "Tell 'Acharneh: rapport préliminaire sur la campagne de fouilles de 2001," in *Tell 'Acharneh 1998-2004. Preliminary Reports on Excavation Campaigns and Study Seaon, Subartu XVIII*. Edited by M. Fortin, pp. 107-120. Turnhout: BREPOLS Publishers.

Fortin, M. 2006c. "Tell 'Acharneh: rapport préliminaire sur la campagne de fouilles 2002," in *Tell 'Acharneh 1998-2004. Preliminary Reports on Excavation Campaigns and Study Seaon, Subartu XVIII*. Edited by M. Fortin, pp. 121-140. Turnhout: BREPOLS Publishers.

Frame, G. 2006. "The Tell Acharneh Stela of Sargon II of Assyria," in *Tell 'Acharneh 1998-2004. Preliminary Reports on Excavation Campaigns and Study Seaon, Subartu XVIII*. Edited by M. Fortin, pp. 49-68. Turnhout: BREPOLS Publishers.

Francis, E. D., and M. Vickers. 1985. Greek Geometric Pottery at Hama and its Implications for Near Eastern Chronology. *Levant* 17:131-138.

Frankenstein, S. 1979. "The Phoenicians in the Far West: A Function of Neo-Assyrian Imperialism," in *Power and Propaganda A Symposium on Ancient Empires, Copenhagen Studies in Assyriology 7*. Edited by M. T. Larsen, pp. 263-298. Copenhagen: Akademisk Forlag.

Frankfort, H. 1952. The Origin of the Bit-Hilani. *Iraq* 14:120-131.

Franklin, N. 2006. Revealing Stratum V at Megiddo. *Bulletin of the American Schools of Oriental Research* 342:95-111.

Frantzen, A. J. 1990. *Desire for Origins. New Language, Old English, and Teaching the Tradition.* New Brunswick: Rutgers University Press.

Freeth, Z., and H. V. F. Winstone. 1978. *Explorers of Arabia: From the Renaissance to the End of the Victorian Era*. London.

French, D. H., J. Moore, and H. F. Russell. 1982. Excavations at Tille 1979-1982. An Interim Report. *Anatolian Studies* 32:161-187.

Frenkel, Y. a. 2001. "Muslim Travellers to *Bilad al-Sham* (Syria and Palestine) from the thirteenth to the sixteenth centuries: Maghribi travel accounts," in *Travellers in the Levant: Voyagers and Visionaries*. Edited by S. Searight and M. Wagstaff, pp. 109-120. Durham: Astene.

Fritz, V. 2000. "Bronze Age Features in Cities of the Early Iron Age in the Southern Levant," in *Proceedings of the First International Congress on the Archaeology of the Ancient Near East, Rome, May 18-23, 1998. Volume I*. Edited by P. Matthiae, A. Enea, L. Peyronel, and F. Pinnock, pp. 507-512. Rome: Universita degli Studi di Roma 'La Sapienza'.

Frost, H. 2004. "Byblos and the Sea," in *Decade: A Decade of Archaeology and History in the Lebanon*. Edited by C. Doumet-Serhal, pp. 316-347. Beirut: The Lebanese British Friends of the National Museum.

Frye, R. N. 1992. Asyria and Syria: synonyms. *Journal of Near Eastern Studies* 51:281-285.

Fuchs, A. 1994. *Die Inschriften Sargons II aus Khorsabad*. Göttingen: Cuvillier Verlag

Fugmann, E. 1958. *Hama: fouilles et recherches, 1931-1938: L'architecture des périodes pré-hellénistiques*. Vol. 2.1. Copenhague: NationalMuséet.

Funk, R. W., and H. N. Richardson. 1958. The 1958 Sounding at Pella. *Biblical Archaeologist* 21:82-96.

Gaal, E. 1976. "The King Parrattarnna died and was cremated?," in *Wirtschaft und Gesellschaft im Alten Vorderasien*. Edited by J. Harmatta and G. Komorocy, pp. 281-286. Budapest.

Gadd, C. J. 1954. Inscribed Prisms of Sargon II from Nimrud. *Iraq* 16:173-201.

Gal, Z. 1988. The Late Bronze Age in Galilee: A Reassessment. *Bulletin of the American Schools of Oriental Research* 272:79-84.

Gal, Z. 1992a. Hurbat Rosh Zayit and the Early Phoenician Pottery. *Levant* 24:173-186.

Gal, Z. 1992b. *Lower Galilee During the Iron Age. Dissertation series (American Schools of Oriental Research) ; no. 9.* Winona Lake, IN: Eisenbrauns.

Gal, Z. 2003. The Iron Age 'Low Chronology' in Light of the Excavations at Horvat Rosh Zayit. *Israel Exploration Journal* 53:147-150.

Galaty, M. L. 1999. *Nestor's Wine Cups. Investigating Ceramic Manufacture and Exchange in a Late Bronze Age "Mycenaean" State. BAR International Series 766*. Oxford: BAR Publishing.

Gallagher, W. R. 1999. *Sennacherib's Campaign to Judah. New Studies. Studies in the History and Culture of the Ancient Near East XVIII*. Leiden: Brill.

Garnsey, P. 1999. *Food and Society in Classical Antiquity*. Cambridge: Cambridge University Press.

Garstang, J. 1931. *Joshua-Judges: The Foundations of Bible History*. London.

Geertz, C. 1973. *The Interpretation of Cultures*. New York.

Gelb, I. J. 1939. *Hittite Hieroglyphic Monuments. Oriental Institute Publications 45*. Chicago: University of Chicago Press.

Gelin, M. 2004. *l'Archéologie en Syrie et au Liban à l'Époque du Mandat (1919-1946): Histoire & Organisation*. Paris: Geuthner.

George, A. R. 1997. "Assyria and the Western World," in *Assyria 1995: Proceedings of the 10th Anniversary Symposium of the Neo-Assyrian Text Corpus Project Helsinki, September 7-11, 1995*. Edited by S. Parpola and R. M. Whiting, pp. 69-76. Helsinki.

Gershuny, L. 1985. *Bronze Vessels from Israel and Jordan. Prähistorische Bronzefunde II.6*. München: C.H. Beck'sche Verlagsbuchhandlung.

Geva, S. 1971. A Reassessment of the Chronology of Beth Shean Strata V and IV. *Israel Exploration Journal* 29:6-10.

Geva, S. 1982. Archaeological Evidence for the Trade Between Israel and Tyre? *Bulletin of the American Schools of Oriental Research* 248:69-72.

Giddens, A. 1984. *The Constitution of Society: Outline of the Theory of Structuration*. Berkeley and Los Angeles: University of California Press.

Gilboa, A. 1989. New Finds at Tell Dor and the Beginning of Cypro-Geometric Pottery Import to Palestine. *Israel Exploration Journal* 39:204-218.

Gilboa, A. 1996. Assyrian Type Pottery at Dor and the Status of the Town during the Assyrian Occupation Period. *Eretz Israel* 25:122-135 (Hebrew).

Gilboa, A. 1999. The Dynamics of Phoenician Bichrome Pottery: A View from Tel Dor. *Bulletin of the American Schools of Oriental Research* 316:1-22.

Gilboa, A., and I. Sharon. 2001. Early Iron Age Radiometric Dates from Tel Dor: Preliminary Implications for Phoenicia and Beyond. *Radiocarbon* 43:1343-1351.

Gilboa, A., and I. Sharon. 2003. An Archaeological Contribution to the Early Iron Age Chronological Debate: Alternative Chronologies for Phoenicia and Their Effects on the Levant, Cyprus and Greece. *Bulletin of the American Schools of Oriental Research* 332:7-80.

Gilmour, G. 1995. "Aegean Influence in Late Bronze Age Funerary Practices in the Southern Levant," in *The Archaeology of Death in the Ancient Near East, Oxbow Monographs 51*. Edited by S. Campbell and A. Green, pp. 155-170. Oxford: Oxbow Books.

Ginsberg, H. L. 1969a. "The Legend of King Keret," in *Ancient Near Eastern Texts Relating to the Old Testament*, Third Edition with Supplement edition. Edited by J. B. Pritchard, pp. 142-149. Princeton, New Jersey: Princeton University Press.

Ginsberg, H. L. 1969b. "The Tale of Aqhat," in *Ancient Near Eastern Texts Relating to the Old Testament*, Third Edition with Supplement edition. Edited by J. B. Pritchard, pp. 149-155. Princeton, New Jersey: Princeton University Press.

Gitin, S. 2002. The House That Albright Built. *Near Eastern Archaeology* 65:5-10.

Gitin, S. 2003. "The Philistines: Neighbours of the Canaanites, Phoenicians and Israelites," in *One Hundred Years of American Archaeology in the Middle East*. Edited by D. R. Clark and V. H. Matthews, pp. 57-86. Boston: American Schools of Oriental Research.

Gjerstad, E. 1944. The Initial Date of the Cypriote Iron Age. *Opuscula Archaeologica* 3:73-106.

Gjerstad, E. 1948. *The Swedish Cyprus Expedition IV.2: The Cypro-Geometric, Cypro-Archaic and Cypro-Classical Periods*. Stockholm: The Swedish Cyprus Expedition.

Gjerstad, E. 1953. Reply to Albright's "Correspondence with Professor Einar Gjerstad on the Chronology of 'Cypriote' Pottery from Early Iron Levels in Palestine". *Bulletin of the American Schools of Oriental Research* 130:22-26.

Glatz, C., and R. Matthews. 2005. Anthropology of a Frontier Zone: Hittite-Kaska Relations in Late Bronze Age North-Central Anatolia. *Bulletin of the American Schools of Oriental Research* 339:47-65.

Glueck, N. 1938. Archaeological Exploration and Excavation in Palestine, Transjordan, and Syria during 1937. *American Journal of Archaeology* 42(1):165-176.

Gonnella, J., W. Khayyata, and K. Kohlmeyer. 2005. *Die Zitadelle Von Aleppo und der Tempel des Wettergottes: Neue Forschungen und Entdeckungen*. Munster: Rhema.

Goodby, R. G. 1998. "Technological Patterning and Social Boundaries: Ceramic Variability in Southern New England, A.D. 1000-1675," in *The Archaeology of Social Boundaries*. Edited by M. T. Stark, pp. 161-182. Washington and London: Smithsonian Institution Press.

Grabois, A. 1988. "Medieval Pilgrims, the Holy Land and its Image in European Civilisation," in *The Holy Land in History and Thought: Papers Submitted to the International Conference on the Relations Between the Holy Land and the World Outside it, Johannesburg 1986*. Edited by M. Sharon, pp. 65-79. Leiden.

Gras, M., Rouillard, P. and J. Teixidor. 1991. The Phoenicians and Death. *Berytus* 39: 127-176.

Grayson, A. K. 1991a. *Assyrian Rulers of the Early First Millennia BC I (1114-859 BC). The Royal Inscriptions of Mesopotamia Assyrian Periods 2*. Toronto: University of Toronto Press.

Grayson, A. K. 1991b. "Assyria: Tiglath-pileser III to Sargon II (744-705 B.C.)," in *The Assyrian and Babylonian Empires and other States of the Near East, from the Eighth to the Sixth Centuries B.C., Cambridge Ancient History III.2 (Second Edition)*. Edited by J. Boardman, I. E. S. Edwards, N. G. L. Hammond, E. Sollberger, and C. B. F. Walker, pp. 71-102. Cambridge: Cambridge University Press.

Grayson, A. K. 1996. *Assyrian Rulers of the Early First Millennia BC II (858-745 BC). The Royal Inscriptions of Mesopotamia Assyrian Periods 3*. Toronto: University of Toronto Press.

Green, A., and A. Hausleiter. 2000. Tell Ahmar Survey. Preliminary Report on the Work of the British-German Team in 1999. *Orient-Express* 2000:79-81.

Green, A., and A. Hausleiter. 2001. Tell Ahmar Survey. Preliminary Report on the Work in 2000. *Orient-Express* 2001:103-106.

Greenacre, M. J. 1984. *Theory and Applications of Correspondence Analysis*. London: Academic Press, Inc.

Griffiths, D. 2003. Petrographic Analysis on Pottery from Tomb IV at Tell el Rachidieh. *Archaeology and History in the Lebanon* 17:58-59.

Gröndahl, F. 1967. *Die Personennamen der Texte aus Ugarit*. Rome: Päpstliches Bibelinstitut.

Gubel, E. 1990a. Tell Kazel, Syria: Excavations of the AUB Museum, 1985-1987, Preliminary Reports: Area I. *Berytus* 38:23-54.

Gubel, E. 1990b. Tell Kazel (Sumur/Simyra) a l'epoque perse. *Transeuphraténe* 2:37-49.

Gubel, E. 1994. "Phoenician Foundations in Archaeological Perspective," in *Nuove fondazioni nel Vicino Oriente Antico: realta e ideologia*. Edited by S. Mazzoni, pp. 341-355. Pisa.

Gubel, E. 1995. "The AUB Excavations at Tell Kazel, Syria: the Oriental Material from Area I," in *Actes du IIIe Congrès International des Études Phéniciennes et Puniques. Tunis, 11-16 Novembre 1991*, pp. 118-127. Tunis: Institut National du Patrimoine.

Gubel, E. 2000. "Amurru, Sumur (Tell Kazel) and the Mitanni. New Perspectives," in *Proceedings of the First International Congress on the Archaeology of the Ancient Near East, Rome, May 18-23, 1998. Volume I.* Edited by P. Matthiae, A. Enea, L. Peyronel, and F. Pinnock, pp. 573-586. Rome: Universita degli Studi di Roma 'La Sapienza'.

Gubel, E., and E. Lipinski. Editors. 1985. *Phoenicia and Its Neighbours. Studia Phoenicia III.* Leuven: Peeters.

Guigues, P. E. 1937. Lébé'a, Kafer-Ğarra, Qrayé, Nécropoles de la Région Sidonienne. *Bulletin du Musée de Beyrouth* 1:35-76.

Guigues, P. E. 1938. Lébé'a, Kafer-Ğarra, Qrayé, Nécropoles de la Région Sidonienne. *Bulletin du Musée de Beyrouth* 2:27-72.

Guigues, P. E. 1939. Lébé'a, Kafer-Ğarra, Qrayé, Nécropoles de la Région Sidonienne. *Bulletin du Musée de Beyrouth* 3:53-63.

Guy, P. L. O. 1931. *New Light from Armageddon: Second Provisional Report (1927-29) on the Excavations at Megiddo in Palestine. Oriental Institute Publications 9.* Chicago: University of Chicago Press.

Guy, P. L. O. 1938. *Megiddo Tombs. Oriental Institute Publications 33.* Chicago: University of Chicago Press.

Haarer, P. 2001. "Problematising the Transition from Bronze to Iron," in *The Social Context of Technological Change. Egypt and the Near East, 1650-1550 B.C.* Edited by A. J. Shortland, pp. 255-273. Oxford: Oxbow Books.

Hachmann, R. 1966. "Verlauf und Ergebnisse der Grabung des Jahres 1964," in *Bericht uber die Ergebnisse der Ausgrabungen in Kamid el-Loz (Libanon) in den Jahren 1963 und 1964, Saarbrucker Beitrage zur Altertumskunde Band 3.* Edited by R. Hachmann and A. Kuschke, pp. 43-68. Bonn: Rudolf Habelt Verlag.

Hachmann, R. 1969a. Rapport Preliminaires sur les Fouilles au Tell de Kamid el-Loz de 1966 a 1968: I. Introduction. *Bulletin du Musée de Beyrouth* 22:49-51.

Hachmann, R. 1969b. Rapport Preliminaires sur les Fouilles au Tell de Kamid el-Loz de 1966 a 1968: II. Historique des Quartre Campagnes de Fouilles. *Bulletin du Musée de Beyrouth* 22:53-60.

Hachmann, R. 1969c. Rapport Preliminaires sur les Fouilles au Tell de Kamid el-Loz de 1966 a 1968: III. Nouvelles decouvertes provenant du cimetiere de l'epoque Perse. *Bulletin du Musée de Beyrouth* 22:61-65.

Hachmann, R. 1969d. Rapport Preliminaires sur les Fouilles au Tell de Kamid el-Loz de 1966 a 1968: V. Le cemetiere de l'age du Bronze Moyen sur la Pente Nord du Tell. *Bulletin du Musée de Beyrouth* 22:77-84.

Hachmann, R. (ed.) 1970a. *Bericht uber die Ergebnisse der Ausgrabungen in Kamid el-Loz (Libanon) in den Jahren 1966 und 1967. Saarbrucker Beitrage zur Altertumskunde Band 4.* Bonn: Rudolf Habelt Verlag.

Hachmann, R. 1970b. Grabungen auf dem Tell Kamid el-Loz (Libanon). *Archiv fur Orientforschung* 23:135-140.

Hachmann, R. 1978a. Rapport preliminaire sur les fouilles au Tell de Kamid el-Loz de 1969 a 1972, avec un appendice concernant des decouvertes epigraphiques par G. Wilhelm. *Bulletin du Musée de Beyrouth* 30:7-26.

Hachmann, R. 1978b. Rapport preliminaire sur les fouilles au Tell de Kamid el-Loz en 1973. *Bulletin du Musée de Beyrouth* 30:27-41.

Hachmann, R. (ed.) 1983. *Fruhe Phoniker im Libanon: 20 Jahre deutsche Ausgrabungen in Kamid el-Loz.* Mainz am Rhein: Verlag Philipp von Zabern.

Hachmann, R. 1989. Kamid el-Loz 1963-1981. German Excavations in Lebanon (Part 1). *Berytus* 37:5-187.

Hachmann, R., and A. Kuschke. 1966. Rapport Preliminaires sur les travaux au Tell Kamid el-Loz durant les annees 1963 et 1964. *Bulletin du Musée de Beyrouth* 19:107-134.

Hachmann, R., and M. Metzger. 1973. Arbeiten auf dem Tell Kamid el-Loz (Libanon) 1970 und 1971. *Archiv fur Orientforschung* 24:176-180.

Haggi, A. 2006. Phoenician Atlit and its Newly-Excavated Harbour: A Reassessment. *Tel Aviv* 33(1):43-60.

Haines, R. C. 1971. *Excavations in the Plain of Antioch. vol. 2. The Structural Remains of the Later Phases. Oriental Institute Publications 95.* Chicago: University of Chicago.

Hakker-Orion, D. 1984. "The Role of the Camel in Israel's Early History," in *Animals and Archaeology: 3. Early Herders and their Flocks, BAR International Series 202.* Edited by J. Clutton-Brock and C. Grigson, pp. 207-212. Oxford: BAR.

Hall, J. M. 1997. *Ethnic Identity in Greek Antiquity.* Cambridge: Cambridge University Press.

Hallo, W. W. 1992. "From Bronze Age to Iron Age in Western Asia: Defining the Problem," in *The Crisis Years: The 12th Century B.C. From Beyond the Danube to the Tigris.* Edited by W. A. Ward and M. S. Joukowsky, pp. 1-9. Iowa: Kendall Hunt.

Hamilakis, Y. 1996. Wine, Oil and the Dialectics of Power in Bronze Age Crete: A Review of the Evidence. *Oxford Journal of Archaeology* 15(1):1-32.

Hamilakis, Y. 1999. Food Technologies/Technologies of the Body: The Social Context of Wine and Oil Production and Consumption in Bronze Age Crete. *World Archaeology* 31(1):38-54.

Hamilton, R. W. 1934. Tell Abu Hawam; Interim Report. *Quarterly of the Department of Antiquities in Palestine* 3:74-80.

Hamilton, R. W. 1935. Excavations at Tell Abu Hawam. *Quarterly of the Department of Antiquities in Palestine* 4:1-69.

Hammade, H. 1999. The Cuneiform Inscription from Aleppo Citadel. *Les Annales Archeologiques Arabes Syriennes* 43: 251-252.

Hanfmann, G. M. A. 1951. Review: The Swedish Cyprus Expedition, Vol. IV, Part 2; The Cypro-Geometric, Cypro-Archaic and Cypro-Classical Periods by Einar Gjerstad. *American Journal of Archaeology* 55(4):424-428.

Hankey, V. 1967. Mycenaean Pottery in the Middle East: Notes on Finds Since 1951. *Annual of the British School at Athens* 62:107-147.

Hankey, V. 1993. "Pottery as Evidence of Trade: the Levant from the Mouth of the River Orontes to the Egyptian Border," in *Wace and Blegen: Pottery as Evidence for Trade in the Aegean Bronze Age. 1939-1989*. Edited by C. Zerner, P. Zerner, and J. Winder, pp. 101-108. Amsterdam: J C Gieben.

Hannestad, L. 1996. "Absolute Chronology: Greece and the Near East c. 1000-500 BCE," in *Absolute Chronology: Archaeological Europe 2500-500 BC, Acta Archaeologica Vol. 67*. Edited by K. Randsborg, pp. 39-49. Kobenhavn: Munksgaard.

Harrison, T. P. 2001a. "Tell Ta'yinat and the Kingdom of Unqi," in *The World of the Aramaeans II: Studies in History and Archaeology in Honour of Paul-Eugene Dion, Journal for the Study of the Old Testament Supplement Series 325*. Edited by P. M. M. Daviau, J. W. Wevers, and M. Weigl, pp. 115-132. Sheffield: Sheffield Academic Press.

Harrison, T. P. 2001b. The Evidence for Aramaean Cultural Expansion in the Amuq Plain. *Canadian Society for Mesopotamian Studies Bulletin* 36:135-144.

Harrison, T. P. 2004a. "Stratigraphy and Chronology," in *Megiddo 3. Final Report on the Stratum VI Excavations, Oriental Insititute Publications Volume 127*. Edited by T. P. Harrison, pp. 8-14. Chicago: University of Chicago.

Harrison, T. P. 2004b. "The Pottery," in *Megiddo 3. Final Report on the Stratum VI Excavations, Oriental Insititute Publications Volume 127*. Edited by T. P. Harrison, pp. 23-42. Chicago: University of Chicago.

Harrison, T. P. 2004c. "Cultural and Historical Synthesis The Stratum VI Settlement," in *Megiddo 3. Final Report on the Stratum VI Excavations, Oriental Insititute Publications Volume 127*. Edited by T. P. Harrison, pp. 105-108. Chicago: University of Chicago.

Harrison, T. P., S. Batiuk, and L. Pavlish. 2004. The 2002 Ta'yinat Archaeological Project Geomagnetic Survey. *Arkeometri Sonuclari Toplantisi* 19:121-130.

Harrison, T. P., S. Batiuk, and L. Pavlish. 2005. 2003 Yili Tayinat Jeomanyetik Arastirmasi. *Arkeometri Sonuclari Toplantisi* 20:153-158.

Harrison, T. P., S. Batiuk, and H. Snow. 2006. 2004 Yili Tayinat Hoyuk Kazilari. *Kazi Sonuclari Toplantisi* 27:353-362.

Harvey, F. D. 1976. Sostratos of Aegina. *Parola del Popolo* 31:206-214.

Hawkins, J. D. 1972. Building Inscriptions of Carchemish. The Long Wall of Sculpture and Great Staircase. *Anatolian Studies* 22:87-114.

Hawkins, J. D. 1972-1975. "Hamath," in *Reallexikon der Assyriologie und Vorderasiatischen Archäologie*. Edited by E. Ebeling and B. Meissner, pp. 67-70. Berlin: Walter de Gruyter.

Hawkins, J. D. 1974. Assyrians and Hittites. *Iraq* 36:67-83.

Hawkins, J. D. 1976-1980. "Karkamiş," in *Reallexikon der Assyriologie und Vorderasiatischen Archäologie*. Edited by E. Ebeling and B. Meissner, pp. 426-446. Berlin: Walter de Gruyter.

Hawkins, J. D. 1980. The 'Autobiography of Ariyahina's Son': An Edition of the Hieroglyphic Luwian Stelae Tell Ahmar 1 and Aleppo 2. *Anatolian Studies* 30:139-156.

Hawkins, J. D. 1982. "The Neo-Hittite States in Syria and Anatolia," in *The Prehistory of the Balkans; and the Middle East and the Aegean World, Tenth to Eighth Centuries B.C., Cambridge Ancient History III.1 (Second Edition)*. Edited by J. Boardman, I. E. S. Edwards, N. G. L. Hammond, and E. Sollberger, pp. 372-442. Cambridge: Cambridge University Press.

Hawkins, J. D. 1983. The Hittite Name of Til Barsip: Evidence From a New Hieroglyphic Fragment from Tell Ahmar. *Anatolian Studies* 33:131-136.

Hawkins, J. D. 1988. Kuzi-Teshub and the 'Great Kings' of Karkamish. *Anatolian Studies* 38:99-108.

Hawkins, J. D. 2002. "Anatolia: The End of the Hittite Empire and After," in *Die Nahöstlichen Kulturen und Griechenland an der Wende vom 2. zum 1. Jarhtausend v. Chr.* Edited by E. A. Braun-Holzinger and H. Matthaus, pp. 143-152. Mohnesee-Wamel: Bibliopolis.

Hawkins, J. D. 2003. "Scripts and Texts," in *The Luwians, Handbuch der Orientalistik 68*. Edited by H. C. Melchert, pp. 128-169. Leiden: Brill.

Healey, J. F. 1983. "Phoenician and the Spread of Aramaic," in *Atti del I Congresso Internazionale di Studi Fenici e Punici, Roma, 5-10 Novembre 1979, Volume Terzo*. Edited by S. Moscati, pp. 663-666. Roma: Consiglio Nazionale Delle Ricerche.

Healey, J. F. 1995. "Death in West Semitic Texts: Ugarit and Nabataea," in *The Archaeology of Death in the Ancient Near East, Oxbow Monographs 51*. Edited by S. Campbell and A. Green, pp. 188-191. Oxford: Oxbow Books.

Heffner, E. H., E. P. Blegen, and M. Burrows. 1932. Archaeological News. *American Journal of Archaeology* 36(1):43-73.

Heinz, M. 2004. "Kamid el-Loz: From Village to City and Back to Village. 3000 Years of Settlement History in the Beqa'a Plain," in *Decade: A Decade of Archaeology and History in the Lebanon*. Edited by C. Doumet-Serhal, pp. 562-581. Beirut: The Lebanese British Friends of the National Museum.

Heinz, M., and K. Bartl. 1997. Bey 024 'Place Debbas' Preliminary Report. *Baal* 2:236-257.

Heinz, M., D. Bonatz, A. Gilibert, H. Heckmann, I. Holzer, C. Jauss, J. Kirchhofer, P. Knötzele, L. Petersen, M. Sommer, and T. Zaven. 2001. Kamid el-Loz in the Beqa'a plain, Lebanon: Continuity and Change in the Settlement of a Region. *Baal* 5:5-92.

Heinz, M., A. Gross, E. John, L. Kirsch, S. Kulemann-Ossen, L. von Lengerich, C. Leschke, J. Nieling, C. von Rüden, and E. Wagner. 2004. Kamid el-Loz in the Beqa'a plain, Lebanon: Excavations in 2001, 2002 and 2004. *Baal* 8:83-118.

Heltzer, M. 1969. "Problems of the Social History of Syria in the Late Bronze Age," in *Syria nel Tardo Bronzo*. Edited by M. Liverani, pp. 31-46. Rome: Centro per le antichita e la storia dell'arte del Vicino Oriente.

Heltzer, M. 1976. *The Rural Community in Ancient Ugarit*. Wiesbaden: Dr Ludwig Reichert.

Heltzer, M. 1978. *Goods, Prices and the Organisation of Trade in Ugarit (Marketing and Transportation in the Eastern Mediterranean in the Second Half of the Second Millennium BCE*. Wiesbaden: Dr Ludwig Reichert Verlag.

Heltzer, M. 1982. *The Internal Organisation of the Kingdom of Ugarit*. Wiesbaden: Dr Ludwig Reichert Verlag.

Heltzer, M. 1995. "Phoenician Trade and Phoenicians in Hamath," in *Immigration and Emigration within the Ancient Near East, Orientalia Lovaniensia Analecta 65*. Edited by K. van Lerberghe and A. Schoors, pp. 101-105. Leuven: Peeters.

Heltzer, M. 1996. "The Symbiosis of the Public and Private Sectors in Ugarit, Phoenicia, and Palestine," in *Privatization in the Ancient Near East and Classical World: a Colloquium held at New York University, November 17-18, 1994*. Edited by M. Hudson and B. Levine, pp. 177-196. Cambridge, Mass.: Peabody Museum of Archaeology and Ethnology.

Heltzer, M. 2001. "A New Weight from Hamath and Trade Relations with the South in the Ninth-Eighth Centuries BCE," in *The World of the Aramaeans II*. Edited by P. M. M. Daviau, J. W. Wevers, and M. Weigl, pp. 133-135. Sheffield.

Herodotus. 2003 edition. *The Histories*. London: Penguin.

Herbert, S. 2003. "Excavating Ethnic Strata: The Search for Hellenistic Phoenicians in the Upper Galilee of Israel," in *The Politics of Archaeology and Identity in a Global Context*. Edited by S. Kane, pp. 101-113. Boston, MA: Archaeological Institute of America.

Herzog, Z. 1992. "Settlement and Fortification Planning in the Iron Age," in *The Architecture of Ancient Israel*. Edited by A. Kempinski and R. Reich, pp. 231-274. Jerusalem: Israel Exploration Society.

Herzog, Z., and L. Singer-Avitz. 2004. Redefining the Centre: The Emergence of State in Judah. *Tel Aviv* 31(2):209-244.

Herzog, Z., and L. Singer-Avitz. 2006. Sub-Dividing the Iron Age IIA in Northern Israel: A Suggested Solution to the Chronological Debate. *Tel Aviv* 33(2):163-195.

Hesiod. *Works and Days: 2006 translation by C. M. Schlegel and H. Weinfield*. Ann Arbor: University of Michigan Press.

Hestrin, R., and E. Stern. 1973. Two "Assyrian" Bowls from Israel. *Israel Exploration Journal* 23:152-155.

Hillers, D. R. 1964. An Alphabetic Cuneiform Tablet from Taanach. *Bulletin of the American Schools of Oriental Research* 173:45-50.

Hitti, P. K. 1951. *History of Syria, Including Lebanon and Palestine*. London.

Hodder, I. 1982. *Symbols in Action*. New York: Cambridge University Press.

Hodder, I. 1986. *Reading the Past*, Cambridge.

Hodder, I. 1987. *Archaeology as Long-Term History*. Cambridge: Cambridge University Press.

Hodder, I. 1992. *Theory and Practice in Archaeology*. London.

Hogarth, D. G. 1909. Carcemish and its Neighbourhood. *Liverpool Annals of Archaeology and Anthropology* 2:165-184.

Hogarth, D. G. 1911-1912. Hittite Problems and the Excavation of Carchemish. *Proceedings of the British Academy* 1911-1912:361-374.

Hogarth, D. G. 1914a. *Carchemish: Report on the Excavations at Djerabis on Behalf of the British Museum (conducted by C. L. Woolley and T. E. Lawrence); Part 1 - Introductory*. London: British Museum.

Hogarth, D. G. 1914b. Revealing the Civilisation of the Hittites of Syria: Excavations at Carchemish. *Illustrated London News*:132-135.

Holladay, J. S. 1990. Red Slip, Burnish, and the Solomonic Gateway at Gezer. *Bulletin of the American Schools of Oriental Research* 277/278:23-70.

Holladay, J. S. 2001. "Toward a New Paradigmatic Understanding of Long-Distance Trade in the Ancient Near East: From the Middle Bronze II to Early Iron II - A Sketch," in *The World of the Aramaeans II: Studies in History and Archaeology in Honour of Paul-Eugene Dion, Journal for the Study of the Old Testament Supplement Series 325*. Edited by P. M. M. Daviau, J. W. Wevers, and M. Weigl, pp. 136-198. Sheffield: Sheffield Academic Press.

Homsy, G. 2003. Des céramiques de l'Âge du Fer provenant de Byblos-Jbeil. *Baal* 7:245-279.

Horden, P., and N. Purcell. 2000. *The Corrupting Sea: A Study of Mediterranean History*. Oxford: Blackwell.

Hornung, E. 1964. *Untersuchungen zur Chronolgie und Geschichte des Neuen Reiches*. Wiesbaden: Otto Harrassowitz.

Horowitz, D. 1992. "Ethnic Identity," in *Ethnicity: Theory and Experience*. Edited by N. Gazer and D. Moynihan, pp. 111-140. Cambridge, MA.

Howes, D. 1996. "Introduction: commodities and cultural borders," in *Cross-Cultural Consumption: Global Markets, Local Realities*. Edited by D. Howes, pp. 1-16. London: Routledge.

Hourani, A. 1991. *A History of the Arab Peoples*. London: Faber and Faber.

Hrouda, B. 1962. *Tell Halaf IV. Die Kleinfunde aus Historischer Zeit*. Berlin: Walter de Gruyter & Co.

Hrouda, B. 1967. Forshungs- und Ausgrabungsergebnisse in Syrien seil 1960. *Mitteilungen der Deutschen Orient-Gesellschaft zu Berlin* 98:46-55.

Hughes, J. 1990. *Secrets of the Times. Journal for the Study of the Old Testament Supplement Series*. Sheffield: JSOT Press.

Humann, C. 1898. "Bericht über die erste Ausgrabungen von Sindjirli 1888," in *Ausgrabungen in Sendschirli II: Ausgrabungsbericht und Architektur, Mittheilungen aus den Orientalischen Sammlungen Heft 12*. Edited by F. von Luschan, pp. 88-102. Berlin: W. Spemann.

Humbert, J.-B. 1993. "Keisan, Tell," in *The New Encyclopaedia of Archaeological Excavations in the*

Holy Land. Edited by E. Stern, pp. 862-867. New York: Simon and Schuster.

Hunt, L. 1989. "Introduction: history, culture, and text," in *The New Cultural History.* Edited by L. Hunt, pp. 1-22. Berkeley.

Iggers, G. G. 1995. Historicism: The History and Meaning of the Term. *Journal of the History of Ideas* 56:129-152.

Ingholt, H. 1934. *Rapport Preliminaire sur la Premiere Campagne de Hama.* Kobenhavn: Ejnar Munksgaard.

Ingholt, H. 1940. *Rapport Preliminaire sur Sept Campagnes de Fouilles a Hama en Syrie.* Kobenhavn: Ejnar Munksgaard.

Ingholt, H. 1942. The Danish Excavations at Hama on the Orontes. *American Journal of Archaeology* 46(4):469-476.

Isajiw, W. W. 1974. Definition of Ethnicity. *Ethnicity* 1:111-124.

Isserlin, B. S. J. 1982. "The Earliest Alphabetic Writing," in *The Prehistory of the Balkans; and the Middle East and the Aegean World, Tenth to Eighth Centuries B.C., Cambridge Ancient History III.1 (Second Edition).* Edited by J. Boardman, I. E. S. Edwards, N. G. L. Hammond, and E. Sollberger, pp. 794-833. Cambridge: Cambridge University Press.

Isserlin, B. S. J. 1983. "Phoenician and Punic Rural Settlement and Agriculture: Some Archaeological Considerations," in *Atti del I Congresso Internazionale di Studi Fenici e Punici, Roma, 5-10 Novembre 1979, Volume Primo.* Edited by S. Moscati, pp. 157-164. Roma: Consiglio Nazionale Delle Ricerche.

Jabak, S., J. Sapin, J. Elayi, and H. Sayegh. 1998. "Les Jarres et Amphores du Locus 130," in *Un Quartier du Port Phenicien de Beyrouth au Fer III/Perse. Les Objets, Supplement no. 6 a Transeuphraténe.* Edited by J. Elayi and H. Sayegh, pp. 25-43. Paris: Gabalda.

James, P. 1987. Southeastern Anatolia and Northern Syria: Dating Neo-Hittite Art and Architecture. *Journal of the Ancient Chronology Forum* 1:36-40.

James, P., I. J. Thorpe, N. Kokkinos, R. Morkot, and J. A. Frankish. 1991. *Centuries of Darkness. A Challenge to the Conventional Chronology of Old World Archaeology.* London: Jonathan Cape.

James, W. F. 1966. *The Iron Age at Beth Shean: A Study of Levels VI - IV.* Philadelphia: University Museum.

Jamieson, A. S. 1990. "Area A - Pottery," in *Tell Ahmar: 1988 Season, Abr-Nahrain Supplement Series 2.* Edited by G. Bunnens, pp. 25-105. Leuven: Orientaliste.

Jamieson, A. S. 1999. "Neo-Assyrian Pottery from Tell Ahmar," in *Iron Age Pottery in Northern Mesopotamia, Northern Syria and South-Eastern Anatolia : papers presented at the meetings of the international "table ronde" at Heidelberg (1995) and Nieborów (1997) and other contributions, Altertumskunde des vorderen Orients ; Bd. 10.* Edited by A. Hausleiter and A. Reiche, pp. 287-308. Münster: Ugarit-Verlag.

Jamieson, A. S. 2000. "Identifying Room Use and Vessel Function: A Case Study of Iron Age Pottery from Building C2 at Tell Ahmar, North Syria," in *Essays on Syria in the Iron Age, Ancient Near Eastern studies. Supplement 7.* Edited by G. Bunnens, pp. 259-303. Louvain: Peeters Press.

Jamieson, A. S. 2005. PhD Thesis, University of Melbourne.

Janeway, B. 2006-2007. The Nature and Extent of Aegean Contact at Tell Ta'yinat and Vicinity in the Early Iron Age: Evidence of the Sea Peoples? *Scripta Mediterranea* XXVII-XXVIII: 123-146.

Jenkins, I. 1992. *Archaeologists and Aesthetes.* London.

Jensen, J. 1996. "Report of the Excavation at the Southern Harbour of Sukas August 27th - November 9th 1960," in *Sukas X: The Bronze and Early Iron Age Remains at the Southern Harbour, Publications of the Carlsberg Expedition to Phoenicia 12. Historisk-filosofiske skrifter (Kongelige Danske Videnskabernes Selskab) 17.* Edited by P. J. Riis, J. Jensen, M.-L. Buhl, and B. Otzen, pp. 11-24. Copenhagen: Munksgaard.

Jidejian, N. 1971. *Byblos Through the Ages.* Beirut: Dar el-Marcheq Publishers.

Jidejian, N. 1992. "An Introduction to Tyre," in *The Heritage of Tyre. Essays on the History, Archaeology, and Preservation of Tyre.* Edited by M. S. Joukowsky, pp. 7-12. Dubuque, Iowa: Kendall Hunt Publishing Company.

Johansen, K. F. 1923. *Les vases sicyoniens: étude archéologie.* Paris: E. Champion.

Johns, C. N. 1931. Excavations at Pilgrim's Castle ('Atlit). *Quarterly of the Department of Antiquities in Palestine* 1:111-129.

Johns, C. N. 1932. Excavations at 'Atlit (1930-1931): The South-Eastern Cemetery. *Quarterly of the Department of Antiquities in Palestine* 2:41-104, Pl XIV-XXXVII.

Johns, C. N. 1934. Excavations at Pilgrim's Castle, 'Atlit, (1932); The Ancient Tell and the Outer Defences of the Castle. *Quarterly of the Department of Antiquities in Palestine* 3:145-164, Pls LVIII-LXV.

Johns, C. N. 1935. Excavations at Pilgrim's Castle, 'Atlit, (1931-2); An Unfinished Church in the Suburb. *Quarterly of the Department of Antiquities in Palestine* 4:122-137; Pls LXXI-LXXV.

Johns, C. N. 1938. Excavations at Pilgrims' Castle, 'Atlit (1933): Cremated Burials of Phoenician Origin. *Quarterly of the Department of Antiquities in Palestine* 6:121-152.

Johns, C. N. 1993. "'Athlit," in *The New Encyclopedia of Archaeological Excavations in the Holy Land.* Edited by E. Stern, pp. 112-117. New York: Simon & Schuster.

Jones, A. 1999. The World on a Plate: Ceramics, Food Technology and Cosmology in Neolithic Orkney. *World Archaeology* 31(1):55-77.

Jones, S. 1997. *The Archaeology of Ethnicity: Constructing Identities in the Past and Present.* London.

Joukowsky, M. S. 1972. The Pottery of Tell el Ghassil in the Beqa'a: A comparative study and analysis of the Iron Age and Bronze Age wares (Unpublished Masters Thesis), American University of Beirut.

Joukowsky, M. S. 1997. "Byblos," in *The Oxford Encyclopedia of Archaeology in the Near East. Vol. 1.* Edited by E. M. Meyers, pp. 390-394. Oxford: Oxford University Press.

Kadioğlu, A. 1996. The Paradox of Turkish Nationalism and the Construction of Official Identity. *Middle Eastern Studies* 32:177-193.

Kamp, K., and N. Yoffee. 1980. Ethnicity in Ancient Western Asia During the Early Second Millennium BC: Archaeological Assessments and Ethnoarchaeological Prospectives. *Bulletin of the American Schools of Oriental Research* 237:85-104.

Karageorghis, V., N. Coldstream, P. M. Bikai, A. W. Johnston, M. Robertson, and L. Jehasse. 1981. *Excavations at Kition, 4. The Non-Cypriote Pottery.* Nicosia: Department of Antiquities.

Karageorghis, V., and M. Demas. 1985. *Excavations at Kition, 5. The Pre-Phoenician Levels: areas I and II.* Nicosia: Department of Antiquities.

Karam, N. 1997. Bey 013 Rapport préliminaire. *Baal* 2:95-113.

Kenyon, K. M. 1964. Megiddo, Hazor, Samaria and Chronology. *Bulletin of the Institute of Archaeology, The University of London* 4:143-156.

Kepecs, S. 1997a. Introduction to New Approaches to Combining the Archaeological and Historical Records. *Journal of Archaeological Method and Theory* 4(3/4):193-198.

Kestemont, G. 1985. "Les Pheniciens en Syrie du Nord," in *Phoenicia and Its Neighbours, Studia Phoenicia 3.* Edited by E. Gubel and E. Lipinski, pp. 135-161. Leuven: Uitgeverij Peeters.

Keswani, P. 1989. Dimensions of Social Hierarchy in Late Bronze Age Cyprus: An Analysis of the Mortuary Data from Enkomi. *Journal of Mediterranean Archaeology* 2:49-86.

Keswani, P. 2004. *Mortuary Ritual and Society in Bronze Age Cyprus. Monographs in Mediterranean Archaeology.* London: Equinox.

Khalifeh, I. A. 1988. *Sarepta II. The Late Bronze and Iron Age Periods of Area II, X.* Beruit.

Khalifeh, I. A. 1997a. "Sarepta," in *The Oxford Encyclopedia of Archaeology in the Near East. Vol. 4.* Edited by E. M. Meyers, pp. 488-491. Oxford: Oxford University Press.

Khalifeh, I. A. 1997b. "Beirut," in *The Oxford Encyclopedia of Archaeology in the Near East. Vol. 1.* Edited by E. M. Meyers, pp. 292-295. Oxford: Oxford University Press.

Khayyata, W., and K. Kohlmeyer. 1998. Die Zitadelle von Aleppo. *Damaszener Mitteilungen* 10:69-96.

Khayyata, W., and K. Kohlmeyer. 2000. "Syro-German Excavations on the Citadel of Aleppo, 1996 and 1997," in *Proceedings of the First International Congress on the Archaeology of the Ancient Near East, Rome, May 18-23, 1998. Volume I.* Edited by P. Matthiae, A. Enea, L. Peyronel, and F. Pinnock, pp. 733-742. Rome: Universita degli Studi di Roma 'La Sapienza'.

Killebrew, A. E. 2005. *Biblical Peoples and Ethnicity; An Archaeological Study of Egyptians, Canaanites, Philistines, and Early Israel 1300-1100 B.C.E.* Atlanta: Society of Biblical Literature.

Kinet, D. 1981. *Ugarit - Geschichte und Kultur einer Stadt in der Umwelt des Alten Testamentes. Stuttgarter Bibelstudien 104.* Stuttgart: Verlag Katholisches Bibelwerk GmbH.

King, L. W. 1909. Note on the Inscription upon the Eastern Lion at Tell-Ahmar. *Liverpool Annals of Archaeology and Anthropology* 2:185-186.

King, L. W. 1915. *Bronze Reliefs from The Gates of Shalmaneser, King of Assyria B.C. 860-825.* London: British Museum.

Kirk, G. S. 1975. "The Homeric Poems as History," in *History of the Middle East and the Aegean Region c. 1380-1000 B.C., Cambridge Ancient History II.2 (First Edition).* Edited by I. E. S. Edwards, C. J. Gadd, N. G. L. Hammond, and E. Sollberger, pp. 820-850. Cambridge: Cambridge University Press.

Kitchen, K. A. 1986. *The Third Intermediate Period in Egypt.* Warminster: Aris & Philips.

Kitchen, K. A. 1991. The Chronology of Ancient Egypt. *World Archaeology* 23:201-208.

Kitchen, K.A. 2003. "Egyptian Interventions in the Levant in Iron Age II" in *Symbiosis, Symbolism, and the Power of the Past. Canaan, Ancient Israel, and Their Neighbours from the Late Bronze Age through Roman Palaestina.* Edited by W. G. Dever and S. Gitin, pp. 113-132. Winona Lake, IN: Eisenbrauns.

Kitchen, K. A. 2007. "Egyptian and Related Chronologies - Look, no Science, no Pots!," in *The Synchronisation of Civilisations in the Eastern Mediterranean in the Second Millennium B.C. III.* Edited by M. Bietak and E. Czerny, pp. 163-171. Vienna: Verlag der Österreichischen Akademie der Wissenschaften.

Klengel, H. 1984. Sumur/Simyra und die Eleutheros Ebene in der Geschichte Syriens. *Klio* 66:5-18.

Klengel, H. 1992. *Syria 3000 to 300 B.C.: A Handbook of Political History.* Berlin: Akademie Verlag.

Klengel, H. 2000a. "The 'Crisis Years' and the New Political System in Early Iron Age Syria: Some Introductory Remarks," in *Essays on Syria in the Iron Age., Ancient Near Eastern studies. Supplement 7.* Edited by G. Bunnens, pp. 21-30. Louvain: Peeters Press.

Klengel, H. 2000b. Qatna 1999-Ein historischer Uberblick. *Mitteilungen der Deutschen Orient-Gesellschaft zu Berlin* 132:239-252.

Klengel-Brandt, E. 1990. "Kleinfunde aus Deve Hoyuk bei Karkamisch im Vorderasiatischen Museum Berlin," in *Resurrecting the Past: A Joint Tribute to Adnan Bounni.* Edited by P. Matthiae, M. Van Loon, and H. Weiss, pp. 133-151. Istanbul: Nederlands Historisch-Archaeologisch Instituut te Istanbul.

Kletter, R. 2006. *Just Past? The Making of Israeli Archaeology.* London: Equinox.

Knapp, A. B. 1992a. "Archaeology and *Annales*: time, space, and change," in *Archaeology, Annales, and Ethnohistory.* Edited by A. B. Knapp, pp. 1-21. Cambridge: Cambridge University Press.

Knapp, A. B. 1992b. "Independence and Imperialism: Politico-Economic Structures in the Bronze Age

Levant," in *Archaeology, Annales, and Ethnohistory*. Edited by A. B. Knapp, pp. 83-98. Cambridge: Cambridge University Press.

Knapp, A. B. 1996. Archaeology Without Gravity: Postmodernism and the Past. *Journal of Archaeological Method and Theory* 3:127-158.

Knapp, A. B. 1998. "Mediterranean Bronze Age trade: distance, power and place," in *The Aegean and the Orient in the Second Millennium. Proceedings of the 50th Anniversary Symposium, Cincinnati, 18-20 April 1997*. Edited by E. H. Cline and D. Harris-Cline, pp. 193-205. Liege: Annales d'archéologie égéenne de l'Université de Liège et UT-PASP.

Koehl, R. B. 1985. *Sarepta III: The Imported Bronze and Iron Age Wares from Area II, X*. Beirut.

Kohl, P. L. 1989. "The Material Culture of the Modern Era in the Ancient Orient: Suggestions for Future Work," in *Domination and Resistance, One World Archaeology 3*. Edited by D. Miller, M. J. Rowlands, and C. Tilley, pp. 240-245. London: Unwin Hyman Ltd.

Kohl, P. L., and C. Fawcett. Editors. 1995a. *Nationalism, Politics, and the Practice of Archaeology*. Cambridge: Cambridge University Press.

Kohl, P. L., and C. Fawcett. 1995b. "Archaeology in the service of the state: theoretical considerations," in *Nationalism, Politics, and the Practice of Archaeology*. Edited by P. L. Kohl and C. Fawcett, pp. 3-18. Cambridge: Cambridge University Press.

Kohl, P. L., and G. R. Tsetskhladze. 1995. "Nationalism, politics, and the practice of archaeology in the Caucasus," in *Nationalism, Politics, and the Practice of Archaeology*. Edited by P. L. Kohl and C. Fawcett, pp. 149-174. Cambridge: Cambridge University Press.

Kohlmeyer, K. 2000. *Der Tempel des Wettergottes von Aleppo*. Munster: Rhema-Verlag.

Koldewey, R. 1898. "Die Architektur von Sendschirli," in *Ausgrabungen in Sendschirli II: Ausgrabungsbericht und Architektur, Mittheilungen aus den Orientalischen Sammlungen Heft 12*. Edited by F. von Luschan, pp. 103-199. Berlin: W. Spemann.

Koşay, H. Z. 1951. *Alaca Höyük Kazisi = Les Fouilles d'Alaca Höyük: Entreprises par la Societe d'Histoire Turque. Rapport Preliminaire sur les Travaux en 1937-1939*. Ankara: Türk Tarih Kurumu Basimevi.

Kraeling, E. G. H. 1918. *Aram and Israel*. New York.

Kramer, C. 1977. "Pots and People," in *Mountains and Lowlands*. Edited by L. Levine and T. C. Young, pp. 91-112. Malibu: Undena.

Krogman, W. M. 1949. Ancient Cranial Types at Chatal Hüyük and Tell al-Judaidah, Syria, from the Late Fifth Millennium B. C. to the mid-Seventh Century A. D. *Belleten* 13:407-478.

Kuhne, H. 2002. "Thoughts About Assyria After 612 BC," in *Of Pots and Plans: Papers on the Archaeology and History of Mesopotamia and Syria Presented to David Oates in Honour of His 75th Birthday*. Edited by L. al-Galiani Werr, J. Curtis, H. Martin, A. McMahon, J. Oates, and J. Reade, pp. 171-175: Nabu Publications.

Kuhrt, A. 1995. *The Ancient Near East c. 3000-330 BC*. Vol. II London and New York: Routledge.

Kuniholm, P. I., S. L. Tarter, and C. B. Griggs. 1993. "Appendix 2: Dendrochronological Report," in *Tille Höyük 4. The Late Bronze Age and the Iron Age Transition, British Institute of Archaeology at Ankara Monograph 15*. Edited by G. D. Summers, pp. 179-190. London: The British Institute of Archaeology at Ankara.

Kuschke, A. 1966. "Die Biqa', ihre altorientalischen Siedlungen und Verkherswege," in *Bericht uber die Ergebnisse der Ausgrabungen in Kamid el-Loz (Libanon) in den Jahren 1963 und 1964, Saarbrucker Beitrage zur Altertumskunde Band 3*. Edited by R. Hachmann and A. Kuschke, pp. 15-30. Bonn: Rudolf Habelt Verlag.

Kuschke, A. 1978. Preliminary Remarks on an Archaeological Survey in the Northern Biqa'. *Bulletin du Musée de Beyrouth* 30:43-45.

Lagarce, J. 1983. "Rapports de Ras Ibn Hani avec la Phénicie et la Méditerranée Orientale à l'Âge du Fer," in *Atti del I Congresso Internazionale di Studi Fenici e Punici, Roma, 5-10 Novembre 1979, Volume Primo*. Edited by S. Moscati, pp. 223-226. Roma: Consiglio Nazionale Delle Ricerche.

Lagarce, E., and J. Lagarce. 2000. "Coastal Syria and Lebanon," in *The Synchronisation of Civilisations in the Eastern Mediterranean in the Second Millennium B.C.* Edited by M. Bietak, pp. 140-146. Vienna: Verlag der Österreichischen Akademie der Wissenschaften.

Lagarce, J., and E. Lagarce. 1992. Ras Ibn Hani. *Chronique Archeologique en Syrie* 1:117-120.

Lagarce, J., and E. Lagarce. 1995. "Ras Ibn Hani au Bronze Recent," in *Les Pays d'Ougarit autour de 1200 av. J.-C. Histoire et archéologie. Actes du Colloque International Paris, 28 juin - 1er juillet 1993, Ras Shamra-Ougarit XI*. Edited by M. Yon, M. Sznycer, and P. Bordreuil, pp. 141-154. Paris: Éditions Recherche sur les Civilisations.

Lamon, R. S. 1948. "The City Gate," in *Megiddo II. Seasons of 1935-1939*. Edited by G. Loud, pp. 46-57. Chicago: University of Chicago.

Lamon, R. S., and G. M. Shipton. 1939. *Megiddo I. Seasons of 1925-1934 Strata I-IV*. Chicago: University of Chicago.

Landsberger, B. 1948. *Sam'al: Studien zur Entdeckung der Ruinenstaette Karatepe. Veroffentlichungen der Turkischen Historischen Gesellschaft VII Serie Nr. 16*. Ankara: Druckerei der Turkischen Historischen Gesellschaft.

Lanfranchi, G. B., and S. Parpola. 1990. *The Correspondence of Sargon II. Part II Letters from the Northern and Northeastern Provinces. State Archives of Assyria V*. Helsinki: Helsinki University Press

Lapointe, R. 1970. Etymologie sémitique de Syrie. *Vetus Testamentus* 20(2):233-236.

Lapp, P. 1970. "The Pottery of Palestine in the Persian Period," in *Archäeologie und Altres Testament, Festschrift für Kurt Galling zum 8 Jan 1970*. Edited by A. Kuschke and E. Kutsch, pp. 179-198. Tübingen: Mohr.

Larsen, M. T. 1987a. Orientalism and the Ancient Near East. *Culture and History* 2:96-115.
Larsen, M. T. 1987b. "Commercial Networks in the Ancient Near East," in *Centre and Periphery in the Ancient World, New directions in archaeology*. Edited by M. J. Rowlands, M. T. Larsen, and K. Kristiansen, pp. 47-56. Cambridge.
Larsen, M. T. 1989. "Orientalism and Near Eastern Archaeology," in *Domination and Resistance, One World Archaeology 3*. Edited by D. Miller, M. J. Rowlands, and C. Tilley, pp. 229-239. London.
Larsen, M. T. 1996. *The Conquest of Assyria: Excavations in an Antique Land 1840-1860*. London.
Lebeau, M. 1983. *La Ceramique de l'Age du Fer II-III a Tell Abou Danne: et ses rapports avec la ceramique contemporaine en Syrie*. Paris.
Lehmann, G. 1996. *Untersuchungen zur späten Eisenzeit in Syrien und Libanon : Stratigraphie und Keramikformen zwischen ca. 720 bis 300 v. Chr. Altertumskunde des vorderen Orients ; Bd. 5.* Münster: Ugarit-Verlag.
Lehmann, G. 1998. Trends in the Local Pottery Development of the Late Iron Age and Persian Period in Syria and Lebanon, ca. 700 to 300 B.C. *Bulletin of the American Schools of Oriental Research* 311:35-54.
Lehmann, G. 2001. "Phoenicians in Western Galilee: First Results of an Archaeological Survey in the Hinterland of Akko," in *Studies in the Archaeology of the Iron Age in Israel and Jordan, Journal for the Study of the Old Testament Supplement Series 331*. Edited by A. Mazar and V. T. Mathias, pp. 65-112. Sheffield: Sheffield Academic Press.
Lehmann, G. 2002a. *Bibliographie der archäologischen Fundstellen und Surveys in Syrien und Libanon. Orient-Archäologie ; Bd. 9.* Rahden/Westf.: M. Leidorf.
Lehmann, G. 2002b. "Zur Siedlungsgeschichte des Hinterlandes von Akko (Israel) in der Eisenzeit. Erste Ergebnisse einer archaologischen Landesaufnahme.," in *Ausgrabungen und Surveys im Vorderen Orient I, Orient-Archaologie 5*. Edited by R. Eichmann, pp. 49-78. Rahden/Westf.: Verlag Marie Leidorf GmbH.
Levi-Strauss, C. 1966. The Culinary Triangle. *Partisan Review* 33:586-595.
Levi-Strauss, C. 1968. *Structural Anthropology vol. I*: Penguin Press.
Levi-Strauss, C. 1970. *The Raw and the Cooked*. New York: Harper Torchbooks.
Levine, P. 1986. *The Amateur and the Professional: Antiquarians, Historians and Archaeologists in Victorian England, 1838-1886*. Cambridge.
Levy, T. E., and T. Higham. 2005. "Introduction: Radiocarbon Dating and the Iron Age of the Southern Levant: Problems and potentials for the Oxford conference," in *The Bible and Radiocarbon Dating: Archaeology, Text and Science*. Edited by T. E. Levy and T. Higham, pp. 3-14. London: Equinox.
Lie, A. G. 1929. *The Inscriptions of Sargon II, King of Assyria, Part 1. The Annals*. Paris.
Lillios, K. T. 1995. "Nationalism and Copper Age Research in Portugal During the Salazar Regime (1932-1974)," in *Nationalism, Politics, and the Practice of Archaeology*. Edited by P. L. Kohl and C. Fawcett, pp. 57-69. Cambridge: Cambridge University Press.
Lipinski, E. 1991. "The Territory of Tyre and the Tribe of Asher," in *Phoenicia and the Bible: Proceedings of the Conference held at the University of Leuven on the 15th and 16th of March 1990, Studi Phoenicia 9.* Edited by E. Lipinski, pp. 153-166. Leuven: Uitgeverij Peeters.
Lipinski, E. 2000a. *The Aramaeans Their Ancient History, Culture, Religion*. Leuven: Peeters Press..
Lipinski, E. 2000b. "The Linguistic Geography of Syria in Iron Age II (c. 1000-600 B.C.)," in *Essays on Syria in the Iron Age, Ancient Near Eastern studies. Supplement 7*. Edited by G. Bunnens, pp. 125-142. Louvain: Peeters Press.
Liverani, M. 1983. "Dall'acculturazione alla deculturazione: considerazioni sul ruolo dei contatti politici ed economici nella storia siro-palestinese pre-ellenistica," in *Modes de contact et processus de transformation dans les societes anciennes*, pp. 503-522. Pisa and Rome.
Liverani, M. 1987. "The Collapse of the Near Eastern Regional System at the End of the Bronze Age: The Case of Syria," in *Centre and Periphery in the Ancient World, New directions in archaeology*. Edited by M. J. Rowlands, M. T. Larsen, and K. Kristiansen, pp. 66-73. Cambridge: Cambridge University Press.
Liverani, M. 1988. The Growth of the Assyrian Empire in the Habur/Middle Euphrates Area: A New Paradigm. *State Archives of Assyria Bulletin* 2:81-98.
Liverani, M. Editor. 1990. *Prestige and Interest: International Relations in the Near East ca. 1600-1000 B.C.* Padova: Sargon Press.
Liverani, M. 1994. History as a War Game. *Journal of Mediterranean Archaeology* 7:241-248.
Liverani, M. 1996. "Ancient Propaganda and Historical Criticism," in *The Study of the Ancient Near East in the Twenty-First Century*. Edited by J. S. Cooper and G. M. Schwartz, pp. 283-289. Winona Lake, Indiana: Eisenbrauns.
Liverani, M. 1997a. "Ancient Near Eastern Cities and Modern Ideologies," in *Die Orientalische Stadt: Kontinuitat, Wandel, Bruch*. Edited by G. Wilhelm, pp. 85-107. Saarbrucken.
Liverani, M. 1997b. "Beyond Deserts, Beyond Oceans," in *Profumi d'Arabia, Saggi di Storia Antica 11*. Edited by A. Avanzini, pp. 557-564. Rome: L'Erma di Bretschneider.
Liverani, M. 1999. "History and Archaeology in the Ancient Near East: 150 Years of a Difficult Relationship," in *Fluchtpunkt Uruk. Archaologische Einheit aus Methodischer Vielfalt. Schriften fur Hans Jorg Nissen, Internationale Archaologie Studia Honoria 6*. Edited by H. Kuhne, R. Bernbeck, and K. Bartl, pp. 1-11. Rahden.
Liverani, M. 2003. "The Influence of Political Institutions on Trade in the Ancient Near East (Late Bronze to Early Iron Age)," in *Mercanti e Politica Nel Mondo Antico, Saggi di Storia Antica 21*. Edited by C.

Zaccagnini, pp. 119-137. Roma: L'Erma di Bretschneider.
Liverani, M. 2004. *Myth and Politics in Ancient Near Eastern Historiography*. Ithaca, New York: Cornell University Press.
Liverani, M. 2005a. "Imperialism," in *Archaeologies of the Middle East: Critical Perspectives*. Edited by S. Pollock and R. Bernbeck, pp. 223-243. Oxford: Blackwell.
Liverani, M. 2005b. *Israel's History and the History of Israel*. London: Equinox.
Lloyd, S. 1947. *Foundations in the Dust: A Story of Mesopotamian Exploration*. London.
London, G. A. 1989. A Comparison of Two Contemporaneous Lifestyles of the Late Second Millennium BC. *Bulletin of the American Schools of Oriental Research* 273:37-55.
London, G. A. 2003. "Ethnicity and Material Culture," in *Near Eastern Archaeology: A Reader*. Edited by S. Richard, pp. 146-149. Winona Lake, Indian: Eisenbrauns.
Long, B. O. 1993. Mythic Trope in the Autobiography of William Foxwell Albright. *Biblical Archaeologist* 56:36-45.
Long, B. O. 1997. "Historical Imaginings, Ideological Gestures: W.F. Albright and the 'Reasoning Faculties of Man'," in *The Archaeology of Israel: Constructing the Past, Interpreting the Present, Journal for the Study of the Old Testament Supplement Series 237*. Edited by N. A. Silberman and D. Small, pp. 82-94. Sheffied: Sheffield Academic Press.
Loud, G. 1948. *Megiddo II. Seasons of 1935-1939*. Chicago: University of Chicago.
Lowenthal, D. 1985. *The Past is a Foreign Country*. Cambridge: Cambridge University Press.
Luciani, M. 2000. "Iron Age Graves in Northern Syria: The Tell Shiukh Fawqani Evidence," in *Proceedings of the First International Congress on the Archaeology of the Ancient Near East, Rome, May 18-23, 1998. Volume I*. Edited by P. Matthiae, A. Enea, L. Peyronel, and F. Pinnock, pp. 803-814. Rome: Universita degli Studi di Roma 'La Sapienza'.
Luciani, M. 2002. "Operation K," in *Excavating Qatna, I. Preliminary Report on the 1999 and 2000 Campaigns of the Joint Syrian-Italian-German Archaeological Research Project at Tell Mishrifeh, Documents d'Archeologie Syrienne IV*. Edited by M. al-Maqdissi, M. Luciani, D. Morandi Bonacossi, M. Novák, and P. Pfälzner, pp. 145-168. Damascus.
Luciani, M. 2005. "Area G: The Iron Age Productive Area (Period IX) and the Inhumation Cemetery (Period X)," in *Tell Shiyukh Fawqani 1994-1998*. Edited by L. Bachelot and F. M. Fales, pp. 719-993. Padova: Sargon Editrice e Libreria.
Luckenbill, D. D. 1926. *Ancient Records of Assyria and Babylonia, Vol. I*. Chicago.
Luckenbill, D. D. 1927. *Ancient Records of Assyria and Babylonia, Vol. II: From Sargon to the End*. Chicago.
Luke, J. 2003. *Ports of Trade, Al Mina and Geometric Greek Pottery in the Levant. BAR International Series 1100*. Oxford: BAR Publishing.
Lund, J. 1986. *Sukas VIII: The Habitation Quarters. Publications of the Carlsberg Expedition to Phoenicia 10. Historisk-filosofiske skrifter (Kongelige Danske Videnskabernes Selskab) 12*. Copenhagen: Munksgaard.
Lundquist, J. M. 1983. Tell Qarqur. The 1983 Season. *Les Annales Archeologiques Arabes Syriennes* 33:272-288.
MacClancy, J. 1992. *Consuming Culture*. London: Chapmans.
MacGilivray, J. A. 2004. "A Middle Minoan Cup from Sidon," in *Decade: A Decade of Archaeology and History in the Lebanon*. Edited by C. Doumet-Serhal, pp. 124-131. Beirut: The Lebanese British Friends of the National Museum.
Macridi-Bey, T. 1904a. A travers les nécropoles sidoniennes. *Revue Biblique* 1:547-572.
Macridi-Bey, T. 1904b. Le temple d'Echmoun à Sidon, fouilles exécutées par le Musée impérial ottoman. *Revue Biblique* 1:390-403.
Magazzù, G. 2002. Afis 2000-2001. Area N: le fortificazioni orientali dell acropoli. *Egitto e Vicino Oriente* 25:54-58.
Maggioli, A. 2002. Afis 2000-2001. Area B: indagini sulla porta urbica del Ferro II-III. *Egitto e Vicino Oriente* 25:23-24.
Mahdi, M. 1964. *Ibn Khaldun's Philosophy of History*. Chicago: University of Chicago Press.
Maisels, C. K. 1993. *The Near East: Archaeology in the 'Cradle of Civilization'. The Experience of Archaeology*. London.
Maisels, C. K. 1999. *Early civilisations of the Old World: The Formative Histories of Egypt, the Levant, Mesopotamia, India and China*. London.
Makinson, M. 1999. "La culture materielle du moyen Euphrate au premier millenaire avant J.C.," in *Archaeology of the Upper Syrian Euphrates, The Tishrin Dam Area: Proceedings of the International Symposium Held at Barcelona, January 28th-30th 1998*. Edited by G. del Olmo Lete and J.-L. Montero Fenollos, pp. 363-391. Barcelona: Editorial AUSA.
Makinson, M. 2005. "La stratigraphie générale et l'occupation de l'Âge du Fer (architecture et matériel)," in *Tell Shiyukh Fawqani 1994-1998*. Edited by L. Bachelot and F. M. Fales, pp. 411-580. Padova: Sargon Editrice e Libreria.
Malamat, A. 1973. "The Aramaeans," in *Peoples of Old Testament Times*. Edited by D. J. Wiseman, pp. 134-155. Oxford: Clarendon Press.
Mallowan, M. E. L. 1937. The Syrian City of Til Barsip. *Antiquity* 11:328-339.
Mallowan, M. E. L. 1972. Carchemish: Reflections on the Chronology of the Sculpture. *Anatolian Studies* 22:63-85.
al-Maqdissi, M. 1990. Tell Kazel, Syria: Excavations of the AUB Museum, 1985-1987, Preliminary Reports: Area III. *Berytus* 38:87-93.
al-Maqdissi, M. 1993. Chronique des Activites Archeologiques en Syria (I). *Syria* 70:443-560.
al-Maqdissi, M. 2001. Kurzbericht uber die syrischen Ausgrabungen in Misrife-Qatna. *Mitteilungen der*

Deutschen Orient-Gesellschaft zu Berlin 133:141-155.

al-Maqdissi, M. 2003. Ergebnisse der siebten und achten syrischen Grabungskampagne 2001 und 2002 in Mišrife - Qatna. *Mitteilungen der Deutschen Orient-Gesellschaft zu Berlin* 135:219-245.

al-Maqdissi, M. and M. Badawi. 2002. "Rapport Préliminaire sur la sixième campagne des fouilles syriennes à Mishrifeh/Qatna," in *Excavating Qatna, I. Preliminary Report on the 1999 and 2000 Campaigns of the Joint Syrian-Italian-German Archaeological Research Project at Tell Mishrifeh, Documents d'Archeologie Syrienne IV*. Edited by M. al-Maqdissi, M. Luciani, D. Morandi Bonacossi, M. Novák, and P. Pfälzner, pp. 25-62. Damascus.

al-Maqdissi, M., M. Luciani, D. Morandi Bonacossi, M. Novák, and P. Pfälzner. 2002a. "Introduction," in *Excavating Qatna, I. Preliminary Report on the 1999 and 2000 Campaigns of the Joint Syrian-Italian-German Archaeological Research Project at Tell Mishrifeh, Documents d'Archeologie Syrienne IV*. Edited by M. al-Maqdissi, M. Luciani, D. Morandi Bonacossi, M. Novák, and P. Pfälzner, pp. 7-16. Damascus.

al Maqdissi, M., M. Luciani, D. Morandi Bonacossi, M. Novák, and P. Pfälzner. Editors. 2002b. *Excavating Qatna, I. Preliminary Report on the 1999 and 2000 Campaigns of the Joint Syrian-Italian-German Archaeological Research Project at Tell Mishrifeh*. Damascus.

Marchand, S. 1996. "Orientalism as *Kulturpolitik*: German Archaeology and Cultural Imperialism in Asia Minor," in *Volksgeist as Method and Ethic: Essays on Boasian Ethnocgraphy and the German Anthropological Tradition, History of Anthropology 8*. Edited by G. W. Stocking, pp. 298-336. Madison, WI: University of Wisconsin Press.

Marfoe, L. 1995. *Kamid el-Loz 13: The Prehistoric and Early Historic Context of the Site. Saarbrucker Beitrage zur Alterumskunde Bd. 41*. Bonn.

Marfoe, L. 1998. *Kamid el-Loz 14: Settlement History of the Biq'a up to the Iron Age. Saarbrücker Beiträge zur Altertumskunde Bd. 53*. Bonn: Rudolph Habelt Verlag GMBH.

Margueron, J. 1979. "Un <hilani> á Emar," in *Archaeological Reports From the Tabqa Dam Project - Euphrates Valley, Syria*. Edited by D. N. Freedman, pp. 153-176. Cambridge, MA: American Schools of Oriental Research.

Margueron, J. 1985. "Les palais syriens à l'Âge du Bronze," in *Le système palatial en Orient, en Grèce et à Rome, Travaux du Centre de Recherche sur le Proche-Orient et la Grèce Antiques*. Edited by E. Levy, pp. 127-158. Strasbourg: Université des Sciences Humaines.

Marquis, P. 2004. "Les fouilles du Centre-Ville de Beyrouth. Dix ans après," in *Decade: A Decade of Archaeology and History in the Lebanon*. Edited by C. Doumet-Serhal, pp. 266-279. Beirut: The Lebanese British Friends of the National Museum.

Marquis, P., and R. Ortali-Tarazi. 1996. Bey 009 L'immeuble de la Banco di Roma; Nouveaux éléments pour l'étude du forum de Béryte. *Baal* 1:148-175.

Maspero, G. 1896. *The Struggle of the Nations. Translated from the French by M. L. McClure*. New York: Appleton.

Masry, A. H. 1981. Traditions of Archaeological Research in the Near East. *World Archaeology* 13:222-239.

Matermawi, M. 2005. Tell Afis, Siria (2002-2004) Area A3. *Egitto e Vicino Oriente* 28:30-31.

Mathias, V. T., and P. J. Parr. 1989. The Early Phases at Tell Nebi Mend: A Preliminary Account. *Levant* 21:13-32.

Matthäus, H. 1980. *Die Bronzegefässe der kretisch-mykenischen Kultur. Prähistorische Bronzefunde II.1*. Munchen: C.H. Beck'sche Verlagsbuchhandlung.

Matthäus, H. 1985. *Metallgefässe und Gefässuntersätze der Bronzezeit, der geometrichen und archaischen Periode auf Cypern, mit einem Anhang der bronzezeitlichen Schwertfunde auf Cypern. Prähistorische Bronzefunde II.8*. Munchen: C.H. Beck'sche Verlagsbuchhandlung.

Matthers, J. 1978. Tell Rifa'at 1977: Preliminary Report of an Archaeological Survey. *Iraq* 40:119-162.

Matthers, J. 1981a. "Introduction," in *The River Qoueiq, Northern Syria, and its Catchment: Studies arising from the Tell Rifa'at Survey 1977-79. Part i, B.A.R. International Series 98 (i)*. Edited by J. Matthers, pp. 1-10. Oxford: BAR Publishing.

Matthers, J. 1981b. "The Iron Age (c. 1200-700 B.C.)," in *The River Qoueiq, Northern Syria, and its Catchment: Studies arising from the Tell Rifa'at Survey 1977-79. Part ii, B.A.R. International Series 98 (ii)*. Edited by J. Matthers, pp. 415-437. Oxford: BAR Publishing.

Matthiae, G. S. 2002. Afis 2000-2001. Area B: la coroplastica del Bronzo Antico IV A-B e del bronzo Medio II. *Egitto e Vicino Oriente* 25:16-18.

Matthiae, P. 1979. Sondages a Tell Afis (Syrie), 1978. *Akkadica* 14:2-5.

Matthiae, P. 1981. *Ebla: An Empire Rediscovered*. Garden City, NY: Doubleday.

Matthiae, P. 1985. "Ricerche della Missione Archeologia in Siria dell'Universita di Roma: Tell Mardikh, Tell Tuqan, Tell Afis," in *Scavi e ricerche archeologiche degli anne 1976-1979*, pp. 163-219. Roma.

Matthiae, P. 2000. "Studies in the Relative and Absolute Chronology of Syria in the IInd Millennium B.C. An Integrated Parallel Project," in *The Synchronisation of Civilisations in the Eastern Mediterranean in the Second Millennium B.C.* Edited by M. Bietak, pp. 136-139. Vienna: Verlag der Österreichischen Akademie der Wissenschaften.

Mazar, A. 1988. "Some Aspects of the "Sea Peoples" Settlement," in *Society and Economy in the Eastern Mediterranean (c. 1500-1000 B.C.)*. Edited by M. Heltzer and E. Lipinski, pp. 251-260. Leuven.

Mazar, A. 1985. Emergence of the Philistine Material Culture. *Israel Exploration Journal* 35(2): 95-107.

Mazar, A. 1992. *Archaeology of the Land of the Bible 10 000-586 B.C.E.* New York: Doubleday.

Mazar, A. 1993a. "Beth Shan," in *The New Encyclopaedia of Archaeological Excavations in the Holy Land.* Edited by E. Stern, pp. 214-223. New York: Simon and Schuster.
Mazar, A. 1993b. Beth Shean in the Iron Age: Preliminary Report and Conclusions of the 1990-1991 Excavations. *Israel Exploration Journal* 43:201-229.
Mazar, A. 1997a. "The Excavations at Tel Beth Shean during the Years 1989-1994," in *The Archaeology of Israel: Constructing the Past, Interpreting the Present, Journal for the Study of the Old Testament Supplement Series 237.* Edited by N. A. Silberman and D. Small, pp. 144-164. Sheffied: Sheffield Academic Press.
Mazar, A. 1997b. Iron Age Chronology - A Reply to Israel Finkelstein. *Levant* 29:155-165.
Mazar, A. 1998. "On the Appearance of Red Slip in the Iron Age I Period in Israel," in *Mediterranean Peoples in Transition: Thirteenth to Early Tenth Centuries BCE-In Honour of Professor Trude Dothan.* Edited by S. Gitin, A. Mazar, and E. Stern, pp. 368-378. Jerusalem: Israel Exploration Society.
Mazar, A. 1999a. The 1997-1998 Excavations at Tel Rehov: Preliminary Report. *Israel Exploration Journal* 49:1-42.
Mazar, A. 1999b. Beth Shean During the Iron Age II: Stratigraphy, Chronology and Hebrew Ostraca. *Eretz Israel* 26 (Frank M. Cross Volume):91-100 (Hebrew).
Mazar, A. 2000. Will Tel Rehov Save the United Monarchy? *Biblical Archaeology Review* 26:38-48, 50-51, 75.
Mazar, A. 2001. "Beth Shean During the Iron Age II: Stratigraphy, Chronology and Hebrew Ostraca," in *Studies in the Archaeology of the Iron Age in Israel and Jordan, Journal for the Study of the Old Testament Supplement Series 331.* Edited by A. Mazar and V. T. Mathias, pp. 289-309. Sheffield: Sheffield Academic Press.
Mazar, A. 2004. Greek and Levantine Iron Age Chronology; A Rejoinder. *Israel Exploration Journal* 54:24-36.
Mazar, A. 2005. "The Debate over the Chronology of the Iron Age in the Southern Levant: Its history, the current situation, and a suggested resolution," in *The Bible and Radiocarbon Dating: Archaeology, Text and Science.* Edited by T. E. Levy and T. Higham, pp. 15-30. London: Equinox.
Mazar, A. 2006. "Beth-Shean from the Late Bronze Age IIB to the Medieval Period: A Summary," in *Excavations at Tel Beth-Shean 1989-1996. Volume I From the Late Bronze Age IIB to the Medieval Period.* Edited by A. Mazar, pp. 26-47. Jerusalem: Israel Exploration Society.
Mazar, A., and I. Carmi. 2001. Radiocarbon Dates from Iron Age Strata at Tel Beth Shean and Tel Rehov. *Radiocarbon* 43:1333-1342.
Mazar, A., H. J. Bruins, N. Panitz-Cohen, and J. van der Plicht. 2005. "Ladder of Time at Tel Rehov: Stratigraphy, archaeological context, pottery and radiocarbon dates," in *The Bible and Radiocarbon Dating: Archaeology, Text and Science.* Edited by T. E. Levy and T. Higham, pp. 193-255. London: Equinox.
Mazar, A., and R. A. Mullins. 2006. "Introduction: The Hebrew University Excavations at Tel Beth-Shean," in *Excavations at Tel Beth-Shean 1989-1996. Volume I From the Late Bronze Age IIB to the Medieval Period.* Edited by A. Mazar, pp. 3-25. Jerusalem: Israel Exploration Society.
Mazar, A., and N. Panitz-Cohen. 2006. "Area S: The Stratigraphy and Architecture of the Iron Age IIA and Later Structures," in *Excavations at Tel Beth-Shean 1989-1996. Volume I From the Late Bronze Age IIB to the Medieval Period.* Edited by A. Mazar, pp. 173-201. Jerusalem: Israel Exploration Society.
Mazar (Maisler), B. 1951. The Stratification of Tell Abu Hawam on the Bay of Acre. *Bulletin of the American Schools of Oriental Research* 124:21-25.
Mazar (Maisler), B. 1957. The Campaign of Pharaoh Shishak to Palestine. *Vetus Testamentum* Sup. 4:57-66.
Mazar, E. 1993. "Achzib," in *The New Encyclopaedia of Archaeological Excavations in the Holy Land.* Edited by E. Stern, pp. 32-36. New York: Simon and Schuster.
Mazar, E. 2000. "Phoenician Family Tombs at Achziv: A Chronological Typology (1000-400)," in *Fenicios Y Territorio.* Edited by A. Gonzalez-Pratz, pp. 89-226. Alicante.
Mazar, E. 2001. The Phoenicians in Achziv, The Southern Cemetery: Jerome L. Joss Expedition Final Report of the Excavations 1988-1990. *Cuadernos de Arqueologia Mediterranea* 7.
Mazar, E. 2004. The Phoenician Family Tomb N. 1 at the Northern Cemetery of Achziv (10th-6th Centuries BCE): Sam Turner Expedition Final Report of the Excavations. *Cuadernos de Arqueologia Mediterranea* 10.
Mazzoni, S. 1984. L' insediamento persiano-ellenistico di Tell Mardikh. *Studi Eblaiti* 7:87-132.
Mazzoni, S. 1988a. Lo scavo dell'edificio del Settore D. *Egitto e Vicino Oriente* 10/2:25-83.
Mazzoni, S. 1988b. Tell Afis: Seminario 1987: Passato e Presente. *Egitto e Vicino Oriente* 10/2:1-3.
Mazzoni, 1990a. Tell Afis and the Chronology of Iron Age in Syria. *Les Annales Archeologiques Arabes Syriennes* 40:76-92.
Mazzoni, S. 1990b. La periode perse a Tell Mardikh dans le cadre de l'evolution de l'age du fer en Syrie. *Transeuphratene* 2:187-199.
Mazzoni, S. 1990c. "Observations about Iron Age Glyptics from Tell Afis and Tell Mardikh," in *Resurrecting the Past: A Joint Tribute to Adnan Bounni.* Edited by P. Matthiae, M. Van Loon, and H. Weiss, pp. 215-226. Istanbul: Nederlands Historisch-Archaeologisch Instituut te Istanbul.
Mazzoni, S. 1990d. Tell Afis e L'eta del Ferro. *Seminari di Orientalistica* 2:157-196.
Mazzoni, S. 1991-1992. Lo Sviluppo Degli Insediamenti in Siria in eta Persiana. *Egitto e Vicino Oriente* 14-15:55-72.
Mazzoni, S. 1992a. "L'eta del Ferro a Tell Mardikh e nella sue regione," in *Tell Afis e l'età del ferro,*

Seminari di orientalistica 2. Edited by S. Mazzoni, pp. 99-155. Pisa: Giardini editori e stampatori in Pisa.

Mazzoni, S. 1992b. "Tell Afis e il Ferro I in Siria," in *Tell Afis e l'età del ferro, Seminari di orientalistica 2*. Edited by S. Mazzoni, pp. 157-196. Pisa: Giardini editori e stampatori in Pisa.

Mazzoni, S. 1994a. "Aramaean and Luwian New Foundations," in *Nuove fondazioni nel Vicino Oriente Antico: realta e ideologia*. Edited by S. Mazzoni, pp. 319-340. Pisa.

Mazzoni, S. 1994b. "Drinking Vessels in Syria: Ebla and the Early Bronze Age," in *Drinking in Ancient Societies: History and Culture of Drinks in the Ancient Near East*. Edited by L. Milano, pp. 245-276. Padova: Sargon srl.

Mazzoni, S. 1995. "Settlement Pattern and New Urbanisation in Syria at the Time of the Assyrian Conquest," in *Neo-Assyrian Geography: Papers Presented at the International Conference Held in the University of Rome 'La Sapienza', November 10th to 12th, 1993., Quaderni di geografia storica ; 5*. Edited by M. Liverani, pp. 181-191. Roma: Università di Roma 'La Sapienza' Dipartimento di Scienze storiche archeologiche e antropologiche dell'Antichità.

Mazzoni, S. 1997a. "The Gate and the City: Change and Continuity in Syro-Hittite Urban Ideology," in *Die orientalische stadt: Kontinuitat, Wandel, Bruch*. Edited by G. Wilhelm, pp. 307-338. Saarbrucken: saarbrucker Druckerei und Verlag.

Mazzoni, S. 1997b. L'arte siro-ittita nel suo contesto archeologico. *Contributi e Materiali di Archeologia Orientale* 7:287-327.

Mazzoni, S. 1998a. "The Late Iron I and Early Iron II Levels (Area E1)," in *Tell Afis, Siria: scavi sull'acropoli 1988-1992 = The 1988-1992 excavations on the Acropolis, Ricerche di archeologia del Vicino Oriente ; 1*. Edited by S. M. Cecchini and S. Mazzoni, pp. 163-199. Pisa: Edizioni ETS.

Mazzoni, S. 1998b. *The Italian Excavations of Tell Afis (Syria): From Chiefdom to an Aramaean state*. Piza: Edizioni ETS.

Mazzoni, S. 1998c. Unc nouvcllc stèlc d'époque aranéenne de Tell Afis (Syrie). *Transeuphratene* 16:9-18.

Mazzoni, S. 2000a. "Syria and the Periodization of the Iron Age: A Cross-Cultural Perspective," in *Essays on Syria in the Iron Age., Ancient Near Eastern studies. Supplement 7*. Edited by G. Bunnens, pp. 31-59. Louvain: Peeters Press.

Mazzoni, S. 2000b. Tell Afis Siria 1999, Summary. *Egitto e Vicino Oriente* 22:55-60.

Mazzoni, S. 2000c. "Pots, People and Cultural Borders in Syria," in *Landscapes: Territories, Frontiers and Horizons in the Ancient Near East: Papers Presented to the XLIV Recontre Assyriologique Internationale, Venezia, 7-11 July 1997; Part II Geography and Cultural Landscapes, History of the Ancient Near East/ Monographs - III/2*. Edited by L. Milano, S. de Martino, F. M. Fales, and G. B. Lanfranchi, pp. 139-152. Padova: Sargon srl.

Mazzoni, S. 2000d. "Crisis and Change: the beginning of Iron Age in Syria," in *Proceedings of the First International Congress on the Archaeology of the Ancient Near East, Rome, May 18-23, 1998. Volume II*. Edited by P. Matthiae, A. Enea, L. Peyronel, and F. Pinnock, pp. 1043-1060. Rome: Universita degli Studi di Roma 'La Sapienza'.

Mazzoni, S. 2001. "Tell Afis and the Lu'ash in the Aramaean Period," in *The World of the Aramaeans II: Studies in History and Archaeology in Honour of Paul-Eugene Dion, Journal for the Study of the Old Testament Supplement Series 325*. Edited by P. M. M. Daviau, J. W. Wevers, and M. Weigl, pp. 99-114. Sheffield: Sheffield Academic Press.

Mazzoni, S. 2002a. Afis 2000-2001. Summary (English). *Egitto e Vicino Oriente* 25:68-72.

Mazzoni, S. 2002b. Afis 2000-2001. Area B: l'occupazione dell'Eta del Ferro. *Egitto e Vicino Oriente* 25:18-19.

Mazzoni, S. 2002c. Afis 2000-2001. Campagne XVIII-XIX: Bilancio e Prospettive. *Egitto e Vicino Oriente* 25:5-7.

Mazzoni, S. 2002d. Afis 2000-2001. Area B: la ceramica del Bronzo Medio. *Egitto e Vicino Oriente* 25:14-16.

Mazzoni, S. 2002e. "Late Bronze Age Pottery Production in Northwestern Central Syria," in *Céramique de l'Âge du Bronze en Syrie, I. La Syrie du Sud et la Vallée l'Oronte, BAH 161*. Edited by M. al Maqdissi, V. Matoïan, and C. Nicolle, pp. 129-151. Damas: IFAPO.

Mazzoni, S. 2005a. Tell Afis, Siria, 2002-2004: Summary. *Egitto e Vicino Oriente* 28:113-122.

Mazzoni, S. 2005b. Tell Afis, the Survey and the Regional Sequence. *Egitto e Vicino Oriente* 28:5-13.

Mazzoni, S., and S. M. Cecchini. 1995. Tell Afis (Siria) 1994: Rapporto preliminare. *Egitto e Vicino Oriente* 18:243-306.

McCall, H. 1998. "Rediscovery and Aftermath," in *The Legacy of Mesopotamia*. Edited by S. Dalley, pp. 183-213. Oxford.

McClellan, T. L. 1992. "Twelfth Century B.C. Syria: Comments on H. Sader's paper: The 12th century B.C. in Syria: The Problem of the Rise of the Aramaeans," in *The Crisis Years: The 12th Century B.C. From Beyond the Danube to the Tigris*. Edited by W. A. Ward and M. S. Joukowsky, pp. 164-173. Iowa: Kendall Hunt.

McEwan, C. W. 1937. The Syrian Expedition of the Oriental Institute of the University of Chicago. *American Journal of Archaeology* 41(1):8-16.

McFadden, G. H. 1954. A Late Cypriote III Tomb from Kourion Kaloriziki No. 40. *American Journal of Archaeology* 58(2):131-142.

McNicoll, A. W., R. H. Smith, and J. B. Hennessy. 1982a. *Pella in Jordan 1: An interim report on the joint University of Sydney and the College of Wooster Excavations at Pella 1979-1981 (Text)*. Canberra: Australian National Gallery.

McNicoll, A. W., R. H. Smith, and J. B. Hennessy. 1982b. *Pella in Jordan 1: An interim report on the joint University of Sydney and the College of Wooster Excavations at Pella 1979-1981 (Plates and Illustrations)*. Canberra: Australian National Gallery.

McNicoll, A. W., P. C. Edwards, J. Hanbury-Tenison, J. B. Hennessy, T. F. Potts, R. H. Smith, A. Walmsley, and P. Watson. Editors. 1992. *Pella in Jordan 2: The second interim report of the joint University of Sydney and College of Wooster Excavations at Pella 1982-1985. Mediterranean Archaeology Supplement 2.* Sydney: MEDITARCH.

Melchert, H. C. 2003a. "Introduction," in *The Luwians, Handbuch der Orientalistik 68.* Edited by H. C. Melchert, pp. 1-7. Leiden: Brill.

Melchert, H. C. 2003b. "Prehistory," in *The Luwians, Handbuch der Orientalistik 68.* Edited by H. C. Melchert, pp. 8-26. Leiden: Brill.

Mellink, M. J. 1974. "Hittite friezes and gate sculptures," in *Anatolian Studies Presented to H.G. Güterbock on the Occasion of his 65th Birthday.* Edited by K. Bittel, P. H. J. Houwink ten Cate, and E. Reiner, pp. 201-214. Istanbul.

Merrillees, R. S. 2003. "The Greek Emergence in the Eastern Mediterranean - Cyprus and Phoenicia," in *One Hundred Years of American Archaeology in the Middle East.* Edited by D. R. Clark and V. H. Matthews, pp. 87-96. Boston: American Schools of Oriental Research.

Meskell, L. 1996. The Somatization of Archaeology. *Norwegian Archaeological Review* 29:1-16.

Metcalf, P., and R. Huntington. 1995. *Celebrations of Death: The Anthropology of Mortuary Ritual.* Cambridge: Cambridge University Press.

du Mesnil du Buisson, R. 1926. Les Ruines d'el-Mishrife au NORD-EST de Homs (Emese). *Syria* 7:289-325.

du Mesnil du Buisson, L. C. 1927a. Les Ruines d'el-Mishrifé au Nord-East de Homs (Émèse). *Syria* 8:13-33.

du Mesnil du Buisson, R. 1927b. L'Ancienne Qatna: ou les ruines d'el-Mishrife au N.-E. de Homs (Emese), Deuxieme Campagne de Fouilles, 1927. *Syria* 8: 13-33, 277-301.

du Mesnil du Buisson, R. 1928. L'Ancienne Qatna: ou les ruines d'el-Mishrife au N.-E. de Homs (Emese), Deuxieme Campagne de Fouilles, 1927. *Syria* 9:6-24, 81-89.

du Mesnil du Buisson, R. 1930. Compte rendu de la quatrième campagne de fouilles a Mishrifé-Qatna. *Syria* 11:146-163.

du Mesnil du Buisson, L. C. 1934. Le Site de Qadesh (Tell Nebi Mend). *Melanges Maspero: Memoires de l'Institut Francais d'Archeologie Orientale, Cairo*:909-927.

du Mesnil du Buisson, R. 1935. *Le Site archeologique de Mishrife-Qatna.* Paris.

Meurdrac, M., and L. Albanèse. 1938. A travers les nécropoles gréco-romaines de Sidon. *Bulletin du Musée de Beyrouth* 2:73-98.

Meurdrac, M., and L. Albanèse. 1939. A travers les nécropoles gréco-romaines de Sidon. *Bulletin du Musée de Beyrouth* 3:37-51.

Millard, A. R. 1976. A Text in Shorter Cuneiform Alphabet from Tell Nebi Mend. *Ugarit-Forschungen* 8:459-460.

Miller, G. F. 1958. *Aleppo, Past and Present: Its history, its citadel, its museum and its antique monuments.* Archaeological Society of Aleppo.

Miller, J. M. 1988. "Antecedents to Palestinian Archaeology," in *Benchmarks in Time and Culture. An Introduction to Palestinian Archaeology. Essays in honor of Joseph A. Callaway.* Edited by J. F. Drinkard, G. L. Mattingly, and J. M. Miller, pp. 3-14. Atlanta: Scholars Press.

Mongne, P., F. Stephan, and S. Zarazir. 2003. Opération Bey 048-Nord; Rapport Préliminaire des fouilles de la Place des Martyrs à Beyrouth. *Baal* 7:49-82.

Montet, P. 1928. *Byblos et l'Égypte. Quatre Campagnes de Fouilles a Gebeil 1921 -1922 - 1923 - 1924 (Texte). Bibliothèque Archéologique et Historique XI.* Paris: Librairie Orientaliste Paul Geuthner.

Montet, P. 1929. *Byblos et l'Égypte. Quatre Campagnes de Fouilles a Gebeil 1921 -1922 - 1923 - 1924 (Atlas). Bibliothèque Archéologique et Historique XI.* Paris: Librairie Orientaliste Paul Geuthner.

Moorey, P. R. S. 1975. Iranian Troops at Deve Hüyük in Syria in the Earlier Fifth Century B.C. *Levant* 7:108-117.

Moorey, P. R. S. 1980. *Cemeteries of the First Millennium B.C. at Deve Hüyük, Near Carchemish, Salvaged by T. E. Lawrence and C. L. Woolley in 1913. B.A.R. International.* Oxford: B.A.R.

Moorey, P. R. S. 1988. "Bronzeworking centres of Western Asia c. 1000-539 BC: problems and perspectives," in *Bronzeworking centres of Western Asia c. 1000-539 BC.* Edited by J. Curtis, pp. 23-32. London: Keegan Paul International.

Moorey, P. R. S. 1991. *A Century of Biblical Archaeology.* Westminster: John Knox Press.

Moorey, P. R. S. 1994. *Ancient Mesopotamian Materials and Industries: The Archeological Evidence.* Oxford: Clarendon Press.

Morandi Bonacossi, D. 1999. "Die eisenzietliche Keramik der TAVO-Gelandebegehung des Unteren Habur-Gebiets," in *Iron Age Pottery in Northern Mesopotamia, Northern Syria and South-Eastern Anatolia : papers presented at the meetings of the international "table ronde" at Heidelberg (1995) and Nieborów (1997) and other contributions, Altertumskunde des vorderen Orients ; Bd. 10.* Edited by A. Hausleiter and A. Reiche, pp. 193-229. Münster: Ugarit-Verlag.

Morandi Bonacossi, D. 2000a. "The Beginning of the Early Bronze Age at Tell Shiukh Fawqani (Tishrin Dam Area) and in the Upper Syrian Euphrates Valley," in *Proceedings of the First International Congress on the Archaeology of the Ancient Near East, Rome, May 18-23, 1998. Volume II.* Edited by P. Matthiae, A. Enea, L. Peyronel, and F. Pinnock, pp. 1105-1124. Rome: Universita degli Studi di Roma 'La Sapienza'.

Morandi Bonacossi, D. 2000b. "The Syrian Jezireh in the Late Assyrian Period: A View from the Countryside," in *Essays on Syria in the Iron Age., Ancient Near Eastern studies. Supplement 7.* Edited by G. Bunnens, pp. 349-396. Louvain: Peeters Press.

Morandi Bonacossi, D. 2000c. "Marina sa sade, Marina, Burmarina: The Many Names of an Assyrian Town on the Euphrates," in *Patavina Orientalia Selecta, History of the Ancient Near East Monographs IV.* Edited by E. Rova, pp. 221-230. Padova: Sargon srl.

Morandi Bonacossi, D. 2002. "Operation J," in *Excavating Qatna, I. Preliminary Report on the 1999 and 2000 Campaigns of the Joint Syrian-Italian-German Archaeological Research Project at Tell Mishrifeh, Documents d'Archeologie Syrienne IV.* Edited by M. al-Maqdissi, M. Luciani, D. Morandi Bonacossi, M. Novák, and P. Pfälzner, pp. 123-144. Damascus.

Morandi Bonacossi, D. 2008a. "The Acropolis of Tell Mishrifeh During the Second and First Millennia BC. Preliminary Results of the Work of the Italian Component of the Syrian-Italian-German Project at Tell Mishrifeh/Qatna," in *Proceedings of the 4th International Congress of the Archaeology of the Ancient Near East, 29 March-3April 2004, Freie Universität Berlin. Volume 2.* Edited by H. Kuhne, R. M. Czichon, and F. J. Kreppner, pp. 316-376. Wiesbaden: Harrassowitz Verlag.

Morandi Bonacossi, D. 2008b. Excavations on the Acropolis of Mishrifeh, Operation J. A New Early Bronze Age III-Iron Age III Sequence for Central Inner Syria. Part I: Stratigraphy, Chronology and Architecture. *Akkadica* 129(1): 55-127.

Morris, E. L. 1994. Production and Distribution of Pottery and Salt in Iron Age Britain: a Review. *Proceedings of the Prehistoric Society* 60:371-393.

Morris, I. 1987. *Burial and Ancient Society. The rise of the Greek city-state.* Cambridge: Cambridge University Press.

Morris, I. 1992. *Death-Ritual and Social Structure in Classical Antiquity.* Cambridge.

Morris, I. 1997. Archaeology as Cultural History. *Archaeological Review from Cambridge* 14:3-16.

Morris, I. 2000. *Archaeology as Cultural History: Words and Things in Iron Age Greece.* Oxford.

Morrison, K. D., and M. T. Lycett. 1997. Inscriptions as Artifacts: Precolonial South India and the Analysis of Texts. *Journal of Archaeological Method and Theory* 4:215-238.

Moscati, S. 1973. *The World of the Phoenicians.* London: Cardinal.

Moscati, S. 1988. *The Phoenicians.* New York: Abbeville Press Publishers.

Moscrop, J. J. 2000. *Measuring Jerusalem. The Palestine Exploration Fund and British Interests in the Holy Land.* London: Leicester University Press.

Mountjoy, P. A. 1993. *Mycenaean Pottery: An Introduction. Oxford University Committe for Archaeology Monograph 36.* Oxford: Oxbow.

Muhesen, S. 1985. Chronique Archeologique. *Syria* 62.

Muhesen, S. 1999. "Salvage Excavations at the Tishrin Dam Area," in *Archaeology of the Upper Syrian Euphrates, The Tishrin Dam Area: Proceedings of the International Symposium Held at Barcelona, January 28th-30th 1998.* Edited by G. del Olmo Lete and J.-L. Montero Fenollos, pp. 11-12. Barcelona: Editorial AUSA.

Muhly, J. D. 1970. Homer and the Phoenicians; The Relations Between Greece and the Near East in the Late Bronze and Early Iron Ages. *Berytus* 19:19-64.

Muhly, J. D. 1980. "The Bronze Age Setting," in *The Coming of the Age of Iron.* Edited by T. A. Wertime and J. D. Muhly, pp. 25-68. New Haven and London: Yale University Press.

Muhly, J. D. 1984. "The Role of the Sea Peoples in Cyprus during the LC III Period," in *Cyprus at the Close of the Late Bronze Age.* Edited by V. Karageorghis and J. D. Muhly, pp. 39-56. Nicosia.

Muhly, J. D. 1986. "The Role of Cyprus in the Economy of the Eastern Mediterranean During the Second Millennium B.C.," in *Acts of the International Archaeological Symposium 'Cyprus Between the Orient and the Occident'.* Edited by V. Karageorghis. Nicosia: Department of Antiquities, Cyprus.

Muhly, J. D. 1992. "The Crisis Years in the Mediterranean World: Transition or Cultural Disintegration?," in *The Crisis Years: The 12th Century B.C. From Beyond the Danube to the Tigris.* Edited by W. A. Ward and M. S. Joukowsky, pp. 10-26. Dubuque: Kendall/ Hunt.

Muhly, J. D. 2003. "Archaeology in the Eastern Mediterranean: Where We Went Wrong in the Twentieth Century and How We Can Do Better in the Twenty-first," in *One Hundred Years of American Archaeology in the Middle East.* Edited by D. R. Clark and V. H. Matthews, pp. 19-32. Boston: American Schools of Oriental Research.

Na'aman, N. 1985. Arad in the Topographical List of Shishak. *Tel Aviv* 12:91-92.

Na'aman, N. 1992. Israel, Edom and Egypt in the 10th Century BCE. *Tel Aviv* 19:71-93.

Na'aman, N. 1995a. "Province System and Settlement Pattern in Southern Syria and Palestine in the Neo-Assyrian Period," in *Neo-Assyrian Geography: Papers Presented at the International Conference Held in the University of Rome 'La Sapienza', November 10th to 12th, 1993., Quaderni di geografia storica ; 5.* Edited by M. Liverani, pp. 103-115. Roma: Università di Roma 'La Sapienza' Dipartimento di Scienze storiche archeologiche e antropologiche dell'Antichità.

Na'aman, N. 1995b. Tiglath-pileser III's Campaigns Against Tyre and Israel (734-732 B.C.E.). *Tel Aviv* 22:268-278.

Na'aman, N. 1997. Historical and Literary Notes on the Excavations of Tel Jezreel. *Tel Aviv* 24:122-166.

Na'aman, N., and Y. Thareani-Sussely. 2006. Dating the Appearance of Imitations of Assyrian Ware in Southern Palestine. *Tel Aviv* 33(1):61-82.

Naumann, R. 1971. *Architektur Kleinasiens.* Tübingen.

Nibbi, A. 1975. *The Sea Peoples and Egypt.* New Jersey: Noyes Press.

Niemann, H. M. 2000. Megiddo and Solomon: A Biblical Investigation in Relation to Archaeology. *Tel Aviv* 27:61-74.

Niemeier, W.-D. 2001. Archaic Greeks in the Orient: Textual and Archaeological Evidence. *Bulletin of the American Schools of Oriental Research* 322:11-32.

Novák, M., and P. Pfälzner. 2000. Ausgrabungen in Tall Mishrife-Qatna 1999; Vorbericht der deutschen Komponente des internationalen Kooperationsprojekts. *Mitteilungen der Deutschen Orient-Gesellschaft zu Berlin* 132:253-296.

Novák, M., and P. Pfalzner. 2001. Ausgrabungen in Tall Mishrife-Qatna 2000; Vorbericht der deutschen Komponente des internationalen Kooperationsprojekts. *Mitteilungen der Deutschen Orient-Gesellschaft zu Berlin* 133:157-198.

Novák, M., and P. Pfälzner. 2002. "Excavations in the Western Part of the Bronze Age Palace (Operation G)," in *Excavating Qatna, I. Preliminary Report on the 1999 and 2000 Campaigns of the Joint Syrian-Italian-German Archaeological Research Project at Tell Mishrifeh, Documents d'Archeologie Syrienne IV*. Edited by M. al-Maqdissi, M. Luciani, D. Morandi Bonacossi, M. Novak, and P. Pfalzner, pp. 63-110. Damascus.

Núñez, F. J. 2001-2002. An Approach to Exchange Relations in Iron Age Beirut: The Ceramic Evidence. *Aram* 13-14:37-49.

Núñez, F. J. 2004a. "Catalogue of Urns," in *The Phoenician Cemetery at Tyre El-Bass, Excavations 1997-1999, Baal, Hors Serie 1*. Edited by M. E. Aubet, pp. 63-203. Beirut: Ministère de la Culture Direction Générale des Antiquités.

Núñez, F. J. 2004b. "Preliminary Report on Pottery," in *The Phoenician Cemetery at Tyre El-Bass, Excavations 1997-1999, Baal, Hors Serie 1*. Edited by M. E. Aubet, pp. 281-373. Beirut: Ministère de la Culture Direction Générale des Antiquités.

Oates, D., and J. Oates. 2001. *Nimrud. An Assyrian Imperial City Revealed*. London: British School of Archaeology in Iraq.

Oates, J. 1959. Late Assyrian Pottery from Fort Shalmaneser. *Iraq* 21:130-147.

O'Callaghan, R. T. 1948. *Aram Naharaim. A contribution to the history of upper Mesopotamia in the second millennium B.C.* Roma: Pontificium Institutum Biblicum.

Oded, B. 1974. The Phoenician Cities and the Assyrian Empire in the Time of Tiglath-Pileser III. *Zeitschrift des Deutschen Palästina-Vereins* 90:38-49.

Oded, B. 1979. *Mass Deportations and Deportees in the Neo-Assyrian Empire*. Wiesbaden: Dr. Ludwig Reichert Verlag.

Oelmann, F. 1921. Zur Baugeschichte von Sendschirli. *Jahrbuch des Deutschen Archäologischen Instituts* 36:85-98.

Oggiano, I. 1997. The Pottery of Iron Age II from Tell Afis. *Contributi della Scuola di Specializzazione di Archeologia di Pisa* 1:185-211.

Oggiano, I. 1999. "Cataloguistics: The Experience at Tell Afis, Syria," in *Iron Age pottery in Northern Mesopotamia, Northern Syria and South-Eastern Anatolia : papers presented at the meetings of the international "table ronde" at Heidelberg (1995) and Nieborów (1997) and other contributions, Altertumskunde des vorderen Orients ; Bd. 10*. Edited by A. Hausleiter and A. Reiche, pp. 377-402. Münster: Ugarit-Verlag.

Oggiano, I. 2002. Afis 2000-2001. Area J. *Egitto e Vicino Oriente* 25:41-45.

Olmstead, A. T. 1931. *History of Palestine and Syria to the Macedonian Conquest*. New York and London: Charles Scribner's Sons.

Oppenheim, , A. L. 1964. *Ancient Mesopotamia*. Chicago: University of Chicago Press.

Oren, E. D. 1973. *The Northern Cemetery of Beth Shan*. Leiden: E. J. Brill.

Orthmann, W. 1964. Zu den Ausgrabungen in Tell Ain Dara. *Archäologischer Anzeiger* 1964:137-143.

Orthmann, W. 1971. *Untersuchungen zur späthethitischen Kunst. Saarbrucker Beitrage zur Altertumskunde 8*. Bonn.

Orton, C., P. Tyers, and A. Vince. 1993. *Pottery in Archaeology*. Cambridge: Cambridge University Press.

Otto, A. 2006. *Alltag und Gesellschaft zur Spätbronzezeit: Eine Fallstudie aus Tall Bazi (Syrien). Subartu XIX*. Turnhout: BREPOLS Publishers.

Otzen, B. 1979. "Israel Under the Assyrians," in *Power and Propaganda A Symposium on Ancient Empires, Copenhagen Studies in Assyriology 7*. Edited by M. T. Larsen, pp. 251-262. Copenhagen: Akademisk Forlag.

Özduğan, M. 1998. "Ideology and Archaeology in Turkey," in *Archaeology Under Fire: Nationalism, Politicss and Heritage in the Eastern Mediterranean and Middle East*. Edited by L. Meskell, pp. 111-123. London: Routledge.

Pakman, D. 1992. Late Iron Age Vessels at Tel Dan. *Eretz Israel* 23:230-240 (Hebrew).

Pamir, H. 2005. "The Orontes Delta Survey," in *The Amuq Valley Regional Projects Volume 1: Surveys in the Plain of Antioch and Orontes Delta, Turkey, 1995-2002, Oriental Institute Publications 131*. Edited by K. A. Yener, pp. 67-98. Chicago: The Oriental Institute of the University of Chicago.

Papasavvas, G. 2003. Writing on Cyprus: Some Silent Witnesses. *Report of the Department of Antiquities Cyprus*:79-94.

Parker Pearson, M. 2000. *The Archaeology of Death and Burial*. College Station: Texas A & M University.

Parker Pearson, M. 2003. "Food, Culture and Identity: An Approach and Overview," in *Food, Culture and Identity in the Neolithic and Early Bronze Age, BAR International Series 1117*. Edited by M. Parker Pearson, pp. 1-30. Oxford: Archaeopess.

Parpola, S. 1987. *The Correspondence of Sargon II. Part I Letters from Assyria and the West. State Archives of Assyria I*. Helsinki: Helsinki University Press.

Parr, P. J. 1983. The Tell Nebi Mend Project. *Les Annales Archeologiques Arabes Syriennes* 33:99-117.

Parr, P. J. 1991. The Tell Nebi Mend Project: A Progress Report. *Journal of the Ancient Chronology Forum* 4:78-85.

Peckham, B. 2001. "Phoenicians and Aramaeans: The Literary and Epigraphic Evidence," in *The World of the Aramaeans II: Studies in History and Archaeology in Honour of Paul-Eugene Dion, Journal for the Study of the Old Testament Supplement Series 325*. Edited by P. M. M. Daviau, J. W. Wevers, and M.

Weigl, pp. 19-44. Sheffield: Sheffield Academic Press.

Pedrazzi, T. 2002. Afis 2000-2001. Area E: l'occupazione del Bronzo Tardo II - Ferro I. Il settore Nord. *Egitto e Vicino Oriente* 25:33-38.

Perreault, J. Y. 1986. Ceramique et echanges: les importations Attiques au Proche-Orient du VIe au Milieu du Ve Siecle Avant J.-C. les Donnees Archeologiques. *Bulletin de Correspondence Hellenitique* 110:145-175.

Perring, D., H. Seeden, P. Sheehan, and T. Williams. 1996. Bey 006, 1994-1995 The Souks Area; Interim Report of the AUB Project. *Baal* 1:176-206.

Perrot, J. 1979. *Syria-Palestine I. From the Origins to the Bronze Age*. Geneva: Nagel Publishers.

Petrie, W. M. F. 1891. *Tell el Hesy (Lachish)*. London: Committee of the Palestine Exploration Fund.

Petrie, W. M. F. 1904. *Methods and Aims in Archaeology*. London: MacMillan & Co.

Petrie, W. M. F. 1930. *Beth Pelet I (Tell Fara)*. London.

Petrie, W. M. F. 1932. *Seventy Years in Archaeology*. London.

Pettinato, G. 1981. *The Archives of Ebla*. New York: Doubleday & Company, Inc.

Pézard, M. 1922. Mission Archeologique a Tell Nebi Mend (1921) Rapport Sommaire. *Syria* 3:89-115.

Pézard, M. 1931. *Qadesh: Mission Archeologique a Tell Nebi Mend 1921-1922. Bibliotheque Archeologique et Historique XV*. Paris: Geuthner.

Philip, G. 1991. Cypriot Bronzework in the Levantine World: Conservatism, Innovation and Social Change. *Journal of Mediterranean Archaeology* 4:59-108.

Philip, G., and B. S. Ottaway. 1983. Mixed Data Cluster Analysis: an Illustration Using Cypriot Hooked-Tang Weapons. *Archaeometry* 25(2):119-133.

Philip, G., P. W. Clogg, and D. Dungworth. 2003. Copper Metallurgy in the Jordan Valley from the Third to the First Millennium BC: Chemical, Metallographic and Lead Isotope Analyses of Artefacts from Pella. *Levant* 35:71-100.

Philip, G., M. Abdulkarim, P. G. Newson, D. Bridgland, M. Bshesh, A. D. Shaw, R. Westaway, and K. Wilkinson. 2005. Settlement and Landscape Development in the Homs Region, Syria. Report on Work Undertaken during 2001-2003. *Levant* 37:21-42.

Phythian Adams, W. J. 1923. Report on the Stratification of Askalon. *Palestine Exploration Fund Quarterly Statement* April 1923:60-84.

Pinnock, F. 1994. "Considerations on the <Banquet Theme> in the Figurative Art of Mesopotamia and Syria," in *Drinking in Ancient Societies: History and Culture of Drinks in the Ancient Near East*. Edited by L. Milano, pp. 15-26. Padova: Sargon srl.

Pitard, W. T. 1987. *Ancient Damascus: A Historical Study of the Syrian City-State from Earliest Times until its Fall to the Assyrians in 732 BCE*. Winona Lake, Indiana: Eisenbrauns.

Pitard, W. T. 1997. "Aleppo" in *The Oxford Encyclopedia of Archaeology in the Near East. Vol. 1*. Edited by E. M. Meyers, pp. 63-65. Oxford: Oxford University Press.

du Plat Taylor, J. 1959. The Cypriot and Syrian Pottery from Al Mina, Syria. *Iraq* 21:69-92.

Ploix de Rotrou, G. 1931a. Les Fouilles de la Citadelle d'Alep. *Revue Archeologique Syrienne* 1:17-18, 43-44, 66, 102, 131.

Ploix de Rotrou, G. 1931b. Le nouveau relief de la Citadelle d'Alep. *Revue Archeologique Syrienne* 1:8-10.

Ploix de Rotrou, G. 1932a. Les orthostates a dessin geometrique de la Citadelle d'Alep. *Revue Archeologique Syrienne* 2:7-9.

Ploix de Rotrou, G. 1932b. Fouilles de la Citadelle d'Alep. *Revue Archeologique Syrienne* 2:20, 111-112.

Ploug, G. 1973. *Sukas II: The Aegean, Corinthian and Eastern Greek Pottery and Terracottas. Publications of the Carlsberg Expedition to Phoenicia 2*. Copenhagen: Munksgaard.

Poidebard, A. 1939a. *Un Grand Port Disparu Tyr: Recherches Aeriennes et Sous-Marines 1934-1936 (Text). Bibliotheque Archeologique et Historique Tome XXIX*. Paris: Librairie Orientaliste Paul Geuthner.

Poidebard, A. 1939b. *Un Grand Port Disparu Tyr: Recherches Aeriennes et Sous-Marines 1934-1936 (Atlas). Bibliotheque Archeologique et Historique Tome XXIX*. Paris: Librairie Orientaliste Paul Geuthner.

Popham, M., A. M. Pollard, and H. Hatcher. 1983. Euboean Exports to Al Mina, Cyprus, and Crete: a Reassessment. *Annual of the British School at Athens* 78:281-290.

Poppa, R. 1978. *Kamid el-Loz 2: Der Eisenzeitliche Friedhof Befunde unde Funde. Saarbrucker Beitrage zur Alterumskunde 18*. Bonn: Rudolph Habelt Verlag GMBH.

Potts, T. F., S. M. Colledge, and P. C. Edwards. 1985. Preliminary Report on a Sixth Season of Excavation by the University of Sydney at Pella in Jordan 1983/1984. *Annual of the Department of Antiquities of Jordan* 29:181-210.

Potts, T. F., S. J. Bourke, P. C. Edwards, F. Richards, and G. J. Wightman. 1988. Preliminary Report on the Eighth and Ninth Seasons of Excavation by the University of Sydney at Pella (Tabaqat Fahl) 1986 and 1987. *Annual of the Department of Antiquities of Jordan* 32:115-150.

Poulsen, F. 1912. *Der Orient und die Frühgriechische Kunst*. Berlin: Teubner.

Prausnitz, M. W. 1959. Akhziv, Notes and News. *Israel Exploration Journal* 9:271.

Prausnitz, M. W. 1960a. Akhziv. *Revue Biblique* 67:398.

Prausnitz, M. W. 1960b. Akhziv, Notes and News. *Israel Exploration Journal* 10:260-261.

Prausnitz, M. W. 1962. Akhziv. *Revue Biblique* 69:404-405.

Prausnitz, M. W. 1963. Akhziv, Notes and News. *Israel Exploration Journal* 13:337-338.

Prausnitz, M. W. 1965. Akhziv, Notes and News. *Israel Exploration Journal* 15:256-258.

Prausnitz, M. W. 1982. "Die Nekropolen von Akhziv und die Entwicklung der Keramik von 10. bis zum 7. Jahrhundert v. Chr. in Akhziv, Samaria und Ashdod," in *Phönizier im Westen : die Beiträge des Internationalen Symposiums über "Die Phönizische Expansion im Westlichen Mittelmeerraum" in Köln vom 24. bis 27. April, 1979, Madrider Beiträge ; Bd. 8*. Edited by H. G. Niemeyer, pp. 45-82. Mainz am Rhein: P. von Zabern.

Pritchard, J. B. 1968. "New Evidence on the Role of the Sea Peoples in Canaan at the Beginning of the Iron Age," in *The Role of the Phoenicians in the Interaction of Mediterranean Civilisations*. Edited by W. A. Ward, pp. 99-112. Beirut: American University of Beirut.

Pritchard, J. B. 1969. *The Ancient Near East in Pictures Relating to the Old Testament. Second Edition with Supplement*. Princeton, NJ: Princeton University Press.

Pritchard, J. B. 1975. *Sarepta: A Preliminary Report on the Iron Age*. Philadelphia.

Pritchard, J. B. 1978. *Recovering Sarepta, A Phoenician City: Excavations at Sarafand, Lebanon, 1969-1974, by the University Museum of the University of Pennsylvania*. Princeton: Princeton University Press.

Pritchard, J. B. 1988. *Sarepta IV: The Objects from Area II, X*. Beirut: Librairie Orientale.

Pucci, M. 2008a. "Visual Communication of Architecture: The Syro-Hittite Town of Zincirli" in *Proceedings of the 4th International Congress of the Archaeology of the Ancient Near East, 29 March-3April 2004, Freie Universität Berlin. Volume 1*. Edited by H. Kuhne, R. M. Czichon, and F. J. Kreppner, pp. 545-555. Wiesbaden: Harrassowitz Verlag.

Pucci, M. 2008b. *Functional Analysis of Space in Syro-Hittite Architecture, BAR International Series 1738*. Oxford: BAR Publishing.

Ranke, L. 1885. *Geschichten der Romanischen und Germanischen Völker von 1494 bis 1514*. Leipzig (originally published 1824).

Rast, W. E. 1978. *Taanach I, Studies in the Iron Age Pottery*. Cambridge, MA: American Schools of Oriental Research.

Rcadc, J. 1976. Sargon's Campaigns of 720, 716, and 715 B.C.: evidence from the sculptures. *Journal of Near Eastern Studies* 35:95-104.

Reade, J. 1978. Studies in Assyrian Geography. *Revue D'Assyriologie* 72:47-72, 175-180.

Reade, J. 1987. "Reflections on Layard's Archaeological Career," in *Austen Henry Layard Tra l'Oriente e Venezia*. Edited by F. M. Fales and B. J. Hickey, pp. 47-54. Roma.

Reade, J. 1993. Hormuzd Rassam and His Discoveries. *Iraq* 55:39-62.

Reisner, G. A., C. S. Fisher, and D. G. Lyon. 1924. *Harvard Excavations at Samaria, 1908-1910*. Cambridge, MA: Harvard University Press.

Renan, E. 1864. *Mission to Phénicie*. Paris.

Renfrew, C., and P. Bahn. 2004. *Archaeology: Theories, Methods and Practice*, Fourth edition. London: Thames and Hudson.

Repiccioli, M., D. Giannessi, and G. Aletta. 2002. Afis 2000-2001. Area B: le fortificazioni del Bronzo Medio II. *Egitto e Vicino Oriente* 25:12-14.

Restö, J. 1991. The Domestication of the Camel and the Establishment of the Frankincense Road from South Arabia. *Orientalia Suecana* 40:187-219.

Rey-Coquais, J.-P. 1976. Inscription Greque Decouverte a Ras Ibn Hani. *Les Annales Archeologiques Arabes Syriennes* 26:51-61.

Riis, P. J. 1948. *Hama: fouilles et recherches 1931-1938: Les cimetieres a cremation. NationalMuséets skrifter. Storre beretninger ; 1*. Copenhagen: NationalMuséet.

Riis, P. J. 1960. l'Activité de la Mission Archéologique Danoise sur la Côte Phenicienne en 1959. *Les Annales Archeologiques Arabes Syriennes* 10:111-132.

Riis, P. J. 1961. l'Activité de la Mission Archéologique Danoise sur la Côte Phenicienne en 1960. *Les Annales Archeologiques Arabes Syriennes* 11:133-144.

Riis, P. J. 1965. L'activité de la Mission Archéologique Danoise sur la Côte Phénicienne en 1963. *Les Annales Archeologiques Arabes Syriennes* 15(2):57-82.

Riis, P. J. 1970. *Sukas I. The North-East Sanctuary and the First Settling of Greeks in Syria and Palestine. Publications of the Carlsberg Expedition to Phoenicia 1*. Kobenhavn: Munksgaard.

Riis, P. J. 1979. *Sukas VI: The Graeco-Phoenician Cemetery and Sanctuary at the Southern Harbour. Publications of the Carlsberg Expedition to Phoenicia 7. Historisk-filosofiske skrifter (Kongelige Danske Videnskabernes Selskab) 10.2*. Copenhagen: Munksgaard.

Riis, P. J. 1983. "La ville phénicienne de Soukas de la fin de l'âge du bronze à la conquête romaine," in *Atti del I Congresso Internazionale di Studi Fenici e Punici, Roma, 5-10 Novembre 1979, Volume Secondo*. Edited by S. Moscati, pp. 509-514. Roma: Consiglio Nazionale Delle Ricerche.

Riis, P. J. 1996. "Introduction," in *Sukas X: The Bronze and Early Iron Age Remains at the Southern Harbour, Publications of the Carlsberg Expedition to Phoenicia 12. Historisk-filosofiske skrifter (Kongelige Danske Videnskabernes Selskab) 17*. Edited by P. J. Riis, J. Jensen, M.-L. Buhl, and B. Otzen, pp. 5-10. Copenhagen: Munksgaard.

Riis, P. J., and M.-L. Buhl. 1990. *Hama: fouilles et recherches, 1931-1938: Les objets de la periode dite syro-hittite: (Age du Fer). NationalMuséets skrifter. Større Beretninger; 12*. Kobenhavn: NationalMuséet.

Ripinsky, M. 1975. The Camel in Ancient Arabia. *Antiquity* 49:295-298.

Robertson, M. 1940. The Excavations at Al Mina, Sueidia (Continued). *Journal of Hellenic Studies* 60:2-21.

Robertson, M. 1946. The Excavations of Al Mina, Sueidia. IV. The Early Greek Vases. A Postscript. *Journal of Hellenic Studies* 66:125.

Robinson, E. 1841. *Biblical Researches in Palestine, Mount Sinai, and Arabia Petraea*. Boston: Crocker and Brewster.

Robinson, E., and E. Smith. 1867 (original 1841). *Biblical Researches in Palestine and Adjacent Regions: A Journal of Travels in the Years 1838 & 1852*, Third edition. London: John Murray.

Roobaert, A. 1996. A Neo-Assyran Statue from Til Barsip. *Iraq* 59:79-87.

Roobaert, A. 1990. "The City Gate Lions," in *Tell Ahmar: 1988 Season, Abr-Nahrain Supplement Series 2*. Edited by G. Bunnens, pp. 126-135. Leuven: Orientaliste.

Roobaert, A., and G. Bunnens. 1999. "Excavations at Tell Ahmar - Til Barsib," in *Archaeology of the Upper Syrian Euphrates, The Tishrin Dam Area: Proceedings of the International Symposium Held at Barcelona, January 28th-30th 1998*. Edited by G. del Olmo Lete and J.-L. Montero Fenollos, pp. 163-178. Barcelona: Editorial AUSA.

Rosenthal, F. 1968. *A History of Muslim Historiography*. Leiden: E. J. Brill.

Sader, H. 1987. *Les Etats arameens de Syrie depuis leur fondation jusqu' a leur transformation en provinces assyriennes*. Beirut.

Sader, H. 1990a. Tell Kazel and Ancient Simyra. *Berytus* 38:16-22.

Sader, H. 1990b. Tell Kazel, Syria: Excavations of the AUB Museum, 1985-1987, Preliminary Reports: Appendix I An Epigraphic Note on a Phoenician Inscription from Tell Kazel. *Berytus* 38:94-97.

Sader, H. 1991. Phoenician stelae from Tyre. *Berytus* 39:101-126.

Sader, H. 1992. "The 12th Century B.C. in Syria: The Problem of the Rise of the Aramaeans," in *The Crisis Years: The 12th Century B.C. From Beyond the Danube to the Tigris*. Edited by W. A. Ward and M. S. Joukowsky, pp. 157-163. Iowa: Kendall Hunt.

Sader, H. 2000. "The Aramaean Kingdoms of Syria: Origin and Formation Processes," in *Essays on Syria in the Iron Age*. Edited by G. Bunnens, pp. 61-76. Leuven: Peeters Press.

Saggs, H. W. F. 1955a. The Nimrud Letters, 1952-I: The Ukin-Zer Rebellion and Related Texts. *Iraq* 17:21-56.

Saggs, H. W. F. 1955b. The Nimrud Letters-II: Relations with the West. *Iraq* 17:126-160.

Saghieh, M. S. 1983. *Byblos in the Third Millennium BC: A reconstruction of the stratigraphy and a study of the cultural connections*. Warminster: Aris & Phillips LTD.

Saghieh, M. S. 1996. Bey 001 & 004 Preliminary Report. *Baal* 1:23-59.

Sagona, A. G., and C. Haymes. 1992-1993. Excavations at Tell Nebi Mend, Syria. *Mediterranean Archaeology* 5/6:168-170.

Sagona, A. G. 1982. Levantine Storage Jars of the 13th to 4th Century B.C. *Opuscula Atheniensia* 14:7:73-110.

Said, E. 1978. *Orientalism: Western Conceptions of the Orient*. London: Routledge & Kegan Paul.

Said, E. W. 1996. *Peace and its Discontents*. New York: Vintage Books.

Saidah, R. 1966. Fouilles de Khaldé. Rapport Préliminaire sur la Première et Deuxième Campagnes (1961-1962). *Bulletin du Musée de Beyrouth* 19:51-90.

Saidah, R. 1969. Archaeology in the Lebanon 1968-1969. *Berytus* 18:119-142.

Saidah, R. 1971. Objets Grecs d'époque géomètrique découverts récemment sur le littoral Libanais (à Khaldé près de Beyrouth). *Les Annales Archeologiques Arabes Syriennes* 21:193-198.

Saidah, R. 1977. Une tombe de l'Age du Fer a Tambourit (Region de Sidon). *Berytus* 25:135-146.

Saidah, R. 1983. "Nouveaux éléments de datation de la céramique de l'Âge du Fer au Levant," in *Atti del I Congresso Internazionale di Studi Fenici e Punici, Roma, 5-10 Novembre 1979, Volume Primo*. Edited by S. Moscati, pp. 213-216. Roma: Consiglio Nazionale Delle Ricerche.

Saliège, J.-F., and H. Pessin. 2005. "Datation 14C de Tell Shiukh Fawqani: La Méthode et Ses Limites," in *Tell Shiyukh Fawqani 1994-1998*. Edited by L. Bachelot and F. M. Fales, pp. 1075-1079. Padova: Sargon Editrice e Libreria.

Salles, J.-F. 1980. *La Necropole 'K' de Byblos. Recherche sur les grandes civlisations*. Boulogne: Maison de l'Orient.

Salles, J.-F. 1995. "Rituel Mortuaire et Rituel Social à Ras Shamra/Ougarit," in *The Archaeology of Death in the Ancient Near East, Oxbow Monographs 51*. Edited by S. Campbell and A. Green, pp. 171-184. Oxford: Oxbow Books.

Salvensen, A. 1998. "The Legacy of Babylon and Nineveh in Aramaic Sources," in *The Legacy of Mesopotamia*. Edited by S. Dalley, pp. 139-161. Oxford: Oxford University Press.

Sapin, J., and H. Sayegh. 1998. "La Ceramique Commune," in *Un Quartier du Port Phénicien de Beyrouth au Fer III/Perse. Les Objets, Supplement no. 6 a Transeuphraténe*. Edited by J. Elayi and H. Sayegh, pp. 53-110. Paris: Gabalda.

Sayegh, H. 1996. Bey 010 Les Souks, Secteur nord/est. *Baal* 1:235-269.

Schachermeyr, F. 1982. *Die Levante im Zeitalter der Wanderungen: vom 13. bis zum 11. Jahrhundert v. Chr.* Vienna: Österreichische Akademie der Wissenschaft.

Schaeffer, C. F.-A. 1938. Les Fouilles des Ras Shamra-Ugarit Neuvième Campagne (Printemps 1937). Rapport Sommaire. *Syria* 19:193-255.

Schaeffer, C. F.-A. 1949. *Ugaritica II: Nouvelles études relatives aux découvertes de Ras Shamra. Mission de Ras Shamra Tome V. Bibliotheque Archeologique et Historique XLVII.* Paris: Librairie Orientaliste Paul Geuthner.

Schiffer, S. 1930-1931. Tell Ahmar. *Archiv fur Orientforschung* 6:133-134.

Schloen, J. D. 2001. *The House of the Father as Fact and Symbol Patrimonialism in Ugarit and the Ancient Near East. Studies in the Archaeology and History of the Levant 2*. Winona Lake, IN: Eisenbrauns.

Schloen, J. D. 2002. W. F. Albright and the Origins of Israel. *Near Eastern Archaeology* 65:56-62.

Schloen, D. 2003. "The Iron Age State as a State of Mind: A Response," in *One Hundred Years of American Archaeology in the Middle East*. Edited by D. R. Clark and V. H. Matthews, pp. 283-292. Boston: American Schools of Oriental Research.

Schnapp, A. 1996. *The Discovery of the Past. The Origins of Archaeology*. London: British Museum Press.

Schneider, E. 1999a. "Die eisenzietliche Keramik von Tell Sheikh Hassan (Syrien)," in *Iron Age Pottery in*

Northern Mesopotamia, Northern Syria and South-Eastern Anatolia : papers presented at the meetings of the international "table ronde" at Heidelberg (1995) and Nieborów (1997) and other contributions, Altertumskunde des vorderen Orients ; Bd. 10. Edited by A. Hausleiter and A. Reiche, pp. 325-346. Münster: Ugarit-Verlag.

Schneider, E. 1999b. "'Assyrische' Schalen aus Tell Sheikh Hassan (Syrien) und ihre Stellung innerhalb der Keramik des assyrischen Einflussgebietes," in *Iron Age Pottery in Northern Mesopotamia, Northern Syria and South-Eastern Anatolia: papers presented at the meetings of the international "table ronde" at Heidelberg (1995) and Nieborów (1997) and other contributions, Altertumskunde des vorderen Orients ; Bd. 10.* Edited by A. Hausleiter and A. Reiche, pp. 347-375. Münster: Ugarit-Verlag.

Schreiber, N. 2003. *The Cypro-Phoenician Pottery of the Iron Age. Culture and history of the ancient Near East; v. 13.* Leiden: Brill.

Schumacher, G. 1908. *Tell el-Mutesellim I.* Leipzig.

Schwartz, G. M. 1989. "The Origins of the Aramaeans in Syria and Northern Mesopotamia: Research Problems and Potential Strategies," in *To the Euphrates and Beyond: Archaeological Studies in Honour of Maurits N. van Loon.* Edited by O. M. C. Haex, H. H. Curvers, and P. M. M. G. Akkermans, pp. 275-291. Rotterdam: A. A. Balkema.

Schwartz, S. 1995. Language, Power and Identity in Ancient Palestine. *Past and Present* 148:3-47.

Scigliuzzo, E. 2002. Afis 2000-2001. Area B: la ceramica. *Egitto e Vicino Oriente* 25:20-22.

Scigliuzzo, E. 2005. Tell Afis, Siria (2002-2004) Area B3: occupazione di Ferro II-III. *Egitto e Vicino Oriente* 28:41-45.

Seeden, H. 1989. Bulldozers Destroy What Scientific Archaeology Exposed. *Berytus* 37:3-4.

Seeden, H. 1990. Editorial. *Berytus* 38:3-5.

Seeden, H. 1991. A *Tophet* in Tyre? *Berytus* 39:39-82.

Seeden, H., and R. Thorpe. 1997-1998. Beirut from Ottoman Sea Walls and Landfill to a Twelfth Century BC Burial: Interim Report on the Archaeological Excavations of the Souks Northern Area (BEY 007). *Berytus* 43:221-254.

Seirafi, F., A. Kirichian, and M. Dunand. 1965. Recherches archéologiques à Ayin Dara au nord-ouest d'Alep. *Les Annales Archeologiques Arabes Syriennes* 15(2):3-20.

Seton-Williams, M. V. 1961. Preliminary Report on the Excavations at Tell Rifa'at. *Iraq* 23:68-87.

Seton-Williams, M. V. 1967. The Excavations at Tell Rifa'at, 1964. Second Preliminary Report. *Iraq* 29:16-33.

Seton-Williams, M. V. 1980. 'Tell Keisan, 1935-1936' in *Tell Keisan, 1971-1976: Une cite phenicienne en Galilee.* Edited by J. Briend and J.-B. Humbert, pp. 381-392; Pls 107-111. Paris: J Gabalda.

Shaath, S. 1996. *Qal'at Halab.* Aleppo.

Shanks, M., and C. Tilley. 1987a. *Re-Constructing Archaeology.* Cambridge.

Shanks, M., and C. Tilley. 1987b. *Social Theory and Archaeology.* Cambridge: Polity Press.

Sharon, I. 2001. "Philistine Bichrome Painted Pottery: Scholarly Ideology and Ceramic Typology," in *Studies in the Archaeology of Israel and Neighbouring Lands in Memory of Douglas L Esse, ASOR Books 5: Studies in Ancient OrientalCcivilisation 59.* Edited by S. R. Wolff, pp. 135-142. Chicago and Atlanta.

Shefton, B. B., J. Elayi, H. Sayegh, and Z. Ghadar. 1998. "La Ceramique Grecque Importee," in *Un Quartier du Port Phénicien de Beyrouth au Fer III/Perse. Les Objets, Supplement no. 6 a Transeuphraténe.* Edited by J. Elayi and H. Sayegh, pp. 111-149. Paris: Gabalda.

Shennan, S. 1989. "Introduction: Archaeological Approaches to Cultural Identity," in *Archaeological Approaches to Cultural Identity.* Edited by S. Shennan, pp. 1-31. London: Unwin Hyman.

Shennan, S. 1997. *Quantifying Archaeology*, Second edition. Edinburgh: Edinburgh University Press.

Sherratt, A., and S. Sherratt. 1991. "From Luxuries to Commodities: The Nature of Mediterranean Bronze Age Trading Systems," in *Bronze Age Trade in the Mediterranean, Studies in Mediterranean Archaeology 90.* Edited by N. H. Gale, pp. 351-386. Jonsered: Aströms Förlag.

Sherratt, A., and S. Sherratt. 1998. "Small Worlds: Interaction and Identity in the Ancient Mediterranean," in *The Aegean and the Orient in the Second Millennium. Proceedings of the 50th Anniversary Symposium, Cincinnati, 18-20 April 1997.* Edited by E. H. Cline and D. Harris-Cline, pp. 329-343. Liege: Annales d'archéologie égéenne de l'Université de Liège et UT-PASP.

Sherratt, S. 1990. Reading the Texts: Archaeology and the Homeric Question. *Antiquity* 64:807-824.

Sherratt, S. 1994a. "Commerce, Iron and Ideology: Metallurgical Innovation in 12th-11th Century Cyprus," in *Proceedings of the International Symposium, Cyprus in the Eleventh Century B.C.* Edited by V. Karageorghis, pp. 59-107. Nicosia.

Sherratt, S. 1994b. Comment on Ora Negbi, 'The "Libyan Landscape" from Thera: A Review of Aegean Enterprises Overseas in the Late Minoan IA Period'. *Journal of Mediterranean Archaeology* 7:237-240.

Sherratt, S. 1998. "'Sea Peoples' and the Economic Structure of the Late Second Millennium in the Eastern Mediterranean," in *Mediterranean Peoples in Transition: Thirteenth to Early Tenth Centuries BCE-In Honour of Professor Trude Dothan.* Edited by S. Gitin, A. Mazar, and E. Stern, pp. 292-313. Jerusalem: Israel Exploration Society.

Sherratt, S. 1999. *"E pur si muove*: pots, markets and values in the second millennium Mediterranean" in *The Complex Past of Pottery: Production, Circulation and Consumption of Mycenaean and Greek Pottery (sixteenth to early fifth centuries BC).* Edited by J.P. Crieelard, V. Stissi and G.J. van Wijngaarden, pp. 163-211. Amsterdam.

Sherratt, S. 2000. "Circulation of Metals and the End of the Bronze Age in the Eastern Mediterranean," in *Metals Make the World Go Round. The Supply and*

Circulation of Metals in Bronze Age Europe. Proceedings of a conference held at the University of Birmingham in June 1997. Edited by C. F. E. Pare, pp. 82-98. Oxford: Oxbow Books.

Sherratt, S. 2003a. Visible Writing: Questions of Script and Identity in Early Iron Age Greece and Cyprus. *Oxford Journal of Archaeology* 22(3):225-242.

Sherratt, S. 2003b. "The Mediterranean Economy: 'Globalisation' at the End of the Second Millennium BCE," in *Symbiosis, Symbolism, and the Power of the Past. Canaan, Ancient Israel, and Their Neighbours from the Late Bronze Age through Roman Palaestina*. Edited by W. G. Dever and S. Gitin, pp. 37-61. Winona Lake, IN: Eisenbrauns.

Sherratt, S. 2005a. "'Ethnicities', 'ethnonyms' and archaeological labels. Whose ideologies and whose identities?," in *Archaeological Perspectives on the Transmission and Transformation of Culture in the Eastern Mediterranean, Levant Supplementary Series 2*. Edited by J. Clarke, pp. 25-38. Oxford: Oxbow Books.

Sherratt, S. 2005b. "High Precision Dating and Archaeological Chronologies: Revisiting an old problem," in *The Bible and Radiocarbon Dating: Archaeology, Text and Science*. Edited by T. E. Levy and T. Higham, pp. 114-128. London: Equinox.

Sherratt, S., and A. Sherratt. 1993. The Growth of the Mediterranean Economy in the Early First Millennium B.C. *World Archaeology* 24:361-378.

Shipton, G. M. 1939. *Notes on the Megiddo Pottery of Strata VI-XX. Studies in Ancient Oriental Civlisation 17*. Chicago: University of Chicago Press.

Shortland, A. J. 2005. "Shishak, King of Egypt: The challenges of Egyptian calendrical chronology," in *The Bible and Radiocarbon Dating: Archaeology, Text and Science*. Edited by T. E. Levy and T. Higham, pp. 43-56. London: Equinox.

Silberman, N. A. 1982. *Digging for God and Country. Exploration, Archaeology, and the Secret Struggle for the Holy Land*. New York: Alfred A. Knopf.

Silberman, N. A. 1989. *Between Past and Present: Archaeology, Ideology and Nationalism in the Modern World*. New York: Henry Holt and Company.

Silberman, N. A. 1991. Desolation and Restoration: The Impact of a Biblical Concept on Near Eastern Archaeology. *Biblical Archaeologist* 54:76-87.

Silberman, N. A. 1993a. *'A Prophet From Amongst You' - The Life of Yigael Yadin: Soldier, Scholar and Mythmaker of Modern Israel*. Reading, MA: Addison-Wesley.

Silberman, N. A. 1993b. Visions of the Future: Albright in Jerusalem, 1919-1929. *Biblical Archaeologist* 56:8-17.

Silberman, N. A. 1993c. "Petrie and the Founding Fathers," in *Biblical Archaeology Today*. Edited by A. Biran and J. Aviram, pp. 545-554. Jerusalem: Israel Exploration Journal.

Silberman, N. A. 1995. "Promised Lands and Chosen Peoples: The Politics and Poetics of Archaeological Narrative," in *Nationalism, Politics, and the Practice of Archaeology*. Edited by P. L. Kohl and C. Fawcett, pp. 249-262. Cambridge: Cambridge University Press.

Silberman, N. A. 1997. "Structuring the Past: Israelis, Palestinians, and the Symbolic Authority of Archaeological Monuments," in *The Archaeology of Israel: Constructing the Past, Interpreting the Present, Journal for the Study of the Old Testament Supplement Series 237*. Edited by N. A. Silberman and D. Small, pp. 62-81. Sheffied.

Silberman, N. A. 1998a. "Whose Game is it Anyway? The Political and Social Transformations of American Biblical Archaeology," in *Archaeology Under Fire: Nationalism, Politicss and Heritage in the Eastern Mediterranean and Middle East*. Edited by L. Meskell, pp. 175-188. London: Routledge.

Silberman, N. A. 1998b. "The Sea Peoples, the Victorians and Us: Modern Social Ideology and Changing Archaeological Interpretation of the Late Bronze Age Collapse," in *Mediterranean Peoples in Transition: Thirteenth to Early Tenth Centuries BCE-In Honour of Professor Trude Dothan*. Edited by S. Gitin, A. Mazar, and E. Stern, pp. 268-275. Jerusalem: Israel Exploration Society.

Silberman, N. A. 2003. "A Century of American Archaeology in the Middle East: Looking Back and Looking Ahead," in *One Hundred Years of American Archaeology in the Middle East*. Edited by D. R. Clark and V. H. Matthews, pp. 7-18. Boston: American Schools of Oriental Research.

Singer, I. 1985. The Beginning of Philistine Settlement in Canaan and the Northern Boundary of Philistia. *Tel Aviv* 12:109-122.

Smith, A. T. 2001. Review: On Landscapes in the Ancient Near East. *Journal of the Economic and Social History of the Orient* 44(3):363-371.

Smith, G. A. 1892. *The Historical Geography of the Holy Land, Especially in Relation to the History of Israel and of the Early Church*. London: Hodder and Stoughton.

Smith, R. H. 1973. *Pella of the Decapolis, I: The 1967 Season of the Wooster Expedition to Pella*.

Smith, R. H. 1993. "Pella," in *The New Encyclopaedia of Archaeological Excavations in the Holy Land*. Edited by E. Stern, pp. 1174-1180. New York: Simon and Schuster.

Smith, R. H., and T. F. Potts. 1992. "The Iron Age," in *Pella in Jordan 2: The Second Interim Report of the Joint University of Sydney and College of Wooster Excavations at Pella 1982-1985, Mediterranean Archaeology Supplement 2*. Edited by A. W. McNicoll, P. C. Edwards, J. Hanbury-Tenison, J. B. Hennessy, T. F. Potts, R. H. Smith, A. Walmsley, and P. Watson, pp. 83-102. Sydney: MEDITARCH.

Smith, S. 1938. *Assyrian Sculptures in the British Museum: From Shalmaneser III to Sennacherib*. London: The British Museum.

Smith, S. 1942. The Greek Trade at al Mina: A Footnote to Oriental History. *The Antiquaries Journal* 22:87-112.

Snodgrass, A. M. 1980a. "Iron and Early Metallurgy in the Mediterranean," in *The Coming of the Age of Iron*.

Edited by T. A. Wertime and J. D. Muhly, pp. 335-374. New Haven and London: Yale University Press.

Snodgrass, A. M. 1980b. *Archaic Greece. The Age of Experiment*. London: Dent.

Soldi, S. 2005. Tell Afis, Siria (2002-2004) Area A1: il settore occidentale. *Egitto e Vicino Oriente* 28:24-29.

Sørensen, M. L. S. 1997. Material Culture and Typology. *Current Swedish Archaeology* 5:179-192.

Sørensen, M. L. S. 2000. *Gender Archaeology*. Cambridge: Blackwell Publishers.

Stager, L. E. 1985. Merenptah, Israel and Sea Peoples: New Light on an Old Relief. *Eretz Israel* 18 (Nahman Avigad Volume):56-64.

Stager, L. E. 1995. "The Impact of the Sea Peoples in Canaan (1185-1050 BCE)," in *The Archaeology of Society in the Holy Land*. Edited by T. E. Levy, pp. 332-348. London.

Stager, L. E. 2001. "Port Power in the Early and the Middle Bronze Age: The Organization of Maritime Trade and Hinterland Production," in *Studies in the Archaeology of Israel and Neighboring Lands in Memory of Douglas L. Esse*. Edited by S. R. Wolff, pp. 625-638. Chicago.

Stark, M. T. 1998. "Technical Choices and Social Boundaries in Material Culture Patterning: An Introduction," in *The Archaeology of Social Boundaries*. Edited by M. T. Stark, pp. 1-11. Washington and London: Smithsonian Institution Press.

Stark, M. T., R. L. Bishop, and E. Miksa. 2000. Ceramic Technology and Social Boundaries: Cultural Practices in Kalinga Clay Selection and Use. *Journal of Archaeological Method and Theory* 7:295-331.

Steegman, J. S. 1986. *The Rule of Taste from George I to George IV*. London.

Steel, L. 1998. The Social Impact of Mycenaean Imported Pottery in Cyprus. *Annual of the British School at Athens* 93:285-296.

Steel, L. 2002. Consuming Passions: A Contextual Study of the Local Consumption of Mycenaean Pottery at Tell el'Ajjul. *Journal of Mediterranean Archaeology* 15(1):25-51.

Steel, L. 2004. "A reappraisal of the distribution, context and function of Mycenaean pottery in Cyprus," in *La Ceramique Mycenienne de l'Egee au Levant. Hommage a Vronwy Hankey, Travaux de la Maison de l'Orient et de la Mediterranee*. Edited by J. Balensi, J.-Y. Monchambert, and S. Muller Celka, pp. 69-86. Lyon: Jean Pouilloux.

Steiner, M. 2001-2002. The Hellenistic to Byzantine Souk; Results of the Excavations at Bey 011. *Aram* 13-14:113-127.

Stern, E. 1982. *Material Culture of the Land of the Bible in the Persian Period 538-332 B.C.* Jerusalem.

Stern, E. 1990a. New Evidence from Dor for the First Appearance of the Phoenicians Along the Northern Coast of Israel. *Bulletin of the American Schools of Oriental Research* 279:27-34.

Stern, E. 1990b. Hazor, Dor and Megiddo in the Time of Ahab and Under Assyrian Rule. *Israel Exploration Journal* 40:12-30.

Stern, E. 1991. "Phoenicians, Sikils, and Israelites in the Light of Recent Excavations at Tel Dor," in *Phoenicia and the Bible: Proceedings of the Conference held at the University of Leuven on the 15th and 16th of March 1990, Studi Phoenicia 9*. Edited by E. Lipinski, pp. 85-94. Leuven: Uitgeverij Peeters.

Stieglitz, R. R. 1990. The Geopolitics of the Phoenician Littoral in the Early Iron Age. *Bulletin of the American Schools of Oriental Research* 279:9-12.

Stieglitz, R. R. 1991. The City of Amurru. *Journal of Near Eastern Studies* 50:45-48.

Stone, B. J. 1995. The Philistines and Acculturation: Culture Change and Ethnic Continuity in the Iron Age. *Bulletin of the American Schools of Oriental Research* 298:7-32.

Stone, E. C., and P. E. Zimansky. 1999. *The Iron Age Settlement at 'Ain Dara, Syria. Survey and Soundings. BAR International Series 786*. Oxford: BAR Publishing.

Strange, J. 2000. "A Proposal for a New Periodization for the Bronze and Iron Ages in the Southern Levant (Jordan, Palestine and Israel)," in *Proceedings of the First International Congress on the Archaeology of the Ancient Near East, Rome, May 18-23, 1998. Volume II*. Edited by P. Matthiae, A. Enea, L. Peyronel, and F. Pinnock, pp. 1571-1577. Rome: Universita degli Studi di Roma 'La Sapienza'.

Strika, V. 2000. "The Perception of the Past in the Near East. Two Case-Studies: Iraq and Saudi Arabia," in *Proceedings of the First International Congress on the Archaeology of the Ancient Near East, Rome, May 18th-23rd 1998*. Edited by P. Matthiae, A. Enea, L. Peyronel, and F. Pinnock, pp. 1579-1586. Rome.

Strommenger, E. 1985a. "Hadatu (Arslantash)," in *Ebla to Damascus: Art and Archaeology of Ancient Syria: An Exhibition from the Director-General of Antiquities and Museums, Syrian Arab Republic*. Edited by H. Weiss, pp. 330-332. Washington DC: Smithsonian Institution Travelling Exhibition Service.

Strommenger, E. 1985b. "Til Barsip/Kar Shalmaneser (Tell Ahmar)," in *Ebla to Damascus: Art and Archaeology of Ancient Syria: An Exhibition from the Director-General of Antiquities and Museums, Syrian Arab Republic*. Edited by H. Weiss, pp. 330. Washington DC: Smithsonian Institution Travelling Exhibition Service.

Stronach, D. 1997. "Notes on the Fall of Nineveh," in *Assyria 1995: Proceedings of the 10th Anniversary Symposium of the Neo-Assyrian Text Corpus Project Helsinki, September 7-11, 1995*. Edited by S. Parpola and R. M. Whiting, pp. 307-324. Helsinki: The Neo-Assyrian Text Corpus Project.

Stuart, B. 2001-2002. Cemeteries in Beirut. *Aram* 13-14:37-49.

Stubbings, F. H. 1951. *Mycenaean Pottery from the Levant*. Cambridge.

Stucky, R. 2000. Le Sanctuaire Sidonien d'Echmoun Aperçu historique du site, des fouilles et des découvertes faites à Bostan ech-Cheikh. *Baal* 4:123-148.

Stucky, R. 2004. "Eléments de construction en marbre à décor animalier de Sidon et la culture phénicienne aux époques perse et hellénistique ancienne," in *Decade:*

A Decade of Archaeology and History in the Lebanon. Edited by C. Doumet-Serhal, pp. 214-223. Beirut: The Lebanese British Friends of the National Museum.

Suleiman, A. 1984. Excavations at Ansari-Aleppo for the Seasons 1973-1980. Early and Middle Bronze Ages. *Akkadica* 40:1-16.

Suleiman, A. 1999. Excavations at Ansari Near Aleppo. *Les Annales Archéologiques Arabes Syriennes* 43:247-248.

Summers, G. D. 1993. *Tille Höyük 4. The Late Bronze Age and the Iron Age Transition. British Institute of Archaeology at Ankara Monograph 15*. London: The British Institute of Archaeology at Ankara.

Swift, G. F. Jr 1958. The Pottery of the 'Amuq Phase K to O, and its Historical Relationship, Unpublished Dissertation, University of Chicago.

Sznycer, M. 1975. Quelques remarques à propos de la formation de l'alphabet phénicien. *Semitica* 24:4-12.

Tadmor, H. 1958. The Campaigns of Sargon II of Assur: A Chronological-Historical Study. *Journal of Cuneiform Studies* 12(1):22-40.

Tadmor, H. 1975. "Assyria and the West: the Ninth Century and its Aftermath," in *Unity and Diversity. Essays in the History, Literature, and Religion of the Ancient Near East*. Edited by H. Goedicke and J. J. M. Roberts, pp. 36-48. London: The Johns Hopkins University Press.

Tadmor, H. 1981. "History and Ideology in the Assyrian Royal Inscriptions," in *Assyrian Royal Inscriptions: New Horizons in literary, ideological, and historical analysis. Papers of a Symposium held in Cetona (Siena) June 26-28, 1980*. Edited by F. M. Fales, pp. 13-34. Roma: Instituo per l'Oriente.

Tadmor, H. 1997. "Propaganda, Literature, Historiography: Cracking the Code of the Assyrian Royal Inscriptions," in *Assyria 1995: Proceedings of the 10th Anniversary Symposium of the Neo-Assyrian Text Corpus Project Helsinki, September 7-11, 1995*. Edited by S. Parpola and R. M. Whiting, pp. 325-338. Helsinki: The Neo-Assyrian Text Corpus Project.

Tainter, J. 1988. *The Collapse of Complex Societies*. Cambridge: Cambridge University Press.

Taylor, W. 1948. *A Study of Archaeology*. Menasha, WI.

Taylor, M. B. 1984. The Pottery from the Iron Age Gate at Tell Qarqur, Brigham Young University.

Tefnin, R. 1976. Le Tell Abou Danne: Mission belge en Mesopotamie (Syrie du Nord). *Archeologia* 95:62-63.

Tefnin , R. 1978-1979. Tall Abu Danna. *Archiv fur Orientforschung* 26:145-147.

Tefnin, R. 1980a. "Deux campagnes de fouilles au Tell Abou Danne (1975-1976)," in *Le Moyen Euphrate: Actes du Colloque de Strasbourg 10-12 mars 1977*. Edited by J.-C. Margueron, pp. 179-199. Strasbourg.

Tefnin, R. 1980b. Les niveaux supérieurs du Tell Abou Danné: Chantier A 1977-78. *Syro-Mesopotamian Studies* 3(3):110-168.

Tefnin, R. 1981-1982. Tall Abu Danna. *Archiv fur Orientforschung* 28:201-202.

Tefnin, R. 1983. Apercu sur Neuf Campagnes de Fouilies Belges aux Tells Abou Danne et Oumm el-Marra (1975-1983). *Les Annales Archeologiques Arabes Syriennes* 33:141-151.

Tenu, A., and L. Bachelot. 2005. Tell Shiukh Fawqani (Syrie): la campagne de sondages 2003 dans la nécropole à incinération. *Akkadica* 126:159-168.

Thalmann, J.-P. 1978a. Tell 'Arqa 1978-1979 Rapport Provisoire. *Bulletin du Musée de Beyrouth* 30:61-75.

Thalmann, J.-P. 1978b. Tell 'Arqa (Liban Nord): Campagnes I-III (1972-1974): Chantier I: Rapport Preliminaire. *Syria* 55:1-151.

Thalmann, J.-P. 1981-1982. Tall 'Arqa. *Archiv fur Orientforschung* 28:206-208.

Thalmann, J.-P. 1983. "Les Niveaux de l'Age de Bronze et de l'Age du Fer a Tell 'Arqa (Liban)," in *Atti del I Congresso Internazionale di Studi Fenici e Punici, Roma, 5-10 Novembre 1979, Volume Primo*. Edited by S. Moscati, pp. 217-222. Roma: Consiglio Nazionale Delle Ricerche.

Thalmann, J.-P. 1989-1990. Prospection de la plaine du Akkar syrien, 1986-7. *Archiv fur Orientforschung* 36/37:344-346.

Thalmann, J.-P. 1990. Tell 'Arqa, de la conquête assyrienne à l'époque perse. *Transeuphratène* 2:51-56.

Thalmann, J.-P. 1991a. L'age du Bronze a Tell Arqa (Liban-Nord). *Orient-Express* 1991:20-21.

Thalmann, J.-P. 1991b. L'age du Bronze a Tell 'Arqa: Bilan et Perspectives (1981-1991). *Berytus* 39:21-38.

Thalmann, J.-P. 2000a. "Le peuplement de la Plaine du Akkar a l'age du Bronze," in *Proceedings of the First International Congress on the Archaeology of the Ancient Near East, Rome, May 18-23, 1998. Volume II*. Edited by P. Matthiae, A. Enea, L. Peyronel, and F. Pinnock, pp. 1615-1636. Rome: Universita degli Studi di Roma 'La Sapienza'.

Thalmann, J.-P. 2000b. Tell Arqa. *Baal* 4:5-74.

Thalmann, J.-P. 2003. Transporter et Conserver: Jarres de l'age du Bronze a Tell Arqa. *Archaeology and History in the Lebanon* 17:25-37.

Thalmann, J.-P. 2006a. *Tell Arqa I. Les niveaux de l'âge du Bronze Volume I: Texte. BAH 177*. Beyrouth: IFAPO.

Thalmann, J.-P. 2006b. *Tell Arqa I. Les niveaux de l'âge du Bronze Volume II: Planches. BAH 177*. Beyrouth: IFAPO.

Thalmann, J.-P., and M. al-Maqdissi. 1989. "Prospection de la trouée de Homs," in *Contribution française à l'archéologie syrienne*, pp. 98-101. Damascus: Institut français d'archéologie du Proche-Orient, Centre de Damas.

Thomas, H. L. 1967. *Near Eastern, Mediterranean and European Chronology. The historical, archaeological, radiocarbon, pollen-analytical and geochronological evidence*. Lund: Studies in Mediterranean Archaeology XVII.

Thomsen, P. 1935-1936. Hamath. *Archiv fur Orientforschung* 10:302.

Thomsen, P. 1936-1937a. Hamath. *Archiv fur Orientforschung* 11:267-268.

Thomsen, P. 1936-1937b. Byblos. *Archiv fur Orientforschung* 11:268.

Thomsen, P. 1937-1939. Hamath. *Archiv fur Orientforschung* 12:298.

Thorpe, R. 1998-1999a. Bey 007 The Souks Area, Preliminary Report of the AUB/ACRE Project. *Baal* 3:31-56.
Thorpe, R. 1998-1999b. Bey 045, Preliminary Report on the Excavations. *Baal* 3:57-83.
Thuesen, I. 1988. *Hama: fouilles et recherches, 1931-1938: I. The Pre- and Protohistoric Periods*. Vol. 1. Copenhague: Nationalmuseet.
Thureau-Dangin, F. 1929. Tell Ahmar. *Syria* 10:185-205.
Thureau-Dangin, F. 1930a. L'inscription des lions de Til-Barsip. *Revue D'Assyriologie* 27:11-21.
Thureau-Dangin, F. 1930b. Un specimen des peintures assyriennes de Til Barsip. *Syria* 11:113-132.
Thureau-Dangin, F., Barrois, A., Dossin, G. and M. Dunand. 1931a. *Arslan-Tash (Texte)*. Paris: Libraire Orientaliste Paul Geuthner.
Thureau-Dangin, F., Barrois, A., Dossin, G. and M. Dunand. 1931b. *Arslan-Tash (Atlas)*. Paris: Libraire Orientaliste Paul Geuthner.
Thureau-Dangin, F., and M. Dunand. 1936a. *Til-Barsib (Texte). Bibliothèque archéologique et historique (Haut-Commissariat de la République Française en Syrie et au Liban, Service des antiquités) 23*. Paris: Geuthner.
Thureau-Dangin, F., and M. Dunand. 1936b. *Til-Barsib (Album). Bibliothèque archéologique et historique (Haut-Commissariat de la République Française en Syrie et au Liban, Service des antiquités) 23*. Paris: Geuthner.
Thurston, T. L. 1997. Historians, Prehistorians, and the Tyranny of the Historical Record: Danish State Formation Through Documents and Archaeological Data. *Journal of Archaeological Method and Theory* 4:239-264.
Trigger, B. G. 1980. *Gordon Childe: Revolutions in Archaeology*. New York: Columbia University Press.
Trigger, B. G. 1984. Alternative Archaeologies: Nationalist, Colonialist, Imperialist. *Man* 19:355-370.
Trigger, B. G. 1989. *A History of Archaeological Thought*. Cambridge.
Trigger, B. G. 1995. "Romanticism, Nationalism and Archaeology," in *Nationalism, Politics, and the Practice of Archaeology*. Edited by P. L. Kohl and C. Fawcett, pp. 263-279. Cambridge: Cambridge University Press.
Trigger, B. G., and I. Glover. 1981. Editorial. *World Archaeology* 13:133-137.
Tubb, J. N. 1988. "The Role of the Sea Peoples in the bronze industry of Palestine/Transjordan in the Late Bronze Age - Eary Iron transition," in *Bronzeworking centres of Western Asia c. 1000-539 BC*. Edited by J. Curtis, pp. 251-270. London: Keegan Paul International.
Turner, G. 1968. The Palace and *Batiment aux Ivoires* at Arslan Tash: A Reappraisal. *Iraq* 30:62-68.
Tvedtnes, J. A. 1981. The Origin of the Name 'Syria'. *Journal of Near Eastern Studies* 40(2):139-140.
Tylecote, R. F. 1980. "Furnaces, Crucibles, and Slags," in *The Coming of the Age of Iron*. Edited by T. A. Wertime and J. D. Muhly, pp. 183-228. New Haven and London: Yale University Press.
Ussishkin, D. 1966. Building IV in Hamath and the Temples of Solomon and Tell Tayanat. *Israel Exploration Journal* 16:104-110.
Ussishkin, D. 1968. 'Der Alt Bau' in Zincirli. *Bulletin of the American Schools of Oriental Research* 189:50-53.
Ussishkin, D. 1967a. On the Dating of Some Groups of Reliefs from Carchemish and Til Barsip. *Anatolian Studies* 17:181-192.
Ussishkin, D. 1971. Was Bit-Adani a Neo-Hittite or Aramaean State? *Orientalia* 40:431-437.
Ussishkin, D. 1976. The Monuments of the Lower Palace Area in Carchemish: A Rejoinder. *Anatolian Studies* 26:105-112.
Ussishkin, D. 1982. *The Conquest of Lachish by Sennacherib*. Tel Aviv: Tel Aviv University.
Ussishkin, D. 1990a. Notes on Megiddo, Gezer, Ashdod and Tel Batash in the Tenth to Ninth Centuries B.C. *Bulletin of the American Schools of Oriental Research* 277/278:71-91.
Ussishkin, D. 1990b. The Assyrian Attack on Lachish: The Archaeological Evidence from the Southwest Corner of the Site. *Tel Aviv* 17:53-86.
Ussishkin, D. 1997. "Megiddo," in *The Oxford Encyclopedia of Archaeology in the Near East. Vol. 3*. Edited by E. M. Meyers, pp. 460-469. Oxford: Oxford University Press.
Ussishkin, D., and J. Woodhead. 1992. Excavations at Tel Jezreel 1990-1991: Preliminary Report. *Tel Aviv* 19:3-56.
Ussishkin, D., and J. Woodhead. 1994. Excavations at Tel Jezreel 1992-1993: Second Preliminary Report. *Levant* 26:1-48.
Ussishkin, D., and J. Woodhead. 1997. Excavations at Tel Jezreel 1994-1996: Third Preliminary Report. *Tel Aviv* 24:6-72.
van Beek, G. W. 1951. Cypriote Chronology and the Dating of Iron I Sites in Palestine. *Bulletin of the American Schools of Oriental Research* 124:26-29.
van Beek, G. W. 1955. The Date of Tell Abu Hawam Stratum III. *Bulletin of the American Schools of Oriental Research* 138:34-38.
van Beek, G. W. 1987. Digging up Tell Jemmeh. *Archaeology* 36:12-19.
van Loon, M. 1995. "A Neo-Hittite Relief in Aleppo," in *Studio historiae ardens: Ancient Near Eastern Studies Presented to Philo H. J. Houwink ten Cate on the Occasion of his 65th birthday*. Edited by T. P. J. van den Hout and J. De Roos, pp. 181-186. Leiden: Uitgaven vat het nederlands historisch-archaeologisch intituut te Istanbul 74.
van Seters, J. 1997. *In Search of History: Historiography in the Ancient World and the Origins of Biblical History*. Winona Lake: Eisenbrauns.
van Wijngaarden, G. J. 1999. "Production, Circulation and Consumption of Mycenaean Pottery (Sixteenth to Twelfth Centuries BC)," in *The Complex Past of Pottery. Production, Circulation and Consumption of Mycenaean and Greek Pottery (Sixteenth to Early Fifteenth Centuries B.C.)*. Edited by J. P. Crielaard, V. Stissi, and G. J. van Wijngaarden, pp. 21-47. Amsterdam: J. C. Gieben.

van Wijngaarden, G. J. 2002. *Use and Appreciation of Mycenaean Pottery in the Levant, Cyprus and Italy (1600-1200 BC)*. Amsterdam: Amsterdam University Press.

Veit, U. 1989. "Ethnic Concepts in German Prehistory: a case study on the relationship between identity and objectivity," in *Archaeological Approaches to Cultural Identity*. Edited by S. Shennan, pp. 35-56. London: Unwin Hyman.

Venturi, F. 1997. Mission archeologique de Tell Afis (Syrie), campagne 1996. *Orient-Express* 1997:11-13.

Venturi, F. 1998a. "The Late Bronze II and Early Iron I Levels (Area E1)," in *Tell Afis, Siria : scavi sull'acropoli 1988-1992 = The 1988-1992 excavations on the Acropolis, Ricerche di archeologia del Vicino Oriente ; 1*. Edited by S. M. Cecchini and S. Mazzoni, pp. 123-162. Pisa: Edizioni ETS.

Venturi, F. 1998b. Mission archeologique de Tell Afis (Siria). Campagne1997. *Orient-Express* 1998/1:13-16.

Venturi, F. 2000a. "Le Premier Age du Fer a Tell Afis et in Syrie Septentrionale," in *Essays on Syria in the Iron Age, Ancient Near Eastern studies. Supplement 7*. Edited by G. Bunnens, pp. 505-536. Louvain: Peeters Press.

Venturi, F. 2000b. Il Bronzo Tardo II ed il Ferro I (livelli 10-5). *Egitto e Vicino Oriente* 22.

Venturi, F. 2000c. "Tell Afis et la 'Dark Age' (les niveaux de la fin du Bronze recent II et de l'age du Fer I)," in *Proceedings of the First International Congress on the Archaeology of the Ancient Near East, Rome, May 18-23, 1998. Volume II*. Edited by P. Matthiae, A. Enea, L. Peyronel, and F. Pinnock, pp. 1715-1730. Rome: Universita degli Studi di Roma 'La Sapienza'.

Venturi, F. 2002. Afis 2000-2001. Area E: l'occupazione del Bronzo Tardo II - Ferro I. Il settore Sud. *Egitto e Vicino Oriente* 25:29-33.

Venturi, F. 2005. Tell Afis, Siria (2002-2004) Area E4b sud: Bronzo Tardo II - Ferro I. *Egitto e Vicino Oriente* 28:69-76.

Venturi, F. 2008. "Elements de continuite dans l'architecture domestique du Levant à travers l'âge obscure; Le cas des maisons aux piliers" in *Proceedings of the 4th International Congress of the Archaeology of the Ancient Near East, 29 March-3April 2004, Freie Universität Berlin. Volume 1*. Edited by H. Kuhne, R. M. Czichon, and F. J. Kreppner, pp. 587-600. Wiesbaden: Harrassowitz Verlag.

Vesco, D., and S. Soldi. 2002. L'informatizzazione della documentazione grafica. *Egitto e Vicino Oriente* 25:7-9.

Vickers, M., and D. Gill. 1994. *Artful Crafts: Ancient Greek Silverware and Pottery*. Oxford: Clarendon.

Virgilio, F. 2002. Afis 2000-2001. Area B: le mura del Ferro II-III. *Egitto e Vicino Oriente* 25:19-20.

Virgilio, F. 2005. Tell Afis, Siria (2002-2004) Area B2: fortificazione e zona abitative del Ferro II-III. *Egitto e Vicino Oriente* 28:37-41.

Volney, M. C.-F. 1972 (original 1787). *Travels through Syria and Egypt, in the Years 1783, 1784, and 1785*. London: G G J and J Robinson.

von Luschan, F. 1902. *Ausgrabungen in Sendschirli III, Thorsculpturen*. Berlin: W. Spemann.

von Oppenheim, M. 1933. *Tell Halaf: A New Culture in Oldest Mesopotamia*. London and New York: G P Putnam's Sons.

von Oppenheim, M. 1962. *Tell Halaf. Vierter Band: Die Kleinfunde aus Historischer Zeit*. Berlin: Walter de Gruyter & Co.

Wachtsmuth, F. 1923-1924. Die Baugeschichte von Sendschirli (Samal). *Jahrbuch des Deutschen Archäologischen Instituts* 38-39:158-169.

Wada, H. 1994. Pottery Vessels in the Iron Age II in the South Area of Tell Mastuma in 1993 Season. *Bulletin of the Ancient Orient Museum* 15:51-76.

Wailes, B., and A. L. Zoll. 1995. "Civilization, barbarism, and nationalism in European archaeology," in *Nationalism, Politics, and the Practice of Archaeology*. Edited by P. L. Kohl and C. Fawcett, pp. 21-38. Cambridge: Cambridge University Press.

Wainwright, G. A. 1936. The Coming of Iron. *Antiquity* 10:5-24.

Wakita, S., I. Asano, H. Wada, A. Tsuneki, and T. Nakamura. 1994. Tell Mastuma. A Preliminary Report of the Excavations in Idlib, Syria 1993. *Bulletin of the Ancient Orient Museum* 15:23-50.

Wakita, S., I. Asano, H. Wada, T. Adachi, S. Nishiyama, Y. Okada, and Y. Ikeda. 1995. Tell Mastuma. A Preliminary Report of the Excavations in Idlib, Syria 1994 and 1995. *Bulletin of the Ancient Orient Museum* 16:1-73.

Wakita, S., H. Wada, and S. Nishiyama. 2000. "Tell Mastuma: Change in Settlement Plans and Historical Context During the First Quarter of the First Millennium B.C.," in *Essays on Syria in the Iron Age, Ancient Near Eastern studies. Supplement 7*. Edited by G. Bunnens, pp. 537-557. Louvain: Peeters Press.

Waldbaum, J. C. 1978. *From Bronze to Iron. The Transition from the Bronze Age to the Iron Age in the Eastern Mediterranean. Studies in Mediterranean Archaeology 54*. Göteburg: Paul Åströms Förlag.

Waldbaum, J. C. 1980. "The First Archaeological Appearance of Iron and the Transition to the Iron Age," in *The Coming of the Age of Iron*. Edited by T. A. Wertime and J. D. Muhly, pp. 69-98. New Haven and London: Yale University Press.

Waldbaum, J. C. 1994. Early Greek Contacts with the Southern Levant, ca. 1000-600 B.C.: The Eastern Perspective. *Bulletin of the American Schools of Oriental Research* 293:53-66.

Waldbaum, J. C. 1997. Greeks in the East or Greek and the East? Problems in the Definition and Recognition of Presence. *Bulletin of the American Schools of Oriental Research* 305:1-18.

Waldbaum, J. C. 1999. "The Coming of Iron in the Eastern Mediterranean: Thirty Years of Archaeological and Technological Research," in *The Archaeometallurgy of the Asian Old World*. Edited by V. Pigott, pp. 27-58. Philadelphia: University Museum.

Waldbaum, J. C., and J. Magness. 1997. The Chronology of Early Greek Pottery: New Evidence from Seventh Century B.C. Destruction Levels in Israel. *American Journal of Archaeology* 101(1):23-40.

Walmsley, A., P. G. Macumber, P. C. Edwards, S. J. Bourke, and P. M. Watson. 1993. The Eleventh and Twelfth Seasons of Excavation at Pella (Tabaqat Fahl) 1989-1990. *Annual of the Department of Antiquities of Jordan* 37:165-240.

Wapnish, P. 1984. "The Dromedary and Bactrian Camel in Levantine Historical Settings: The Evidence from Tell Jemmeh," in *Animals and Archaeology: 3. Early Herders and their Flocks, BAR International Series 202*. Edited by J. Clutton-Brock and C. Grigson, pp. 171-200. Oxford: BAR.

Ward, W. A., and M. S. Joukowsky. Editors. 1992. *The Crisis Years: The 12th Century B.C. From Beyond the Danube to the Tigris*. Dubuque, Iowa: Kendall/ Hunt.

Ward, W. H. 1888. Unpublished or Imperfectly Published Hittite Monuments III: Reliefs at Carchemish=Jerablus. *The American Journal of Archaeology and of the History of the Fine Arts* 4:172-174.

Ward, W. H., and A. L. Frothingham. 1887. Unpublished or Imperfectly Published Hittite Monuments II: Sculptures Near Sindjirli. *The American Journal of Archaeology and of the History of the Fine Arts* 3:62-69.

Wartke, R.-B. 2005. *Sam'al: Ein aramäischer Stadtstaat des 10. bis 8. Jhs v. Chr. Und die Geschichte seiner Erforschung*. Mainz am Rhein: Verlag Philipp von Zabern.

Waterfield, G. 1963. *Layard of Nineveh*. London: John Murray.

Waterfield, G. 1987. "Reflections on A.H. Layard's Character and Early Career," in *Austen Henry Layard Tra l'Oriente e Venezia*. Edited by F. M. Fales and B. J. Hickey, pp. 7-18. Roma.

Watson, P., and J. Tidmarsh. 1996. Pella/Tall al-Husn Excavations 1993: The University of Sydney's 15th Season. *Annual of the Department of Antiquities of Jordan* 40:293-314.

Watzinger, C. 1929. *Tell el-Mutesellim: Bericht überdie 1903 bis 1905 mit Unterstützung Sr. Majestät des Deustchen Kaisers und der Orient-Gesellschaft vom Deutschen verein zur Erforschung Palästinas veranstaltelen Ausgrabungen, Band 2. Die Funde*. Leipzig: J. C. Hinrichs.

Webb, J. M., and D. Frankel. 1994. Making an Impression: Storage and Surplus Finance in Late Bronze Age Cyprus. *Journal of Mediterranean Archaeology* 7(1):5-26.

Weiss, H. Editor. 1985. *Ebla to Damascus: Art and Archaeology of Ancient Syria: An Exhibition from the Director-General of Antiquities and Museums, Syrian Arab Republic*. Washington DC: Smithsonian Institution Travelling Exhibition Service.

Weiss, H. 1991. Archaeology in Syria. *American Journal of Archaeology* 95(4):683-740.

Weiss, H. 1994. Archaeology in Syria. *American Journal of Archaeology* 98(1):101-158.

Weiss, H. 1997. Archaeology in Syria. *American Journal of Archaeology* 101(1):97-148.

Wengrow, D. 2006. "The Idea of Prehistory in the Middle East," in *A Future for Archaeology: The Past in the Present*. Edited by R. Layton, S. Shennan, and P. Stone, pp. 187-197. London: UCL Press.

Wertime, T. A. 1980. "The Pyrotechnologic Background," in *The Coming of the Age of Iron*. Edited by T. A. Wertime and J. D. Muhly, pp. 1-24. New Haven and London: Yale University Press.

West, M. L. 1988. The Rise of the Greek Epic. *Journal of Hellenic Studies* 108:151-172.

Whincop, M. R. 2007. The Iron Age II at Tell Nebi Mend: Towards an Explanation of Ceramic Regions. *Levant* 39:185-212.

Whitelam, K. 1996. *The Invention of Ancient Israel, the silencing of Palestinian history*. London: Routledge.

Whiting, C. 2002. Complexity and Diversity in the Late Iron Age Southern Levant: The Investigation of 'Edomite' Archaeology and Scholarly Discourse. Unpublished PhD Thesis, Durham University.

Whiting, C. 2007a. *Complexity and Diversity in the Late Iron Age Southern Levant: The investigation of 'Edomite' archaeology and scholarly discourse. BAR International Series 1672*. Oxford: BAR Publishing.

Whiting, C. 2007b. Telling it as it was in the southern Levant: a task of biblical proportions? *Antiquity* 81:210-213.

Wightman, G. J. 1985. Megiddo VIA-III: Associated Structures and Chronology. *Levant* 17:117-129.

Wightman, G. J. 1990a. The Myth of Solomon. *Bulletin of the American Schools of Oriental Research* 277/278:5-22.

Wightman, G. J. 1990b. "Area B - Sounding in the Middle City," in *Tell Ahmar: 1988 Season, Abr-Nahrain Supplement Series 2*. Edited by G. Bunnens, pp. 106-120. Leuven: Orientaliste.

Wilhelm, G. 1973. Eine Krughenkelschrift in Alphabetischer Keilschrift aus Kamid el-Loz. *Ugarit-Forschungen* 5:284-285.

Wilken, R. L. 1992. *The Land Called Holy: Palestine in Christian History and Thought*. New Haven and London.

Wilkens, B. 1998. "The Faunal Remains," in *Tell Afis, Siria: scavi sull'acropoli 1988-1992 = The 1988-1992 excavations on the Acropolis, Ricerche di archeologia del Vicino Oriente ; 1*. Edited by S. M. Cecchini and S. Mazzoni, pp. 433-450. Pisa: Edizioni ETS.

Wilkens, B. 2002. Afis 2000-2001. I resti faunistici. *Egitto e Vicino Oriente* 25:58-67.

Wilkinson, T. J. 2003. *Archaeological Landscapes of the Near East*. Tucson: University of Arizona Press.

Will, E., J. M. Dentzer, and J.-P. Thalmann. 1973. La Premiere Campagne de Fouilles a Tell Arqa (Liban Nord). *Bulletin du Musée de Beyrouth* 26:61-79.

Williamson, H. G. M. 1991. Jezreel in the Biblical Texts. *Tel Aviv* 18:72-92.

Wilson, J. A. 1968. "A Century of Near Eastern Archaeology and the Future," in *The Role of the Phoenicians in the Interaction of Mediterranean*

Civilisations. Edited by W. A. Ward, pp. 113-122. Beirut: American University of Beirut.

Wilson, J. A. 1969a. "The War Against the Peoples of the Sea," in *Ancient Near Eastern Texts Relating to the Old Testament*, Third Edition with Supplement edition. Edited by J. B. Pritchard, pp. 262-263. Princeton, New Jersey: Princeton University Press.

Wilson, J. A. 1969b. "The Campaign of Sheshonk I," in *Ancient Near Eastern Texts Relating to the Old Testament*, Third Edition with Supplement edition. Edited by J. B. Pritchard, pp. 263-264. Princeton, New Jersey: Princeton University Press.

Wilson, J. A. 1969c. "Tiglath-Pileser III (744-727): Campaigns Against Syria and Palestine," in *Ancient Near Eastern Texts Relating to the Old Testament*, Third Edition with Supplement edition. Edited by J. B. Pritchard, pp. 282-284. Princeton, New Jersey: Princeton University Press.

Wilson, J. A. 1969d. "Sargon II (721-705): The Fall of Samaria," in *Ancient Near Eastern Texts Relating to the Old Testament*, Third Edition with Supplement edition. Edited by J. B. Pritchard, pp. 284-287. Princeton, New Jersey: Princeton University Press.

Wilson, J. A. 1969e. "The Journey of Wen-Amon to Phoenicia," in *Ancient Near Eastern Texts Relating to the Old Testament*, Third Edition with Supplement edition. Edited by J. B. Pritchard, pp. 25-29. Princeton, New Jersey: Princeton University Press.

Wilson, J. A. 1969f. "Papyrus Harris I," in *Ancient Near Eastern Texts Relating to the Old Testament*, Third Edition with Supplement edition. Edited by J. B. Pritchard, pp. 262. Princeton, New Jersey: Princeton University Press.

Wilson, K. A. 2005. *The Campaign of Pharaoh Shoshenq I into Palestine. Forschungen zum Alten Testament 2 Reihe 9*. Tübingen: Mohr Siebeck.

Winstone, H. V. F. 1982. *The Illicit Adventure: The Story of Political and Military Intelligence in the Middle East from 1898 to 1926*. London: Jonathan Cape.

Winstone, H. V. F. 1990. *Woolley of Ur: The Life of Sir Leonard Woolley*. London: Secker & Warburg.

Winter, I. J. 1973. North Syria in the Early First Millennium B.C., with Special Reference to Ivory Carving. (Unpublished Dissertation), Columbia University, New York.

Winter, I. J. 1976. Phoenician and North Syrian Ivory Carving in Historical Context: Questions of style and distribution. *Iraq* 38:1-22.

Winter, I. J. 1988. "North Syria as a bronzeworking centre in the early first millennium BC: luxury commodities at home and abroad," in *Bronzeworking centres of Western Asia c. 1000-539 BC*. Edited by J. Curtis, pp. 193-226. London: Keegan Paul International.

Winter, I. J. 1995. "Homer's Phoenicians: History, ethnography, or literary trope? A perspective on early orientalism," in *The Age of Homer: A Tribute to Emily Townsend Vermeule*. Edited by J. B. Carter and S. P. Morris, pp. 247-271. Austin: University of Texas Press.

Wiwjorra, I. 1996. "German Archaeology and its relation to nationalism and racism," in *Nationalism and Archaeology in Europe*. Edited by M. Diaz-Andreu and T. C. Champion, pp. 164-188. London: University College London Press.

Woolley, C. L. 1914. Hittite Burial Customs. *Liverpool Annals of Archaeology and Anthropology* 6:87-98.

Woolley, C. L. 1914-1916. A North Syrian Cemetery of the Persian Period. *Liverpool Annals of Archaeology and Anthropology* 7:115-129.

Woolley, C. L. 1921a. *Carchemish: Report on the Excavations at Jerablus on Behalf of the British Museum (conducted by C. L. Woolley, T. E. Lawrence and P. L. O. Guy); Part II - The Town Defences*. London: British Museum.

Woolley, C. L. 1921b. La Phenicie et les peuples egeens. *Syria* 2:177-194.

Woolley, C. L. 1927. Classification of the Pottery of Central and Northern Syria. *Classifications des Ceramiques Antiques* 2:1-6.

Woolley, C. L. 1934. The Prehistoric Pottery of Carchemish. *Iraq* 1:146-162.

Woolley, C. L. 1937. Excavations near Antioch in 1936. *The Antiquaries Journal* 17:1-15.

Woolley, C. L. 1938a. Excavations at Al Mina, Sueidia: The Archaeological Report. *Journal of Hellenic Studies* 58:1-30.

Woolley, C. L. 1938b. Excavations at Al Mina, Sueidia II. *Journal of Hellenic Studies* 58:133-170.

Woolley, C. L. 1939a. Excavations at Atchana-Alalakh, 1938. *The Antiquaries Journal* 19:1-37, 133-170.

Woolley, C. L. 1939b. The Iron-Age Graves of Carchemish. *Liverpool Annals of Archaeology and Anthropology* 26:11-37.

Woolley, C. L. 1948. The Date of al Mina. *Journal of Hellenic Studies* 68:148.

Woolley, C. L. 1952. *Carchemish: Report on the Excavations at Jerablus on Behalf of the British Museum (conducted by D G Hogarth, R Campbell Thompson and Sir Leonard Woolley, with T E Lawrence, P L O Guy and H Reitlinger); Part III - The Excavations in the Inner Town*. London: Trustees of the British Museum.

Woolley, C. L. 1955. *Alalakh: An Account of the Excavations at Tell Atchana in the Hatay, 1937-1949. Reports of the Research Committee of the Society of Antiquaries of London, 18*. London: The Society of Antiquaries.

Wriedt Sorenson, L. 1997. "Travelling Pottery Connections Between Cyprus, the Levant, and the Greek World in the Iron Age," in *Res Maritimae. Cyprus and the Eastern Mediterranean from Prehistory to Late Antiquity. Proceedings of the Second International Symposium 'Cities on the Sea', Nicosia, Cyprus, October 18-22, 1994*. Edited by S. Swiny, R. L. Hohlfelder, and H. Wylde Swiny, pp. 285-298. Atlanta, GA: Scholars Press.

Wright, G. E. 1937. *The Pottery of Palestine From the Earliest Times to the End of the Early Bronze Age. American Schools of Oriental Research Publications of the Jerusalem School I*. New Haven: American Schools of Oriental Research.

Wright, G. E. 1939. Iron: The Date of its Introduction into Common Use in Palestine. *American Journal of Archaeology* 43:458-463.

Wright, G. E. 1950. The Discoveries at Megiddo 1935-39. *Biblical Archaeologist* 13:28-46.

Wright, G. E. 1959. Philistine Coffins and Mercenaries. *Biblical Archaeologist* 22:54-67.

Wright, G. E. 1961. "The Archaeology of Palestine," in *The Bible and the Ancient Near East. Essays in honor of William Foxwell Albright.* Edited by G. E. Wright, pp. 73-112. London: Routledge and Kegan Paul.

Wright, G. R. H. 1966. A Method of Excavation Common in Palestine. *Zeitschrift des Deutschen Palästina-Vereins* 82:113-124.

Wright, W. 1882. Kadesh on the Orontes. *Quarterly Statement: Palestine Exploration Fund* 1882:132.

Wylie, M. A. 1982. "Epistemological Issues Raised by a Structuralist Archaeology," in *Symbolic and Structural Archaeology.* Edited by I. Hodder, pp. 39-46. Cambridge.

Yadin, Y. 1960. New Light on Solomon's Megiddo. *Biblical Archaeologist* 23:62-68.

Yadin, Y. 1966. *Masada. Herod's Fortress and the Zealot's Last Stand.* London: Weidenfeld & Nicolson.

Yadin, Y. 1970. Megiddo of the Kings of Israel. *Biblical Archaeologist* 33:66-96.

Yadin, Y. 1972. *Hazor. The Head of all these Kingdoms.* London: Oxford University Press.

Yadin, Y. 1975. *Hazor: The Rediscovery of a Great Citadel of the Bible.* London: Weidenfeld and Nicolson.

Yadin, Y., Y. Aharoni, R. Amiran, T. Dothan, I. Dunayevski, and J. Perrot. 1958. *Hazor I: An Account of the First Season of Excavations, 1955.* Jerusalem: Magnes Press Hebrew University

Yadin, Y., Y. Aharoni, R. Amiran, T. Dothan, M. Dothan, and J. Perrot. 1961. *Hazor III-IV: An Account of the Third and Fourth Seasons of Excavation, 1957-1958 (Plates).* Jerusalem: Magnes Press Hebrew University of Jerusalem : Israel Exploration Society.

Yadin, Y., Y. Aharoni, R. Amiran, T. Dothan, M. Dothan, and J. Perrot. 1989. *Hazor III-IV: An Account of the Third and Fourth Seasons of Excavation, 1957-1958 (Text).* Jerusalem: Magnes Press Hebrew University of Jerusalem : Israel Exploration Society.

Yadin, Y., Y. Aharoni, R. Amiran, T. Dothan, I. Dunayevski, and J. Perrot. 1960. *Hazor II: An Account of the Second Season of Excavations, 1956.* Jerusalem: Magnes Press Hebrew University.

Yadin, Y., and S. Geva. 1986. *Investigations at Beth Shean: The Early Iron Age Strata. Qedem 23.* Jerusalem: Hebrew University.

Yahya, A. H. 2005. "Archaeology and Nationalism in the Holy Land," in *Archaeologies of the Middle East: Critical Perspectives, Blackwell Studies in Global Archaeology.* Edited by S. Pollock and R. Bernbeck, pp. 66-77. Oxford: Blackwell Publishing.

Yamada, S. 2000. *The Construction of the Assyrian Empire. A Historical Study of the Inscriptions of Shalmeneser III (859-824 B.C.) Relating to His Campaigns to the West. Culture & History of the Ancient Near East 3.* Leiden: Brill.

Yasur-Landau, A. 2003. "The Absolute Chronology of the Late Helladic III C Period: A View from the Levant," in *LH III C Chronology and Synchronisms: Proceedings of the International Workshop held at teh Austrian Academy of Sciences at Vienna, May 7th and 8th, 2001.* Edited by S. Deger-Jalkotzy and M. Zavadil, pp. 235-244. Wien: Österreichischen Akademie der Wissenschaften.

Yasur-Landau, A. 2004. The Chronological Use of Imported Mycenaean Pottery in the Levant: Towards a Methodological Common Ground. *Ägypten und Levante* 14:339-346.

Yener, K. A. 2000a. "Between the Tigris-Euphrates and the Mediterranean Sea: The Oriental Institute Amuq Valley Regional Projects, Turkey," in *Proceedings of the First International Congress on the Archaeology of the Ancient Near East, Rome, May 18-23, 1998. Volume II.* Edited by P. Matthiae, A. Enea, L. Peyronel, and F. Pinnock, pp. 1801-1812. Rome: Universita degli Studi di Roma 'La Sapienza'.

Yener, K. A. 2000b. *The Domestication of Metals. The Rise of Complex Metal Industries in Anatolia. Culture and History of the Ancient Near East 4.* Leiden: Brill.

Yener, K. A. 2005. "Introduction," in *The Amuq Valley Regional Projects Volume 1: Surveys in the Plain of Antioch and Orontes Delta, Turkey, 1995-2002, Oriental Institute Publications 131.* Edited by K. A. Yener, pp. 1-23. Chicago: The Oriental Institute of the University of Chicago.

Yener, K. A., C. Edens, T. P. Harrison, J. Verstraete, and T. J. Wilkinson. 2000a. The Amuq Valley Regional Projects 1995-1998. *American Journal of Archaeology* 104(2):163-220.

Yener, K. A., C. Edens, J. Casana, B. Diebold, H. Ekstrom, M. Loyet, and R. Özbal. 2000b. Tell Kurdu Excavations 1999. *Anatolica* 26:31-117.

Yogev, O. 1988-1989. Tel Yizre'el - October 1987-January 1988. *Excavations and Surveys in Israel* 7-8:191-195.

Yon, M. 1987. "Les rhytons du sanctuaire," in *Ras Shamra-Ougarit 3. Le Centre de la Ville. 38e-44e campagnes 1978-1984*, pp. 343-350. Lyon: Editions Recherche sur les Civilisations.

Yon, M. 1992. "The End of the Kingdom of Ugarit," in *The Crisis Years: The 12th Century B.C. From Beyond the Danube to the Tigris.* Edited by W. A. Ward and M. S. Joukowsky, pp. 111-122. Iowa: Kendall Hunt.

Yon, M., and A. Caubet. 1990. Tell Kazel, Syria: Excavations of the AUB Museum, 1985-1987, Preliminary Reports: Appendix II Les Ceramiques Importees de l'Ouest. *Berytus* 38:98-118.

Yon, M., V. Karageorghis, and N. Hirschfeld. 2000. *Céramiques Mycéniennes d'Ougarit. Ras Shamra-Ougarit 13.* Nicosia: Leventis Foundation.

Zarins, J. 1978. The Camel in Ancient Arabia: A Further Note. *Antiquity* 52:44-46.

Zarzeki- Peleg, A. 1997. Hazor, Jokneam and Megiddo in the 10th Century BCE. *Tel Aviv* 24:258-288.

Zettler, R. L. 1996. "Written Documents as Excavated Artifacts and the Holistic Interpretation of the Mesopotamian Archaeological Record," in *The Study of the Ancient Near East in the Twenty-First Century*. Edited by J. S. Cooper and G. M. Schwartz, pp. 81-101. Winona Lake, Indiana: Eisenbrauns.

Zimhoni, O. 1992. The Iron Age Pottery from Tel Jezreel: An Interim Report. *Tel Aviv* 19:57-71.

Zimhoni, O. 1997a. "The Iron Age Pottery from Tel Jezreel: An Interim Report," in *Studies in the Iron Age Pottery of Israel: Typological, Archaeological, and Chronological Aspects, Tel Aviv Occasional publications no. 2*, pp. 13-28. Tel Aviv: Tel Aviv University, Institute of Archaeology.

Zimhoni, O. 1997b. "Clues from the Enclosure Fills: Pre-Omride Settlement at Tel Jezreel," in *Studies in the Iron Age Pottery of Israel: Typological, Archaeological, and Chronological Aspects, Tel Aviv Occasional publications no. 2*, pp. 29-56. Tel Aviv: Tel Aviv University, Institute of Archaeology.

Zimhoni, O. 1997c. Clues from the Enclosure Fills: Pre-Omride Settlement at Tel Jezreel. *Tel Aviv* 24:83-109.

Zuckerman, S. 2007a. Anatomy of a Destruction: Crisis Architecture, Termination Ritual and the Fall of Canaanite Hazor. *Journal of Mediterranean Archaeology* 20(1):3-32.

Zuckerman, S. 2007b. "Dating the Destruction of Canaanite Hazor without Mycenaean Pottery?," in *The Synchronisation of Civilisations in the Eastern Mediterranean in the Second Millennium B.C. III*. Edited by M. Bietak and E. Czerny, pp. 621-627. Vienna: Verlag der Österreichischen Akademie der Wissenschaften.

Internet resources

http://www.pef.org.uk (accessed 03/2006)
http://www.pef.org.uk/EarlySyriaPages/Carchemish.htm (accessed 13/04/2007)
http://www.syriatourism.org (accessed 10/08/2006)
http://www.uni-freiburg.de/aorient/aohome.html (accessed 05/05/2006)
http://www.utoronto.ca/tap/2005.htm (accessed 15/03/2007)
www.reshafim.org.il/ad/egypt/shoshenqi.htm (accessed 04/04/2007)
http://www.qatna.org/en-index.html (accessed 07/05/2006)
http://www.sidonexcavation.org (accessed 16/03/2006)

APPENDICES

APPENDIX A

Gazetteer of North Levantine Iron Age Sites

This gazetteer draws together a base level of information from North Levantine sites that have published a reasonable amount of Iron Age pottery or have contributed significantly to our understanding of this period. While sites like Zincirli, Aleppo and Arslan Tash have published little or no pottery, they have made a significant contribution to knowledge of the architecture and urban planning of the Iron Age, and for that reason are included. The site data is presented in a systematic fashion: the location and setting; historical connection; list of excavation; short list of published Iron Age ceramics.

A.1 Abou Danne, Tell (Syria)

Located on the Raqqa–Aleppo road, Tell Abou Danne lies 25 km due east of Aleppo and 50 km west of the Euphrates. The site is situated in the remarkably fertile basin of the *nahr Toman*, which flows into the large Djabboul (Jabbul) Lake 10 km to the south-east. The regular mound has a diameter of 300 m (c. 7 ha.) and height of 25 m (Lebeau 1983, 11).

Ancient Iron Age Name:
The ancient name of the site is currently not known.

Seasons of Excavation
R. Tefnin: 1974-1978, 1981

Published Iron Age Ceramics
Lebeau 1983; Tefnin 1980a

A.2 Abu Hawam, Tell (Israel)

Tell Abu Hawam is composed of a small (c. 4 ha.), low mound (7 m above the plain) surrounded by an extensive settlement (c. 10 ha.). There is also evidence for two inhumation necropolises situated less than one kilometre from the tell, and three silted-up anchorage facilities to the west (Balensi 1993, 7). The site is today situated on the coastal plain near Mount Carmel (within the city limits of Haifa) and lies around 2 km from the Mediterranean, though the site originally stood on the coast at the point where the Kishon River emptied into the bay of Haifa - the river mouth has long since silted up and moved away from the site. Tell Abu Hawam lies approximately midway between Cyprus and the Nile Delta, which might have facilitated participation in maritime trade.

Ancient Iron Age Name:
The ancient name of Tell Abu Hawam is unknown, though Dever (1997c) suggests it is the biblical city of Shihor-Libnah mentioned in Joshua 19:26.

Seasons of Excavation
P. L. O. Guy and G. M. Fitzgerald: 1922
L. A. Mayer and N. Makhouly: 1929
N. Makhouly, D. Baramki and A. Vilensky: 1930
R. W. Hamilton: 1932-1933
E. Anati, Y. Olami and M. Prausnitz: 1952, 1963
J. Balensi: 1984-1989
M. Artzy: 2005?

Published Iron Age Ceramics
Hamilton 1934; 1935

A.3 Afis, Tell (Syria)

Tell Afis is located to the immediate west of the modern Hama-Aleppo highway c. 3 km north of Saraqab, in the district (*Mohafazeh*) of Idlib. The region is fertile, favourable to both rainfall and irrigation agriculture, and is still renowned for olive oil production. The site itself consists of a circular lower mound (c. 28 ha), crowned with a steep acropolis toward its northern edge (Mazzoni and Cecchini eds 1998, 5).

Ancient Iron Age Name:
Apsuna (Ugaritic, Late Bronze Age)
Hazrak/Hazrek (Aramaic)
Hatarikka (Neo-Assyrian)
Hadrach (Hebrew, Old Testament)
Apish? (Aramaic)

In 1903, H. Pognon, the French Consul in Aleppo, passed through Saraqeb and discovered fragments of a stele bearing an Aramaic inscription. His account of the discovery does not explicitly mention Tell Afis, but the site has come to be identified with the city mentioned in the inscription, ancient Hazrek (Oggiano 1997, 186). The stele is dedicated by Zakkur, king of Hamath and Lu'ash, to god Iluwer for protecting Hazrek against a siege undertaken by a coalition led by the king of Damascus (Mazzoni 1998b, 7-9). Dornemann (2003, 4), however, has suggested an alternate identification with the ancient city of Apish, which also appears in the Zakkur stele.

Seasons of Excavation
P. Matthiae: 1970, 1972, 1978, 1985
S. Mazzoni and S. M. Cecchini: 1986-1989, 1991-1992, 1994, 1996-2006

Published Iron Age Ceramics
Bonatz 1998; Canuti 2000; Cecchini 1988; 1993; 1998; 2000a; 2002; 2005; Chiti 2005; Ciafardoni 1988; D'Amore 1998; 2005; Degli Esposti 1998; Magazzù 2002; Matermawi 2005; Mazzoni 1988; 1990a; 1998a; 2000b; Oggiano 1997; 1999; 2002; Pedrazzi 2002; Scigliuzzo 2002; 2005; Soldi 2005; Venturi 1998a; 2000a; 2000b; 2002; 2005; del Vesco *et al.* 2005; Virgilio 2005

A.4 Ahmar, Tell (Syria)

Located on the east bank of the River Euphrates, Tell Ahmar (literally the 'Red Mound') lies 20 km downstream from Jerablus (Carchemish) and the Turkish border. Now partially submerged by the waters of the Tishrin Dam, Tell Ahmar was once situated at the edge of a terrace overlooking the long alluvial plain of the Euphrates River (Bunnens 1990a, 2). The site stands 1.5 km from the point of confluence of the Euphrates and Sajur Rivers, and overlooks what was once a natural crossing point for the Euphrates. Like the nearby site of Carchemish, the site of Tell Ahmar boasts a tripartite plan: the Acropolis, Middle City and Lower City. The small Acropolis (c. 2 ha.) once stood 25 m over the surrounding plain; the French expedition removed the top seven metres of deposit from the tell and extended its southern edge with a large spoil-dump. The Middle city extends 350 m to the west and lies under much of the modern village and cemetery. The Lower City is extensive; its semi-circular shape has a diameter of around 1200 m, suggesting an area of *c.* 59 hectares (Roobaert and Bunnens 1999, 163; Anastasio *et al.* 2004, 43).

Ancient Iron Age Name
Til Barsip (Aramaic)
Kar-Shalmeneser (Neo-Assyrian)
KTK (Aramaic)
Masuwari (Luwian)

The history of Tell Ahmar in the Iron Age has attracted much attention. The site is widely recognised as the capital of the small 'Syro-Hittite' kingdom of Masuwari (Hawkins 1983), before being transformed into the headquarters of the 'Aramaean' state of Bit-Adini, and then becoming the Assyrian Kar-Shalmeneser of the eighth and seventh centuries BC (Schwartz 1989, 278). Ussishkin (1971), however, has reversed the sequence- he prefers to see the early Iron Age site as Aramaean before becoming Syro-Hittite about a century before Assyrian conquest. While the matter remains unresolved, Hawkins (1980; 1983) has suggested an alternative interpretation: that Tell Ahmar was a small kingdom of Hittite (Luwian) descent that was under regional authority of an Aramaean tribal leader. This, according to Hawkins, explains the Syro-Hittite character of the local art and the mention of Tell Ahmar as part of the Aramaean coalition against the Neo-Assyrians.

Seasons of Excavation
F. Thureau-Dangin: 1928-1931.
G. Bunnens: 1988-2006

Published Iron Age Ceramics
Wightman 1990b; Jamieson 1999; 2000; Green and Hausleiter 2000

A.5 Ain Dara (Syria)

The site of Ain Dara lies in the Afrin valley, northern Syria, on the left bank of the Afrin River, upstream from the Amuq plain and 67 km north-west of Aleppo. The name of the site derives from the spring which lies 700 m to the east of the mound. The site consists of a steep upper mound (acropolis), which rises 30 m above the plain, and an extensive lower mound (24 ha.), which surrounds the acropolis on the northern and eastern sides (Abou Assaf 1997 and Stone and Zimansky 1999, 2). The many visitors to the site, due to its proximity to both St Simeon and Cyrrhus, approach the acropolis via a modern road which winds up through the lower city's northern gate, one of four in the Iron Age settlement (Abou Assaf 1990, Fig. 2). The lower tell was settled primarily during the Late Bronze and Iron Age (Stone and Zimansky 1999, 3-7). The acropolis, however, appears to be much older, with evidence of occupation as early as the fourth millennium BC (Abou Assaf 1997).

Ancient Iron Age Name
Matthiae (1981, 33) has suggested Ain Dara be identified with ancient Kinalua, the capital of the kingdom of Hattina, though Harrison *et al.* (2004, 122) prefers to associate Kinalua with Tell Ta'yinat. Regardless, Ain Dara and Tel Ta'yinat were probably part of this same kingdom.

Seasons of Excavation
M. Dunand and F. Seirafi: 1956, 1962, 1964
A. Abou Assaf: 16 seasons between 1976 and 1992
E. C. Stone and P. E. Zimansky: 1982-1984

Published Iron Age Ceramics
Stone and Zimansky 1999

Little has been published regarding the Iron Age ceramics of Ain Dara. Even the final report of the American excavation of the lower town presents a limited corpus of ceramic profiles despite over 129,000 sherds having been catalogued during the course of two seasons (Stone and Zimansky 1999, 59). The ceramics from the Syrian expedition to the acropolis are currently in preparation for publication.

A.6 Akhziv (Israel)

Akhziv is located on the northern coast of Palestine 15 km north of 'Akko and 25 km south of Tyre (Dayagi-Mendels 2002, 1). The site itself is unremarkable, except for the associated 'Phoenician' cemeteries. The southern cemetery (*Minet ez-Zib*) is situated on the Mediterranean shore less than a kilometre south of the double tell and extends over a great area, about 5000 square metres according to Ben Dor (Dayagi-Mendels 2002, 1). The tombs are on a sandstone (*kurkar*) ridge below the sand dunes, with nothing on the surface to indicate their presence. Other cemeteries are located to the east of the tell (*er-Ras*), on a hard limestone spur, and to the north (a large ashlar-built chamber tomb and cremation field).

Ancient Iron Age Name
Achzib (Hebrew, Old Testament)
Ez-Zib (Aramaic)
Akzibi/Accipu (Neo-Assyrian)

The city of Achzib is mentioned twice in the biblical text as an undefeated Canaanite city. It appears in Assyrian records as a fortified city conquered by Sennacherib in 701 BC, as part of the same campaign that saw Sidon, Sarepta and 'Akko subjugated (E. Mazar 1993, 32). The identification of the site with ancient Akhzib is due to the etymologically similar modern name; further excavation of the tell would be needed to confirm this identification.

Seasons of Excavation
I. Ben Dor: 1941-1944
Z. Goldmann: 1954
M. Prausnitz: nine seasons between 1957-1984
E. Mazar: 1988-1990, 1994, 2002

Published Iron Age Ceramics
Culican 1970; 1982; E. Mazar 1993; 2001; 2004; Dayagi-Mendels 2002

A.7 Aleppo (Syria)

The second largest city within modern Syria, Aleppo is located in the northern part of the country on the banks of the River Quweiq, equidistance from the Mediterranean Sea and Euphrates River. Due to the extensive size of the modern city, archaeological investigation has been somewhat limited, though excavations have recently returned to the ancient tell situated within the Aleppo citadel. Determining the size and extent of ancient Aleppo remains difficult, though the listing of Aleppo as a UNESCO World Heritage site will protect the region for possible future investigation.

Ancient Iron Age Name
HRB (Akkadian, Late Bronze Age)
Halpa (Hittite, Bronze Age)
Halman/Halwan (Neo Assyrian)

Aleppo is known as the administrative and cultic centre of ancient Halab, the capital of the independent 'Syro-Hittite' kingdom of Yamhad (Pitard 1997). Following repeated conquests by the Assyrians (853 and 743 BC) and Urartians, Aleppo was reduced to political insignificance during much of the Iron Age, though texts suggest it remained a major centre for the worship of Hadad, the storm god (Miller 1958, 2).

Seasons of Excavation
G. Ploix de Rotrou: 1929-1936
R. Saba: 1981, 1983
W. Khayyata and K. Kohlmeyer: 1995-1997, 1999-2005

Published Iron Age Ceramics
None

A.8 Amuq Plain (Turkey)

The Amuq Valley (alternatively known as the Hatay or Plain of Antioch), which was annexed to the Turkish Republic in 1939, constitutes the southernmost province of Turkey. The large (1,400 sq. km) fertile plain is extremely well-watered, with three large rivers flowing into the basin before conjoining and flowing west to the Mediterranean (Pamir 2005, Fig. 3.1). The Orontes River enters the Amuq from the south and bends westward. The *Kara Su* (literally black water) and Afrin River enter the Amuq from the north and east respectively, before joining the Orontes' flow toward the sea. The fertile plain is framed by mountains on all sides, except where the three rivers enter and exit the plain; the Taurus Mountains to the north rise up toward the Anatolian plateau. Of the sites surveyed by the Oriental Institute in the 1930s, three yielded significant Iron Age assemblages: Chatal Höyük,

Tell Judeideh and Tell Ta'yinat. The first two sites are located near the town of Rihaniyyah, while the third lies immediately east of the modern village of *Jisr el-Hadid.* Excavation was also undertaken at Tell Atchana (ancient Alalakh) and Tell Kurdu, though neither yielded significant (if any) Iron Age material.

Ancient Iron Age Name
Amuq Region
Paddan-Aram (Hebrew, Old Testament)
Patina (Luwian)
Unqi (Akkadian)
'MQ (Aramaic)
Umqa (Syriac)

Tell Ta'yinat
Kullani (Neo-Assyrian)
Kunalua (Luwian)

The territory of the Amuq was occupied by the 'Syro-Hittite' kingdom of Patina (Harrison 2001a, 116-117), though it appears to also have been known by the name Unqi (Braidwood 1937, 1; McEwan 1937, 8). Fragments of a small Iron Age bowl were found at Tell Ta'yinat inscribed with the word KNLH, linguistically very similar to Kunulua, capital city of the kingdom of Patina/Unqi (Harrison *et al.* 2004, 122; Swift 1958, 191-192). This identification appears supported by the military annals of Ashurnasirpal II (c. 870 BC), who locates the city at the southern edge of the Amuq (Hawkins 1982, 389). The later use of the term 'MQ (the Aramaic equivalent of the Akkadian Unqi) has been interpreted as representing Aramaic ascendancy in the area. The OI excavations at Tell Ta'yinat produced an extensive corpus of Akkadian, Aramaic and Luwian inscriptions; Luwian hieroglyphic inscriptions account for the largest number.

Seasons of Excavation
Oriental Insitute: C. W. McEwan and R. J. Braidwood:
Chatal Höyük: 1932-1935
Tell al-Judeideh: 1934-1936
Tell Ta'yinat: 1935-1938

University of Toronto: T. Harrison
Tell Ta'yinat: 2004-present

Published Iron Age Ceramics
Batiuk *et al.* 2005; Casana and Wilkinson 2005; Harrison *et al.* 2006; Janeway 2006-2007

A.9 Arqa, Tell (Lebanon)

Tell Arqa is a high (40 m), multi-period tell situated in northern Lebanon. It lies twenty kilometres north of Tripoli and c. 5 km inland, near where the Homs-Tripoli gap debouches into the Mediterranean. The tell looms over the surrounding Akkar Plain, the vast fertile plain watered by the *Nahr el-Kebir*, which is today the Syrian border. Found at the very southern edge of the plain, Tell Arqa is one of three large tells in the Akkar (the other two are Tell Kazel and Tell Jamous). Despite its imposing height, the modest area of the mound (*c.* 4 ha.) belies a settlement of intermittent occupation and sporadic prosperity (Thalmann 1978, 3).

Ancient Iron Age Name
Irqata (Akkadian, Amarna Letters)
RQT (Akkadian, Late Bronze Age)
Arka (Hebrew, Old Testament)
'Arqa (Phoenician)
'Arqu (Aramaic)

Seasons of Excavation
J.P. Thalmann: 1972-1974, 1978-1981, 1992-2005

Published Iron Age Ceramics
Will *et al.* 1973; Thalmann 1978a; 1978b; 1983; 1990; Chaaya 2000

A.10 Arslan Tash (Syria)

Arslan Tash is located in northern Syria approximately 30 km northeast of Tell Ahmar and the Euphrates River and 6 km south of the modern Turkish border. The large oval mound, c. 30.5 ha. (728 m x 560 m), lies in the middle of the fertile Plain of Saruj (Anastasio *et al.* 2004, 58). The name of the site, which literally means 'lion stone' in Turkish, derives from the large basalt lion statues that were originally found lying on the site's surface.

Ancient Iron Age Name
Hadatu (?)
Dur-Tukulti-apil-Esarra (Neo-Assyrian)

Seasons of Excavation
G. Paşa: 1886
F. Thureau-Dangin: 1927-1929
L. Delaporte: 1938

Published Iron Age Ceramics
None identified, despite the two volume publication of the French project (Thureau-Dangin *et al.* 1931a; 1931b).

A.11 'Athlit (Israel)

'Athlit is located on the northern coast of Palestine, 12 km south of Haifa. It is situated on a *kurkar* ridge adjacent to a natural bay. Evidence for Iron Age occupation initially consisted of the remains of fourteen chamber tombs exposed beneath the medieval castle (Johns 1931, 111) on a natural promontory (Johns 1931, Pl. 40; 1932, Fig. 1). The remains of an Iron Age settlement were found to the east and south-east of the promontory within a low lying (7 m above sea level), medium-sized (c. 8 ha) tell (Johns 1934, 145-146).

Ancient Iron Age Name
Although 'Athlit is not mentioned explicitly in ancient documents, the site has been tentatively identified with Kartah of Zebulun mentioned in Joshua 21:34 (Johns 1993, 112).

Seasons of Excavation
C. N. Johns: 1930-1933, 1935
A. Raban and A. Haggi: 2002-2003

Published Iron Age Ceramics
Johns 1932; 1934; 1938; Haggi 2006

A.12 Beirut (Lebanon)

While Lebanon's capital, Beirut is today a thriving metropolis that covers a large area, the ancient settlement of Beirut was initially located on the northern side of the natural promontory that today forms the Beirut Town Centre. Occupation was originally restricted to a small tell (2.5 ha.) that lies to the south of the modern harbour of Beirut (Badre 1998, 73). Ancient Biruta (of the Amarna texts) took the form of a small arc facing the sea, before expanding during the Classical period to cover much of the promontory.

Ancient Iron Age Name
Beruta/Biruti (Amarna letters, 14th century BC)
Kur $_{\text{bi-ru-ù-ti}}$ / Kur $_{\text{bi-ru-ut-ti}}$ (Ugaritic texts, 14th-13th centuries BC)
Bi'ru (Neo-Assyrian, Esarhaddon Prism inscription)
Beerot (Phoenician)
Berytos (Greek)

The oldest confirmed references to Beirut are found in the Amarna Letters of the fourteenth century BC (Curvers and Stuart 1996, 228). During the Iron Age, however, there is only one possible mention of the site, a reference that lists Bi'ru amongst the cities within the kingdom of Sidon (Badre 1997b, 12). The conspicuous scarcity of Iron Age references has brought into question the importance of this city during the first millennium BC, though recent excavations have begun to reverse that image.

Seasons of Excavation
R. du Mesnil du Buisson and T. Homolle: 1924
L. Brossé, L. Albanèse, F. Anus and M. Chéhab: 1924-1932
R. du Mesnil du Buisson: 1927-1928
H. Fleisch: 1947
H. Kalayan: 1968-1969
C. Forest and J.D. Forest: 1977
BCD Archaeology Project: 1993-2003

Published Iron Age Ceramics
Badre 1997b; 1998; 2000b; Doumet-Serhal and Sayegh 1998; Jabak *et al.* 1998; Sapin and Sayegh 1998; Shefton *et al.* 1998; Elayi and Sayegh 2000

Despite a good Iron Age ceramic sequence being obtained from the excavations under the BCD 'umbrella', no comprehensive typology or catalogue has been published (e.g. Núñez 2001-2002; Curvers and Stuart 1997, 182-183).

A.13 Beth Shan, Tel (Israel)

Tel Beth Shan (*Tell el-Husn* in Arabic) is located on the western edge of the Jordan Valley, c. 20 km south of the Sea of Galilee (*Yam Kinneret*). The site guards the ford of the Jordan River 5 km to the east and lies near the junction of the Jezreel and Jordan Valleys. The steep but small (c. 4 ha) conical tell is situated on a high hill, though still 300 m below sea level. Excavations have revealed an almost continuous occupation of the site from the Neolithic to the Early Medieval period. The classical settlement (Scythopolis) outgrew the tell and a large urban centre developed around the foot of the mound.

Ancient Iron Age Name
Beth Shan (Hebrew, Old Testament)
Bit Sani (Amarna Letters)
Beit She'al (Egyptian)

Beth Shan is mentioned in the biblical text as an unconquered Canaanite city (Joshua 17:11), and where the Philistines impaled the body of Saul and his sons on the city-wall (1 Samuel 31:10).

Seasons of Excavation
C.S. Fisher: 1921-1923
A. Rowe: 1925-1928
G.M. Fitzgerald: 1930, 1933
Y. Yadin and S. Geva: 1983
A. Mazar: 1989-1996

Published Iron Age Ceramics
Beth Shan: James 1966; Oren 1973; Yadin and Geva 1986; A. Mazar 1993b; 2001

A.14 Byblos (Lebanon)

The ancient site of Byblos is located on a short headland known locally as *Djebeil* (or *Gabayl*). It lies on the Lebanese coast roughly equidistant (c. 40 km) between Beirut and Tripoli. The short promontory provided two natural harbours for the site, a feature common amongst coastal settlements of the region. Before excavation the large mound (c. 10 ha.) stood 12 m. above the surrounding coastline.

Ancient Iron Age Name
Gebal (Hebrew, Old Testament)
KBN (Akkadian, Late Bronze Age)
KPNJ (Akkadian, Late Bronze Age)
Gubla (Akkadian, Amarna Letters, Late Bronze Age)
Gublu (Neo-Assyrian)
Byblos (Greek, Achaemenid and Hellenistic Periods)

Seasons of Excavation
E. Renan: 1860-1861
J. Rouvier: 1899
Freres Maristes: 1914
P. Montet: 1919, 1921-1924
M. Dunand: 1926, 1928-1933, 1935-1941, 1943-1944, 1948-1952, 1954-1962, 1964-1965, 1971-1973
J. Lauffray: 1940
Direction Générale des Antiquities: 1946
M. Chéhab: 1960

Published Iron Age Ceramics
Dunand 1939a; 1939b; 1950; 1954; Culican 1970; Salles 1980; Homsy 2003

Chatal Höyük (Turkey) - see A.8 Amuq

A.15 Dan, Tel (Israel)

Tel Dan (Arabic *Tell el-Qadi*, literally 'mound of the judge') lies at the foot of Mount Hermon (Biran 1994, 21). It is situated in the heart of a rich and fertile region; the spring flowing in and around the mound is one of three sources of the Jordan River. The edge of the large (c. 20 ha) tell rises 20 m above the plain, while the central area descends 10 m in height, giving the impression of a shallow crater.

Ancient Iron Age Name
Dan (Hebrew, Old Testament)
Laish/Leshem (Canaanite)

The identification of biblical Dan with *Tell el-Qadi* was first suggested in the mid-nineteenth century by Robinson, but wasn't confirmed until 1976, when excavations discovered a bilingual (Greek and Aramaic) inscription; "To the god who is in Dan" (Biran 1994, 25). The city is mentioned repeatedly in biblical texts as the northern-most point of ancient Israel. The city also appears in Egyptian Execration texts and the Mari Letters, though under the city's former name, Laish or Leshem. According to the biblical account, the name was changed to Dan when that Israelite tribe conquered Laish.

Seasons of Excavation
Z. Yeiven: 1963
A. Biran: 1966-1972, 1974-2005

Published Iron Age Ceramics
Biran 1982; 1989a; 1989b; 1994; 1996b; 1999; Pakman 1992

A.16 Deve Höyük (Turkey)

Deve Höyük is a small village in the Sajur Valley, about 30 km west of Jerablus and the Euphrates River. The village consists of a cluster of houses surrounding the southern slope of a small tell. Little more is known about the site, except that an ancient cemetery was located on the north eastern edge of a shallow depression surrounding the mound.

Ancient Iron Age Name
Not known

Seasons of Excavation
No systematic excavation at Deve Höyük was undertaken, only the illicit plundering of "some fifty

peasants... in full view of the Baghdad railway" (Woolley 1914-1916, 115).

Published Iron Age Ceramics
Woolley 1914; 1914-1916; Moorey 1975; 1980; Klengel-Brandt 1990

A.17 Ghassil, Tell el- (Lebanon)

The small mound (c. 3 ha.) of Tell el-Ghassil is located in the northern reaches of the Beqa' Valley of Lebanon. The small mound lies 18 km southwest of Baalbek and to the immediate east of the Litani River (*Nahr al-Litani*). The site derives its name from the nearby spring, Ayn Ghassil, which emerges from the 1,000 m contour of the Lebanese Range.

Ancient Iron Age Name
Not known.

Seasons of Excavation
D. C. Baramki: 1956-1961, 1963, 1968-1969
L. Badre: 1972-1974.

Published Iron Age Ceramics
Baramki 1961; 1964; 1966; Joukowsky 1972 (unpublished); Marfoe 1995

A.18 Hama (Syria)

Situated in west-central Syria, the ancient mound of Hama is located on the left bank of the River Orontes at the point where a major north-south road crossed the river. The modern city of the same name, however, has significantly grown to encompass both banks of the river, and the site itself is now on its north-western outskirts. The steep tell still dominates the city skyline as it towers 46 m above the river (Ingholt 1942, 469).

Ancient Iron Age Name
Hamath (Hebrew, Old Testament)
Hamati (Akkadian, Late Bronze Age)
'MT/HMT (Akkadian, Late Bronze Age)
Tunip (Amarna Letters, Late Bronze Age)
Amatu (Luwian)
Hamatu (Neo-Assyrian)

Neo-Assyrian records reveal that during the middle of the 9th century BC, Urhilina, king of Hamatu, played an important role in a coalition that was able to temporarily arrest Assyrian advances toward the west. By the late 8th century BC, however, many of the provinces belonging to Hamatu had to be surrendered and a large number of inhabitants were deported by the rampaging Assyrians (Buhl 1992, 34).

Seasons of Excavation
H. Ingholt: 1931-1938
K. Chehade: 1960

Published Iron Age Ceramics
Riis 1948; 1970; Fugmann 1958; Buhl 1983; Francis and Vickers 1985; Riis and Buhl 1990

A.19 Hazor (Israel)

Biblical Hazor has long been associated with *Tell el-Qedah* located 14 km north of the Sea of Galilee and 8 km southwest of Lake Hula. The site comprises two distinct areas: the 40 m high acropolis mound (c. 10 ha); and the extensive (c. 80 ha) rectangular mound of the lower city (Ben Tor 1997, 107; Yadin *et al.* 1958, Pl. I).

Ancient Iron Age Name
Hazor (Hebrew, Old Testament and Armana Letters)

Hazor and Laish (Tel Dan) are the only two Canaanite cities mentioned in the Mari archives of the Middle Bronze Age. The frequency with which Hazor appears in these documents is indicative of its importance during the 2nd millennium BC; indeed, the site covers its largest area during this period (Ben Tor 1993a, 603). Hazor appears in the biblical record foremost as the chief Canaanite city of the region, which is conquered and settled by the invading Israelite tribes (Joshua 11:10-13).

Seasons of Excavation
J. Garstang: 1928
Y. Yadin: 1955-1958, 1968-1969
A. Ben Tor: 1990-2007

Published Iron Age Ceramics
Yadin *et al.* 1958; 1960; 1961; Ben Tor *et al.* 1997; Ben Tor and Ben Ami 1998; Finkelstein 1999a; Ben Ami 2001; 2006

A.20 Jerablus (Turkey) including Kefrik and Merj Khamis

The site of Barak, more commonly known as Jerablus (or *Jerabis* in Arabic), lies at the extreme north of the plain of Jerablus (Anastasio *et al.* 2004, 83). It is a large tell site on the right bank of the Euphrates River, situated by the modern border of the Turkish and Syrian Arab Republics, c. 100 km north east of Aleppo and 60 km south of Gaziantep. The somewhat tense, though generally cordial relations between these two nations means the site, which is actually cut by the border, is unavailable for further study. The eastern half of the tell is well protected by the deeply incised Euphrates River, while the western portion of the site is protected by long mud-brick city walls, still visible as large earth mounds at the time of the British Expedition. In ancient times the extensive city commanded the main ford across the Euphrates.

Ancient Iron Age Name
KRKMS (Akkadian, Late Bronze Age)
Kargamiş (Hittite, Late Bronze Age)

The city of Carchemish, known from biblical, Assyrian and Egyptian sources, was first identified with Jerablus by George Smith in 1876 (www.pef.org.uk/EarlySyriaPages/Carchemish.htm).

Seasons of Excavation
A. E. Henderson: 1876-1879
D.G. Hogarth, C.L. Woolley and T.E. Lawrence: 1911-1914
C.L. Woolley: 1920

Published Iron Age Ceramics
Woolley 1914; 1921a; 1939b; 1952; Moorey 1980

A.21 Jezreel, Tel (Israel)

The site of Tel Jezreel (or *Yizre'el*, literally 'God sows') is situated at the eastern part of the Valley of Jezreel, and lies almost equidistant between the two large mounds of Beth Shan and Megiddo. Situated on the very edge of a 100 m high ridge extending along the southern side of the valley, the view from the site is a commanding and strategic one. The relatively small (c. 6 ha.) mound is roughly rectangular in shape, indicative of the regular casemate wall exposed in the excavations (Ussishkin and Woodhead 1992).

Ancient Iron Age Name
Jezreel (Hebrew, Old Testament)

Seasons of Excavation
O. Yogev: 1987-1988
J. Woodhead and D. Ussishkin: 1990-1996.

Published Iron Age Ceramics
Zimhoni 1992 (reprinted as Zimhoni 1997a); Zimhoni 1997c (reprinted as Zimhoni 1997b)

A.22 Joya (Lebanon)

The cemetery of Joya lies about 10 km inland (east) of Tyre in the south of Lebanon, approximately equidistant between Khirbet Silm and Tyre (Chapman 1972, 56). The nature of the site is not known.

Ancient Iron Age Name
Not known

Seasons of Excavation
Clandestine excavation only

Published Iron Age Ceramics
S. V. Chapman 1972

Judeideh, Tell (Turkey) - see A.8 Amuq

A.23 Jurn Kabir, Tell (Syria)

Located on the right bank of the Upper Syrian Euphrates, Tell Jurn Kabir is situated 2.5 km south of the Islamic fortress *Qal'at Najm* (Eidem 1999a, 149-153). Aleppo lies 80 km to the west. The small (c. 3 ha.), low mound of Tell Jurn Kabir is heavily eroded. The site is surrounded by the village of Jurn Kabir, founded sometime in the late nineteenth century (Eidem 1999a, 152; Anastasio *et al.* 2004, 188).

Ancient Iron Age Name
The ancient name of the site is currently not known.

Seasons of Excavation
J. Eidem: 1993-1998

Published Iron Age Ceramics
Eidem1999b; Eidem and Ackermann 1999; Eidem and Putt 1999; 2001

A.24 Kamid el-Loz (Lebanon)

Kamid el-Loz is a steep tell site within the southern reaches of the Beqa' Valley, where the alluvial plain rises toward the Anti-Lebanon Range (Kushcke 1966, 15; Hachmann 1989, Fig. 6). A modern town of the same name is located to the immediate south of the oval mound and a modern cemetery covers the south eastern quarter of the site. The tell of Kamid el-Loz is one of the largest (c. 7 ha) and tallest (26 m) within the Beqa' Valley, though it is relatively small when compared to the size of the large Syrian and Mesopotamian tells (Hachmann 1989, 13). The Litani River lies 3 km to the west, though there is a spring to the immediate north.

Ancient Iron Age Name
Kumidi (Akkadian, Amarna Letters, Late Bronze Age)
KMT (Egyptian texts)

Cuneiform texts found within the Late Bronze Age temple at Kamid el-Loz ha ve confirmed the identification of the site with the Bronze Age city of Kumidi (Heinz 2004, 570), an identification that was originally postulated in the late nineteenth century (Hachmann 1989, 89).

Seasons of Excavation
R. Mouterde and A. Beaulieu: 1934
R. Hachmann and A. Kuschke: 1963-1964
R. Hachmann: 1966-1971, 1973-1974, 1977-1981
M. Heinz: 1997, 1999-2002, 2004-2005

Published Iron Age Ceramics
Hachmann 1966; Hachmann and Kuschke 1966; Poppa 1978; Marfoe 1995; 1998; Heinz *et al.* 2004

A.25 Kazel, Tell (Syria)

Tell Kazel is situated 3.5 km from the Syrian coast, within the northern limits of the Akkar Plain and 18 km south of Tartous (Badre 2006, 65). The tell is oval in shape with particularly well-defined upper edges, suggesting the presence of significant fortifications on the edge of the acropolis; three depressions along the edge are probably the three city gates. The height of the large (11 ha.) tell reaches 25 m above the level of the surrounding Akkar Plain (Badre 1990a, 13). Tell Kazel is the largest site in the Akkar (Thalmann and al-Maqdissi 1989, 98) and is the only one boasting evidence of continuous occupation from the 4th to 1st millennium BC.

Ancient Iron Age Name
Zemar (Hebrew, Old Testament)
D̲MR (Akkadian, Late Bronze Age)
Sumur (Akkadian, Amarna Letters, Late Bronze Age)
Samuri/Simirra (Neo-Assyrian)
Simyra (Greek)

The site is thought to be historical Simyra/Sumur, an identification first presented by M. Dunand (Dunand and Saliby 1957, 1-10), developed by H. Klengel (1984, 6) and today widely accepted (Gubel 2000; Sader 1990a; cf. Stieglitz 1991). However, this interpretation is based on a number of assumptions; foremost being that the historical importance of Simyra is comparable with the size of Tell Kazel. The identification of Tell Kazel with Sumur has not yet been established by any inscription found on the site (Badre 2006, 67).

Seasons of Excavation
M. Dunand and N. Saliby: 1960-1962
A. Bounni: 1985
L. Badre: 1985-2001

Published Iron Age Ceramics
Dunand and Saliby 1957; Dunand *et al.* 1964; Gubel 1990a; Badre 1990b; 2006; al Maqdissi 1990; Yon and Caubet 1990; Badre *et al.* 1994; 2005; Badre and Gubel 1999-2000; Capet and Gubel 2000; Capet 2003

A.26 Keisan, Tell (Israel)

Situated between Haifa and Akko, the high conical mound of Tell Keisan is prominent within the landscape of the Akko (Acre) Plain (Briend and Humbert 1980, Pl. 2). Although the site today lies 8 km inland, it originally stood much closer to the coast, effectively linking the hills of the Lower Galilee with the Mediterranean Sea (Briend and Humbert 1980, Pl. 1). The surviving mound rises to a height of at least 25 m above the alluvial plain and covers an area of c. 6 hectares (Briend and Humbert 1980, 3; Seton-Williams 1980, 381).

Ancient Iron Age Name
The ancient name is unknown, despite a number of suggested biblical links: Kishion; Allammelech; Mishal; Akshaph (Garstang 1931, 354; Humbert 1993, 862).

Seasons of Excavation
J. Garstang and A. Rowe: 1935-1936
R. de Vaux, J. Briend and J.-B. Humbert: 1971-1976

Published Iron Age Ceramics
Briend and Humbert 1980; Seton-Williams 1980

A.27 Khalde (Lebanon)

About 12 km south of Beirut, Khalde (*Kobbet Choueifat*) is situated beside the Beirut International Airport, cut by the Sidon highway (Saidah 1966, 51). The large necropolis (500 m in length) lies atop two promontories sloping down to the Mediterranean. Khalde's proximity to Beirut ensured its importance, as indicated by the large number (over 400) of Iron Age burials found *in situ* (Badre 1997a). The extensive cemetery implies a large settlement nearby, though not found.

Ancient Iron Age Name
Hi-il-du-ya (Akkadian, Neo-Assyrian)
Heldua? (Greek)

Khalde is identified with ancient Hi-il-du-ya, a city of the kingdom of Sidon annexed by Esarhaddon in the first half of the seventh century BC.

Seasons of Excavation
H. Kalayan: 1960
R. Saidah: 1961-1966, 1974-1975

Published Iron Age Ceramics
Saidah 1966; 1971; Culican 1982

A.28 Khirbet Silm (Lebanon)

Khirbet Silm (also Khirbet Selim) lies several kilometres inland from Tyre in southern Lebanon. Nothing further has been published regarding the nature of the site.

Ancient Iron Age Name
The ancient name of Khirbet Silm remains unknown.

Seasons of Excavation
Albanèse: 1924-1926

Published Iron Age Ceramics
Chapman 1972; Culican 1982

A.29 Mastuma, Tell (Syria)

Located in northwest Syria approximately 60 km southwest of Aleppo and 5 km directly south of Idlib, Tell Mastuma consists of a relatively small tell (diameter of around 200 m; c. 3 ha.) that rises little more than 20m above the surrounding plains. At 450 m above sea-level, the site lies a few kilometres west of the Syrian steppe.

Ancient Iron Age Name
Astamaku (Neo-Assyrian)

Seasons of Excavation
N. Egami: 1980-1982, 1984, 1986, 1988, 1993- 1995

Published Iron Age Ceramics
Egami & Masuda 1982; 1984; Egami 1983; 1988; Wada 1994; Wakita *et al.* 1995; Wakita *et al.* 2000.

A.30 Megiddo (Israel)

The mound of *Tell al Mutesellim* has been associated with the biblical site of Megiddo since the end of the nineteenth century (Smith 1892, 385). The medium-sized (c. 15 ha) conical tell rises 40-60 m above the surrounding plain. The position of the mound at the point where the *Nahal 'Iron* enters the Jezreel Valley gave it strategic control over the narrow 'Iron Valley, which passes through the formidable coastal mountains (Aharoni *et al.* 1993, 1003).

Ancient Iron Age Name
Megiddo (Hebrew, Old Testament)

Seasons of Excavation
G. Schumacher: 1903-1905
C. Fisher: 1925-1927
P. L. O. Guy: 1928-1934
G. Loud: 1935-1939
Y. Yadin: 1960, 1966-1967, 1971/2.
I. Finkelstein and D. Ussishkin: 1992, 1994, 1996, 1998, 2000, 2002, 2004, 2006

Published Iron Age Ceramics
Guy 1938; Lamon and Shipton 1939; Loud 1948; Culican 1982; Finkelstein *et al.* 2000b; Harrison 2004b; Finkelstein 2006; Arie 2006

A.31 Mina, al (Turkey)

The site of al Mina is located in the Turkish province of the Hatay on the northern bank of the Orontes River, at the point where the river empties into the Mediterranean Sea. It lies in the gulf of Suwaidia, which is protected on all sides by mountains; Mount Casius, Ras al Khanzir and the Amanus to the south, north and east, respectively. Access to the site is granted either from the sea or along the course of the Orontes River (S. Smith 1942, 88). Apart from a good defensive position, al Mina, and the Amuq Plain to the east, was located in an ideal position for linking the Mediterranean and Mesopotamian worlds. The size of the tell itself is difficult to determine because a large portion of the site has been lost to erosion or obscured by thick alluvial deposits.

Ancient Iron Age Name
Al Mina was originally identified with classical Poseideion, but subsequent excavation of nearby Ras al Bassit has also shown strong connections with the Greek world whilst also, significantly, retaining an echo of the classical name (S. Smith 1942, 96-98). As a result, al Mina has been left 'nameless' (Boardman 1990, 170).

Seasons of Excavation
C. L. Woolley: 1935-1936

Published Iron Age Ceramics
Woolley 1937; 1938a; Robertson 1940; Boardman 1959; Du Plat Taylor 1959; Peltenberg 1969; Luke 2003

A.32 Mishrifeh, Tell (Syria)

Tell Mishrifeh lies 18 km north east of the modern city of Homs. It is located in a large fertile plain between the desert steppe and the alluvial Orontes Valley, an area crossed by a seasonal net of *wadis*. The site consists of a large, gently sloping tell, which is enclosed by massive Bronze Age ramparts. The fortifications, which still stand to heights of 15-20 m, protect and enclose a roughly square area of c. 110 ha (al-Maqdissi *et al.* 2002a, 9).

Ancient Iron Age Name:
Qaṭna (Amarna Letters, Late Bronze Age)

King Akizzi of Qatna wrote a number of letters to Pharaoh Akhenaton requesting help against the aggressive Hittite king Šuppiluliuma I. In addition to the Amarna letters, Hittite sources also mention the siege and conquest of Qatna. The site then fades from the textual record (Klengel 2000b).

Seasons of Excavation
R. du Mesnil du Buisson: 1924, 1927-1929
M. al-Maqdissi: 1994-1998
M. al-Maqdissi, D. Morandi-Bonacossi and P. Pfälzner: 1999-2006.

Published Iron Age Ceramics
du Mesnil du Buisson 1926; 1927a; 1927b; 1928; al-Maqdissi 2003; al-Maqdissi and Badawi 2002; Barro 2002; Luciani 2002; Morandi Bonacossi 2002.

A.33 Nayrab (Syria)

The small tell of Nayrab lies c. 5 km to the south-east of the Aleppo citadel. The low mound, which rises less than 10 m above the surrounding plain, is today completely enclosed by modern dwellings, though at the time of excavation the south-east corner of the tell was exposed (Carrière and Barrois 1927, Pl. XXXV.2).

Ancient Iron Age Name
nrb (Aramaic, funerary stelae)
ni-ri-ib (Neo-Babylonian, Nayrab tablets)

The identification of Nayrab comes from textual evidence recovered from the site.

Seasons of Excavation
B. Carrière, A, Barrois and A. Parrot: 1926
F. M. Abel, A. Barrois, A. Parrot and J. Darrous: 1927

Published Iron Age Ceramics
Carrière and Barrois 1927; Abel and Barrois 1928

A.34 Nebi Mend, Tell (Syria)

Tell Nebi Mend is located in west central Syria approximately 30 km south west of Homs (ancient Emessa) and only 4-5 km directly south of the Lake of Homs (also Lake Qattineh). The modern village of the same name surrounds the large mound, which lies between the Orontes River, to the east, and its tributary, the Mukadiyeh, to the west: the confluence of the two lies to the north. The large mound (10 ha.) rises 30 m above the surrounding plain, though thick alluvial deposits along the Orontes River imply an even more imposing sequence of occupation.

Ancient Iron Age Name
Qadeš (Old Babylonian, Bronze Age)

QDS/Kinza (Akkadian, Late Bronze Age)
Qidša (Akkadian, Amarna Letters, Late Bronze Age)
Qi-di-si (Neo-Assyrian)

Tell Nebi Mend was first identified with ancient Kadesh-on-the-Orontes in 1848 by W. M. Thompson (1848, 691-692). Subsequent geographical, textual and archaeological investigations have since confirmed the identification (Parr 1983, 99-100; 1991, 78; Conder 1881; 1882; 1883; W. Wright 1882). While no direct evidence for the Battle of Kadesh has been located, the position of the large tell between two rivers recalls the depiction of the city upon the Egyptian monuments of Ramesses II.

Seasons of Excavation
M. Pézard and C. L. Brossé: 1921-1922
DGAM: 1950
P. J. Parr: 1975-1982, 1986, 1990, 1992, 1995

Published Iron Age Ceramics
Pezard 1931; Sagona and Haymes 1992-1993; Whincop 2007

A.35 Pella (Jordan)

The ruins at *Tabaqat Fahl* have been associated with ancient Pella since the mid-nineteenth century. Located approximately 30 km south of the Sea of Galilee, the site is positioned amid the foothills of the eastern side of the Jordan Valley (biblical Gilead) and enjoys a commanding view of the Jordan River and surrounding plains. Despite this 'elevated' position, the site only lies at *c.* sea level (McNicoll *et al.* 1982a, 12). Nevertheless, summers are much cooler here than in the valley, and winters much milder. A perennial spring also provides sufficient water for settlement, which is supplemented by winter rains sufficient for spring crops. Parts of the site, which includes a tell (*Tell el-Husn*) and small settled tableland (*tabaqah* in Arabic), have been occupied for at least eight millennia. Pella is remembered as one of the Decapolis cities of Roman Syria (R. Smith 1993).

Ancient Iron Age Name
Pihilum (Semitic?, Late Bronze Age)
Pehel (Egyptian, Late Bronze Age)

The name of the site is absent from Iron Age literary sources (McNicoll *et al.* 1982a, 55). The Hellenised name, Pella, was introduced after the conquests of Alexander the Great, who was born at Pella in Macedonia, though the similarity between Pella and the Late Bronze Age name (Pihilum) has not been adequately explained (R. Smith 1993, 1174).

Seasons of Excavation
R.W. Funk and H.N. Richardson: 1958
Jordanian Department of Antiquities: 1964
R. Smith: 1967
R. Smith, A.W. McNicoll and J.B. Hennessy: 1979-1985
A. W. McNicoll and J. B. Hennessy: 1986-1991
S.J. Bourke: 1994-2007

Published Iron Age Ceramics
McNicoll *et al.* 1982a; 1982b; Potts *et al.* 1985; 1988; R. Smith and Potts 1992; Bourke *et al.* 2003; Bourke 1997

A.36 Qarqur, Tell (Syria)

Tell Qarqur is located within the northern reaches of the Gharb Valley, a fertile strip of Western Syria enclosed by the Jebel Anasariya to the west and the Idlib plateau to the east. The heavily occupied Amuq Plain lies less than 50 km (30 miles) to the north. The site itself consists of a double mound; a high southern tell rising *c.* 30 m above the plain, and an extensive (22 ha) lower city mound stretching to the north. While today Tell Qarqur lies 2 km east of the Orontes River, the course of the river in pre-Roman times would have run close to the western edge of the mound.

Ancient Iron Age Name
Karkara (Neo-Assyrian, Shalmeneser III's inscriptions)
Qarqar (Neo-Assyrian, Sargon II's inscriptions)

Tell Qarqur has long been identified with ancient Qarqar (Pritchard 1955; Yamada 2000, 154), site of two major battles between the Neo-Assyrians and Hama-led coalitions. The Neo-Assyrian records mention a battle at the time of Shalmeneser III (853 BC) and later during the reign of Sargon II (720 BC); both battles were fought against a coalition of Syrian and Palestinian states. The focus of Assyrian attack was against the northern border of the important Kingdom of Hamath, and more specifically the royal city of Qarqar (Dornemann 2003a, 4). The site of Tell Qarqur today lies at the very northern end of the Hama district, a fact that lends support to its equation with ancient Qarqar, though no epigraphic evidence has been found at the site to confirm the identification.

Seasons of Excavation
J M Lundquist: 1983-1984
R. H. Dornemann: 1993-2006

Published Iron Age Ceramics
Dornemann 1999; 2000; 2003a

A.37 Qasmieh (Lebanon)

Little has been published regarding the nature of Qasmieh, which lies near the southern Lebanese coast just north of the Litani River, *c.* 10 km north of Tyre.

Ancient Iron Age Name
Bit-Gisimeia (Neo-Assyrian)

Seasons of Excavation
Unknown

Published Iron Age Ceramics
S. V. Chapman 1972

A.38 Qrayé (Lebanon)

Situated in the foothills of the Lebanons, the village of Qrayé lies 8 km southeast of Sidon (Guigues 1939, 53). Along the village's western edge a Bronze and Iron Age cemetery was exposed. Little else has been published regarding the nature of the site.

Ancient Iron Age Name
The ancient name of Qraye is currently not known.

Seasons of Excavation
G. Contenau: 1914
P.E. Guigues: 1924

Published Iron Age Ceramics
S. V. Chapman 1972

A.39 Rachidieh, Tell (Lebanon)

Tell Rachidieh (also *Tell Habish*) lies 5 km south of Tyre (Dunand 1940, 123); close enough to once be considered part of Tyre. The tell lies 1 km from the Ras al-Ayn spring, a key source for fresh water (Doumet-Serhal 2004a).

Ancient Iron Age Name
Uzu (Papyrus Anastasi, New Kingdom)
'ITW (Akkadian)
Usu/Ushu (Neo-Assyrian)
Dur Bel-Harransadua (Neo-Assyrian)
Hosa (Hebrew, Old Testament)

Seasons of Excavation
T. Macridi-Bey:1901, 1903 (7 Iron Age tombs)
D. Le Lasseur: 1921
M. Chéhab: 1942-1943
B. Kaoukabani and H. Chéhab: 1974

Published Iron Age Ceramics
Macridi-Bey 1904a; Culican 1982; Doumet 1982; Doumet and Kawkabani 1995; Bordreuil 2003; 2004; Doumet-Serhal 2003; 2004b; Griffiths 2003

A.40 Ras al Bassit (Syria)

Ras al-Bassit lies on the Syrian coast between Lattakia (55 km to the south) and the mouth of the Orontes River (25 km to the north). The tell lies on a natural terrace overlooking the cove formed by a natural headland jutting out from the coast. An ancient necropolis is situated c. 75 m to the south west, between the tell and a large classical acropolis. The shore line has changed over the centuries, and the buildings on the tell were once situated directly on the cove, though today the tell lies 100 m from the water (Courbin 1993a, Fig. 1). The tell itself is of a low height (the highest point is 6.40 m above sea-level) and modest size (< 1 ha.), though the late Iron Age settlement was apparently not restricted to the tell (Courbin 1986, 182).

Ancient Iron Age Name
R'S(Y) (Ugaritic, Late Bronze Age)
SNR(Y) (Ugaritic, Late Bronze Age)
Sinuru (Late Bronze Age)
Posideion (Greek, Herodotus)

Ras al-Bassit has strong connections with the Greek world and, significantly, retains an etymological echo of the classical Posideion (Boardman 1990, 170).

Seasons of Excavation
P. Courbin: 1971-1986, 1990, 1992

Published Iron Age Ceramics
Courbin 1973; 1974-1975; 1978; 1983; 1986; 1990a; 1990b; 1993a; Braemer 1986

A.41 Ras Ibn Hani (Syria)

Ras Ibn Hani is situated on the north Syrian coast 9 km north of Lattakia (Bounni *et al.* 1976, 233, 265). Sited on a 2.5 km long limestone promontory, the widely spread settlement (c. 15 ha) was protected on three sides by the sea, leaving only its eastern side needing fortifications (Bounni *et al.* 1978, Fig. 1). The natural topography provided unimpeded access to the Mediterranean, which encouraged participation in maritime trade whilst maintaining visible communication with Ugarit 4.5 km to the north-east (Bounni *et al.* 1998, 7-8). The moderate height (9 m) of the tell is indicative of limited occupation.

Ancient Name
B'IR/Biruti (Ugaritic, Late Bronze Age)
PU/Bi-i-ri (Akkadian, Late Bronze Age)

Since the discovery of a Late Bronze Age archive at Ras Ibn Hani, a number of proposals have been advanced regarding the identification of the ancient site; Biruti being the most popular candidate (Bounni *et al.* 1998, 92-93).

Seasons of Excavation
K. Toueir: 1974
A. Bounni and J. Lagarce: 1975-1984, 1986-1995

Published Iron Age Ceramics
Badre 1983; Lagarce 1983; Bounni *et al.* 1976; 1978; 1979; 1981; 1998

A.42 Rehov, Tel (Israel)

Tel Rehov (*Tell es-Sarem* in Arabic) is one of the largest mounds (c. 10 ha) in Palestine. The upper tell rises 20 m above the surrounding alluvium of the Beth Shan Valley and is visible from the site of Pella, 10 km to the east (A. Mazar 1999a, Fig. 1). The lower mound is elongated and rises 8 m above the plain and is heavily eroded on its eastern edge. Tel Rehov is situated at the western edge of the Jordan Valley, 6km from the Jordan River and 5 km south of Beth Shan.

Ancient Iron Age Name
Rehob (Hebrew)
Rehob (Egyptian, Shoshenq's Karnak inscription)

Rehov (literally 'street' in Hebrew) was the name of several cities mentioned in ancient texts, and is identified with *Tell es-Sarem* due to the survival of the name Rehov at the nearby Islamic holy tomb (*esh-Sheikh er-Rihab*).

Seasons of Excavation
F. Vitto: 1974-1980 (five seasons)
A. Mazar: 1997, 1998, 1999, 2000, 2001, 2002, 2003, 2004, 2005, 2006, 2007

Published Iron Age Ceramics
A. Mazar 1999a

A.43 Rifa'at, Tell (Syria)

Tell Rifa'at occupies a commanding position between the Qoueiq River and the Jebel Sam'an, 35 km north of Aleppo. The red soil and high water table offer an extremely fertile environment for the large site (c. 5 ha) (Seton-Williams 1961, 69-70). The imposing tell rises 36 m above the surrounding plain. The mound itself, with its heavily eroded edges and steep slopes, is completely surrounded by the town of the same name, which, since the 1870s, has occupied an area roughly equivalent to the lower settlement of the Iron Age (Seton-Williams 1967, Pl. V). Extensive quarrying of the site by local villagers, combined with heavy winter rains, has led to extensive damage of the tell.

Ancient Iron Age Name
Arpad (Aramaic, Old Testament)
Arpadda (Neo-Assyrian)

The identification of Tell Rifa'at with ancient Arpad, capital of the Aramaean kingdom of Bit-Agusi, was proposed on etymological and geographic grounds prior to excavation (Kraeling 1918, 65). According to texts, the city and district of Arpad were seized by Aramaeans under their leader, Agusi, sometime in the first half of the 9th century BC. From that point on Arpad played a significant role in the shifting politics of a region in dangerous proximity to the re-invigorated Assyrians.

Seasons of Excavation
B. Hrozny: 1924-1925
M. V. Seton-Williams: 1956, 1960, 1964

Published Iron Age Ceramics
Seton-Williams 1961; 1967; Matthers 1981b; Lehmann 1996

A.44 Sarepta (Lebanon)

Located on the southern coast of Lebanon, Sarepta lies 22 km north of Tyre and 13 km south of Sidon. The site conforms to the classic pattern of settlement within Iron Age Lebanon, utilising either an island or promontory that gives access to at least two natural harbours. The modern village of Sarafand, approximately 2 km to the south, still makes use of the dual harbours. The low Iron Age tell (*Ras el-Qantara*) rises abruptly from the shore and extends in a gradual incline to the south and east, reaching a modest height of 15 m (Pritchard 1978, 71).

Ancient Iron Age Name
DRPT (Akkadian, Late Bronze Age)
Zarpat/Zarephath (Hebrew, Old Testament)
Sa-ri-ip-tu/Za-ri-ib-tu (Neo-Assyrian)
Sarepta (Greek)

The site of Sarafand has long been identified with ancient Sarepta, primarily based on etymological similarity (Pritchard 1975, 7-9). The 1969 discovery at Sarafand of a Greek inscription dedicated to "the holy god of Sarepta," supports this identification (Pritchard 1978, Fig. 16). When in 1971 an inscribed scarab was discovered with the four letters ***s r p t*** (Sarepta) the identification appears confirmed (Anderson 1988, 36). While literary sources depict Sarepta as a major city of the first millennium BC, it is always coupled with Sidon and appears to have been one of many satellite fortresses within the Sidonian kingdom. The city also held significant fame within the early Christian period; remembered for the two miracles performed there by the biblical prophet Elijah (see 1 Kings 17:7 *ff* and Luke 4:24-26).

Seasons of Excavation
H. Ingolt, E. Fugmann: 1932
R. Saidah: 1962, 1964, 1968
J. B. Pritchard: 1969-1974

Published Iron Age Ceramics
Culican 1970; Pritchard 1975; 1978; 1988; Koehl 1985; Anderson 1988; 1990; Khalifeh 1988; Bettles 2003

A.45 Sheikh Hassan, Tell (Syria)

Tell Sheikh Hassan is situated in northern Syria within the elbow of the Euphrates River, about half-way between Tawi and Mumbaqat. Originally situated on the left bank of the Euphrates River, the tell is now an island within Lake Assad as a result of the inundation of the Tabqa Dam (Anastasio *et al.* 2004, 300). The tell originally rose 14 m above the Euphrates River and possessed a diameter of over 250 m (c. 5 ha.), but today is a small (c. 0.5 ha.) island situated 500 m from land. Needless-to-say, that the inundation of the site has greatly damaged the mound.

Ancient Iron Age Name
The ancient name of the site is currently not known.

Seasons of Excavation
A. Bounni: 1972-1974
J. Cauvin: 1976
W. Orthmann: 1981
J. Boese: 1984-1990, 1992-1994

Published Iron Age Ceramics
Boese 1995; Schneider 1999a; 1999b

A.46 Shiyukh Fawqani, Tell (Syria)

One of a few Iron Age sites excavated as part of Euphrates dam salvage projects, Tell Shiyukh Fawqani is located on the left bank of the Euphrates equidistant between Jerablus, a few kilometres to the north, and Tell Shiyukh Tahtani, to the south (Bachelot and Fales 2005). While the lower town is not extensive (c. 4 ha.), the upper acropolis is particularly small (c. 1.3 ha.). From its position on the river terrace, the tell rises 25 m above the surrounding village, and 30 m above the alluvial plain (Luciani 2005, 719).

Ancient Iron Age Name
Bur-mar'ina (Neo-Assyrian)
^{Uru} Marina sa sade (Middle Assyrian)

A text excavated from the mound of Tell Shiyukh Fawqani mentions the city of Burmarina (Fales *et al.* 2005; Fales 1999), which is also mentioned in the Assyrian annals as a city conquered by Shalmeneser III. Bur-mar'ana is mentioned as belonging to Ahuni, man of Adini (Pritchard 1955, 277; Morandi Bonacossi 2000c).

Seasons of Excavation
L. Bachelot (Centre National de la Recherche Scientifique) and M. Fales (University of Udine, Italy): 1994-1998, 2003

Published Iron Age Ceramics
Bachelot 1999; Luciani 2000; 2005; al-Bahloul *et al.* 2005; Makinson 2005

A.47 Sidon (Lebanon)

The ancient site of Sidon is identified with the modern city of Saïda located on the southern Lebanese coast 20 km south of Beirut. The ancient city occupied two distinct areas; the oldest town ('little Sidon') was built on a promontory between two bays, while 'Greater Sidon' was situated slightly inland from the coast (Doumet-Serhal 2004d, 102). It is impossible to determine the size and height of the tell, as the modern city of Saïda covers and obscures the ancient settlement.

Ancient Iron Age Name
Sidon (Hebrew, Old Testament)
DDN/Sidna (Akkadian)
Siduna (Amarna letters)
Sidunu (Neo-Assyrian)
SDN (Phoenician)

Ancient documents (e.g. Homer, Hebrew Bible) depict Sidon as the most important and ancient of the Levantine coastal cities, a portrait yet to be fully explored through archaeological means.

Seasons of Excavation
O. Hamdy Bey: 1887
T. Macridi-Bey: 1900-1904, 1909
C. C. Torrey: 1902
P. Perdrizet: 1903
G. Contenau: 1914, 1920
C. L. Brossé: 1923
M. Dunand: 1924
B. E. Guigues: 1925
M. Meurdrac and L. Albanèse: 1929
A. Poidebard: 1946-1950
H. Frost: 1963-1964
M. Dunand: 1964-1965, 1967-1968
C. Ghadban: 1960s
R. Saidah: 1967-1972
C. Doumet-Serhal: 1998, 2000-2006

Published Iron Age Ceramics
Contenau 1920; Culican 1975; Riis and Buhl 1990; Doumet-Serhal 2000; 2004d; 2004g; 2006

A.48 Sukas, Tell (Syria)

Tell Sukas is a site on the north Syrian coast bordering the Gabala plain, 26 km south of Lattakia. This highly fertile plain, c. 40 km in length, is surrounded by mountains to the north, east and south (Riis 1970, 7). The relatively small (c. 2 ha) mound rises 20 m above the natural promontory upon which it resides. The site is flanked by two small bays, to the north and south (Riis 1970, Fig. 3).

Ancient Name
Su-uk-su/Suksi (Ugaritic – Late Bronze Age)

No find from Tell Sukas has confirmed the name of the settlement, though Sukas corresponds well with Suksu, the name of a town at the southern frontier of the kingdom of Ugarit, a position that accords well with the location of Tell Sukas.

Seasons of Excavation
E. O. Forrer: 1934
P. J. Riis: 1958-1961, 1963

Published Iron Age Ceramics
Riis 1960; 1970; 1979; Ploug 1973; Buhl 1983; 1996; Lund 1986; Jensen 1996.

A.49 Tambourit (Lebanon)

The village of Tambourit lies on the Lebanese coast 6 km south east of Sidon.

Ancient Iron Age Name
The ancient name of Tambourit is not known.

Seasons of Excavation
R. Saidah: 1971

Published Iron Age Ceramics
Saidah 1977

Ta'yinat, Tell (Turkey) - see A.8 Amuq

A.50 Tille Höyük (Turkey)

Situated in the Upper Euphrates Valley of south-eastern Turkey, Tille Höyük lies 60 km east of Adiyaman and 130 km north of the Syrian-Turkish border. The site consists of a small mound (c. 0.25 ha.) surrounded by an extensive (c. 12 ha.) lower town (Anastasio *et al.* 2004, 326). While the site once stood on the right bank of the Euphrates River, since 1990 it has lain beneath the waters of the Ataturk Dam.

Ancient Iron Age Name
The ancient name of the site is currently not known.

Seasons of Excavation
D. French: 1979-1990

Published Iron Age Ceramics
Summers 1993; Blaylock 1999.

A.51 Tyre (Lebanon)

Ancient Tyre is located opposite the plain of Tyre, which stretches 25 km south from the Litani River. The ancient site of Tyre was situated on an island opposite the plain. When Alexander laid siege to the city, he built a causeway joining the island to the mainland, thus creating today's peninsula. Tyre lies 94 km south of Beirut, and midway between Sidon and Akko. At this point the coastal plain reaches its maximum width of 2 km. The ancient city of Tyre is today covered by the modern city of Sour, leaving much of the past settlement beyond reach.

Ancient Iron Age Name
Surru (Akkadian)
Sor/Zor/Tsor (Hebrew, Old Testament)
Tor/Tur (Aramaic)
SR (Phoenician)
Tyros (Greek)

Seasons of Excavation
J. de Bertou: 1837 (northern harbour)
E. Renan: 1860 (northern harbour)
J. Nepomuk Sepp: 1874 (Emperor Barbarossa's tomb)
D. Le Lasseur: 1921-1923
L. Albanese: 1923-1928
A. Poidebard: 1934-1936 (underwater survey)
M. Dunand: 1937 (Roman tomb)
M. Chéhab: 1946, 1950-1951, 1961-1974
M. Simson: 1947-1948
A. Bounni: 1957-1959
R. Saidah: 1960 (Roman tombs)
H. Frost: 1963, 1966
C. Asmar: 1966-1967 (Roman baths)
P. Bikai: 1973, 1974
M.E. Aubet: 1997, 1999, 2002-2005 (Iron Age cemetery)

Published Ceramics
Patricia Bikai 1978a; 1978b; Seeden 1991; Aubet 2004a; Aubet *et al.* 1998-1999; Núñez 2004a; 2004b

A.52 Zincirli (Turkey)

Zincirli is situated in the Valley of the Kara-Su of the anti-Taurus Mountains of south central Turkey. It is located immediately east of the Amanus Mountain range in Gaziantep province, approximately 10 km north of Islahiye. The site consists of a large (40 ha.) double mound, with a fortified acropolis and extensive lower town.

Ancient Iron Age Name
Sam'al (Luwian)

Sam'al was the capital of the Neo-Hittite kingdom of Ja'dija.

Seasons of Excavation
F. von Luschan: 1888, 1890-1891, 1894, 1902

Published Iron Age Ceramics
Andrae 1943; Lehmann 1996

FIGURES

001a/1

001a/2

001b/1

001b/2

001c

001d

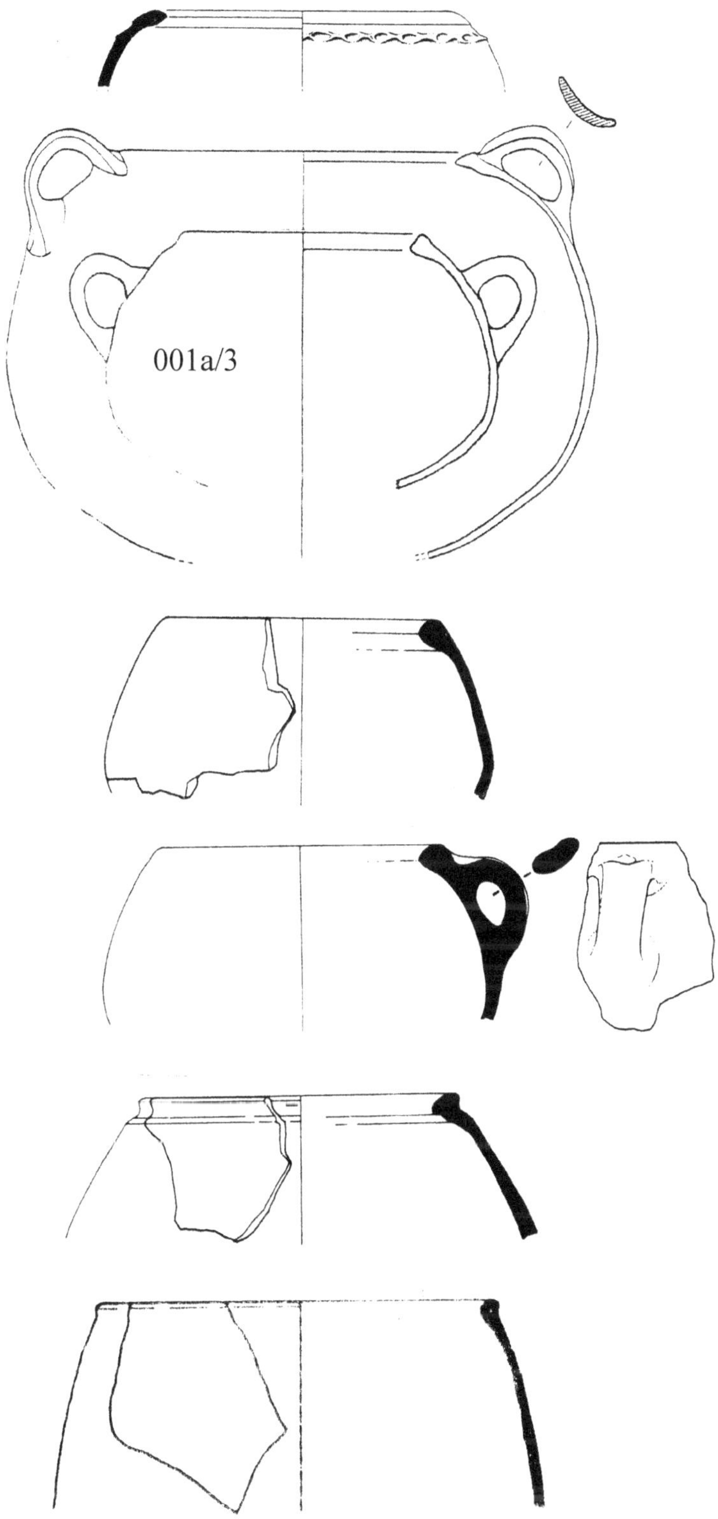

Fig. 1 Cooking Pot CLASSES Used in This Study (Scale1:4)

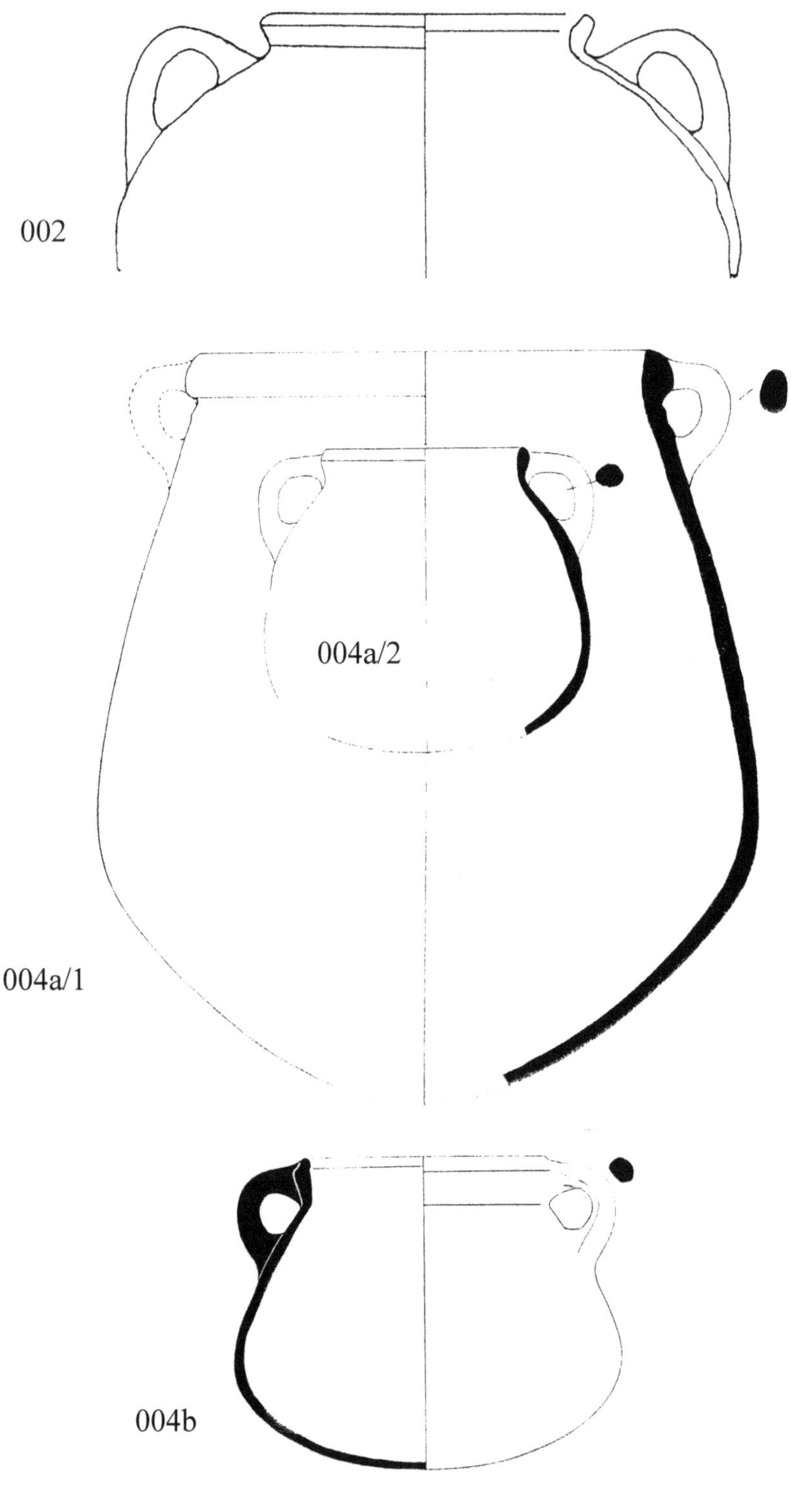

Fig. 2 Cooking Pot CLASSES Used in This Study (Scale 1:4)

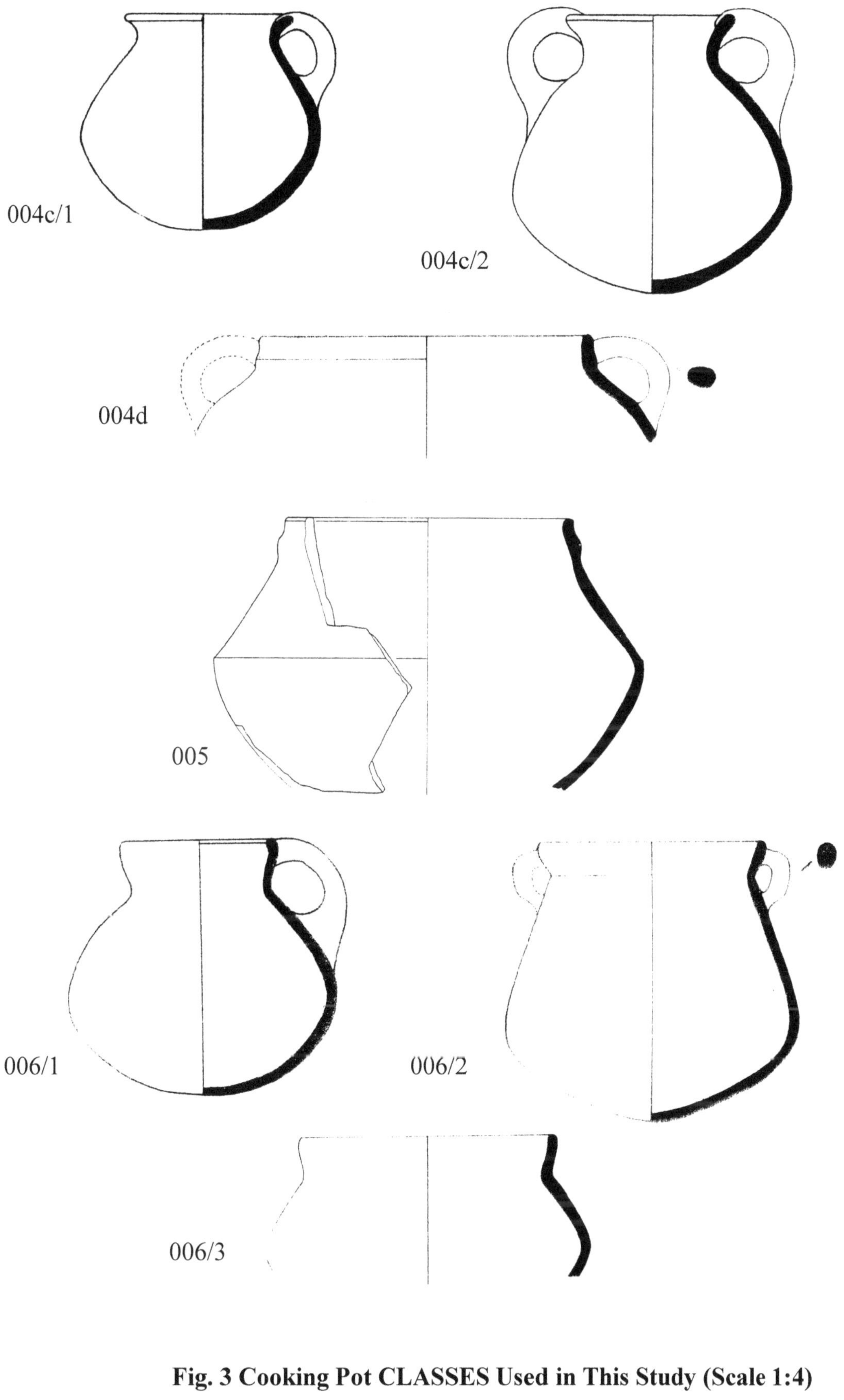

Fig. 3 Cooking Pot CLASSES Used in This Study (Scale 1:4)

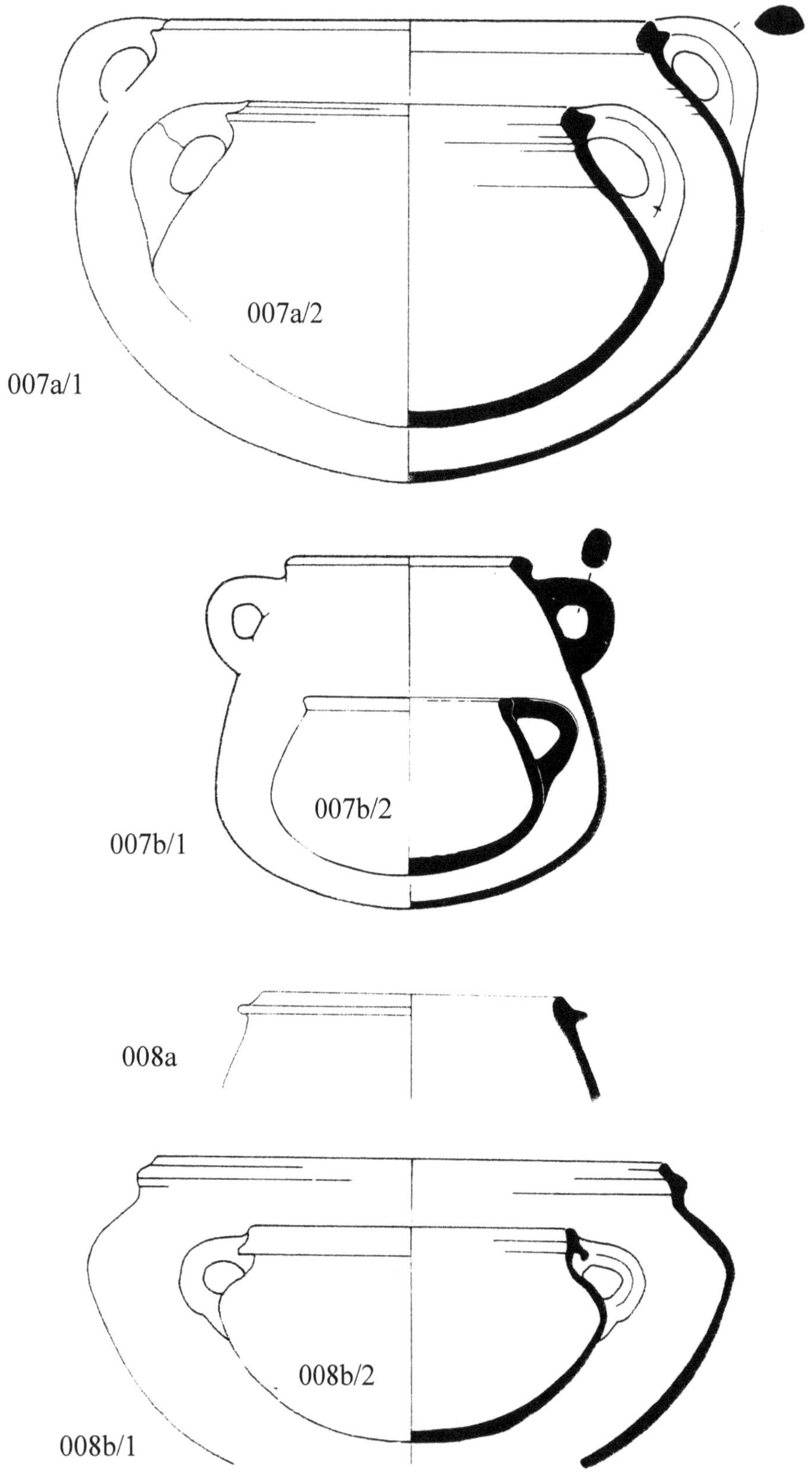

Fig. 4 Cooking Pot CLASSES Used in This Study (Scale 1:4)

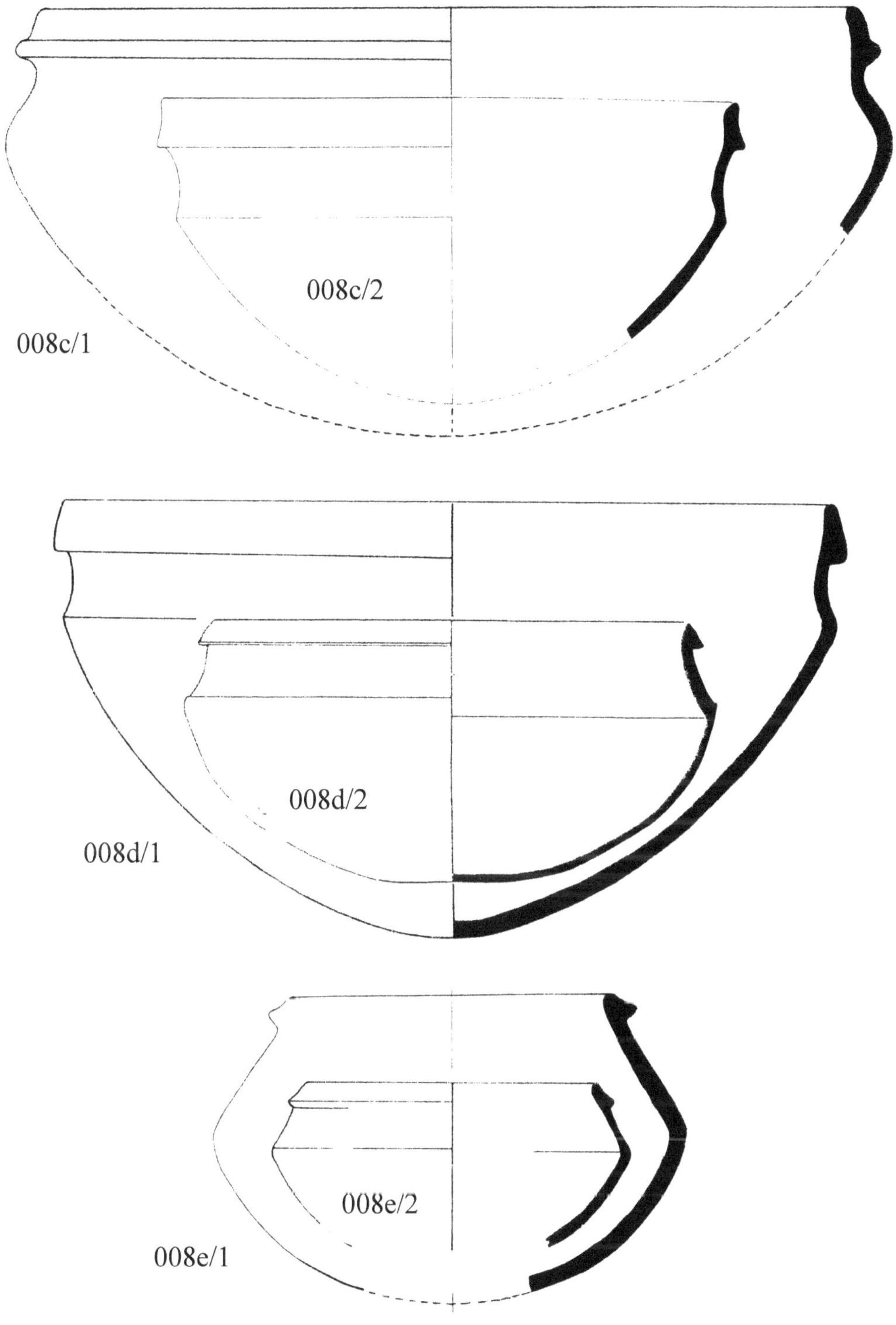

Fig. 5 Cooking Pot CLASSES Used in This Study (Scale 1:4)

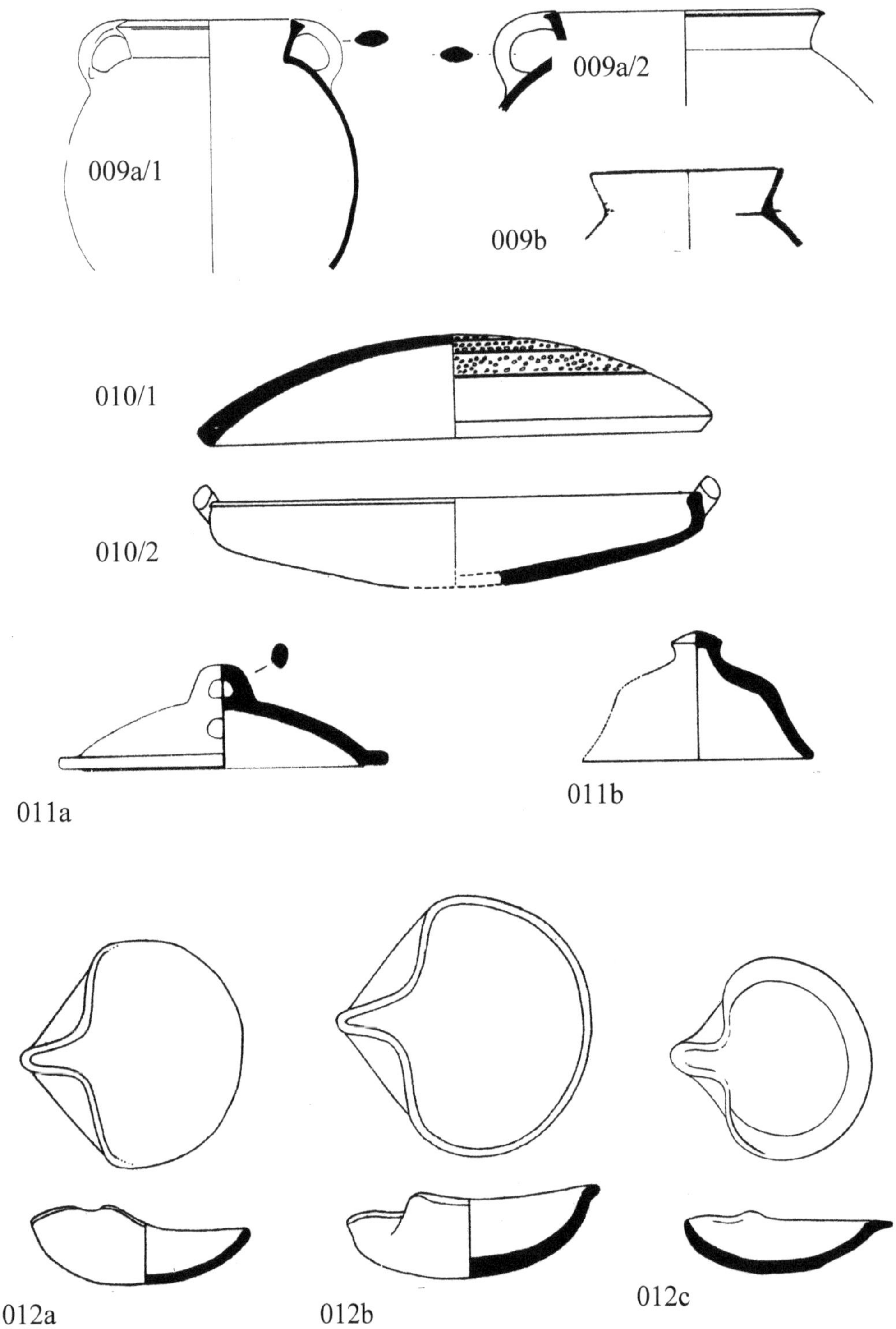

Fig. 6 Cooking Pot and Lid CLASSES Used in This Study (Scale 1:4)

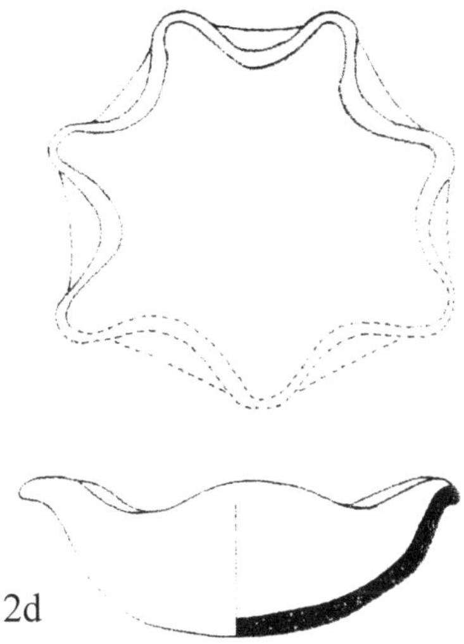

012d

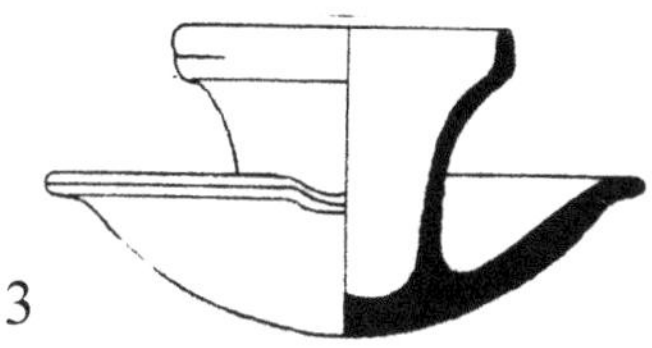

013

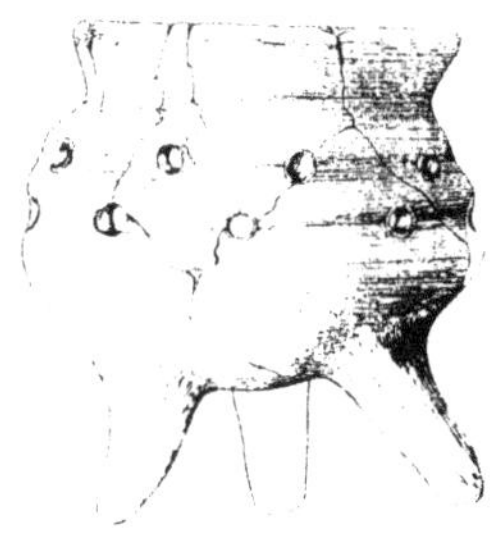

014

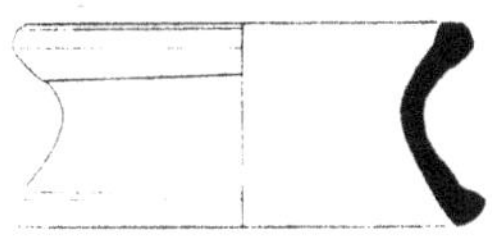

015/1

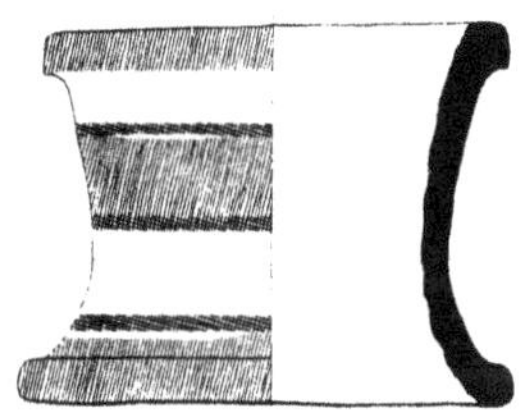

015/2

Fig. 7 Lid and Miscellaneous CLASSES Used in This Study (Scale 1:4)

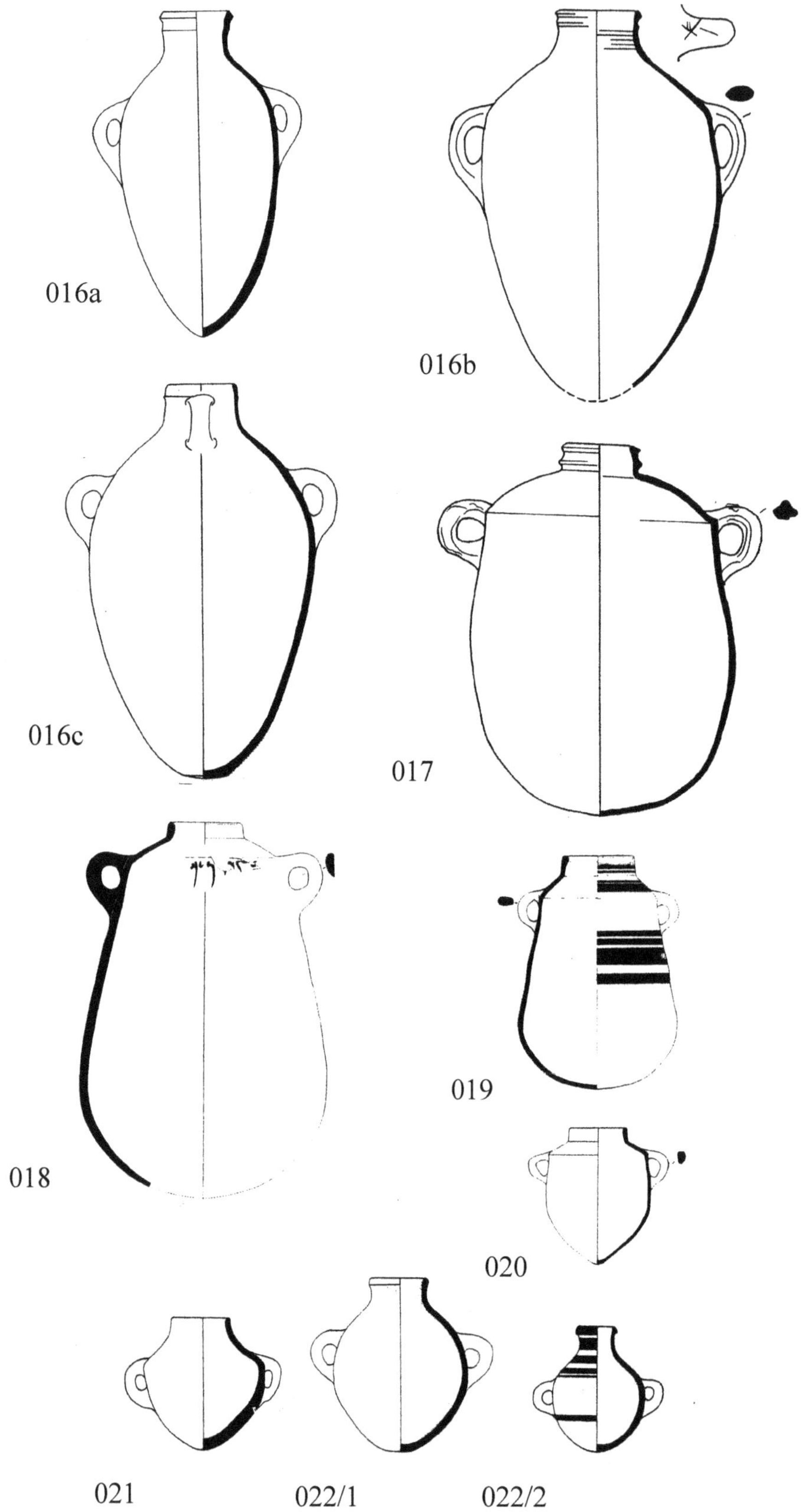

Fig. 8 Transport Amphora CLASSES Used in This Study (Scale 1:10)

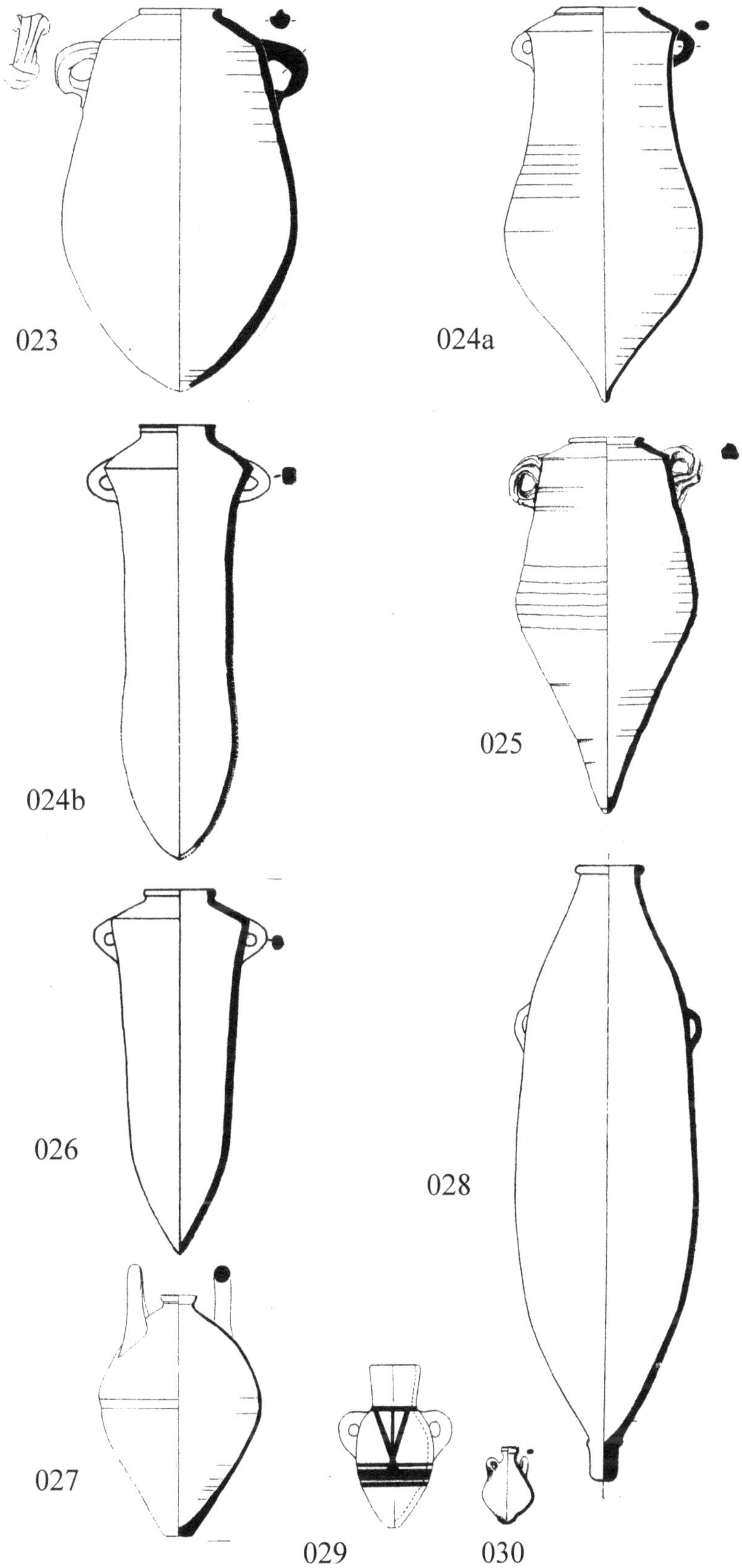

Fig. 9 Transport Amphora CLASSES Used in This Study (Scale 1:10)

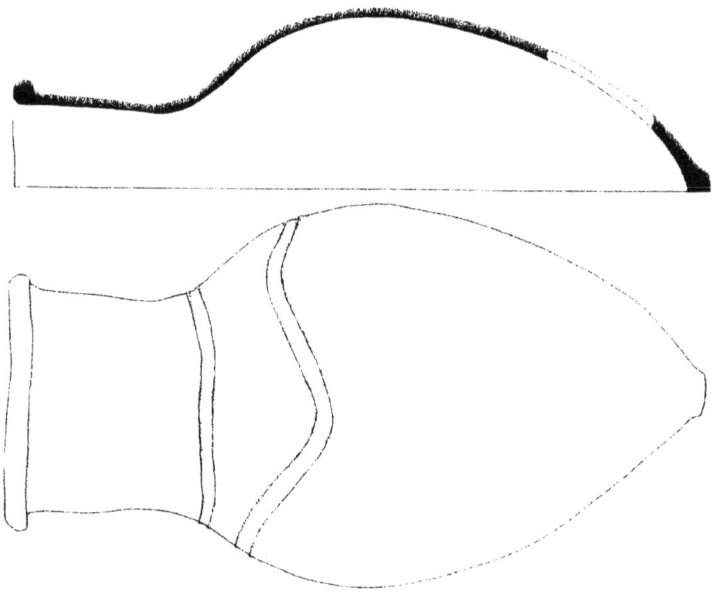

032

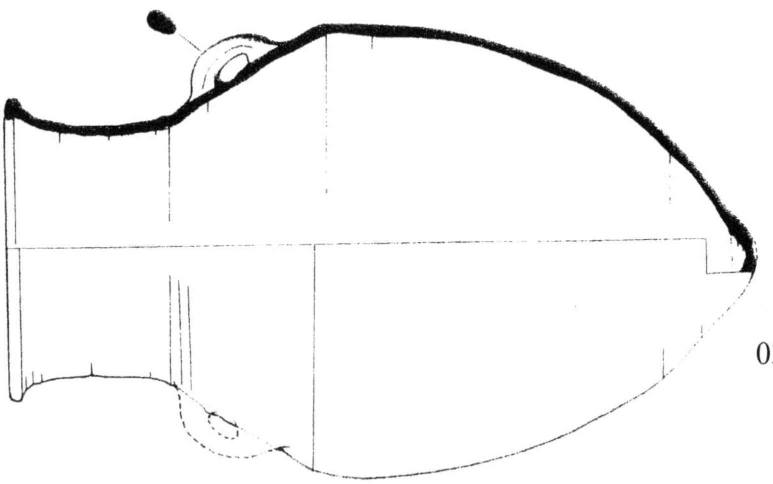

031

Fig. 10 Pithoi CLASSES Used in This Study (Scale 1:10)

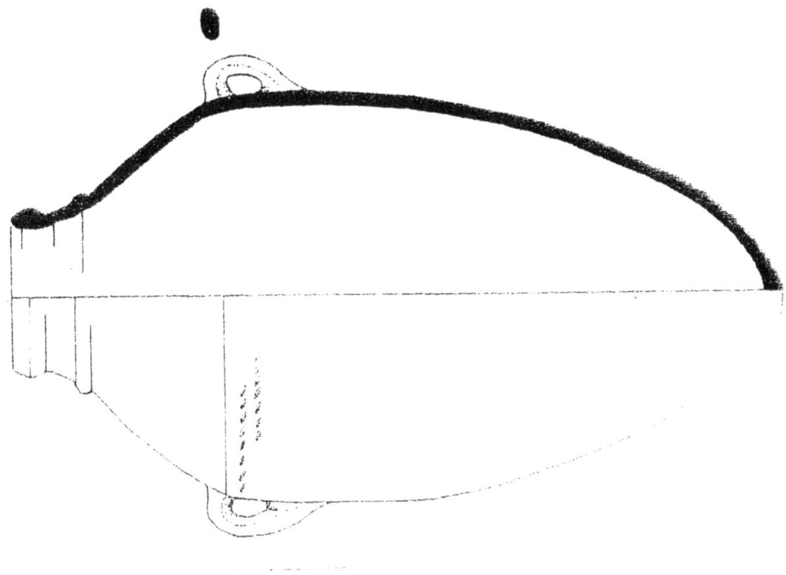

034

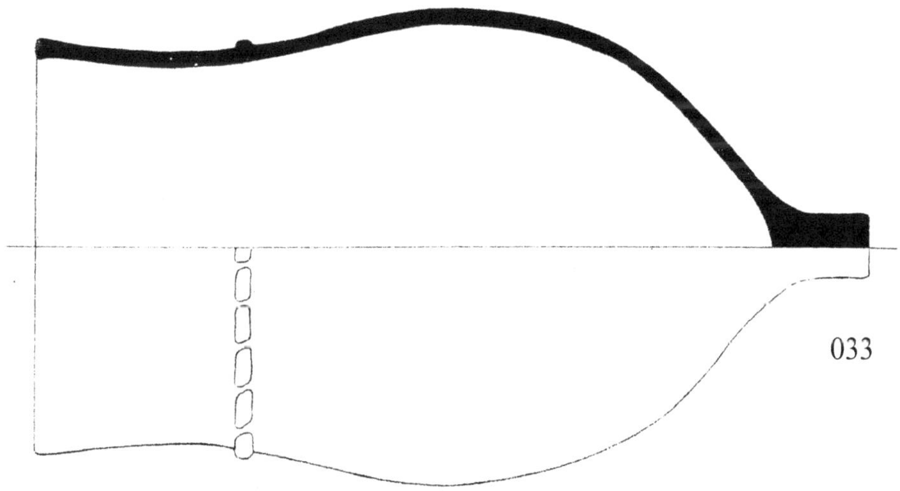

033

Fig. 11 Pithoi CLASSES Used in This Study (Scale 1:10)

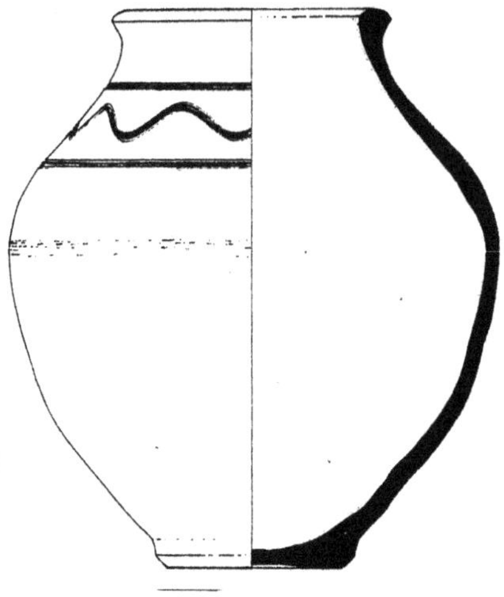

035

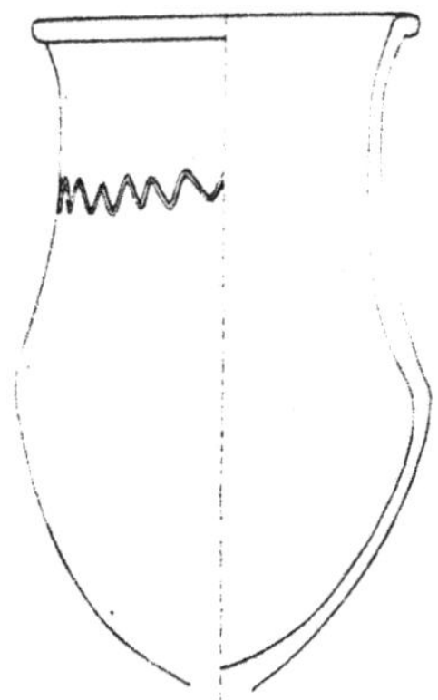

036/1

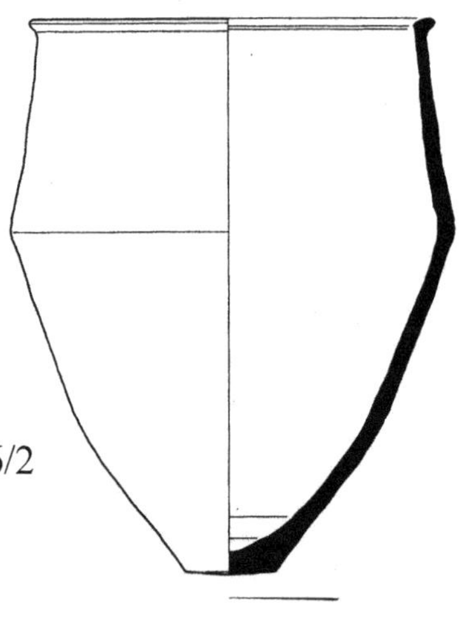

036/2

Fig. 12 Pithoi CLASSES Used in This Study (Scale 1:10)

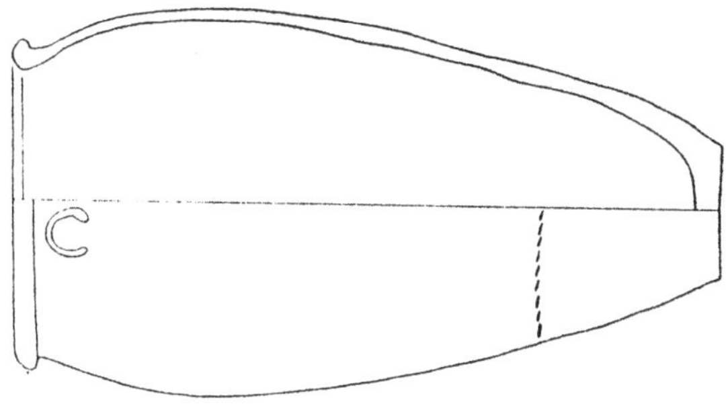

037b/2

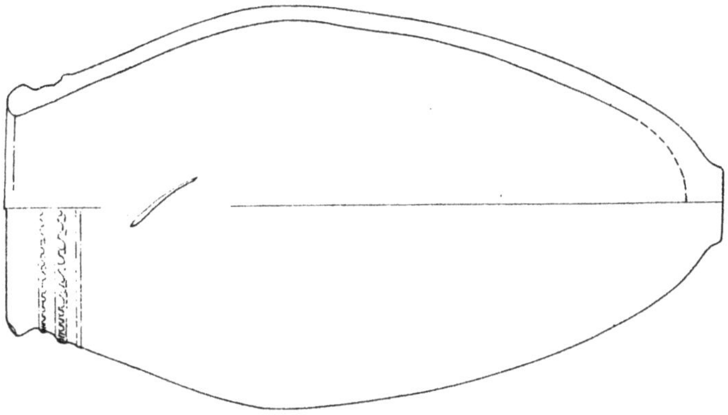

037b/1

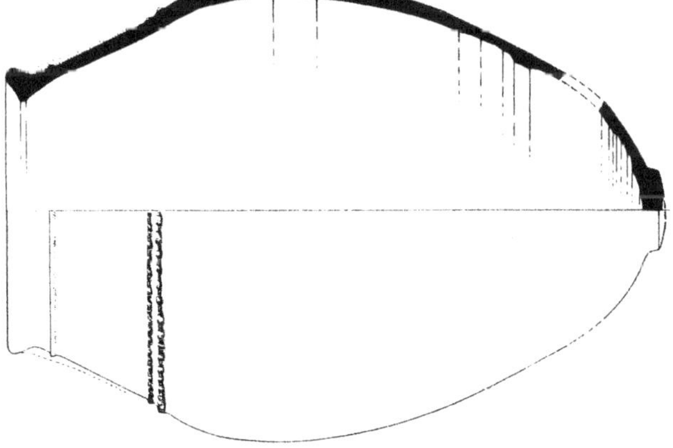

037a

Fig. 13 Pithoi CLASSES Used in This Study (Scale 1:10)

038a

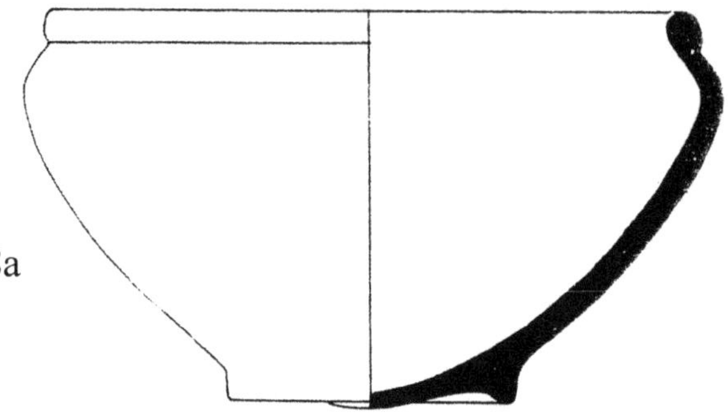

038b

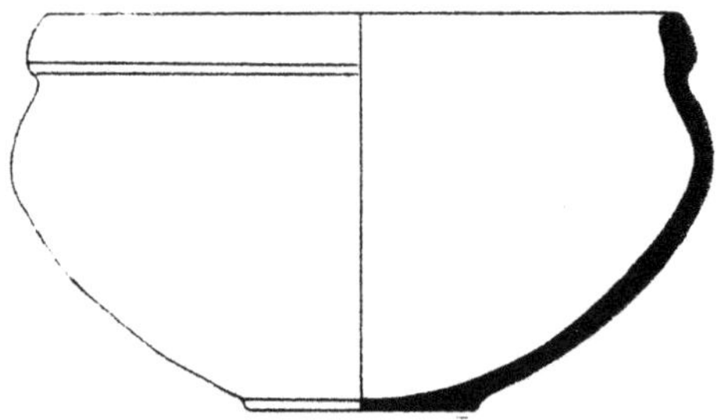

039

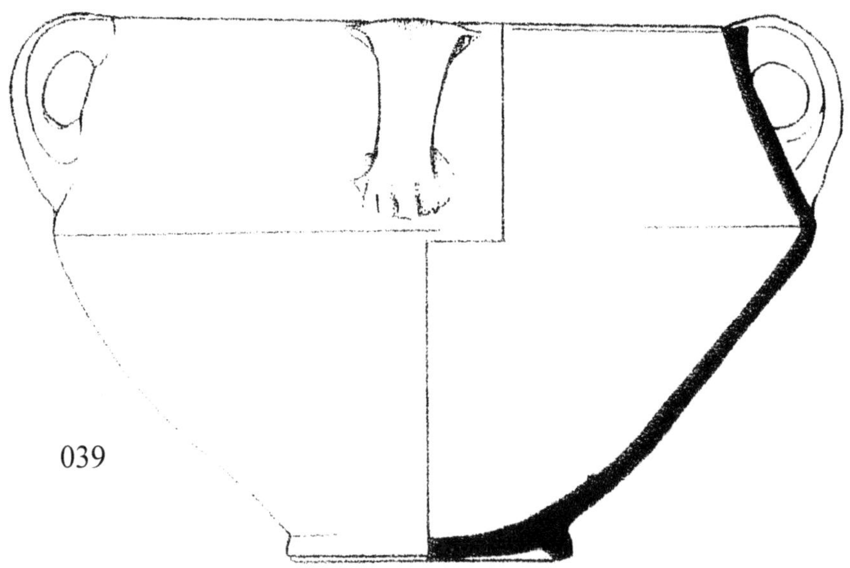

Fig. 14 Krater CLASSES Used in This Study (Scale 1:4)

040a

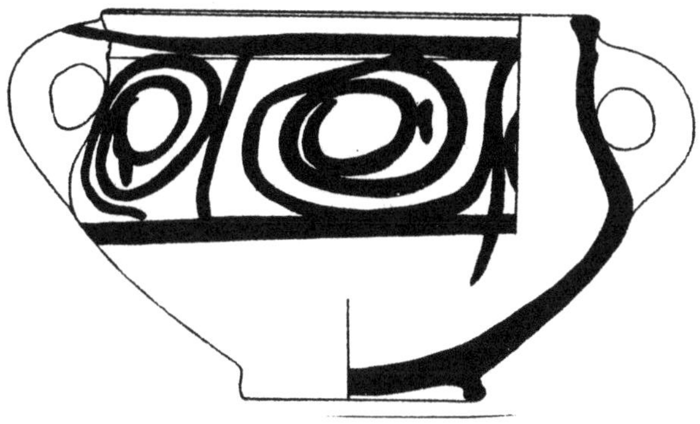

040b

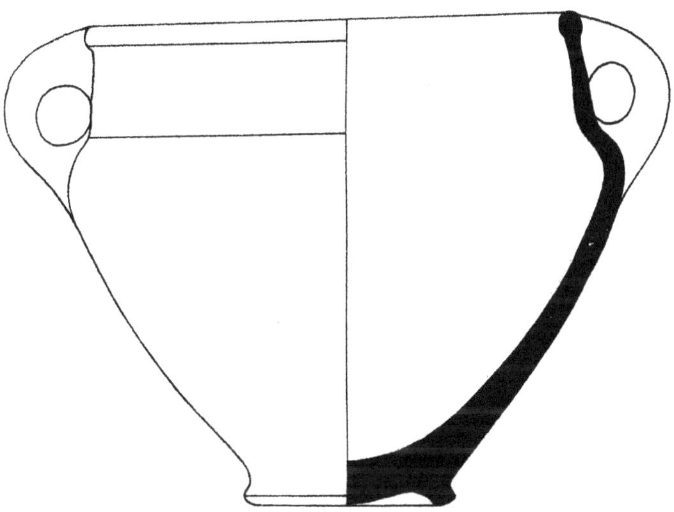

040c

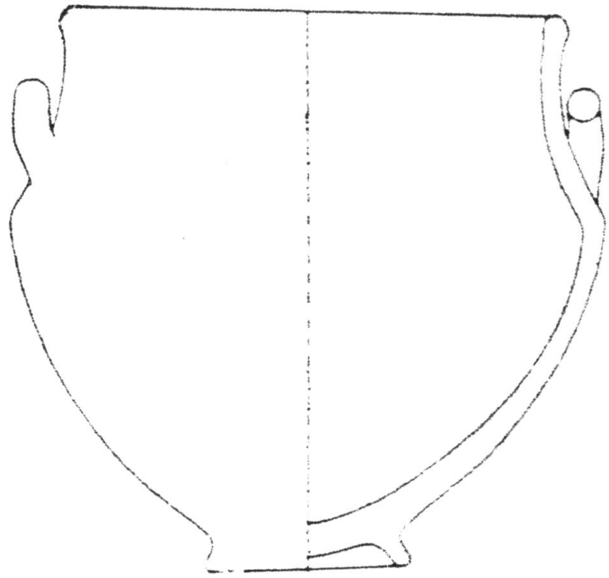

Fig. 15 Krater CLASSES Used in This Study (Scale 1:4)

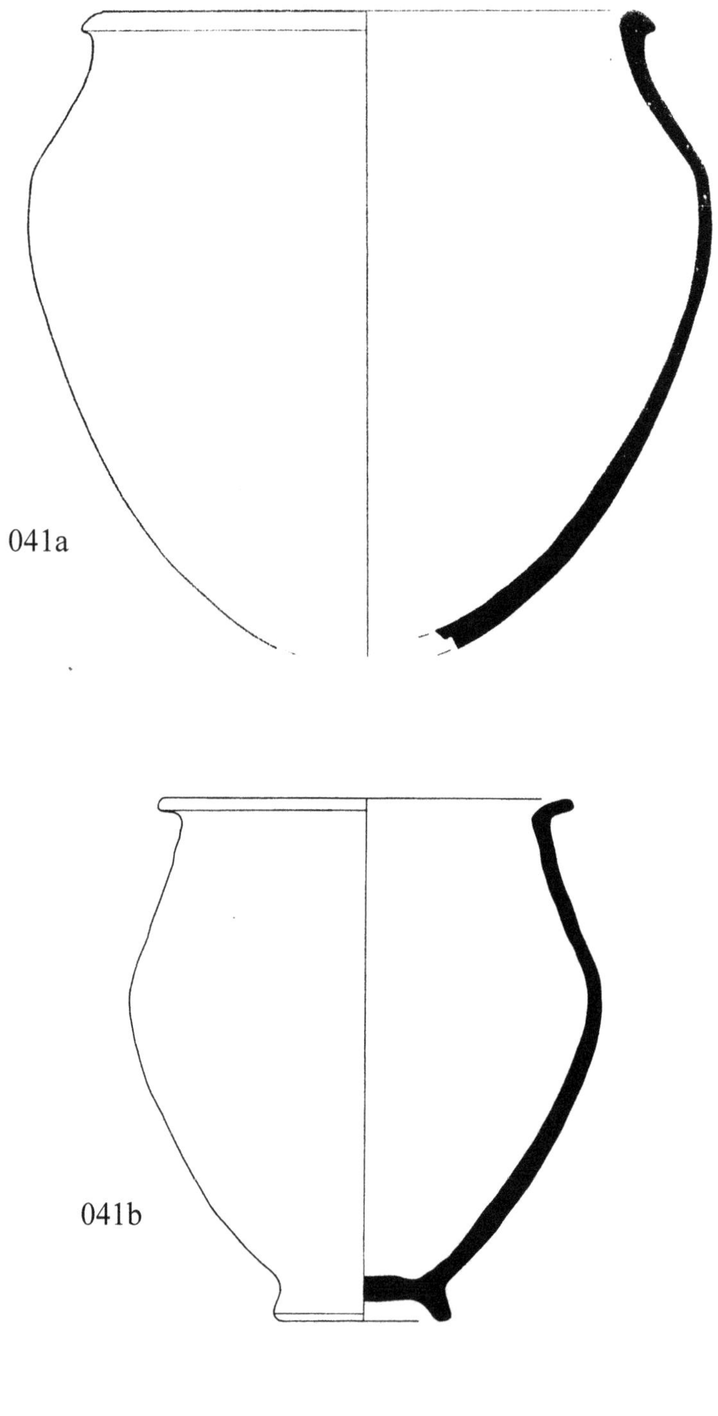

Fig. 16 Krater CLASSES Used in This Study (Scale 1:4)

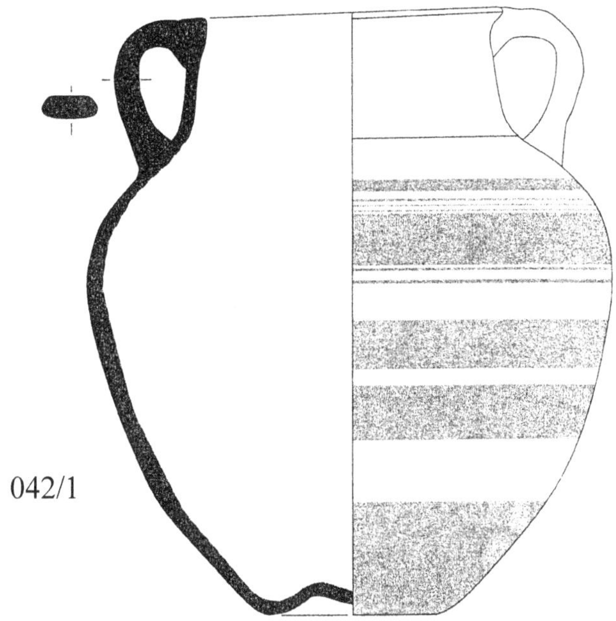

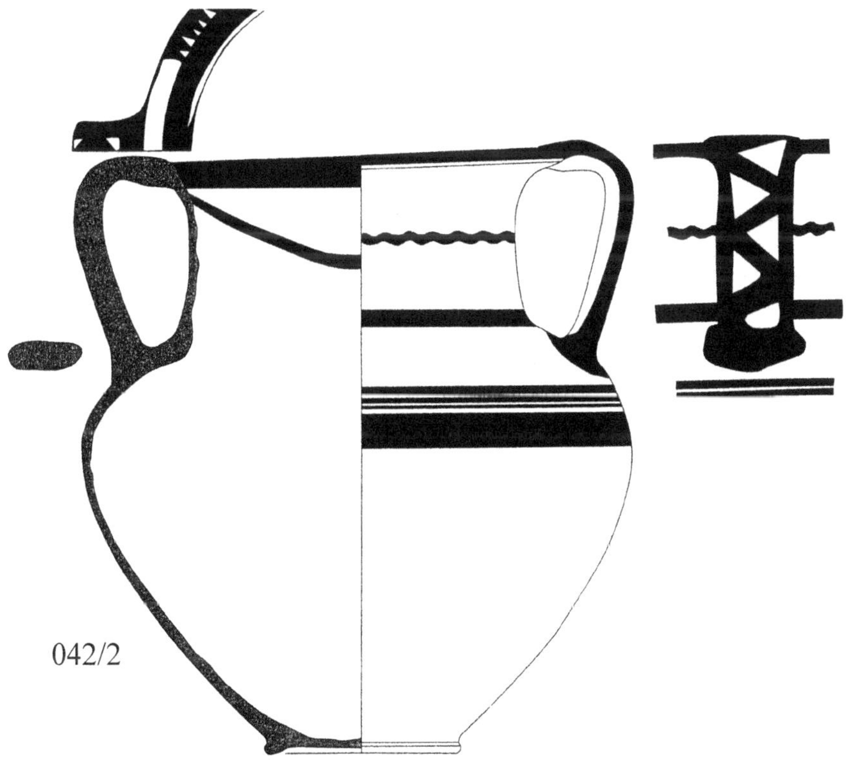

Fig. 17 Krater CLASSES Used in This Study (Scale 1:4)

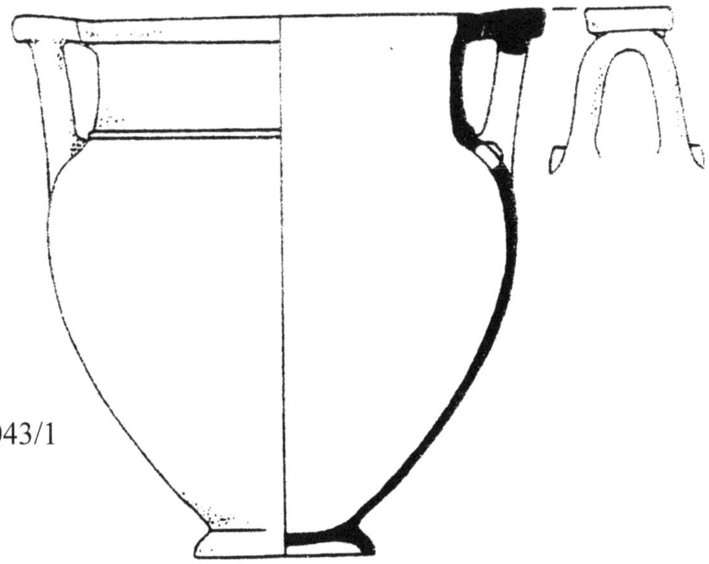

043/1

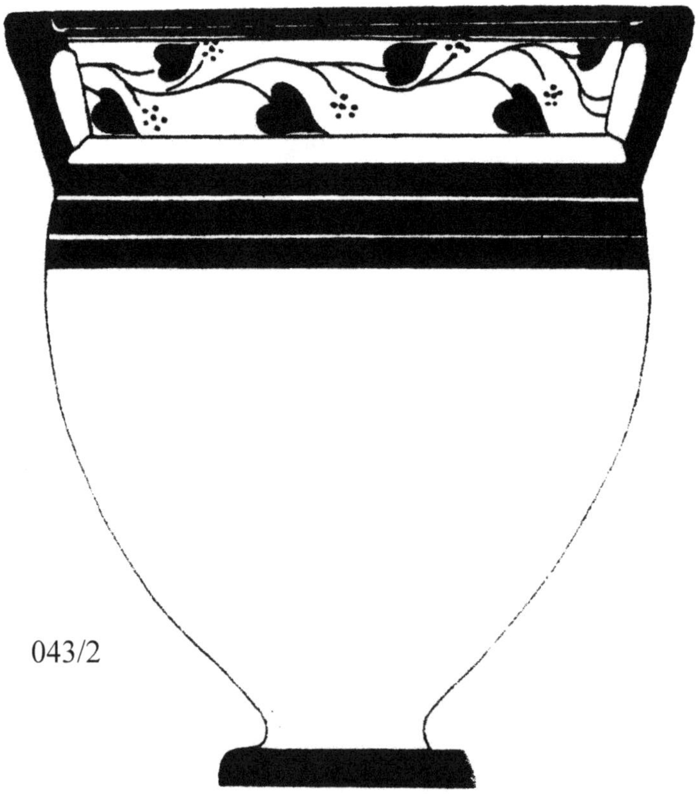

043/2

Fig. 18 Krater CLASSES Used in This Study (Scale 1:4)

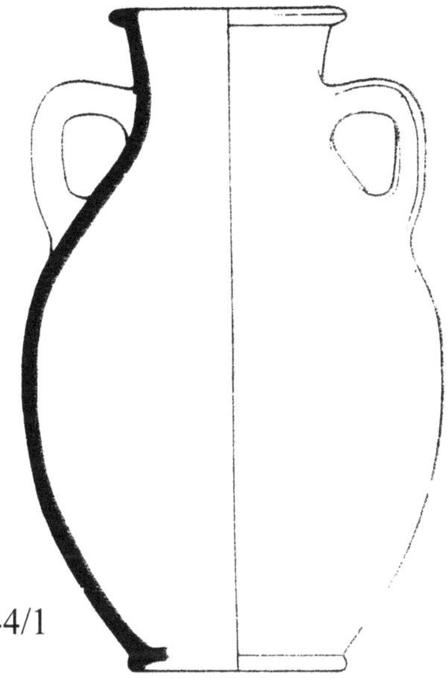

044/1

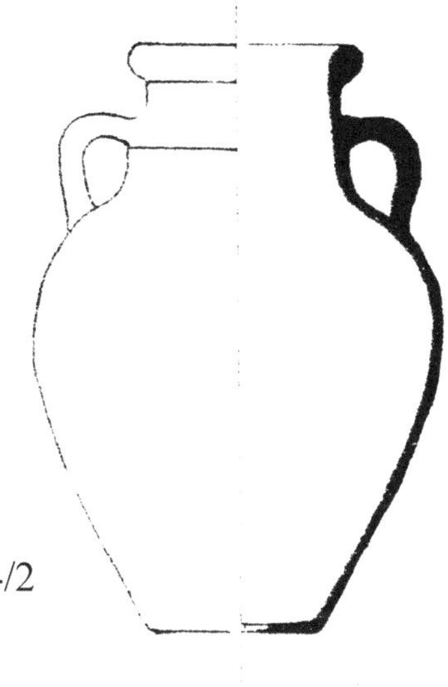

044/2

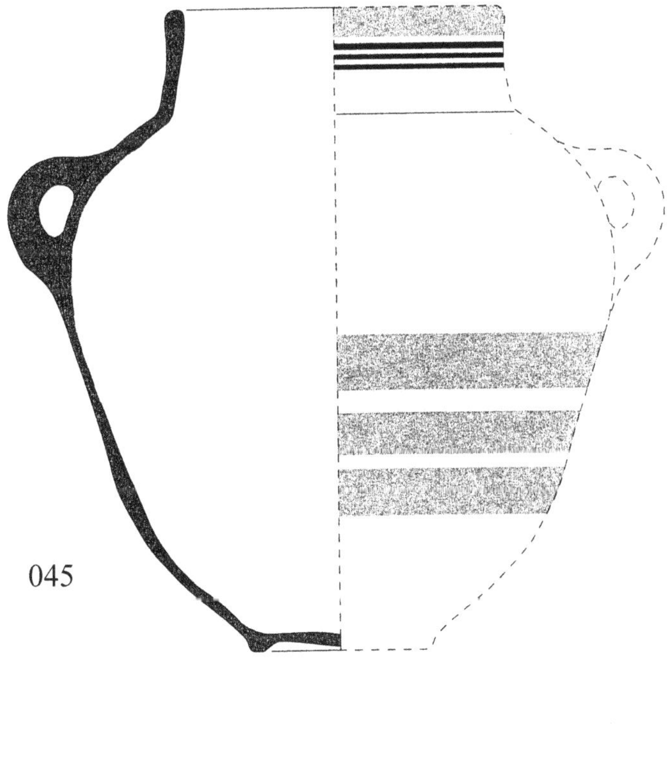

045

Fig. 19 Krater CLASSES Used in This Study (Scale 1:4)

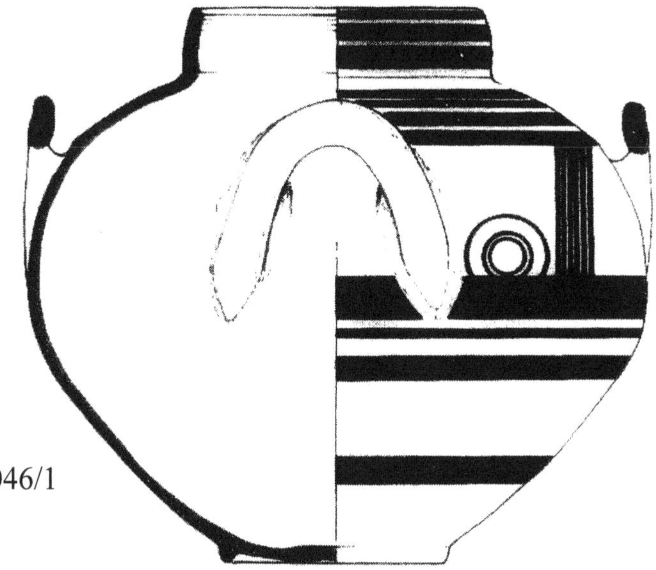

046/1

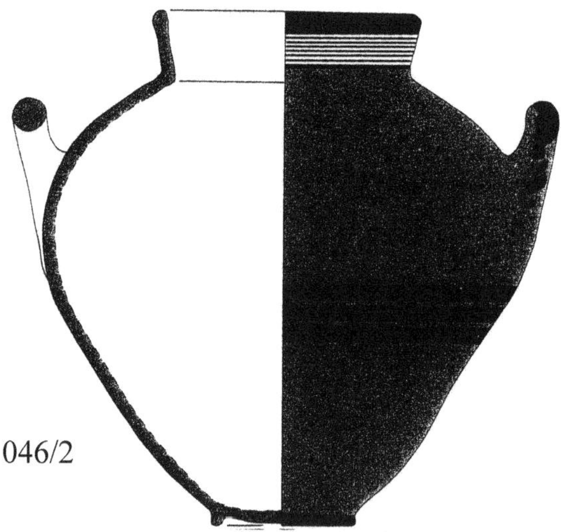

046/2

Fig. 20 Krater CLASSES Used in This Study (Scale 1:4)

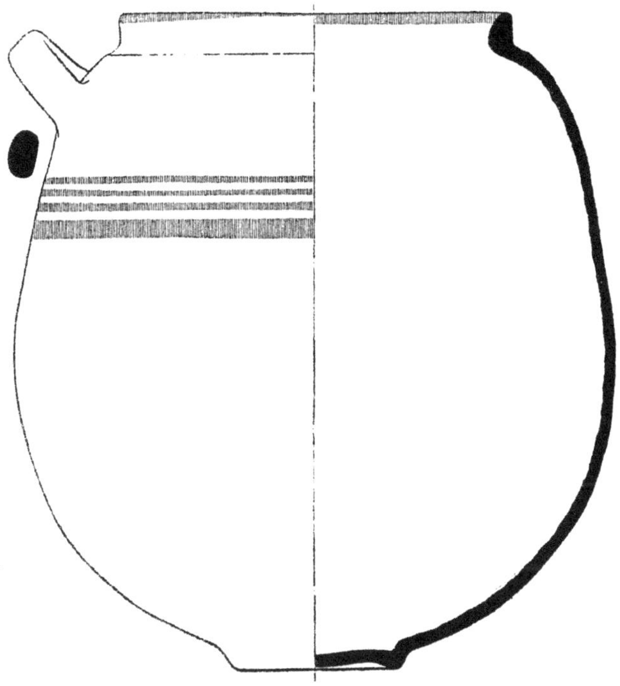

047

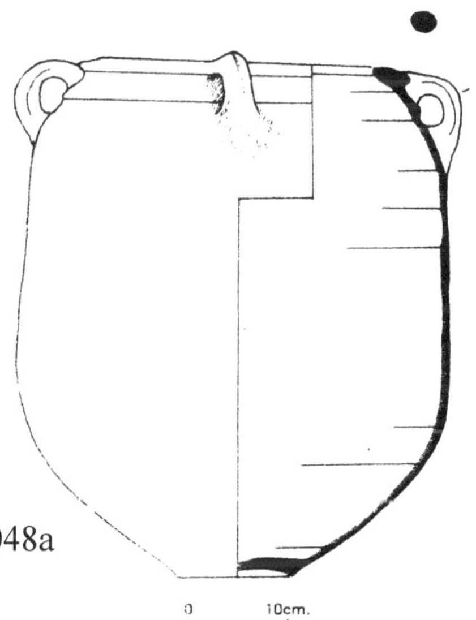

048a

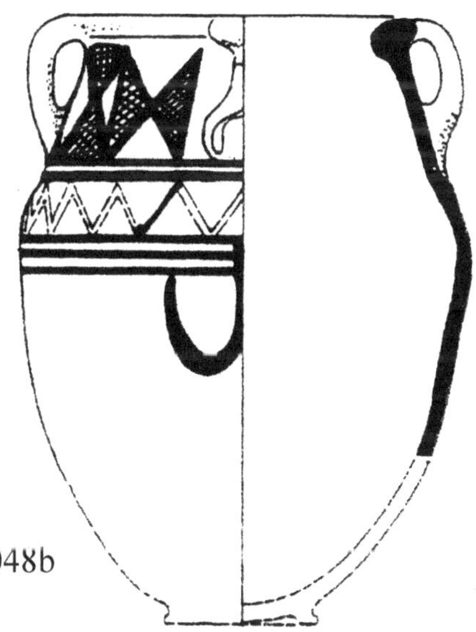

048b

Fig. 21 Krater CLASSES Used in This Study (Scale 1:4; except 048a 1:10)

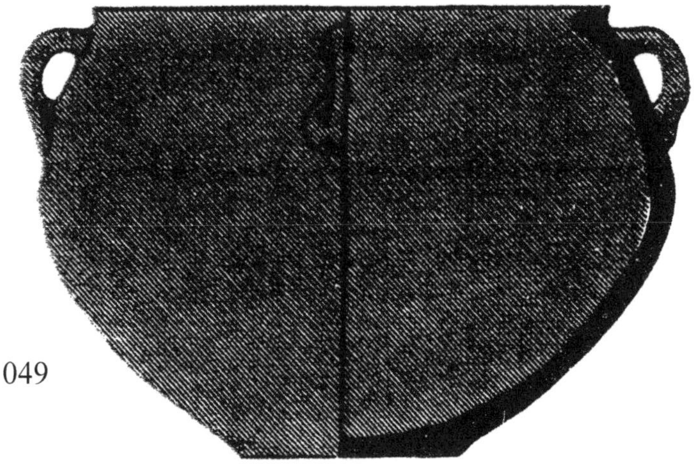

049

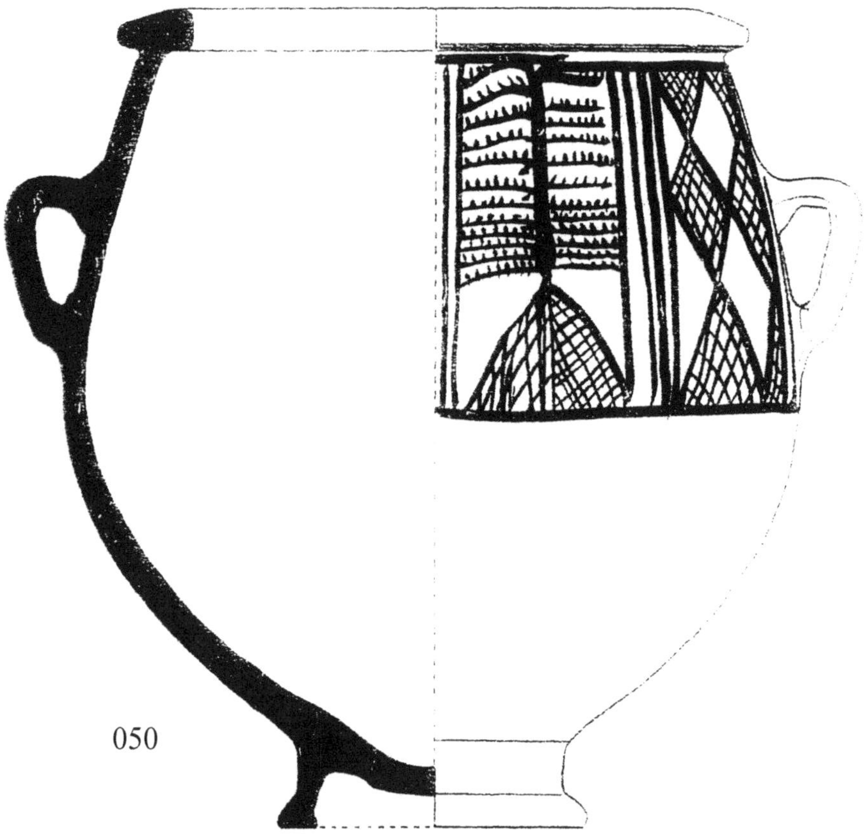

050

Fig. 22 Krater CLASSES Used in This Study (Scale 1:4)

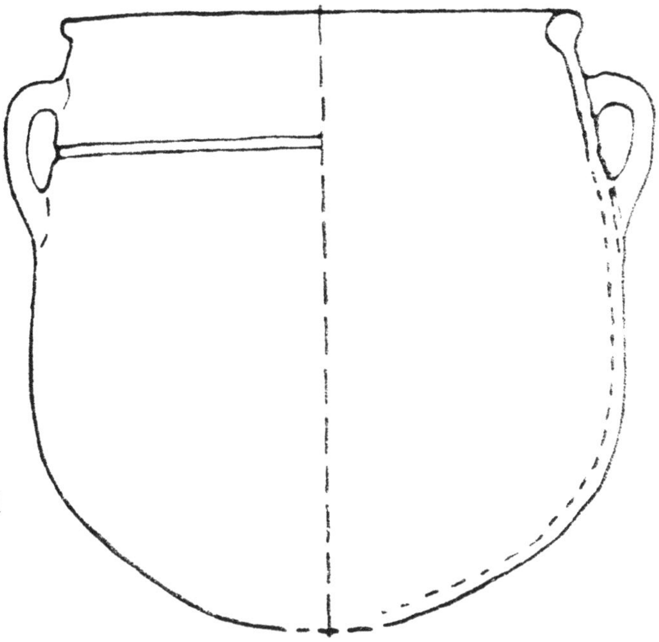

051

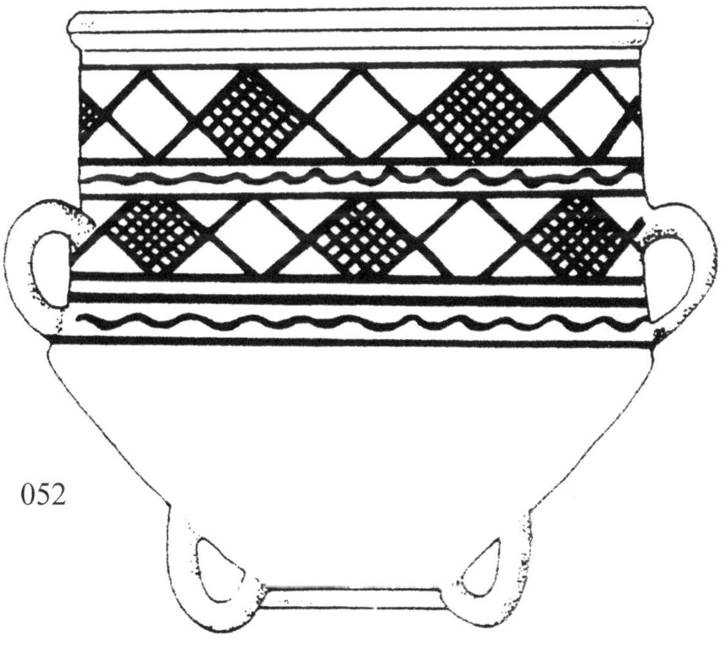

052

Fig. 23 Krater CLASSES Used in This Study (Scale 1:4)

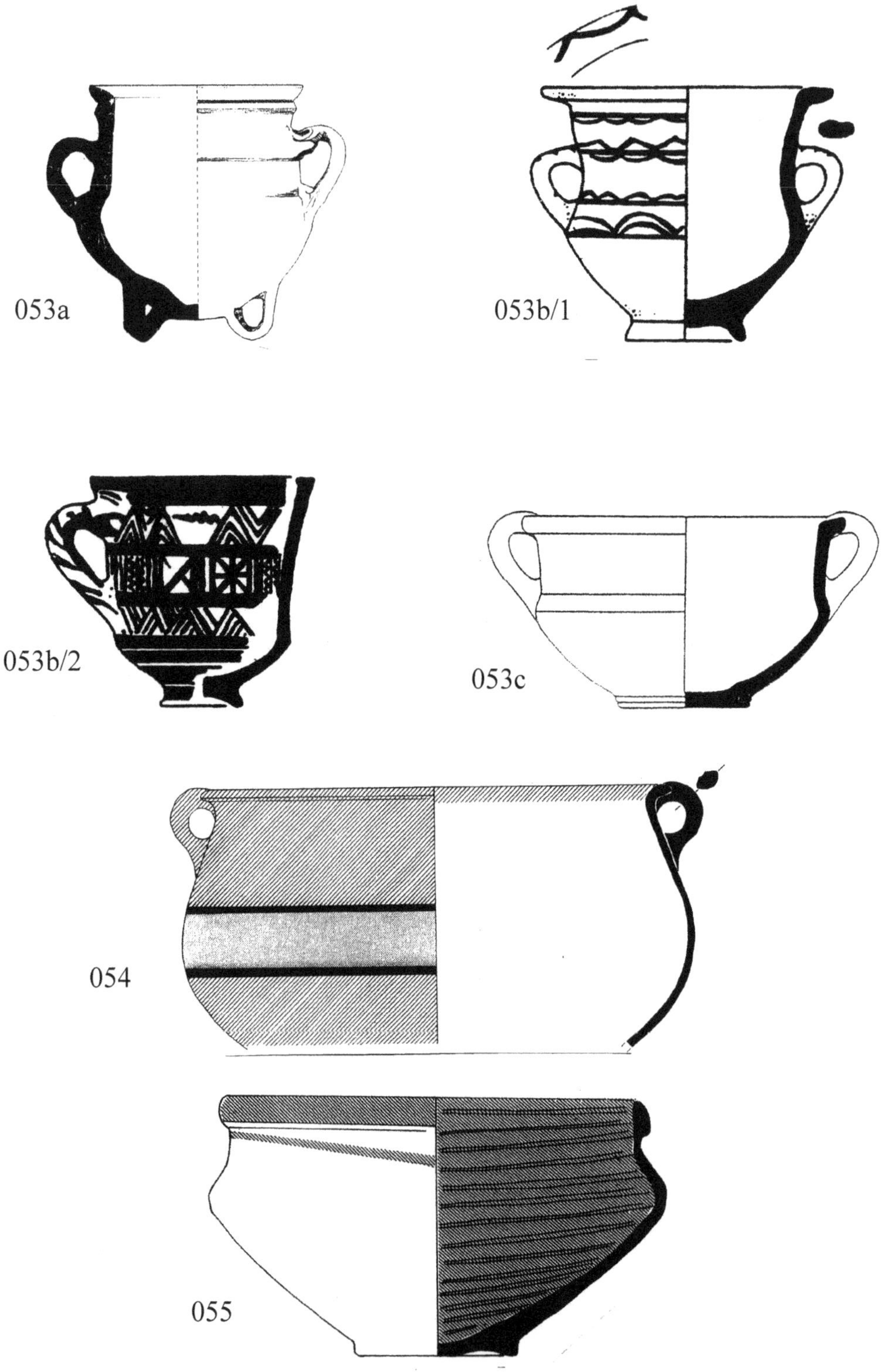

Fig. 24 Krater CLASSES Used in This Study (Scale 1:4)

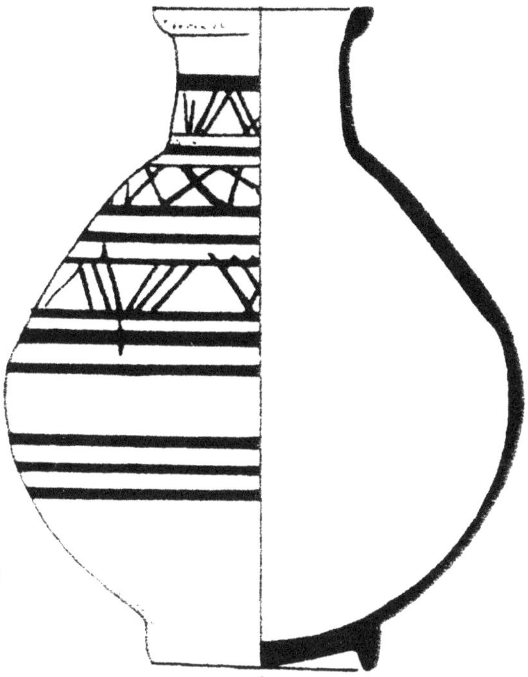

057a

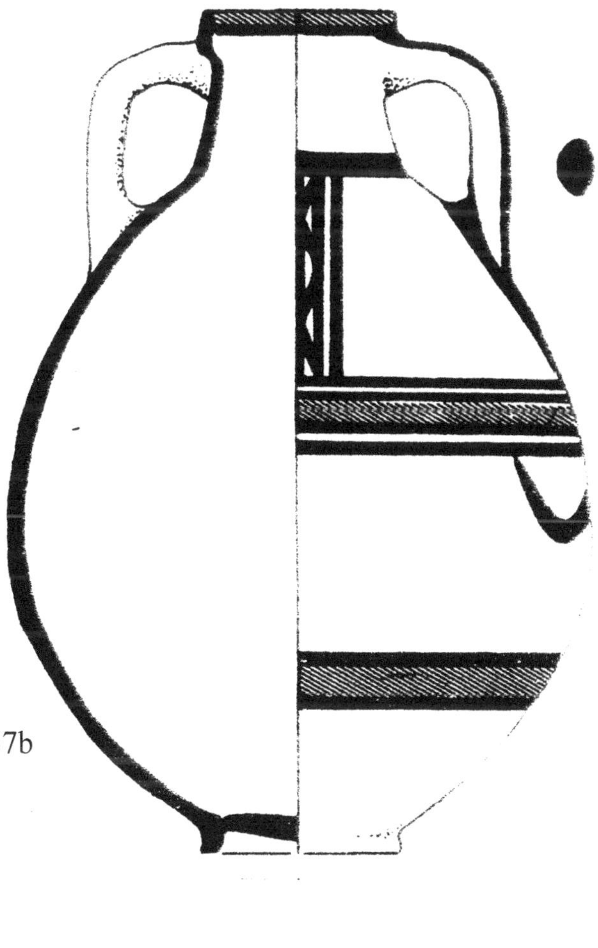

057b

Fig. 25 Urn (Storage Amphora) CLASSES Used in This Study (Scale 1:4)

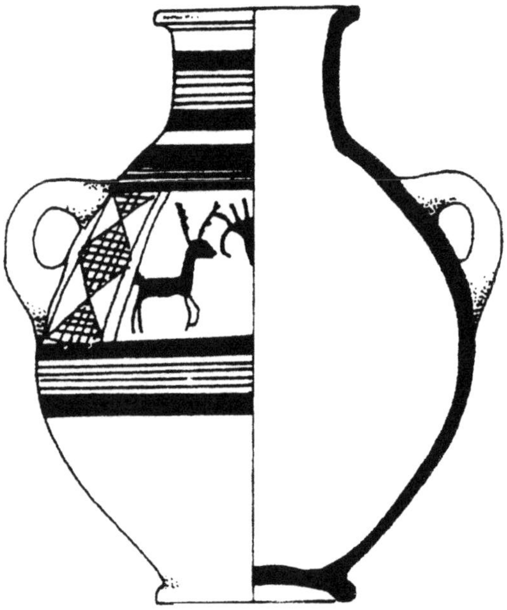

057c/1

057c/2

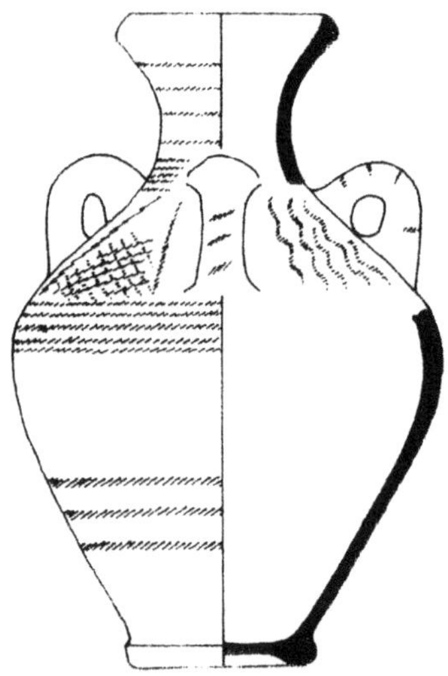

057d

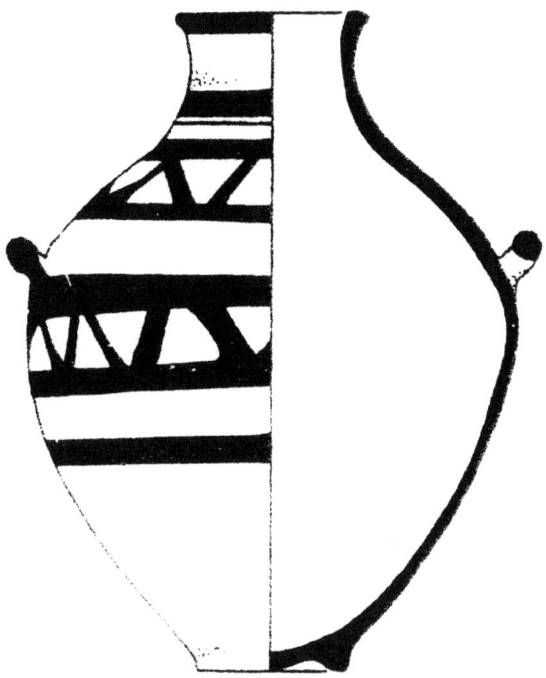

057e

Fig. 26 Urn (Storage Amphora) CLASSES Used in This Study (Scale 1:4)

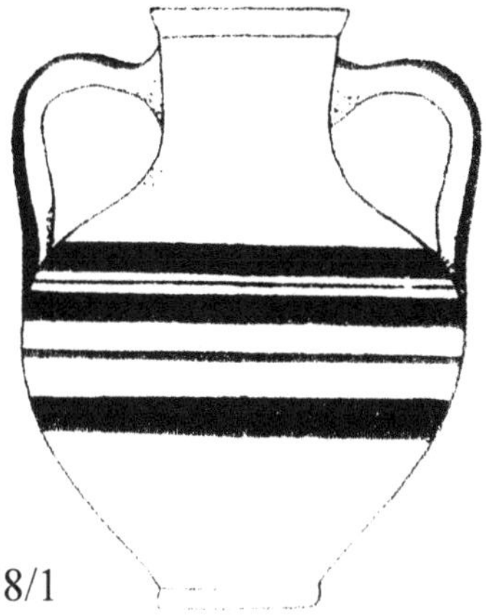

058/1

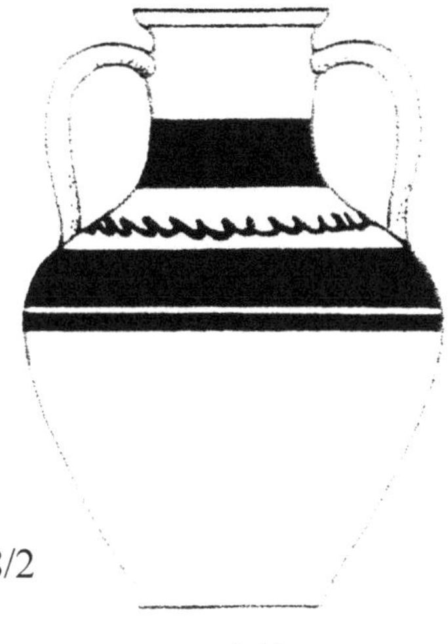

058/2

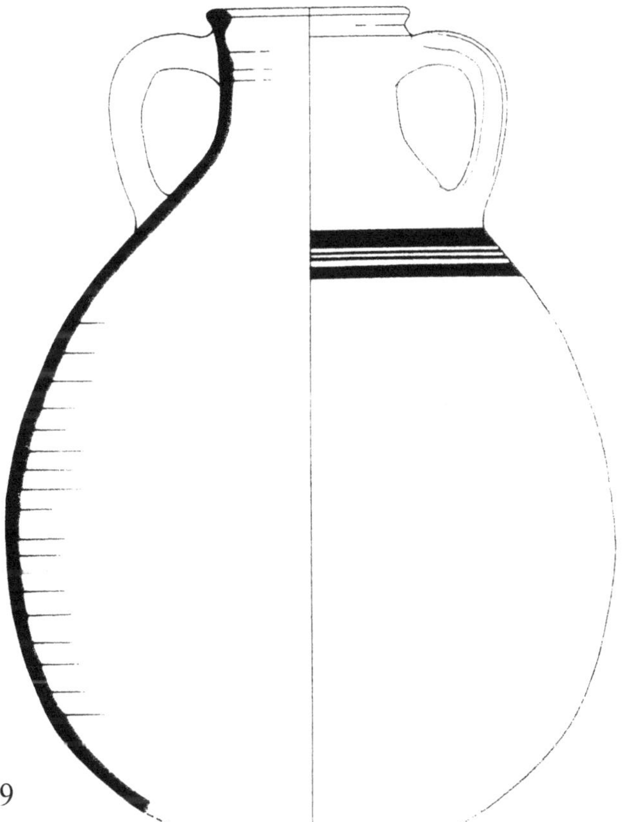

059

Fig. 27 Urn (Storage Amphora) CLASSES Used in This Study (Scale 1:4)

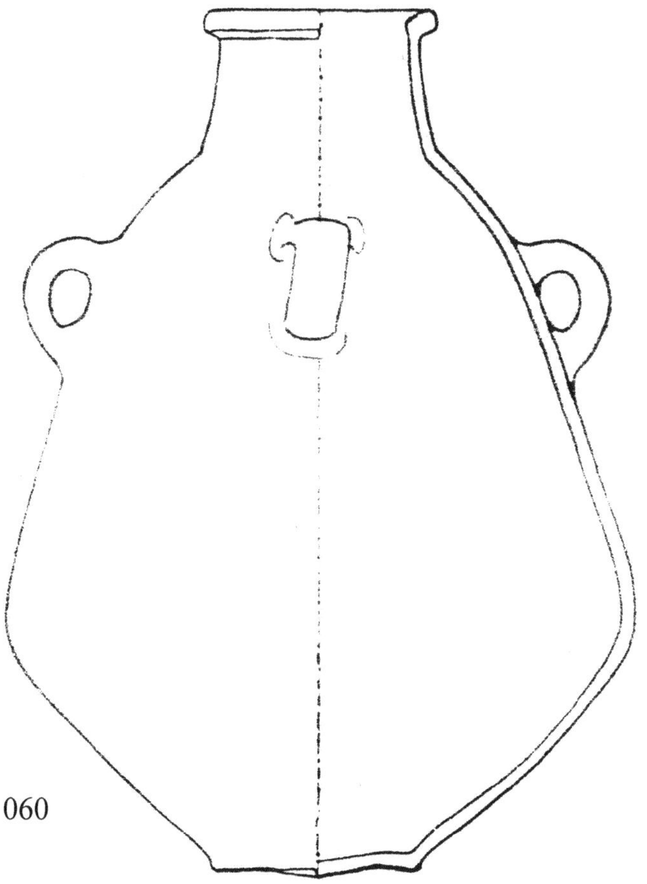

060

061

Fig. 28 Urn (Storage Amphora) CLASSES Used in This Study (Scale 1:4)

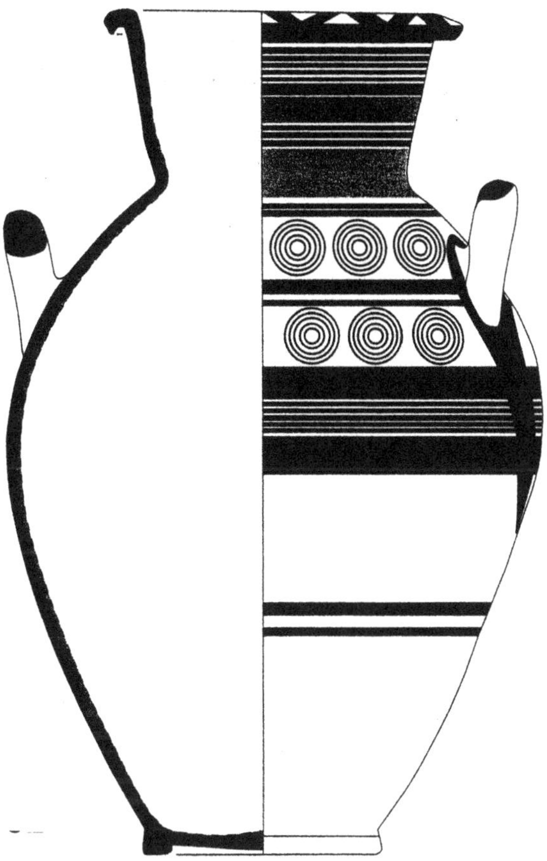

062

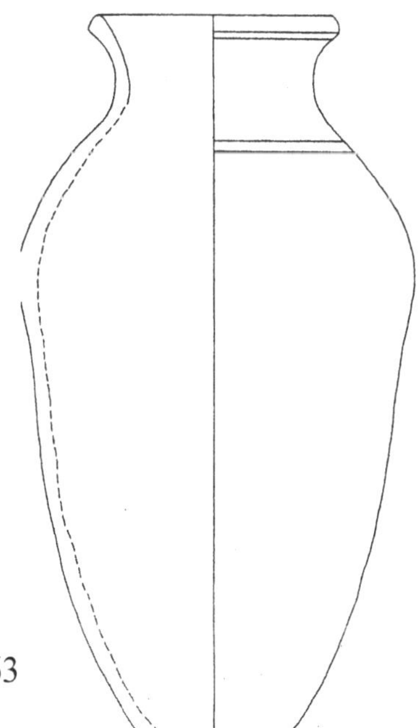

063

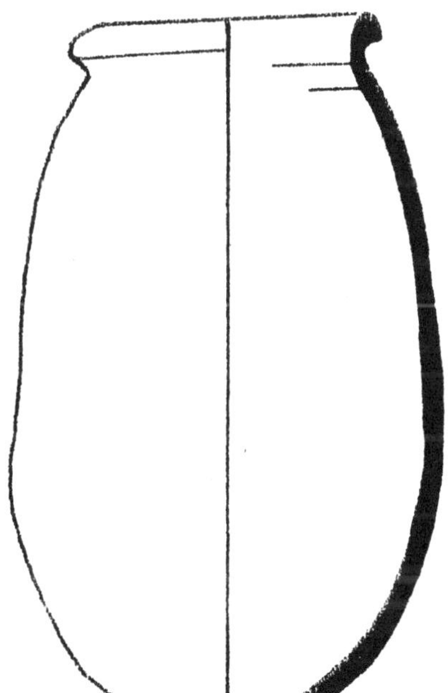

064

Fig. 29 Urn (Storage Amphora) CLASSES Used in This Study (Scale 1:4)

065

066

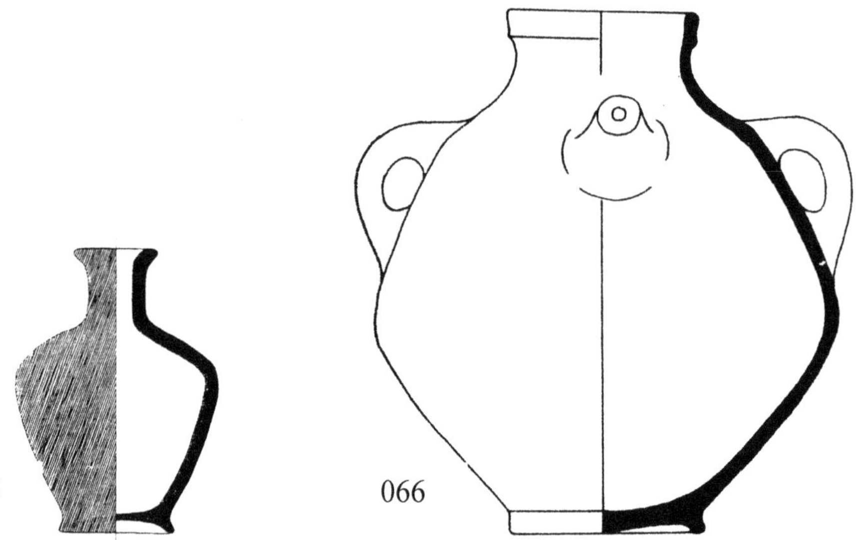

067

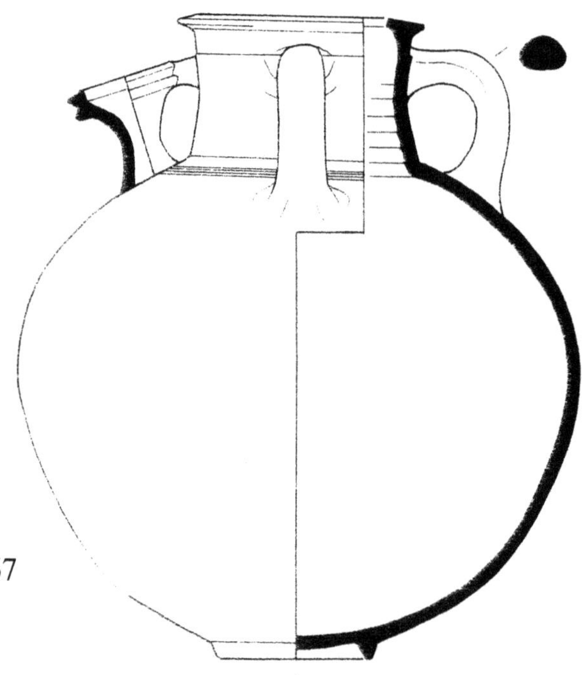

Fig. 30 Urn (Storage Amphora) CLASSES Used in This Study (Scale 1:4)

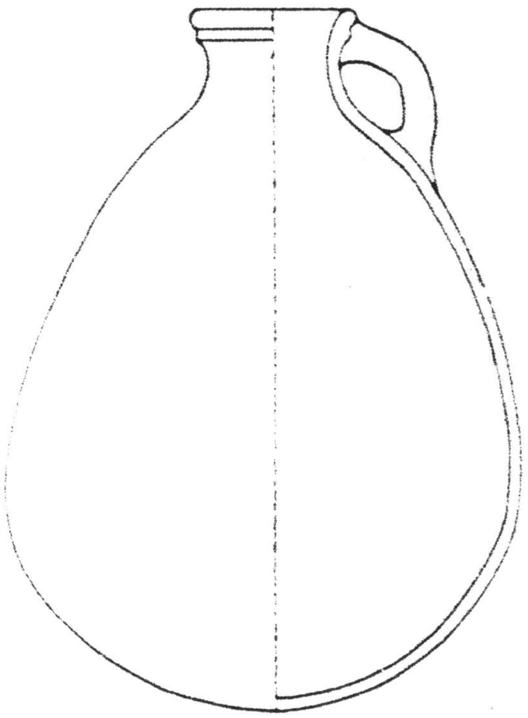

068

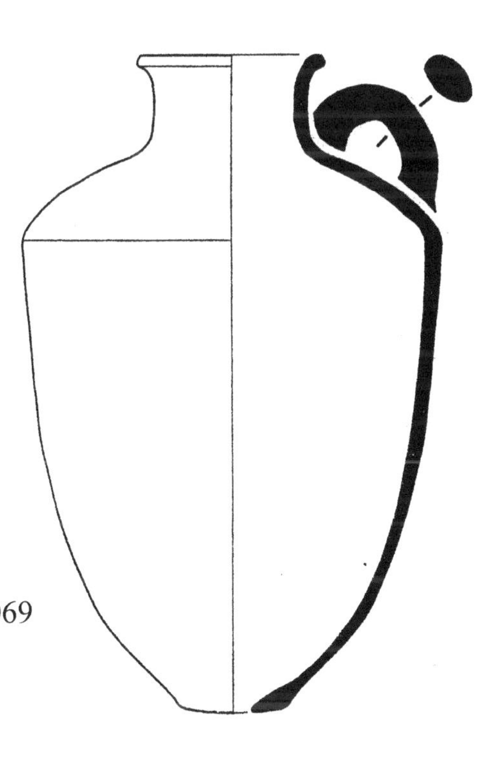

069

Fig. 31 Jug CLASSES Used in This Study (Scale 1:4)

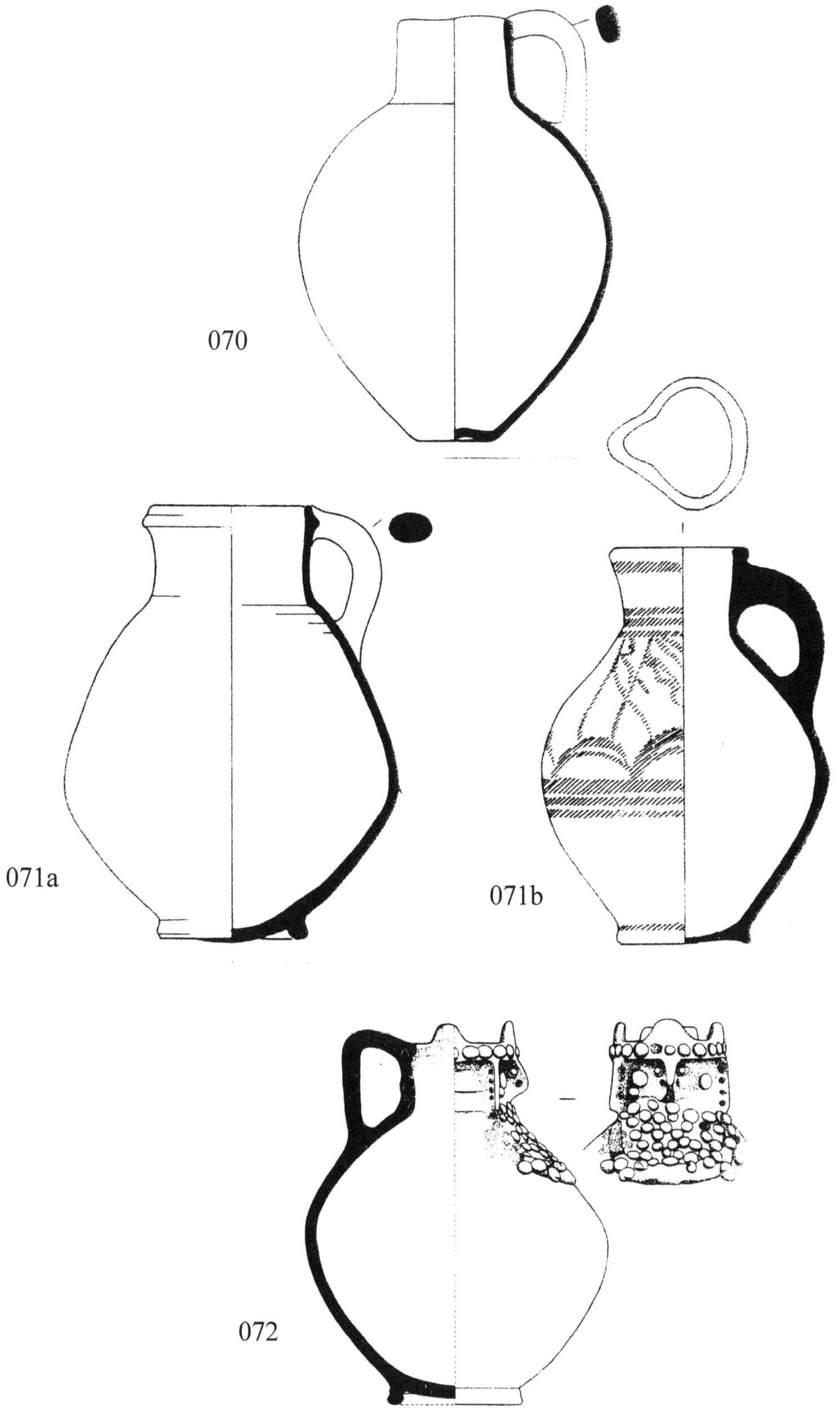

Fig. 32 Jug CLASSES Used in This Study (Scale 1:4)

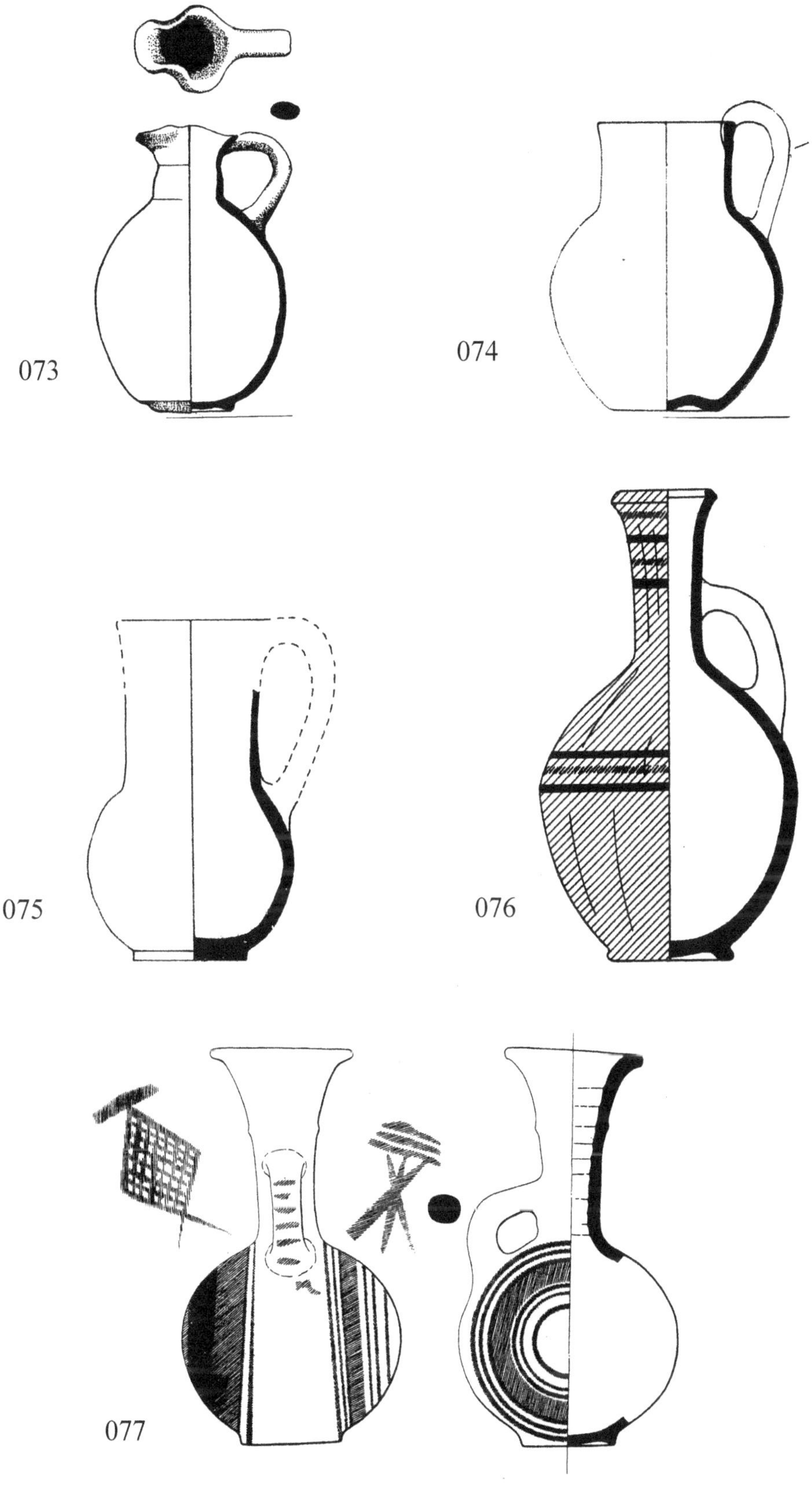

073

074

075

076

077

Fig. 33 Jug CLASSES Used in This Study (Scale 1:4)

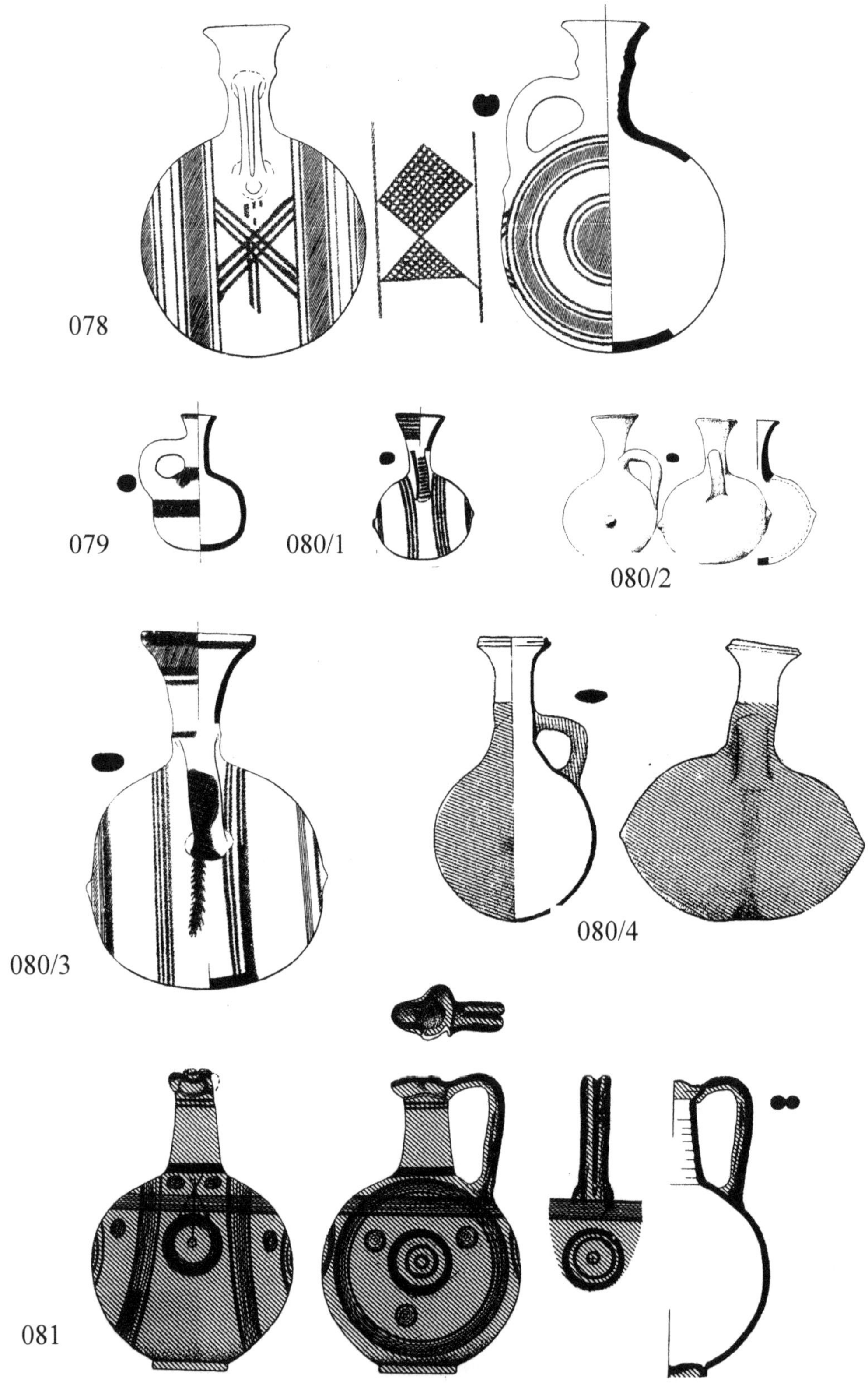

Fig. 34 Jug CLASSES Used in This Study (Scale 1:4)

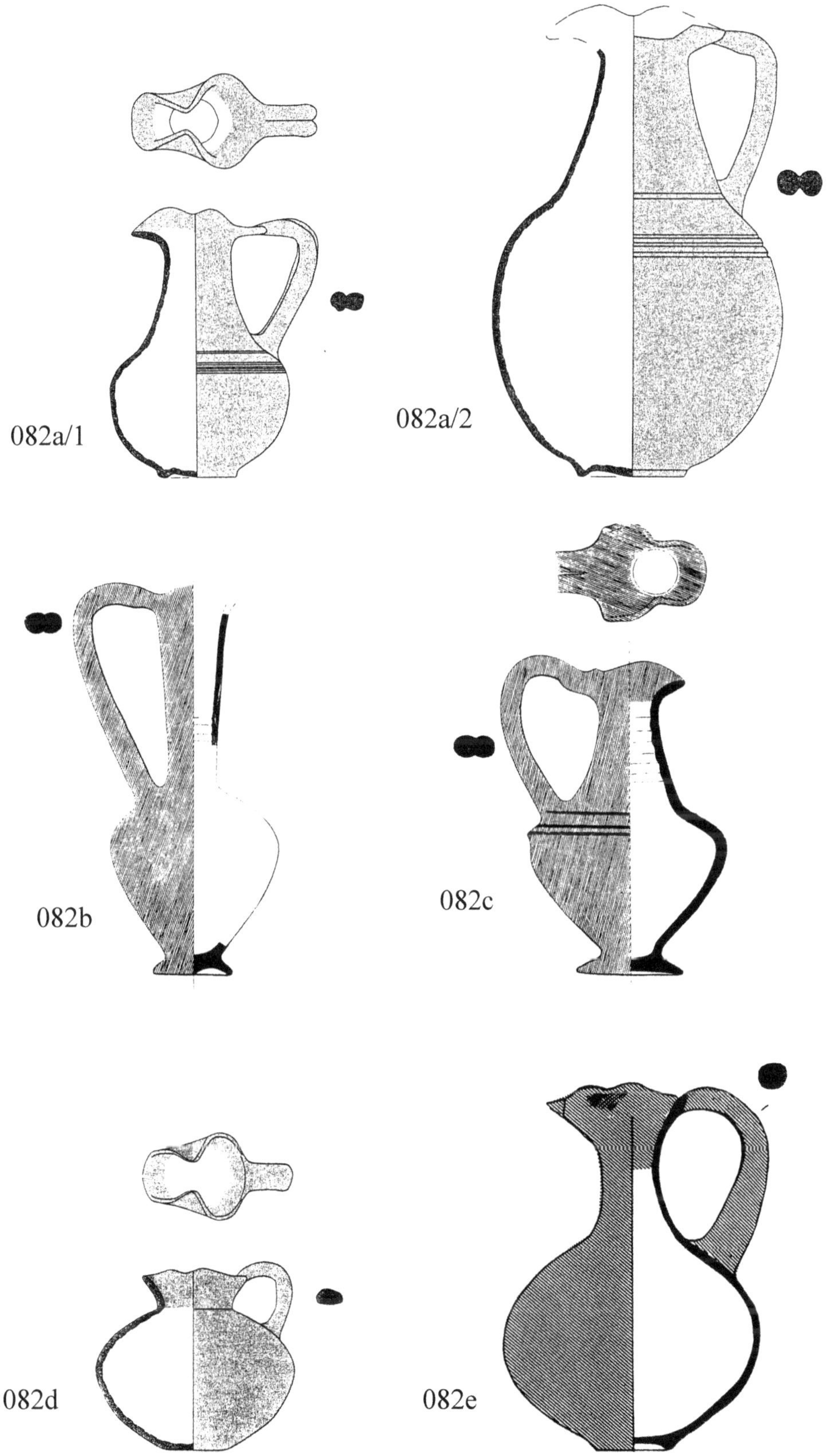

Fig. 35 Jug CLASSES Used in This Study (Scale 1:4)

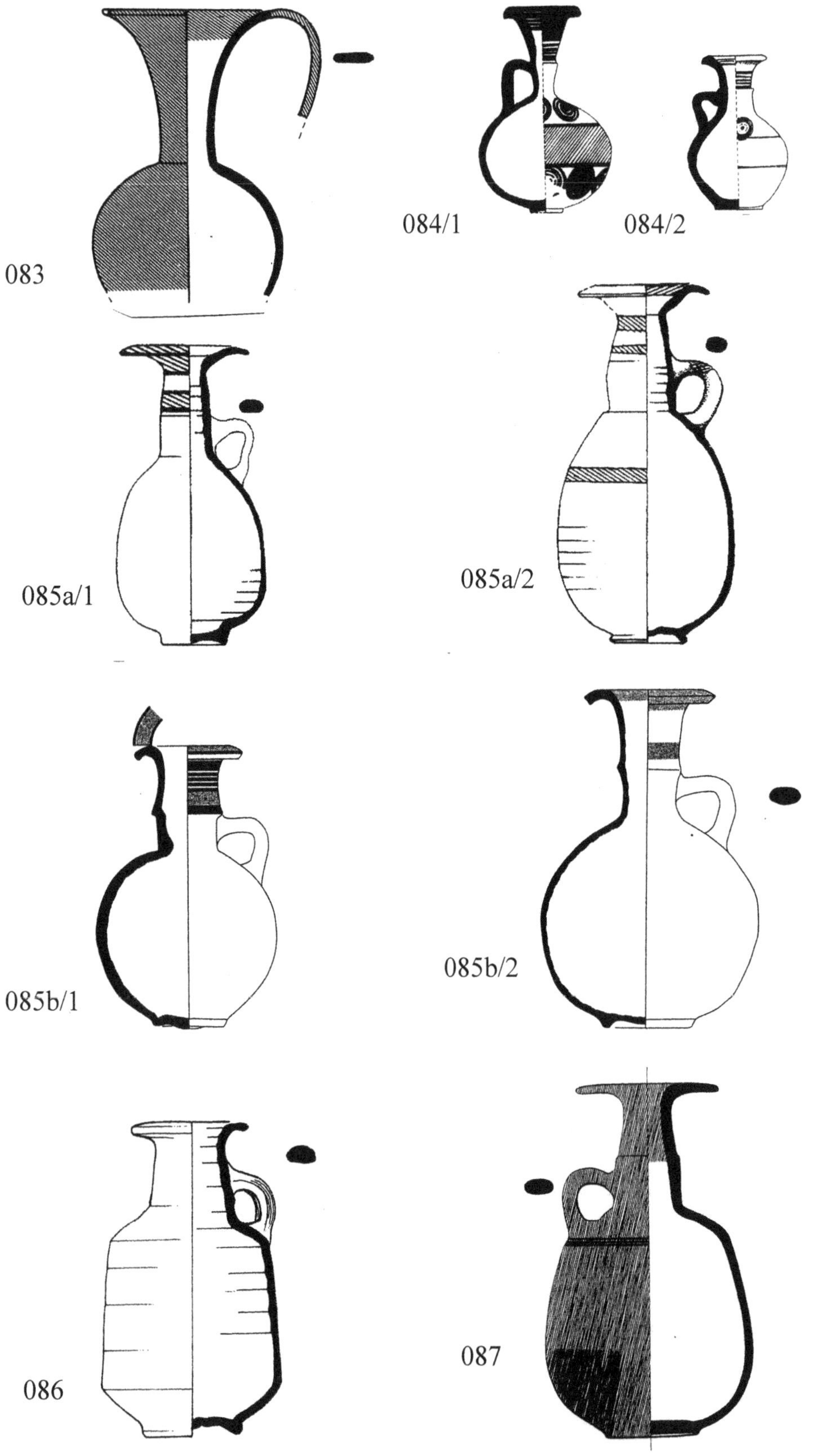

Fig. 36 Jug CLASSES Used in This Study (Scale 1:4)

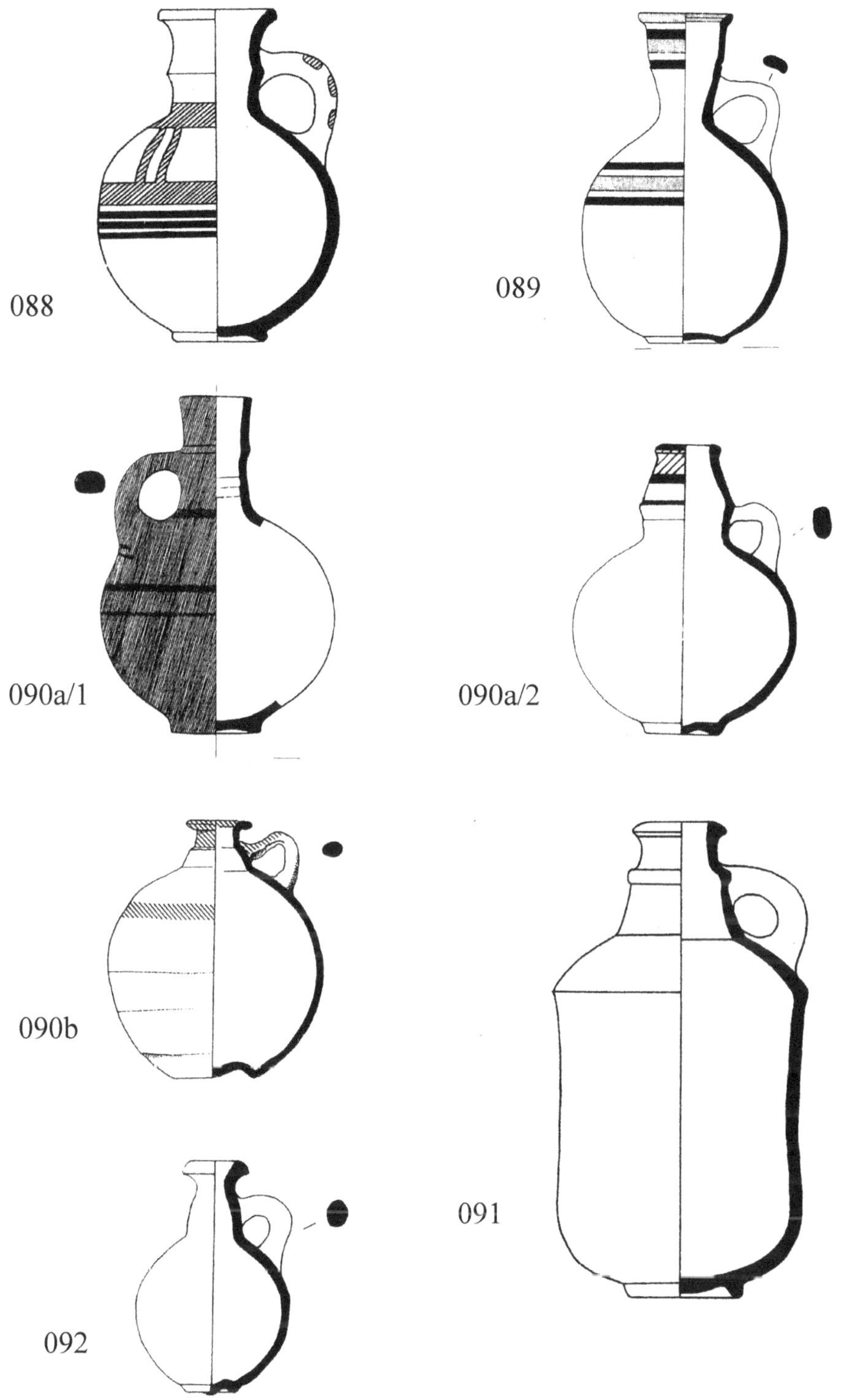

Fig. 37 Jug CLASSES Used in This Study (Scale 1:4)

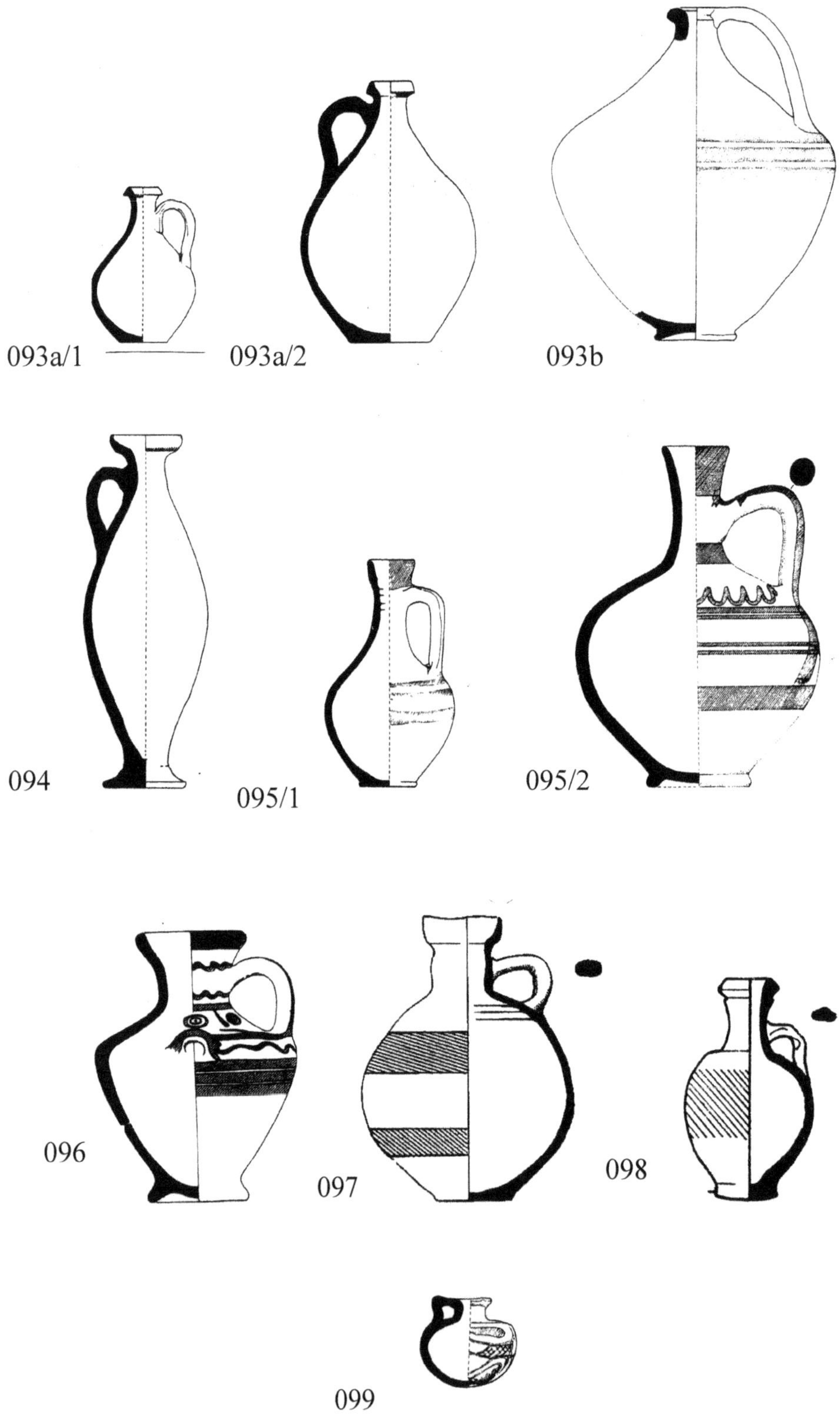

Fig. 38 Jug CLASSES Used in This Study (Scale 1:4)

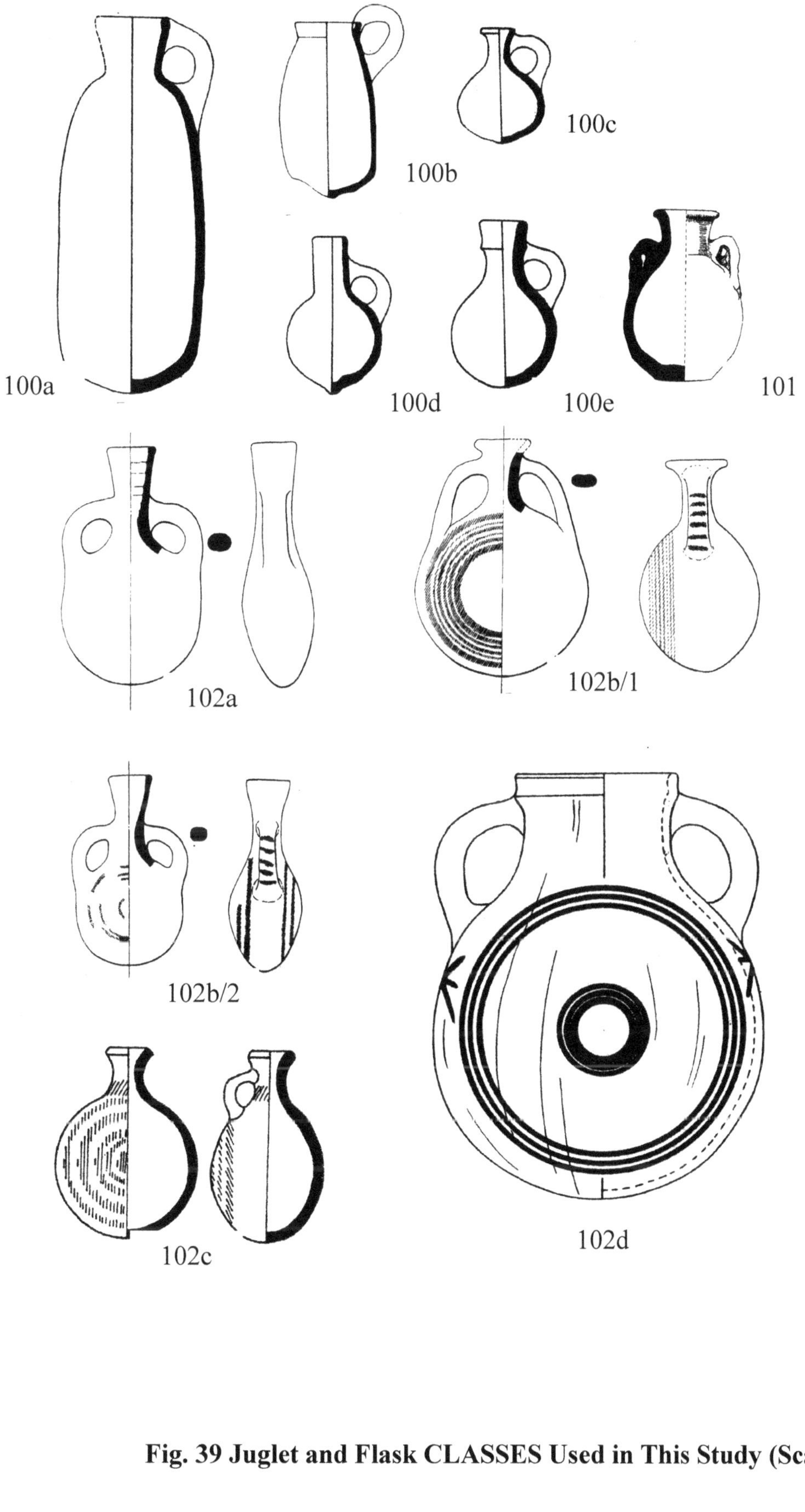

Fig. 39 Juglet and Flask CLASSES Used in This Study (Scale 1:4)

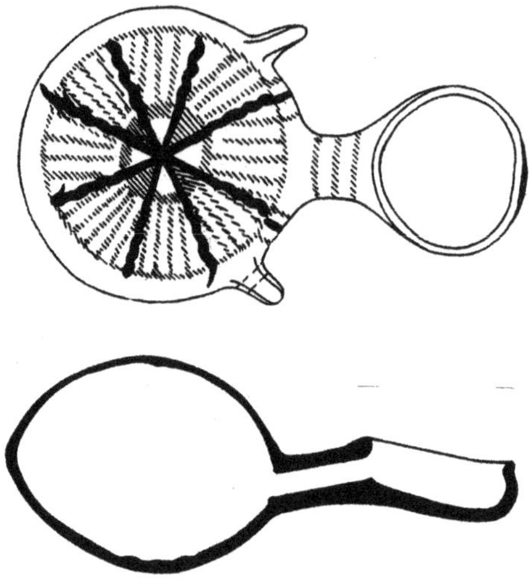

102e

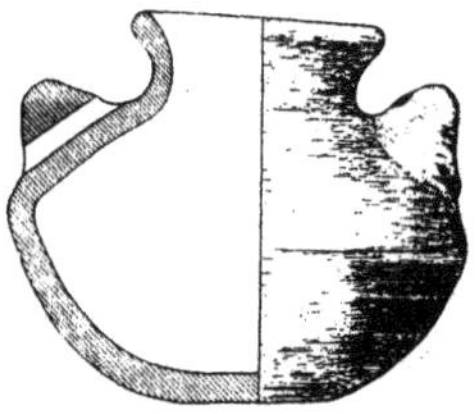

103a

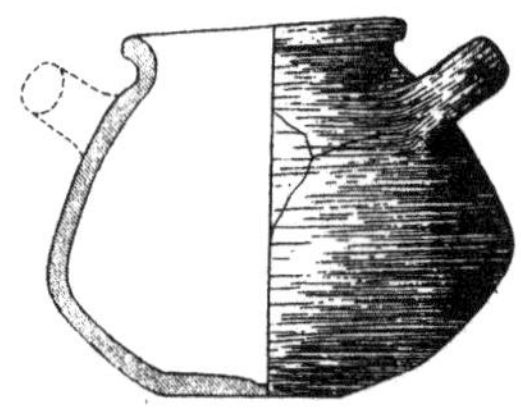

103b

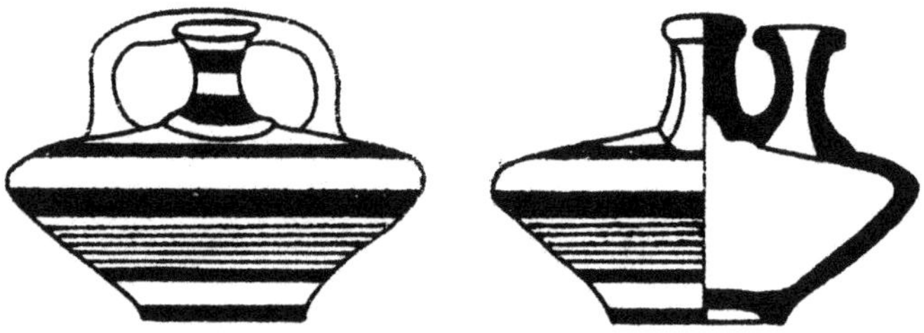

104

105

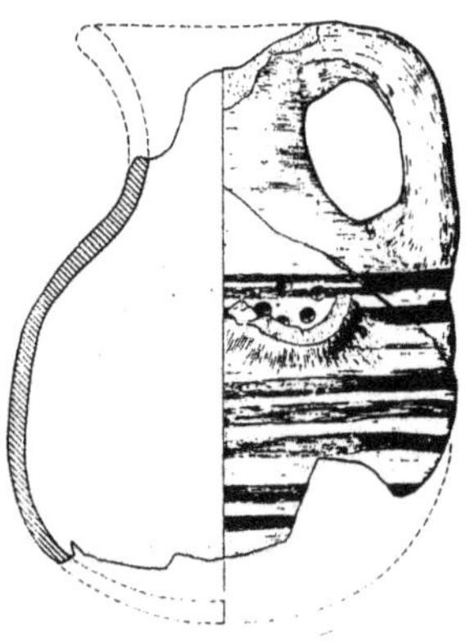

106

Fig. 40 Spouted Vessel CLASSES Used in This Study (Scale 1:4)

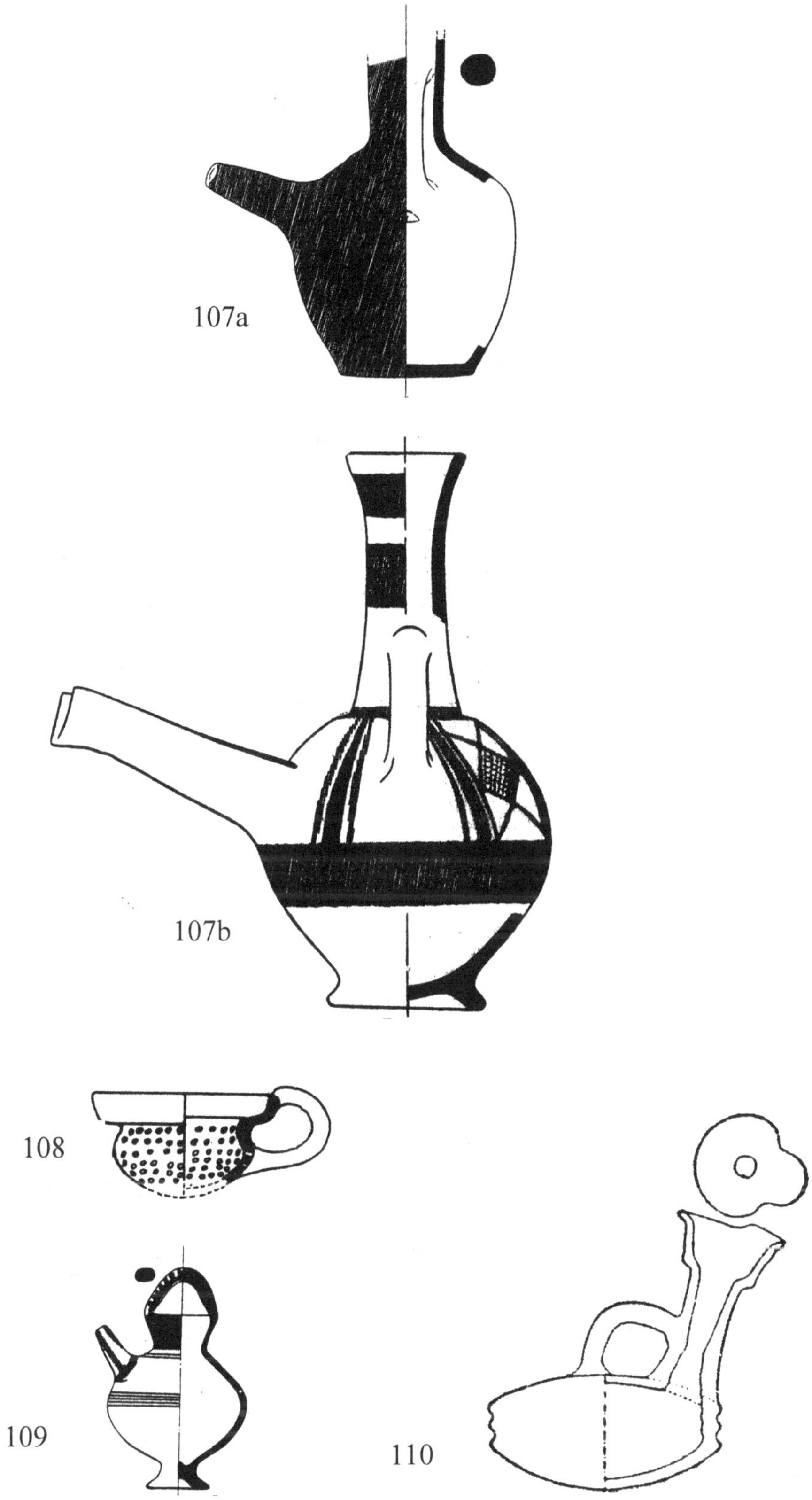

Fig. 41 Spouted Vessel CLASSES Used in This Study (Scale 1:4)

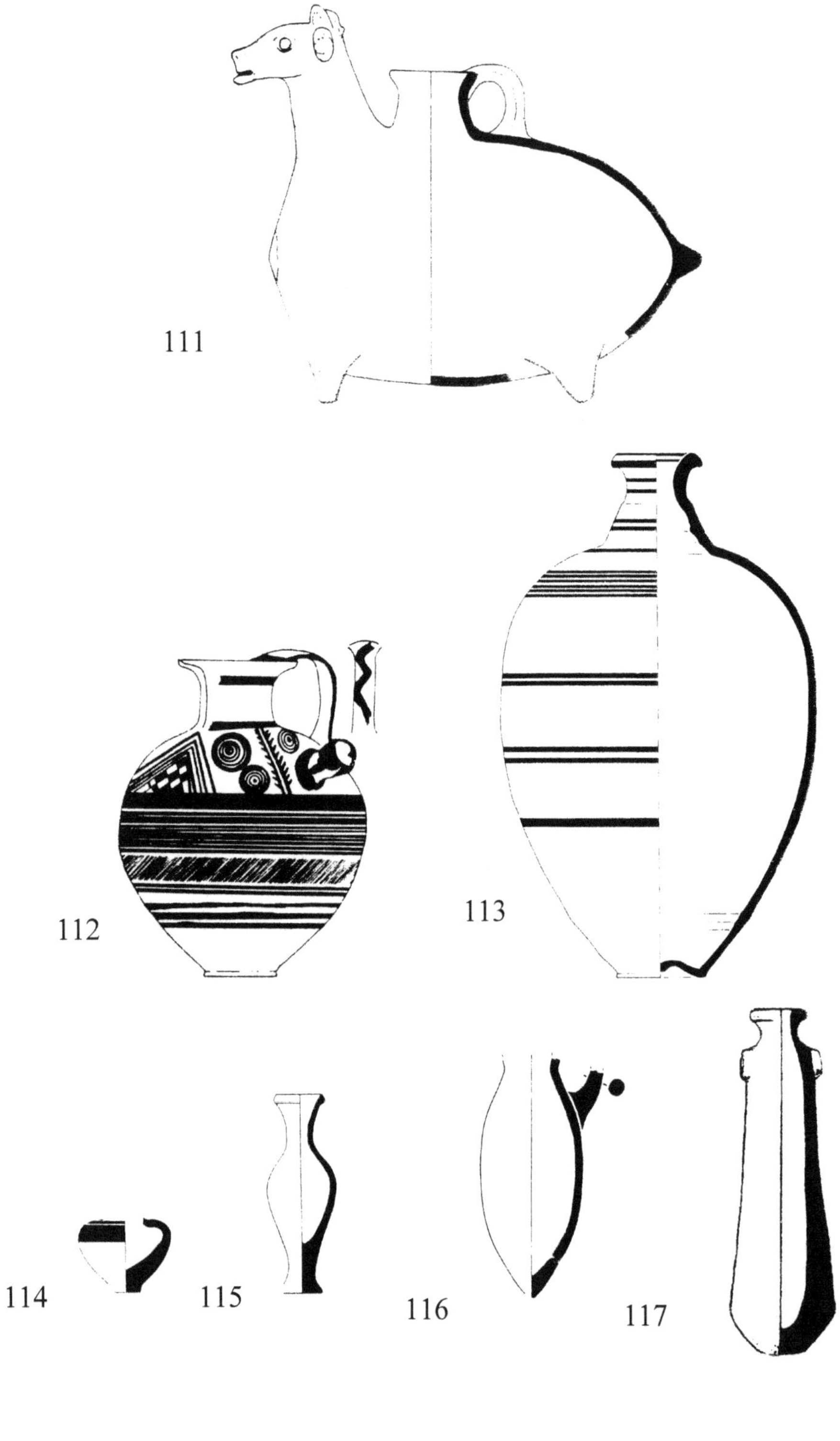

111

112

113

114

115

116

117

Fig. 42 Spouted Vessel and Bottle CLASSES Used in This Study (Scale 1:4)

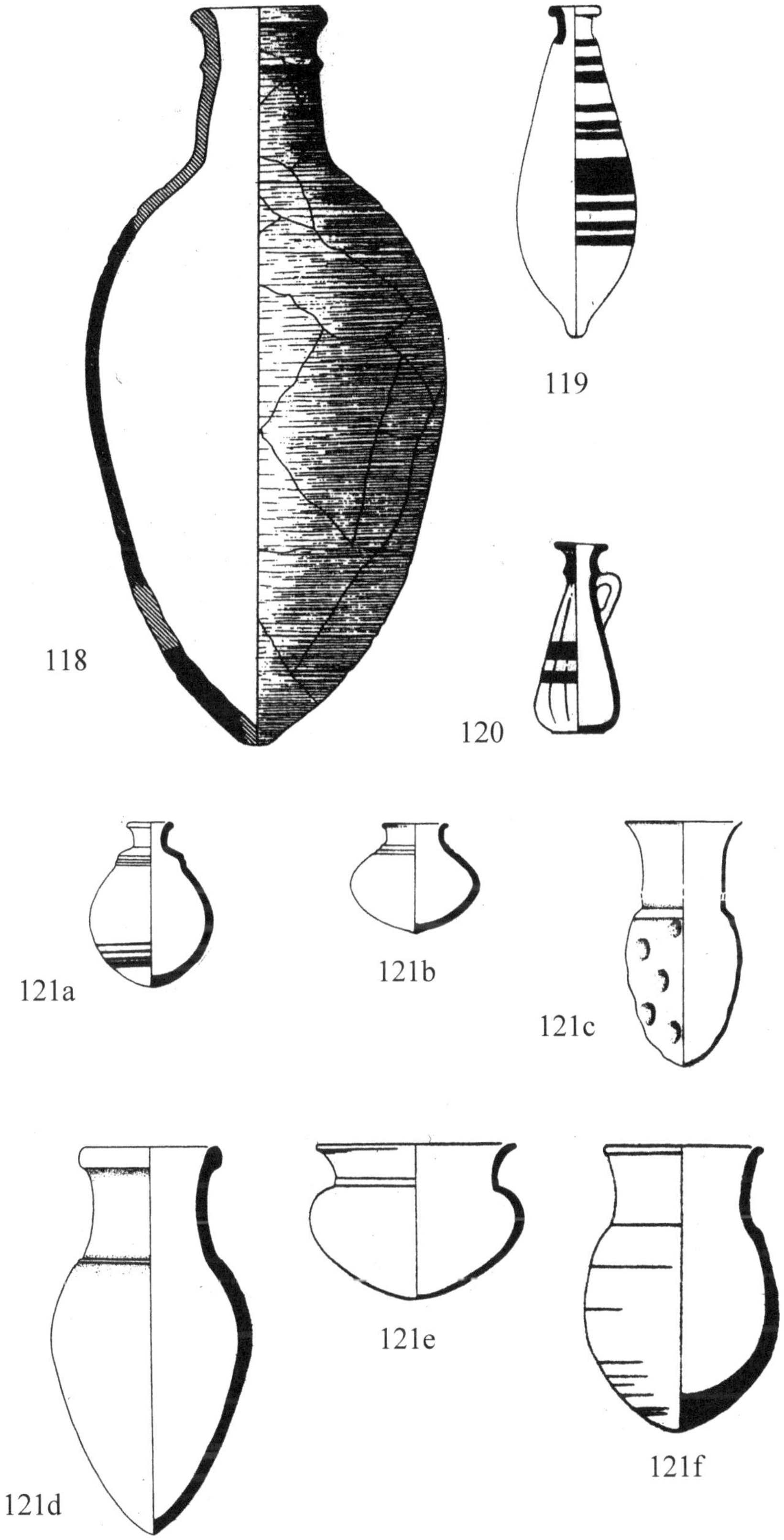

Fig. 43 Bottle CLASSES Used in This Study (Scale 1:4)

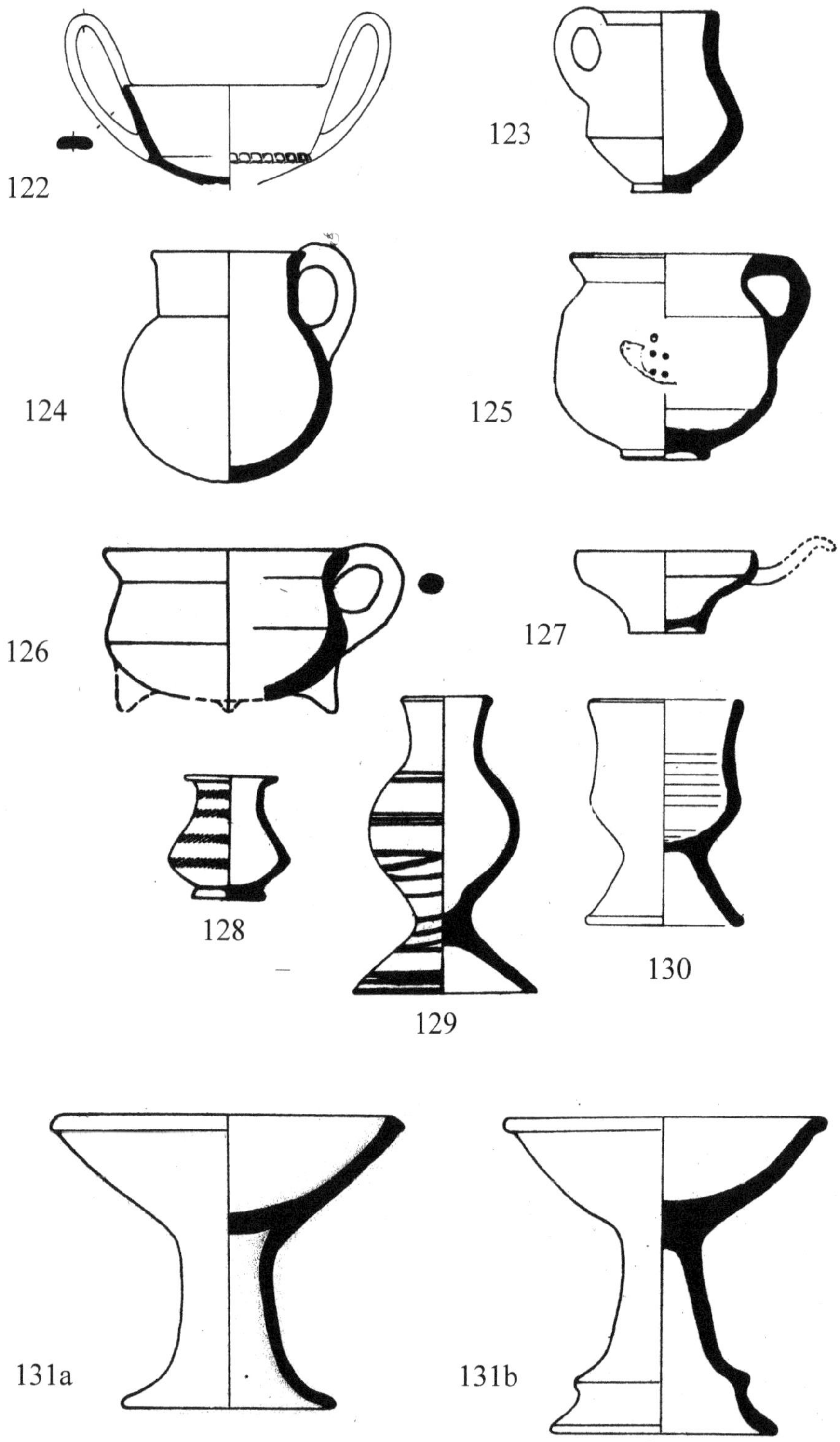

Fig. 44 Cup and Chalice CLASSES Used in This Study (Scale 1:4; except 128 scale unknown)

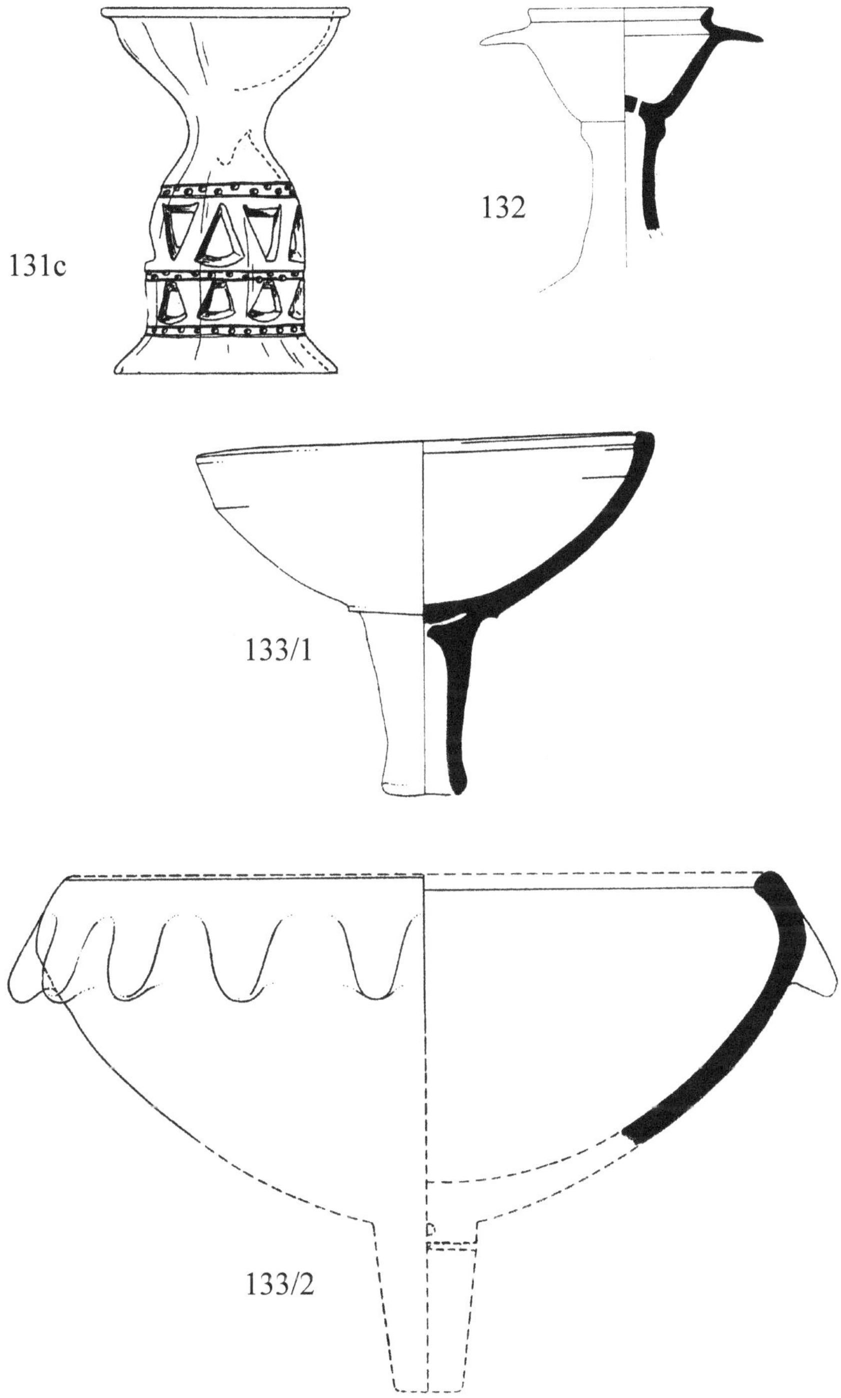

Fig. 45 Chalice CLASSES Used in This Study (Scale 1:4)

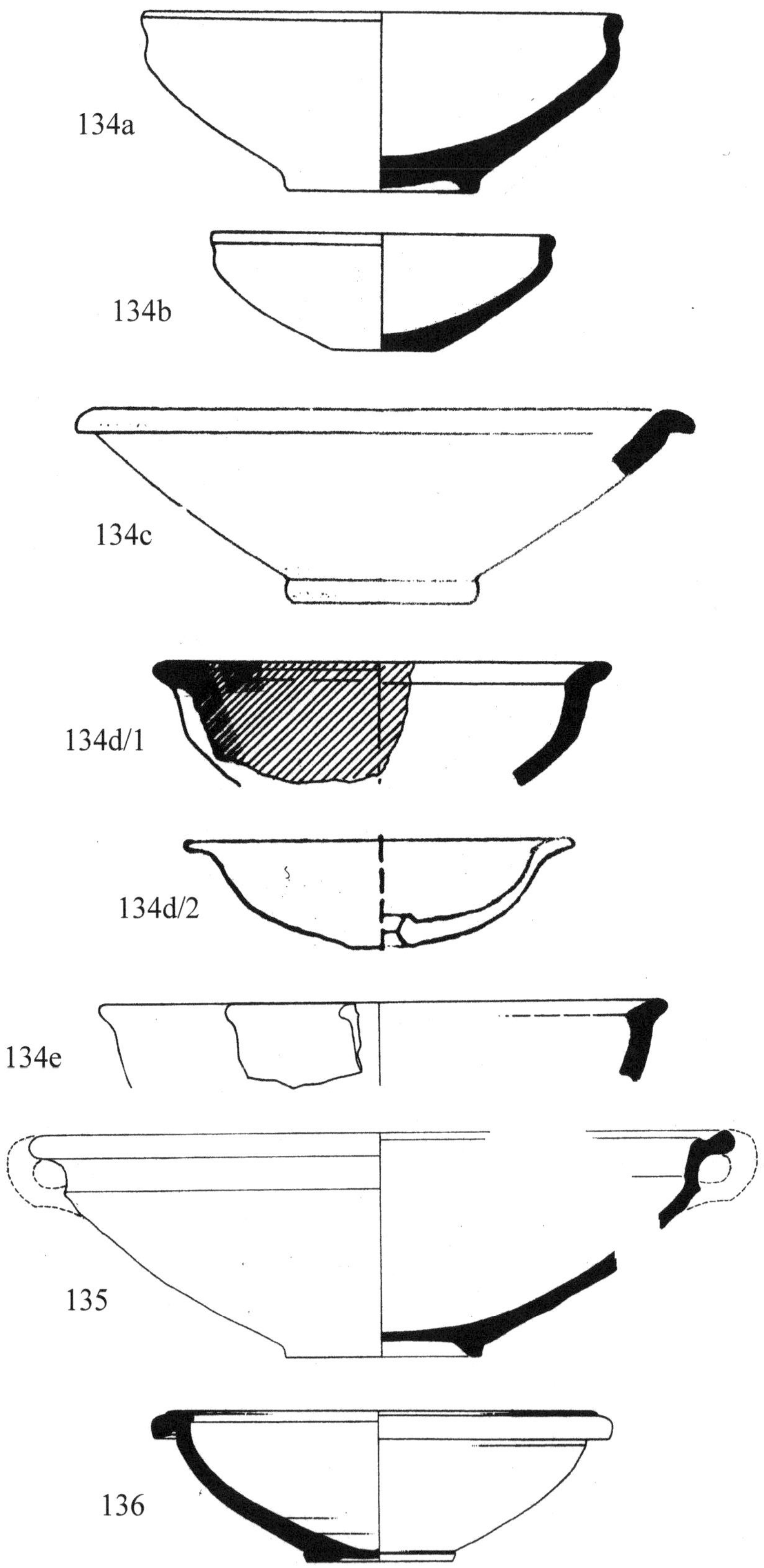

Fig. 46 Bowl CLASSES Used in This Study (Scale 1:4)

137 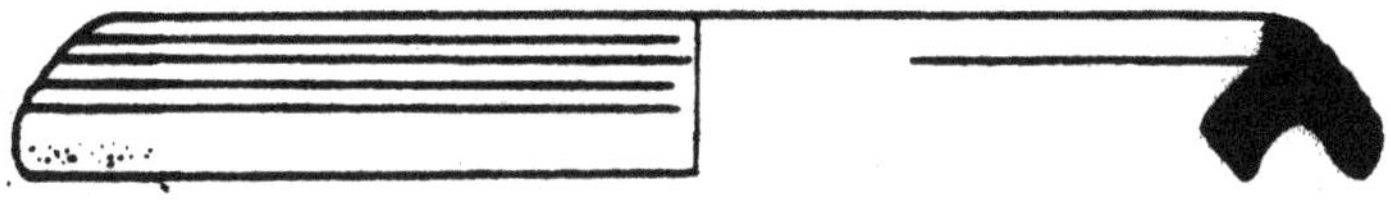

138 

139 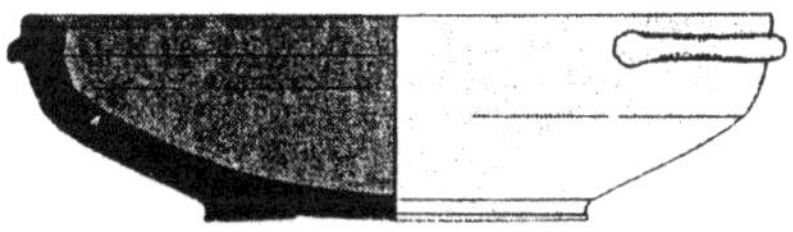

140 

141a 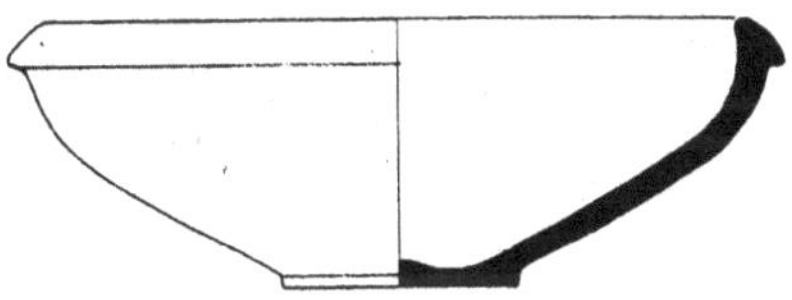

141b 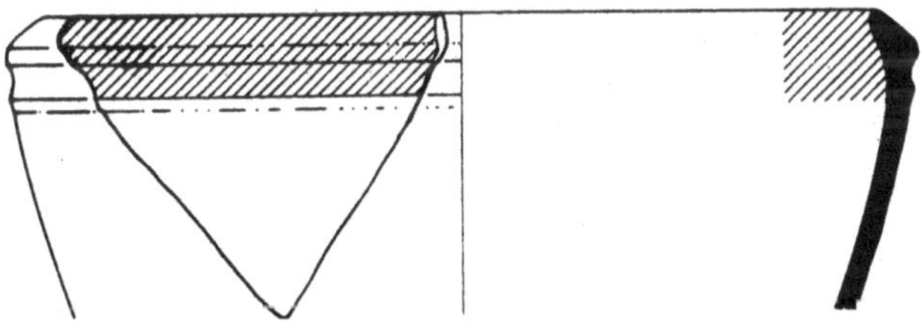

141c 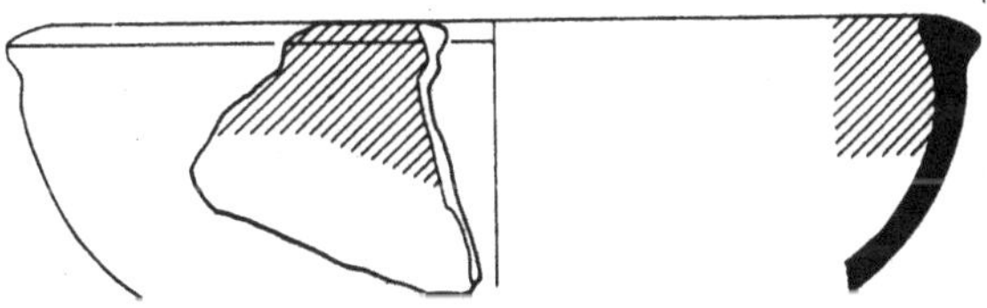

141d 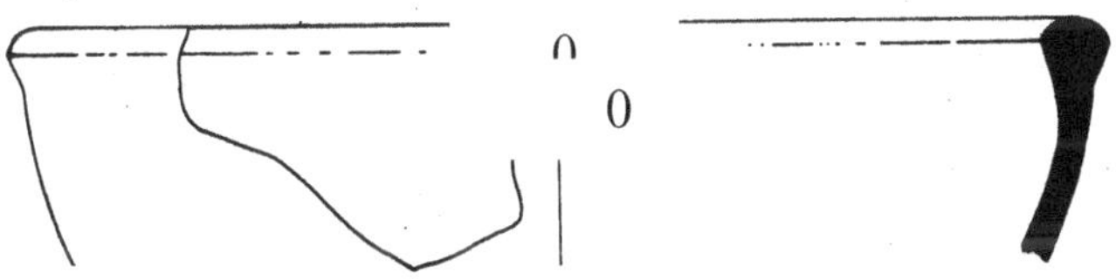

142 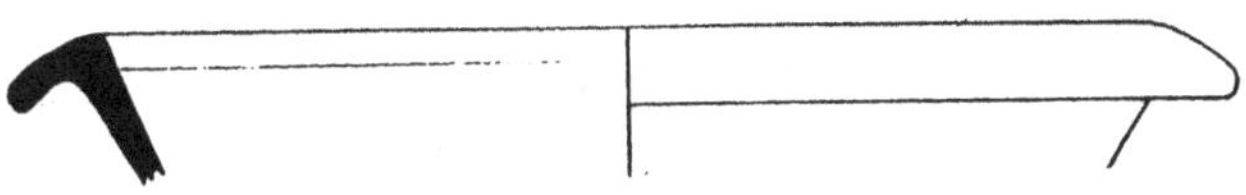

Fig. 47 Bowl CLASSES Used in This Study (Scale 1:4)

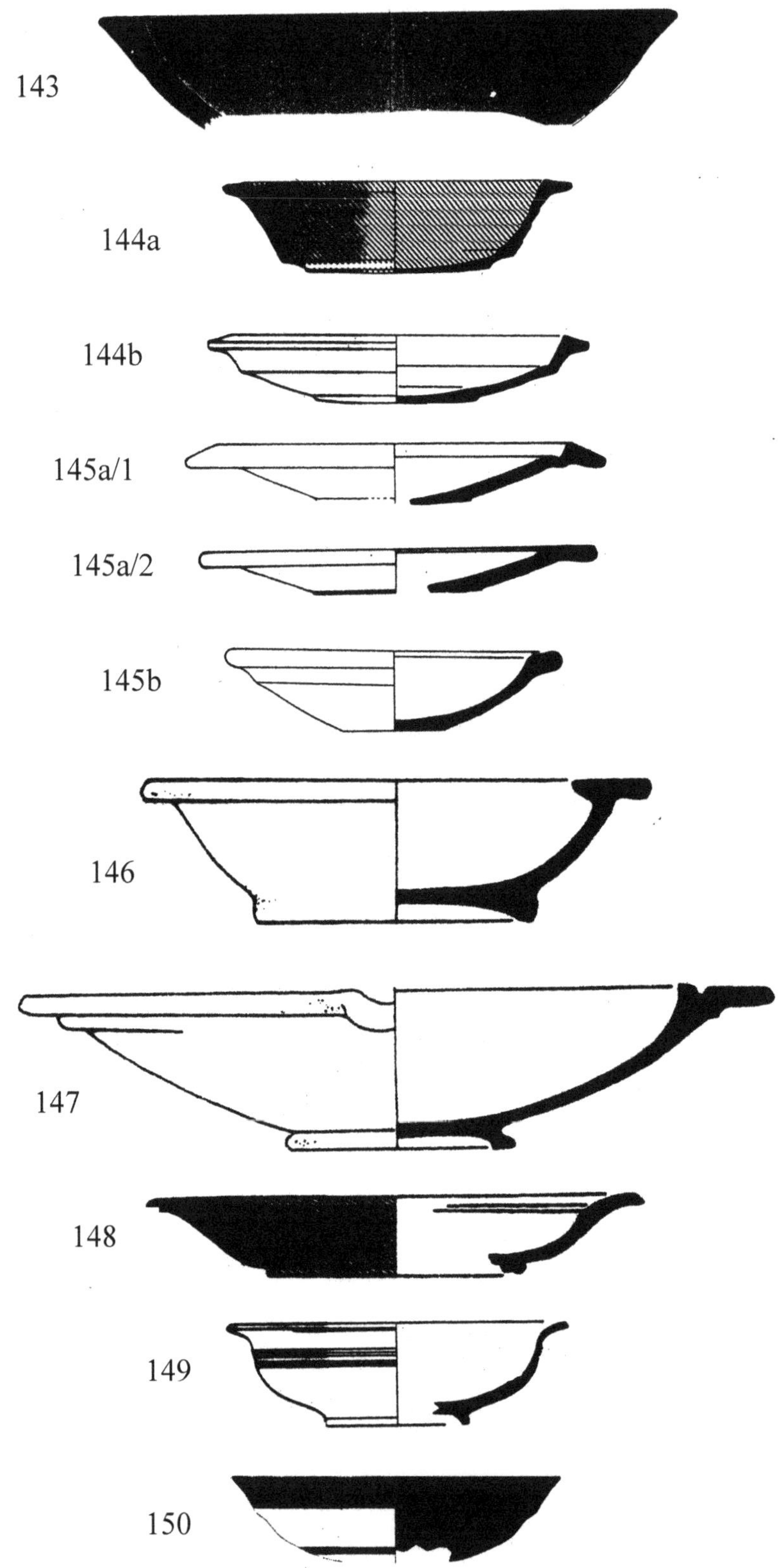

Fig. 48 Bowl CLASSES Used in This Study (Scale 1:4)

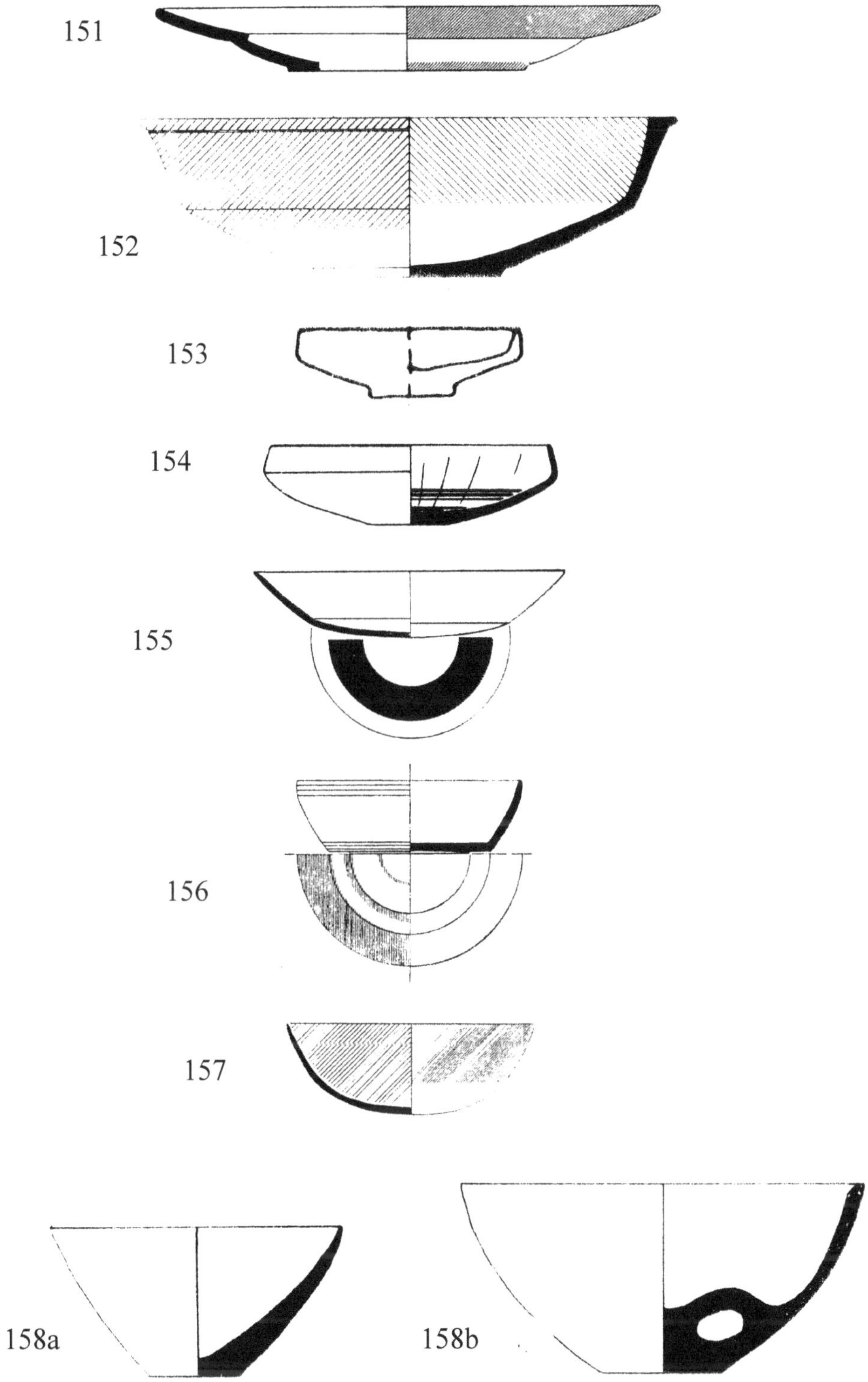

Fig. 49 Bowl CLASSES Used in This Study (Scale 1:4)

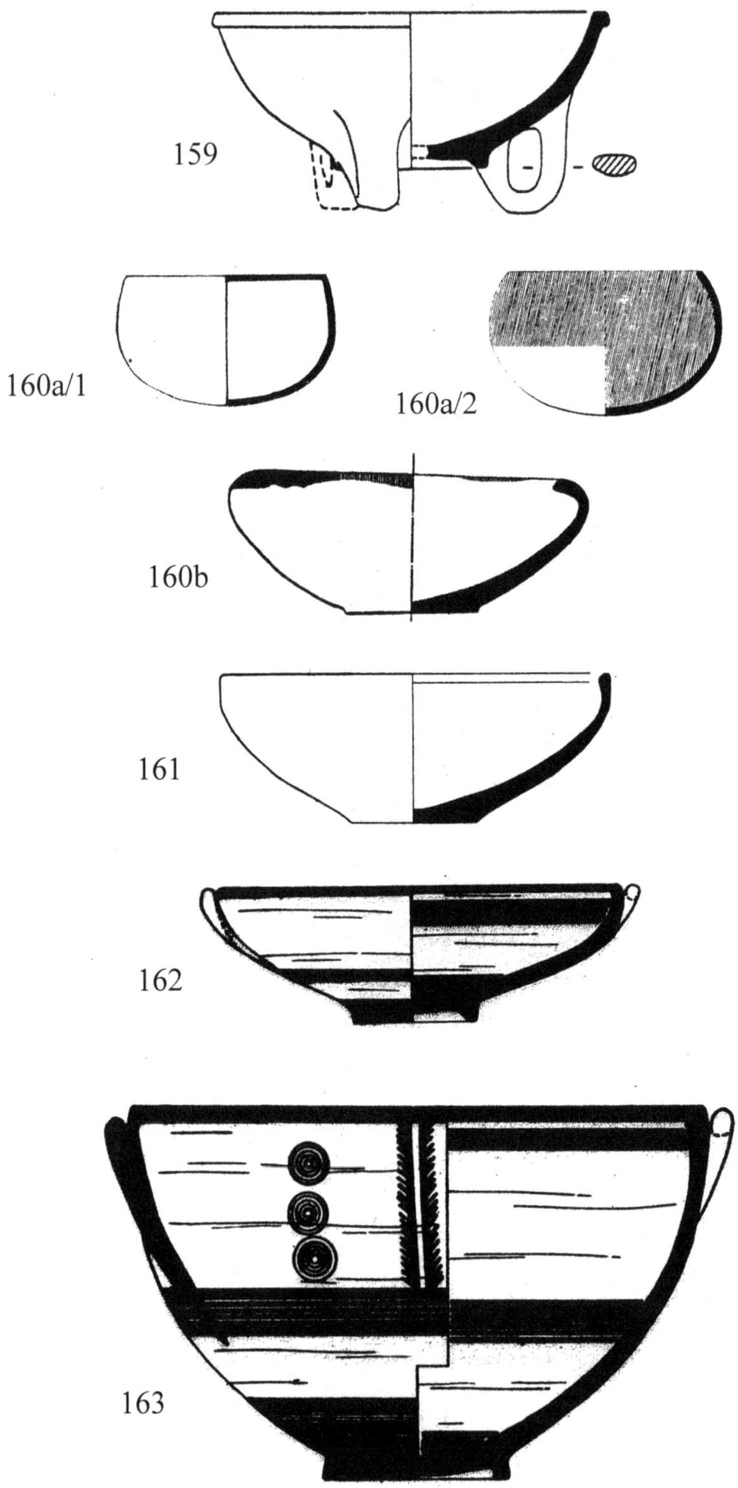

Fig. 50 Bowl CLASSES Used in This Study (Scale 1:4)

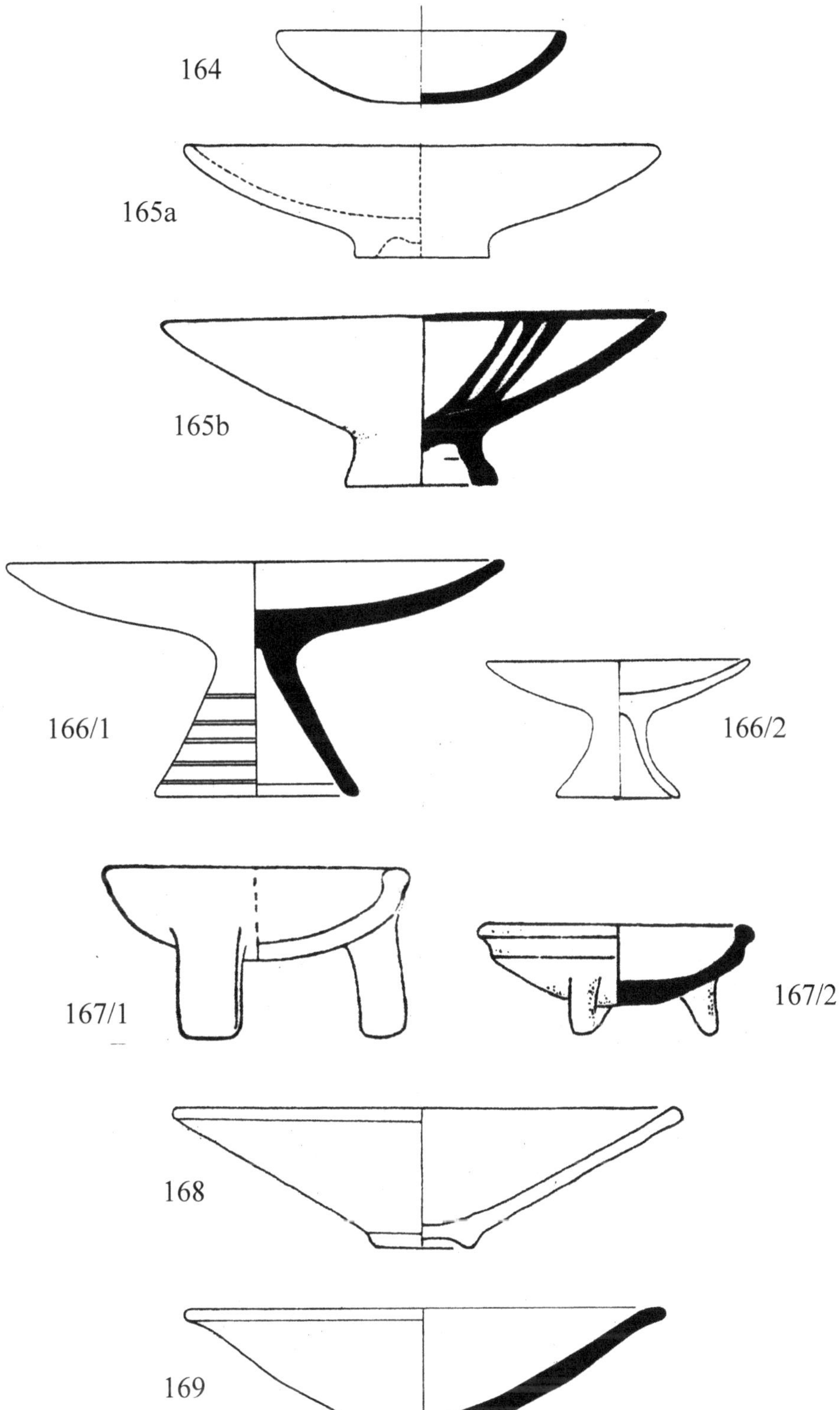

Fig. 51 Bowl CLASSES Used in This Study (Scale 1:4)

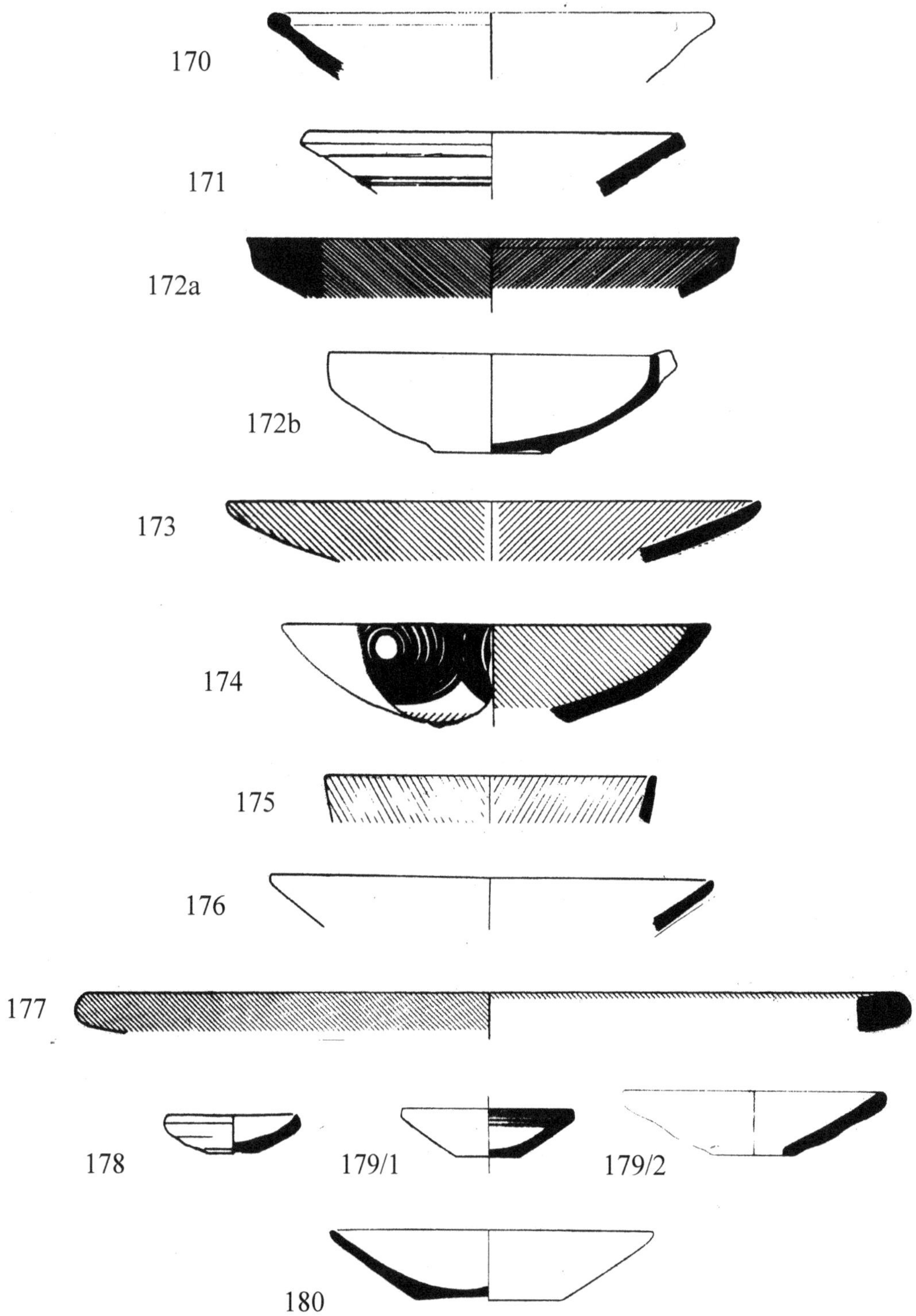

Fig. 52 Bowl CLASSES Used in This Study (Scale 1:4)

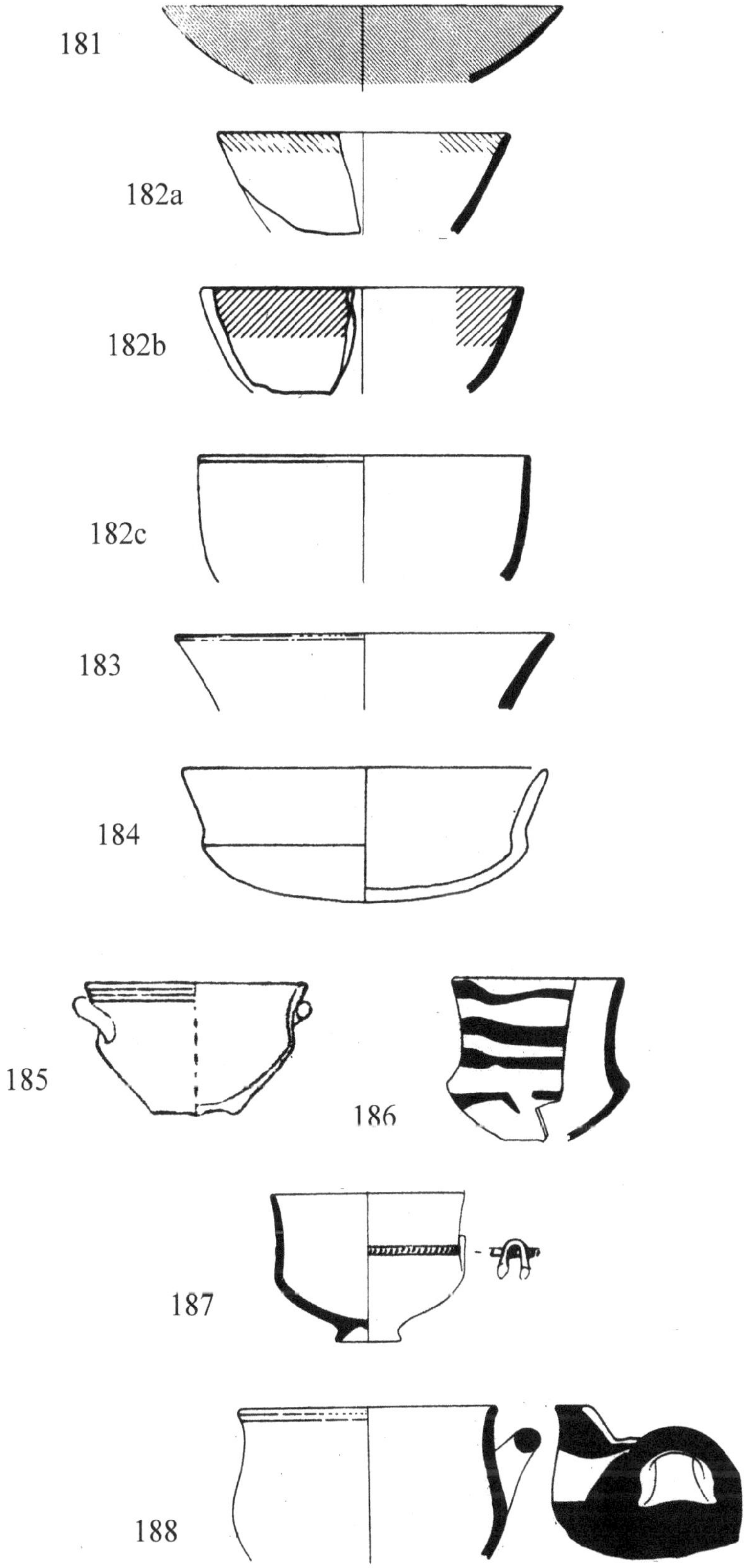

Fig. 53 Bowl CLASSES Used in This Study (Scale 1:4)

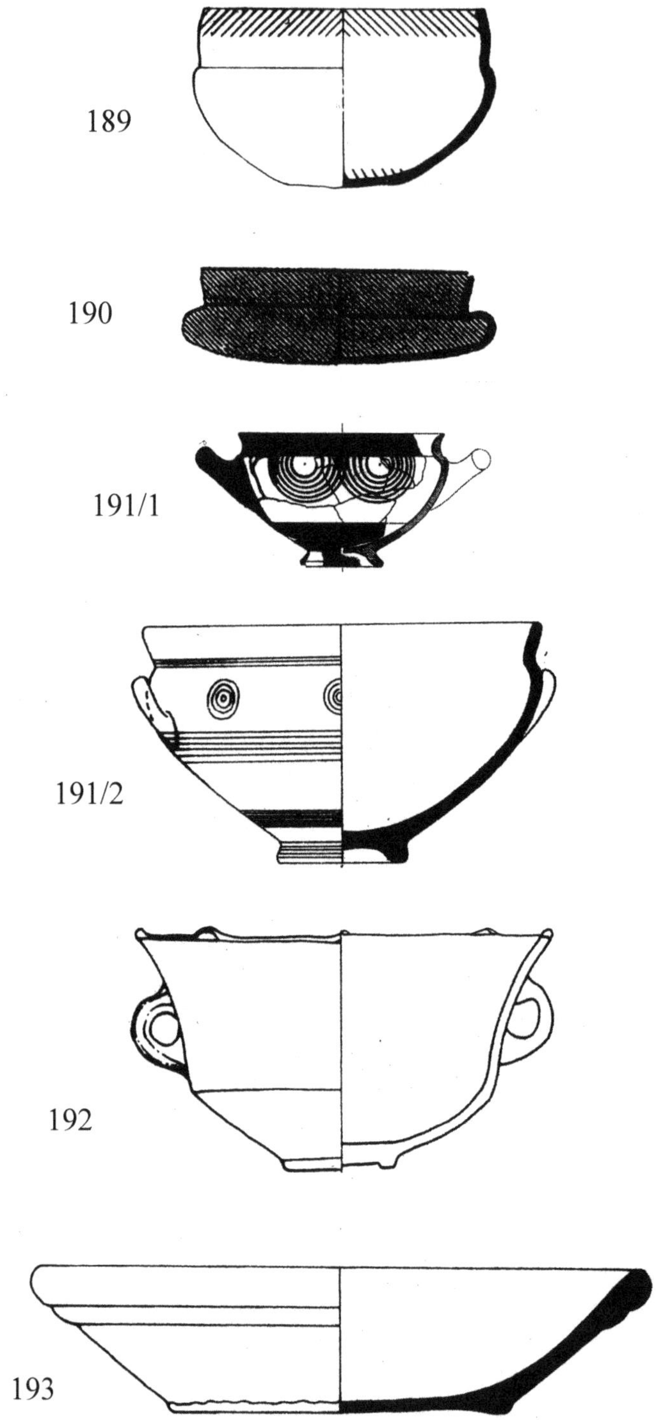

Fig. 54 Bowl CLASSES Used in This Study (Scale 1:4)

MAPS

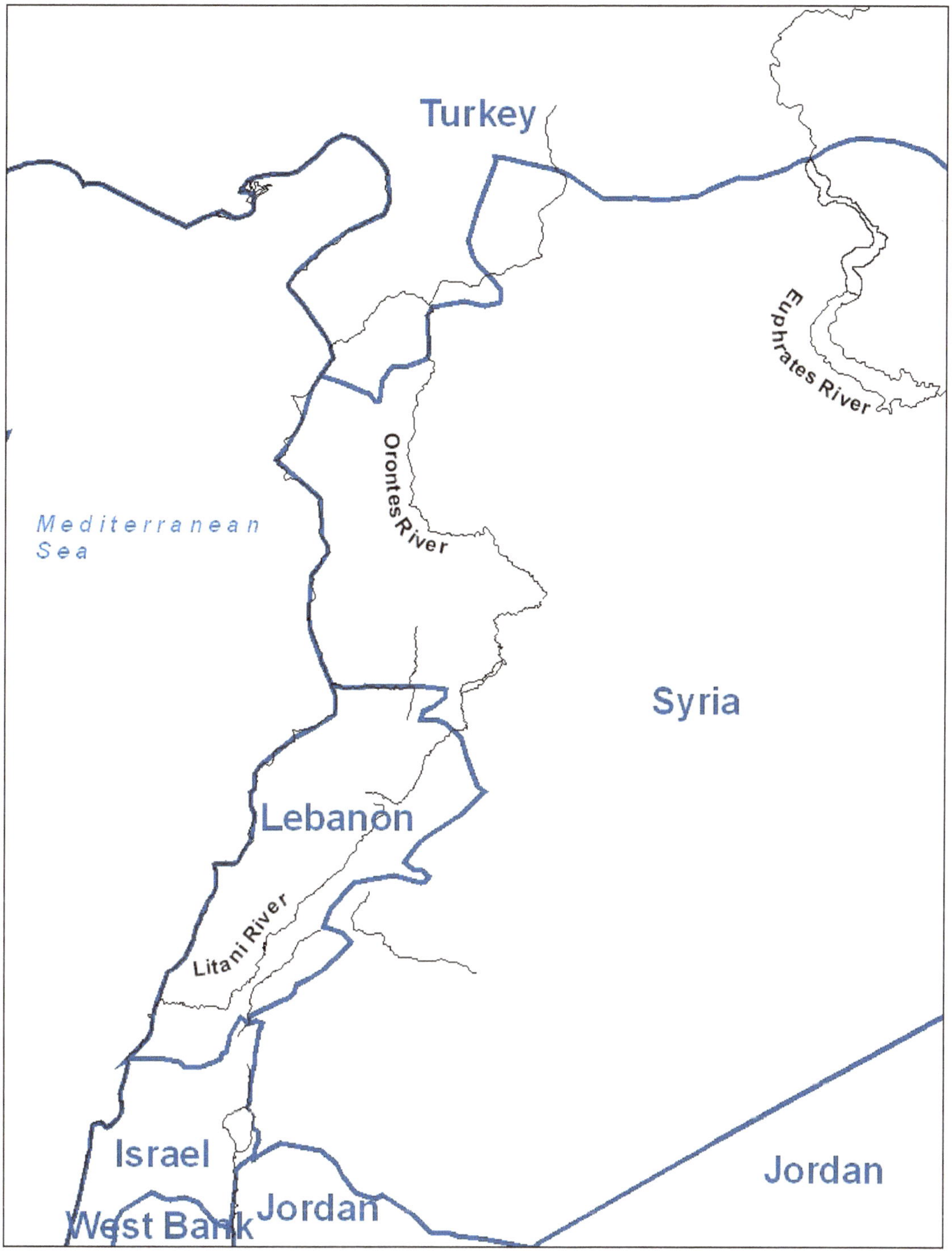

Map 1. Northern Levant with modern national boundaries

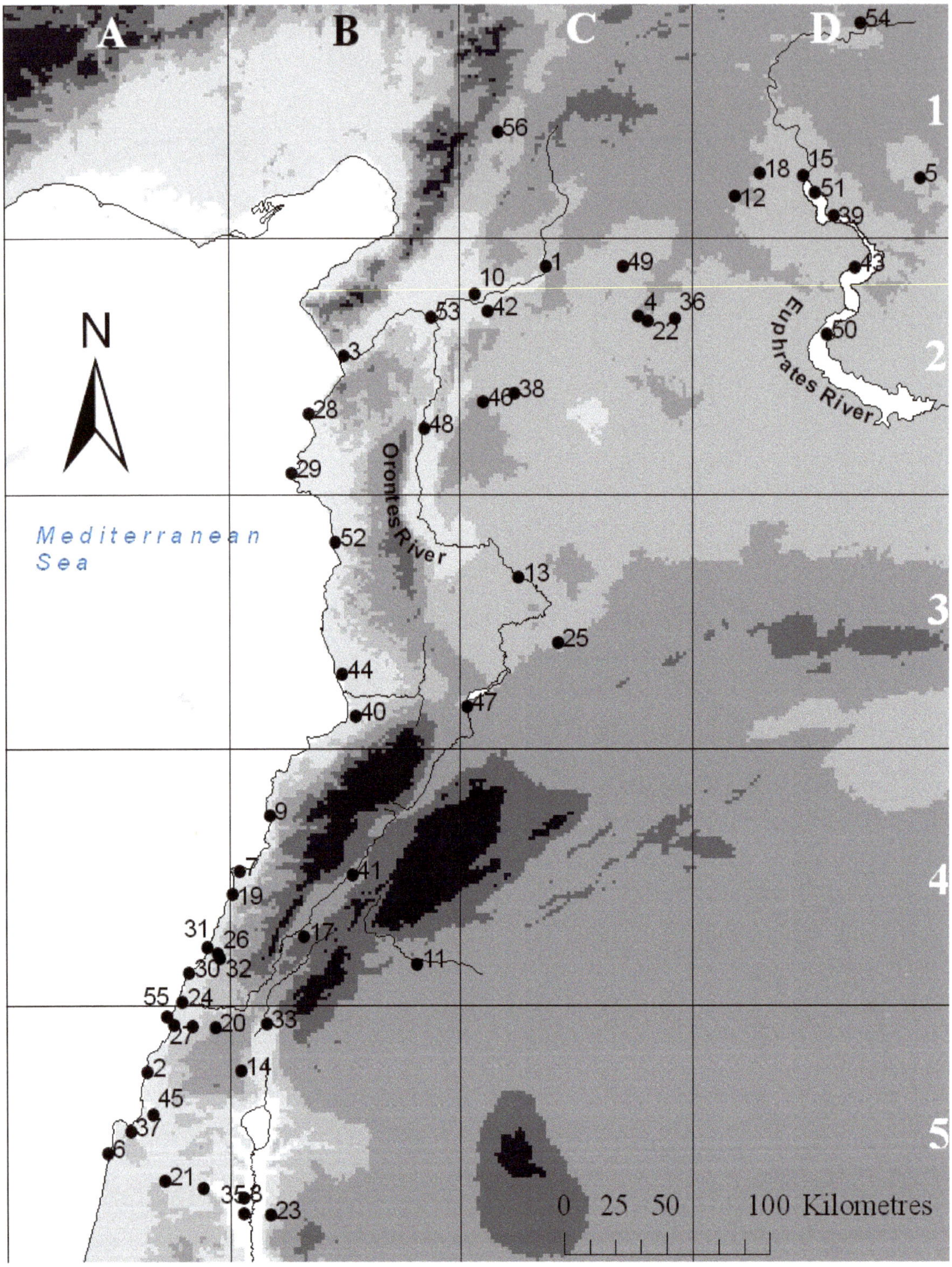

Map 2. Northern Levant with sites used in present study (key overleaf)

Key to Sites in Map 2 (grid reference in parentheses)

1	Ain Dara (C2)	29	Ras Ibn Hani (B2)
2	Akhziv (A5)	30	Sarepta (A4)
3	Al Mina (B2)	31	Sidon (A4)
4	Aleppo (C2)	32	Tambourit (A4)
5	Arslan Tash (D1)	33	Tel Dan (B5)
6	‘Athlit (A5)	34	Tel Jezreel (A5)
7	Beirut (B4)	35	Tel Rehov (B5)
8	Beth Shan (B5)	36	Tell Abou Danne (C2)
9	Byblos (B4)	37	Tell Abu Hawam (A5)
10	Chatal Höyük (C2)	38	Tell Afis (C2)
11	Damascus (B4)	39	Tell Ahmar (D1)
12	Deve Höyük (D1)	40	Tell Arqa (B3)
13	Hama (C3)	41	Tell el Ghassil (B4)
14	Hazor (B5)	42	Tell Judeideh (C2)
15	Jerablus (D1)	43	Tell Jurn Kabir (D2)
16	Joya (A5)	44	Tell Kazel (B3)
17	Kamid el Loz (B4)	45	Tell Keisan (A5)
18	Kefrik (D1)	46	Tell Mastuma (C2)
19	Khalde (B4)	47	Tell Nebi Mend (C3)
20	Khirbet Silm (A5)	48	Tell Qarqur (B2)
21	Megiddo (A5)	49	Tell Rifa’at (C2)
22	Nayrab (C2)	50	Tell Sheikh Hassan (D2)
23	Pella (B5)	51	Tell Shiyukh Fawqani (D1)
24	Qasmieh (A4)	52	Tell Sukas (B3)
25	Qatna/Tell Mishrifeh (C3)	53	Tell Ta’yinat (B2)
26	Qraye (A4)	54	Tille Höyük (D1)
27	Rachidieh (A5)	55	Tyre (A5)
28	Ras al Bassit (B2)	56	Zincirli (C1)

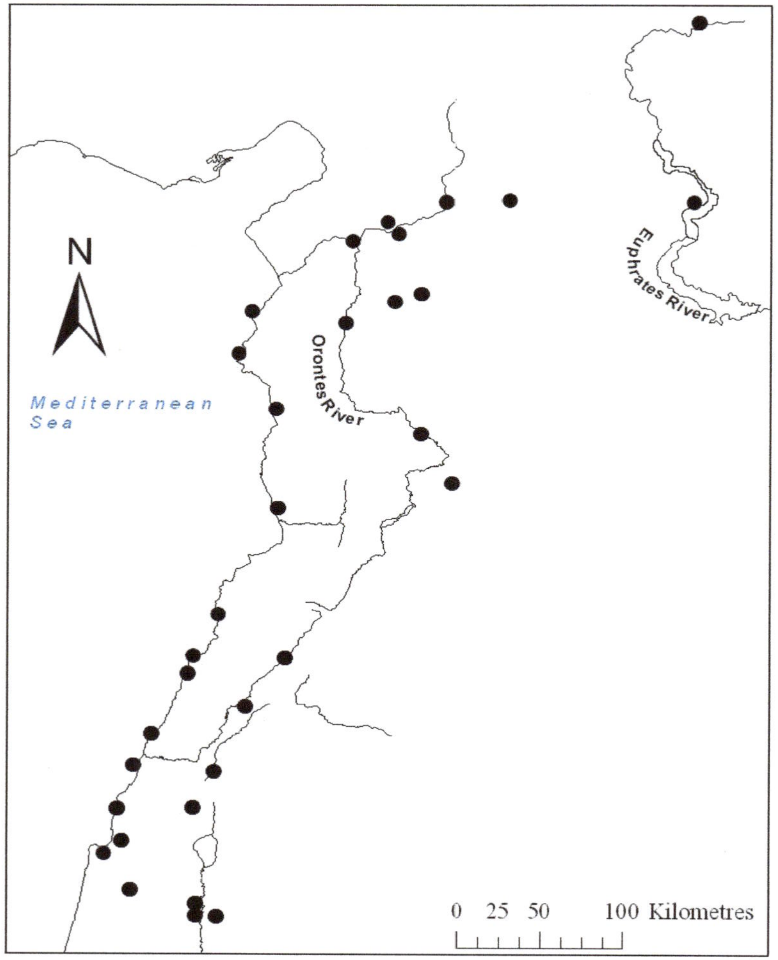

Map 3. Sites from study area with Iron I contexts

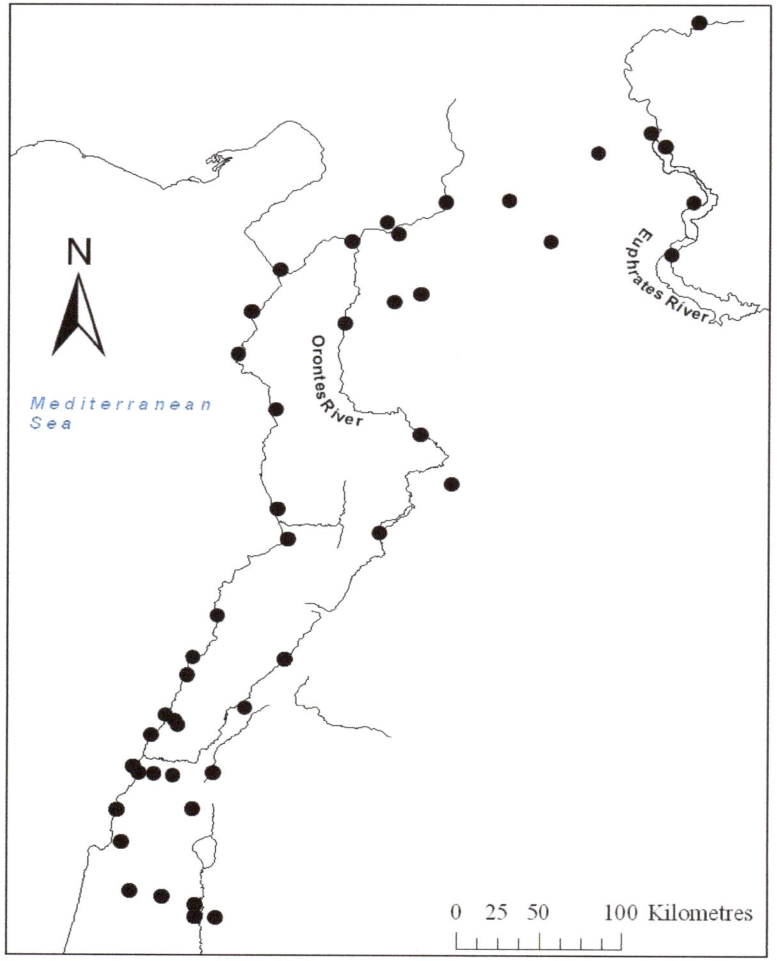

Map 4. Sites from study area with Iron II contexts

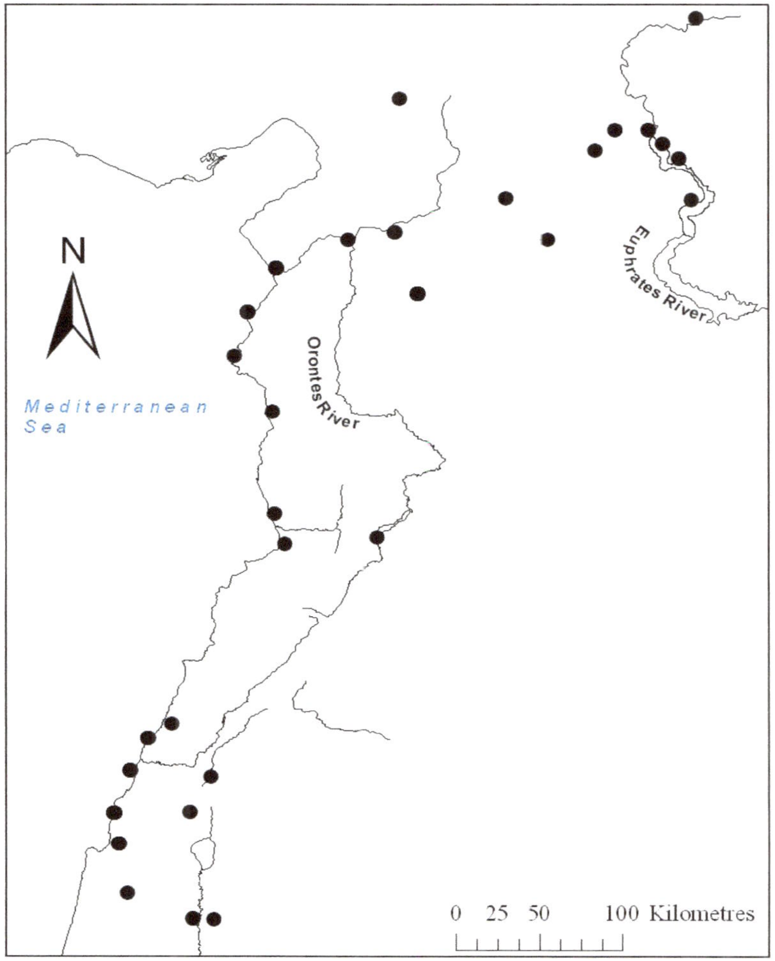

Map 5. Sites from study area with Iron III contexts

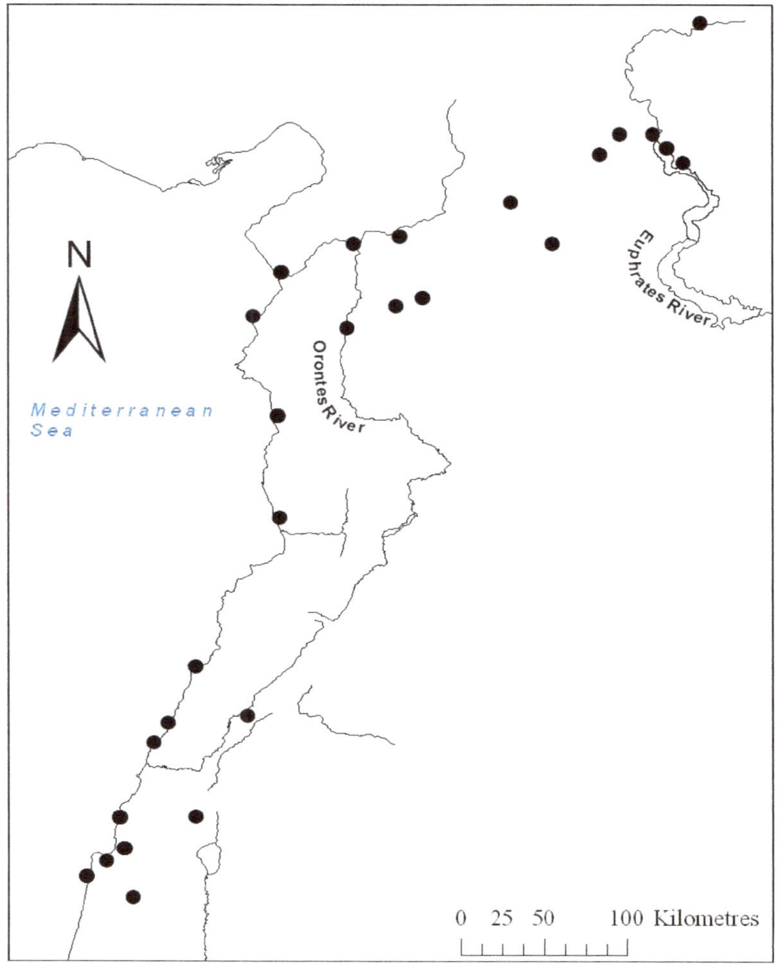

Map 6. Sites from study area with Persian period contexts

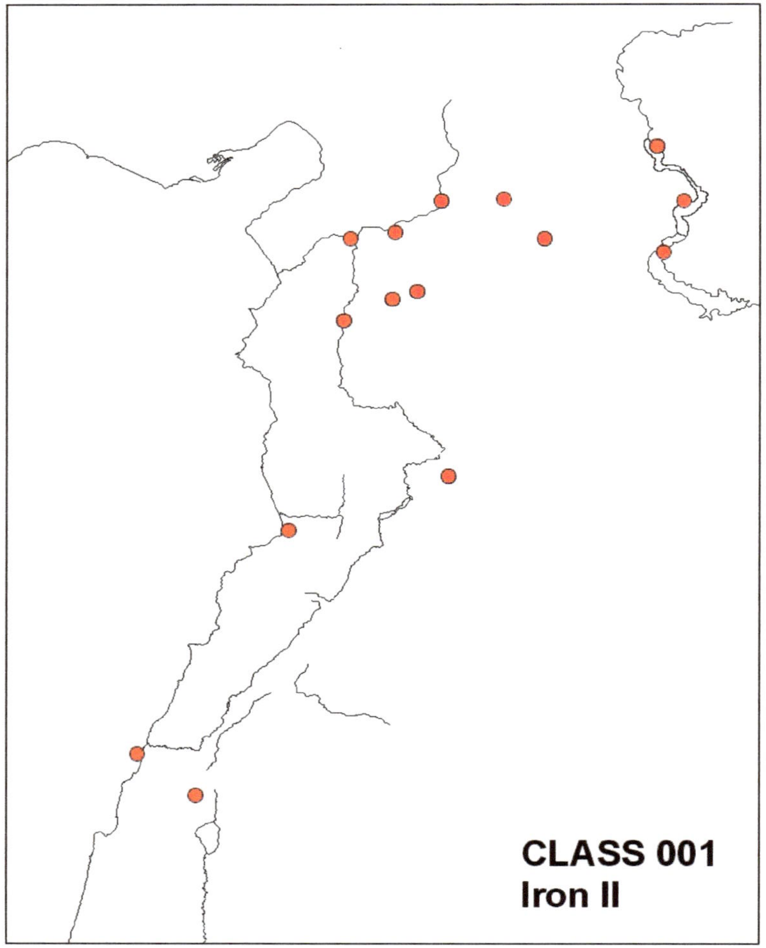

Map 7

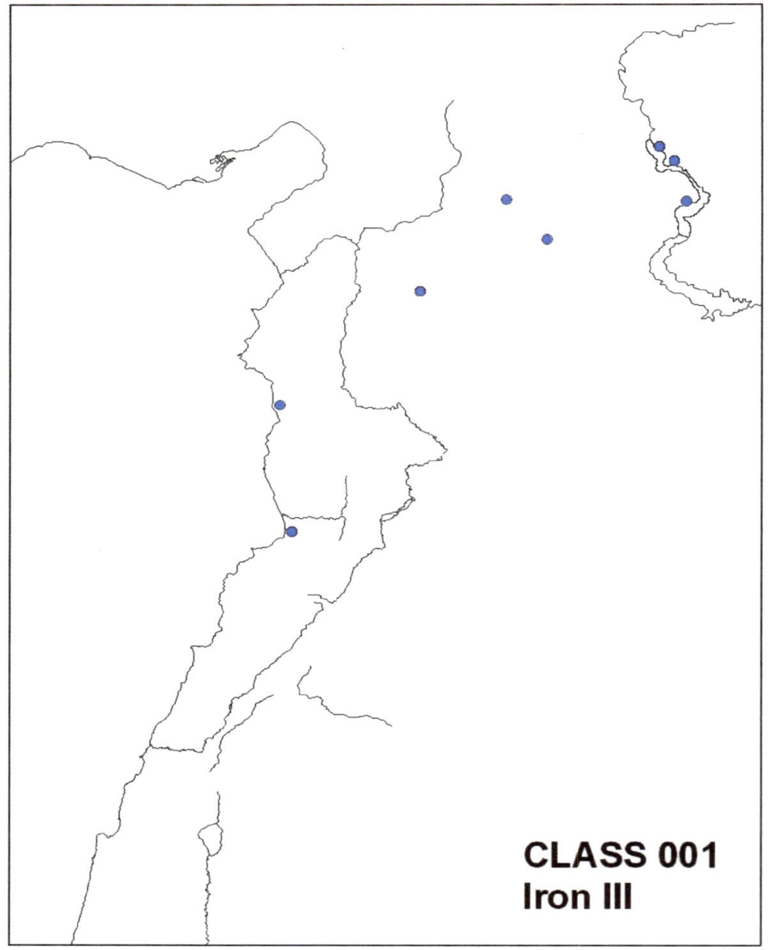

Map 8

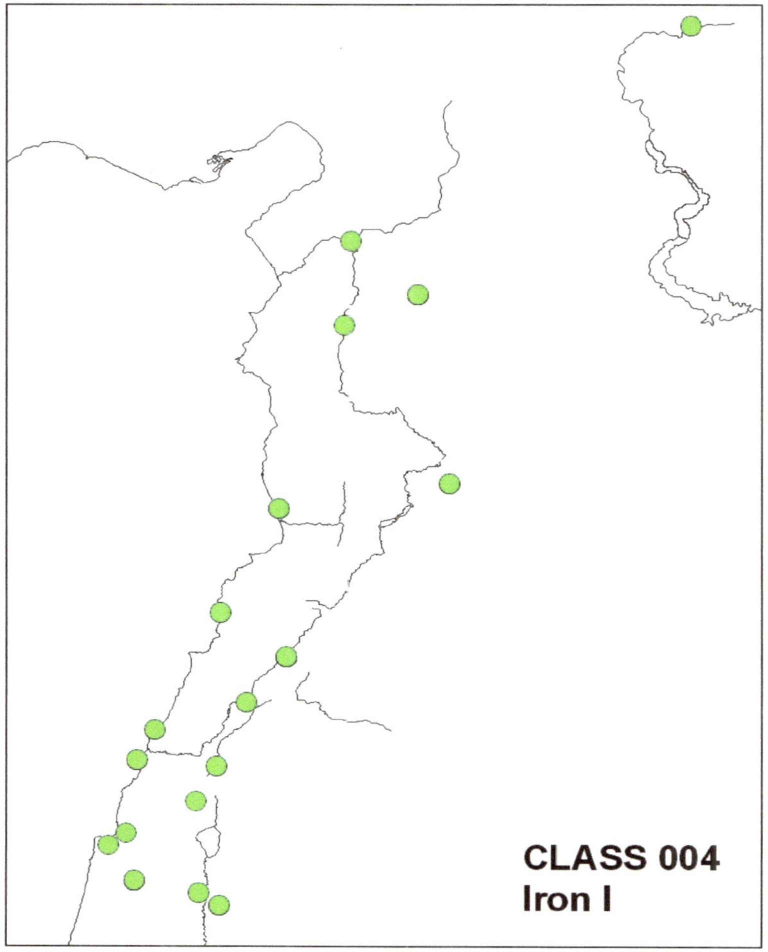

Map 9

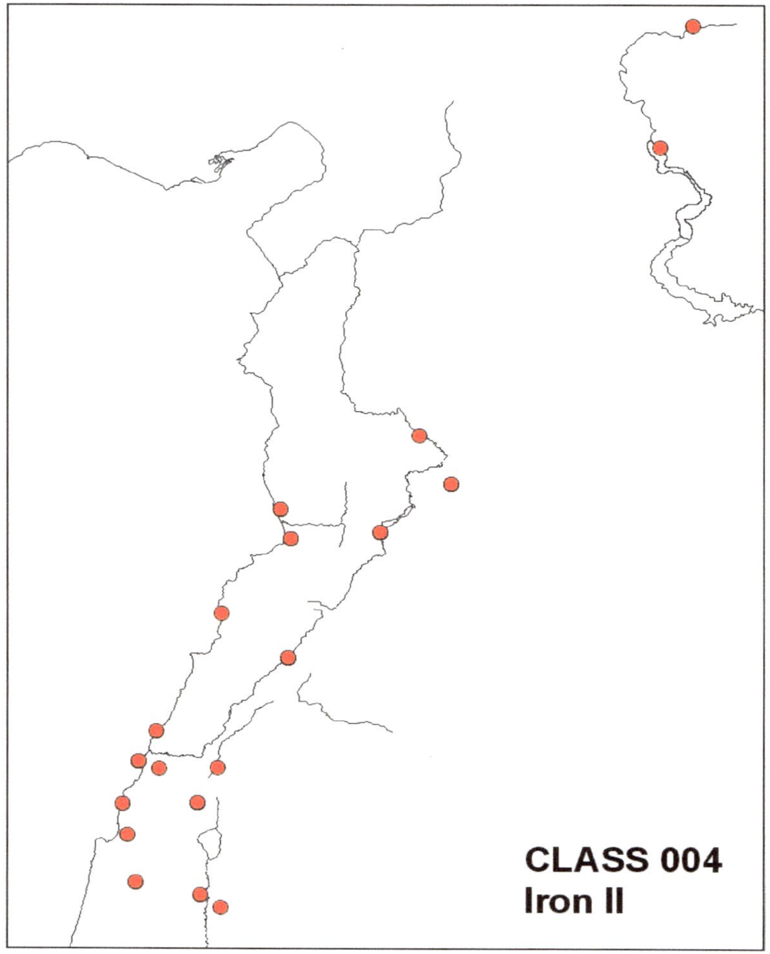

Map 10

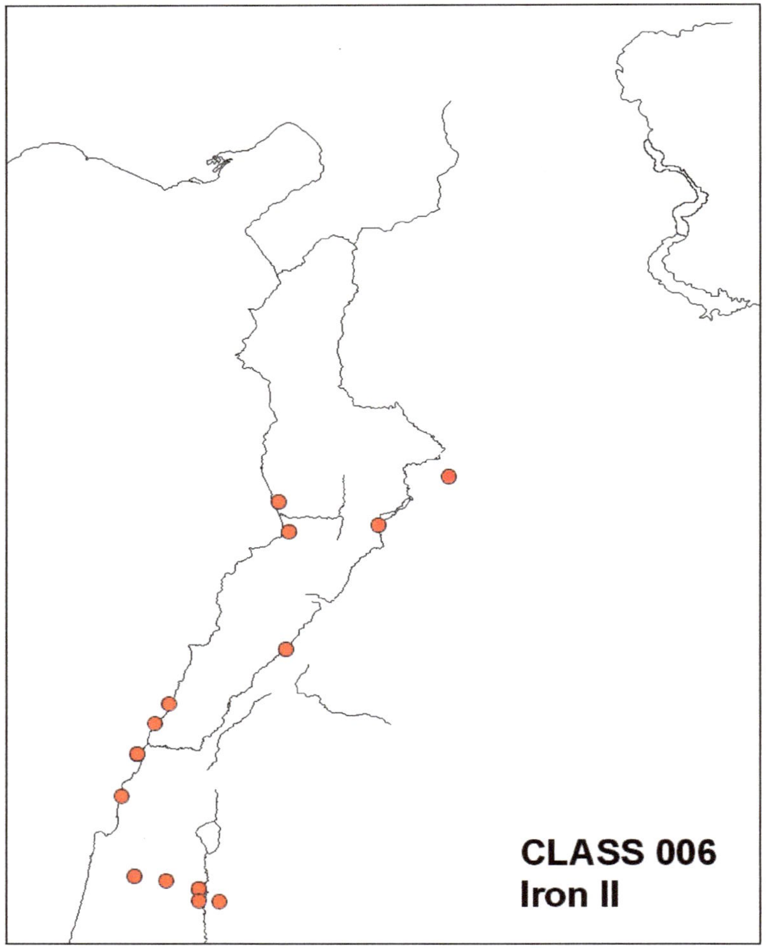

Map 11

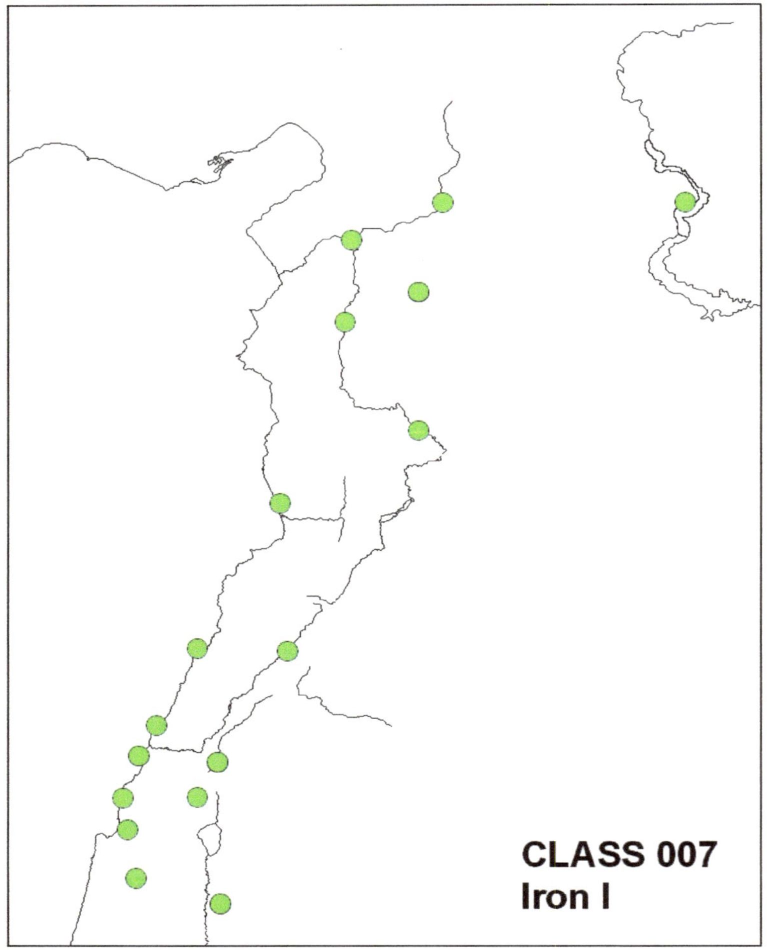

Map 12

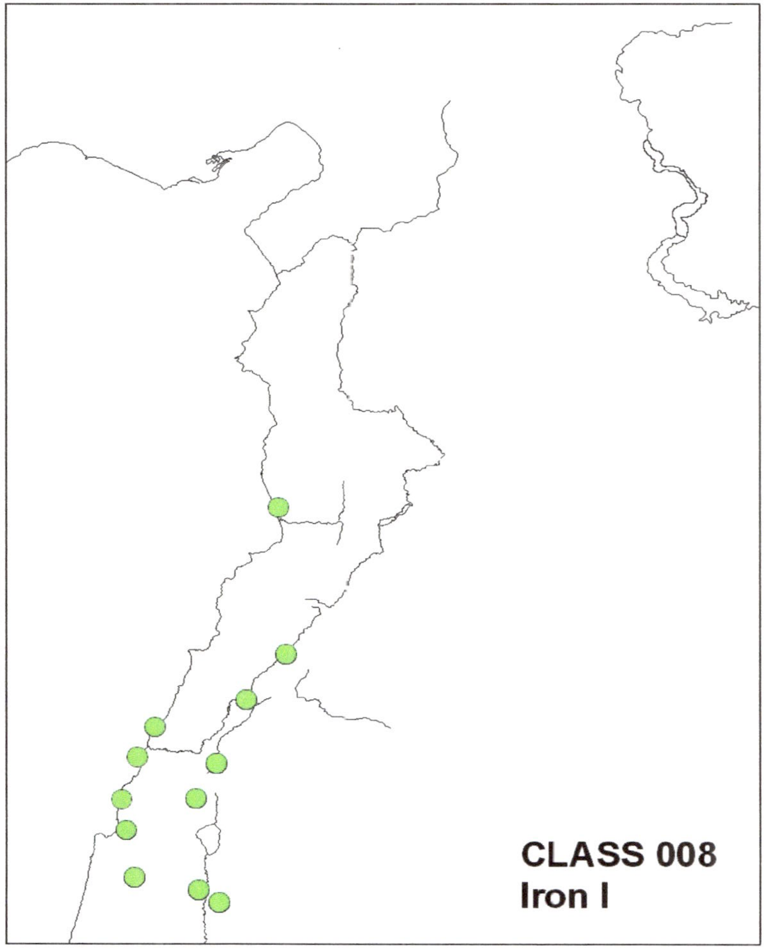

Map 13

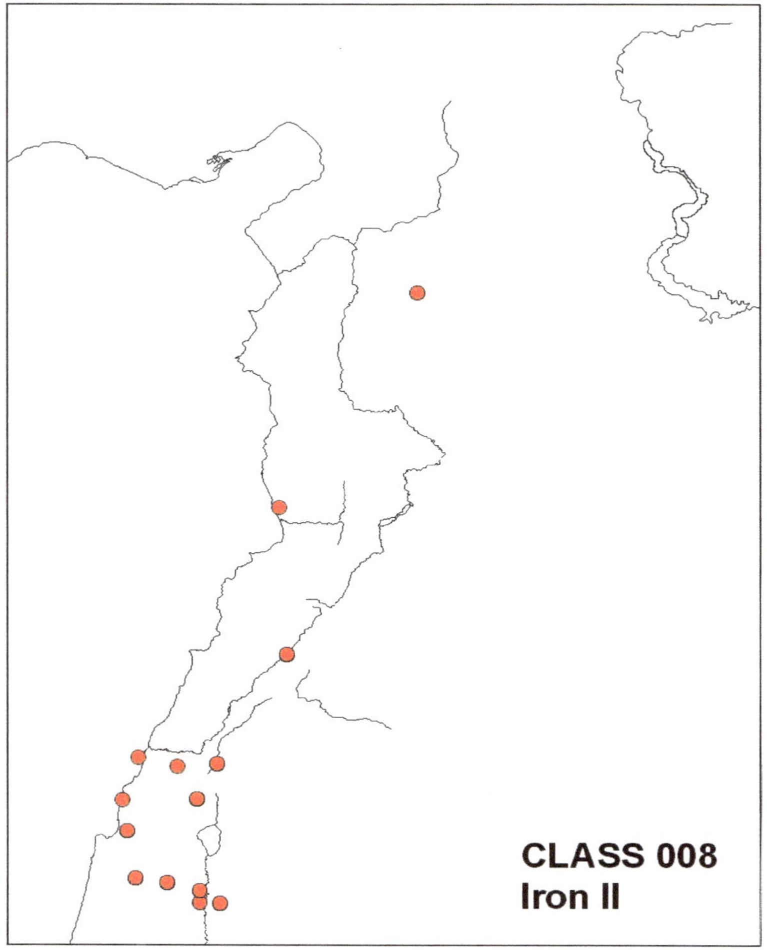

Map 14

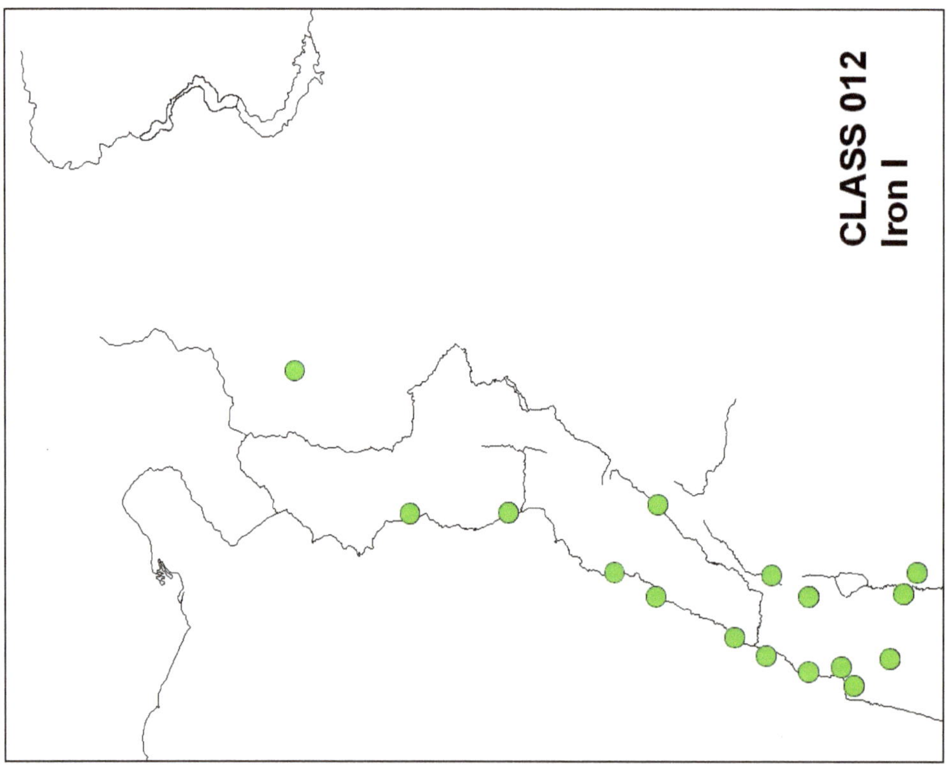

Map 15

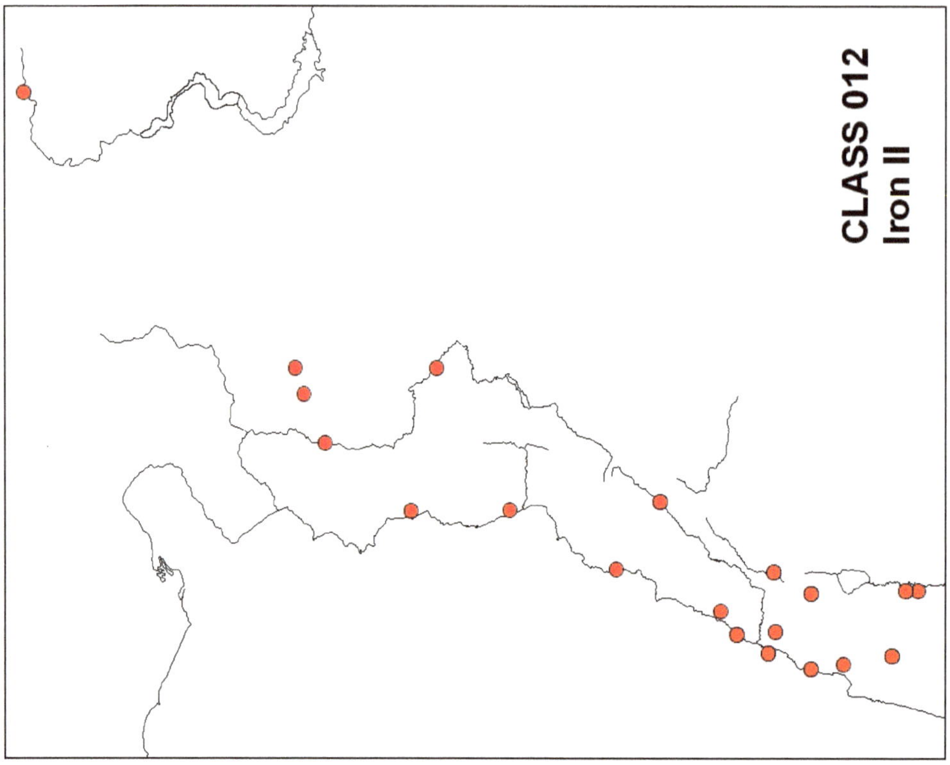

Map 16

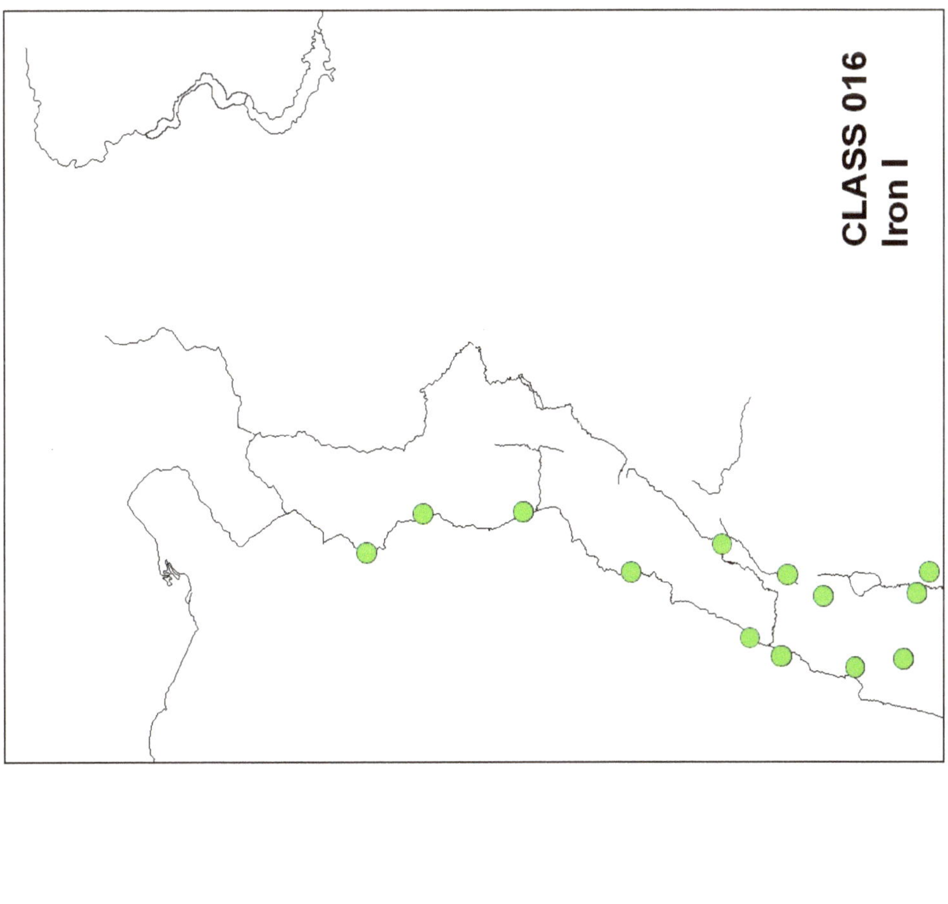

Map 17

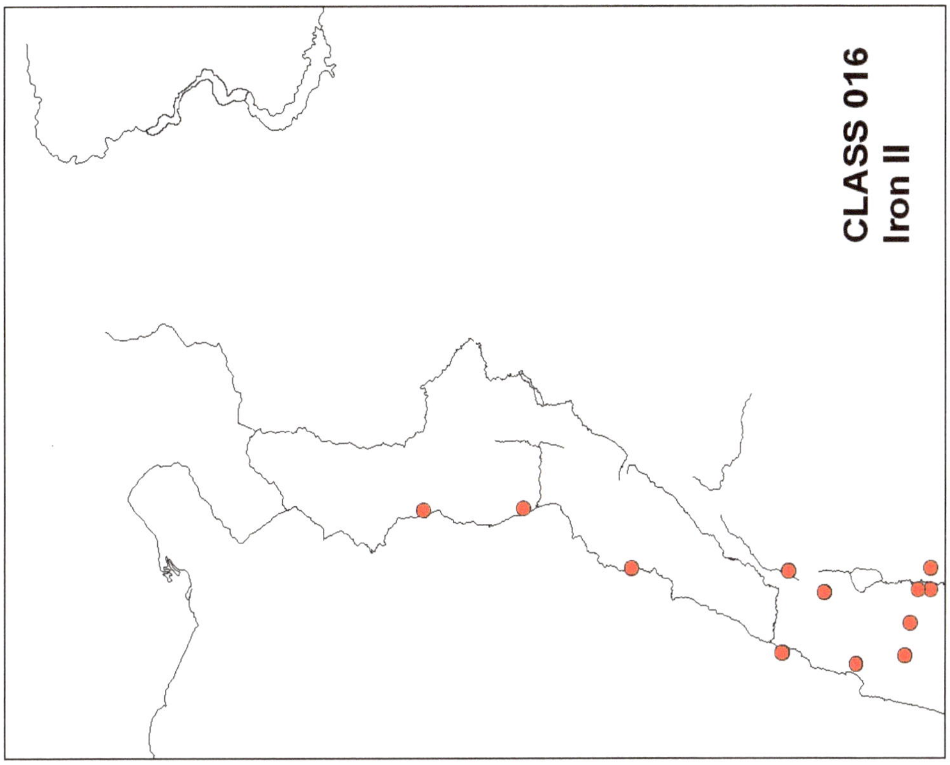

Map 18

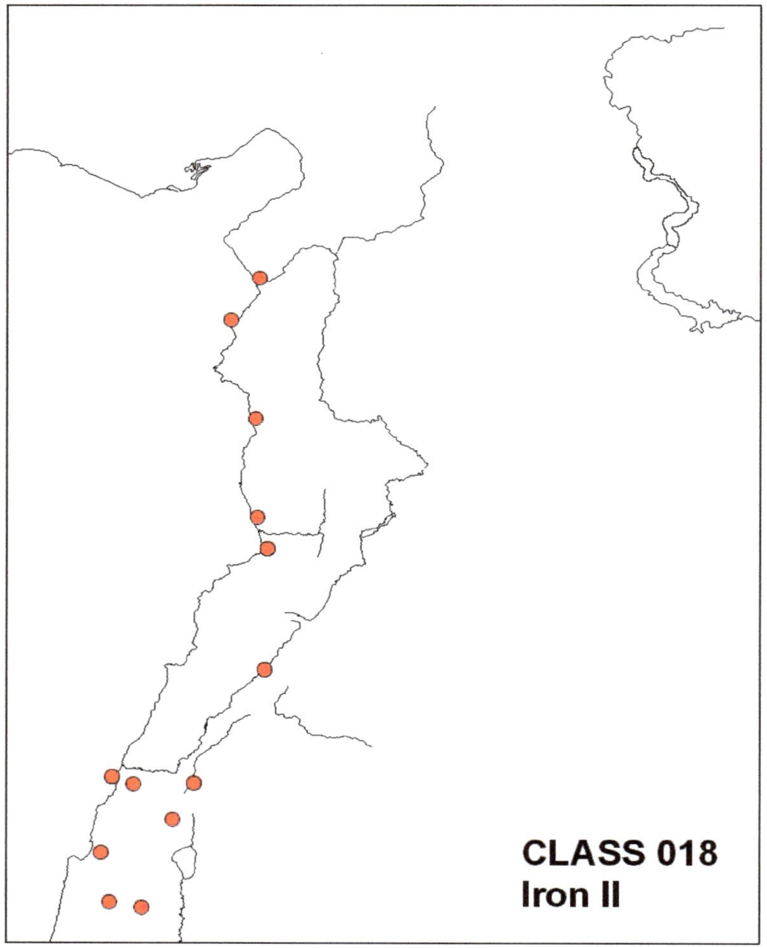

Map 19

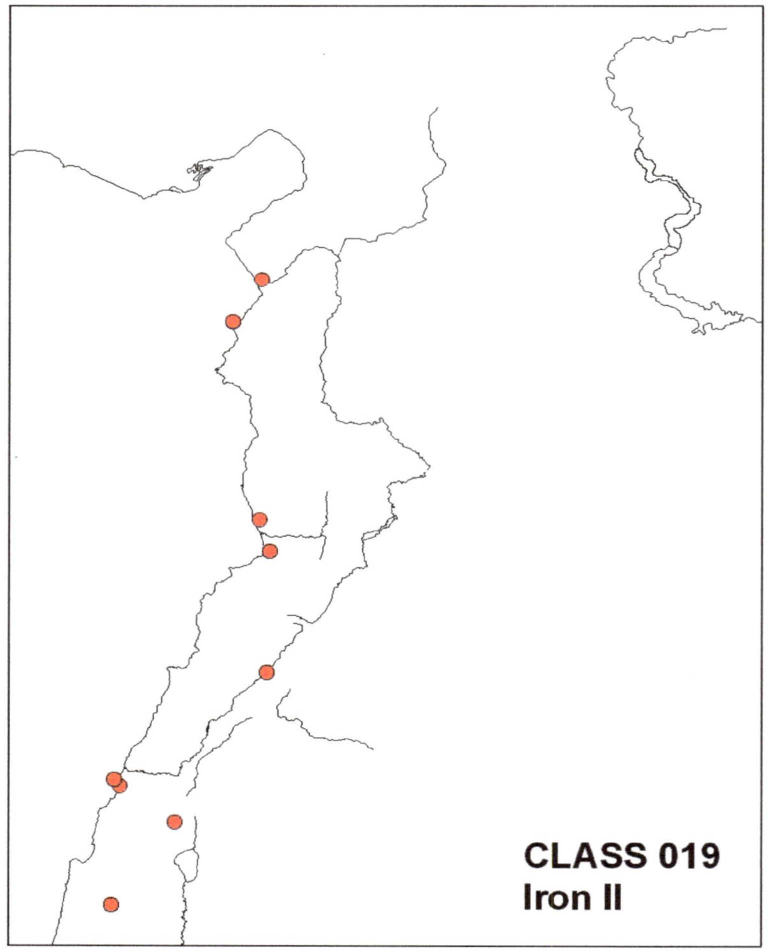

Map 20

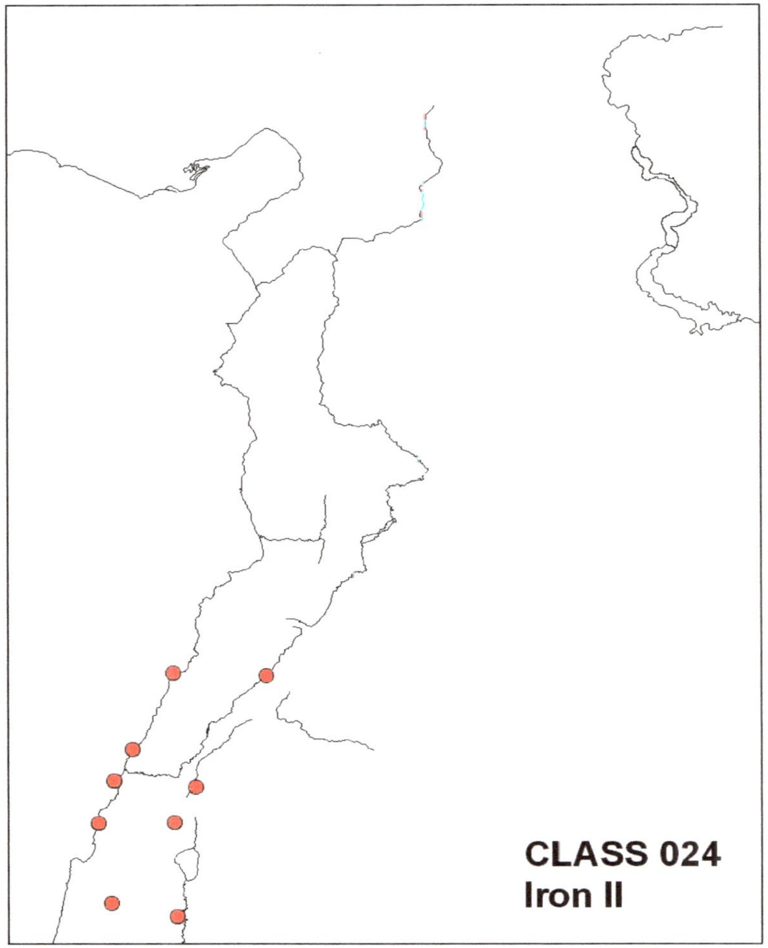

Map 21

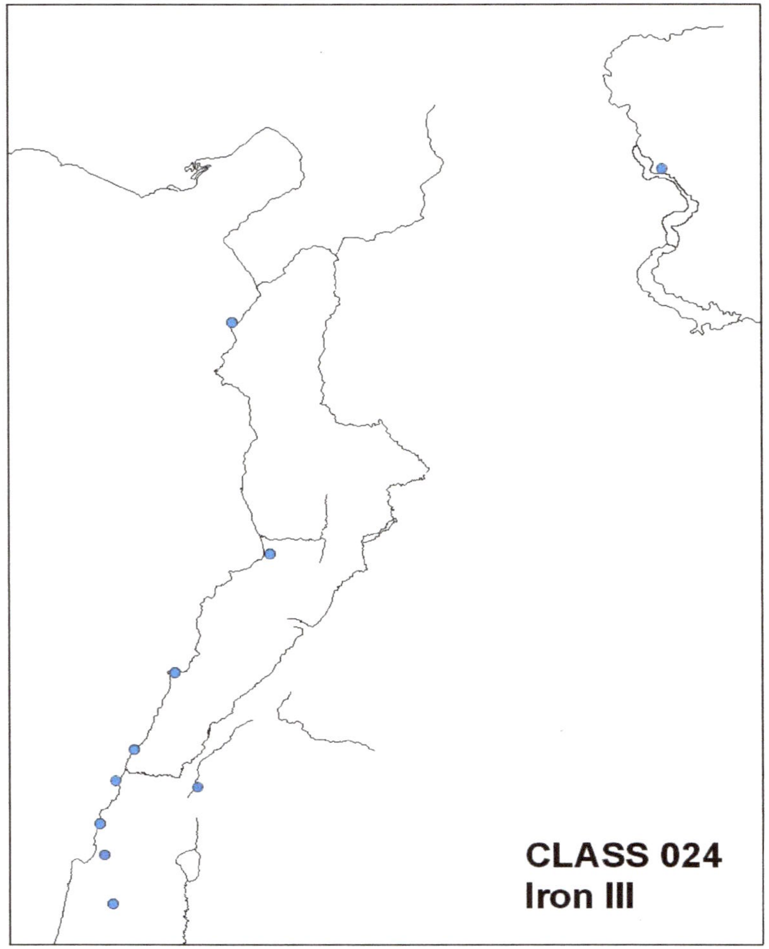

Map 22

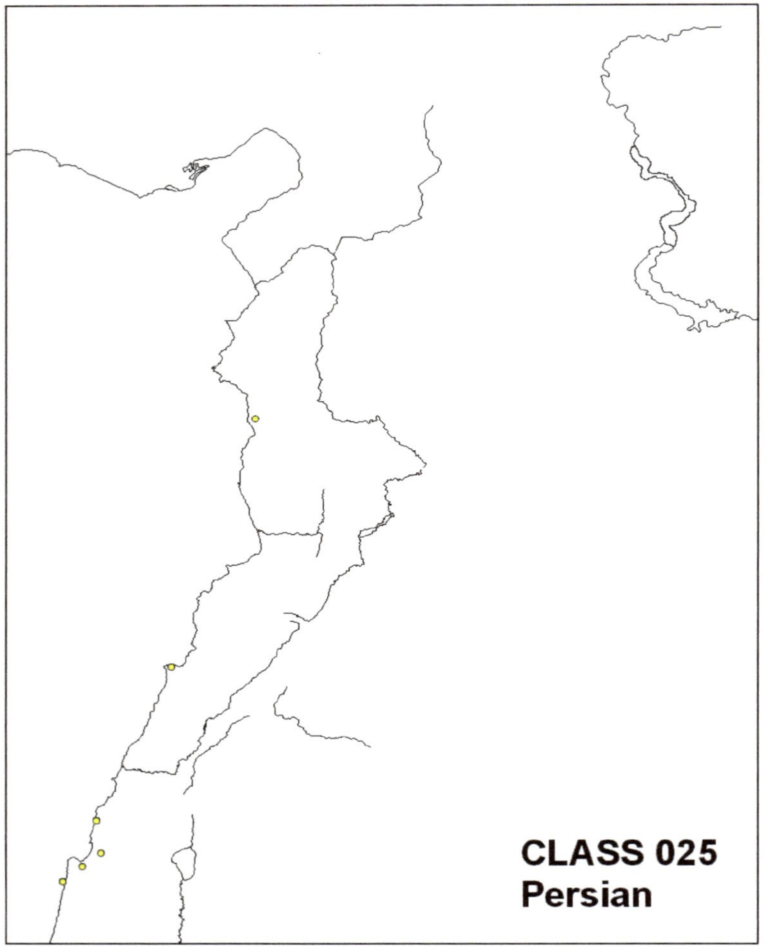

Map 23

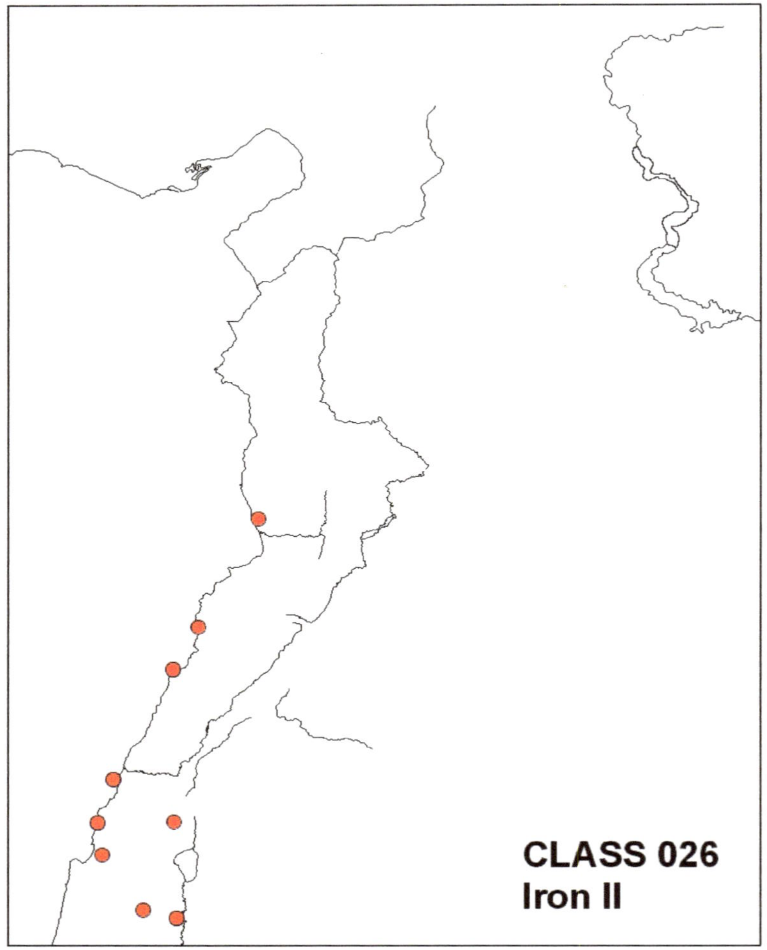

Map 24

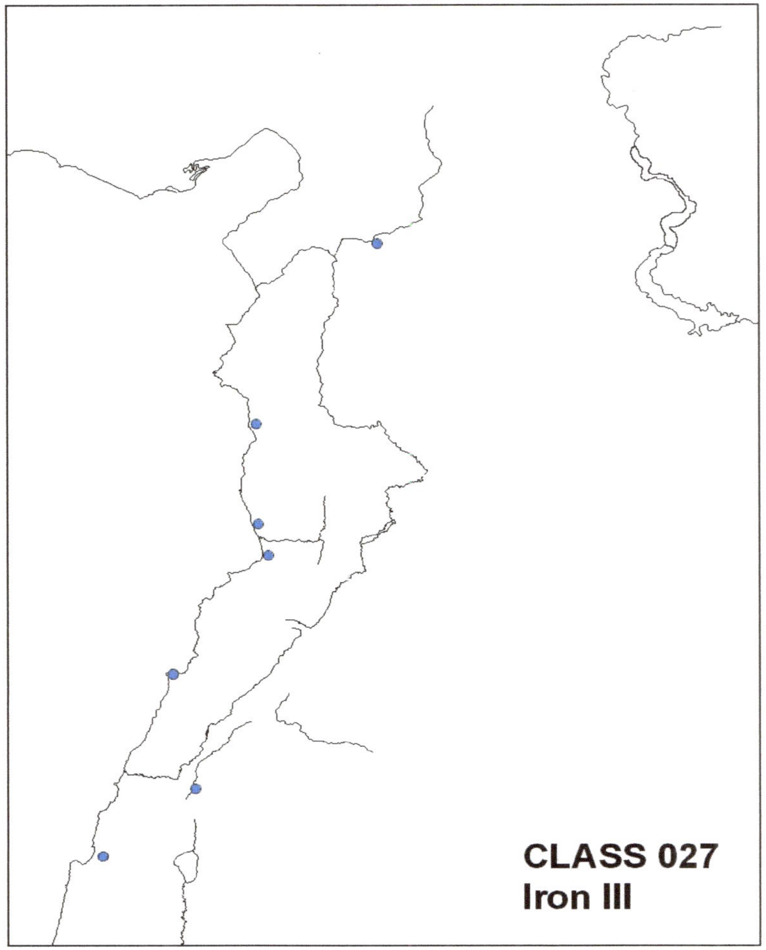

Map 25

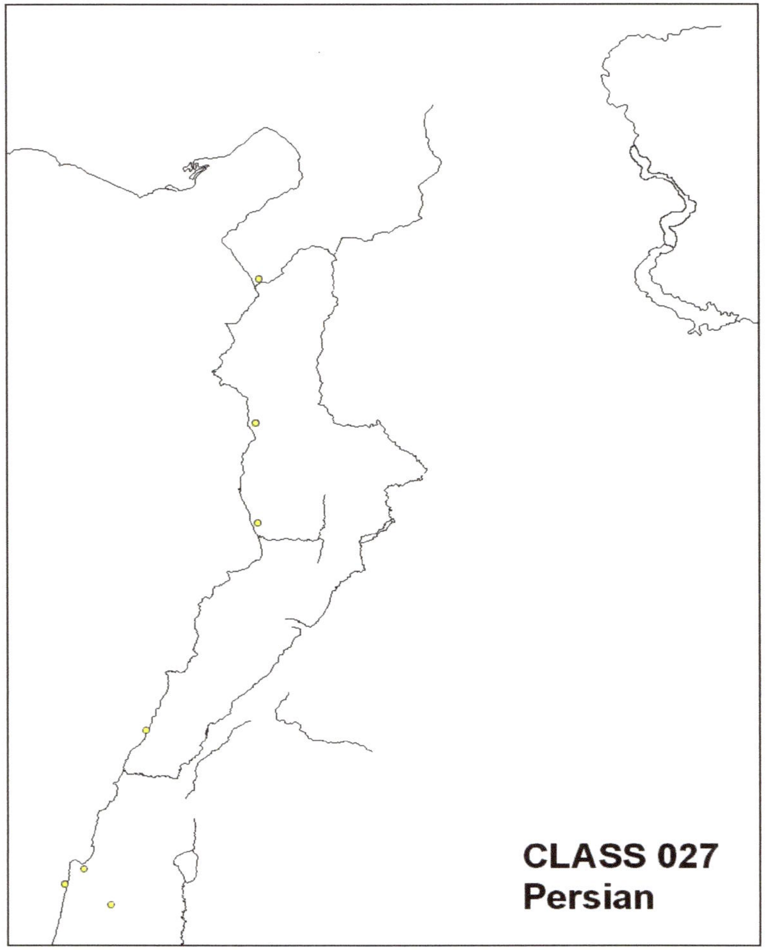

Map 26

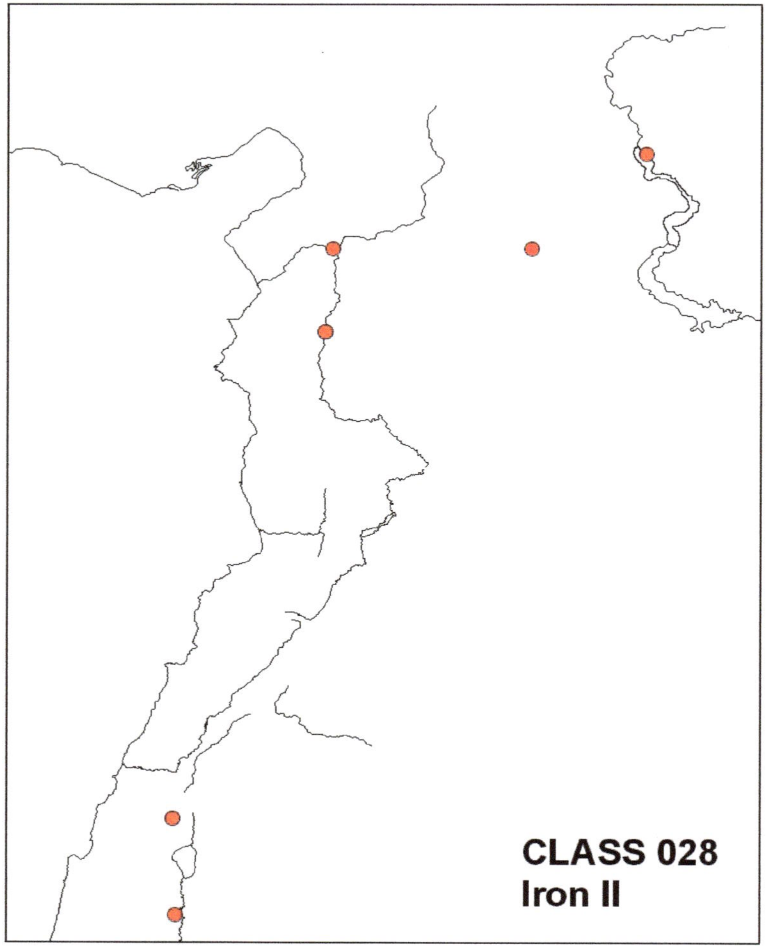

Map 27

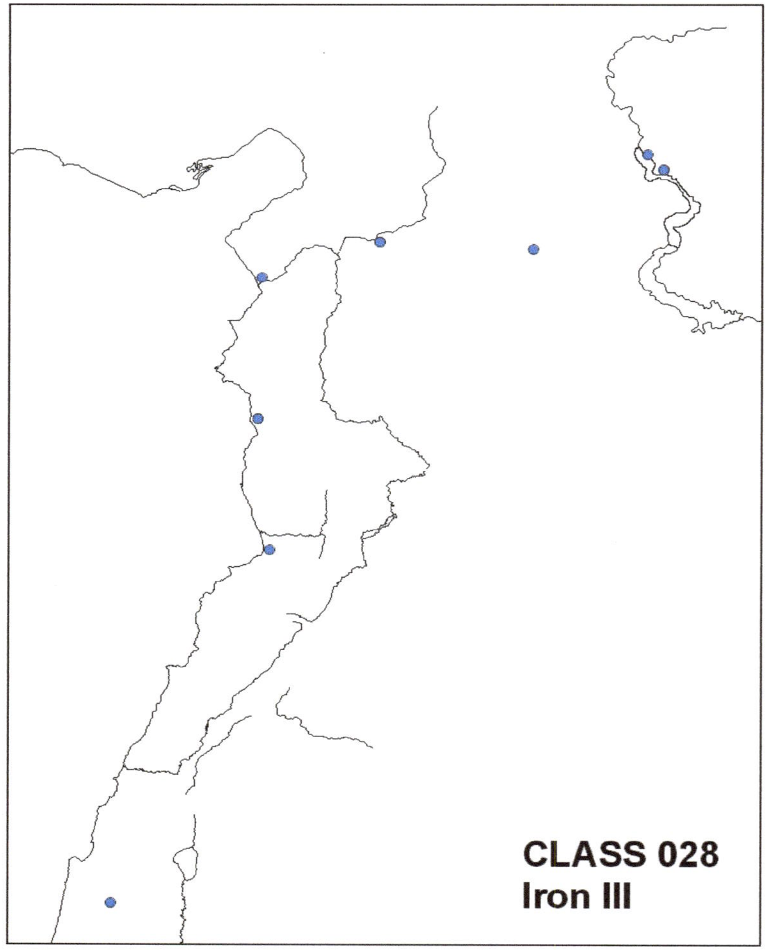

Map 28

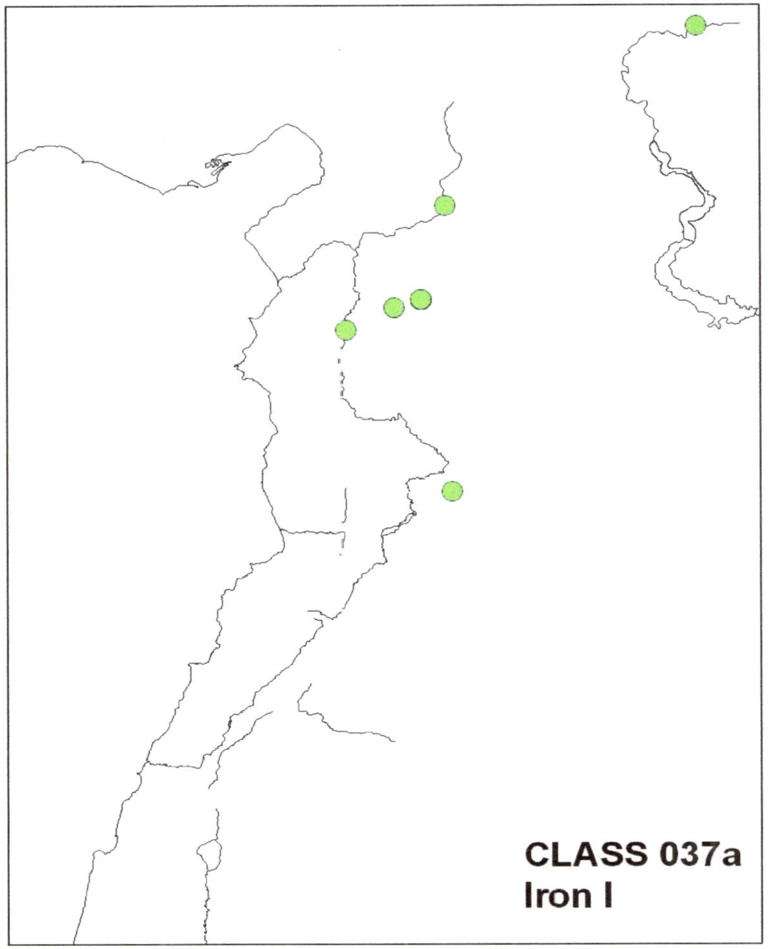

Map 29

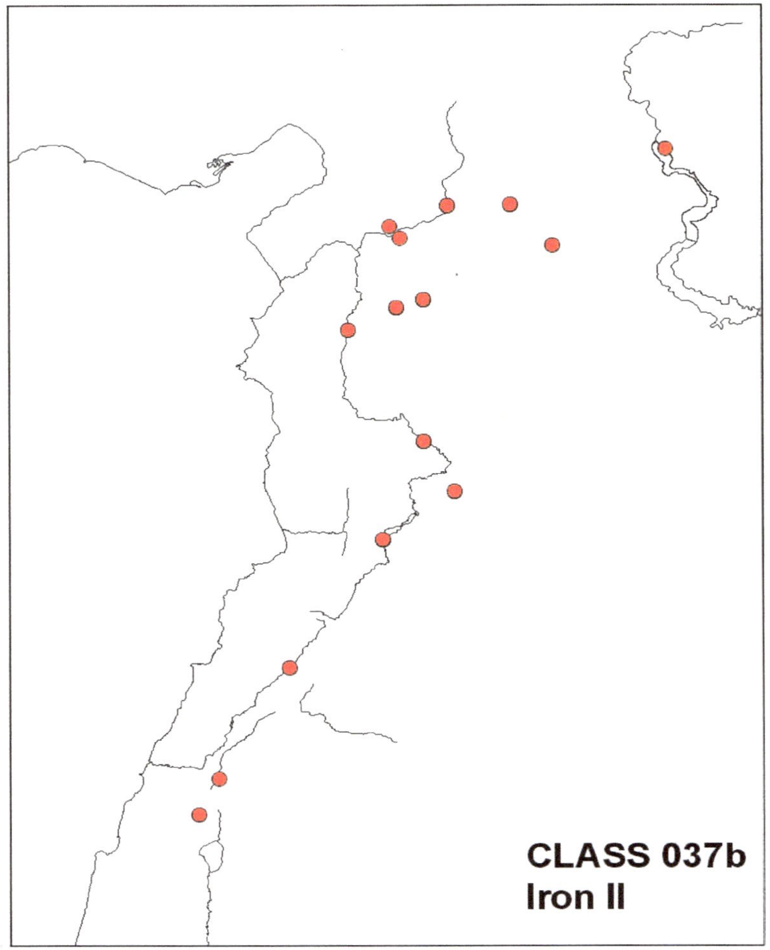

Map 30

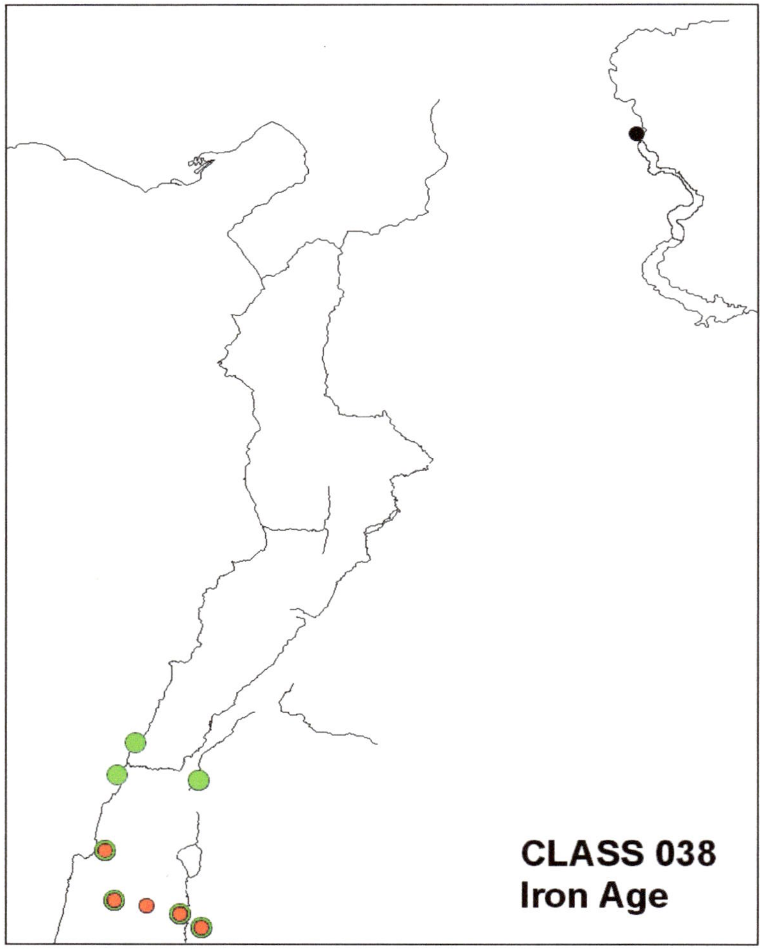

Map 31

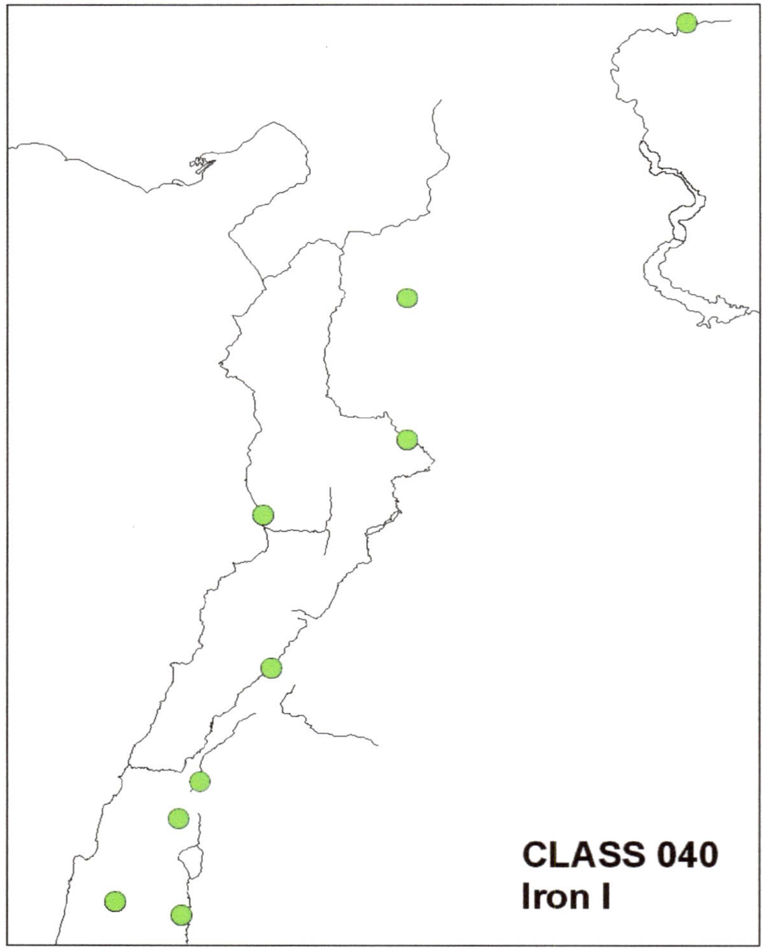

Map 32

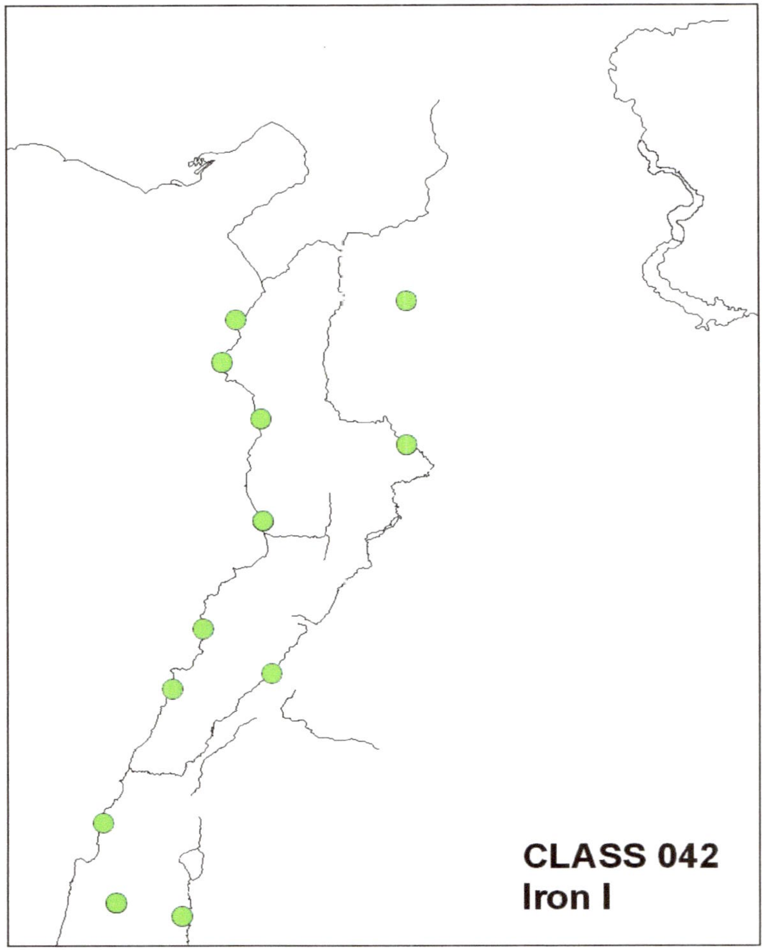

Map 33

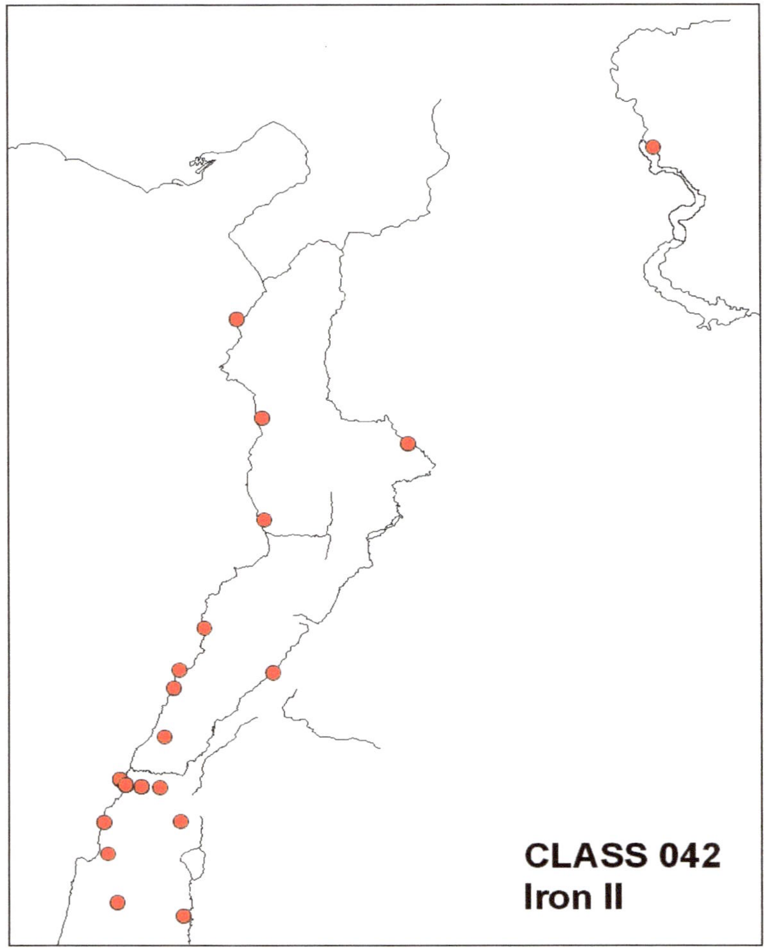

Map 34

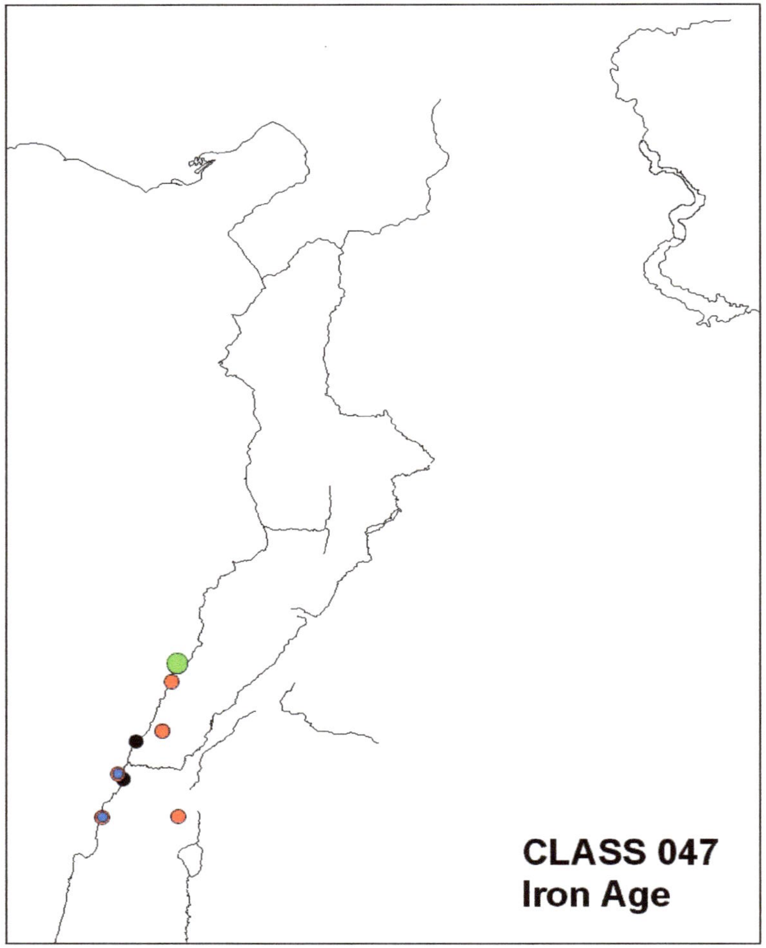

Map 35

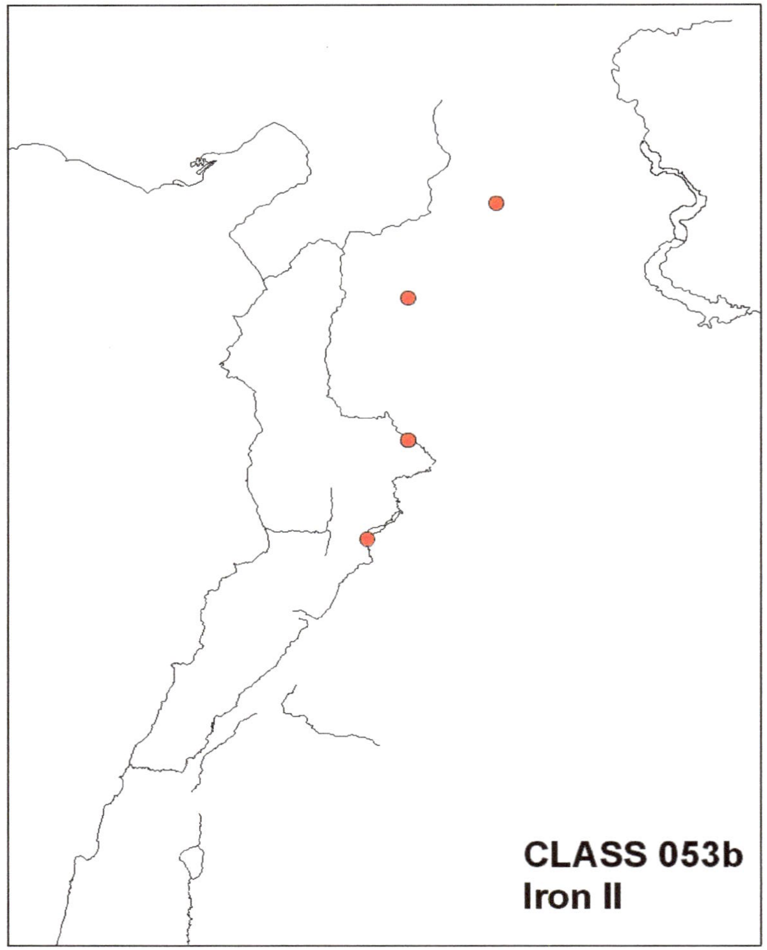

Map 36

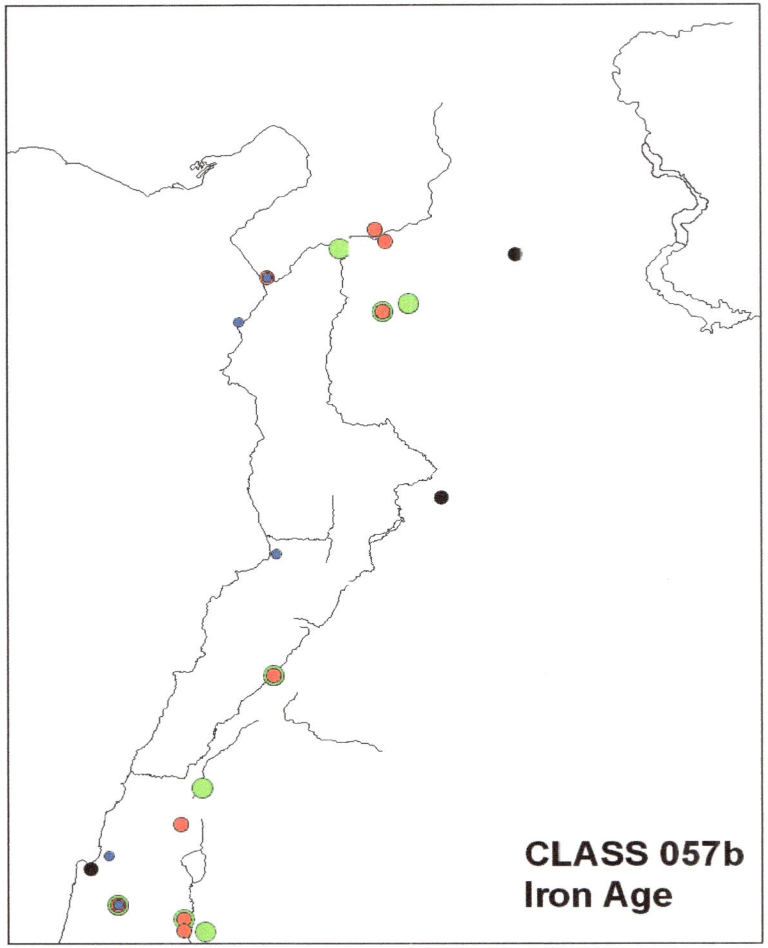

Map 37

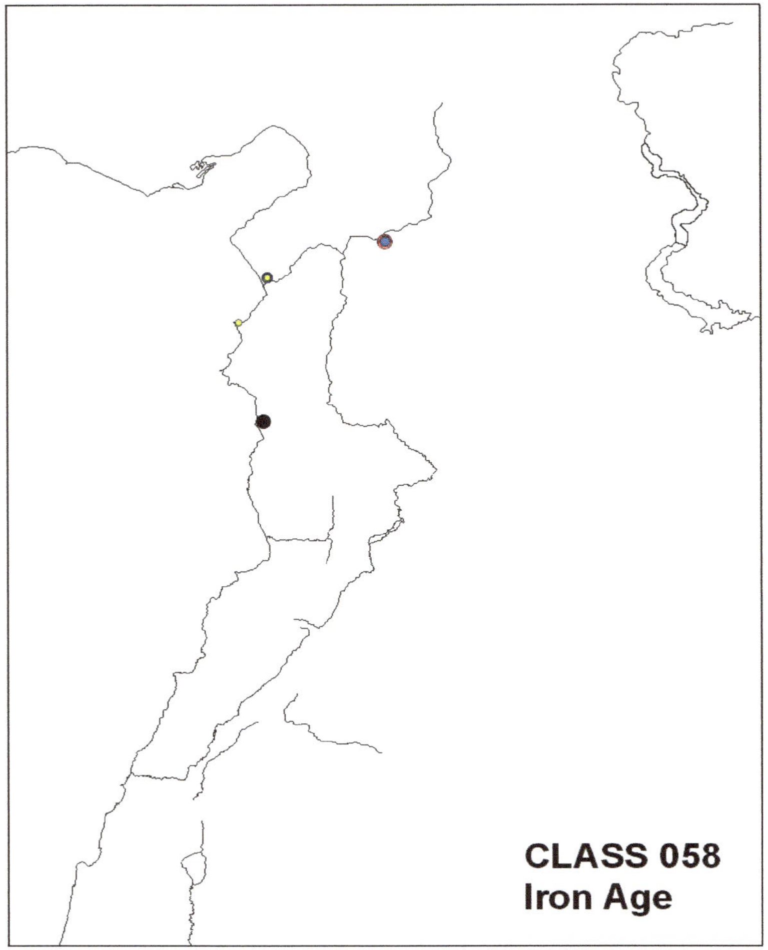

Map 38

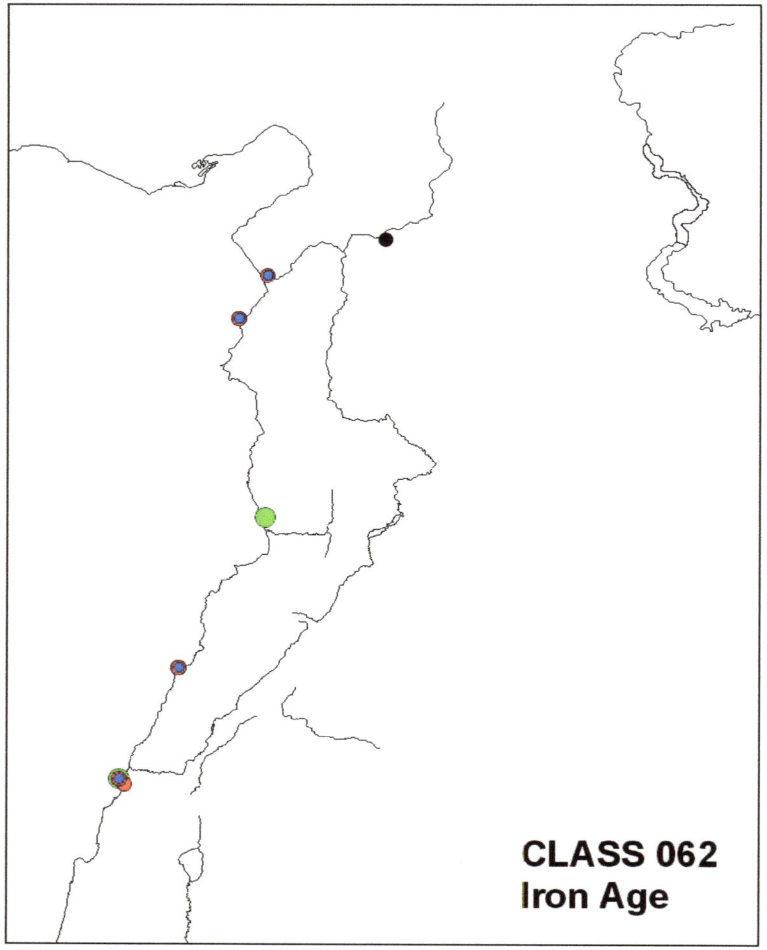

Map 39

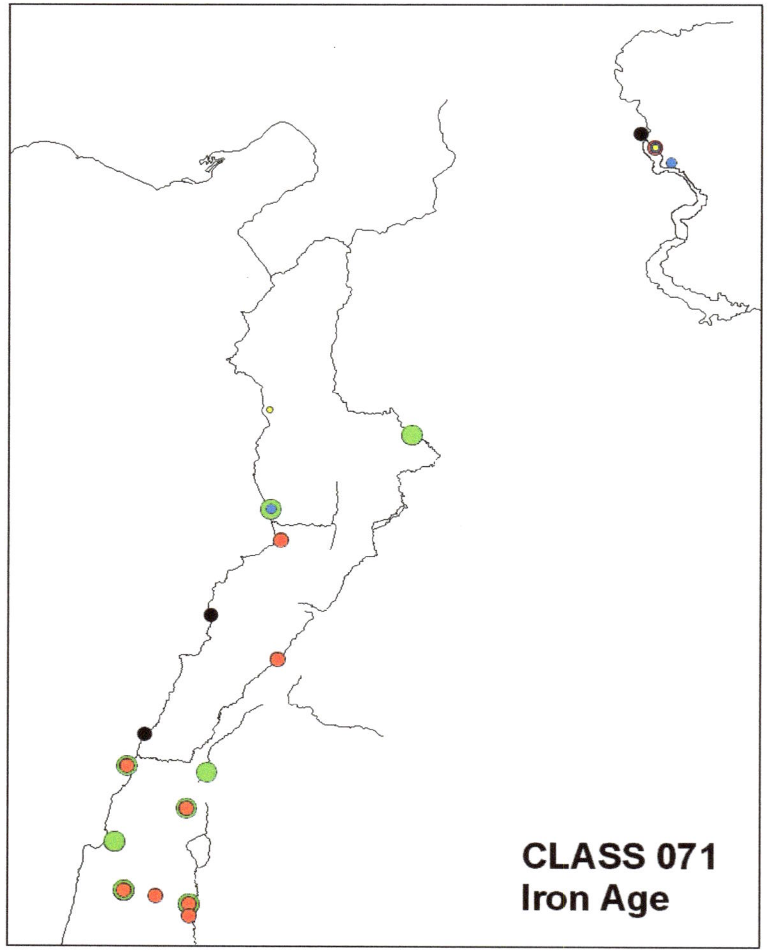

Map 40

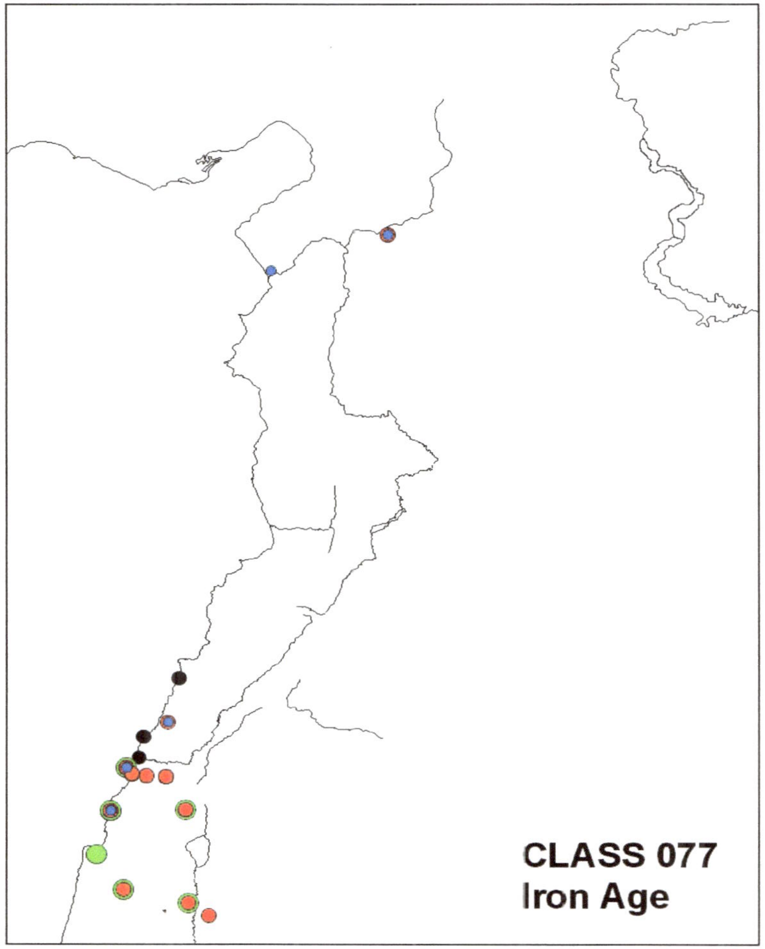

Map 41

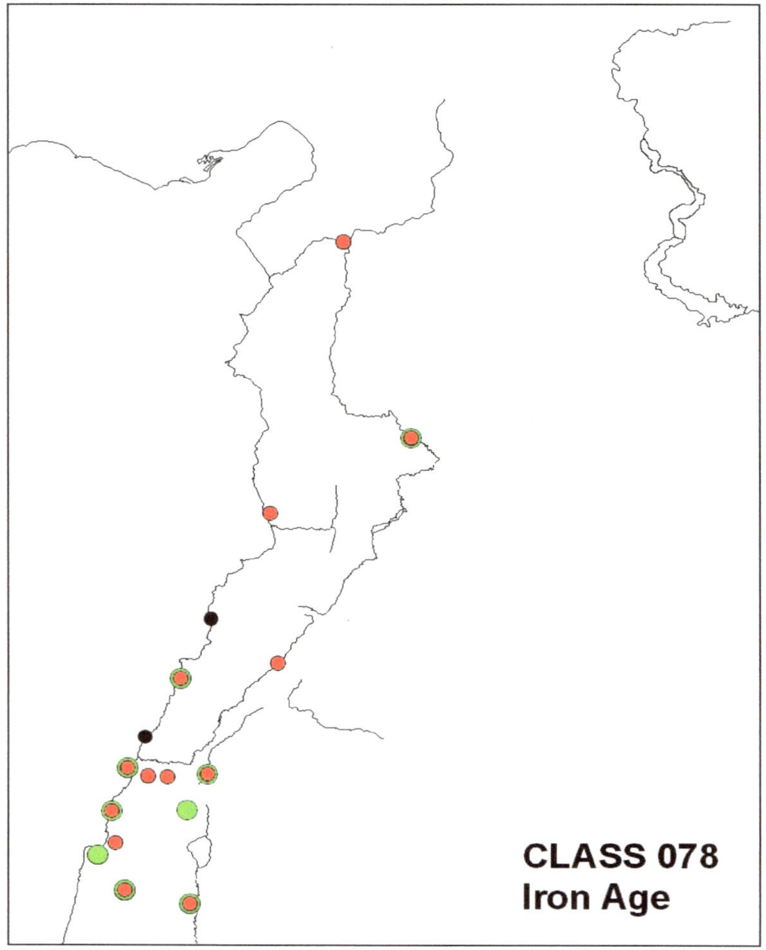

Map 42

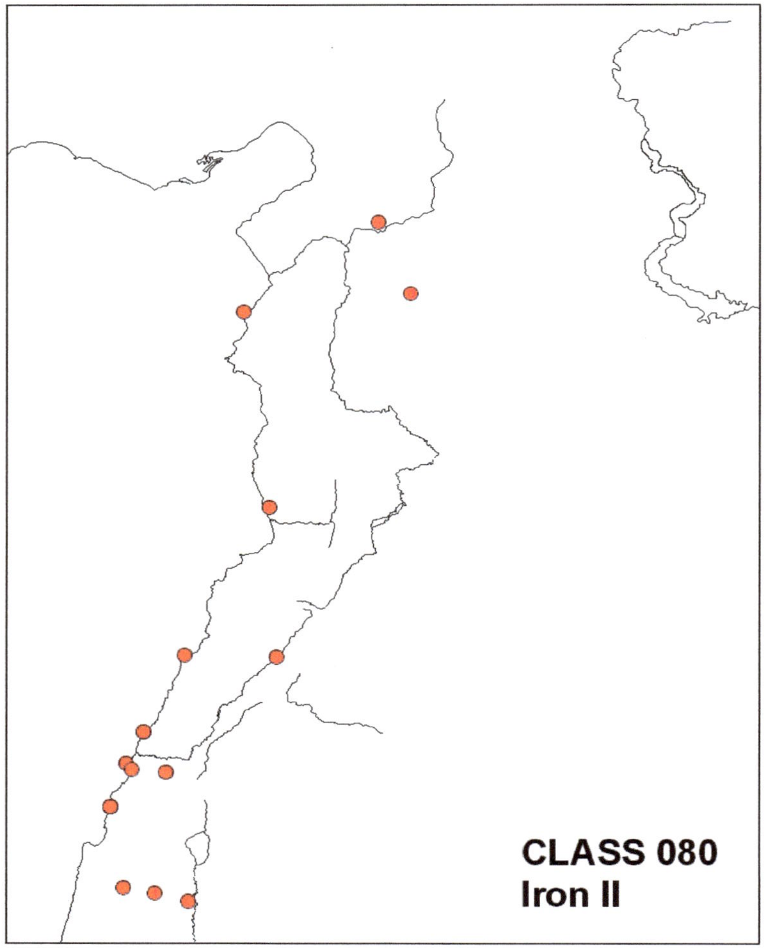

Map 43

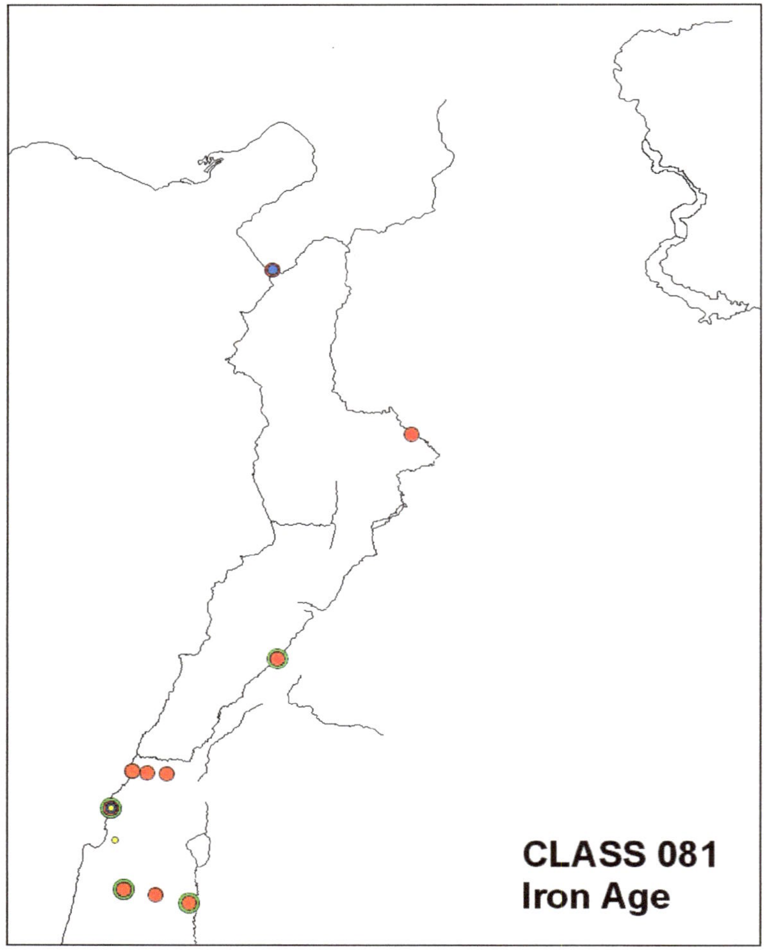

Map 44

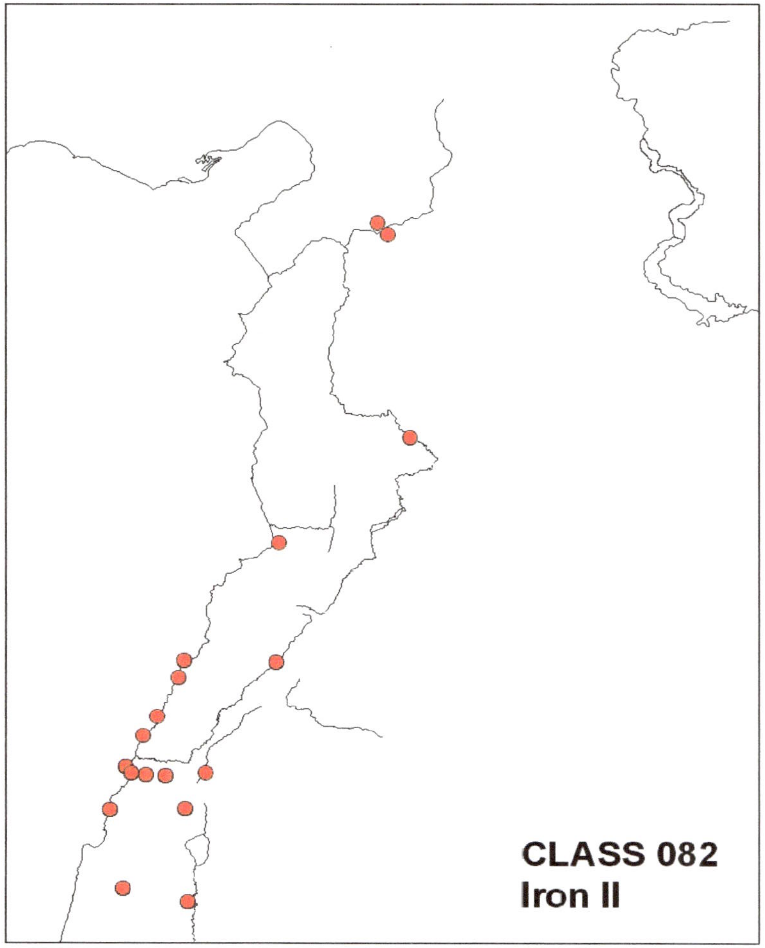

Map 45

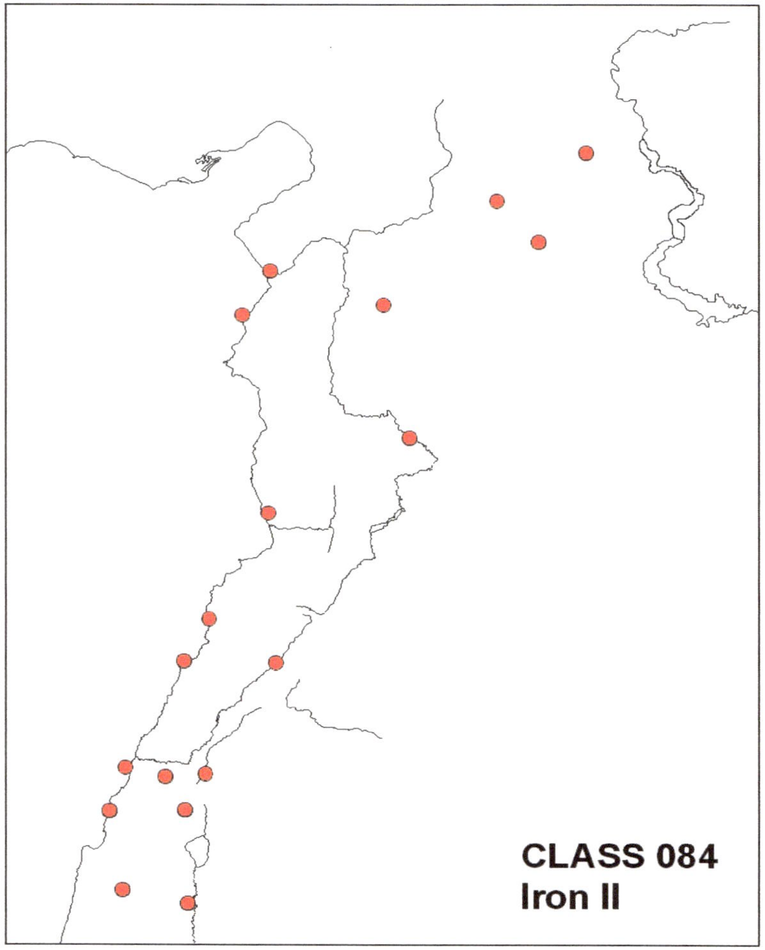

Map 46

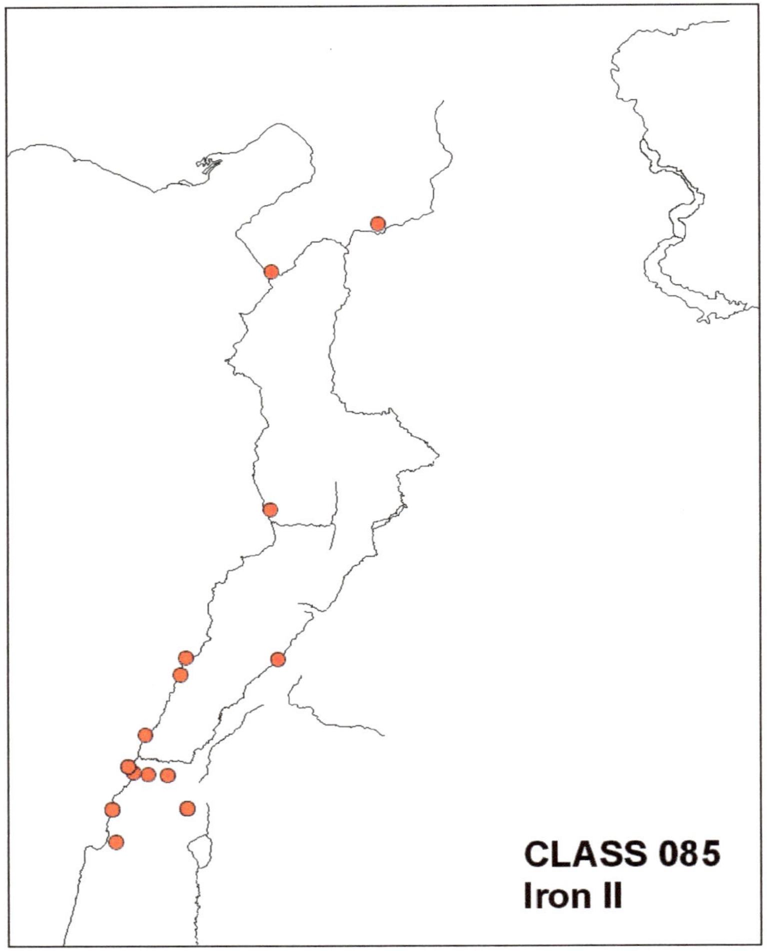

Map 47

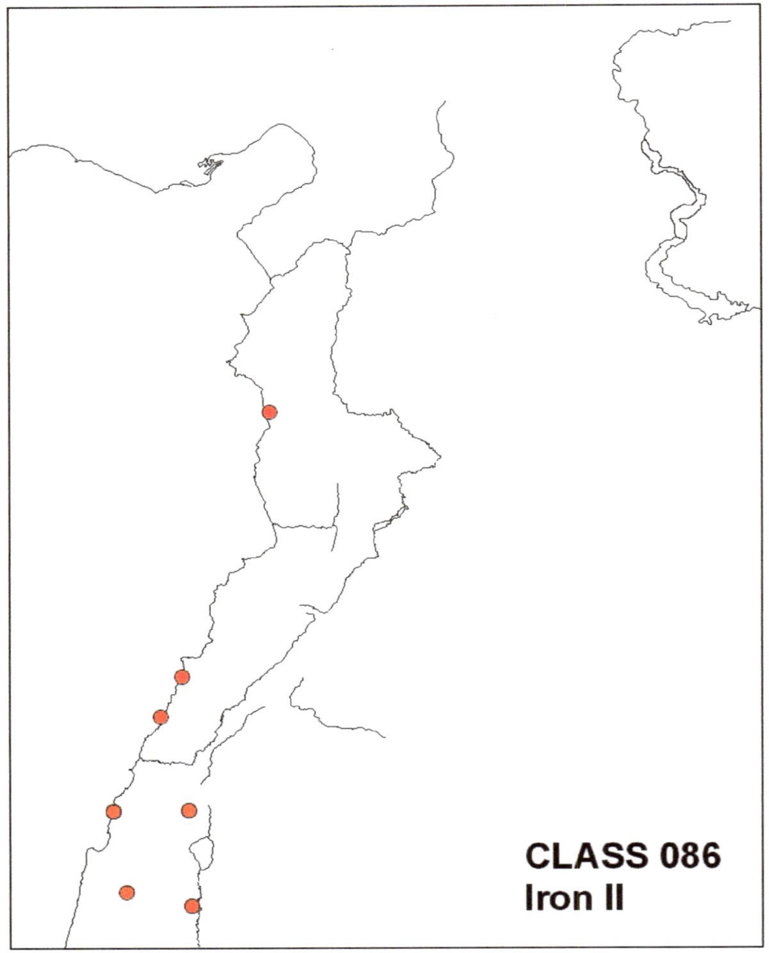

Map 48

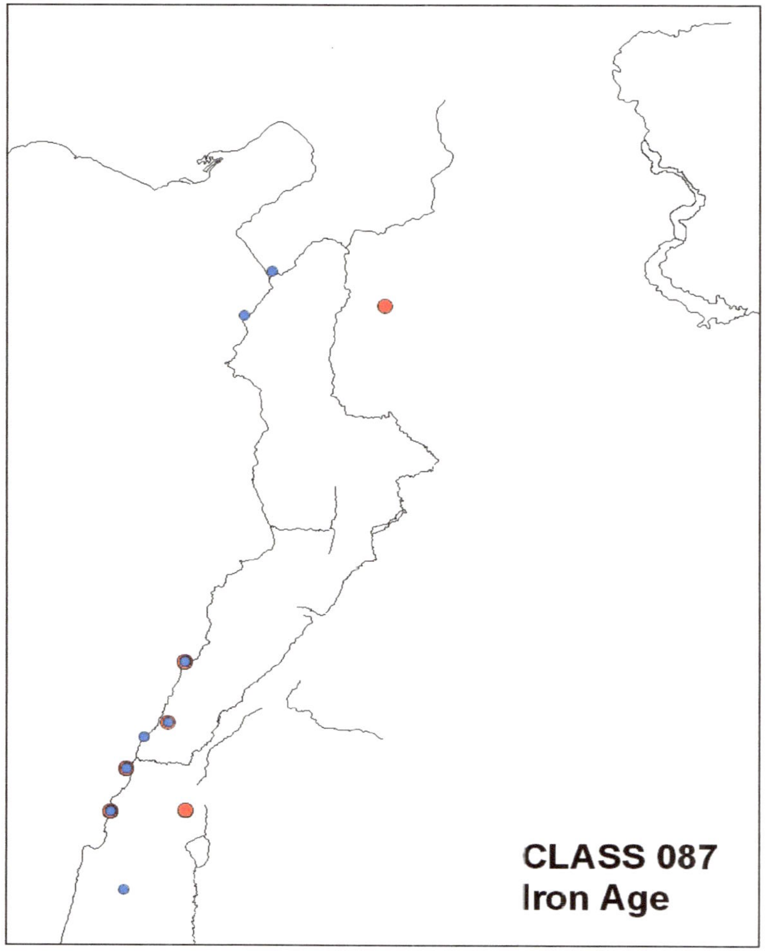

Map 49

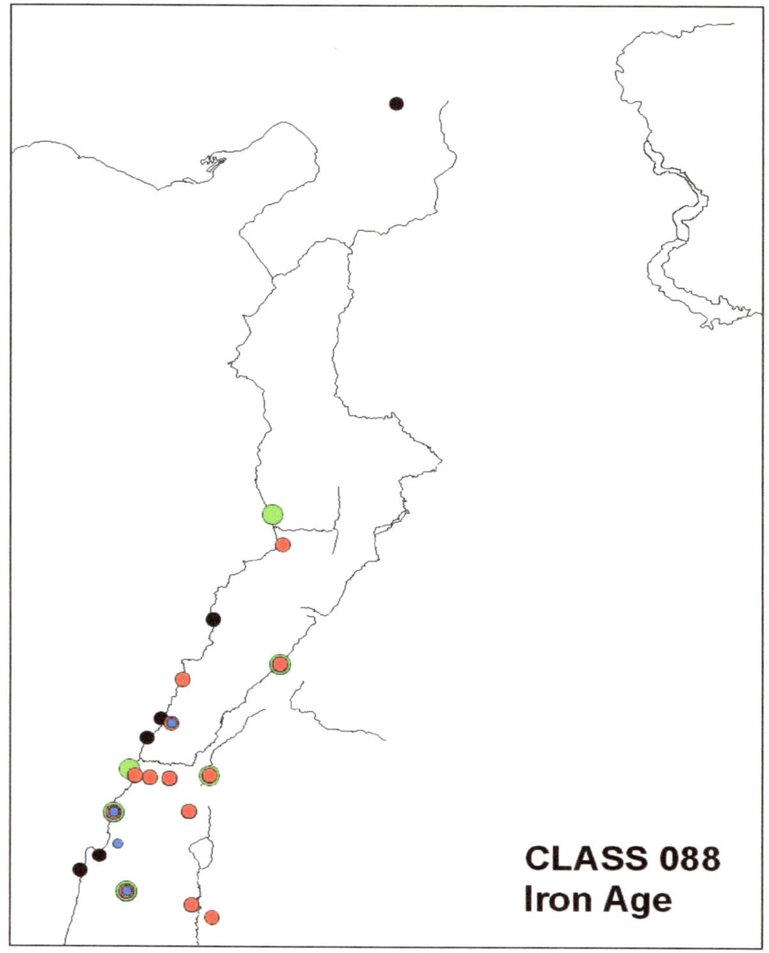

Map 50

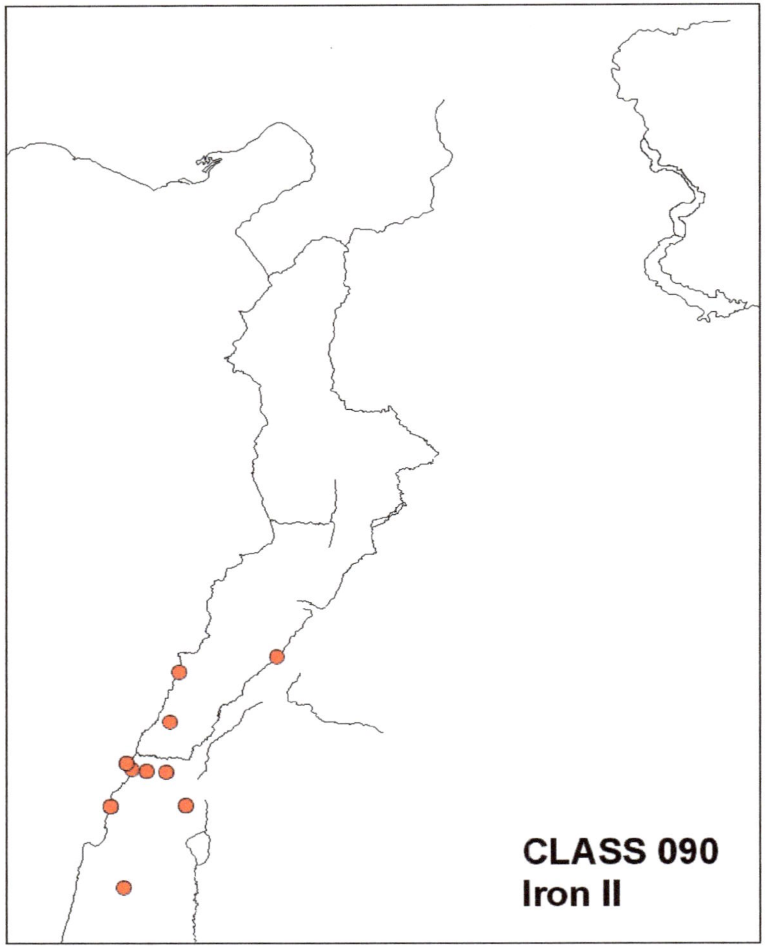

Map 51

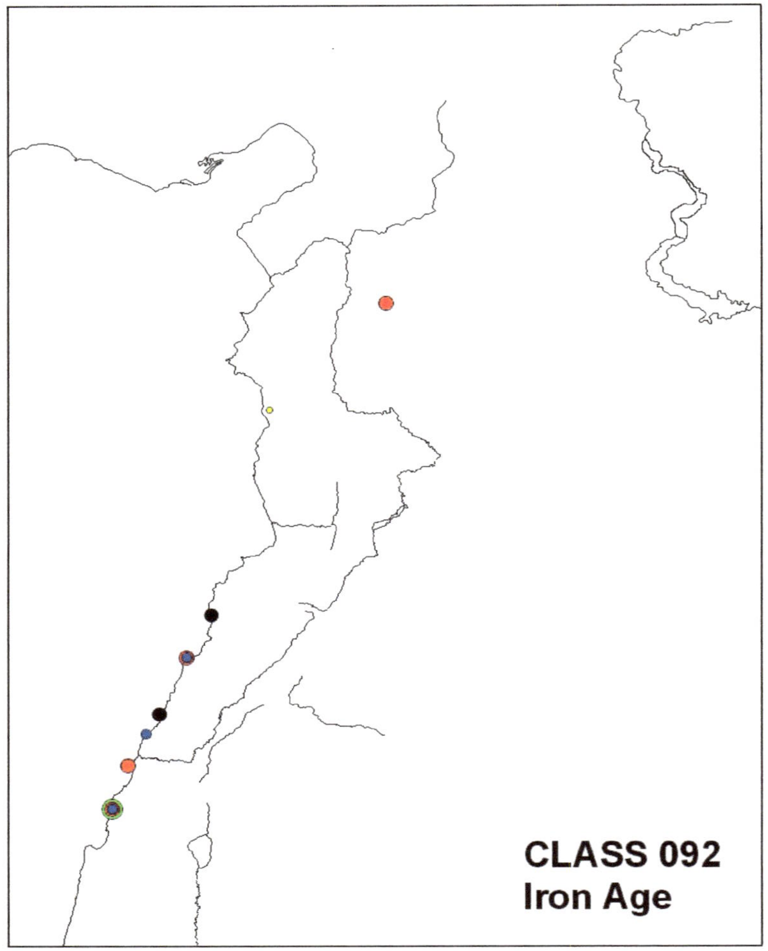

Map 52

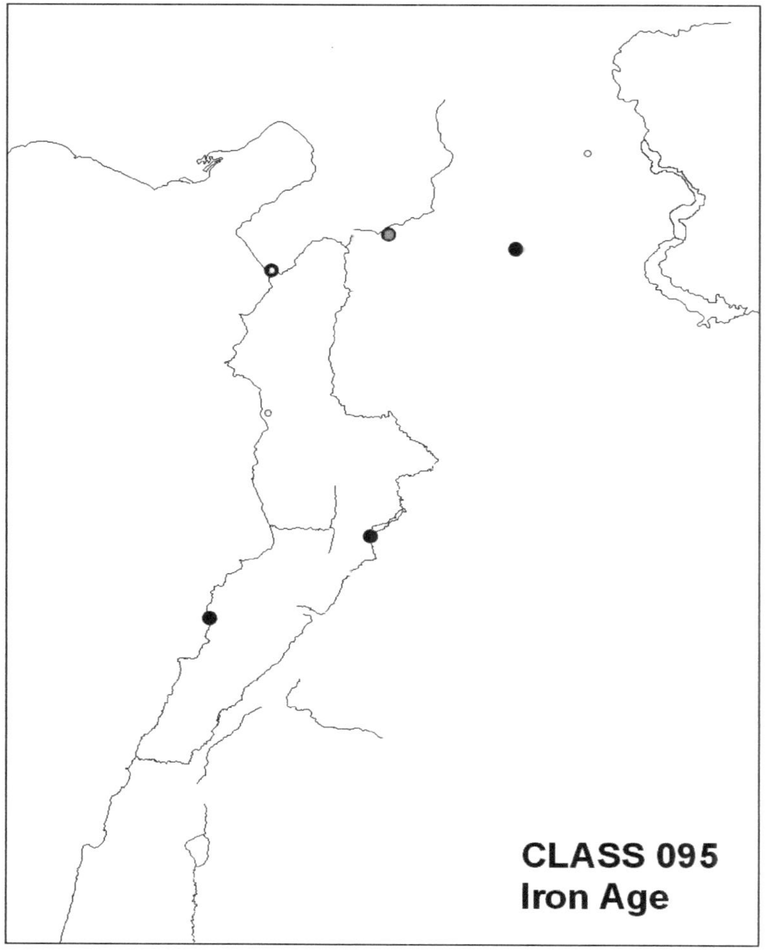

Map 53

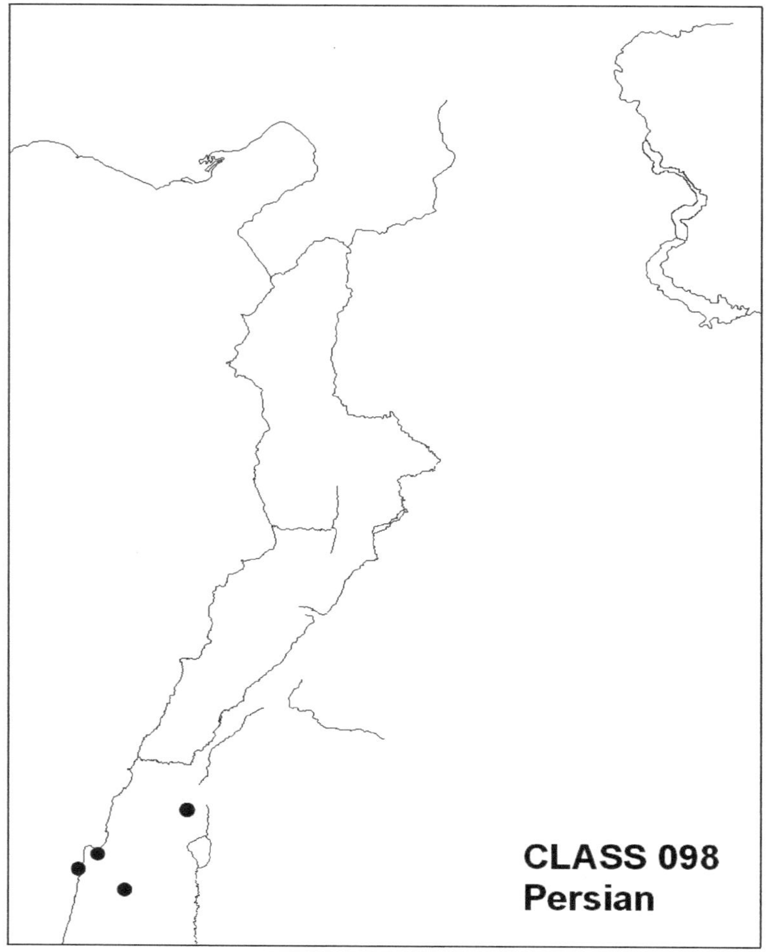

Map 54

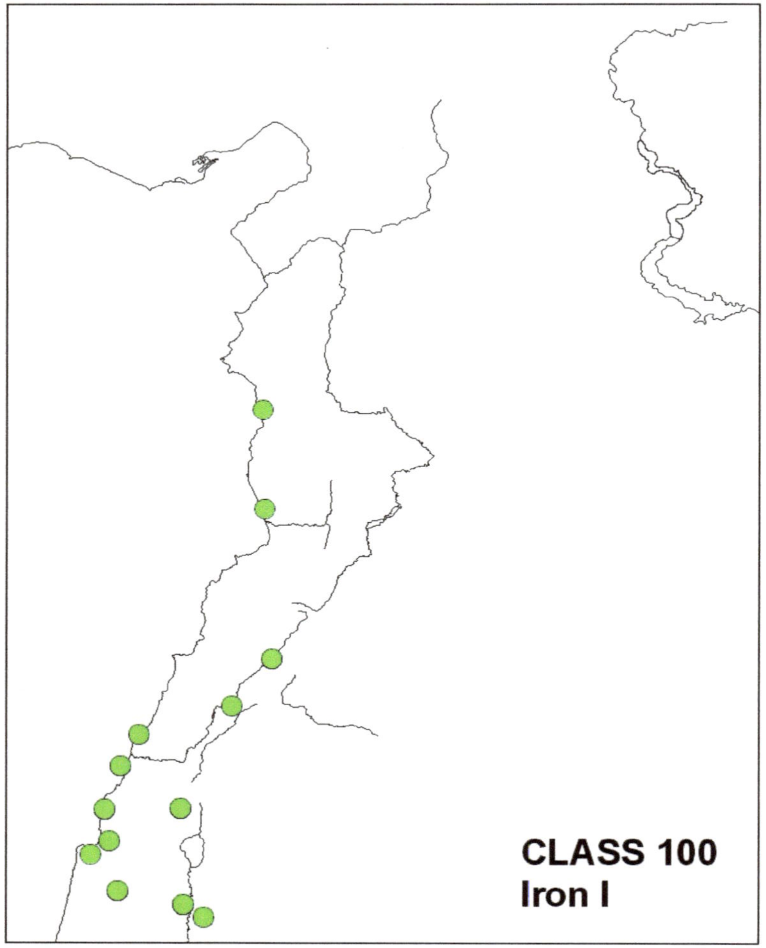

Map 55

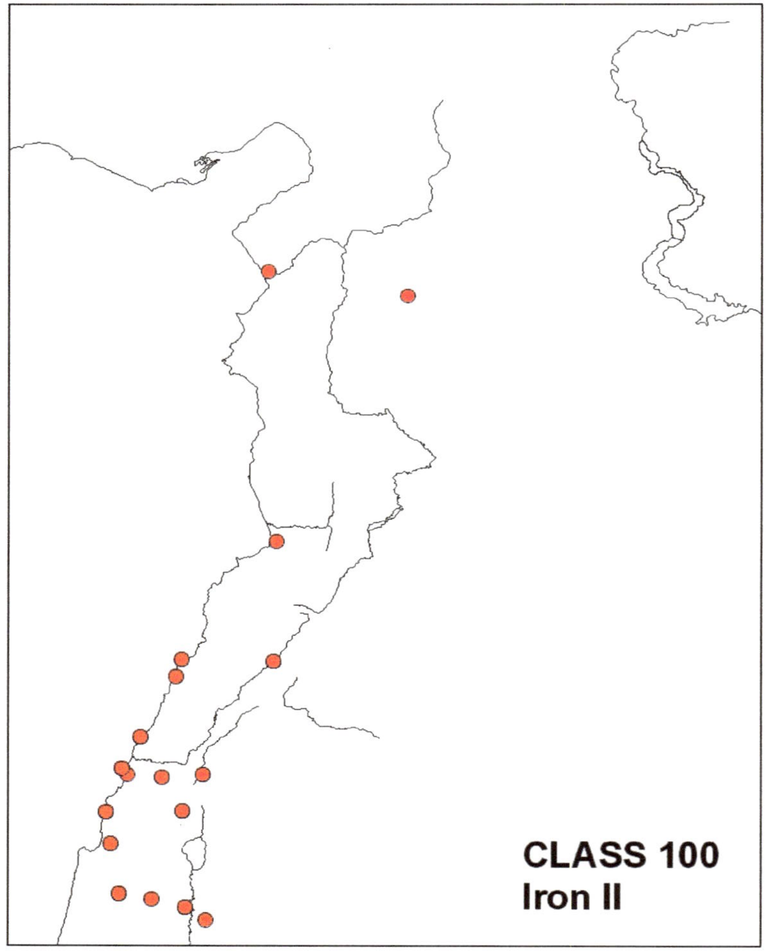

Map 56

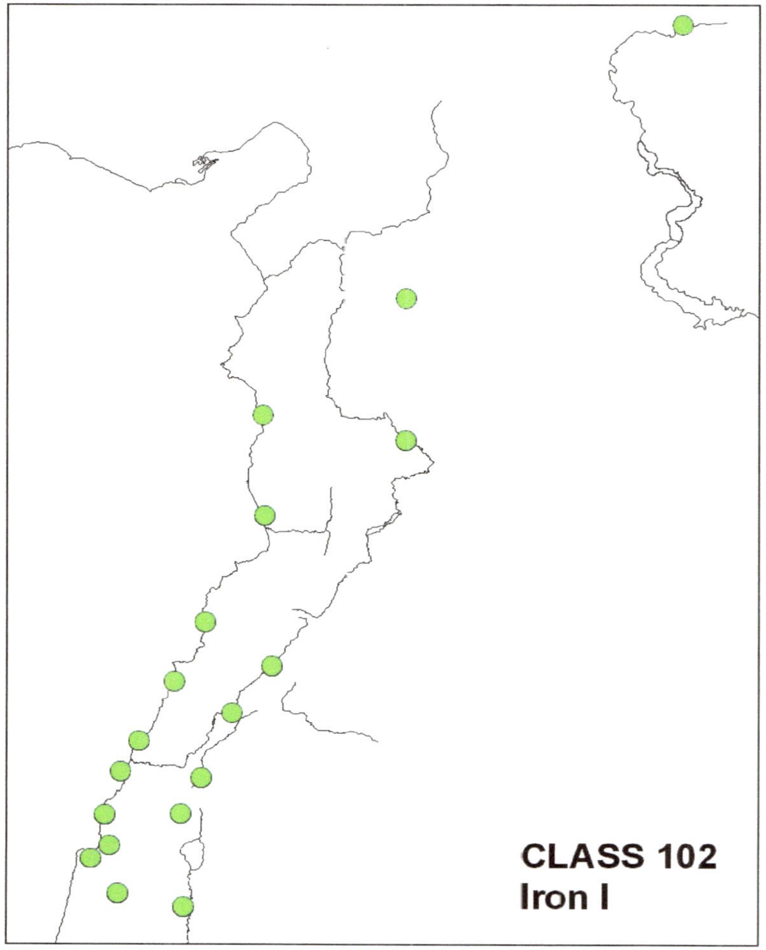

Map 57

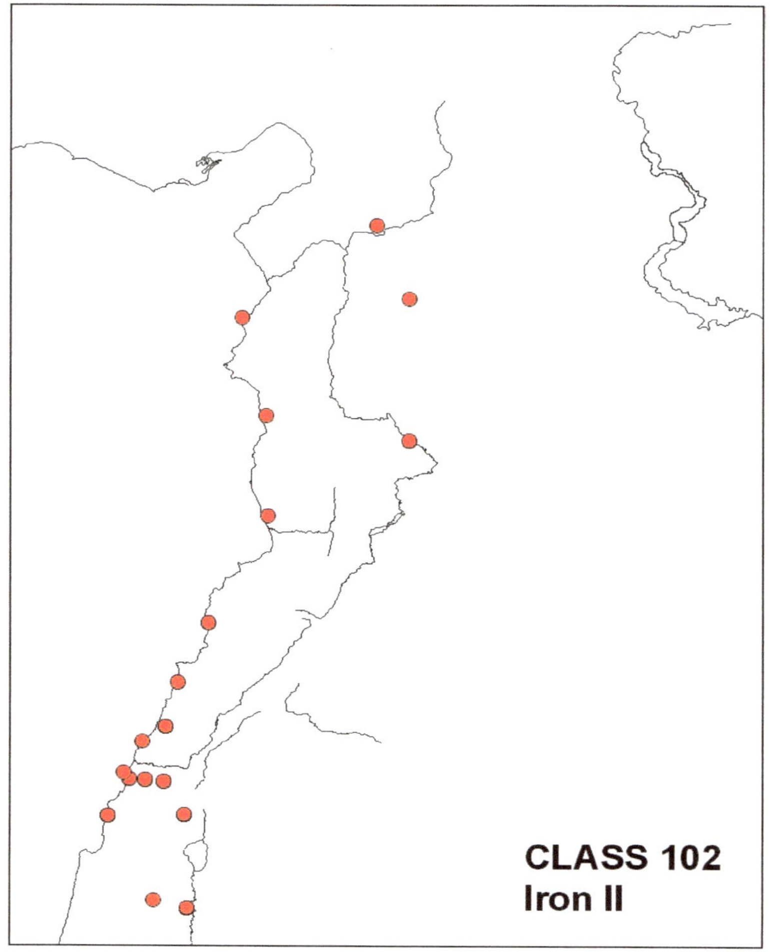

Map 58

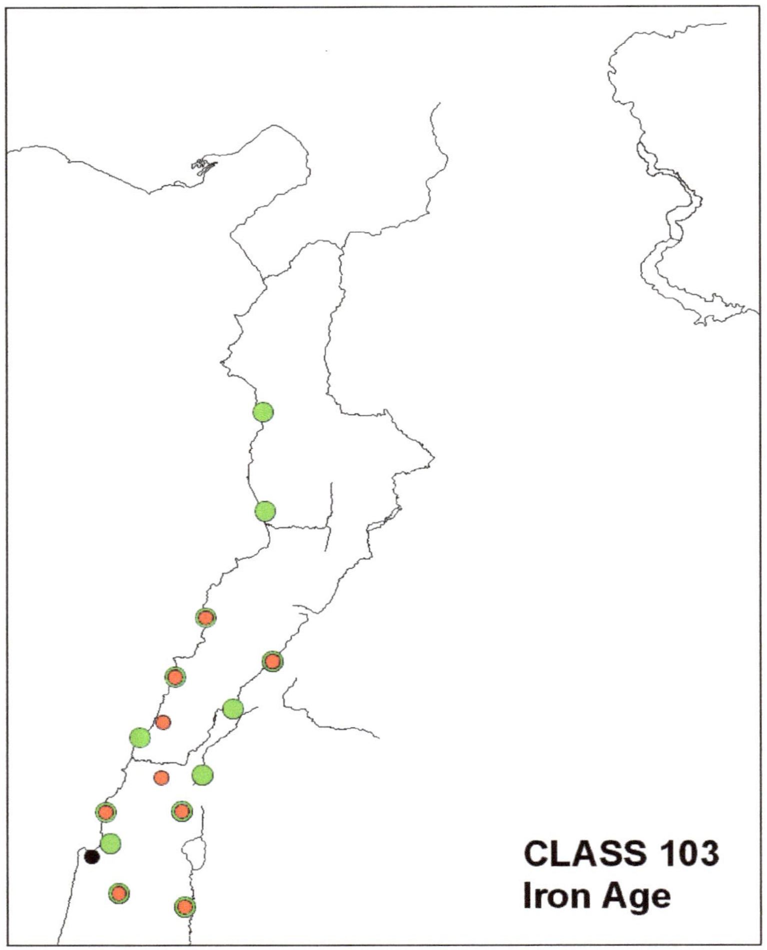

Map 59

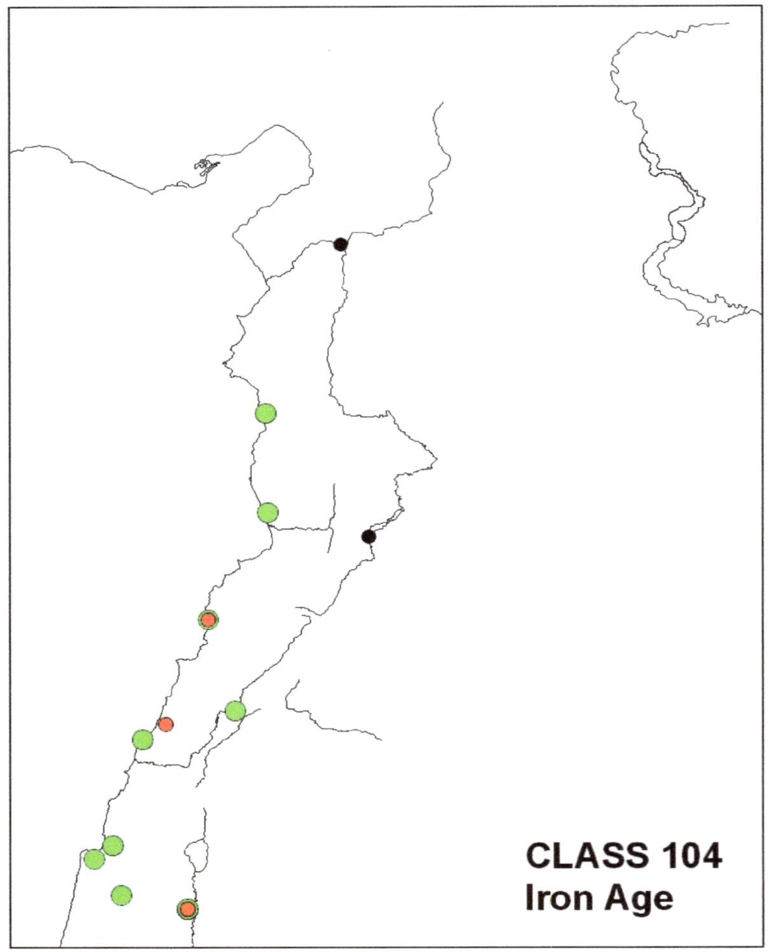

Map 60

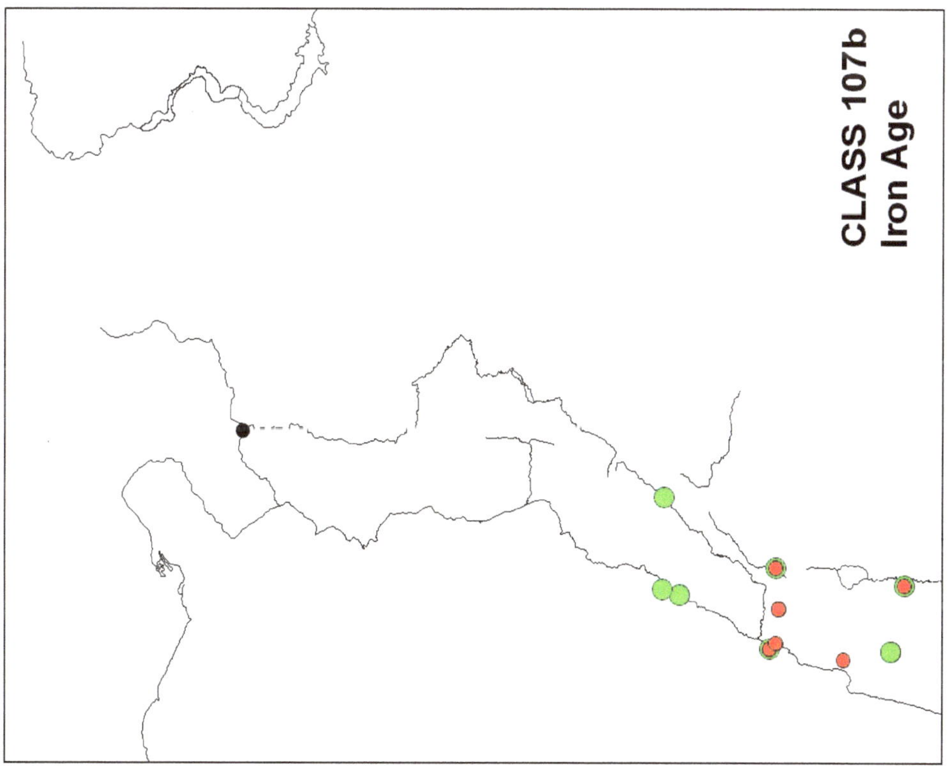

Map 61

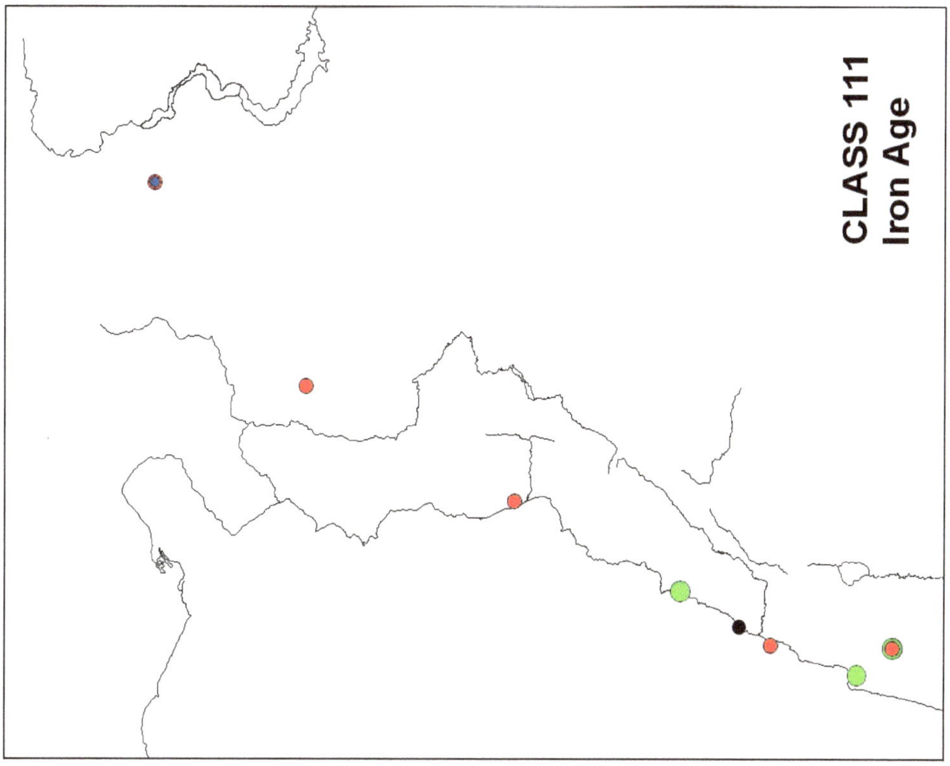

Map 62

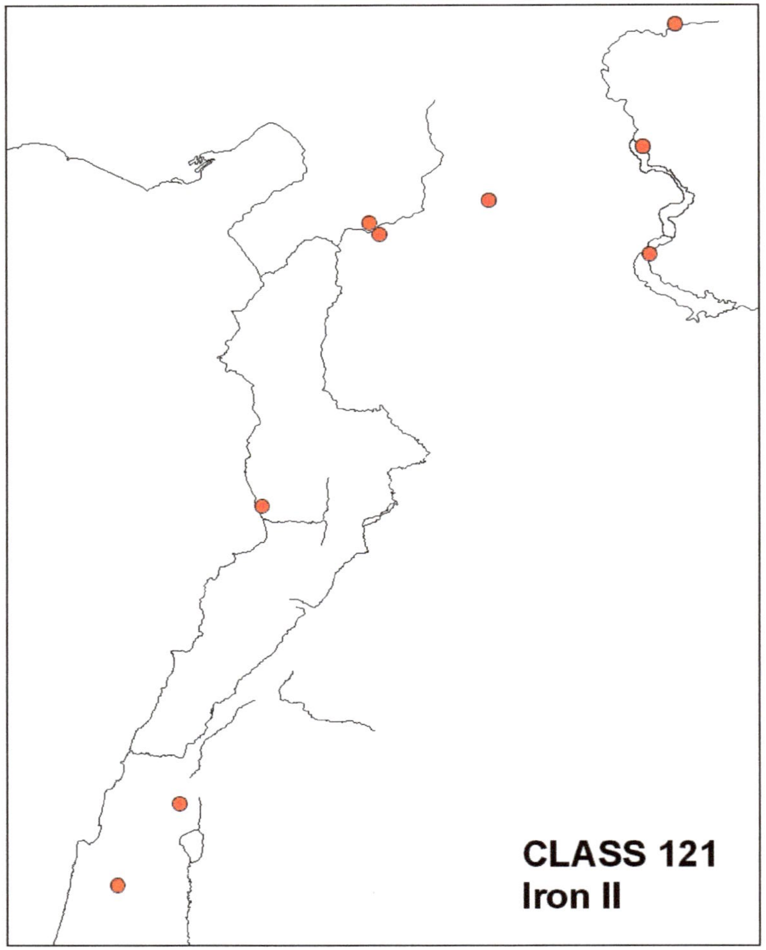

Map 63

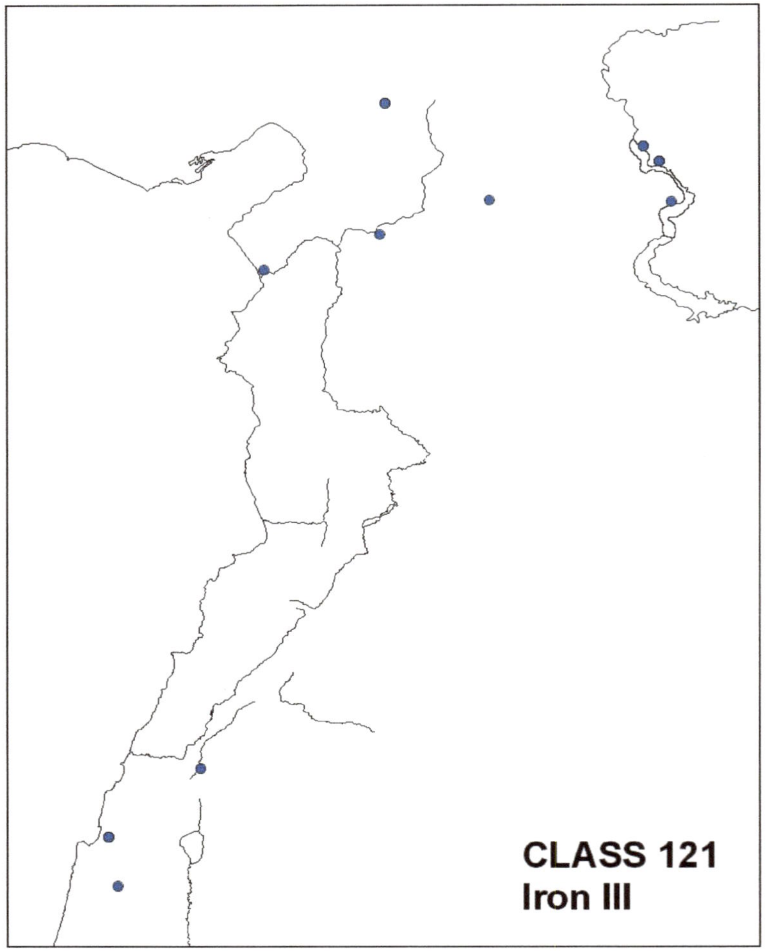

Map 64

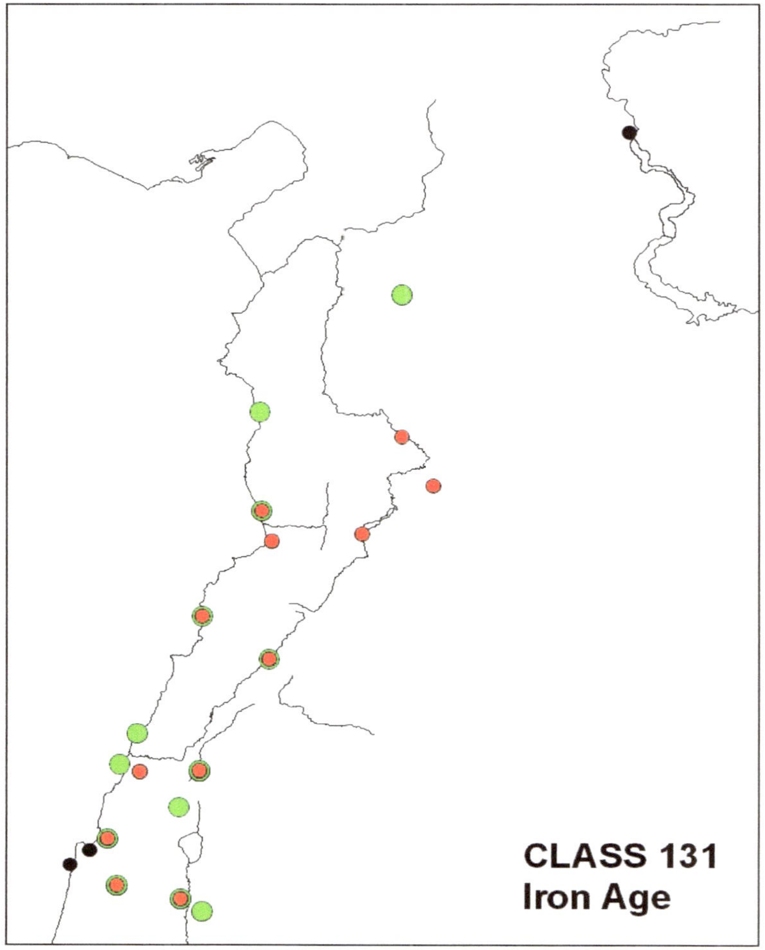

Map 65

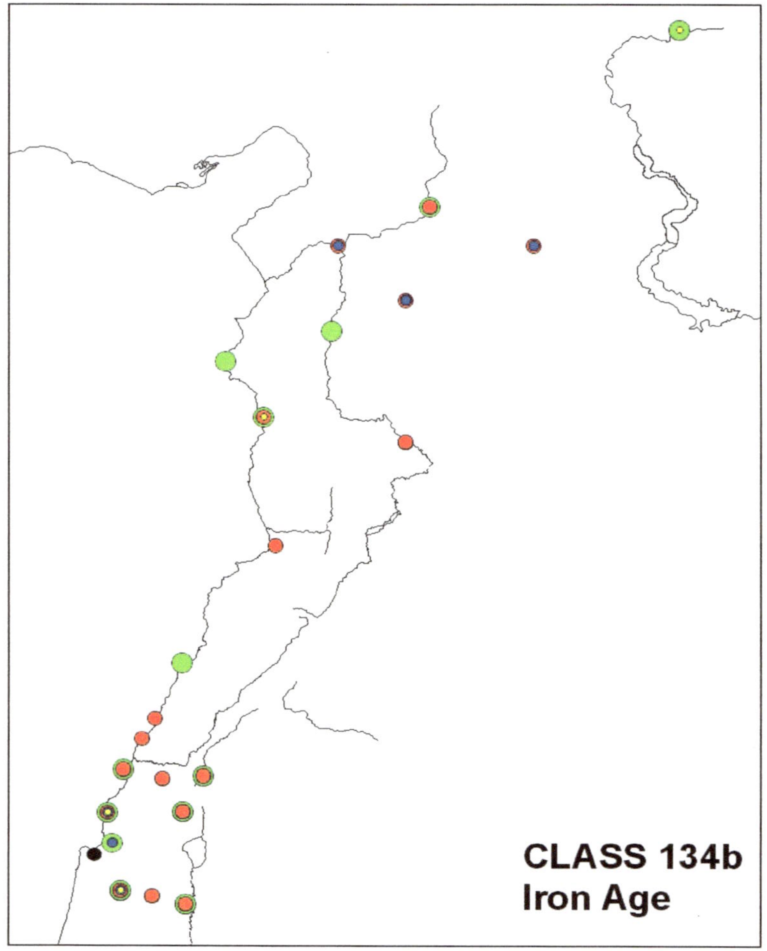

Map 66

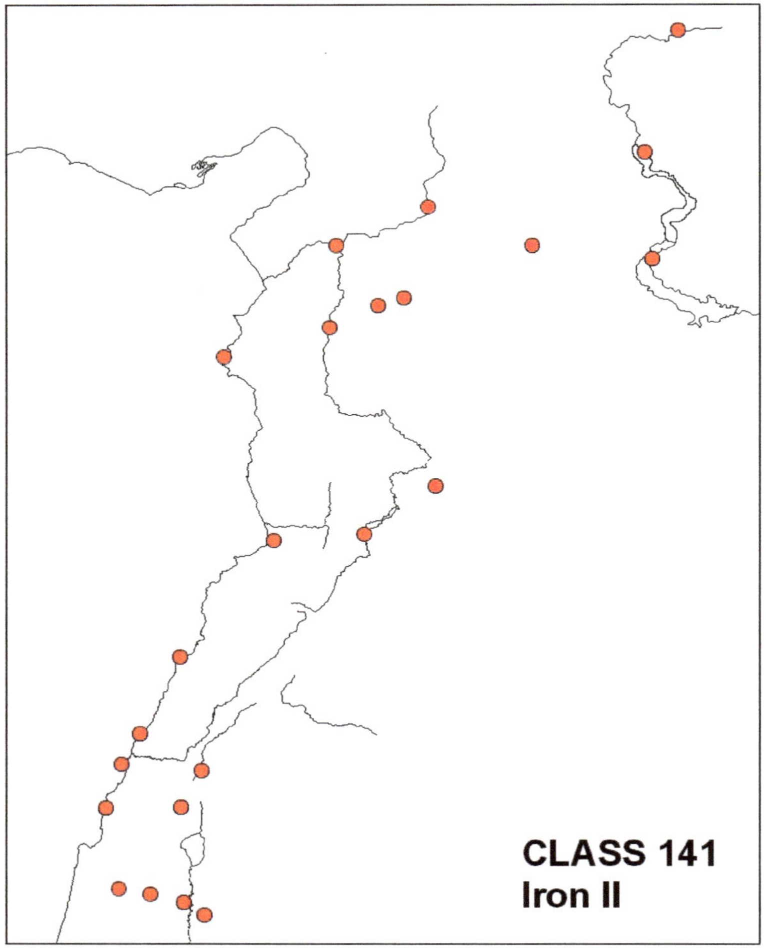

Map 67

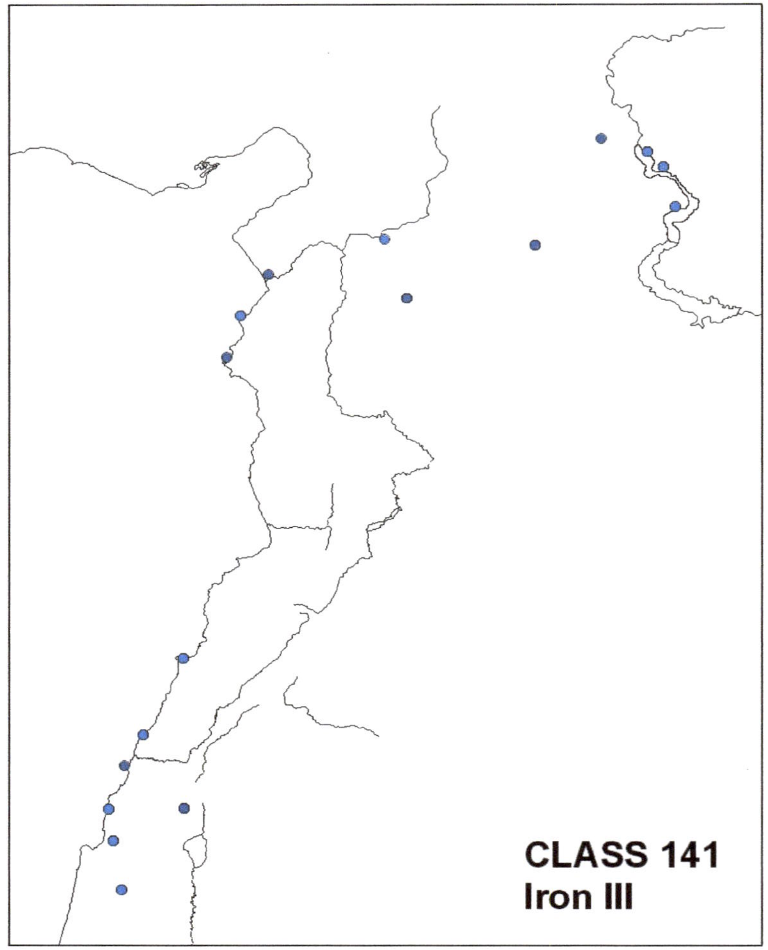

Map 68

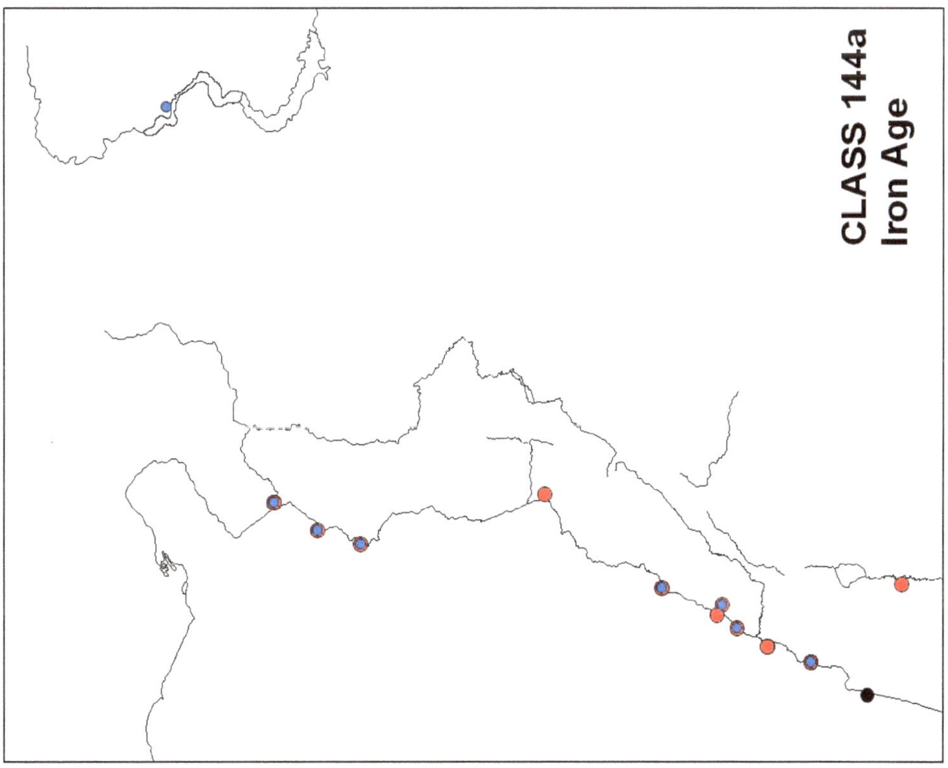

Map 69

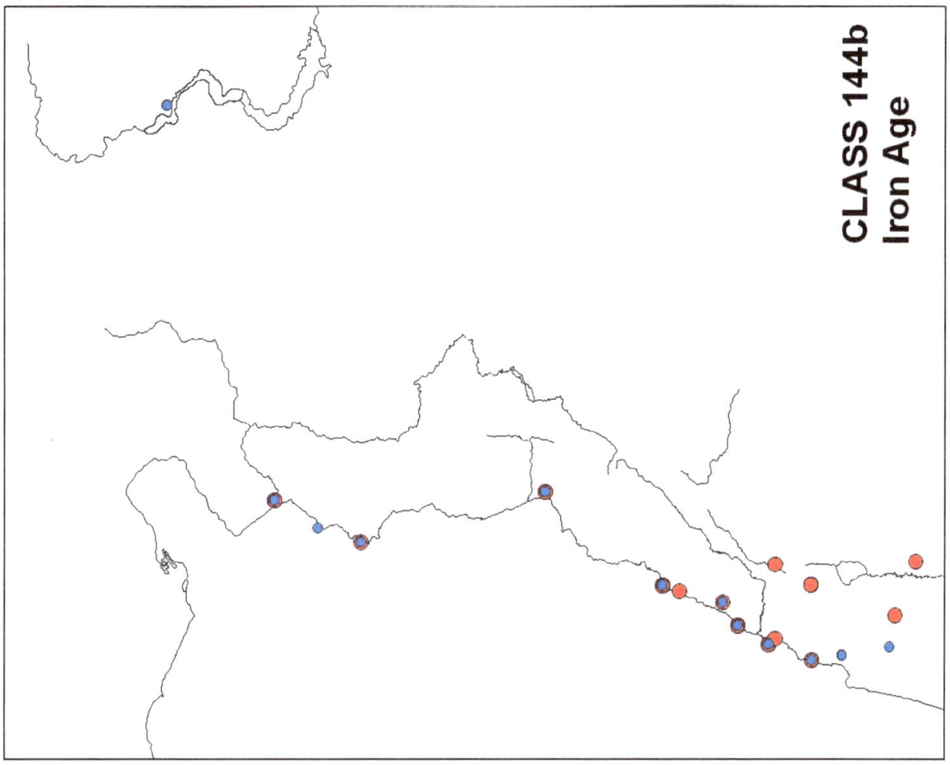

Map 70

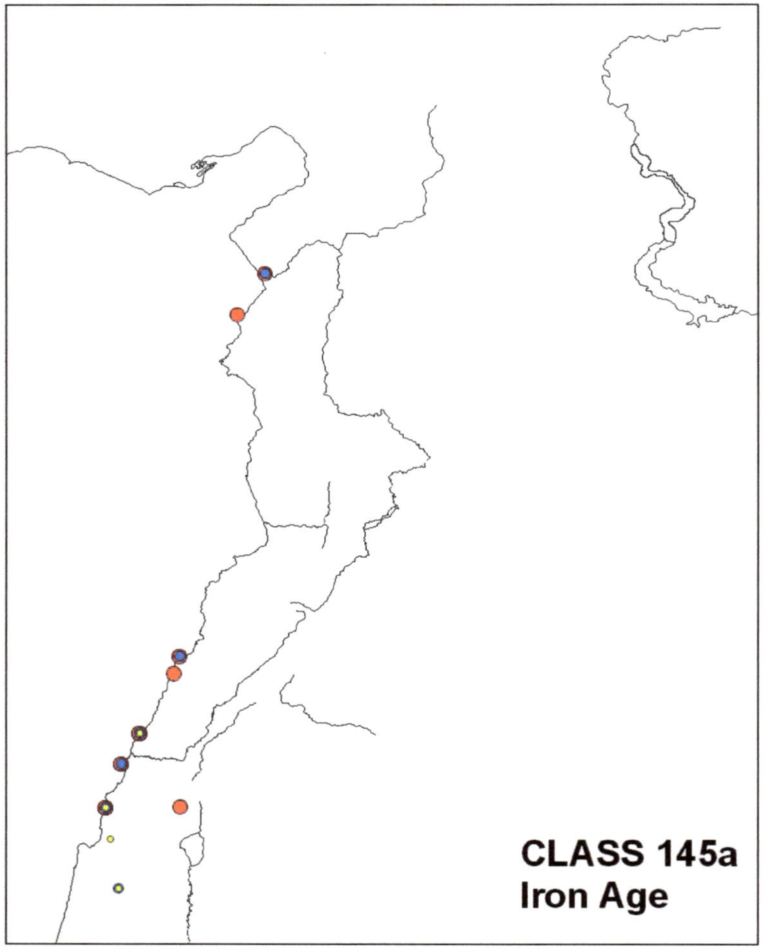

Map 71

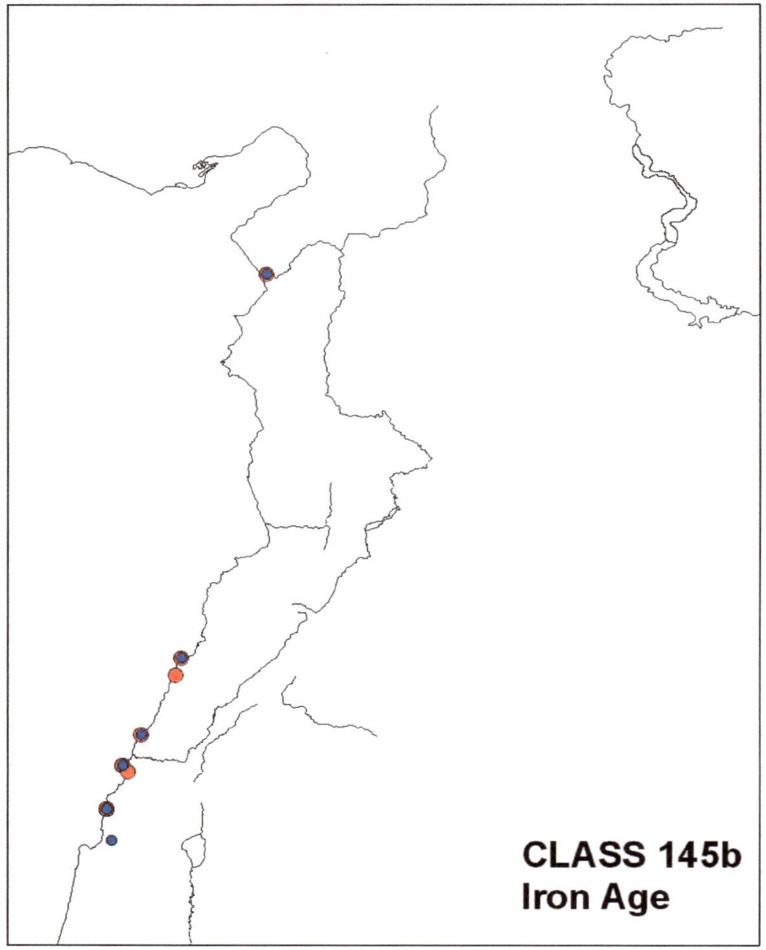

Map 72

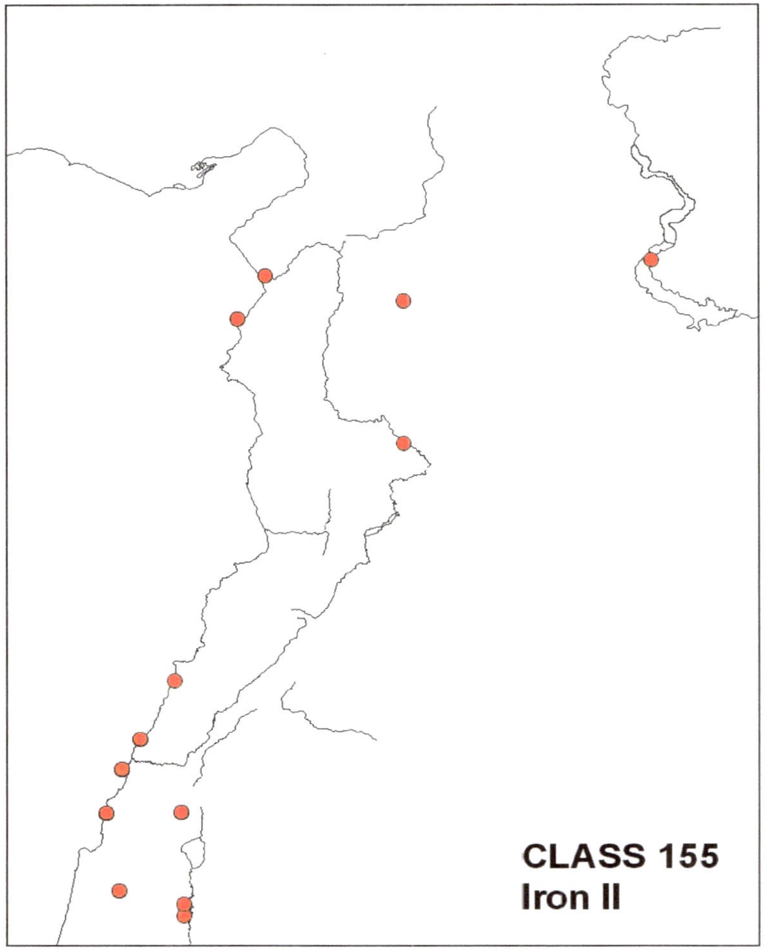

Map 73

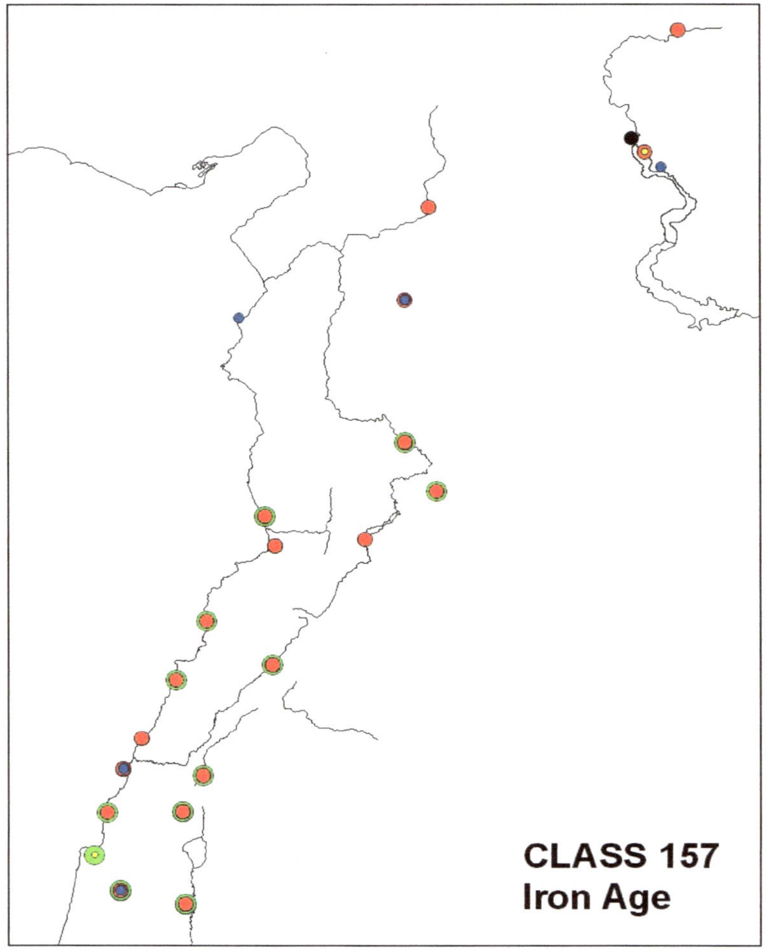

Map 74

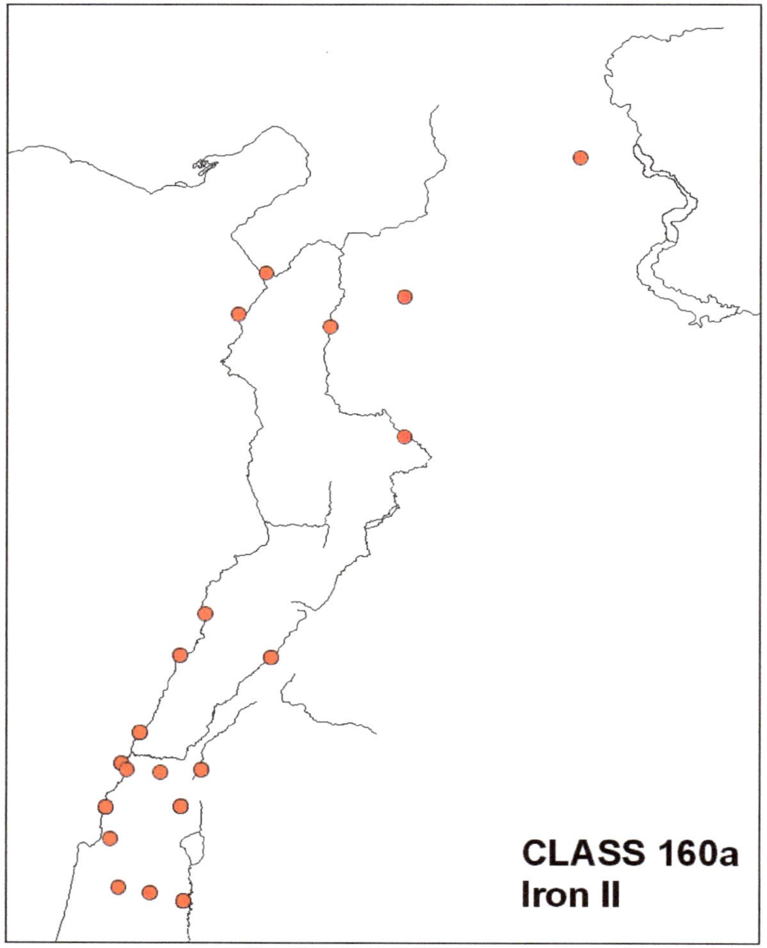

Map 75

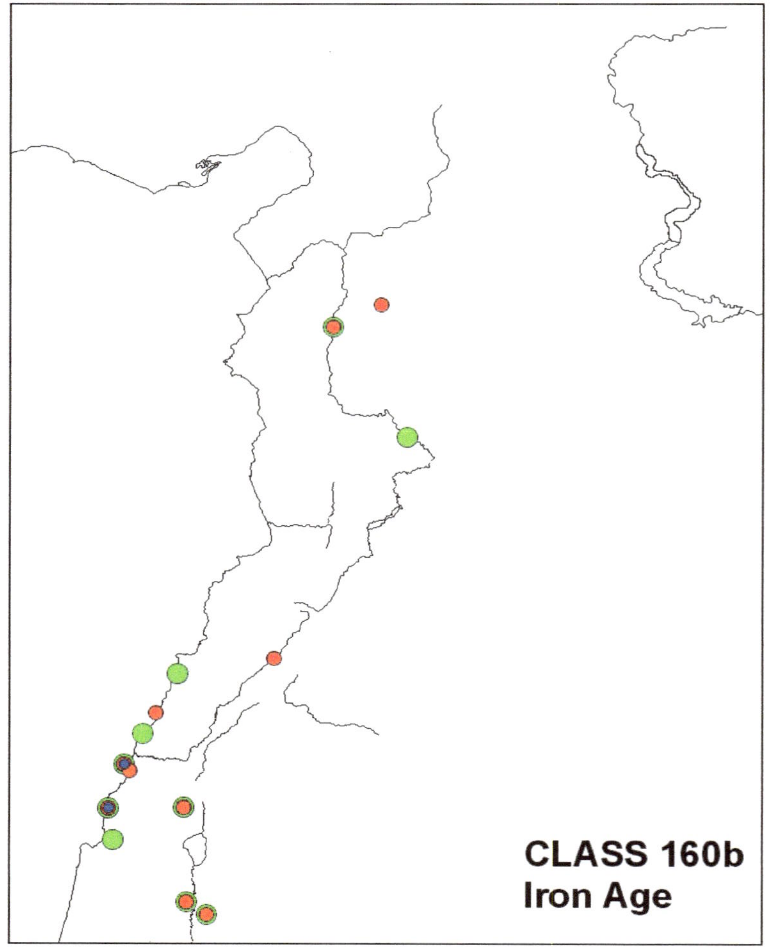

Map 76

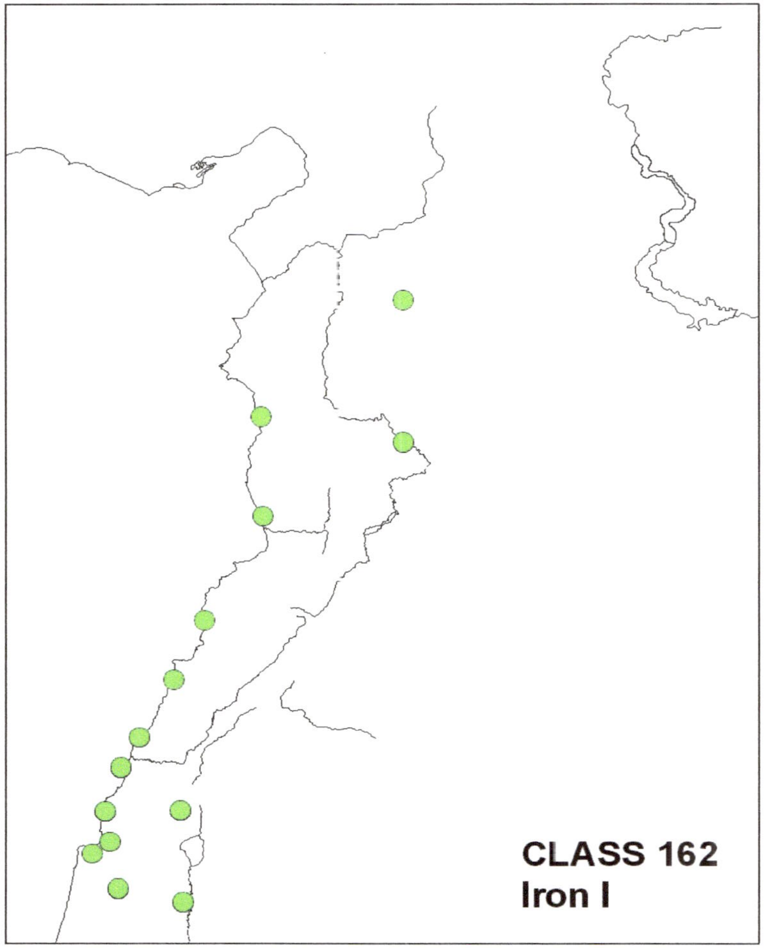

Map 77

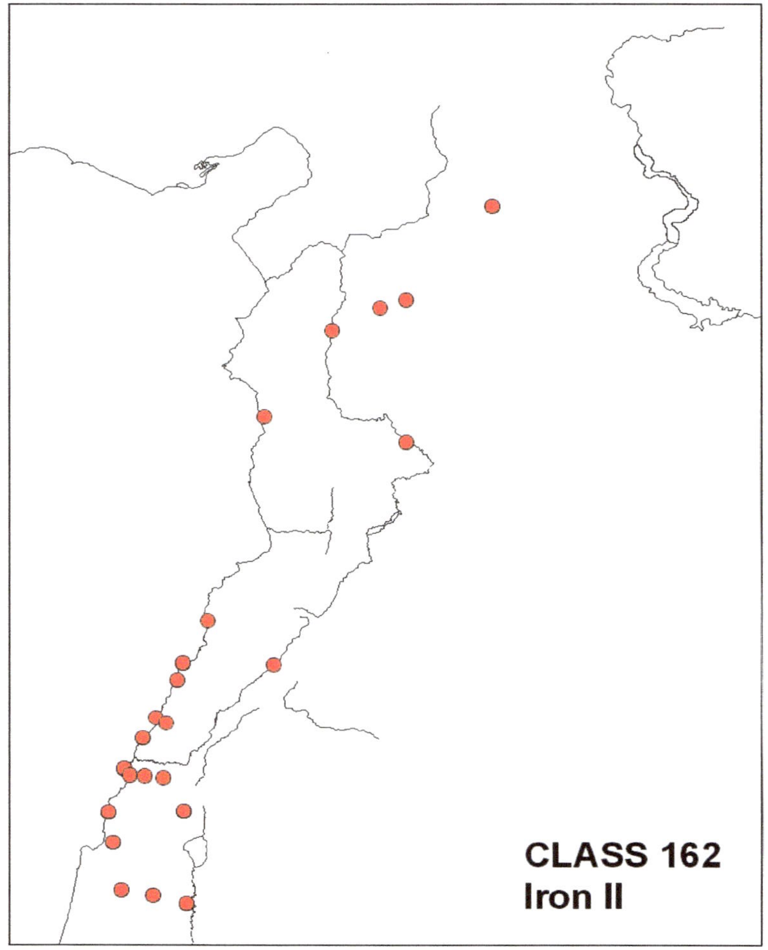

Map 78

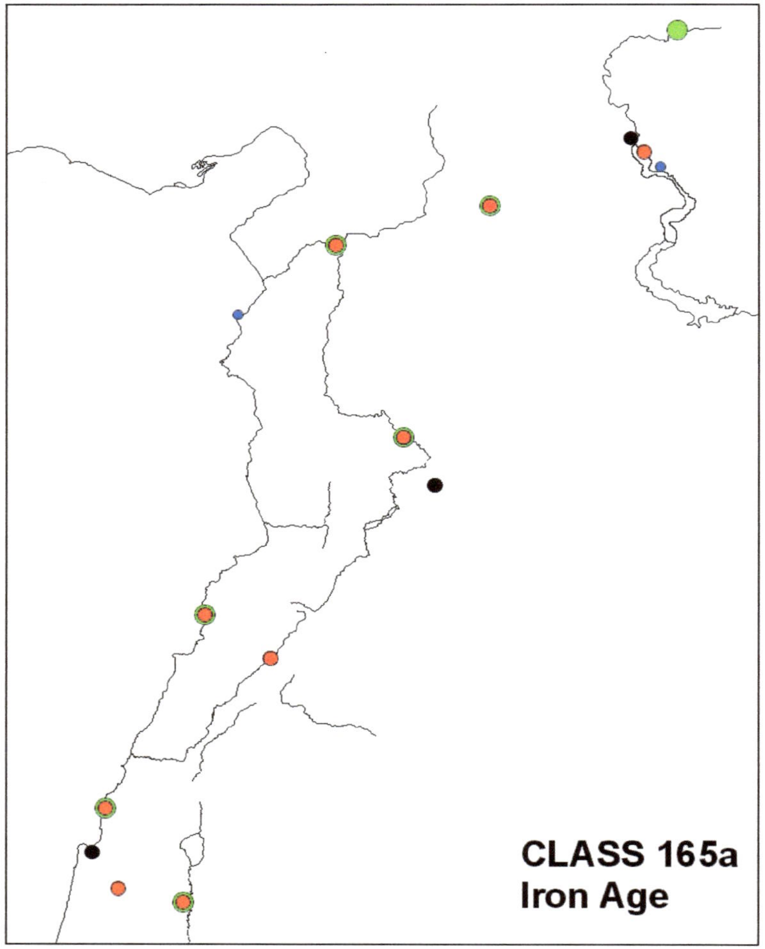

Map 79

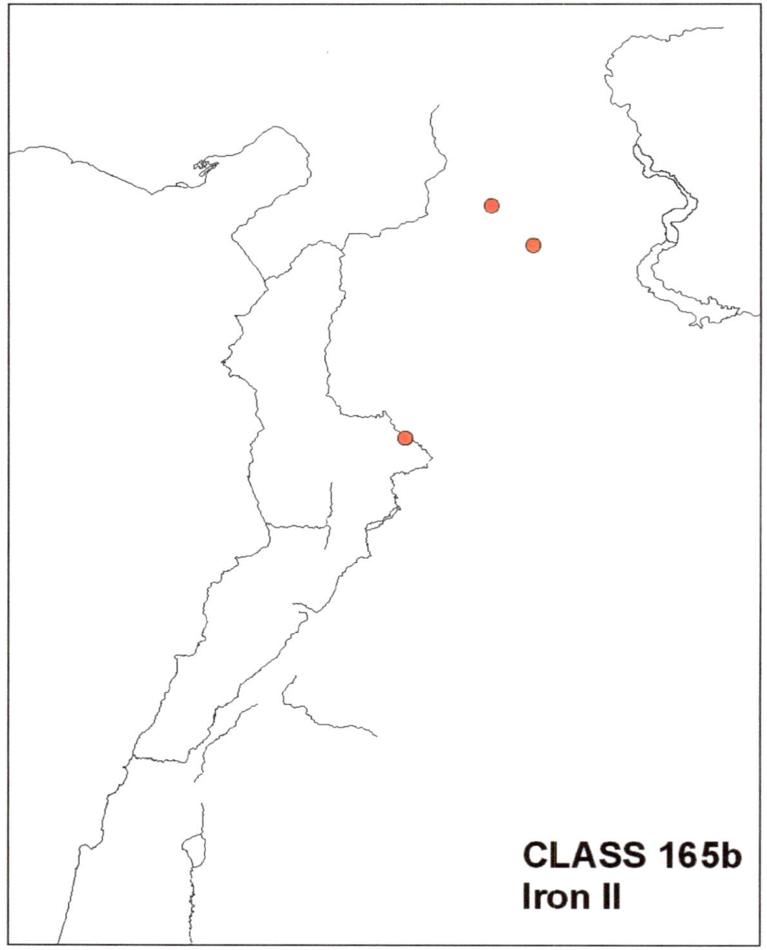

Map 80

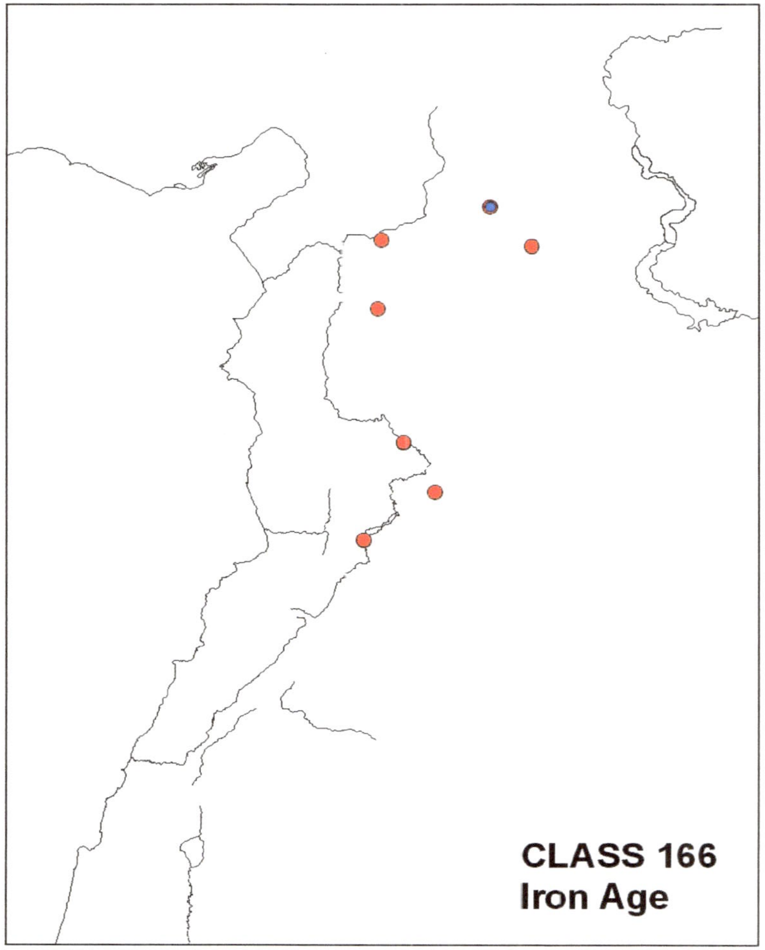

Map 81

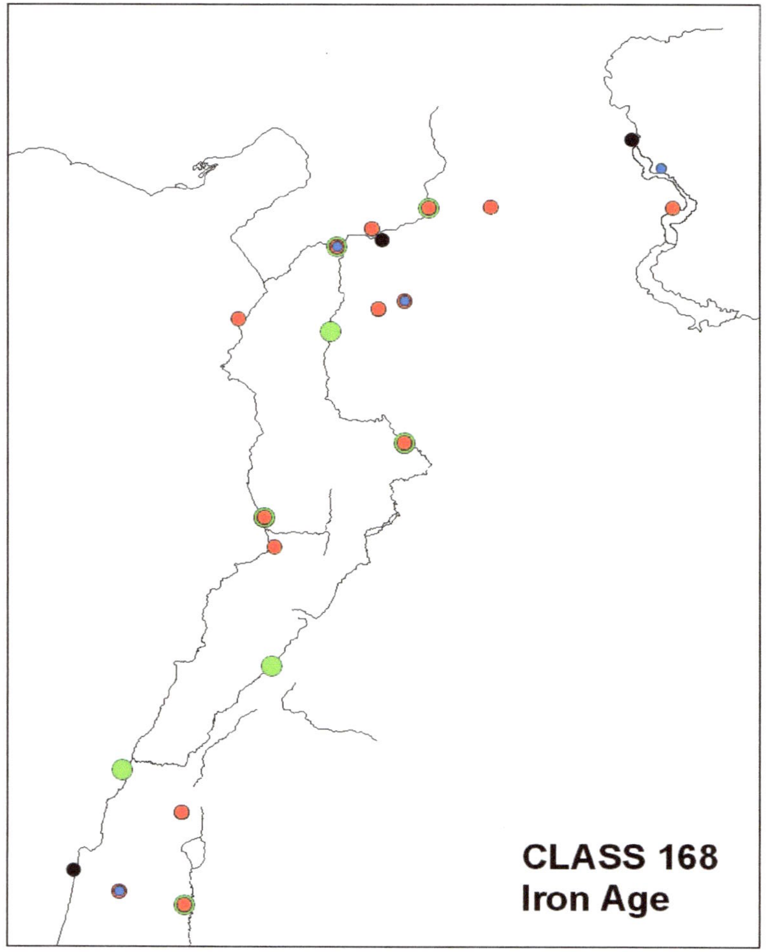

Map 82

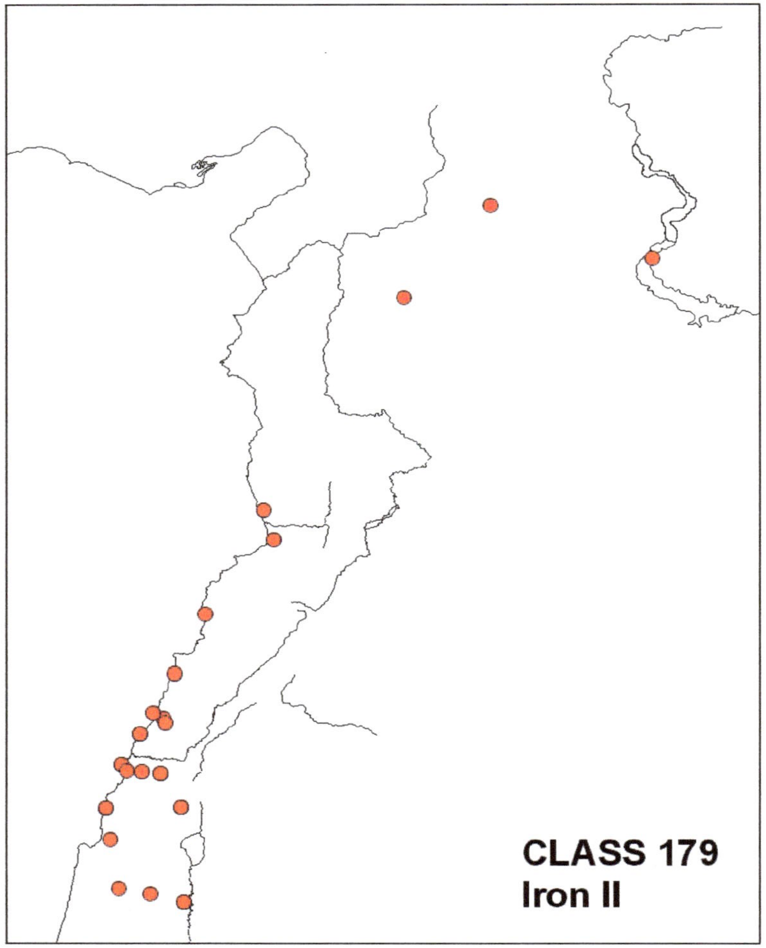

Map 83

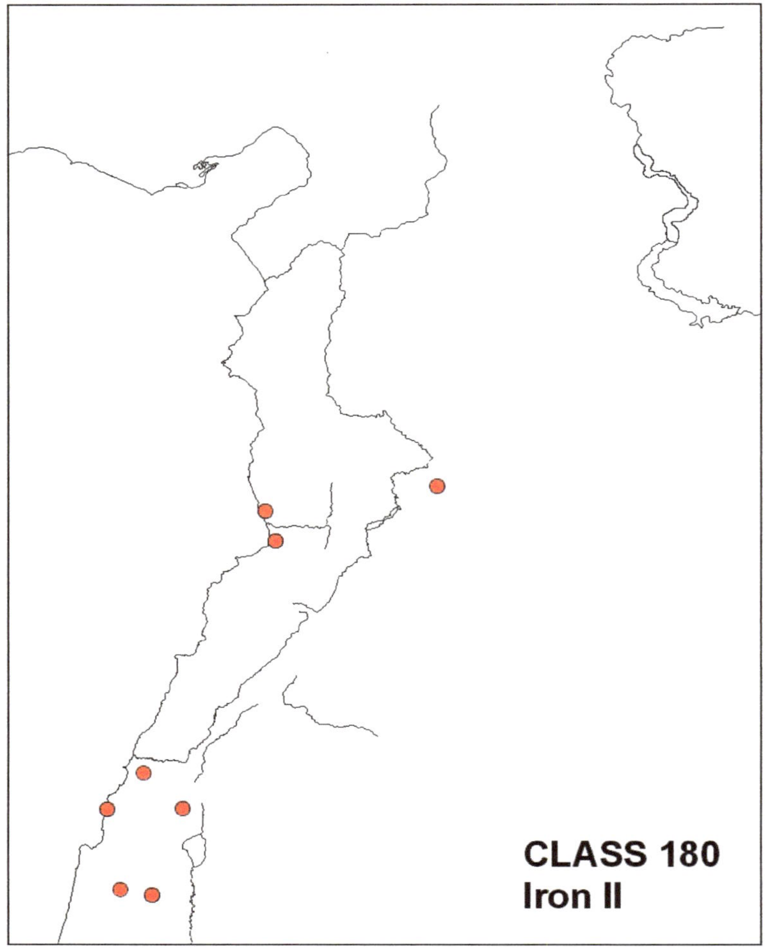

Map 84

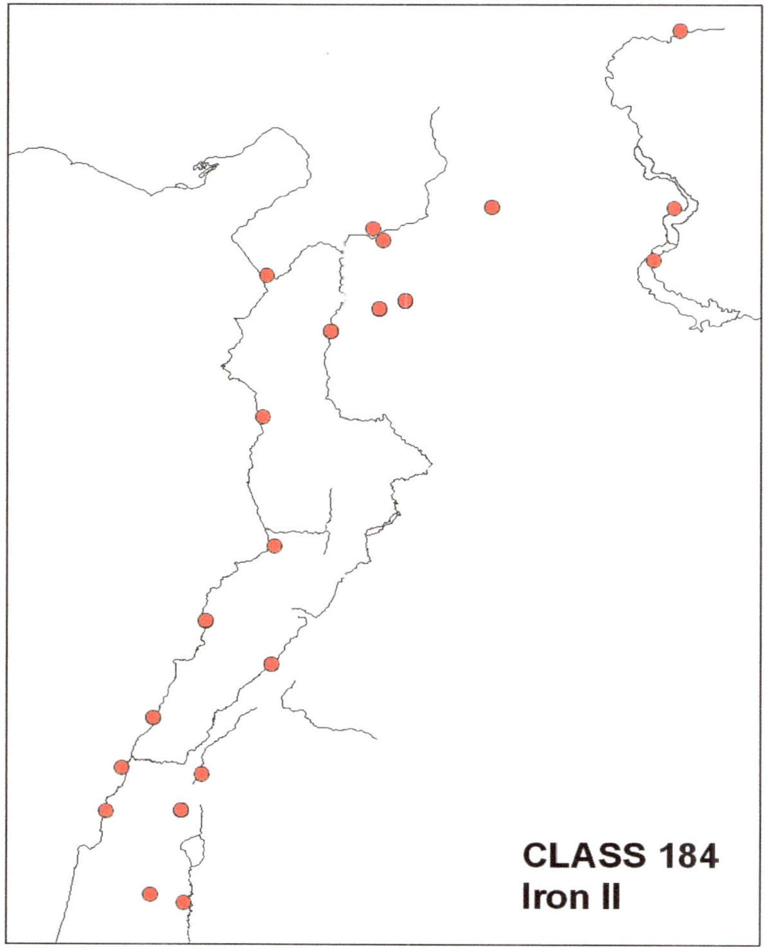

Map 85

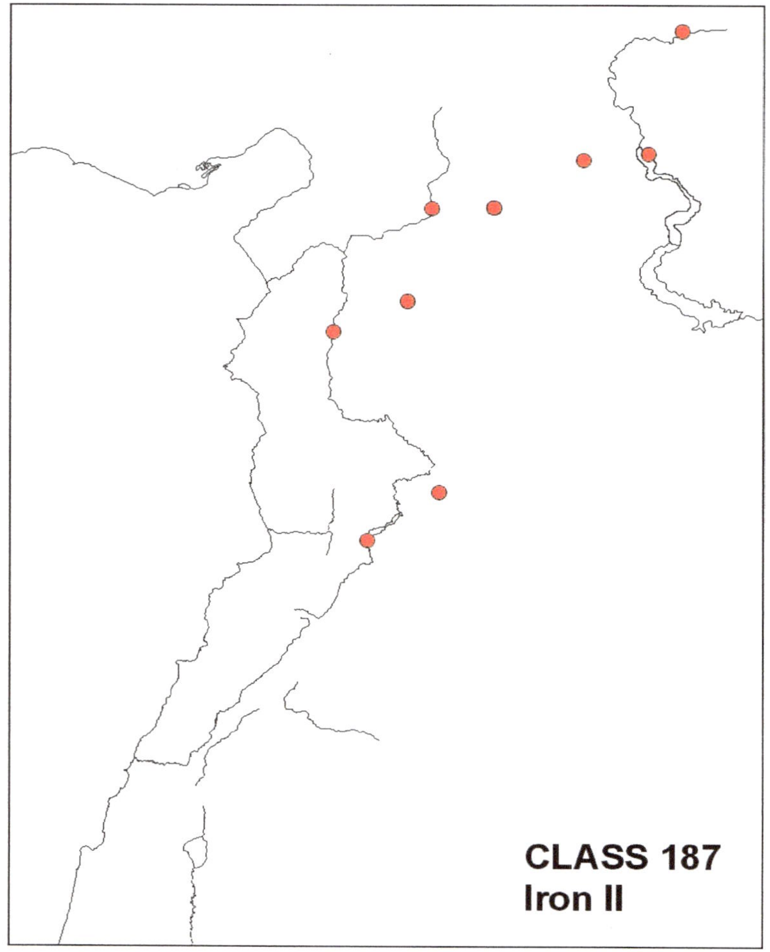

Map 86

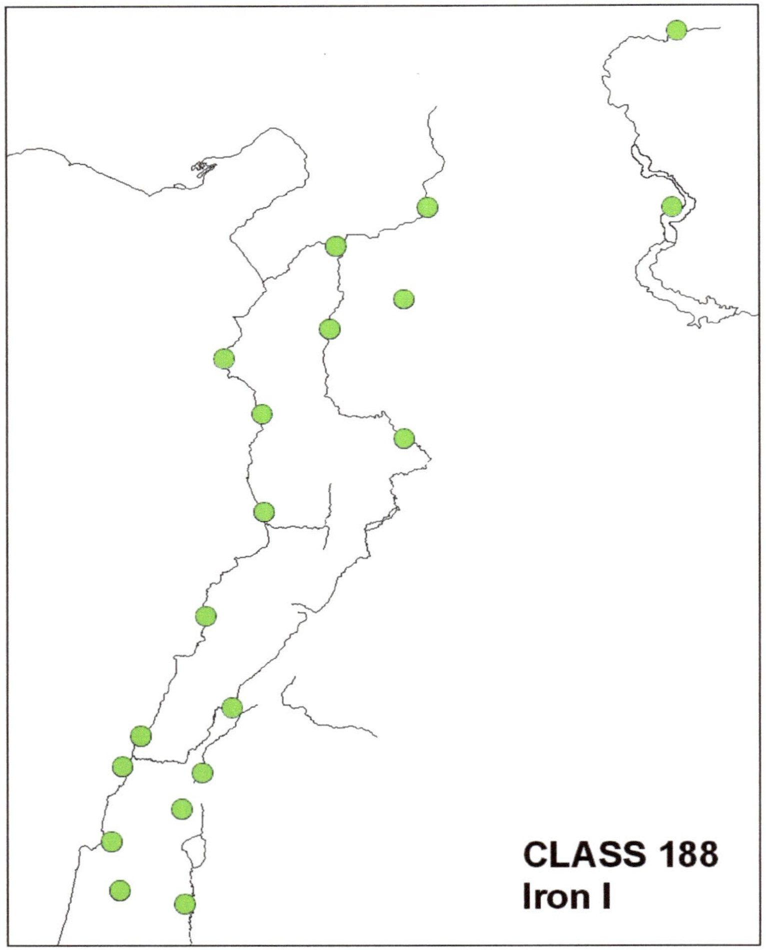

Map 87

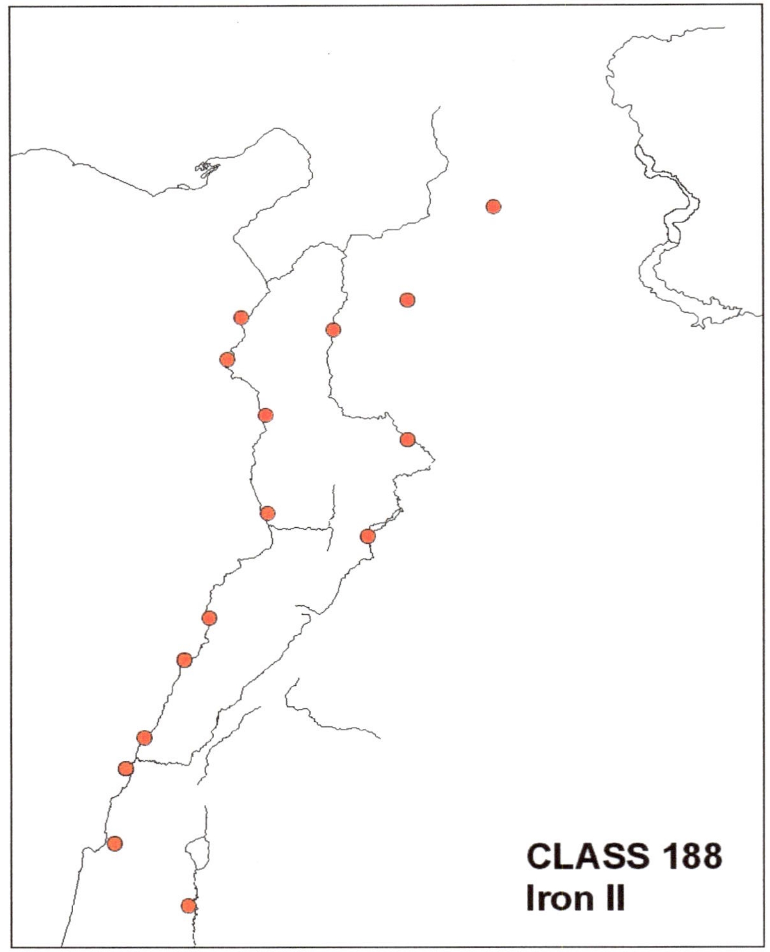

Map 88

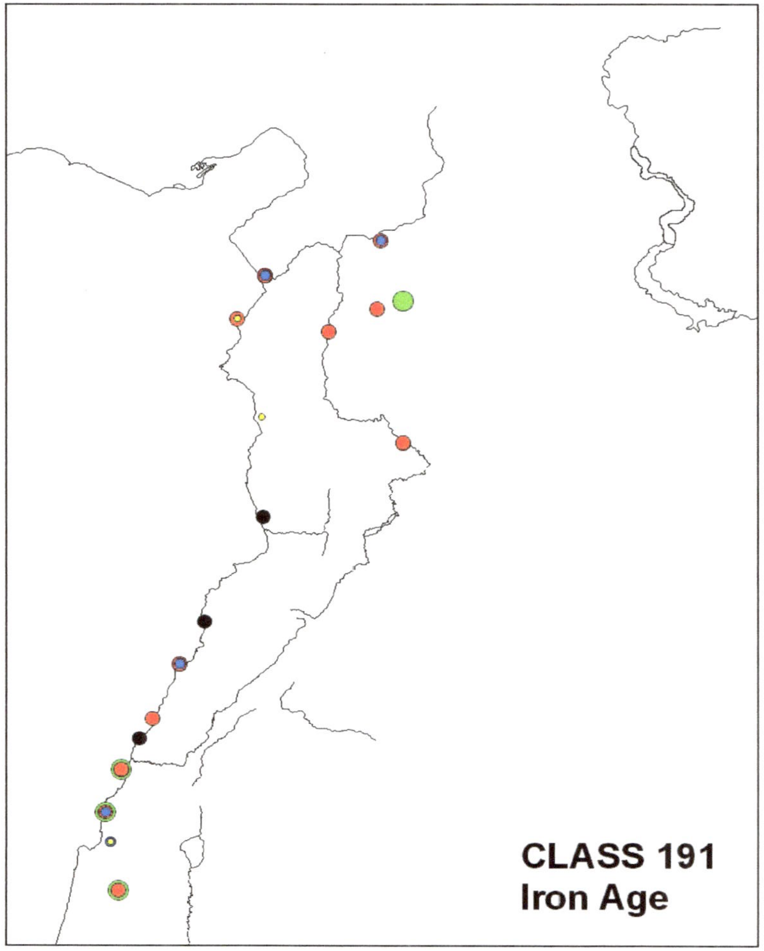

Map 89

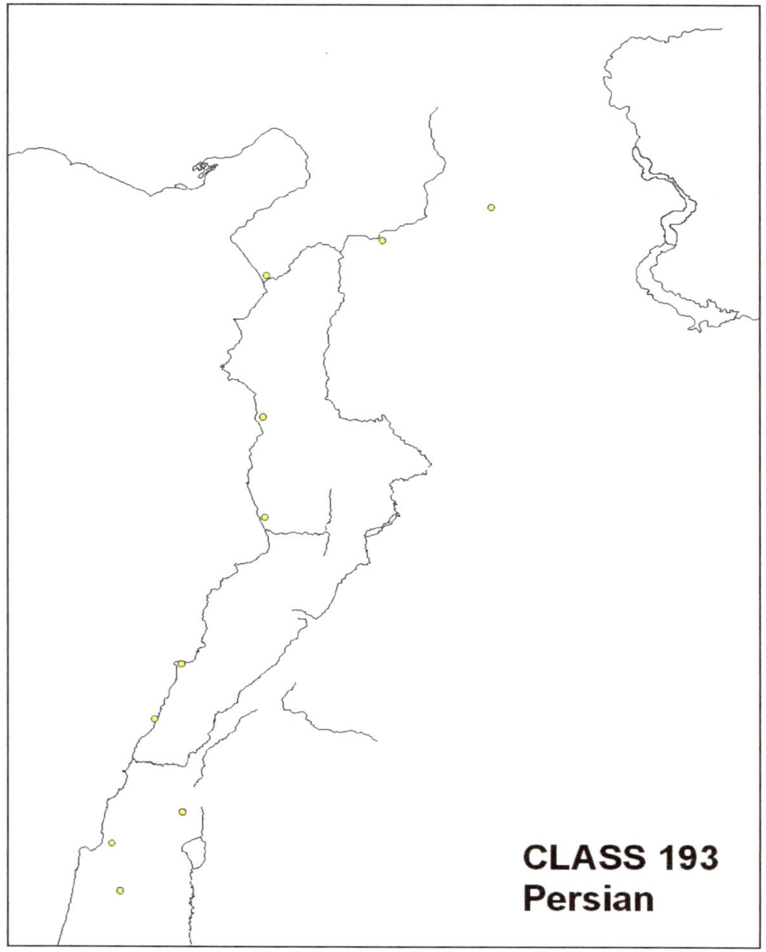

Map 90

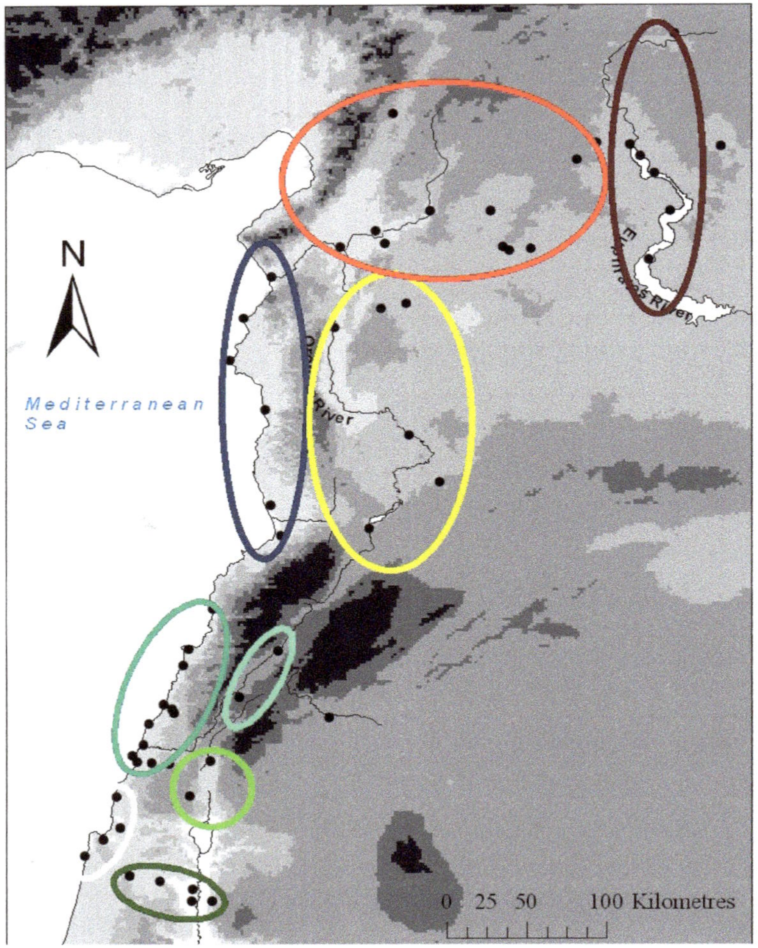

Map 91. Geographic Zones used within present study

Key to Map 91

Colour	Zone	Approx. Modern Equiv.
Blue	1	Syrian Coast
Teal	2	Lebanese Coast
Turquoise	3	Beqa' Valley
Dark Green	4	Jezreel Valley
Green	5	Northern Palestine
White	6	Palestinian Coast
Yellow	7	Orontes Syria
Red	8	North-west Syria (Amuq)
Burgundy	9	Euphrates Syria

See **Table 7.3** for sites within each Zone

www.ingramcontent.com/pod-product-compliance
Lightning Source LLC
LaVergne TN
LVHW072247300926
849160LV00018B/1609

* 9 7 8 1 4 0 7 3 0 3 8 4 0 *